TODAY'S TECHNICIAN™

CLASSROOM MANUAL
FOR AUTOMOTIVE HEATING & AIR CONDITIONING

TODAY'S TECHNICIAN™

CLASSROOM MANUAL
FOR AUTOMOTIVE HEATING & AIR CONDITIONING

MARK SCHNUBEL

NAUGATUCK VALLEY COMMUNITY COLLEGE

WATERBURY, CONNECTICUT

SIXTH EDITION

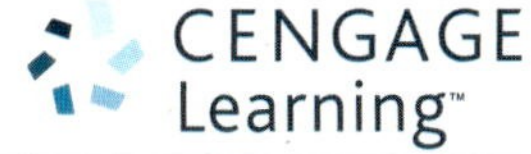

Australia • Canada • Mexico • Singapore • Spain • United Kingdom • United States

Today's Technician™: Classroom Manual for Automotive Heating & Air Conditioning, Sixth Edition

Mark Schnubel

SVP, GM Skills & Global Product Management: Dawn Gerrain

Product Director: Matthew Seeley

Product Team Manager: Erin Brennan

Senior Director, Development: Marah Bellegarde

Senior Product Development Manager: Larry Main

Senior Content Developer: Meaghan Tomaso

Product Assistant: Maria Garguilo

Vice President, Marketing Services: Jennifer Ann Baker

Production Service/Compositor: SPi Global

Marketing Manager: Jonathon Sheehan

Senior Production Director: Wendy Troeger

Production Director: Andrew Crouth

Senior Content Project Manager: Cheri Plasse

Senior Art Director: Benjamin Gleeksman

Cover image(s): © cla78/Shutterstock

Library of Congress Control Number: 2015956682

Book Only ISBN: 978-1-305-49760-3
Package ISBN: 978-1-305-49762-7

Cengage Learning
20 Channel Center Street
Boston, MA 02210
USA

Cengage Learning is a leading provider of customized learning solutions with office locations around the globe, including Singapore, the United Kingdom, Australia, Mexico, Brazil, and Japan. Locate your local office at: **www.cengage.com/global**

Cengage Learning products are represented in Canada by Nelson Education, Ltd.

To learn more about Cengage Learning, visit **www.cengage.com**

Purchase any of our products at your local college store or at our preferred online store **www.cengagebrain.com**

Printed in the United States of America
Print Number: 01 Print Year: 2016

CONTENTS

CONTENTS

Preface

Thanks to the support the *Today's Technician*™ series has received from those who teach automotive technology. Cengage Learning, the leader in automotive-related textbooks, is able to live up to its promise to provide new editions regularly. We have listened and responded to our critics and our fans and present this new updated and revised sixth edition. By revising our series regularly, we can and will respond to changes in the industry, changes in technology, changes in the certification process, and to the ever-changing needs of those who teach automotive technology.

The *Today's Technician*™ series features textbooks that cover all mechanical and electrical systems of automobiles and light trucks (while the heavy-duty trucks portion of the series does the same for heavy-duty vehicles). Principally, the individual titles correspond to the main areas of ASE (National Institute for Automotive Service Excellence) certification. Additional titles include remedial skills and theories common to all of the certification areas and advanced or specific subject areas that reflect the latest technological trends. Each text is divided into two volumes: a Classroom Manual and a Shop Manual.

Unlike yesterday's mechanic, the technician of today and for the future must know the underlying theory of all automotive systems and be able to service and maintain those systems. Dividing the material into two volumes provides the reader with the information needed to begin a successful career as an automotive technician without interrupting the learning process by mixing cognitive and performance learning objectives in one volume.

The design of Cengage's *Today's Technician*™ series was based on features that are known to promote improved student learning. The design was further enhanced by a careful study of survey results in which the respondents were asked to value particular features. Some of these features can be found in other textbooks, whereas others are unique to this series.

Each Classroom Manual contains the principles of operation for each system and subsystem. The Classroom Manual also contains discussions on design variations of key components used by the different vehicle manufacturers. This volume is organized to build upon basic facts and theories. The primary objective of this volume is to allow the reader to gain an understanding of how each system and subsystem operates. This understanding is necessary to diagnose the complex automobiles of today and tomorrow. Although the basics contained in the Classroom Manual provide the knowledge needed for diagnostics, diagnostic procedures appear only in the Shop Manual. An understanding of the basics is also a requirement for competence in the skill areas covered in the Shop Manual.

A spiral-bound Shop Manual covers the "how-to's." This volume includes step-by-step instructions for diagnostic and repair procedures. Photo Sequences are used to illustrate some of the common service procedures. Other common procedures are listed and are accompanied with fine-line drawings and photos that allow the reader to visualize and conceptualize the finest details of the procedure. This volume also contains the reasons for performing the procedures, as well as information on when that particular service is appropriate.

The two volumes are designed to be used together and are arranged in corresponding chapters. Not only are the chapters in the volumes linked together, but the contents of the chapters are also linked. This linking of content is evidenced by marginal callouts that refer the reader to the chapter and page where the same topic is addressed in the other volume. This feature is valuable to instructors. Without this feature, users of other two-volume textbooks must search the index or table of contents to locate supporting information in the other volume. This is not only cumbersome but also creates additional work for an instructor when planning the presentation of material and making reading assignments. This linking feature is also valuable to students; with page references they also know exactly where to look for supportive information.

The art is a vital part of each textbook. Both volumes contain clear and thoughtfully selected illustrations, many of which are original drawings or photos specially prepared for inclusion in this series.

The page layout used in the series is designed to include information that would otherwise break up the flow of information presented to the reader. The main body of the text includes all the "need-to-know" information and illustrations. In the wide side margins of each page are many of the special features of the series: simple examples of concepts just introduced in the text, explanations or definitions of terms that will not be defined in the glossary, examples of common trade jargon used to describe a part or operation, and exceptions to the norm explained in the text. Many textbooks attempt to include this type of information by inserting it in the main body of text; this tends to interrupt the thought process and cannot be pedagogically justified. By placing this information off to the side of the main text, the reader can choose when to refer to it.

Jack Erjavec
Series Advisor

Highlights of this Edition—Classroom Manual

The Classroom Manual of this edition has been updated to include new technology used in the automotive heating and air-conditioning systems of today's vehicles while still retaining information on systems used in older vehicles that are still in use. In addition, an emphasis has been placed on updating images throughout the text with full-color photos. Charts, graphs, and line drawings are now also in full color to be more visually appealing and improve the content comprehension by the reader. Coverage of R-1234yf has been added throughout the text. Chapter 2 covers the basic theories required to fully understand the operation and diagnosis of the complete HVAC system. A chapter on electricity and electronic fundamentals has been added covering the application and use of digital multimeters for those readers that have a limited background in electrical applications. This is intended to improve their understanding of electrical applications material covered in later chapters and prepare them with the electrical knowledge needed to complete future job sheets. Chapter 4 covers the automotive heating system and engine cooling system, including systems used on today's hybrid electric vehicles. The electronic thermostat used on some of today's vehicles is thoroughly explained along with the rationale behind its use.

The rest of the text is laid out in a logical order, beginning with basic air-conditioning system operating principles and progressing to diagnosis of the refrigerant system. The end-of-chapter questions in all chapters have been revised and updated. Updated coverage on advanced electronics has been included, from the operation of electronic variable compressors and electric motor–driven compressors to advanced sensors such as the airborne pollutants sensor. Chapter 11 on HVAC system controls has been updated to include more information on advanced climate control systems while still including a thorough description of CAN system operation.

HIGHLIGHTS OF THIS EDITION—SHOP MANUAL

Safety information remains the first chapter of the Shop Manual and covers general safety issues as well as topics specific to automotive HVAC service. This chapter includes an in-depth discussion on high-voltage safety on today's hybrid and electric vehicles and the equipment necessary to service these vehicles. As with the Classroom Manual, an emphasis has been placed on updating Shop Manual images and photo sequences throughout the text with full-color photos and line art. Chapter 3, "Electricity and Electronic Fundamentals," has been added. This chapter contains detailed information on digital multimeter usage for electrical system diagnosis and troubleshooting.

Chapter 4 and later chapters cover service information related to the system information covered in the corresponding Classroom Manual chapters. Many new job sheets have been added and existing job sheets have been updated with 100 percent of the NATEF tasks covered. The latest use of tools and technology has been integrated into the text, including hybrid electric compressors and the operation and use of SAE standard J2788 refrigerant recovery/recycling/recharging equipment needed to service today's small-capacity refrigerant systems. Added coverage of today's automatic climate control system service and diagnosis has been updated and includes specific examples. This edition of the Shop Manual will guide the student/technician through all the basic tasks related to automive heating and air-conditioning service and repair.

Classroom Manual

Features of this manual include:

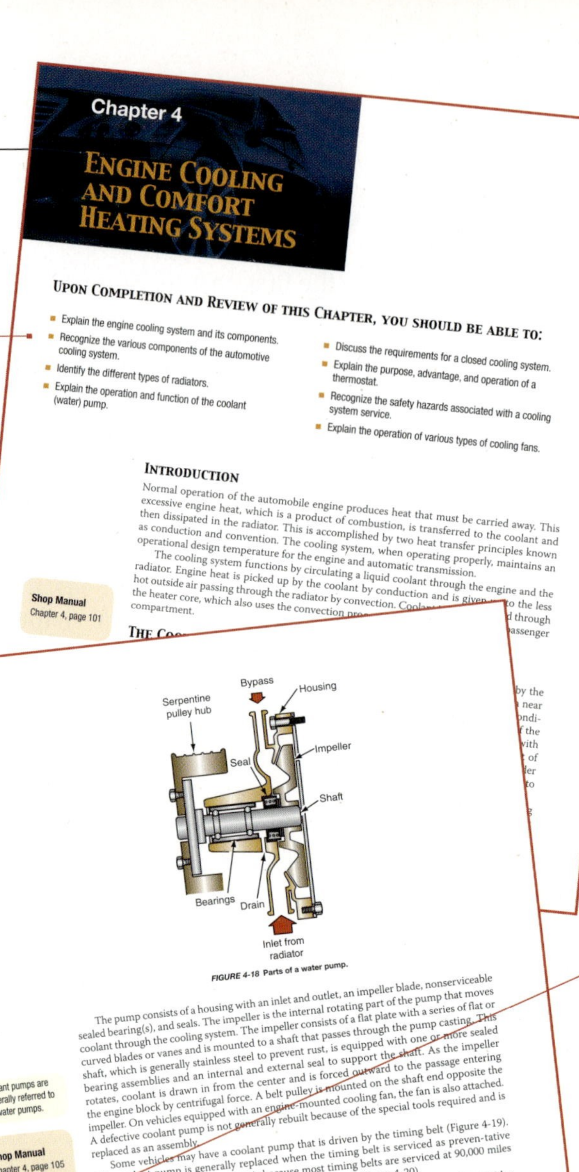

COGNITIVE OBJECTIVES

These objectives outline the contents of the chapter and define what the student should have learned upon completion of the chapter.

Each topic is divided into small units to promote easier understanding and learning.

MARGINAL NOTES

These notes add "nice-to-know" information to the discussion. They may include examples or exceptions, or may give the common trade jargon for a component.

TERMS TO KNOW DEFINITIONS

Many of the new terms are pulled out into the margins and defined.

CROSS-REFERENCES TO THE SHOP MANUAL

Reference to the appropriate page in the Shop Manual is given whenever necessary. Although the chapters of the two manuals are synchronized, material covered in other chapters of the Shop Manual may be fundamental to the topic discussed in the Classroom Manual.

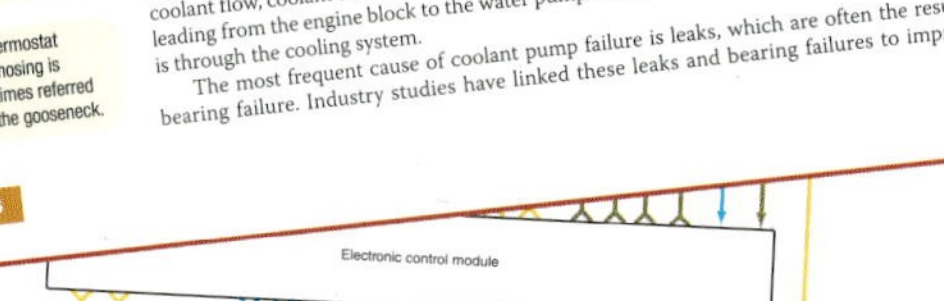

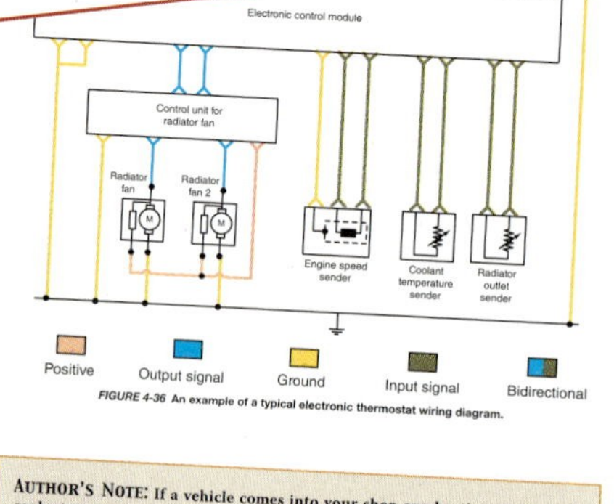

AUTHOR'S NOTES

This feature includes simple explanations, stories, or examples of complex topics. These are included to help students understand difficult concepts.

A BIT OF HISTORY

This feature gives the student a sense of the evolution of the automobile. It not only contains nice-to-know information, but should also spark some interest in the subject matter.

REVIEW QUESTIONS

Short-answer essay, fill-in-the-blank, and multiple-choice questions are found at the end of each chapter. These questions are designed to accurately assess the student's competence in the stated objectives at the beginning of the chapter.

FIGURE 4-66 Ethylene glycol antifreeze.

FIGURE 4-67 Propylene glycol antifreeze.

FIGURE 4-68 Distilled water.

TERMS TO KNOW LIST

A list of new terms appears after the Summary.

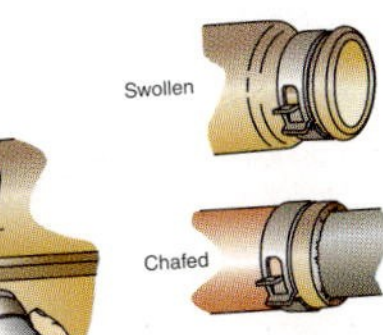
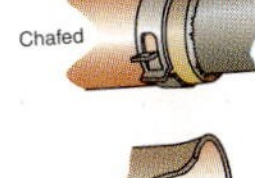

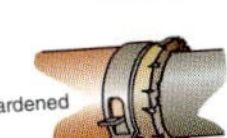

FIGURE 4-70 Typical hose defects.

SUMMARIES

Each chapter concludes with a summary of key points from the chapter. These are designed to help the reader review the contents.

Shop Manual

To stress the importance of safe work habits, the Shop Manual also dedicates one full chapter to safety. Other important features of this manual include:

PERFORMANCE OBJECTIVES

These objectives outline the contents of the chapter and identify what the student should have learned upon completion of the chapter. These objectives also correspond to the list of required tasks for NATEF certification.

Although this textbook is not designed to simply prepare someone for the certification exams, it is organized around the NATEF task list. These tasks are defined generically when the procedure is commonly followed and specifically when the procedure is unique for specific vehicle models. Imported and domestic model automobiles and light trucks are included in the procedures.

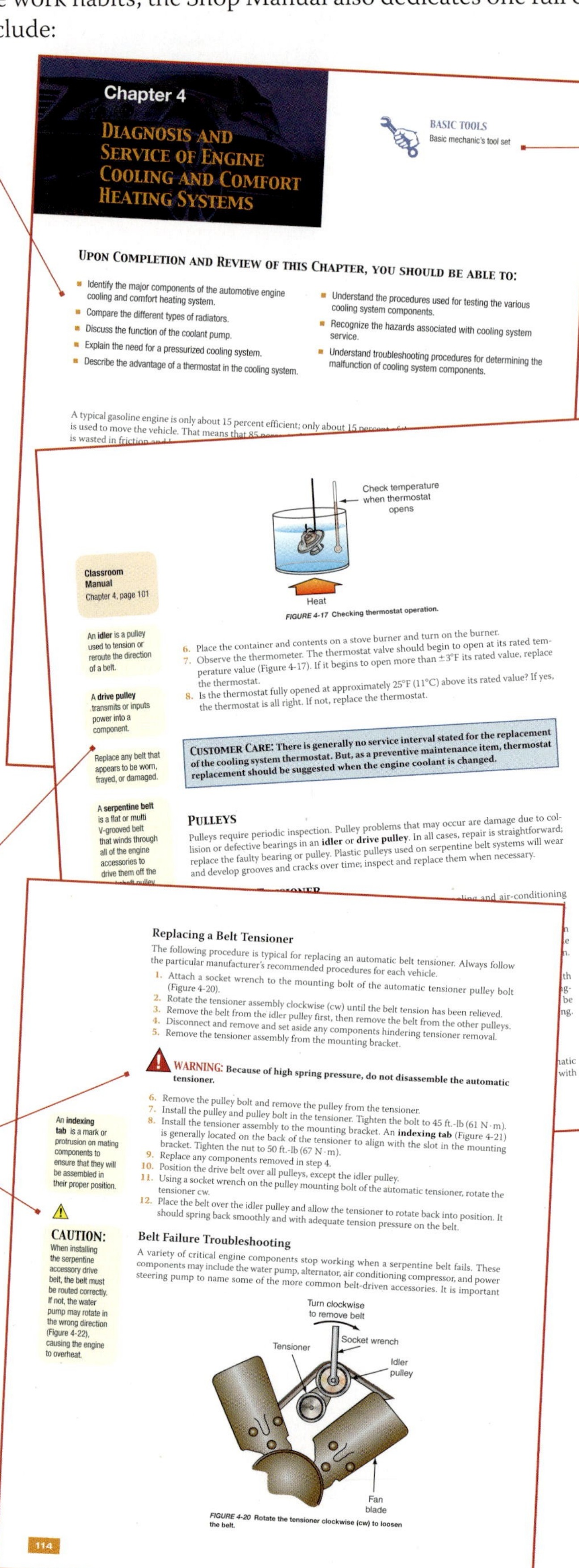

Chapter 4

DIAGNOSIS AND SERVICE OF ENGINE COOLING AND COMFORT HEATING SYSTEMS

BASIC TOOLS
Basic mechanic's tool set

UPON COMPLETION AND REVIEW OF THIS CHAPTER, YOU SHOULD BE ABLE TO:

- Identify the major components of the automotive engine cooling and comfort heating system.
- Compare the different types of radiators.
- Discuss the function of the coolant pump.
- Explain the need for a pressurized cooling system.
- Describe the advantage of a thermostat in the cooling system.
- Understand the procedures used for testing the various cooling system components.
- Recognize the hazards associated with cooling system service.
- Understand troubleshooting procedures for determining the malfunction of cooling system components.

A typical gasoline engine is only about 15 percent efficient; only about 15 percent …
is used to move the vehicle. That means that 85 …
is wasted in friction …

Classroom Manual
Chapter 4, page 101

FIGURE 4-17 Checking thermostat operation.

An **idler** is a pulley used to tension or reroute the direction of a belt.

A **drive pulley** transmits or inputs power into a component.

Replace any belt that appears to be worn, frayed, or damaged.

A **serpentine belt** is a flat or multi V-grooved belt that winds through all of the engine accessories to drive them off the …

6. Place the container and contents on a stove burner and turn on the burner.
7. Observe the thermometer. The thermostat valve should begin to open at its rated temperature value (Figure 4-17). If it begins to open more than ±3°F its rated value, replace the thermostat.
8. Is the thermostat fully opened at approximately 25°F (11°C) above its rated value? If yes, the thermostat is all right. If not, replace the thermostat.

CUSTOMER CARE: **There is generally no service interval stated for the replacement of the cooling system thermostat. But, as a preventive maintenance item, thermostat replacement should be suggested when the engine coolant is changed.**

PULLEYS

Pulleys require periodic inspection. Pulley problems that may occur are damage due to collision or defective bearings in an **idler** or **drive pulley**. In all cases, repair is straightforward; replace the faulty bearing or pulley. Plastic pulleys used on serpentine belt systems will wear and develop grooves and cracks over time; inspect and replace them when necessary.

Replacing a Belt Tensioner

The following procedure is typical for replacing an automatic belt tensioner. Always follow the particular manufacturer's recommended procedures for each vehicle.

1. Attach a socket wrench to the mounting bolt of the automatic tensioner pulley bolt (Figure 4-20).
2. Rotate the tensioner assembly clockwise (cw) until the belt tension has been relieved.
3. Remove the belt from the idler pulley first, then remove the belt from the other pulleys.
4. Disconnect and remove and set aside any components hindering tensioner removal.
5. Remove the tensioner assembly from the mounting bracket.

WARNING: **Because of high spring pressure, do not disassemble the automatic tensioner.**

6. Remove the pulley bolt and remove the pulley from the tensioner.
7. Install the pulley and pulley bolt in the tensioner. Tighten the bolt to 45 ft.-lb (61 N·m).
8. Install the tensioner assembly to the mounting bracket. An **indexing tab** (Figure 4-21) is generally located on the back of the tensioner to align with the slot in the mounting bracket. Tighten the nut to 50 ft.-lb (67 N·m).
9. Replace any components removed in step 4.
10. Position the drive belt over all pulleys, except the idler pulley.
11. Using a socket wrench on the pulley mounting bolt of the automatic tensioner, rotate the tensioner cw.
12. Place the belt over the idler pulley and allow the tensioner to rotate back into position. It should spring back smoothly and with adequate tension pressure on the belt.

Belt Failure Troubleshooting

A variety of critical engine components stop working when a serpentine belt fails. These components may include the water pump, alternator, air conditioning compressor, and power steering pump to name some of the more common belt-driven accessories. It is important

An **indexing tab** is a mark or protrusion on mating components to ensure that they will be assembled in their proper position.

CAUTION: When installing the serpentine accessory drive belt, the belt must be routed correctly. If not, the water pump may rotate in the wrong direction (Figure 4-22), causing the engine to overheat.

FIGURE 4-20 Rotate the tensioner clockwise (cw) to loosen the belt.

114

TOOLS LISTS

Each chapter begins with a list of the Basic Tools needed to perform the tasks included in the chapter. Whenever a Special Tool is required to complete a task, it is listed in the margin next to the procedure.

CUSTOMER CARE

This feature highlights those little things a technician can do or say to enhance customer relations.

TERMS TO KNOW DEFINITIONS

Many of the new terms are pulled out into the margin and defined.

CAUTIONS AND WARNINGS

Throughout the text, cautions are given to alert the reader to potentially hazardous materials or unsafe conditions. Warnings are also given to advise the student of what can go wrong if instructions are not followed or if a nonacceptable part or tool is used.

PHOTO SEQUENCES

Many procedures are illustrated in detailed Photo Sequences. These detailed photographs show the students what to expect when they perform particular procedures. They can also provide a student a familiarity with a system or type of equipment that the school may not have.

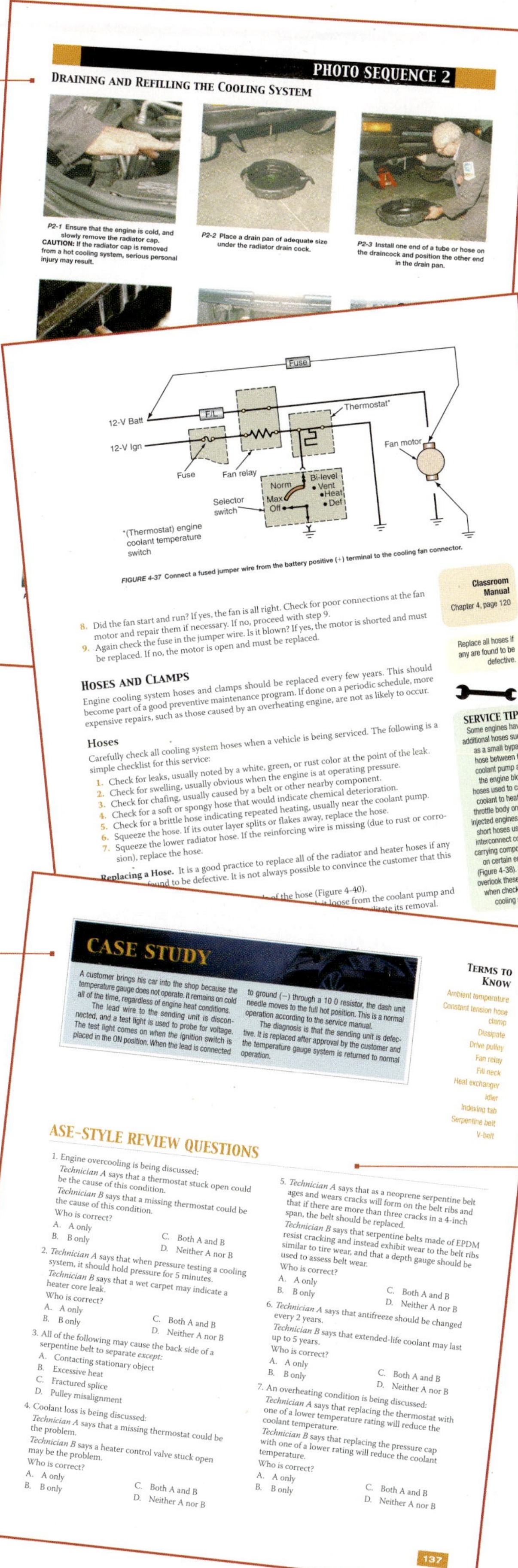

PHOTO SEQUENCE 2

DRAINING AND REFILLING THE COOLING SYSTEM

P2-1 Ensure that the engine is cold, and slowly remove the radiator cap. CAUTION: If the radiator cap is removed from a hot cooling system, serious personal injury may result.

P2-2 Place a drain pan of adequate size under the radiator drain cock.

P2-3 Install one end of a tube or hose on the draincock and position the other end in the drain pan.

FIGURE 4-37 Connect a fused jumper wire from the battery positive (+) terminal to the cooling fan connector.

Classroom Manual Chapter 4, page 120

8. Did the fan start and run? If yes, the fan is all right. Check for poor connections at the fan motor and repair them if necessary. If no, proceed with step 9.
9. Again check the fuse in the jumper wire. Is it blown? If yes, the motor is shorted and must be replaced. If no, the motor is open and must be replaced.

Replace all hoses if any are found to be defective.

HOSES AND CLAMPS

Engine cooling system hoses and clamps should be replaced every few years. This should become part of a good preventive maintenance program. If done on a periodic schedule, more expensive repairs, such as those caused by an overheating engine, are not as likely to occur.

Hoses

Carefully check all cooling system hoses when a vehicle is being serviced. The following is a simple checklist for this service:

1. Check for leaks, usually noted by a white, green, or rust color at the point of the leak.
2. Check for swelling, usually obvious when the engine is at operating pressure.
3. Check for chafing, usually caused by a belt or other nearby component.
4. Check for a soft or spongy hose that would indicate chemical deterioration.
5. Check for a brittle hose indicating repeated heating, usually near the coolant pump.
6. Squeeze the hose. If its outer layer splits or flakes away, replace the hose.
7. Squeeze the lower radiator hose. If the reinforcing wire is missing (due to rust or corrosion), replace the hose.

Replacing a Hose. It is a good practice to replace all of the radiator and heater hoses if any ... found to be defective. It is not always possible to convince the customer that this ...

SERVICE TIP: Some engines have additional hoses such as a small bypass hose between the coolant pump and the engine block, hoses used to carry coolant to heat the throttle body on fuel injected engines, and short hoses used to interconnect coolant carrying components on certain engines (Figure 4-38). Do not overlook these hoses when checking the cooling system.

CASE STUDY

A customer brings his car into the shop because the temperature gauge does not operate. It remains on cold all of the time, regardless of engine heat conditions.

The lead wire to the sending unit is disconnected, and a test light is used to probe for voltage. The test light comes on when the ignition switch is placed in the ON position. When the lead is connected to ground (−) through a 10 Ω resistor, the dash unit needle moves to the full hot position. This is a normal operation according to the service manual.

The diagnosis is that the sending unit is defective. It is replaced after approval by the customer and the temperature gauge system is returned to normal operation.

TERMS TO KNOW

Ambient temperature
Constant tension hose clamp
Dissipate
Drive pulley
Fan relay
Fill neck
Heat exchanger
Idler
Indexing tab
Serpentine belt
V-belt

ASE-STYLE REVIEW QUESTIONS

1. Engine overcooling is being discussed: *Technician A* says that a thermostat stuck open could be the cause of this condition. *Technician B* says that a missing thermostat could be the cause of this condition. Who is correct?
   A. A only
   B. B only
   C. Both A and B
   D. Neither A nor B
2. *Technician A* says that when pressure testing a cooling system, it should hold pressure for 5 minutes. *Technician B* says that a wet carpet may indicate a heater core leak. Who is correct?
   A. A only
   B. B only
   C. Both A and B
   D. Neither A nor B
3. All of the following may cause the back side of a serpentine belt to separate *except:*
   A. Contacting stationary object
   B. Excessive heat
   C. Fractured splice
   D. Pulley misalignment
4. Coolant loss is being discussed: *Technician A* says that a missing thermostat could be the problem. *Technician B* says a heater control valve stuck open may be the problem. Who is correct?
   A. A only
   B. B only
   C. Both A and B
   D. Neither A nor B
5. *Technician A* says that as a neoprene serpentine belt ages and wears cracks will form on the belt ribs and that if there are more than three cracks in a 4-inch span, the belt should be replaced. *Technician B* says that serpentine belts made of EPDM resist cracking and instead exhibit wear to the belt ribs similar to tire wear, and that a depth gauge should be used to assess belt wear. Who is correct?
   A. A only
   B. B only
   C. Both A and B
   D. Neither A nor B
6. *Technician A* says that antifreeze should be changed every 2 years. *Technician B* says that extended-life coolant may last up to 5 years. Who is correct?
   A. A only
   B. B only
   C. Both A and B
   D. Neither A nor B
7. An overheating condition is being discussed: *Technician A* says that replacing the thermostat with one of a lower temperature rating will reduce the coolant temperature. *Technician B* says that replacing the pressure cap with one of a lower rating will reduce the coolant temperature. Who is correct?
   A. A only
   B. B only
   C. Both A and B
   D. Neither A nor B

137

CROSS-REFERENCES TO THE CLASSROOM MANUAL

Reference to the appropriate page in the Classroom Manual is given whenever necessary. Although the chapters of the two manuals are synchronized, material covered in other chapters of the Classroom Manual may be fundamental to the topic discussed in the Shop Manual.

SERVICE TIPS

Whenever a shortcut or special procedure is appropriate, it is described in the text. These tips are generally those things commonly done by experienced technicians.

CASE STUDIES

Case Studies concentrate on the ability to properly diagnose the systems. Beginning with Chapter 3, each chapter ends with a case study in which a vehicle has a problem, and the logic used by a technician to solve the problem is explained.

TERMS TO KNOW LIST

A list of new terms appears after the case study.

ASE-STYLE REVIEW QUESTIONS

Each chapter contains ASE-style review questions that reflect the performance objectives listed at the beginning of the chapter. These questions can be used to review the chapter as well as to prepare for the ASE certification exam.

JOB SHEETS

Located at the end of each chapter, the Job Sheets provide a format for students to perform procedures covered in the chapter. A reference to the NATEF Task addressed by the procedure is referenced on the Job Sheet.

ASE PRACTICE EXAMINATION

An ASE practice exam, located in the Appendix, is included to test students on the content of the complete Shop Manual.

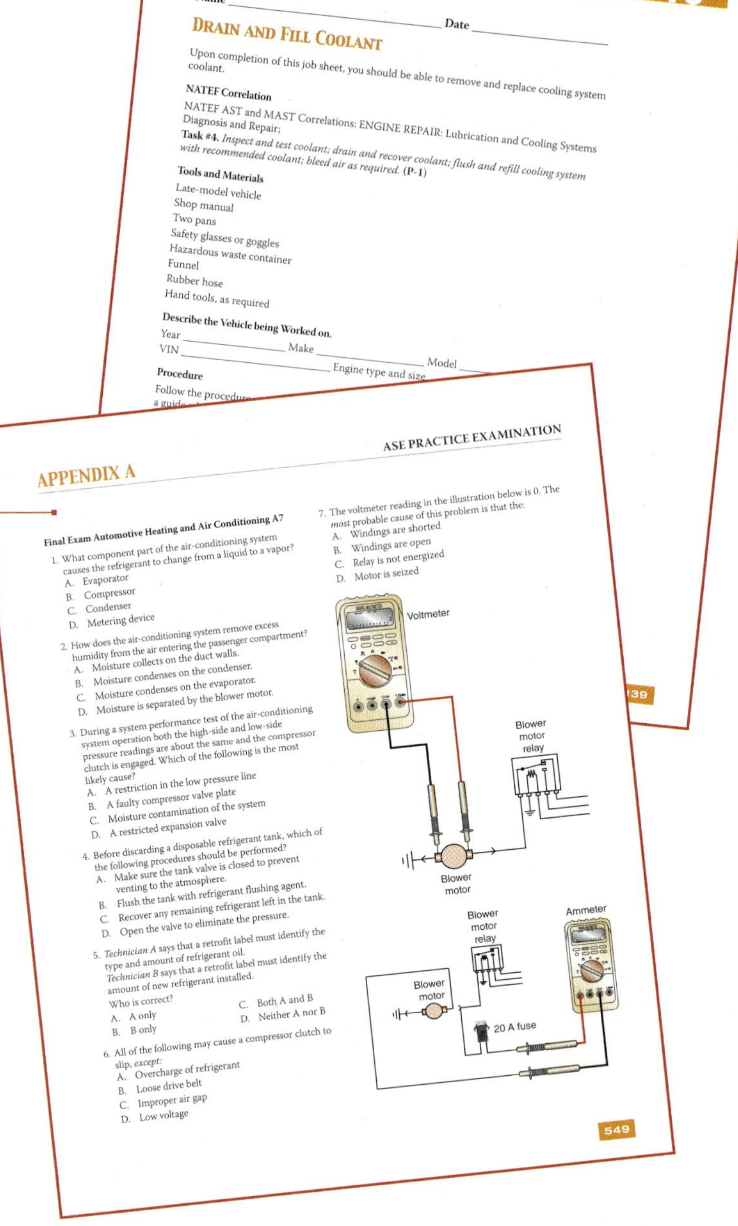

JOB SHEET 10

Name ______________________ Date ______________

DRAIN AND FILL COOLANT

Upon completion of this job sheet, you should be able to remove and replace cooling system coolant.

NATEF Correlation

NATEF AST and MAST Correlations: ENGINE REPAIR: Lubrication and Cooling Systems Diagnosis and Repair;

Task #4. *Inspect and test coolant; drain and recover coolant; flush and refill cooling system with recommended coolant; bleed air as required.* **(P-1)**

Tools and Materials

Late-model vehicle
Shop manual
Two pans
Safety glasses or goggles
Hazardous waste container
Funnel
Rubber hose
Hand tools, as required

Describe the Vehicle being Worked on.

Year ________ Make ____________ Model ________
VIN ______________ Engine type and size

Procedure

Follow the procedu...

139

ASE PRACTICE EXAMINATION

APPENDIX A

Final Exam Automotive Heating and Air Conditioning A7

1. What component part of the air-conditioning system causes the refrigerant to change from a liquid to a vapor?
   A. Evaporator
   B. Compressor
   C. Condenser
   D. Metering device
2. How does the air-conditioning system remove excess humidity from the air entering the passenger compartment?
   A. Moisture collects on the duct walls.
   B. Moisture condenses on the condenser.
   C. Moisture condenses on the evaporator.
   D. Moisture is separated by the blower motor.
3. During a system performance test of the air-conditioning system operation both the high-side and low-side pressure readings are about the same and the compressor clutch is engaged. Which of the following is the most likely cause?
   A. A restriction in the low pressure line
   B. A faulty compressor valve plate
   C. Moisture contamination of the system
   D. A restricted expansion valve
4. Before discarding a disposable refrigerant tank, which of the following procedures should be performed?
   A. Make sure the tank valve is closed to prevent venting to the atmosphere.
   B. Flush the tank with refrigerant flushing agent.
   C. Recover any remaining refrigerant left in the tank.
   D. Open the valve to eliminate the pressure.
5. *Technician A* says that a retrofit label must identify the type and amount of refrigerant oil. *Technician B* says that a retrofit label must identify the amount of new refrigerant installed. Who is correct?
   A. A only
   B. B only
   C. Both A and B
   D. Neither A nor B
6. All of the following may cause a compressor clutch to slip, *except*:
   A. Overcharge of refrigerant
   B. Loose drive belt
   C. Improper air gap
   D. Low voltage
7. The voltmeter reading in the illustration below is 0. The *most* probable cause of this problem is that the:
   A. Windings are shorted
   B. Windings are open
   C. Relay is not energized
   D. Motor is seized

549

Supplements

Instructor Resources

The Instructor Resources, now available both online and on DVD, are a robust ancillary product that contains all preparation tools to meet any instructor's classroom needs. It includes chapter outlines in PowerPoint with images, video clips, and animations that coincide with each chapter's content coverage, chapter tests powered by Cognero with hundreds of test questions, an Image Gallery with all photos and illustrations from the text, theory-based Worksheets in Word that provide homework or in-class assignments, the Job Sheets from the Shop Manual in Word, a NATEF correlation chart, and an Instructor's Guide in electronic format.

To access these Instructor Resources online, go to login.cengagebrain.com, and create an account or log into your existing account.

MindTap

MindTap for ***Today's Technician: Automotive Heating & Air Conditioning, 6th edition***, is a personalized teaching experience with relevant assignments that guide students to analyze, apply, and improve thinking, allowing you to measure skills and outcomes with ease.

- Relevant readings, multimedia, and activities are designed to guide students through progressive levels of learning, from basic knowledge to analysis and application.
- Personalized teaching becomes yours through a Learning Path built with key student objectives and your syllabus in mind. Control what students see and when they see it.
- Analytics and reports provide a snapshot of class progress, time in course, engagement, and completion rates.

MindTap for ***Today's Technician: Automotive Heating & Air Conditioning, 6th edition***, meets the needs of today's automotive classroom, shop, and student. Within the MindTap faculty and students will find editable and submittable job sheets, based on NATEF tasks. MindTap also offers students engaging activities that include videos, matching exercises, and assessments.

Reviewers

The author and publisher wish to thank the instructors who reviewed this text and offered their invaluable feedback:

Dan Cifalia
Mesa Community College
Mesa, AZ

Lance David
College of Lake County
Grayslake, IL

Randy Howarth
Hudson Valley Community College
Troy, NY

Shannon Kies
University of Northwestern Ohio
Lima, OH

John Koehn
Pueblo Community College
Pueblo, CO

Christopher J. Marker
University of Northwestern Ohio
Lima, OH

Gary McDaniel
Metropolitan Community College–Longview
Longview, MO

William McGrath
Moraine Valley Community College
Palos Hills, IL

Rouzbeh "Ross" Oskui
Monroe County Community College
Monroe, MI

Christopher Parrot
Vatterott College
Wichita, KS

Mike Shoebroek
Austin Community College
Austin, TX

Ira Siegel
Moraine Valley Community College
Palos Hills, IL

Stephen Skroch
Mesa Community College
Mesa, AZ

Christopher C. Woods
University of Northwestern Ohio
Lima, OH

Chapter 1

HEATING AND AIR CONDITIONING—HISTORY AND THE ENVIRONMENT

UPON COMPLETION AND REVIEW OF THIS CHAPTER, YOU SHOULD BE ABLE TO:

- Define the term air conditioning.
- Discuss the historic developments of modern refrigeration.
- Discuss the advantages of air conditioning in the automotive industry.
- Explain the importance of the ozone layer.
- Discuss what industry is doing about the ozone depletion problem.
- Discuss what government is doing about the ozone depletion problem.
- Describe how ozone is created.
- Describe how ozone is destroyed.
- Discuss the Clean Air Act.
- Discuss ozone protection regulations.
- Discuss CFC-12 (R-12), HFC-134a (R-134a), and the introduction of HFO-1234yf (R-1234yf) as an automotive refrigerant.
- Describe technician certification.
- Explain special safety precautions.
- Discuss the types of antifreeze/coolant used.
- Discuss the hazardous materials used.
- Describe toxic gases.

INTRODUCTION

Since the dawn of time, humans have been trying to control their environment. It was natural that after the automobile became popular, a passenger comfort heating and cooling system would be required. And so began the quest for vehicle passenger compartment temperature control.

The first year that automotive air conditioning was offered on a production vehicle was 1940, by the Packard Motor Car Company. Cadillac soon followed in 1941. After its initial introduction in the early 1940s, it did not become a popular option until the early 1960s. Since then the popularity of air conditioning has increased annually.

In 1962, just over 11 percent of all cars sold were equipped with air conditioners. This accounted for 756,781 units, including both factory-installed systems and those with add-on systems installed after the purchase, which are referred to as "aftermarket" systems. Just five years later, in 1967, the total number had increased an astounding 469 percent—to 3,546,255 units. Air conditioning is now one of the most popular selections in the entire list of automotive accessories and is often standard equipment on many vehicle models today. At the present time, over 93 percent of all automobiles sold in the United States are equipped with air conditioning units. It is expected that this percentage will remain at approximately 93 percent into the future.

Air conditioning is the process of adjusting and regulating by heating or refrigerating; the quality, quantity, temperature, humidity, and circulation of air in a space or enclosure; to condition the air.

Temperature and *humidity* refer to the quality of the conditioned air.

Volume refers to the quantity of the conditioned air.

Artificial ice, usually cut from frozen ponds, contained many impurities, dirt, and debris.

When mobile air conditioning was first introduced, it was considered a luxury. Its usefulness, however, has made it a necessity. As a matter of fact, if a vehicle is not equipped with air conditioning, its value is dramatically reduced on the resale market.

This text concentrates on the heating and air-conditioning system's function and operation as employed by various automotive manufacturers and the methods they use to improve passenger comfort levels.

Air Conditioning Defined

The definition of air conditioning should be reviewed before tracing its history and its application to the automobile. **Air conditioning**, by definition, is the process by which air is:

- Cooled
- Heated
- Cleaned or filtered
- Humidified or dehumidified
- Circulated or recirculated

In addition, the quantity and quality of the conditioned air are controlled. This means that the temperature, humidity, and volume of air can be controlled at any time in any given situation. Under ideal situations, air conditioning can be expected to accomplish all of these tasks at the same time. It is important to recognize that the air conditioning process includes the process of refrigeration (cooling by removing heat).

Refrigeration

Refrigeration is the term given to a process by which heat is removed from matter—solid, liquid, or vapor. It is the process of lowering the temperature of an enclosure or area by natural, chemical, electrical, or mechanical means. The fluid that circulates through an air-conditioning system is referred to generically as refrigerant. The refrigerants used in automotive air conditioning systems are commonly referred to as R12 and R134a. These refrigerants will be discussed further in this chapter and in Chapter 5.

Historical Development of Refrigeration

Refrigeration, as we know it today, is less than one hundred years old. Some of its principles, however, were known as long ago as 10,000 BC.

The Egyptians developed a method for cooling water. They found that water could be cooled by placing it in porous jugs on the rooftop at sundown. The night breeze evaporated the moisture seeping through the jugs and, in turn, cooled the contents. The Greeks and Romans had snow brought down from mountaintops. They preserved it by placing it in cone-shaped pits lined with straw and covered with a thatched roof. Even earlier, the Chinese learned that ice improved the taste of drinks. They cut it from frozen ponds and lakes in the winter, preserved it in straw, and sold it in the summer.

Domestic Refrigeration

Dr. John Gorrie (1803–1855) of Abbeville, South Carolina, was issued the first U.S. patent for a mechanical refrigeration system in 1851. Gorrie correctly theorized that if air were highly compressed, it would be heated by the energy of compression. If this compressed air were then run through metal pipes that were cooled with water, the air could be cooled to the water temperature. If this air were then expanded back to atmospheric pressure, low temperatures of about 26°F (—33°C—low enough to freeze water in pans in a refrigerator box—could be obtained. The compressor of this system could be powered by horse, water, wind, or steam. Gorrie's original system was installed in the U.S. Marine Hospital in Apalachicola, Florida, where he used it to treat patients suffering from yellow fever. A replica of his system is on display at the John Gorrie State Museum in Apalachicola.

While Dr. Gorrie's mechanism produced ice in quantities, leakage and irregular performance often impaired its operation. Gorrie's basic principle, however, is the one most often used in today's modern refrigeration: cooling caused by the rapid expansion of gases.

Domestic refrigeration systems first appeared in 1910, although in 1896 the Sears, Roebuck and Company catalog offered several refrigerators. Refrigeration, however, was provided by ice. The refrigerator held 25 pounds (11.34 kilograms) of ice and was useful only for short-term storage for the preservation of foods.

In 1899, the first household refrigeration patent was awarded to Albert T. Marshall of Brockton, Massachusetts. A manually-operated refrigerator was produced by J. L. Larsen in 1913. The Kelvinator Company produced the first automatic refrigerator in 1918. The acceptance of this new technology was slow. By 1920, only about 200 refrigerators had been sold.

In 1926, the first hermetic (sealed) refrigerator was introduced by General Electric. The following year, Electrolux introduced an automatic absorption unit. A 4-cubic-foot refrigerator was introduced by Sears, Roebuck, and Company in 1931. The refrigerator cabinet and the refrigeration unit were shipped separately and required assembly.

In terms of the cost per cubic foot of refrigeration, the early refrigerator compares favorably to today's modern machines. In terms of the economy, one worked about four times longer to pay for the 4-cubic-foot refrigerator than one works for today's 16-cubic-foot refrigerator, which is four times larger.

Shortly after the beginning of the twentieth century, T. C. Northcott of Luray, Virginia, became the first person known in history to have a home with central heating and air conditioning. A heating and ventilating engineer, Northcott built his house on a hill above the famous Caverns of Luray. Because of his work, he knew that air filtered through limestone was free of dust and pollen. This fact was important because Northcott and his family suffered from hay fever.

Some distance behind his house he drilled a shaft through the ceiling of the cavern and installed a fan to pull cavern air through the shaft. He then constructed a shed over the shaft and a duct system to the house. The duct system was divided into two chambers, one above the other. The upper duct, which carried air from the cavern, was heated by the sun, providing air to warm the house on cool days. The lower duct, which was unheated, carried air from the cavern to cool the house on warm days.

The moisture content (humidity) of the air was controlled in a chamber in Northcott's basement. Here, air from both ducts could be mixed. Because it is known that warm air contains more moisture than cool air, Northcott was able to direct conditioned air from the mixing chamber to any or all of the rooms in his house through a network of smaller ducts. During the winter season, auxiliary heat was provided by steam coils located in the base of each of the branch ducts.

Cooling accomplished by humidification is only effective in arid (dry) areas of the country.

Mobile Air Conditioning

The first automotive air conditioning unit appeared on the market in 1927. True air conditioning was not to appear in cars for another 13 years. However, air conditioning was advertised as an option in some cars in 1927. At that time, air conditioning meant only that the car could be equipped with a heater, a ventilation system, and a means of filtering the air. In 1938, Nash introduced "air conditioning" heating and ventilation. Fresh outside air was heated and filtered, then circulated around inside the car by fan.

Humidity refers to the amount of moisture in the air.

By 1940, heaters and defrosters were standard equipment on many models. That year Packard offered the first method of cooling a car by means of refrigeration. Actually, these first units were belt-driven commercial air conditioners that were adapted for automotive use and the **evaporators** were usually located in the trunk. Two years earlier, a few passenger buses had been air conditioned by the same method.

Early studies of the effectiveness of vehicles equipped with automotive air conditioning proved that sales and production increased significantly.

Accurate records were not kept in the early days of automotive air conditioning. However, it is known that before World War II between 3,000 and 4,000 units were installed in

Packards. Defense priorities for materials and manufacturing prevented the improvement of automotive air conditioning until the early 1950s. At that time, the demand for air conditioned vehicles began in the Southwest.

Ozone (O_3) is a form of oxygen.

The first of today's modern automotive air-conditioning systems was introduced by Cadillac in 1960. Their bi-level system could cool the top level of the car while heating the lower level. This method provided a means of controlling the in-vehicle humidity.

Ozone (O_3) is an unstable, pale blue gas with a penetrating odor; it is an allotropic form of oxygen (O) that is usually formed by electrical discharge in the air.

Many large firms reported increased sales after air conditioning was installed in the cars of their salespeople. Most commercial passenger-carrying vehicles are now air conditioned. Truck lines realize larger profits because drivers who have air conditioned cabs average more miles per day than those who do not.

In 1967, all of the state police cars on the Florida Turnpike were air conditioned. Since that time, most governmental and law enforcement agencies across the nation have added air conditioning to their vehicles.

Other Applications

Mobile air conditioning is not only found in cars, trucks, and buses. In recent years, mobile air conditioning application has been expanded for use in farm equipment such as tractors, harvesters, and thrashers. Additionally, mobile air-conditioning systems have been developed for use in other off-road equipment, such as backhoes, bulldozers, and graders. Air conditioning may be found in almost any kind of domestic, farm, or commercial equipment that has an enclosed cab and requires an onboard operator.

Oxygen (O) is an odorless, colorless, tasteless element that forms 21 percent of our atmosphere. It is an essential element for plant and animal life.

Refrigerant and the Environment

Since the discovery of the hole in the ozone layer over Antarctica, there has been widespread concern about the consequences for human health and for the environment. Ozone depletion, together with global warming resulting from the greenhouse effect, has attracted widespread media attention and well-founded concern around the world. This chapter will explain:

- The importance of the ozone layer
- How the ozone layer is formed
- How the ozone layer is being depleted
- What is causing ozone depletion
- What industry is doing to correct the damage
- What government is doing to correct the damage

Allotropes are structurally different from elements. For example, though different in structure, the properties of charcoal and diamond are the same as the element carbon; they are both made up of the same element in differing combinations.

What is Ozone?

Ozone is a molecular form of **oxygen**, having a different chemical property. Thus, it is an **allotrope** of oxygen. In large concentrations, ozone is considered to be a poisonous gas. The ozone layer, however, protects life on earth from damaging ultraviolet (UV) radiation.

Ozone has a very pungent odor described by many as irritating. In high concentrations, it has a pale blue color. This is in contrast with oxygen (O), which is colorless, tasteless, and has no odor. Each molecule of ozone (O_3), an allotropic form of oxygen, contains three atoms of oxygen in contrast to the diatomic form, which contains two atoms of oxygen (O_2).

Atmosphere is a general term used to describe the gaseous envelope surrounding the earth to a height of 621 miles (1,000 km); it is 21 percent oxygen, 78 percent nitrogen, and 1 percent other gases.

The Earth's Atmosphere

The earth's **atmosphere** is composed of a thin covering of gases that surround the globe and comprise an enormous mass. This mass is equivalent to about one million tons for every person living on earth. The atmosphere extends skyward for hundreds of miles (Figure 1-1). The lowest part of the atmosphere is the troposphere, which extends from ground level to about seven miles (11 kilometers) depending on the time of year and region of the globe. The troposphere has clouds, wind, storms, and a weather system. Above the troposphere is the

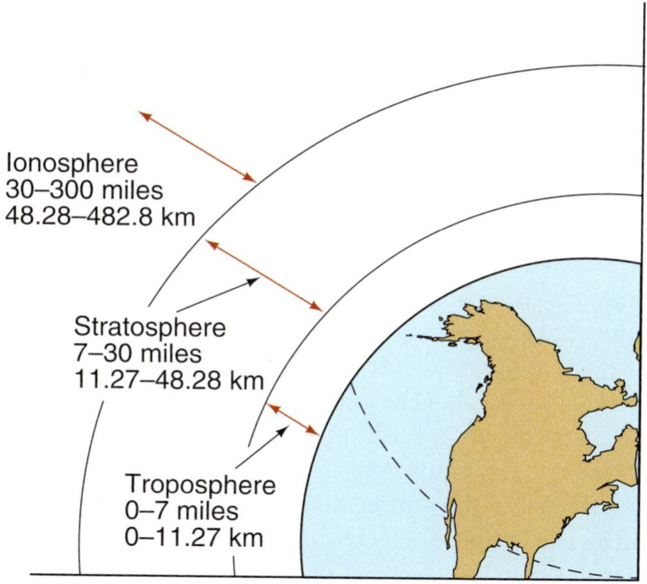

FIGURE 1-1 **The atmosphere extends skyward for hundreds of miles.**

stratosphere which extends to an altitude of about 30 miles (48 kilometers). This is the region where the protective ozone layer resides. The next time when you see a very large anvil shaped thunderstorm cloud (cumulonimbus) with a flat top look at it closely. This flat top is caused by the top of the cloud reaching the highest level of the troposphere and flatting out just below the stratosphere as it is sheared off by high level winds.

The major gases in the atmosphere are **nitrogen (N)**, an inert gas, which comprises 78 percent of the atmosphere by volume, and oxygen, which is vital for life and comprises 21 percent. Water (H_2O) vapor, a portion of which is seen as clouds, accounts for less than 1 percent. The composition of atmospheric gases is as shown in Table 1-1.

Several trace gases are also included in the other 0.00276 percent of the earth's atmosphere. Though very small in volume, they play critical roles in the atmosphere. Carbon dioxide (CO_2), for example, is a trace gas with a concentration of only 350 parts per million by volume (ppmv). This accounts for less than 0.3 percent, but it absorbs infrared radiation, thus warming the atmosphere through the phenomenon of the greenhouse effect.

Nitrogen (N) is an odorless, colorless, tasteless element that forms 78 percent of our atmosphere. It is an essential element for plant and animal life.

The air we breathe contains 1 percent rare gases, such as krypton (Kr).

TABLE 1-1 **COMPOSITION OF THE EARTH'S ATMOSPHERE**

Gas	PPM by Volume	Percentage
Nitrogen (N)	780,840	78
Oxygen (O)	209,460	21
Argon (Ar)	9,340	0.0934
Carbon dioxide (CO_2)	350	0.0035
Neon (Ne)	18.18	0.0002
Helium (He)	5.24	0.00005
Methane CH_4	2.00	0.00002
Krypton (Kr)	1.14	0.00001
Hydrogen (H)	0.50	0.000005
Nitrous oxide N_2O	0.50	0.000005
Ozone O_3	0.40	0.000004
Xenon (Xe)	0.09	0.0000009

Without its ability to retain this heat, the earth would be about 60°F (33°C) colder and could not support life as we know it. An increase of 25 percent in the concentration of carbon dioxide over the past century is one of the primary causes of global warming.

Ozone (O_3), another trace gas, occurs at concentrations of only about 0.4 ppmv (0.000004 percent), but it is also essential for absorbing UV radiation from the sun. Excessive UV radiation is very damaging to life on earth.

Ozone in the Atmosphere

Unlike other gases that are concentrated in the troposphere, about 90 percent of the ozone occurs in the stratosphere, from an altitude of about 9–22 miles (15–35 km). Even at its highest concentration, ozone does not exceed 10 ppmv—equivalent to one ozone molecule in every 100,000 molecules. For example, if all the ozone in the atmosphere were concentrated at sea level, it would form a layer less than 0.125 in. (3 mm) thick. There are about 3,000 million tons of ozone in the atmosphere, equivalent to about 1,600 lb. (726 kg) per person on earth. Compared with the total mass of the atmosphere, however, the amount of ozone is negligible.

Ultraviolet (UV) radiation consists of invisible rays from the sun that have damaging effects on the earth. Ultraviolet radiation causes sunburns.

Ozone is formed by the action of electrical discharges. For this reason, it is sometimes detected by odor near electrical equipment or just after a thunderstorm. More frequently, however, ozone is formed by the action of **ultraviolet (UV) radiation** on oxygen in the stratosphere. The atoms in the oxygen molecules split apart, and the separated atoms recombine with other oxygen molecules to form the triatomic ozone (O_3).

Because sunlight is essential for the formation of stratospheric ozone, it is formed mainly over the equatorial region, where solar radiation is highest. From there, it is distributed throughout the stratosphere by the slight global wind circulation. Stratospheric ozone levels vary throughout the world, being highest at the equator and lowest toward the poles.

Ultraviolet (UV) radiation is to be avoided whenever possible.

Absorption of Ultraviolet Radiation

Incoming radiation from the sun is of various wavelengths, ranging from UV to visible light to infrared. Ultraviolet radiation can cause sunburn, skin cancer, and damage to eyes, including cataracts. It can also cause premature aging and wrinkling of the skin. Ultraviolet radiation breaks down the food chain by destroying minute organisms such as plankton in the ocean, thereby depriving certain species of their natural food. Plant life and crops can also be devastated by excessive UV radiation.

Ozone depletion is the reduction of the ozone layer due to contamination, such as the release of chlorofluorocarbon (CFC) refrigerants into the atmosphere.

Fortunately, the damaging forms of UV radiation are absorbed by ozone in the atmosphere and do not reach the earth. The minute amount of atmospheric ozone is sufficient to absorb this radiation. The ozone layer, then, acts as a giant sunscreen or umbrella enveloping the earth, protecting life from the dangerous UV radiation. **Ozone depletion** results in weakening of this protective shield, however, and allows more UV radiation to strike the earth and living organisms, as shown in Figure 1-2.

Another consequence of the absorption of solar energy by ozone is that the upper stratosphere is somewhat warmer than at lower altitudes, which helps to regulate the earth's temperature. Stratospheric ozone absorbs about 3 percent of incoming solar radiation, thus serving as a heat sink. Loss of this ozone will decrease the temperature of the stratosphere, which will, in turn, affect the troposphere and consequently the weather and climate on the earth's surface.

Ozone is measured in Dobson Units (DU).

Measurement of Ozone

Although the ozone depletion problem was not to be officially addressed for another 50 years or so, the measurement of ozone in the atmosphere began in the 1920s.

Dobson Unit (DU) is a measure of ozone density level named after Gordon Dobson, a British meteorologist who was the inventor of the measuring device (called a spectrophotometer).

The standard term for measuring ozone levels is the **Dobson Unit (DU)**, named after the British meteorologist Gordon Dobson, who was the first to use a spectrophotometer. This device is used to determine the intensity of various wavelengths in a spectrum of light and can thereby measure the density of the ozone layer.

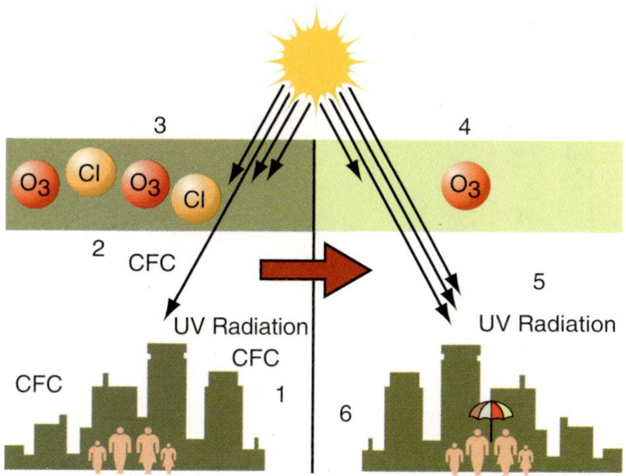

FIGURE 1-2 **Ozone Depletion Process.**

The Ozone Hole

The term *ozone hole* refers to the loss of the blocking effect of ozone against UV radiation. With the depletion of the ozone barrier, a "hole" has been created that allows a much greater amount of UV radiation to penetrate to the earth. Like an umbrella with holes in it that allows the rain through, holes in the ozone layer allow dangerous UV radiation to pass through to the earth's surface.

When ozone measurements were first taken at the British base at Halley Bay in the Antarctic, levels were found to fall drastically in September and October to 150 DU—half the normal level. This is also half the levels measured in the northern hemisphere in the spring. Levels again rose in November to the expected pattern, confirming that the atmosphere over Antarctica differs from elsewhere in the world.

At the center of the depleted area almost all of the ozone had disappeared. At Halley Bay between mid-August and early October, levels fall by 97 percent at a height of 10.25 miles (16.5 km). The hole occurs between 10.6 and 13.7 miles (17 and 22 km) above the earth.

Recent studies by NASA indicate that by the year 2030 climate change may surpass chlorofluorocarbons as the main cause of ozone depletion. Greenhouse gases like methane and carbon dioxide are changing the earth's climate. These effects could delay the recovery of the ozone layer even though most of the industrialized nations have signed international agreements to ban the production and use of CFCs. CFCs once used in the production of refrigerant and other commercial applications will last for decades in the upper stratosphere.

Ozone thinning can also occur when water vapor makes its way to the stratosphere. At these high altitudes, water vapor can be broken down into molecules that attack the ozone molecules. This can occur when methane emissions that migrate to the stratosphere are transformed into water vapor. The greenhouse effect also heats up the lower stratosphere where most of the ozone is concentrated. As it heats up, the chemical reactions that destroy ozone are also accelerated. Computer modeling indicated that the hole in the ozone layer was the largest ever observed by NASA on September 21–30, 2006. Another study indicates that, as the level of CFCs decline and their effect on the ozone layer is taken by itself, the ozone layer will make a full recovery by the year 2065. Unfortunately, the same study indicates that when the other variables such as the greenhouse effect and water vapor in the stratosphere are added back into the equation, the ozone layer will only make a slight improvement by 2040. There is still hope, but CFC reduction is only one piece of the puzzle.

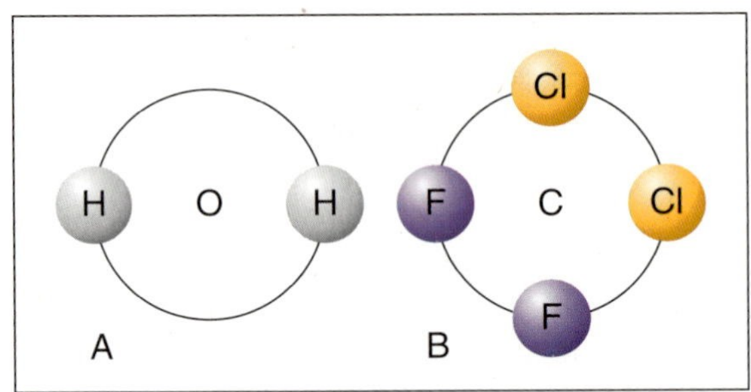

FIGURE 1-3 **Chemical structure of (A) Water; (B) CFC-12.**

How Ozone Is Being Destroyed

Chlorine (Cl) is a poisonous, greenish-yellow gas used in some refrigerants and known to be harmful to the ozone layer.

CFC stands for chlorofluorocarbon, a man-made compound used in refrigerants such as CFC-12 (R-12).

Toxicity refers to the toxic or poisonous quality of a substance.

The chlorine atoms in CFCs are hazardous to the ozone layer.

Ozone is both created and destroyed by the action of UV radiation on oxygen molecules. **Chlorine (Cl)** is the major gas causing the destruction of ozone and starts chain reactions in which a single molecule of chlorine can destroy 100,000 ozone molecules. Such reactions can continue for many years, even a century or more, until the chlorine drifts down into the troposphere or is chemically bound into another compound.

The main sources of chlorine are chlorofluorocarbons (CFCs). **CFCs** (Figure 1-3) are artificially-made chemicals first developed in 1928 and are comprised of:

- Chlorine (Cl)
- Fluorine (F)
- Carbon (C)
- (Often) hydrogen (H)

CFCs are very stable chemicals and are nonflammable, nonirritating, nonexplosive, noncorrosive, odorless, and relatively low in **toxicity**. They vaporize at low temperatures, which makes them very desirable for use as refrigerants in air conditioners and refrigerators. CFCs were also used as solvents for cleaning electronic components, for blowing bubbles in certain types of foam-blown plastics such as sponges and food packaging, in dry cleaning solvent, and as an aerosol propellant. Figure 1-4 depicts the consumption rates of the United States where CFCs were used to produce goods and services prior to the restriction in production under the Clean Air Act. As can be seen, refrigerant made up a large portion of the overall use of CFCs.

During the 1960s and 1970s, aerosol use was widespread due to the stable nature and nonflammability of CFCs. Peak worldwide use of CFCs in the 1970s was on the order of about 700,000 tons (635,460 metric tons) each year. The scheduled phaseout caused drastic reductions, however, and with the decline in the use of aerosols, nonaerosol use has risen.

The consumption of CFCs on a per capita basis in the United States is among the highest in the world (Figure 1-5), a reflection of our affluence and the popularity and use of air conditioners. Although industrialized nations are the major consumers of CFCs, developing nations such as China and India, because of their large populations, have an enormous potential to require CFCs for refrigerators and other uses.

In 1974, two chemists at the University of California, Mario Molina and Sherwood Rowland, asked the simple question: "What has happened to the millions of tons of CFCs released over the previous four decades?" The only "sink" they could suggest was the stratosphere. They hypothesized that the chemical stability of CFCs would enable them to reach the stratosphere, be broken apart by the intense UV radiation, and release chlorine by a process known as photolysis. The chlorine would then react with the ozone, causing its depletion.

It is not the CFCs, as such, that cause the destruction, but rather the chlorine released by the CFCs. The research of the British scientists at Halley Bay, together with international research programs in which samples of stratospheric air are obtained by high-altitude flights

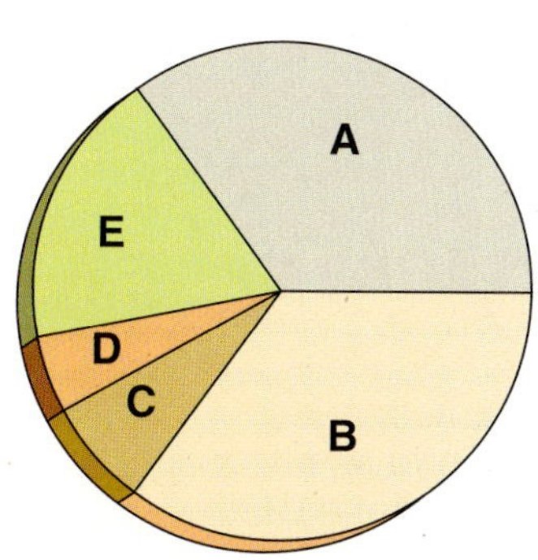

FIGURE 1-4 **United States' consumption rates prior to CFC use being restricted by the Clean Air Act: (A) A/C-Ref 35%, (B) Foam Blowing 35%, (C) Other 7%, (D) Sterilants 5%, (E) Solvents 18%. (Consumption rates are prior to CFC being restricted.)**

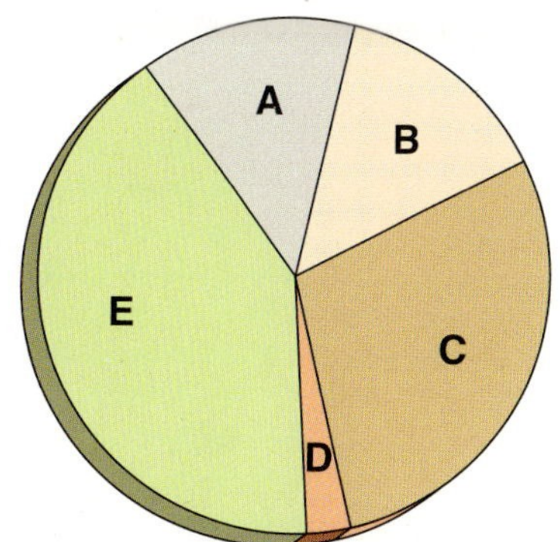

FIGURE 1-5 **Consumption of CFCs at the time of the signing of the Montreal Protocol, by country/ region: (A) The former Soviet Republics 14%, (B) Developing Nations 14%, (C) United States 29%, (D) China and India 2%, (E) Other Developed Nations 41%.**

over Antarctica, have proven the link between CFCs and ozone destruction. A further factor identified as contributing to the loss of ozone is the polar stratospheric clouds that form during the Antarctic winter in the very cold stratospheric air. These comprise tiny particles of frozen water vapor, which condense and form clouds in spring. The clouds act as reservoirs of frozen chlorine during winter until thawed in spring. At that time, the chlorine is released and begins to react with the ozone over the following five to six weeks, then the vortex breaks up and the stratosphere becomes less stable.

A chlorine atom reacts with an ozone molecule by splitting it apart and attaching itself to one of the oxygen atoms to form chlorine monoxide. A free oxygen atom splits the chlorine monoxide molecule to reform a molecule of oxygen (O2), and the chlorine atom is free to attack another ozone molecule (Figure 1-6).

Chlorine atoms split ozone molecules to form chlorine monoxide.

The CFCs take six to eight years to rise up through the atmosphere. Chlorine as used in swimming pools and bleach is unstable and breaks down rapidly without rising into the atmosphere. The concern is that the current hole and depletion that have resulted from CFCs released in early years will only worsen as their full effects are manifested over time in the stratosphere.

Effects of Loss of Ozone on Human Health

As we have seen, ozone protects life on earth from damaging UV radiation. It acts as a giant sunscreen absorbing the UV rays, preventing a certain percentage of them from reaching the earth. As already noted, loss of ozone will only allow more UV radiation to penetrate to the earth and adversely affect human health and the environment. The three areas of our bodies that are adversely affected are the skin, eyes, and the immune system.

Exposure of skin to UV radiation can initially result in sunburn and suntan. If the exposure continues over a long period, as with those who work outdoors, the skin protects itself from UV radiation by gradually thickening and darkening as a pigment called melanin is released in the skin. Continuous exposure of the skin to UV radiation results in its aging and wrinkling and increases the risk of skin cancer.

Excessive UV exposure to the eyes will increase the risk of cataracts, which cause cloudiness in the lens of the eye, limiting vision. Other eye problems such as retina damage, tumors on the cornea, and "snow blindness" may also be caused by exposure to increased levels of UV radiation.

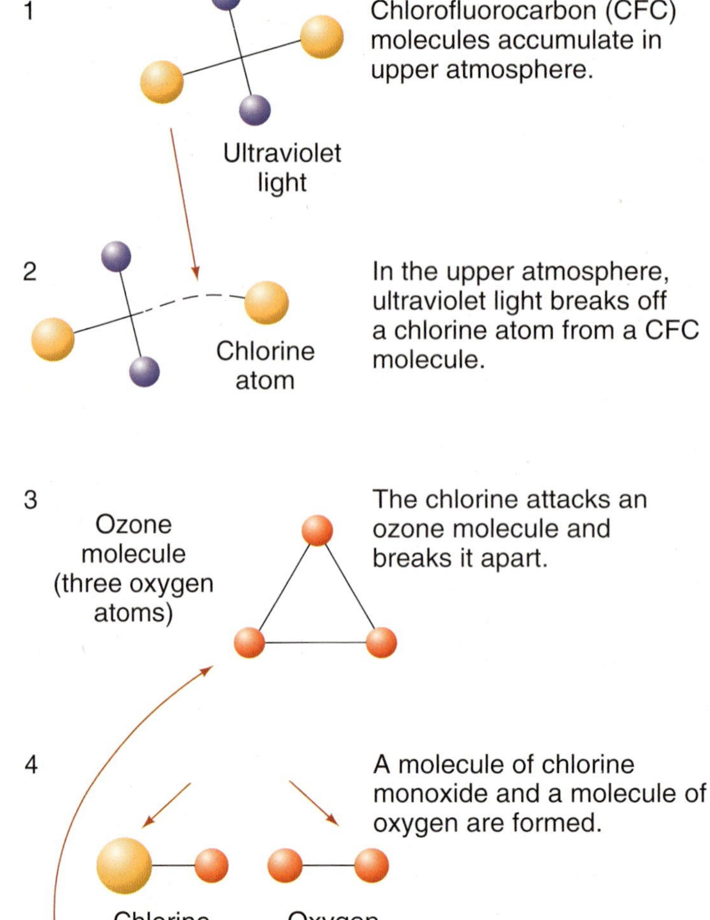

FIGURE 1-6 **How CFCs destroy the ozone.**

The body's immune system protects it from foreign chemicals and infections. If damaged, the immune system cannot protect the body and infections spread more rapidly. Ultraviolet radiation reduces the ability of the immune system to reject cancers, although not much is known about why this happens. Overall, increased UV radiation resulting from ozone depletion has the potential to significantly increase human skin cancers and cataracts and damage the human immune system. It also adversely affects marine and terrestrial plants and animals.

The extent of the damage will depend on the degree to which the earth's ozone layer is depleted. To date, it has been reduced by about 2.5 percent, and it remains to be seen whether the actions taken to control the release of ozone-depleting substances will be sufficient.

Climate Change and the Greenhouse Effect

The loss of ozone and the greenhouse effect are separate phenomena, although CFC's are a common agent in both. Greenhouse gases in effect work as a blanket, warming the lower atmosphere. The earth's atmosphere contains many chemical compounds that act as "greenhouse" gases, and these gases allow sun light to entire the earth's atmosphere freely. As this solar radiation passes through the atmosphere it then strikes and warms the earth. But some of this **infrared** radiation (heat) is re-emitted by the earth's surface and is absorbed and reflected by greenhouse gas molecules in the atmosphere, in all directions (Figure 1-7). The effect of this is to warm the earth's surface and lower atmosphere even further. It is similar to how a greenhouse is warmed by sunlight passing through the glass warming the interior and not allowing the heat to escape unless the vents are open. Think of the glass as the greenhouse gas molecules. Also, think of the interior of a car in the sun on a hot summer day with the windows rolled up. The interior will become much hotter than the outside temperature. Over time the amount of heat energy (infrared radiation) sent from the sun to heat the earth is radiated back into space, thus leaving the earth's temperature relatively constant ideally, if nature is left alone.

Infrared is the invisible light rays just beyond the red end of the visible spectrum and have a penetrating heating effect.

Many gases exhibit greenhouse gas properties, some occur naturally like water vapor, carbon dioxide (CO_2), methane (CH_4), and nitrous oxide. Others are man-made or increased by human activities. The **greenhouse effect** or **global warming** is the result of the release of increasing amounts of so-called "greenhouse gases" into the atmosphere, gases such as CO_2, CH_4, and manmade gases such as CFCs and HFCs. These greenhouse gases act as a blanket around the earth retaining heat. Without any greenhouse effect, the earth would be about 60°F (33°C) colder, too cold to support life as we know it.

Greenhouse effect is a term based on the fact that a greenhouse is warmed because glass allows the sun's radiant heat to enter, but prevents radiant heat from leaving. Likewise, global warming is caused by some gases in the atmosphere that act like greenhouse glass.

Greenhouse gas emissions have increased by about 25 percent over the last 150 years (Figure 1-8). This is also the time period of large scale industrialization in the United States and Europe. In addition, 75 percent of the carbon dioxide emissions produced in the last 20 years was from the burning of fossil fuels.

The earth has a natural processes by which concentrations of carbon dioxide in the atmosphere are regulated known as the "carbon cycle" (Figure 1-9). Through processes like plant photosynthesis carbon is moved from the atmosphere to the land and oceans of the earth. These natural processes are responsible for removing approximately 6.1 billion metric tons of man-made carbon dioxide emissions each year. But this leaves an additional 3.1 billion metric tons of man-made carbon dioxide emissions that are not removed by the carbon cycle and are added to the atmosphere each year. This results in an ever-increasing amount of greenhouse gases accumulating in our atmosphere that are not absorbed naturally.

Global warming is the gradual warming of the earth's atmosphere due to the greenhouse effect.

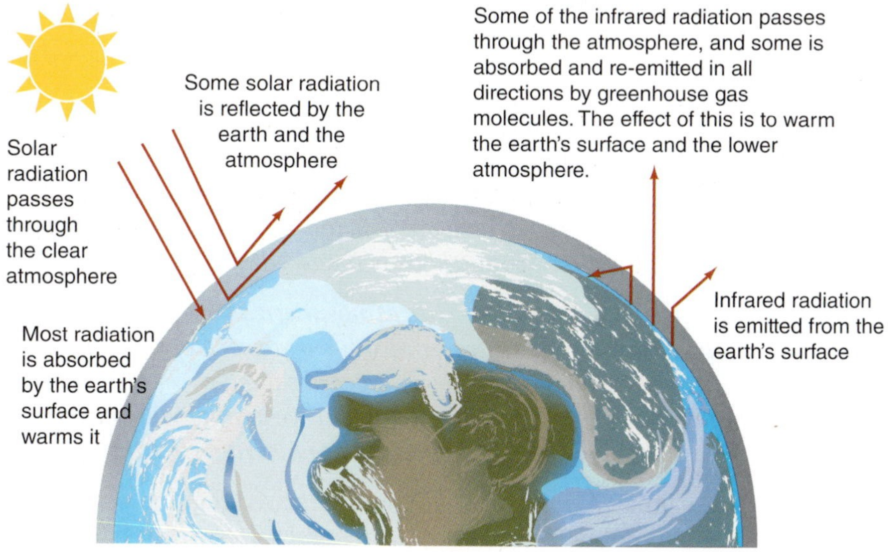

FIGURE 1-7 **The Greenhouse Effect**

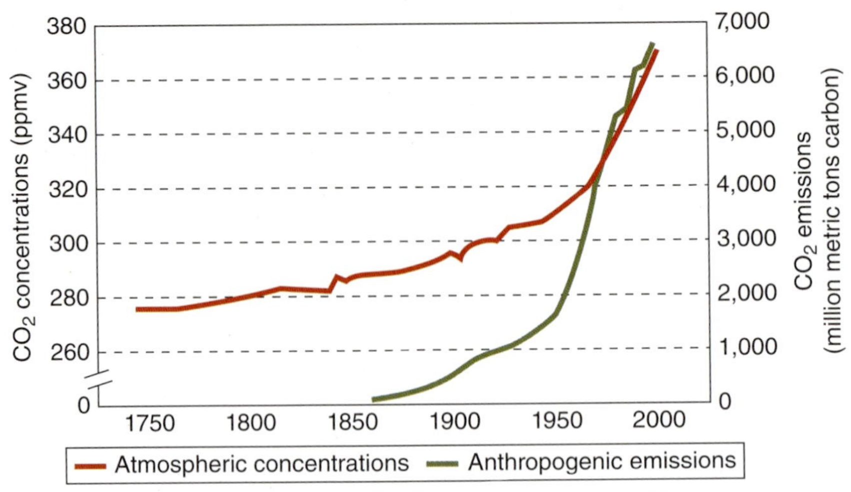

FIGURE 1-8 **Trends in Atmospheric Concentrations and Man-made (Anthropogenic) Emissions of Carbon Dioxide**

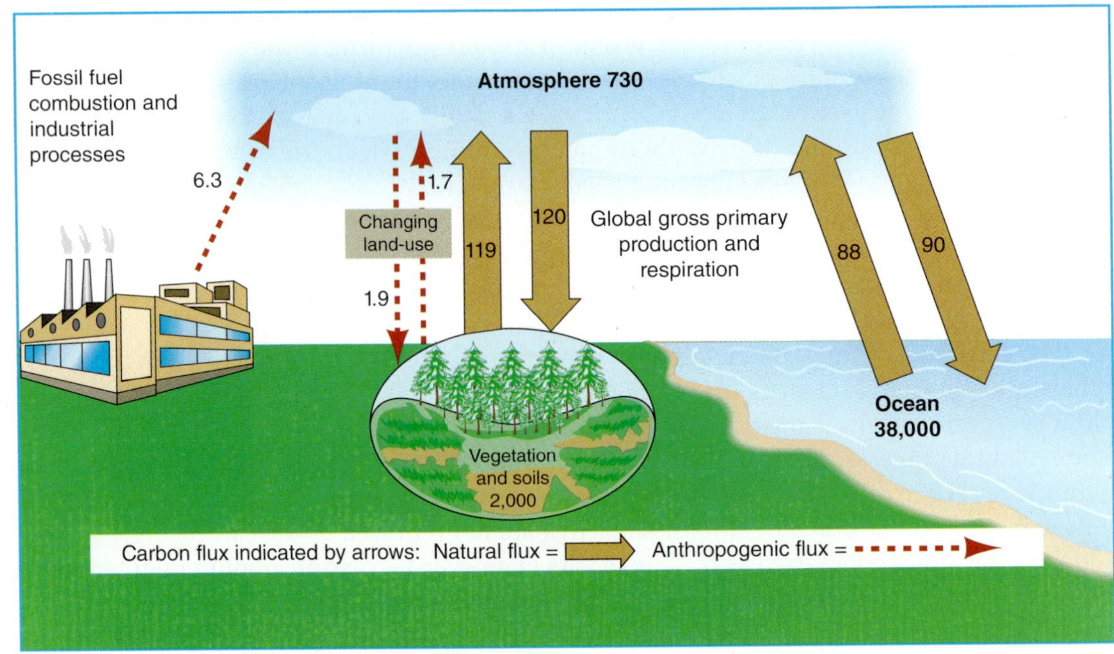

FIGURE 1-9 **The global carbon cycle in billions of metric tons.**

Because the earth's climate is naturally variable, it is difficult to determine the exact extent that human activity has had. Computer analysis has shown that increases in greenhouse gases have been correlated to increases in the earth's temperature. This analysis also indicates that rising temperature may produce changes in sea level, and weather commonly referred to as "climate change." A National Research Council study dated May 2001 stated,

Greenhouse gases are accumulating in Earth's atmosphere as a result of human activities, causing surface air temperatures and sub-surface ocean temperatures to rise. Temperatures are, in fact, rising The changes observed over the last several decades are likely mostly due to human activities, but we cannot rule out that some significant part of these changes is also a reflection of natural variability.

In 1992, the National Climate Data Center (NCDC) declared that the winter of 1991–1992 was the warmest U.S. winter in the 97 years that the federal government had kept a record of climatic conditions. The average temperature was 36.87°F (2.7°C). The previous high average

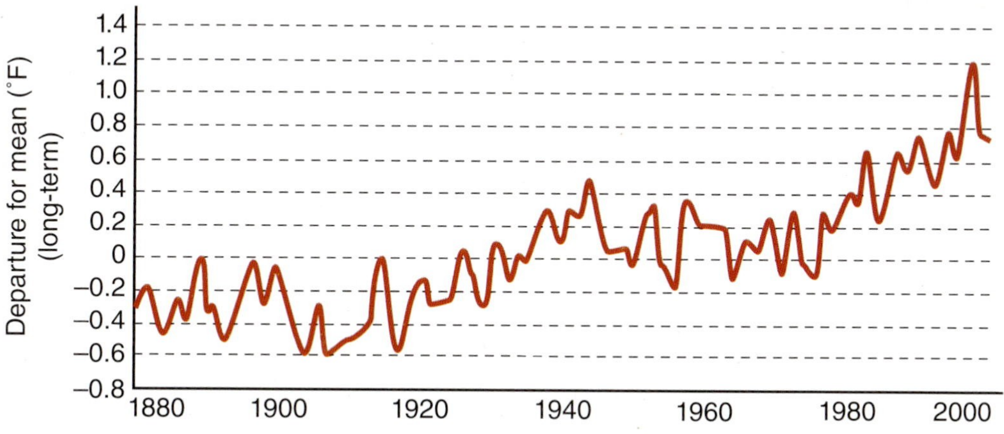

FIGURE 1-10 **U.S. National Climate Data Center 2011 showing an increase in global temperature.**

temperature of 36°F (2.22°C) was recorded in 1953–1954. Then in 1999, the NCDC declared that 1998 was the warmest year on record. In North America, seven of the eight warmest years have occurred since 2001 with the ten warmest years occurring since 1995 (Figure 1-10). This increase in global warming has also increased worldwide rainfall amounts by about 1 percent. Some scientists predict that global surface temperatures could rise 1-4.5 percent over the next 15 years and by 2–10 percent during this century. This could result in sea levels rising by as much as 2 feet along most of the U.S. coastline.

The U.S. is responsible for producing approximately 25 percent of the global CO_2 emissions by burning fossil fuels. Our economy is the largest in the world and 82 percent of our energy needs (i.e., electricity generation and transportation) are derived from the burning of petroleum and natural gas (Figure 1-11). Unfortunately both hybrid electric and battery electric vehicles in the United States still ultimately rely on fossil fuels for the majority of their energy. Man-made gases, which include hydro fluorocarbons (HFCs) used as refrigerants, represent two percent of total emissions. The good news is that the United States is predicted to lower its carbon intensity between 2001 and 2025 by 25 percent (Figure 1-12). The bad news is that worldwide CO_2 emissions levels are expected to increase by 1.9 percent annually over the same time period, primary due to increased levels of emissions by developing countries such as China and India. Developing countries CO_2 emissions levels are expected to increase by 2.7 percent annually between 2001 and 2025 causing industrialized countries gains to be negatively offset.

According to recent statistics automobiles in the United States have leaked approximately 51 thousand tons of R134a refrigerant gas into the atmosphere, which is equivalent to

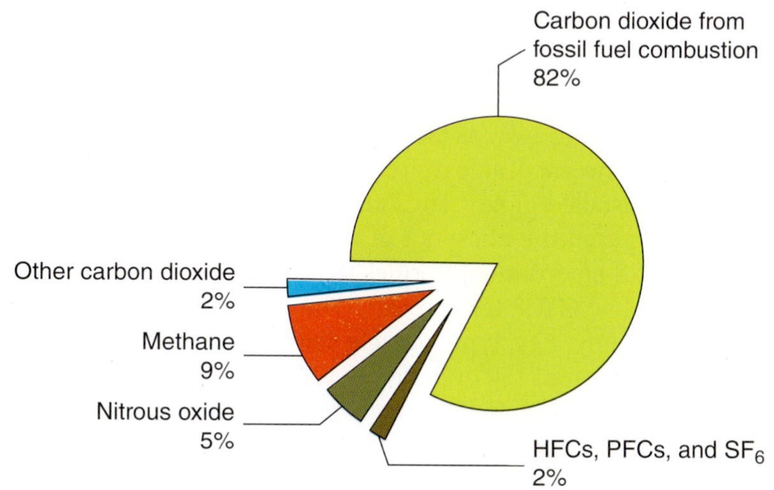

FIGURE 1-11 **United States greenhouse gas emissions by gas.**

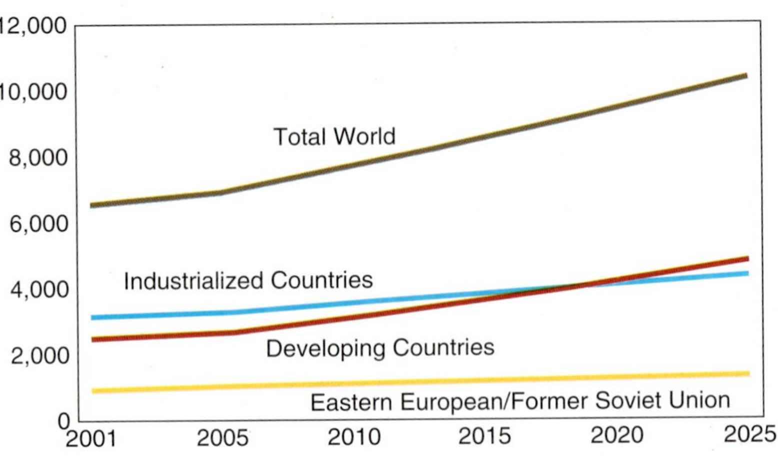

FIGURE 1-12 **Predicted world carbon dioxide emissions by region 2001–2025.**

72.8 million metric tons of greenhouse gases. Though this number seems large, it is a very small percentage when compared to the CO_2 emissions generated from the burning of fossil fuels as was noted in Figure 1-11.

Beginning on January 1, 2011, the European Union EU 2006/40/EC Act provisions went into effect, though the deadline was extended to January 1, 2013, due to lack of a readily available supply of a new refrigerant. This law was designed to phase out global warming refrigerants with complete phaseout of R134a in Europe by January 1, 2017, for new vehicles sold in EU member countries. The act requires all new automotive platform refrigerant systems to use a refrigerant with a **global warming potential (GWP)** that is not to exceed 150. The GWP number is an index number that is an estimate of how much a given mass of a gas will contribute to global warming compared to the same mass of carbon dioxide, where carbon dioxide is given the number 1.

Global warming potential (GWP Global warming potential is an index number which is an estimate of how much a given mass of a gas will contribute to global warming compared to the same mass of carbon dioxide, where carbon dioxide is given the number 1.

The current refrigerant R134a has a GWP number of 1430, meaning it has 1400 times the greenhouse effect of carbon dioxide. The refrigerant that has been chosen to replace R-134A in Europe is R-1234yf, which has a GWP of 4. The environmental life span of CFC refrigerants like R-12 was over 100 years, whereas HFC refrigerants like R-134A have an environmental life span of 10 years compared to R1234yf, which has a much shorter atmospheric life span of only 11 days. Currently the United States has no plan to regulate the use of R134a as a refrigerant gas but is incentivizing manufacturers to choose a lower GWP refrigerant. Some automotive manufacturers such as General Motors began to phase in the use of R-1234yf, which has a lower GWP number beginning in 2013 on some platforms, and other manufacturers began to follow, offering R1234yf on some new platforms. Although higher costs and lack of federal regulation may deter automotive manufacturers from adopting alternative refrigerants in the U.S. market quickly, it is predicted that by 2021, R-134A may be phased out in the United States.

Officially R-134a is not going away as far as the United States and the Environmental Protection Agency (EPA) are concerned, at least for the near term future, and R-134a will continue to be manufactured and installed in new and existing systems. Europe, on the other hand, as was stated earlier, has required the phaseout of R-134a by 2017. The European Union (EU) has enacted some very stiff environmental rules that are now driving the worldwide mobile HVAC industry to adopt low GWP refrigerants. The European Community mandated that beginning January 1, 2011, any "new type" vehicle platform must change over to a low-GWP refrigerant with a GWP number of 150 or below. R-134a's GWP number is 1430. This does not mean a full-scale changeover is required to a new refrigerant, only on completely "new type" platforms. Existing platforms were allowed to continue to be produced using R-134a until full phaseout in 2017 in the EU. So, until a manufacturer developed a new platform (type), they could continue installing the R-134a system in vehicles until 2017.

In the United States and Asia, manufacturers are not required to phase out R-134a, although some manufacturers that sell vehicles in the United States and elsewhere have chosen to switch to a refrigerant with a lower GWP number in an effort to receive EPA "carbon credits" on some of their platforms. A "carbon credit" is a credit that can be used to offset carbon dioxide (CO_2) production (emissions) above the EPA limit of 250 grams per mile (g/mi.) vehicle fleet average beginning in 2016, which also coincides with more stringent corporate average fuel economy (CAFÉ) regulations. Like many EPA regulations, the agency does not tell the industry how to achieve these limits, only what they are. The industry has many options on how they will meet these requirements. As an example, think about vehicle tailpipe emission requirements. The EPA did not tell the industry to abandon carburetors; the EPA set emission limits that required the industry to develop new technology in order to meet more stringent emission regulations. In the early stages, the industry developed electronic ignition systems and electronic feedback carburetors and later implemented more sophisticated fuel injection and computer-controlled subsystems. And today when you open the hood, it is almost unrecognizable from the engine compartments of the 1960s. This is the road we are taking once again, but this time the emission gas that is being regulated is CO_2, and if you remember from basic internal combustion engine theory, the higher the CO_2 levels produced after combustion, the more efficient the engine is running. But these CO_2 limits are not just related to tailpipe emissions, they apply to the entire vehicle. Now the industry must decide how they want to meet these new regulations. If a manufacturer decides to switch over to a low-GWP refrigerant, the EPA will issue carbon credits for a reduced carbon footprint. However, if R-134a systems are improved to provide greater fuel efficiency and reduce lifetime leakage, they may also receive carbon offset credits. Beginning in the 2009 model year, improvements in R-134a systems can receive carbon credits that can be carried forward to when the new regulations are implemented, which in turn reduces the urgency of needing to switch to a new refrigerant. The EPA formula is based in part on the SAE J2727 standard for calculating system improvements. As an example of carbon credits, an R-134a system with system improvements and an electric compressor can receive 9.5 g/mi. credit for cars and 11.7 g/mi. credit for trucks compared to the carbon credit received for switching over to R-1234yf of 13.8 g/mi. credit.

So the question at many automotive manufacturer boardrooms is, "Do we need to switch to a new refrigerant to receive carbon credits or can we improve our current R-134a system in order to meet new EPA requirements?" Like many questions, there will be more than one answer and not every manufacturer will make the same choices. But with a global marketplace and economies of scale, the long-term choices will probably be similar.

The Clean Air Act

The most significant legislation to affect the automotive air conditioning industry in the United States is the **Clean Air Act (CAA)**. The CAA was signed into law by U.S. President George H. W. Bush on November 15, 1990. Most of the rules and regulations of the CAA were a result of the recommendations made at the Montreal Protocol.

Clean Air Act (CAA) is a Title 6 Amendment signed into law in 1990 that established national policy relative to the reduction and elimination of ozone-depleting substances.

The Montreal Protocol and later amendments deal with the environmental problems and issues created by certain refrigerants depleting the ozone on an international level. The CAA deals with this problem on a national level. The Montreal Protocol is structured so that periodic meetings must take place in order to reassess the ozone problem. As new facts about the impact of refrigerants are brought to light, the protocol will be modified accordingly. The majority of protocol modifications will also result in the CAA being modified accordingly.

Language exists in the CAA stating that the Environmental Protection Agency (EPA) can accelerate schedules for the phaseout of refrigerants if it is deemed necessary and practical. The CAA also mandates that phaseout may be accelerated if required by the Montreal Protocol.

The CAA is somewhat more specific than the Montreal Protocol in addressing the ozone depletion problem. The Clean Air Act gives the EPA the authority to establish environmentally safe procedures with respect to the use and reuse of refrigerants. In addition, the EPA will establish standards for certifying those who service refrigeration equipment and for that service itself. These standards will be derived from the information furnished mainly by private sector organizations.

Stratospheric Ozone Protection—Title VI

Title VI of the CAA concerns stratospheric ozone protection. It establishes regulations for the production, use, and phaseout of CFCs, halons, and HCFCs. Other chemicals such as carbon tetrachloride (CCl_4), also covered by Title VI, are not covered in this text. Title VI divides the substances to be regulated into two classes: Class I and Class II.

The chemical that we are primarily concerned with in the automotive industry is CFC-12, a Class I refrigerant. Manufacture of this refrigerant ended in the United States on December 31, 1995.

Hydrofluorocarbons (HFCs) are an alternative to ozone-damaging CFCs in refrigeration systems. The refrigerant currently used in automotive air-conditioning systems is HFC-134a, better known as R-134a. HFCs have been designated greenhouse gases, and HFC-134a has an atmospheric lifetime of about 14 years.

Ozone Protection Regulations

For decades, R-12, or freon more properly known as CFC-12, was used as the refrigerant in motor vehicle air-conditioning systems. However, since the discovery that CFCs damage the ozone layer, the production of ozone-depleting substances has ended. To help ensure that existing CFC-12 is used and reused rather than being wasted and released to the atmosphere, the EPA has issued regulations under Section 609 of the CAA to require that automotive shop technicians use special machines to recover and recycle CFC-12.

On December 31, 1995, the production of CFC-12 in the United States essentially ceased. It is legal, however, to use existing stockpiles of CFC-12, and several companies have also developed several new substitutes. These substitute refrigerants have been reviewed by the EPA's Significant New Alternatives Policy (SNAP) program. It is also illegal to release these substitutes to the atmosphere. As of June 1, 1998, the EPA has allowed refrigerant blends used in motor vehicle air-conditioning systems to be recycled. The EPA stipulates that the equipment used must meet Underwriters Laboratories (UL) standards, and the refrigerant must be returned only to the vehicle from which it was removed.

The California Automotive Repair Bureau (CARB) passed a law that went into effect on January 19, 2001, requiring that every shop in the state that performs mobile air conditioning service have a minimum set of diagnostic equipment.

CFC-12 (R-12)

Those who wish to service or repair motor vehicle air conditioners (MVACs) using CFC-12 as a refrigerant must be trained and certified by an EPA-approved organization. The training program must include pertinent information on the proper care and use of equipment, the regulatory requirements, the importance of refrigerant recovery, and the environmental effects of ozone depletion. To be certified, a technician must pass a test designed to demonstrate his or her knowledge in all of these areas. The supply tank for R-12 is white in color, which applies to both disposable and reusable tanks.

HFC-134a (R-134a)

Any automotive technician who wishes to repair or service HFC-134a MVACs must also be trained and certified by an EPA-approved agency. If, however, a technician is already trained and certified to repair and service CFC-12 systems, he or she does not have to be recertified to service HFC-134a systems.

Characteristics of HFC134a (R-134a). The automotive industry chose R-134a as the replacement refrigerant for CFC12 (R-12). R-134a is an HFC and does not contribute to the depletion of the ozone layer. HFC134a is classified as a contributor to global warming and has a GWP of 1430 though, and as such, is regulated. HFC refrigerants replace the chlorine atom with hydrogen atoms. It is a single-composition refrigerant that changes state at a specified temperature and pressure and has similar performance and vapor pressure characteristics to that of R-12. The power consumption is slightly higher for R-134a, and it has a slightly lower refrigeration capacity of between 3 and 5 percent compared to R-12. The properties of R-134a also require the use of different refrigerant oils to provide proper compressor lubrication, which are not compatible with R-12. Components have been redesigned for the different characteristics of R-134a. The supply tank for R-134a is sky blue in color, which applies to both disposable and reusable tanks. Vehicles with R-134a have unique low- and high-side quick-connect coupler service fittings to avoid system contamination by another refrigerant.

Refrigerant HFO-1234yf (R-1234yf)

The refrigerant that is the preferred replacement for R-134a today is R-1234yf, pronounced R twelve thirty four yf. The new refrigerant was developed by Honeywell and DuPont and is classified as HFO-1234yf (R-1234yf); HFO stands for hydrofluoro olefin and has a chemical structure of CF3CF = CH2 2,3,3,3-tetrafluoropropene. The refrigerant has been determined to be mildly flammable gas. R-1234yf is an environmentally friendly refrigerant that has no ozone depletion potential; a global warming potential (GWP) of 4, which is 99.7% lower than R-134a; and has a vapor pressure of 583 kPa absolute at 20°C and a boiling point of −29.2°C. This new refrigerant received final EPA approval under the Significant New Alternative Program (SNAP) for use in mobile air-conditioning systems (MAC) in early 2011. But the changeover to this new refrigerant may be very slow. General Motors began rolling out a few platforms equipped with R-1234yf in the United States on a very limited basis beginning in 2013. Vehicles with R-1234yf have unique low- and high-side service fittings to avoid system contamination by another refrigerant. The fittings are similar to R-134a fittings, but smaller.

Honeywell developed the commercially viable HFO-1234yf as a low-cost, low-GWP drop-in refrigerant replacement for R134a refrigerant. This new refrigerant is expected to significantly reduce the global warming footprint associated with air-conditioning systems. The GWP of R-1234yf is 4 compared to R-134a, which has a GWP of 1430. Many environmental groups still favor other refrigerant options such as CO_2 (R744), which has a GWP of 1, but for now the higher production costs of these systems has stalled development and production for automotive applications. R-1234yf has an atmospheric lifetime of 11 days and the atmospheric breakdown products are the same as R-134a, which is trifluoroacetic acid (TFA) and does not pose a threat to the environment, based on industry evaluations that took place in the 1990s. In fact, TFA is found in large amounts in the oceans of the world and it has been suggested that TFA is a natural component of saltwater.

Since it does not appear that the United States is going to regulate R-134a out of existence, manufacturers do not have to stop using R-134a in new vehicle platforms like they did with R-12. Because R-134a will still be produced and available, there may not be widespread acceptance of R-1234yf. In addition, it does not look like retrofitting from R-134a to R-1234yf will be approved or allowed by the EPA. In the end, what may keep the U.S. automotive industry from changing over to R-1234yf may be economic—higher product costs and a single supplier. Honeywell-DuPont is retaining sole patent and production rights and ultimately that may be too large an issue for the automotive industry to overlook. The supply tank for R-134a is white with a red band to denote flammability, which applies to both disposable and reusable tanks.

Blend Refrigerants

Automotive technicians who service or repair MVACs that use a blend refrigerant must be trained and certified by an EPA-approved agency. However, a technician that is already trained and certified to handle CFC-12 or HFC-134a does not have to be recertified to handle a blend refrigerant.

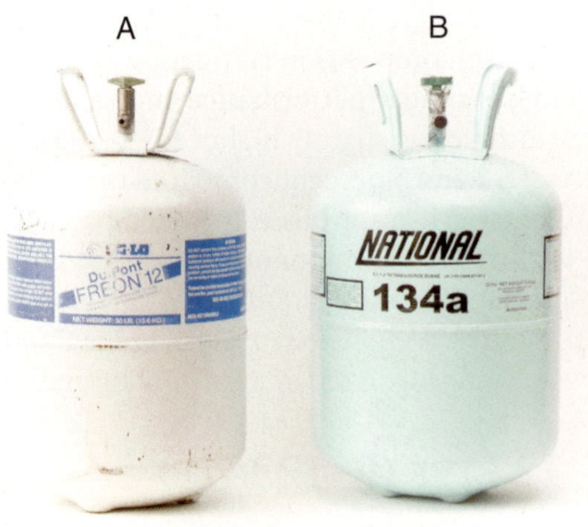

FIGURE 1-13 **Refrigerant cylinders are designated for a particular type refrigerant; (A) pounds (12.08 kg) cylinder R-12; and (B) 30 pounds (12.08 kg) cylinder R-134a.**

Refrigerant Cylinders

Refrigerant cylinders (Figure 1-13) are designed and constructed for definite maximum pressures and for definite quantities of refrigerant that are based on specified maximum temperatures, usually 130°F (54°C). The color of the cylinder indicates the type of refrigerant gas that it contains. A white cylinder indicates R-12, sky blue identifies R-134a, and a white cylinder with a red band around the top indicates R-1234yf refrigerant (Figure 1-14). If the cylinders are subjected to temperatures above those specified, the liquid expands to entirely fill the cylinder; extremely high hydrostatic pressures develop, and the cylinder may burst.

If the cylinders are filled with a greater amount of refrigerant than specified, hydrostatic pressures may develop at ordinary room temperatures, and the cylinder may burst.

Flying pieces of the cylinder may travel at bullet velocity, or in the case of small cylinders or light containers, the container itself may travel like a rocket at projectile speed. Sometimes

FIGURE 1-14 **The R1234yf refrigerant tank is white with and identifying red band since it is a mildly flammable refrigerant.**

of equal or greater danger, the refrigerant itself may burst from the cylinder; technicians have been blinded or suffered freezing injuries from being sprayed with refrigerant.

Many factory cylinders are equipped only with fusible plugs, which offer no protection against overfilling; nor do they offer adequate protection on most cylinders against excessive temperatures. Fusible plugs melt at about 160°F (71°C) and most cylinders are liquid-full at 130°F (54°C), so the fusible plug gives no protection between 130°F and 160°F.

All refrigerant cylinders should be protected by means of pressure-activated relief valves, especially service cylinders, because they are more often abused by overfilling than are factory-filled cylinders. Small combination service valves, with built-in pressure-relief safety valves, were developed by valve manufacturers in cooperation with the Refrigeration Service Engineers Society (RSES) safety and educational department and are available at moderate prices from refrigeration supply wholesalers. Every service cylinder should be equipped with one of the safety valves.

Even with the best of care, cylinders become rusted, damaged, or otherwise weakened after several years of use and should be retested by a hydraulic test approved by the Interstate Commerce Commission (ICC). The ICC requires a retest of all service cylinders and most factory cylinders once every 5 years (Figure 1-15). Do not use cylinders beyond the five-year period without having them retested. It may save your life or prevent serious injury. Your refrigerant supplier should be able to suggest a laboratory for retesting refrigerant cylinders. If not, consult the Yellow Pages of your local telephone directory under Hydrostatic Testing for the nearest facility.

It is a violation of federal law to reuse a disposable refrigerant cylinder.

Corrosion may occur inside a refrigerating system and may also affect external parts. It is commonly due to rusting in damp atmospheres or in areas in which there is a great deal of acidity in the air. As a rule, the parts most likely to be seriously affected are bolts, screws, nuts and rivets, or comparatively thin-walled vessels or tubes, especially those made of iron or steel. Particularly in damp or acid atmospheres, these parts should be inspected occasionally and repaired or replaced if necessary. Keeping parts subject to corrosion properly painted will greatly extend their useful life and lessen the possibility of their suddenly giving way and

FIGURE 1-15 **Service cylinders must be reinspected every 5 years.**

causing an accident. Using protective paints and greases is an inexpensive preventative maintenance that guards against the dangerous and costly breakage of corroded and weakened parts. Water supply lines, gate valves, fittings, and automatic pressure and control valves should be inspected periodically; badly corroded or weakened parts should be replaced.

Technician Certification

Automotive technicians who wish to service mobile air-conditioning systems and refrigeration equipment must be certified by an appropriate testing agency approved by the EPA and receive an EPA section 609 Technician Training Certification. This includes all who work with R-12, R-134a, and R1234yf, or any of the blend refrigerants available and approved for automotive use.

Certificating Agency

If there is any doubt about the integrity of the agency offering training and certification, check with the EPA. The EPA maintains an updated list of all of the approved agencies. The EPA has little mercy for anyone issuing bogus technician certificates to those who have not taken the required exam and can impose prison sentences and fines.

Injuries as a Result of High Pressure

A basic characteristic of a mechanical refrigeration system is the use of a fluid, both gas and liquid, that is at pressures above atmospheric pressure. The fluid must therefore be maintained and transmitted in tanks, pipes, and other vessels that do not allow the fluids to leak and that are strong enough to withstand maximum pressures without splitting or bursting under extreme conditions of use.

It is also a basic characteristic of mechanical refrigeration that these pressures change with fluctuations in temperature or are increased by compressors or pumps. We must, therefore, guard against extra pressures caused by compressors and pumps, as well as the pressures existing in the system because of variations of temperature.

Pressure relief valves are provided to release excess pressure.

Pressure-containing vessels (Figure 1-16) and tubes are designed and constructed to withstand normal pressures caused by normal temperatures, by normal degrees of compression, and normal filling of the vessels. If the vessel or tube is overheated, if an attempt is made to

FIGURE 1-16 **An accumulator is a pressure vessel designed to withstand the normal pressures of an air-conditioning system.**

put too much fluid in it, or if the fluid is compressed above the pressure for which the vessel is designed or constructed, the vessel will "give" somewhat until it reaches its limit of elasticity; then the vessel will burst, often with explosive violence.

Overpressure may cause large parts to be blown out, such as the welded ends of dryers or of receivers. Overpressure may also drive plugs or other small parts out with projectile speed and force.

Explosions or bursting of vessels from overpressure sometimes start with overfilling the vessels with liquids at lower temperature. Then when the completely filled vessel warms up, the liquid expands and exerts tremendous pressure, known as **hydrostatic pressure**. In other words, hydrostatic pressure occurs when a cylinder is full of a liquid and there is no room for expansion as it heats up. As a result, something has to give, and it is the weakest part that gives. Oftentimes it is a hose or hose connection. Sometimes it is the compressor head gasket or a head, which may be most dangerous.

Hydrostatic pressure is the pressure exerted by a fluid.

Special Safety Precautions

Because it is very important that the student be aware of the hazards involved in the use of any refrigerant, the following safety procedures must be observed at all times. Recall that refrigerant is:

- Odorless
- Undetectable in small quantities
- Colorless
- Nonstaining

However, refrigerant is dangerous because of the damage it can cause if allowed to strike the human eye or come into contact with the skin. Suitable eye protection must be worn to protect the eyes from splashing refrigerant (Figure 1-17). If refrigerant does enter the eye, freezing of the eye can occur with resultant blindness. The following procedure is suggested if refrigerant enters the eye(s):

1. Do not rub the eye.
2. Splash large quantities of cool (not hot) water into the eye to raise the temperature.
3. Tape a sterile eye patch over the eye to prevent dirt from entering. Do not use salves or ointments.
4. Go immediately to a doctor or hospital for professional care.

Do not attempt self-treatment for injury.

If liquid refrigerant strikes the skin, frostbite can occur. The same procedure outlined for emergency eye care can be used to combat the effects of refrigerant contact with the skin. Refrigerant in the air is harmless unless it is released in a confined space. In a refrigerated trailer with an evaporator leak, it is possible that the refrigerant could displace the oxygen, resulting in an oxygen-depleted environment. Always exercise caution when a refrigerant leak is suspected in a sealed or confined space and allow for adequate ventilation before performing

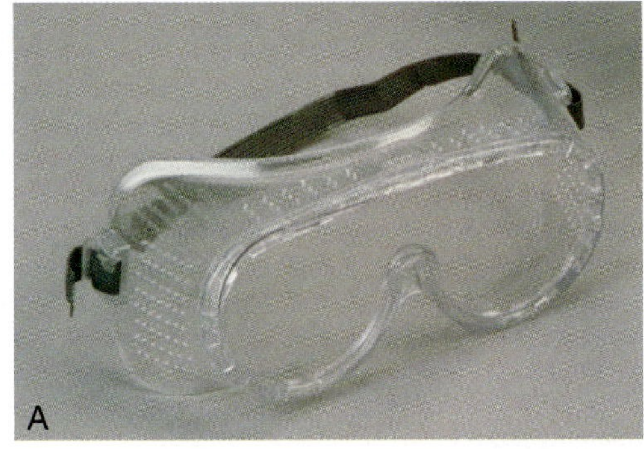

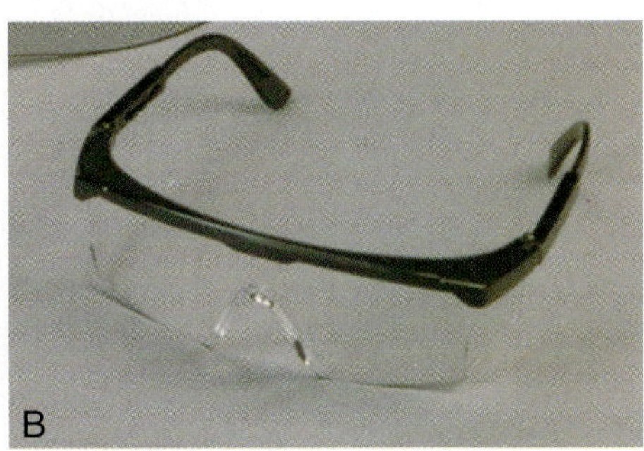

FIGURE 1-17 **Wear suitable eye protection: (A) monogoggle; (B) safety glasses.**

services. Under these conditions, refrigerant displaces oxygen in the air and may cause drowsiness or unconsciousness—even death. However, the automobile owner and the service technician need not be overly concerned about the safety of the automotive air-conditioning system under normal conditions. The small capacity of the system compared to the large area of the car interior or work area minimizes the concentration of any contamination.

Refrigerant must not, however, be allowed to come into contact with an open flame or a very hot metal. Tests made by UL in 1933, shortly after the development of CFC-12, indicated that it produced a highly toxic gas known as phosgene during decomposition. Tests in recent years, however, prove that phosgene gas is not a product of decomposition in this manner. Decomposition does, however, result in the formation of carbonyl fluoride (COF_2) and carbonyl chlorofluoride (COClF) with small amounts of free chlorine ($C1_2$).

Though 20–50 times less toxic than phosgene, as discussed earlier, the decomposed gases of CFC-12 must be avoided. At high concentrations, the lack of oxygen, which results in asphyxiation, is the real hazard. A primary rule, then, is to avoid breathing these or any other fumes. The human body requires oxygen in the quantity found in noncontaminated air. Diluting air with any foreign gas can reduce the available oxygen to a level that may be harmful or, in some cases, fatal.

The following rules must always be observed when handling refrigerants:

1. Never heat a refrigerant cylinder above 125°F (51.7°C) or allow it to reach this temperature. Above 130°F (54.44°C), expanding liquid refrigerant completely fills the container, and hydrostatic pressure builds up rapidly with each degree of temperature rise.
2. Never apply a direct flame to a refrigerant cylinder or container. Never place an electrical resistance heater near or in direct contact with a container of refrigerant.
3. Do not abuse a refrigerant cylinder or container. To avoid damage, use an approved valve wrench for opening and closing the valves. Secure all cylinders in an upright position for storing and withdrawing refrigerant. Carefully invert a refrigerant cylinder to dispense liquid refrigerant (first ensuring that the compressor is not running). Recovery cylinders are not to be inverted; use the liquid valve for dispensing liquid refrigerant (again ensuring that the compressor is not running).
4. Do not handle refrigerant without suitable eye protection.
5. Do not discharge (vent) refrigerant into the atmosphere. Remove refrigerant from a system using approved recovery equipment only.
6. Use only Department of Transportation (DOT) approved refrigerant recovery cylinders (Figure 1-18). Do not fill recovery cylinders beyond 80 percent of their rated capacity.

Do not refill disposable cylinders.

FIGURE 1-18 **Use only DOT-approved recovery cylinders.**

7. Do not mix refrigerants. Cross-contaminated refrigerants must be destroyed or separated by an approved reclamation center.
8. For an automotive air-conditioning system, do not introduce anything but refrigerant acceptable under EPA's SNAP program into the system.
9. Use only lubricant recommended for the refrigerant type. Properly identify, by label and fittings, refrigerant used.
10. Keep refrigerant containers out of direct sunlight.
11. Always work in a well-ventilated area. DO NOT work in a confined area.

ANTIFREEZE/COOLANT

There are four key areas of engine protection. These are:

- Freeze protection
- Boil-over protection
- Corrosion prevention
- Adequate heat transfer

Ethylene Glycol-Based Antifreeze

Ethylene glycol (EG) is the main ingredient of all major antifreeze brands and has long been known to be poisonous. When ingested, EG converts to oxalic acid, which damages the kidneys and may cause kidney failure and death.

Just 2 ounces of undiluted EG antifreeze (Figure 1-19) can kill a dog; 1 teaspoon can be lethal to a cat; and 2 tablespoons can be hazardous to children.

Data compiled by the American Association of Poison Control Centers show that about 3,400 poisonings related to EG occur annually. About 20 percent of these incidents are reported among children under 6 years of age.

Propylene Glycol-Based Antifreeze

A "new" antifreeze, formulated with propylene glycol (PG), is less toxic than EG antifreeze. Therefore, PG antifreeze (Figure 1-20) is much safer for children and animals. Actually, PG is used in specific amounts in the formulation of many consumer products. These products include, but are not limited to, cosmetics, pet food, and certain over-the-counter medications. Nonetheless, PG-based antifreeze should be considered toxic and handled as a hazardous substance.

FIGURE 1-19 **Ethylene glycol antifreeze.**

FIGURE 1-20 **Propylene glycol antifreeze.**

In areas where recycling antifreeze is required, one may locate a facility through a local automotive parts house or in the telephone book's Yellow Pages under *Recycling Centers* or *Hazardous Materials and Waste Contractors.*

Mixing EG and PG

It should be noted that EG-based antifreeze should not be mixed with PG-based antifreeze. Most antifreeze manufacturers caution against mixing the various types and suggest recovery and storage of the different types in separate containers. Also, most vehicle manufacturers require the same type antifreeze be used for top-off or refill that was originally installed in the factory fill to avoid warranty problems. It should be noted that there may be different formulations of antifreeze specified by some vehicle manufacturers. Always refer to manufacturer's specifications before changing or adding antifreeze to a cooling system.

Vehicle Engine Protection

Either EG- or PG-based antifreeze offers excellent protection for vehicle engines against corrosion, freezing, and overheating. A 50/50 blend of ethylene glycol antifreeze and water has a freezing point of −34°F (−36.7°C). If a lower temperature protection is required, it can be attained by increasing the concentration of antifreeze. A 60/40 blend, for example, gives antifreeze protection to −54°F (−47.8°C). It also helps to prevent corrosion in all metals used in automotive cooling systems, including aluminum, brass, copper, cast iron, steel, and the elements contained in solder.

Disposal

Used coolant must be properly disposed of in compliance with local rules and regulations. In areas where recycling is available, both used EG and PG coolants should be offered to recyclers for recycling and reuse.

Hazardous Materials

Refrigerants, refrigeration lubricants, solvents, and other chemicals used in an automotive repair facility may be considered hazardous materials and will include warning and caution labels that should be read and understood by everyone who uses them.

All hazardous materials should be properly labeled, indicating what health, fire, or reactive hazard they pose and what protective equipment is necessary when handling each chemical. The manufacturer of the hazardous material must also provide all warnings and precautionary information that must be read and understood by all users before the material is used. One should pay particular attention to the label information. Using the product according to label directions helps to ensure proper and safe methods, thereby preventing a hazardous condition.

A list of all hazardous materials used in the shop should be posted for all employees to see. Shops must also maintain documented records of the hazardous chemicals in the workplace, training programs, accidents, and spill incidents.

Material Safety Data Sheet

Every employee in a shop is protected by "right-to-know" laws concerning hazardous materials and wastes. The general intent of the law is to ensure that the employer provide a safe work environment. All employees must be trained about their rights under the legislation, the nature of the hazardous chemicals in their workplace, the labeling of chemicals, and the information about each chemical listed and described on Material Safety Data Sheets (MSDS). These sheets (Figure 1-21) are available from the manufacturers and suppliers of the chemicals. They detail the chemical composition and precautionary information for all products that can pose health or safety hazards.

FIGURE 1-21 **Typical Material Safety Data Sheet (MSDS) manual.**

Employees must be familiar with the contents of the MSDS that contain information relative to the intended purposes of the substance, the recommended protective equipment, accident and spill procedures, and any other information regarding safe handling. Training must be provided by the employer annually, and new employees must be trained as a part of their job orientation. When handling any hazardous material, always wear the appropriate safety protection. Always follow the correct procedures while using the material, and be familiar with the information given in the MSDS for that material.

Hazardous Waste

Waste is considered hazardous if it is on the EPA list of known harmful materials. Those materials generally have one or more of the following characteristics:

- Ignitable
- Corrosive
- Reactive
- Toxic

Many service procedures also generate products that may be considered hazardous wastes. Contaminated refrigerants or antifreezes are typical examples of hazardous waste.

Safety Precautions

The following safety precautions in working with hazardous materials should always be observed:

- Do not overfill refrigerant cylinders.
- Do not allow pressure-containing vessels to become overheated.
- Do not put a flame on a refrigerant cylinder, accumulator, receiver, or any other vessel that may contain refrigerant.
- Do not steam clean any vessels that may contain refrigerant.
- Do not change or add refrigerant to any system without first determining system compatibility.
- Always connect both low- and high-pressure gauges before servicing a system. Observe these gauges frequently.
- Before loosening bolts or screws, see that the pressure in the part has been relieved. Gaskets may hold the pressure temporarily, then release suddenly, throwing a full charge of refrigerant in the technician's face.

It is a violation of federal law to intentionally vent refrigerant to the atmosphere.

- Use pressure relief valves on all refrigerant cylinders and other vessels that may be subject to excessive pressures.
- Do not allow a compressor to pump liquid or "slug oil."
- Wear suitable protective gear when handling any materials that may be considered toxic.
- Keep your mind on what you are doing.
- Be vigilant.
- If you are tired, take a break.
- Read and heed all caution labels. Those warning of high pressures, as in the antilock brake systems and the dangers of unexpected air bag deployment, are most important.
- Be aware of under-hood hazards and avoid their danger.

AUTHOR'S NOTE: When you first begin a new job in an automotive shop, you will have many concerns on your first day. Pay particular attention to the location of all exits from the building, fire extinguishers, eye wash stations, the emergency shower, and where the MSDS data book is located in case of emergency. An emergency situation is not the time to try to locate safety equipment or exits.

BREATHING TOXIC GASES

Literally the word *toxic* means "poisonous," so "toxicity" is the condition of being "poisonous." In refrigeration terms, "toxic" is more frequently used with gases that we may breathe and that poison us by being taken into our blood by means of the lungs.

Refrigerants vary a great deal in their degrees of toxicity. Some refrigerants, such as ammonia (NH_3), are so highly toxic that it is dangerous, as well as unpleasant, to breathe air that has only a few parts per million of these gases. Others, such as R-12, may be breathed in large percentages with air without noticeably harmful effects.

It must be remembered, however, that the gas we as humans are suited to breathe is air, which is approximately 21 percent oxygen, 78 percent nitrogen, and 1 percent other inert gases. Any other gas, especially in large concentrations, may be harmful or fatal. Harmful effects depend upon:

- The nature of the gas itself,
- Its concentration in air, and
- How long a time it is breathed.

Decomposition of Gases

Some gases that may have a high safety rating or moderate safety ratings in their natural state become highly toxic if they are exposed to flames or hot surfaces. The heat "decomposes" these relatively safe gases and causes them to form other gases that are very toxic.

Halogen refers to any of the five chemical elements that may be found in some refrigerants: astatine (At), bromine (Br), chlorine (Cl), fluorine (F), and iodine (I).

The refrigerants thus decomposed are those that contain one or more of the **halogens**, a group of elements that includes chlorine (Cl), fluorine (F), bromine (Br), and iodine (I). Any of the refrigerants that have the symbols "Cl" or "F" in their chemical structure may be subject to hazardous decomposition.

Refrigerants should not be allowed to come into contact with an open flame or a very hot metal. Until recently, it was believed that fluorocarbon refrigerants, such as R-12, produce phosgene gas when exposed to hot metal or an open flame. The original tests, made by UL shortly after the development of R-12, indicated that it produced this highly toxic gas during decomposition. Recent tests, however, have shown that phosgene gas is not produced in this manner.

According to a technical specialist for SUVA® Refrigerants at DuPont Chemicals, which was one of the major manufacturers of R-12, commonly known as Freon®, the only products of decomposition of R-12 when in contact with an open flame or glowing metal surface, are hydrofluoric and hydrochloric acids.

Though as much as 50 times less toxic than phosgene gas, the decomposed gases of R-12 must be avoided. At high concentrations, lack of oxygen, which results in asphyxiation, is the real hazard. Avoid breathing these or any other fumes. The human body requires oxygen in the quantity found in noncontaminated air. Diluting air with any foreign gas can reduce the available oxygen to a level that may be harmful or, in some cases, fatal.

These gases of decomposition may not noticeably affect the person breathing them for several hours, so you should vacate the area contaminated by them as soon as you detect them by smell. Also, beware of the gases from burning plastics; one of these is the extremely dangerous phosgene ($COC1_2$).

Precautions. If the nature of the various refrigerants and other gases and fumes are understood and if reasonable care is exercised, a refrigeration service technician need have no fear of possible toxic hazards from refrigerants.

- One must use care, however, and in particular observe the following:
- Do not breathe any gas any more than is absolutely necessary. None of them is harmless under all conditions.
- Do not ignore the possible danger of a gas just because it has very little odor. The odor of a gas is no indication of its toxicity. Do not discharge any gas into any unventilated area.
- Do not discharge any of the hydrocarbon gases into a room in which there is a fire, flame, or electric heating element.
- Do not hesitate to use a gas mask if it is necessary to enter a room that you know or suspect has any of the toxic gases in it.
- Do not leave leaks in refrigerating equipment that may fill the room with gas and pose a danger to someone.
- Do not run an automobile engine in a closed garage; do not sit in a closed car with the engine running.
- Do not allow liquids to boil over on a gas stove; they may put out the flame, but the gas continues to escape.
- Do not use questionable tubing, flexible hose, or connectors.
- Do not work in an unventilated room with a heater having an open flame.
- Do not vent refrigerant. The EPA requires that all refrigerant be recovered.
- Do not breathe fumes from acids, caustics, carbontetrachloride ($CC1_4$), benzol, ketone, xylene, or other toxic cleaning materials. Always keep rooms well ventilated when using cleaning solvents.
- Do not breathe fumes from broken fluorescent lamps; they are poisonous.

The Industry

Automobile air conditioning, once considered a luxury, has become a necessity. Millions enjoy the benefits it produces. Business people are able to drive to appointments in comfort and arrive fresh and alert. People with allergies are able to travel without the fear of coming into contact with excessive dust and airborne pollen and pollution. Because of the extensive use of the automobile, automobile air conditioning is playing an important role in promoting the comfort, health, and safety of travelers throughout the world.

It is easy to understand how automotive air conditioning has become the industry's most sought-after product. In the South and Southwest, many specialty auto repair shops base their entire trade on selling, installing, and servicing automotive air conditioners throughout the year.

FIGURE 1-22 **A typical ASE certificate.**

ASE Certification

The National Institute for Automotive Service Excellence (ASE) has established a certification program for the automotive heating and air conditioning technician (Figure 1-22). This is one of the eight automotive certification areas that lead to certification as a Master Auto Technician (Figure 1-23). ASE also offers other certification programs in other areas, such as heavy-duty truck, collision repair, school bus, engine machine shop technician, parts specialist, alternate fuels, and advanced engine performance.

FIGURE 1-23 **A certified Master Auto Technician.**

ASE's voluntary certification system combines on-the-job experience and tests to confirm that technicians have the necessary skills to work on today's vehicles. The ASE Master Auto Technician status certification is awarded when a technician passes all eight tests that address diagnostic and repair problems in the following areas:

1. Engine repair
2. Automatic transmission/transaxle
3. Manual transmissions and drive axles
4. Suspension and steering
5. Brakes
6. Electrical/electronic systems
7. Heating and air conditioning
8. Engine performance (driveability)

After passing at least one ASE-administered exam and providing proof of two years of hands-on work experience, the technician becomes ASE certified in that particular area. The ASE certification is valid for five years. Retesting is necessary every five years to renew certification.

Work Experience Credit

The technician may be given credit for one of the two years of work experience by substituting relevant formal training in one, or a combination, of the following:

- Secondary training: 3 years of high school training in automotive repair may be substituted for 1 year of work experience.
- Postsecondary training: 2 full years of training after high school in a public or private trade school, vocational-technical institute, community college, or 4-year college may be counted as 1 year of work experience.
- An apprenticeship program: The completion of a state-approved apprenticeship program may be counted as 1 year of work experience. Full credit for the experience requirement is given for satisfactorily completing a 3- or 4-year apprenticeship program. Specialty and short courses: For shorter periods of postsecondary training, one may substitute 1 month of work experience for every 2 months of training.

Test Content

The current heating and air conditioning test consists of 50 multiple-choice questions as follows:

Content Area	Questions	Percent of Test
A/C system diagnosis and repair	17	34
Refrigeration system components diagnosis and repair	10	20
Compressor and clutch (5)		
Evaporator, condenser, and related components (5)		
Heating and engine cooling systems diagnosis and repair	4	8
Operating systems and related controls diagnosis and repair	19	38
Electrical (10)		
Vacuum/mechanical (2)		
Automatic and semiautomatic heating, ventilating, and A/C systems (7)		
Total	50	100%

Additional Questions

The test could contain up to ten additional questions for statistical research purposes that will not affect your score. The 5-year recertification test covers the same content areas as those listed above; however, the number of questions in each content area will be reduced by about 50 percent.

The Questions

The questions are written by a panel of technical service experts, including domestic and import vehicle manufacturers, repair and test equipment and parts manufacturers, working automotive technicians, and automotive instructors. All questions are pretested by a national sample of technicians before they are included in the actual test. Many test questions force the student to choose between two distinct repair methods. Questions similar to the *Technician A, Technician B* format are included in the review questions at the end of each chapter in this text as well as in the Shop Manual.

Why Certify with ASE?

In a word, "recognition." Being an ASE-certified technician provides credentials that attest to your professional abilities to your peers as well as to your prospective employer. As a matter of practice, although ASE certification is voluntary, many employers ask for certified applicants when advertising for employment, or they state "ASE certification preferred." In no small part, certification demonstrates to the employer one's ability to read—an important requirement for technicians of the future.

EPA Certification

To purchase refrigerant or service air conditioning and refrigeration (ACR) systems, one must be certified under section 608 or 609 of the CAA through an agency approved by the EPA. A "609-certified technician" is someone certified by an EPA-approved agency for servicing MVAC and MVAC-like air-conditioning systems. The exam for this certification is open book and is generally available by mail from professional organizations, such as Mobile Air Conditioning Society (MACS), ASE, and others. Testing may also be available in an instructor-led classroom setting.

Under the CAA, a 609-certified technician is not permitted to service domestic or commercial air conditioning or refrigeration equipment, even though the equipment may be similar to an automotive air-conditioning system. A small domestic air conditioner, for example, contains far less refrigerant than the average MVAC; however, a 609-certified technician cannot legally service it. This service requires a 608-certified technician.

A 608-certified technician is someone certified by an EPA-approved agency for servicing particular types of ACR systems. The exam for this certification, with exception, is closed book and is proctored at an approved test site. There are actually four classes of certification, as follows:

1. Type I: One who services high-pressure ACR systems with a capacity of up to 5 pounds of refrigerant.
2. Type II: One who services high-pressure ACR systems with a capacity over 5 pounds of refrigerant.
3. Type III: One who services low-pressure systems, such as centrifugal systems, with up to hundreds of tons of refrigerating capacity.
4. UNIVERSAL: One who is certified in all three types.

The exception to the above is that one may be certified for "small appliances" by taking an open-book exam that is generally administered by mail. This certification, equivalent to Type I, is much more convenient. It is available from trade organizations listed in the Appendix.

For simplicity, some automotive technicians are certified under both sections 608 and 609. For example, one may purchase refrigerant and other supplies at either an automotive parts or refrigeration supply store. When purchasing HFC-134a, for example, the automotive supplier only has cylinders with the "unique" fitting required by EPA. The refrigeration supply store, on the other hand, can supply the cylinder with either fitting: ½-in. Acme for automotive use or ¼-in. SAE for commercial use.

Depending on geographical location, refrigerants may often be less expensive if purchased at a refrigeration supply store. Generally, refrigerant is a "**price leader**" to a refrigeration supply store as engine oil is to an automotive parts store.

Price leader is an item that a merchant may sell at cost or near cost to attract customers.

Cost of Operation

Because the air-conditioning system places an extra load on the engine, it seems apparent that the use of an air conditioner will reduce gasoline mileage. This is only true for stop-and-go driving.

At highway speeds, air conditioned cars, with their windows closed and the air conditioning operating, actually average 2–3 percent better mileage than do cars without air conditioning that have their windows down. The aerodynamic design considerations of today's cars are based upon having the windows closed. When the windows are closed, reduced wind resistance offsets the demand load of the air-conditioning system on the engine.

The modern automobile is designed so it will offer less wind resistance with the windows closed.

SUMMARY

- Refrigeration is the term given to a process by which heat is removed.
- Although the principles were known as long ago as 10,000 B.C., air conditioning and refrigeration were developments of the twentieth century.
- Automotive air conditioning has played a significant and important role in the comfort, health, and safety of the modern motorist.
- The ozone layer protects all life on earth from excess UV radiation.
- Ozone depletion seems to be the greatest during the early winter months.
- The mandatory phaseout in the manufacture and the eventual reduction of use of CFCs has had a positive effect on the ozone layer.
- Increased UV radiation affects the eyes, skin, and the immune system.
- The greenhouse effect is also affected by the release of pollutants.
- Solar radiation passes through the clear atmosphere.
- Most of the radiation is absorbed by the earth's surface to warm it.
- Some of the solar radiation is reflected by the earth and the atmosphere.
- Some of the infrared radiation that passed through the atmosphere is absorbed and readmitted in all directions by *greenhouse gas* molecules. The effect of this is to warm the lower atmosphere.
- All areas of safety should be practiced at all times.
- Antifreeze is either ethylene glycol or propylene glycol based.

Terms to Know

Air conditioning
Allotropes
Atmosphere
CFCs
Chlorine (Cl)
Clean Air Act (CAA)
Dobson Unit (DU)
Global warming
Greenhouse effect
Halogen
Hydrofluorocarbons (HFCs)
Hydrostatic pressure
Infrared
Nitrogen (N)
Oxygen (O)
Ozone depletion
Ozone (O_3)
Price leader
Refrigeration
Toxicity
Ultraviolet (UV) radiation

REVIEW QUESTIONS

Short-Answer Essays

1. How did the early Egyptians cool water?
2. How was humidity controlled in the Northcotts' home?
3. Define the term air conditioning.
4. What is the greenhouse effect?
5. Describe the location and conditions of the troposphere.
6. What does the term ozone hole refer to?
7. Compare the ozone layer to an umbrella.
8. What is the intent of Title VI of the Clean Air Act?
9. Briefly describe the term hydrostatic pressure.
10. What are some of the factors that contribute to the greenhouse effect and global warming?

Fill in the Blanks

1. Most of the earth's protective ozone is found in the _______________.
2. The air we breathe is made up of 21 percent _______________.
3. Increased UV radiation is damaging to the eyes, skin, and _______________.
4. The common name or number used for the hydrofluorocarbon (HFC) refrigerant used in automotive refrigerant systems today is _______________.
5. The element for the chemical symbol O_3 is _______________.
6. The Clean Air Act was signed into law by ___________.
7. The fluid in an air-conditioning system is called _______________.
8. Factory cylinders are equipped with _______________ plugs.
9. Over _______________ percent of all cars produced today are equipped with an air-conditioning system.
10. In the late 1920s and early 1930s, an "air conditioning" option meant that the car was equipped with a (n) _______________ and _______________ system.

Multiple Choice

1. What is the air we breathe made up of?
   A. 21 percent nitrogen and 78 percent oxygen
   B. 12 percent oxygen and 88 percent oxygen
   C. 98 percent oxygen and 2 percent nitrogen
   D. 21 percent oxygen and 78 percent nitrogen
2. What is the main source of ozone-depleting chlorine in the stratosphere?
   A. Chlorine used in swimming pools
   B. Chlorine used in laundry detergent
   C. Chlorine contained in chlorofluorocarbons
   D. All of the above
3. Air conditioning is the process by which air is:
   A. Heated.
   B. Cooled.
   C. Cleaned or filtered.
   D. All of the above.
4. In what year did the first production vehicle automotive air conditioning unit appear on the market?
   A. 1907
   B. 1925
   C. 1940
   D. 1962
5. We are concerned about ozone depletion in what layer of the atmosphere?
   A. Ionosphere
   B. Stratosphere
   C. Troposphere
   D. Ozonosphere
6. Because sunlight is essential for the formation of stratospheric ozone, it is formed mainly over what region of the globe?
   A. The south pole
   B. The north pole
   C. The equator
   D. The United States
7. What can ultraviolet radiation cause?
   A. Skin cancer
   B. Damage to the eyes
   C. Heart disease
   D. Both A and B
8. The hole in the ozone layer was detected over what region of the globe?
   A. The south pole
   B. The north pole
   C. The equator
   D. The United States
9. Technician A says that chlorine (Cl), an ingredient of CFC refrigerants, is harming the ozone layer. Technician B says that the ozone layer is important for pro tection from ultraviolet (UV) radiation. Who is correct?
   A. A only
   B. B only
   C. Both A and B
   D. Neither A nor B
10. In the greenhouse effect what type of radiation is re-emitted by the earth's surface and is absorbed and reflected by greenhouse gas molecules in the atmosphere warming the earth's surface and lower atmosphere?
    A. Gama
    B. Ultraviolet
    C. Infrared
    D. Nuclear

Chapter 2

TEMPERATURE AND PRESSURE FUNDAMENTALS

UPON COMPLETION AND REVIEW OF THIS CHAPTER, YOU SHOULD BE ABLE TO:

- Explain the nature of atoms and molecules.
- Describe the differences between sensible, latent, and specific heat values.
- Discuss the measurement of heat energy.
- Describe how heat flows.
- Explain effects of radiation, conduction, and convection on personal comfort.
- Describe the difference between humidity and relative humidity.
- Discuss the fundamentals of temperature and pressure.
- Explain the term specific gravity.
- Discuss the fundamentals of Boyle's Law, Charles' Law, and Dalton's Law.

INTRODUCTION

Before we can enter into a discussion about the automotive heating and air-conditioning system, you must first have a basic understanding of the chemistry and physics involved in order to properly analyze both systems. The concepts and principles covered in this chapter will form the foundation for understanding heat transfer and the three changes in state of matter, which are the operating principles behind the climate control systems used on today's vehicles. Diagnosing and servicing comfort control systems will become more clear after learning how and why heat transfer takes place. Theories are critical in developing your skills as a technician and, in turn, will increase your productivity and overall worth to the trade.

ELEMENTS AND MATTER

Everything in nature is known as **matter** and is made up of one or more of the 106 known basic elements. Some of the more common and better known of these elements are described here.

Matter is anything that occupies space and possesses mass. All things in nature are composed of matter.

Carbon

Carbon (C) is a nonmetallic element found in many inorganic compounds and in all organic compounds. Carbon is present in many chemical compounds and gases such as Refrigerant-12, a **chlorofluorocarbon (CFC)** refrigerant, and Refrigerant-134a, a hydrofluorocarbon (HFC) refrigerant. At atmospheric pressure and temperatures, carbon normally exists as a solid.

Nearly pure carbon (C), in crystalline form, is known as a diamond.

Chlorine

Chlorine (Cl) is a heavy, greenish-yellow gas used for the purification of water (H_2O) and the manufacture of CFC and HCFC refrigerants. Chlorine, it has been determined, is causing problems with the ozone layer of the atmosphere.

Aluminum

Aluminum (Al) is a lightweight, ductile metal that does not readily tarnish or corrode. Aluminum and aluminum alloys have widespread use in the manufacture of automotive components and parts.

Lead

Lead (Pb) is currently used in the construction of the lead-acid automotive storage battery.

Lead (Pb) is a heavy, soft, blue-gray metal that was once used extensively as a filler material for soldering. It has now been determined that lead is a health hazard and its use is limited and in many cases prohibited.

Nitrogen

Ordinary air that we breathe is about 78 percent nitrogen (N). Nitrogen normally exists as a gas and is a very important element in plant life. Nitrogen is a compound of all living things.

Oxygen

Oxygen (O) makes up 21 percent of the air we breathe. It is absolutely essential to all animal life. Oxygen is a very active element and combines readily with most of the other elements to form oxides or more complex chemicals. Oxygen normally exists as a gas.

Water (H_2O) is formed when hydrogen (H) is burned and it unites with oxygen (O).

Other Gases

The remaining 1 percent of the air we breathe consists of argon (Ar), hydrogen (H), neon (Ne), krypton (Kr), helium (He), xenon (Xe), and other trace nonelement gases, such as carbon dioxide (CO_2) and ozone (O_3).

Hydrogen (H) is a flammable, odorless, colorless gas and is the lightest of all known substances.

Hydrogen (H), an important element in oil, fuels, acids, and many other compounds, and is odorless, tasteless, and colorless. It is normally a gas and is the lightest of the elements. Hydrogen rarely exists alone in nature; its most commonly known mixture is with oxygen to form water (H_2O).

The Atom

The **atom** is the smallest particle of matter.

Each of the elements consists of billions of tiny particles called atoms. An **atom** is so small that it can only be seen with the most powerful microscope. Scientists, however, can measure it and weigh it, and they have learned a great deal about its nature.

An atom is the smallest particle of which an element is composed that still retains the characteristics of that element. For example, an atom of copper (Cu) is copper, and it is different from, say, an atom of aluminum (Al). For our purpose, consider the atom as indivisible and unchangeable. That is, it cannot be divided by ordinary means. Whenever we divide an atom physically and chemically, it retains the characteristics of that element. Atoms of all of the elements are different. Iron (Fe) is composed of iron atoms, lead (Pb) of lead atoms, tin (Sn) of tin atoms, and so on.

An atom (Figure 2-1) is composed of still smaller particles called protons, neutrons, and electrons. The proton has a positive (+) charge; the electron has a negative (−) charge; and the neutron has neither a negative (−) nor a positive (+) charge.

Scientists have been able to split some kinds of elements, such as uranium (U). For our purpose, however, we will focus on the following facts regarding atoms:

- The atom is the very smallest possible particle of matter.
- All of the elements are composed of atoms.
- The atoms of the different elements are different.

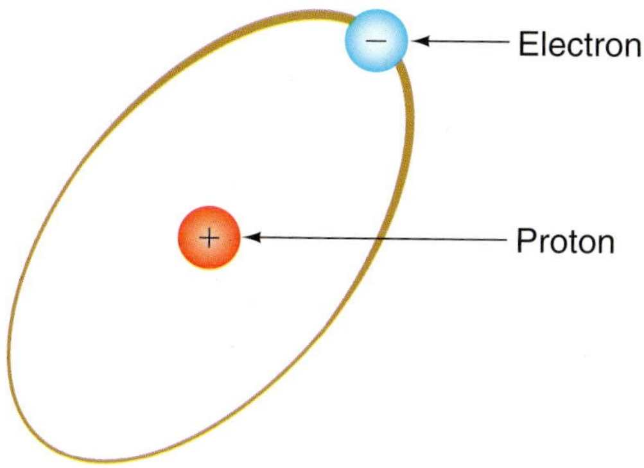

FIGURE 2-1 **A simple hydrogen (H) atom is composed of one proton and one electron.**

THE MOLECULE

The next larger particle of a material is called a **molecule**. If the molecule contains only one kind of atom, the molecule will be a molecule of an element. The molecule of most of the elements has only one atom in it. However, a molecule may have several atoms in it, even though they are all the same kind. For example, a molecule of oxygen (Figure 2-2) contains 2 atoms of oxygen, a molecule of iron has but one atom of iron, and so on for all of the elements. Any substance consists of billions of molecules, and each of those molecules consists of one, two, or several atoms of that element.

A **molecule** is two or more atoms chemically bonded together.

CHEMICAL COMPOUNDS

A molecule may consist of two or more atoms of different elements. In such an instance, the material becomes entirely different and usually does not resemble either of the elements that constitute it. For example, if the molecule consists of one atom of iron (Fe) and one atom of oxygen (O), it becomes iron oxide (FeO), which is quite different from either iron (Fe) or oxygen (O).

How different the compound material itself can be from the elements of which it is composed is illustrated by water (H_2O). The molecule of water, in any form, consists of two atoms of the element hydrogen (H), which is a very light, highly flammable gas, and one atom of the element oxygen (O), which is also a gas that aids combustion. The combination of these two elements, each a gas in its natural state, produces a liquid, water, which is unlike either hydrogen or oxygen.

When hydrogen (H), a flammable gas, and oxygen (O), required for combustion, are combined, water (H_2O) is produced, which is not flammable.

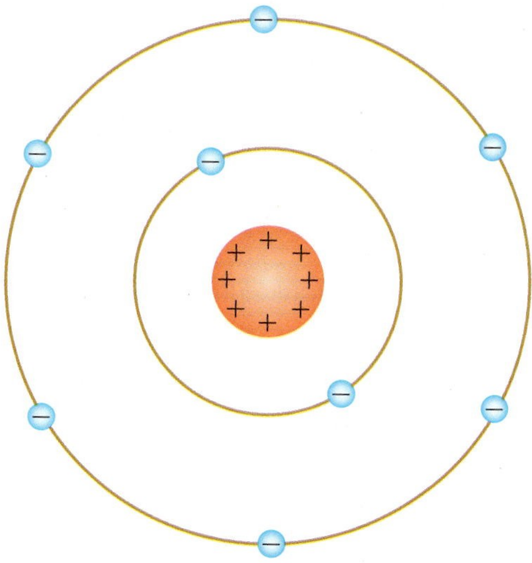

FIGURE 2-2 **An oxygen molecule O_2 has two atoms of oxygen (O).**

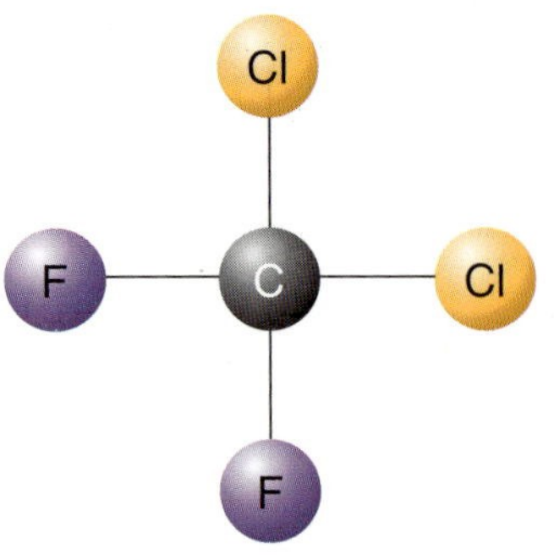

FIGURE 2-3 **Composition of Refrigerant-12.**

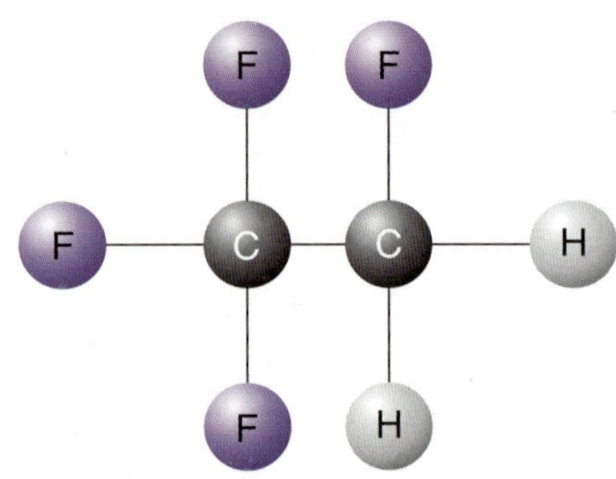

FIGURE 2-4 **Composition of Refrigerant-134a.**

The molecule is often quite complex and may have several kinds of elements in it. Refrigerant-12 (CFC-12), which is normally a colorless gas, is a good example of this. The CFC-12 molecule consists of one atom of carbon (C), normally a black solid; two atoms of chlorine (Cl), normally a yellow-green gas; and two atoms of fluorine (F), normally a pale-yellow gas. The chemical symbol for CFC-12 is CCl_2F_2 (Figure 2-3). HFC-134a, an ozone-friendly refrigerant developed to replace CFC-12, consists of two carbon (C) atoms, four fluorine (F) atoms, and two hydrogen (H) atoms. The chemical symbol for HFC-134a is CF_3CH_2F (Figure 2-4).

Motion of the Molecules

Because mechanical refrigeration is a physical rather than a chemical process, we deal with molecules and their movement. Rarely do we have a need to go into chemical processes that involve breaking down the molecules. It is nonetheless necessary to have an elementary understanding of the composition of matter to more easily understand how gases, liquids, and solids behave under various conditions.

If the matter is a solid, such as copper (Cu) or ice (H_2O), the molecules are held together by their mutual attraction to each other. The mutual attraction of like molecules is called cohesion. The molecules are not tightly bound together, and they are not motionless (Figure 2-5). There are spaces between them, and although their motion is limited, they move somewhat.

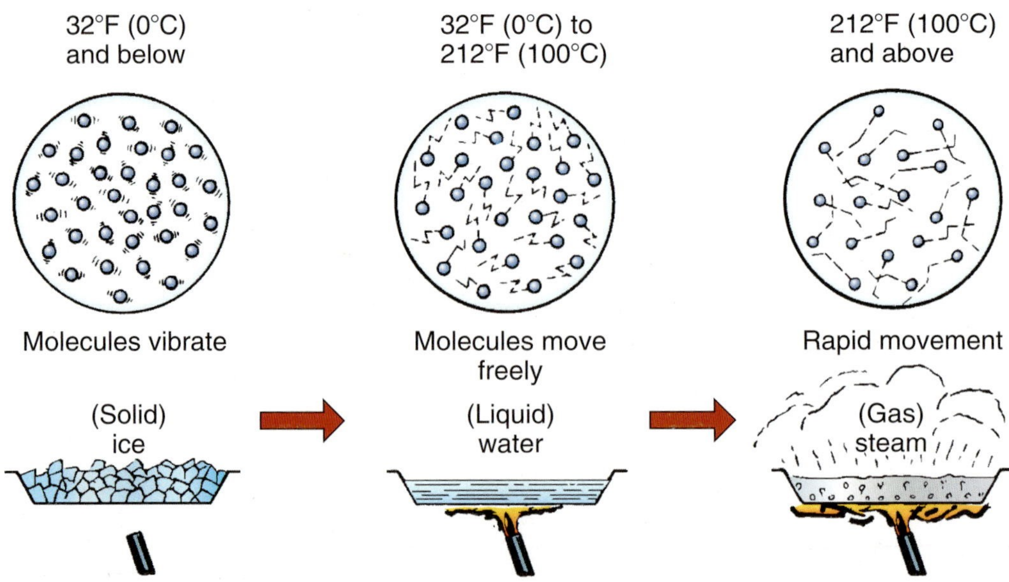

FIGURE 2-5 **Freedom of water (H_2O) molecules in states of matter.**

The colder the solid, the less motion of the molecules. If the matter had absolutely no heat, if it were at a temperature of **absolute zero** (−459.67°F), there would be no motion of the molecules. If the absolute cold matter were heated, the molecules would begin to move. The motion of the molecules becomes greater the warmer the matter becomes.

Absolute zero is the complete absence of heat, believed to be −459°F (−273.15°C).

Heat and Cold

An appropriate definition of **heat** is the sensation of warmth or hotness. The definition of cold, then, is feeling no warmth—uncomfortably chilled. If you sense a temperature above the normal body temperature of 98.6°F (37°C), you tend to feel warm; if the temperature is below normal body temperature, you feel cool. To understand what heat and cold are, you must first understand the law of heat.

Heat is any temperature above absolute zero.

The Law of Heat

Heat is ever-present in all matter. There are three basic terms used to describe the three types of heat: sensible, latent, and specific.

Sensible Heat. Sensible heat is any heat that we can feel and that can be measured with a thermometer. For example, water boils at 212°F (100°C) at sea-level atmospheric pressure. The temperature of water from a spigot may be 58°F (14.4°C). From the spigot temperature to the boiling point, the increase in temperature is 154°F (85.6°C). That increase in temperature is known as sensible heat.

Sensible heat causes a change in the temperature of a substance but does not change the state of the substance.

Latent Heat. Latent heat cannot be measured with a thermometer. To explain, we know that a pan of boiling water does not all turn to steam (gas) as soon as it reaches its boiling point. If the pan is left on the burner and allowed to boil long enough, however, all of the water will boil away. The heat that is added to the boiling water to cause all of it to vaporize is called latent heat. Though it cannot be measured with a thermometer, latent heat is required to cause a change of state in matter.

We cannot heat water at atmospheric pressure (sea level) hotter than 212°F (100°C). The steam form of this water as it boils is also at 212°F (100°C). We will discuss later the fact that water in a sealed container, such as an automobile radiator, may not boil until the temperature is further increased if the system is pressurized. Also, if the pressure is reduced or at a vacuum, water may boil at temperatures much below 212°F (100°C). In fact, we will demonstrate how, under certain conditions, water will boil at 32°F (0°C)—its normal freezing point (Figure 2-6).

Latent heat is the amount of heat required to cause a change of state of a substance without changing its temperature.

As indicated earlier, latent heat must be added to the water to cause it to change to a gas. Because we cannot measure latent heat on a thermometer, we use as a unit of measure the **British thermal unit (Btu)**. One Btu will cause a change of temperature of one degree Fahrenheit (1°F) in one pint (1 lb., or 16 oz.) of water (H_2O). Metrically, one calorie (1 cal) will cause a change of temperature of one degree Celsius (1°C) in one gram (1 g) of water (H_2O).

British thermal unit (Btu) is a measure of heat energy; one Btu is the amount of heat necessary to raise one pound of water 1°F.

One pound (0.4536 kg) of ice taken from the refrigerator at 32°F (0°C) requires 144 Btu of latent heat to cause a change in state to liquid at 32°F (0°C). An additional 180 Btu of sensible heat will bring the liquid to its boiling point of 212°F (100°C), and another 970 Btu of latent heat are required for a change in state to a gas, again at 212°F (100°C).

Specific Heat. Everything in nature has a **specific heat**. We are not to be particularly concerned with this term. It is important, at this time, to know that refrigerant used in automobile air-conditioning systems has an appropriate specific heat value for its application. It is interesting to note that water also has an appropriate specific heat value for its application as an engine coolant.

Specific heat is the quantity of heat required to change one pound of a substance by 1°F.

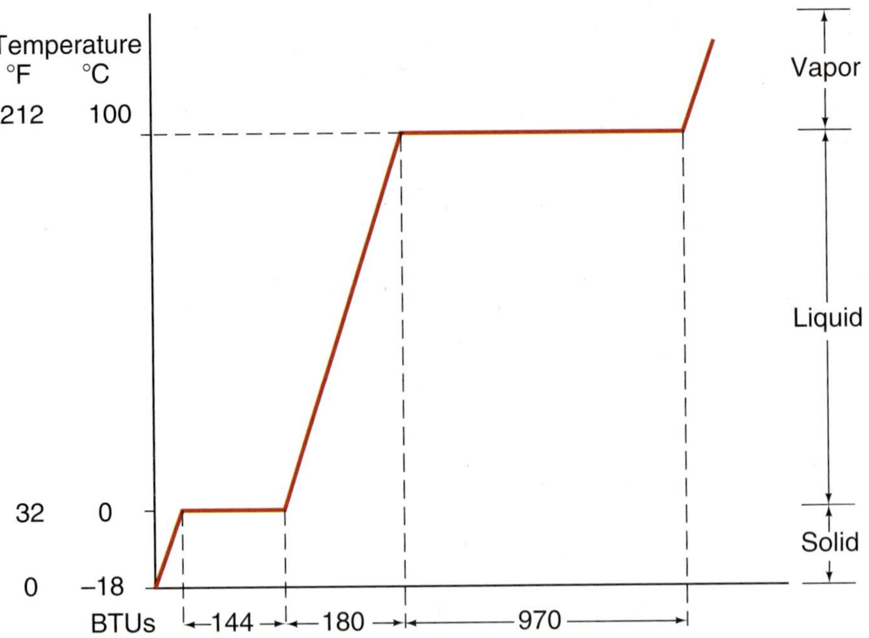

FIGURE 2-6 **Sensible and latent heat values required to effect a temperature or physical change in 1 lb. (0.4536 kg) of water (H_2O) from the freezing to boiling temperature at sea level atmospheric pressure.**

Sensible Heat of a Solid

Energy is required to cause movement or to do work. The molecules must be given heat, a form of energy, in order to give motion. The more heat energy present, the greater the motion and the faster they move.

The heat added to raise the temperature of the solid matter and give the molecules more movement is called sensible heat, for we can tell that heat has been added by one of our senses, the sense of feeling, which tells us that the solid is warmer than before.

If we continue to add heat energy to the solid, it becomes warmer and the molecules move faster, but still within a very limited space, because they are still held to one another by their mutual attraction.

Author's Note: Air-conditioning systems are based on the laws of physics, and it is essential that technicians understand the principles of latent and sensible heat if they are to properly diagnose system function later in this text.

Melting or Fusion

Finally, when sufficient heat is added to the molecules, they receive enough energy to partially overcome their attraction to each other. At that time, the molecules can move about freely and change state to become a liquid. The process of the molecules breaking away from each other and changing state from a solid to a liquid is called melting or fusion.

The attraction of the molecules of a solid to one another is great; therefore, a considerable amount of heat energy is required for a solid to become a liquid. The heat energy required to melt a solid is relatively more than the amount required to warm it by raising its temperature a few degrees.

Latent means hidden.

The heat required to effect a change of state of matter from a solid to a liquid is called the latent heat of melting or, more correctly, the latent heat of fusion. The word latent means

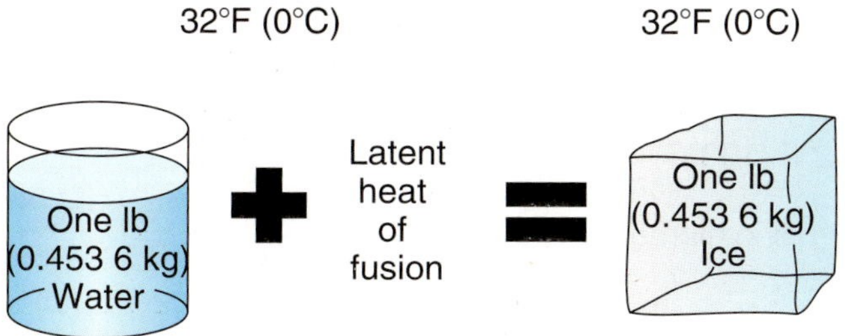

FIGURE 2-7 **The temperature of liquid and solid water (H_2O) has not changed.**

hidden. It is a term that is used to identify something that is present but not visible. The heat that causes a change of state cannot be measured on a thermometer; in this sense, it is hidden heat. Latent heat, then, is heat required to cause a change of state of matter without changing its temperature.

The temperature of the solid immediately before it melts and immediately afterward when it has become a liquid are exactly the same. For example, if the matter (Figure 2-7) were water, its temperature as a solid (ice) and as a liquid would be 32°F.

The molecules, now moving more freely, are not held together. The matter can no longer stand rigidly by itself and must have a container to support it. The speed of the molecules is much greater as a liquid, yet not great enough to overcome the force of gravity. The molecules are therefore held downward in the container. The liquid, however, can be poured from a higher container to a lower one, or pumped from a lower container to a higher container.

Matter in the liquid state has more heat in it than when it is in the solid state; and, except at the exact melting temperature, it will always be warmer.

Because the liquid molecules are much freer in a liquid than in a solid, they are farther apart and require more room to move. Assuming that the same matter and the same weight, the volume of a liquid, then, is greater than the volume of a solid.

Sensible Heat of a Liquid

Because the molecules of a liquid have considerable heat energy, they move about in a lively manner and at a rapid speed. They constantly bump into each other and into the side of their container.

Sensible heat is heat that is added to matter to cause it to become warmer.

As heat energy is added to a liquid and it becomes warmer, the speed of its molecules increases. The heat energy that is added to it is called sensible heat.

Evaporation

Not all of the molecules in the liquid move at the same speed. In fact, some of the molecules at the top of the liquid may attain enough speed to fly out of the liquid and into the space of air above the liquid and escape. Some of them escape from the liquid temporarily but do not have enough speed or energy to entirely escape and eventually fall back into the liquid.

Some molecules, however, do escape to mix with the air or other gas above the liquid. Some of the molecules, then, are constantly escaping. They form a gas or vapor blanket above the liquid and tend to diffuse into and mix with the air. The process of the molecules escaping from the surface of the liquid is called **evaporation**.

Evaporation is the process of forming a vapor.

A good example of evaporation of a liquid is water (H_2O) in an open container (Figure 2-8). When water is placed over heat and its temperature exceeds 212°F (100°C), it slowly evaporates into a gas (water vapor). Eventually all of the water evaporates. The warmer the water, the faster it evaporates. The more heat energy added, the more molecules gain sufficient velocity to escape from the liquid.

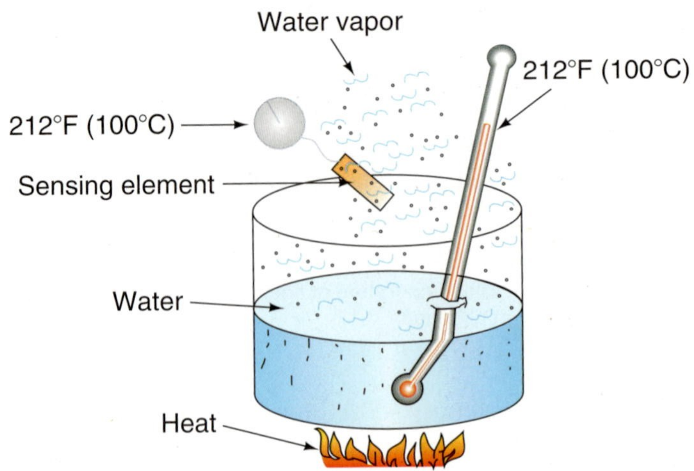

FIGURE 2-8 **Evaporation of liquid water (H_2O) by boiling.**

Boiling or Vaporization

If more heat is added to the liquid above the boiling temperature, it becomes warmer and warmer and the molecules move faster and faster. When enough heat energy has been added, the molecules are moving so rapidly that they lose all restraint and fly out of the liquid, much the same as in evaporation, but in far greater numbers.

At this very high temperature, the liquid disintegrates and breaks loose even from the force of gravity. The molecules fly in all directions. This condition is referred to as a gas or a vapor. As a vapor, the matter requires a great deal more space than when it was a liquid or a solid. The molecules are flying about and are, therefore, widely separated. The volume of the vapor, then, is much greater than when the matter was a liquid.

The process of changing from a liquid to a vapor is called boiling or vaporization. The temperature at which this occurs is called the boiling temperature.

A great amount of heat energy is required to boil a liquid and give the molecules enough energy to escape and form a vapor. This is referred to as the latent heat of boiling or, more correctly, the latent heat of vaporization. Note that this heat too is referred to as latent heat. It is heat that is required for a change of state without a change of temperature.

If the liquid is water (H_2O) in an open container, the molecules that escape into the air form what is known as water vapor. Another term for water vapor is "moisture in the air." A liquid can and does have its own vapor that forms just above its surface. It also diffuses or spreads through the space above the surface of the liquid.

Sensible Heat of a Vapor

Sensible heat can be measured with a thermometer.

Matter in any state can be warmed; a vapor or gas can be warmed just as a solid or liquid can be warmed. If heat energy is added, the speed—or velocity—of the molecules increases and the matter is said to be warmer. The heat that is added to a vapor and causes it to become warmer is called the sensible heat of the vapor. It is also called **superheat**. Superheat is the temperature a substance is heated above its vapor saturation point for a given pressure so that a drop in temperature does not cause a reconversion to a liquid state. It is the added heat intensity given to a gas after the complete evaporation of a liquid. In Chapter 4 there are pressure/temperature charts for both R-134a and R-12 refrigerants. In a refrigerant system, if all the liquid refrigerant in the evaporator core at a given point has gone through a change of state from a liquid to a gas as it picked up heat, and these molecules still have 25 percent of the evaporator core left to travel through, this gas will pick up more heat from the heat load on the evaporator core as it removes heat from the air stream passing over it and even though it is at the same pressure it will become hotter than the pressure/temperature chart indicates it should be. This increase in heat above the normal pressure/temperature relationship is called superheat. This phenomenon only occurs when there are no liquid molecules nearby. Most refrigerant systems are designed to maintain 10°F (−12°C) of superheat in the refrigerant leaving the evaporator so that the gas returning to the compressor is several degrees away from the condensation point of the refrigerant. This is to avoid the risk of liquid refrigerant entering the compressor. The compressor is designed to be a vapor pump and would be damaged if it had to compress liquid refrigerant.

Measuring the Amount of Heat Energy

To understand and apply these principles, we must measure temperature changes and the amounts of heat. If we cannot, for example, measure a material, an action, a process, or an energy, we really cannot properly understand it.

Heat is one of the three forms of energy. Energy does work by causing things to happen. Energy is not a solid, liquid, or gas; and it cannot be measured in traditional terms, such as inches, feet, quarts, or cubic feet. Energy must be measured by what it does and by the effects it produces.

The term *thermal* defines heat.

Adding heat to water raises its temperature. We measure this heat by how much it raises the temperature of the water. For example, 1 lb. (0.45 kg) of water (H_2O), approximately 1 pt. (0.47 L), is at about 63°F (15.6°C). It takes a certain amount of heat (Figure 2-9) to raise its temperature 1°F, from 63°F to 64°F (15.6°C to 17.8°C).

Most countries are on the metric system. The calorie is the heat unit in the metric system. The calorie is now used in the United States only in some scientific laboratories. The calorie is the amount of heat required to raise 1 g (0.035 oz.) of water (H_2O) 1.0°C.

The Btu is the standard measure of the amount of heat, and the degree of temperature is the standard measure of the effect of that heat on 1 lb. (0.45 kg) of water (H_2O). The amount of water must be known. Obviously, it will take two times as much heat (2 Btu) to warm 2 lb. (0.91 kg) of water (H_2O) 1°F (0.56°C) as it would to warm 1 lb. (0.45 kg). It would also take twice as much heat (2 Btu) to warm 1 lb. (0.45 kg) of water (H_2O) 2°F (1.1°C) as it would to warm 1 lb. (0.45 kg) of water (H_2O) 1°F (0.56°C).

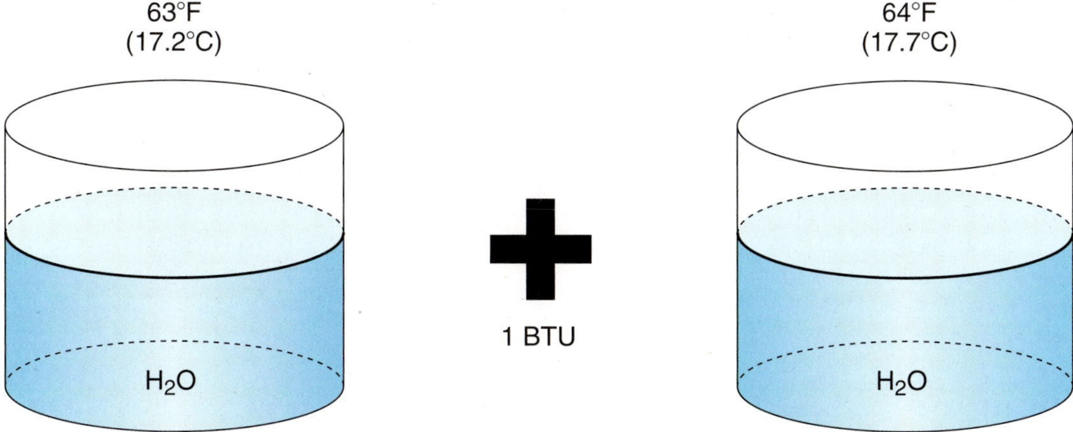

FIGURE 2-9 **One Btu raises 1 lb (0.45 kg) of water (H_2O) 1°F (0.56°C).**

To warm 10 lb. (4.54 kg) of water (H_2O) 10°F (5.56°C) would require 100 Btu of added heat.

$$10 \times 10 = 100$$

The amount of heat required in Btu to warm any amount of water through any known temperature change is found by multiplying the number of pounds of water by the number of degrees Fahrenheit the temperature is to be raised. The answer is the number of Btu of heat that must be added.

Do not forget that the heat energy added to the water results in an increase in the active movement of the molecules to give them more rapidity of motion. This increased rapidity of motion produces the effect of a rise in temperature.

Specific Heat

Materials vary in the amount of heat required to raise their temperature (Figure 2-10). Compared to most other materials—whether solid, liquid, or gas—water requires a great deal of heat energy to raise its temperature. Oil requires only about one-half as much heat energy as water, so it takes only ½ (0.5) Btu to warm 1 lb. (0.45 kg) of oil 1°F (0.56°C). Mercury (Hg) requires only about $\frac{1}{30}$ (0.034) Btu; alcohol about $\frac{3}{5}$ (0.6) Btu; and so on.

Gases vary greatly in the amount of heat required to warm them, depending on their original temperature and their pressure. At atmospheric pressure and room temperature:

- Air: oxygen, nitrogen, and carbon dioxide require $\frac{1}{5}$ to ¼ Btu per pound per degree;
- Sulphur dioxide requires about $\frac{1}{6}$ Btu per pound per degree;
- Ammonia requires about ½ Btu per pound per degree; and
- Refrigerant-12 or Refrigerant 134a requires about $\frac{1}{7}$ Btu per pound per degree.

The amount of heat required to raise 1 lb. (0.45 kg) of matter 1°F is called its specific heat. For water, the specific heat is 1.0. It takes 1 Btu to raise 1 lb. (0.45 kg) of water (H_2O) 1°F, as demonstrated earlier. The specific heat of oil is ½ (0.5) Btu. Tables of specific heats are usually given in decimals, as ice (at 20°F) 0.48, water vapor 0.46, iron 0.13 to 0.17, and so on.

To calculate the amount of heat required to raise matter from one temperature to a higher temperature, first figure out the number of Btu just as if the material were water (H_2O), by multiplying the number of pounds (kg) of the matter by the number of degrees F (C) that it is to be warmed. Next, multiply this amount by the specific heat of that particular matter.

Material	Specific heat	Material	Specific heat
Air	0.240	Nitrogen	0.240
Alcohol	0.600	Oxygen	0.220
Aluminum	0.230	Rubber	0.481
Brass	0.086	Silver	0.055
Carbon dioxide	0.200	Steel	0.118
Carbon tetrachloride	0.200	Tin	0.045
Gasoline	0.700	Water, fresh	1.000
Lead	0.031	Water, sea	0.940

FIGURE 2-10 **Specific heat values of selected solids, liquids, and gases.**

Example

How many Btu are required to raise the temperature of 20 pounds of ice from 20°F to 32°F? If the ice were water, it would require

$$20 \times 12 \times 1 = 240 \text{ Btu}$$

The specific heat of ice, however, is 0.48, so

$$20 \times 12 \times 0.48 = 115.2 \text{ Btu}$$

Latent Heat of Fusion

To change 1 lb. (0.45 kg) of ice at 32°F to water at 32°F requires that the value of the latent heat of fusion of ice be added—144 Btu per pound (0.45 kg):

$$144 \times 20 = 2{,}880 \text{ Btu}$$

Accordingly, 2,880 Btu are required to melt 20 lb. (9.1 kg) of ice.

To warm the 20 lb. (9.1 kg) of water to 50°F (an additional 18°F), we would have to supply another 20 × 18 or 360 Btu.

If, however, we want to heat the water from 32°F to the boiling point of 212°F (through 180°F) instead of just to 50°F, we will have to add 3,600 Btu (20 × 18 = 3,600 Btu).

The latent heat of fusion also varies with the matter. For some other solids, the latent heats of fusion are:

Aluminum (Al)	167.5 Btu per pound
Copper (Cu)	78.0 Btu per pound
Silver (Ag)	43.9 Btu per pound
Gold (Au)	28.7 Btu per pound
Tin (Sn)	25.4 Btu per pound
Lead (Pb)	9.8 Btu per pound

Latent Heat of Vaporization

To boil water at 212°F and turn it into steam, also at 212°F, requires latent heat of vaporization. For water in an open pan, the latent heat of vaporization is 970 Btu per lb. (0.45 kg), so 20 lb. (9.1 kg) at 212°F requires 20 × 970 or 19,400 Btu to turn it into steam or water vapor also at 212°F.

If we want to superheat this steam to 250°F, to raise its temperature above the 212°F, we must add 20 × 38 × 0.46 (the specific heat of steam) or 349.6 Btu.

To warm 20 lb. of ice from 20°F to 32°F, change it to water, heat the water to 212°F, change it to vapor (steam), and then heat the steam to 250°F will require these steps:

- To warm 20 lb. of ice from 20°F to 32°F: 20 × 12 × 0.48 = 115.2 Btu
- To change the 32°F ice to water at 32°F: 20 × 144 = 2,880.0 Btu
- To warm 20 lb. of water from 32°F to 32°F: 20 × 180 × 1 = 3,600.0 Btu
- To change the 212°F water to steam at 212°F: 20 × 970 = 9,400.0 Btu
- To warm 20 lb. of steam from 212°F to 250°F: 20 × 38 × 0.46 = 349.6 Btu
- The total required to change 20 lb. of ice at 20°F to steam at 250°F = 26,344.8 Btu

From this example, it may be seen that the latent heats of fusion and of vaporization are very large compared with the sensible heats. Moreover, the latent heats of fusion and vaporization of water, to change their states from a solid to a liquid and from a liquid to a vapor, are quite large compared to other matter. Most matter has far less heat capacity than ice, water (H_2O), and steam (Figure 2-11).

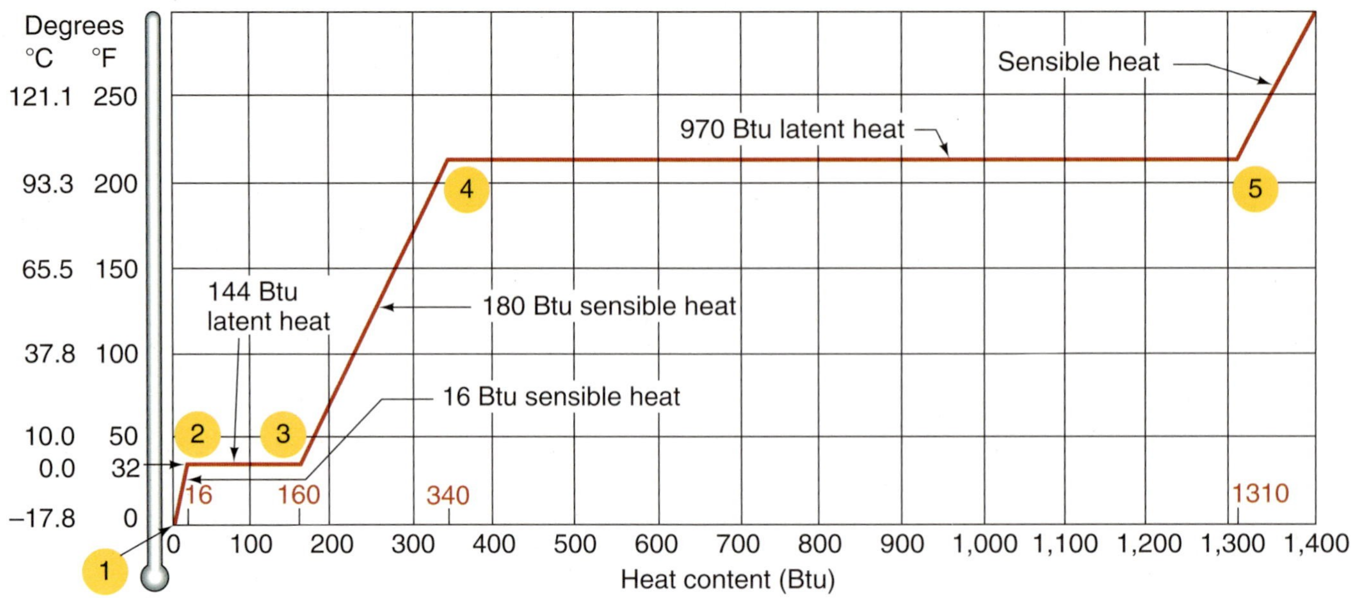

FIGURE 2-11 **Latent and sensible heat values for water (H_2O).**

Heat Flow

Heat has been defined as the energy of the molecules in motion. Motion is transmitted to other molecules that have less motion. Some of the molecules give up some of their energy to other molecules that have less energy. Another way of saying this is that heat flows from the matter at a higher temperature to a matter at a lower temperature.

Natural heat flow from a warm to a less warm area or surface is called gravity.

Heat transfer, therefore, is always downward; from hot to warm, warm to cool, or cool to cold. Heat never flows from low-temperature matter to high-temperature matter. Natural heat flow from a warm to a less warm area or surface is called gravity.

One method of studying heat flow is to place a hot object near, or touching, a colder object. Heat will flow from the hot object to the colder object. Actually, heat is neither added nor removed. It is simply transferred from one place where it is not wanted to another place where it is accepted.

Heat moves in either or all of three ways. Heat moves by:

- Radiation
- Conduction
- Convection

Radiation is the transfer of heat without heating the medium through which it is transmitted.

Radiation

If a hot object is placed near or against a cooler object, heat is transferred by **radiation** (Figure 2-12) across the space between the two objects to warm the cooler object.

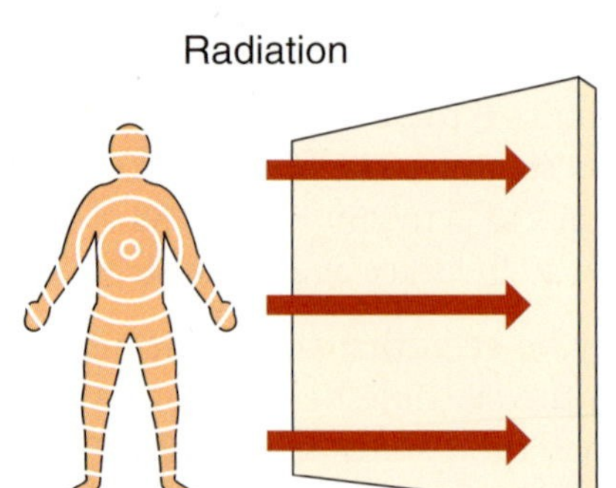

FIGURE 2-12 **Heat is transferred by radiation.**

There does not have to be any gas or other material in the space. An excellent example of radiation is how we receive heat from the sun. This heat radiates through some 92 million miles of vacuum to heat the earth without heating the space in between.

Conduction

If one side of a material is heated, the heat will travel through it from the hot side to the cooler side. In turn, heat may be conducted to another cooler object touching it or radiated to a cooler object some distance away.

The transfer of heat through a material is known as **conduction** (Figure 2-13). The warmer side gives motion to the molecules that, in turn, give motion to nearby molecules, and so through the material. In doing so, some of the heat energy is given up to the molecules and remains as heat energy. All of the heat, therefore, does not pass through the material.

Conduction is the transmission of heat through the direct contact of two objects.

A material that transmits heat easily with little loss is called a conductor of heat. Some of the best conductors are also good conductors of electricity. They are copper (Cu), silver (Ag), and aluminum (Al).

A material that does not conduct heat through itself easily is called an **insulator**. Some of the better insulators are cork, cotton, air, and other materials that are composed of thousands of tiny air cells. A good insulator is a poor conductor.

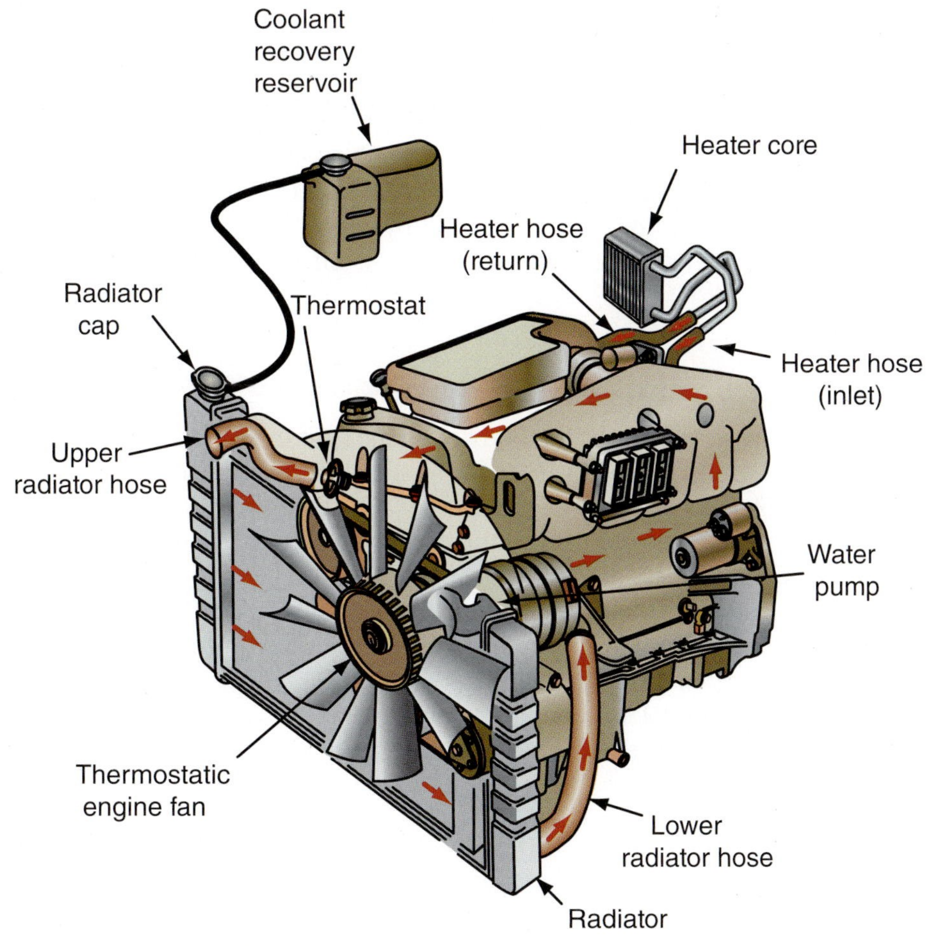

FIGURE 2-13 **Heat is transferred by conduction.**

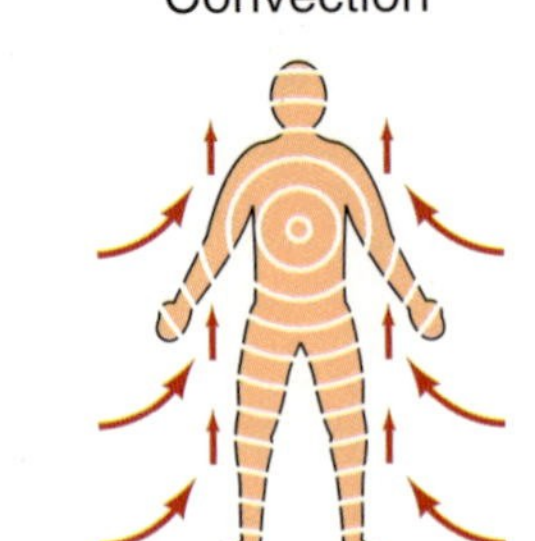

FIGURE 2-14 **Heat is transferred by convection.**

Convection

Convection is the transfer of heat by the circulation of a vapor or liquid.

Convection occurs only in fluids—liquids and gases or vapors. When a material is warmed, it expands in volume and therefore becomes lighter per cubic foot of volume. In the case of fluids, the cooled fluid is heavier and, as a result, crowds out the lighter, warmer fluid. This pushes the warmer fluid upward, setting up a cycle of circulation. This circulation of the fluid carries heat upward on one side and downward on the other side. This means of conducting heat is called convection (Figure 2-14). It is very important in refrigeration and heating, where a large part of the process is cooling or heating fluids. In many applications, the fluids are cooled as a means of carrying heat away from or to foods, human beings, or other objects.

Personal Comfort and Convenience

Twenty-five pounds of ice is equal to 150 Btus in today's terms.

Deep-tissue or **subsurface temperature** is the core temperature of the body.

The normal body temperature of a human adult is 98.6°F (37°C). This temperature is sometimes called **subsurface** or **deep-tissue temperature**, as opposed to surface or skin temperature. An understanding of the process by which the body maintains its temperature is helpful to the student because it explains how air conditioning helps keep the body comfortable.

How the Body Produces Heat

Water (H_2O) weighs 8.3453 pounds (3.79 kilograms) per gallon (3.785 liters). For practical purposes, it is considered that 1 pound (0.4536 kilogram) of water (H_2O) is equal to 1 pint (0.473 liter).

All food and beverage taken into the body contains heat in the form of calories. The calorie is a term used to express the heat value of food. The calorie is the amount of heat required to raise 1 kilogram of water (H_2O) 1 degree Celsius (C). There are 252 calories in 1 Btu.

As calories are taken into the body, they are converted into energy and stored for future use. The conversion process generates heat. All body movements use up the stored energy and, in doing so, add to the heat generated by the conversion process. The body consistently produces more heat than it requires. Therefore, for body comfort all of the excess heat produced must be given off by the body.

The constant removal of body heat takes place through three natural processes (Figure 2-15) discussed earlier, which all occur at the same time. These are:

- Convection
- Radiation
- Evaporation

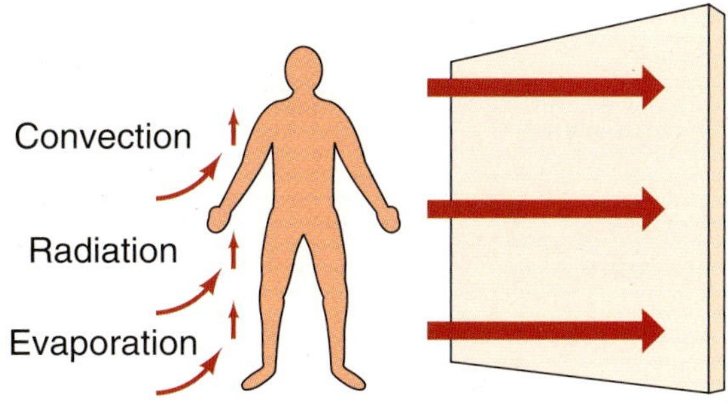

FIGURE 2-15 **The body gives up heat by convection, radiation, and evaporation.**

Convection

The convection process of removing heat is based on two natural phenomena:

- Heat flows from a hot surface to a surface containing less heat. For example, heat flows from the body to the air surrounding the body when the air temperature is lower than the skin temperature (Figure 2-16).
- Heat rises. This is evident by watching the smoke from a burning cigarette or the steam from boiling water.

Gravity causes warm air to rise. Cool air is heavier than warm air, and as warm air rises, cool air falls to take its place.

When these two natural phenomena are applied to the bodily process of removing heat, the following changes occur:

- The body gives off heat to the surrounding air (which has a lower temperature). The surrounding air becomes warmer and moves upward.
- As the warmer air moves upward, air containing less heat takes its place. The convection cycle is then completed.

Radiation

Radiation is the process that moves heat from a heat source to an object by means of heat rays. This principle is based on the phenomenon that heat moves from a hot surface to a surface containing less heat. Radiation takes place independently of convection. The process

Radiation is the transfer of heat without heating the medium through which it is transmitted.

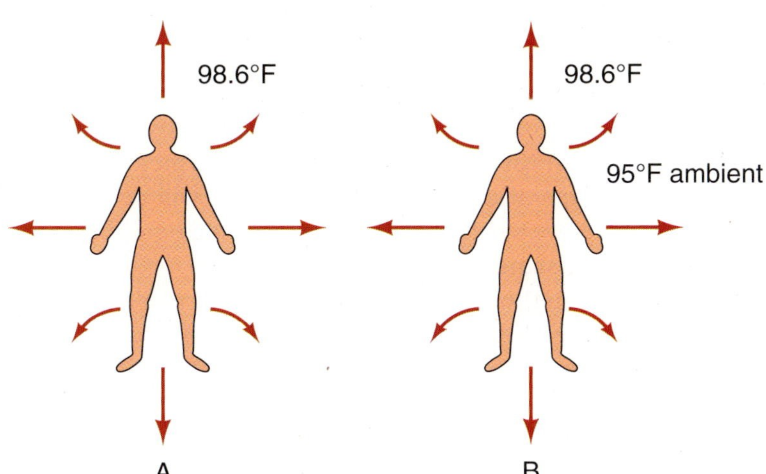

FIGURE 2-16 **The body rapidly gives up heat when the surrounding air temperature is below body temperature (A) and slows as the surrounding air temperature increases (B).**

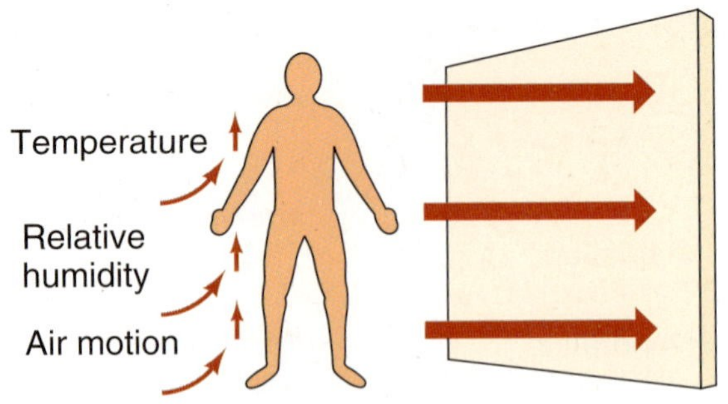

FIGURE 2-17 **Conditions that affect body comfort.**

of radiation does not require air movement to complete the heat transfer. This process is not affected by air temperature, although it is affected by the temperature of the surrounding surfaces.

The body quickly experiences the effects of sun radiation when one moves from a shady area to a sunny area.

Evaporation

Evaporation is the changing of a liquid to a vapor while picking up heat.

Perspiration is the salty fluid secreted by sweat glands through the pores of the skin.

Evaporation is the process by which moisture becomes a vapor. As moisture vaporizes from a warm surface, it removes heat and thus lowers the temperature of the surface. This process takes place constantly on the surface of the body. Moisture is given off through the pores of the skin. As the moisture evaporates, it removes heat from the body.

Perspiration appearing as drops of moisture on the body indicates that the body is producing more heat than is being removed by convection, radiation, and normal evaporation.

The three main factors that affect body comfort (Figure 2-17) are:

- Temperature
- Relative humidity
- Air movement

Temperature

The **temperature** of an object can be described as that which determines the sensation of warmth or coldness when one comes into contact with it. When two objects are placed together, they are said to be in thermal contact. The object with the higher temperature is cooled while the object with the lower temperature is warmed. At some point in time, they will both be the same temperature and no more change will occur. When thermal changes between two objects stop, we say that they are in thermal equilibrium. To our senses, then, both objects would feel the same.

When we say that something is cool or warm, we are speaking in relative terms. For example, consider the following experiment. You will need two pans approximately 7 × 7 × 2 in. (18 × 18 × 5 cm), one pan approximately 9 × 12 × 2 in. (23 × 30 × 5 cm), 9 cups (4.26 liters) tap water, 1 cup (0.47 liter) hot water, and a tray of ice cubes.

SetUp

1. Place the pans on a level surface with the large pan in the middle and the smaller pans on both sides.
2. Put 2 cups (0.473 liter) of tap water in each of the small pans.
3. Put 5 cups (1.18 liter) of water in the large pan.

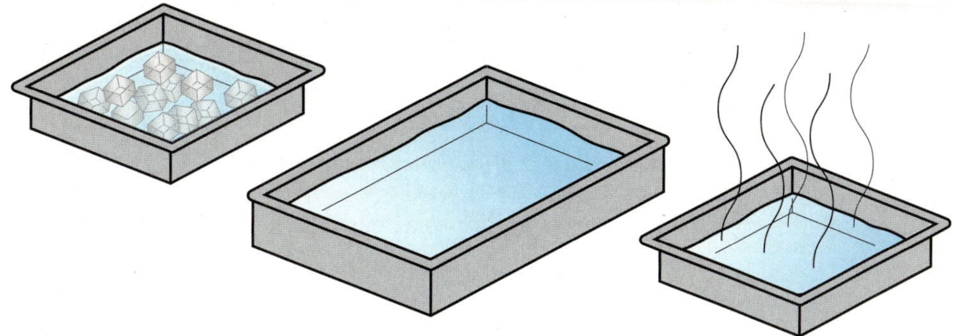

FIGURE 2-18 **Place ice cubes in the left pan.**

4. Allow them to reach thermal equilibrium with the ambient room temperature.
5. Put a tray of ice cubes in the pan on the left (Figure 2-18).
6. Add 1 cup (0.24 liter) of hot water to the pan on the right. Water must not be so hot as to cause personal injury.

The Experiment

1. Place both hands in the center pan for 10–15 seconds (Figure 2-19). What do you feel? (You should experience a feeling of neither warmth nor cold because the water in this pan is in thermal equilibrium with the room temperature.)
2. Place the left hand in the pan with the ice cubes. What do you feel? (You should experience a sensation of cold because this water is below room temperature.)
3. Next, place the right hand in the pan with warm water. What do you feel? (You should experience a sensation of warmth because this water is above room temperature. Remember that we are speaking in relative terms—in this case, relative to the ambient temperature.)
4. Remove your left hand from the left pan and place it in the center pan for a few seconds. What do you feel? (The surface temperature of your left hand was lowered, so the water in this pan should now feel warm.)
5. Remove your right hand from the right pan and place it in the center pan for a few seconds. What do you feel? (The surface temperature of your right hand was raised, so the water should now feel cool.)

Conclusion. In relative terms, the left-hand felt warm because the temperature of the ambient water was higher than the cold water. The right hand felt cool because the temperature of the ambient water was lower than the warm water.

FIGURE 2-19 **Place both hands in pan for 10–15 seconds.**

It should now be obvious that the feeling of cold or warm is relative in relation to the temperature of an object or to the ambient environment.

Cool air increases the rate of convection; warm air slows it down. Cool air lowers the temperature of the surrounding surfaces. Therefore, the rate of radiation increases. Because warm air raises the surrounding surface temperature, the radiation rate decreases. In general, cool air increases the rate of evaporation and warm air slows it down. The evaporation rate also depends on the amount of moisture already in the air and the amount of air movement.

Humidity

Relative humidity (RH) is the actual moisture content of the air in relation to the total moisture that the air can hold at a given temperature.

The moisture of air is measured in terms of **relative humidity (RH)**. The expression "50 percent relative humidity," for example, means that the air contains half the amount of moisture that it is capable of holding at a given temperature.

A low relative humidity permits heat to be taken away from the body by evaporation. Because low humidity means that the air is relatively dry, it can readily absorb moisture. A high relative humidity has the opposite effect. The evaporation process slows down in humid conditions; thus, the speed at which heat can be removed by evaporation decreases. An acceptable comfort range for the human body is 72°F to 80°F (22.2°C to 26.6°C) at 45 to 50 percent relative humidity (RH).

Humid is another term for damp. It is the feeling of dampness when the dew point of the air is close to the actual ambient air temperature, causing water in the air to condense. The closer the dew point temperature of the air is to the actual ambient air temperature, the more humid it feels.

In some areas of the country, the average RH is 50 percent or above most of the time. These areas are said to be "**humid**" regions, and dehumidification is generally required to provide an environment ideal for human comfort. In other parts of the country, average RH is less than 50 percent. These regions are said to be "dry" or "**arid**." Humidification is usually required in arid areas to create an ideal environment.

The recommended relative humidity for an occupied area is between 40 and 60 percent for a healthy environment. Recent studies show that bacteria, fungi, and viruses are more active below 40 percent RH and above 60 percent RH. The evaporator, that part of an air-conditioning system that removes heat from the air, also removes moisture from the air. Some of this moisture, however, tends to cling to the fins and tubes of the evaporator after the system has been turned off. This creates a very high humidity condition within the evaporator, promoting fungi growth from the airborne impurities that were trapped by the moisture. It is not uncommon to hear a complaint of a "musty mildew odor" coming from the automotive air-conditioning system when it is first turned on. This odor is caused by the mildew-type fungi that were formed in the evaporator case during the off period of the air-conditioning system.

Some older people are not comfortable in the normal "comfort range" and require special considerations.

To combat this problem, the blower motor of some systems is turned on for a few minutes after the air conditioner has been turned off for a period of time, generally 30 minutes. There is also a device that can be installed on most vehicles which operates the blower motor some time after the air-conditioning system has been turned off. A delay in this process is intended to provide sufficient time for the moisture to run off the tubes and fins of the evaporator and collect in the bottom of the evaporator case. The fan then forces the water out the drain tube and dries off the evaporator coil.

Arid is another term for dry.

Automotive evaporators may be cleaned of fungi following specific service procedures as given in the appropriate manufacturer's service manual. A typical procedure for correcting this problem may be found in the Shop Manual, Chapter 9.

Air Movement

Another factor that affects the ability of the body to give off heat is the movement of air around the body. As the air movement increases, the following processes occur:

- The evaporation process of removing body heat speeds up because moisture in the air near the body is carried away at a faster rate.

- The convection process increases because the layer of warm air surrounding the body is carried away rapidly.
- The radiation process increases because the heat on the surrounding surfaces is removed at a faster rate. As a result, heat radiates from the body at a faster rate.

As the air movement decreases, the processes of evaporation, convection, and radiation decrease. As the air movement increases, so does evaporation, convection, and radiation, a process known as *windchill factor*.

Windchill Factor

The windchill factor, developed in 1941, is a measure of relative personal discomfort due to combined cold and wind based on physiological studies of the rate of heat loss for various combinations of ambient temperature and wind speed (Table 2-1). The windchill factor is based on the actual air temperature when the wind speed is 4 mph (6.4 km/h) or less. At higher wind speeds, the windchill temperature is lower than the air temperature and measures the increased cold stress and discomfort associated with wind.

The air temperature is not lowered by the windchill factor. Regardless of how strong the wind, the air temperature remains constant. The windchill factor is a measure of how rapidly heat is being removed from a body. If, for example, the air temperature is 40°F(4.4°C) and the wind speed is 20 mph (32 km/h), it feels the same as 19°F(−7°C) with no wind blowing.

A windchill factor near or below 0°F(−17.8°C) is an indication that there is a risk of frostbite or other injury to exposed human flesh. Between 10° and 15°F(−12° and −26°C), there is little danger; between −30° and −70°F (−34° and −57°C), there is danger that human flesh may freeze within 1 minute of exposure. Below −75°F (−59°C), there is great danger that human flesh may freeze within 30 seconds of exposure.

The effects of wind chill, however, depend on many factors such as the amount of clothing worn, health, age, gender, and body weight.

Cold

We have discussed heat and adding or removing heat, but what about "cold?" Actually, there is no such thing as "cold." Remember the definition "feeling no warmth?"

All matter, everything in nature or everything manufactured, contains heat. Some things contain more heat than others, but all contain heat to some degree.

***TABLE 2-1*: THE EFFECT OF THE WINDCHILL FACTOR**

Air Temp (°F)	Wind Speed (mph)							
	5	10	15	20	25	30	35	40*
40	37	28	23	19	16	13	12	11
30	27	16	9	4	1	−2	−4	−5
20	16	3	−5	−10	−15	−18	−20	−21
10	6	−9	−18	−24	−29	−33	−35	−37
0	−5	−22	−31	−39	−44	−49	−52	−53
−10	−15	−34	−45	−53	−59	−64	−67	−69
−20	−26	−46	−58	−67	−74	−79	−82	−84
−30	−38	−58	−72	−81	−88	−93	−97	−100
−40	−47	−71	−85	−95	−103	−109	−113	−115

*Winds above 40 mph (64 km/h) have little additional effect on windchill factor.

Absolute cold is the absence of all heat, or −459.67°F.

Cold, then, refers to an object or matter in which some of its heat has been removed; therefore, an object has more or less of its original heat. In an air-conditioning system, we are not really producing cold air. We are simply transferring heat; removing some of the heat from the car's interior where it is not wanted and transferring it to the outside air. The refrigerant in the air-conditioning system is the medium that is used for the transfer of this heat. The effect is that the car's interior becomes cool.

Specific Gravity

Specific gravity is the ratio of the mass or weight of a given volume of a substance divided by the density of an equal volume of water for solids and liquids. The specific gravity of water is 1.0, which is derived from the density of water divided by itself (62.4 lb/ft / 62.4 lb/ft = 1). We measure the specific gravity of engine coolant to determine the mixture of ethylene glycol and water contained in the system with a refractometer, which gives an exact freeze point temperature of the mixture. The specific gravity of a lead acid batteries' electrolyte solution when fully charged is 1.265.

Gas Laws

To understand how the refrigerant system functions, it is necessary to understand how gases respond to pressure and temperature changes. Several laws will help you to understand the reaction of the refrigerant gas and the pressure-temperature-volume relationship of refrigerant in various parts of the refrigerant system. In addition, these laws help you understand the effect of gas contamination on refrigerant system operation. When using the gas law equations, pressure must be in the absolute pressure scale (psia) and temperature must be measured (1 Kelvin = 1° Celsius), where 0°C is the freezing point of water, or on the Rankine scale, where 0°R is equal to −459.67°F, so water will freeze at 491.67°R. If the correct scale is not used, the solution of the equation will be meaningless. The absolute scales use zero as their starting point.

Gases exert pressure in all directions and will completely fill any container that holds them. Gas molecules have little attraction to each other and have neither definite shape nor volume.

Boyle's Law

Boyle's law was formulated by Robert Boyle (1627–1691) and states that the volume of a gas varies inversely with the absolute pressure at a constant temperature. In other words, as pressure is applied to a volume of gas in a closed container, the volume of gas will decrease and the pressure will increase. An example of this is the piston action of an engine. As the piston travels upward on the compression stroke, the pressure in the cylinder will increase (Figure 2-20). Boyle's law on its own is not practical, because as the gas is compressed, some of the heat generated during compression is transferred to the gas, and when the gas is expanded, some of the heat is given off. Formula for Boyle's Law:

$$P_1 \times V_1 = P_2 \times V_2$$

where P_1 = original absolute pressure

V_1 = original volume

P_2 = new pressure

V_2 = new volume

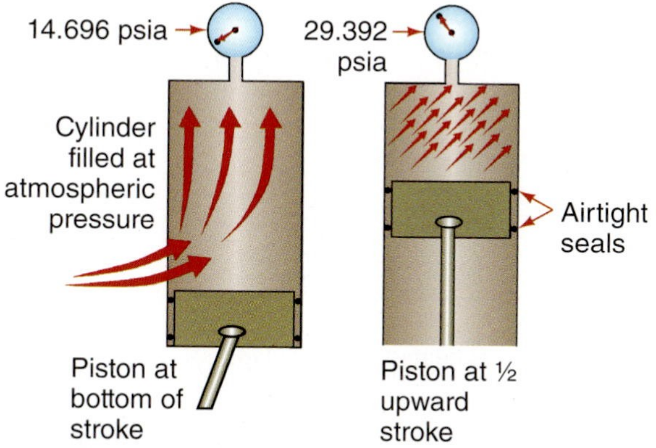

FIGURE 2-20 **If the volume of a container is reduced by half, then the absolute pressure in the container will double.**

As an example, if the original pressure in the container were 25 psia and the volume at the beginning were 25 in, what would the calculated new volume be when the pressure is increased to 50 psia?

$$V_2 = \frac{P_1 \times V_1}{P_2}$$

$$V_2 = \frac{25 \times 25}{50}$$

$$V_2 = 12.5 \text{ in}^3$$

Charles' Law

Charles' Law is also known as Gay-Lussac's Law and states that the volume of a fixed mass of gas held at a constant pressure varies directly with a change in the absolute temperature. And conversely, when a gas is held at a constant volume, its pressure varies directly with the absolute temperature.

$$\frac{V_1}{T_1} = \frac{V_2}{T_2} \text{ or } \frac{V_2}{V_1} = \frac{T_2}{T_1} \text{ or } V_1 \times T_2 = V_2 \times T_1$$

where V_1 = original volume

V_2 = new volume

T_1 = original temperature

T_2 = new temperature

If 100 ft of air were drawn in and passed over a vehicle heater core and were heated from 40°F to 140°F, what would be the volume of the air entering the vehicle passenger compartment? (Figure 2-21).

V_1 = 100 ft

V_2 = unknown volume

$T_1 = 40°F + 460° = 500°R$ (absolute)

$T_2 = 140°F + 460° = 600°R$ (absolute)

$$V_2 = \frac{V_1 \times T_2}{T_1}$$

$$V_2 = \frac{100 \text{ ft}^3 \times 600°R}{500°R}$$

$$V_2 = 120 \text{ ft}^3$$

The above formula shows that air expands as it is heated.

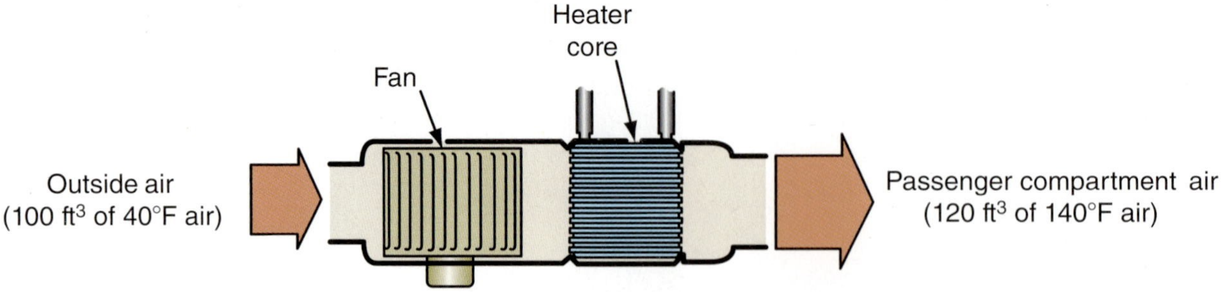

FIGURE 2-21 **As air is heated, it will expand.**

The next formula demonstrates the relationship between pressure and temperature:

$$\frac{P_1}{T_1} = \frac{P_2}{T_2} \text{ or } P_1T_2 = P_2T_1$$

where P_1 = original pressure
P_2 = new pressure
T_1 = original temperature
T_2 = new temperature

If a container containing 100 ft^3 of air were stored at 75°F (24°C) and the container was placed in a vehicle trunk that is 140°F (60°C), what would the pressure be if the original pressure of the air were 30 psig at 75°F (24°C)?

P_1 = 30 psig + 14.696 (absolute pressure) = 44.696 psia
P_2 = unknown volume
T_1 = 75°F + 460° = 535°R (absolute)
T_2 = 140°F + 460° = 600°R (absolute)

$$P_2 = \frac{P_1 \times T_2}{T_1}$$

$$P_2 = \frac{44.696 \times 600°R}{535°R} = 35.43 \text{ psi (absolute)}$$

$$P_2 = 35.43 \text{ psia} - 14.696 = 20.734 \text{ psig}$$

The above formula shows that when a volume of air in a closed container is heated, its pressure will increase.

Dalton' Law

John Dalton discovered in the early 1800s that the atmosphere is made up of several different gases and that each gas creates its own pressure. The total pressure of a confined mixture in a sealed container is equal to the sum of the pressures of each gas in the mixture. As an example, if oxygen is combined with nitrogen in a container of a given size, the resulting pressure will be greater than if either gas stood alone in that same given size container (Figure 2-22). It is assumed in this law that the volume of the mixture is the same as the volume of each individual gas.

AUTHOR'S NOTE: As you will see later in this text, if R134a is contaminated with R12, the resulting pressure at a given temperature will be greater than the individual pressure of the individual gas in its pure form.

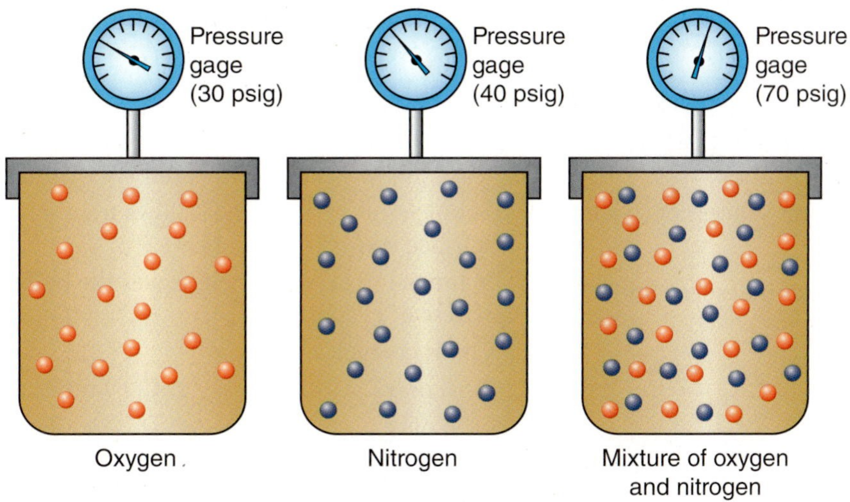

FIGURE 2-22 **The total pressure in a closed container is the sum of the individual pressures of each gas.**

TERMS TO KNOW

Absolute zero
Arid
Atom
British thermal unit (Btu)
Chlorofluorocarbon (CFC)
Conduction
Convection
Deep-tissue (subsurface) temperature
Evaporation
Heat
Humid
Hydrogen (H)
Insulator
Latent heat
Matter
Molecule
Perspiration
Radiation
Relative humidity (RH)
Sensible heat
Specific heat
Superheat
Temperature

SUMMARY

- When the surrounding temperature, known as ambient temperature, is above normal body temperature, one is said to feel warm or hot; when it is below, one is said to feel cool or cold.
- Everything in nature contains heat. This heat is known as specific heat. Some things retain/conduct heat better than other things. The absence of heat is cold.
- Latent heat is hidden heat. It is the heat that is required for a change of state of matter and it cannot be measured with a thermometer.
- Sensible heat can be measured with a thermometer and can be felt (sensed).
- Heat flows from a warm surface or object to a less warm surface or object.

REVIEW QUESTIONS

Short-Answer Essays

1. Explain heat transfer by convection.
2. Briefly define Dalton's Law.
3. Describe the molecular movement of matter based on its temperature.
4. How does liquid water become water vapor?
5. What is meant by the term subsurface temperature?
6. What is the unit of measurement used to measure latent heat?
7. What does the term relative humidity mean?
8. Briefly define latent heat.
9. Describe the transfer of heat by conduction.
10. Briefly define sensible heat.

Fill in the Blanks

1. ______________ Law states that the volume of a fixed mass of gas held at a constant pressure varies directly with a change in absolute temperature.
2. The absence of heat is cold; therefore, ______________ is ever present.
3. Moisture in the air is known as ______________.

4. Latent heat is required to cause a change of ______________.

5. ______________ Law states that the volume of a gas varies inversely with the absolute pressure at a constant temperature.

6. Moisture becomes vapor by the natural process of ______________.

7. ______________ heat cannot be felt or measured with the use of a thermometer.

8. A unit of heat measure is the ______________ thermal unit.

9. ______________exert pressure in all directions and will completely fill any container that holds them.

10. The metric conversion of pounds (weight) is ______________.

Multiple Choice

1. Technician A says that latent heat is hidden heat and cannot be measured on a thermometer.
   Technician B says that latent heat is hidden heat that is required for a change of state of matter.
   Who is correct?
   A. A only
   B. B only
   C. Both A and B
   D. Neither A nor B

2. The heat added to a vapor causing it to become warmer is called what?
   A. Sensible heat
   B. Superheat
   C. Latent heat
   D. Both A and B

3. What is a material called that can block the flow of heat?
   A. A conductor
   B. A convectant
   C. A radiant
   D. An insulator

4. All of the following are factors that affect body comfort except:
   A. Relative humidity.
   B. Air movement.
   C. Concentration.
   D. Temperature.

5. Heat transfer is being discussed:
   Technician A says that heat flows from a hot surface to a surface containing less heat.
   Technician B says that heat leaves the body by the process of evaporation.
   Who is correct?
   A. A only
   B. B only
   C. Both A and B
   D. Neither A nor B

6. The acceptable comfort range for the human body is:
   A. 72° – 80°F
   B. 45–50 percent relative humidity
   C. Both A and B
   D. Neither A nor B

7. Everything in nature contains what type of heat?
   A. Specific heat
   B. Latent heat
   C. Sensible heat
   D. Natural heat

8. What type of heat transfer requires air movement?
   A. Radiation
   B. Convection
   C. Conduction
   D. Evaporation

9. All of the following are processes by which heat moves, except:
   A. Radiation
   B. Latent
   C. Conduction
   D. Convection

10. All of the following are gas laws, except:
    A. Dalton's Law
    B. Boyle's Law
    C. Darwin's Law
    D. Charles' Law

Chapter 3

Electricity and Electronic Fundamentals

Upon Completion and Review of this Chapter, you should be able to:

- Explain the basics of voltage.
- Explain the basics of amperage.
- Explain the basics of resistance.
- Describe how to use Ohm's law.
- Discuss the basic types of electrical circuits.
- Explain the use of digital multimeters.
- Describe the operation of a diode.
- Discuss relay operation.
- Describe the operation of the HVAC blower motor and stepped resistor assemblies.
- Explain how the blower motor power transistor module functions in the blower motor circuit.
- Explain the application and function of an electromagnetic clutch assembly and the role of the clamping diode in the circuit.

Introduction

In order to work on and diagnose electrical elements associated with the automotive heating and air-conditioning system, you must first have a basic understanding of electricity and electronics in order to properly analyze both control and activation of complex systems. The concepts and principles covered in this chapter will form the foundation for understanding basic electricity and electronics, which are the operating principles behind the climate control systems used on today's vehicles. Diagnosing and servicing comfort control systems will become clear after learning basic electrical principles and component operation. Theories are critical in developing your skills as a technician and, in turn, will increase your productivity and overall worth to the trade.

Electrical Principles

Many heating and air-conditioning components are controlled or powered by electricity. Examples of this are the air-conditioning compressor clutch, the blower motor, and the complex climate control system. Therefore, a basic understanding of some of the electrical principles, including voltage, amperage, and resistance, is required. The following section is meant to be a review or overview of these concepts. It is recommended that a technician receive in-depth electrical and electronic training as part of their overall education. Electronics is part of every major system on the automobile and electrical failures have become routine complaints, though not always routine repairs.

The Basics: Voltage, Amperage, and Resistance

For many, it is often easier to think of electricity in terms of water flowing in your home, hydraulic system principles, as there is a visible cause and effect. The flow of electricity is similar to the flow of water through your household plumbing. Where the water pressure is similar to electrical pressure or voltage, the water flow is similar to current flow in a conductor or amperage, and a restriction such as a kink in a hose is similar to the resistance in an electrical system. But, unlike water, electricity does not flow out the end of the wire and poor onto the ground if left open.

Voltage (V) or electromotive force (EMF) is the electrical pressure and is measured in volts (V); it may be either direct current (DC) or alternating current (AC). In the automotive industry, especially on HEV and EV platforms, we deal with both. Current is the flow or rate of flow of electrons under pressure in a conductor between two points having a difference in potential and is measured in **amperes (A)** or amperage. A complete circuit is required for current to flow. A DC circuit requires a complete circuit or loop between positive (+) and negative (−) for current to flow. Resistance is the friction in an electrical circuit that restricts the flow of electrons under pressure and is measured in **ohms (Ω)**. Electrical resistance is a load on the moving current that must be present to do any useful work. Resistance controls the amount of current flow in an electrical circuit. Electrical devices that use electricity to operate have a greater amount of resistance than a conductor (wire) and are considered loads in a circuit. A motor, light bulb, or solenoids are examples of electrical loads in a circuit; a load in a circuit is the electrical device that consumes electricity. Poor connections and corrosion are examples of unwanted electrical loads in a circuit. Resistance in an electrical circuit is measured in ohms (H). If there is resistance (H) in the conductor, electrons will not flow as readily.

For all practical automotive purposes, electricity only flows through a good conductor; in the majority of automotive wiring, this is copper. It takes a high voltage (V) to flow current (A) through a poor conductor. Air, for example, is a poor conductor, but with a high enough voltage even air can flow electricity. Lightning with millions of volts is capable of flowing current through air! High-voltage electrical systems on HEV and EV platforms do not contain voltage as high as lightning, but it is not your 12-volt system either. To jump an inch of air it takes approximately 10,000 volts. Have you ever seen a faulty secondary ignition wire (spark plug wire) arcing to the engine? The voltage on the HEV and EV orange wiring harness is between 200 and 300 volts generally, which is too low to jump through air, but the capacitors and condensers in the system may contain much higher voltage.

To summarize:

- Voltage is the pressure that moves electrons and is measured in either AC or DC volts (V).
- Current or amperage is the flow or volume of electrons flowing in a conductor and is measured in amps (A).
- Electrical resistance is a load or opposition on the moving electrons (current) in a circuit and is measured in ohms (H).

Ohm's Law

The amount of current flow is determined by the amount of resistance in the loop of the circuit. In a fixed voltage circuit, if the amount of resistance in the loop of the circuit is high, the current flow will be low and if the amount of resistance in the loop of the circuit is low, the current flow will be high. This inverse relationship is known as **Ohm's law** and is summarized by the mathematical equation in Figure 3-1. Ohm's law states that it requires one volt of electrical pressure to move one amp through one ohm of resistance. Mathematically, Ohm's

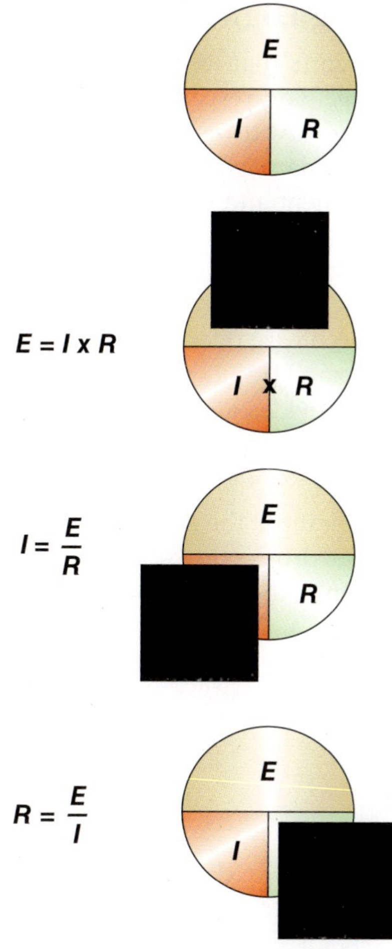

FIGURE 3-1 **With Ohm's law, if you know two of the three electrical factors, you can calculate the third.**

law is expressed by the following equations, using the symbols *E* for voltage, *I* for current, and *R* for resistance:

To find voltage $E = I \times R$

To find current $I = E \div R$

To find resistance $R = E \div I$

For example, if a 12.6-volt circuit has 2 ohms of resistance, you can use Ohm's law to determine the current flow in the circuit as follows:

$I = E \div R$

$I = 12.6 \div 2$

$I = 6.3$ amps

The current flow in the circuit is 6.3 amps.

These equations can be used to calculate the voltage, current, or resistance of a circuit. It is the understanding of this relationship that is most important to you as a technician. You are not generally crunching numbers but instead you are making measurements on a circuit with a digital multimeter (DMM) and trying to understand what this information means. Ohm's law can help to clear up what the meter is telling you for a given reading.

Types of Circuits and Using Digital Multimeters

There are three basic types of circuits that we will be dealing with and a few ways to test these circuits for voltage, amperage, and resistance with a DMM. It is imperative that you understand these basic concepts and how to test a circuit with a DMM to avoid damage to the electrical circuit and components or damage to your expensive DMM. Remember, Snap-On does not warranty misuse of equipment and service managers do not understand expensive mistakes! Take your time and think before you test. Remember the best place to begin your diagnosis is with the electrical diagram and system operation description contained in the vehicle service information database. As stated earlier, this section is meant to be a review and not an in-depth training section, so please refer to *Today's Technician Automotive Electricity & Electronics* for more information and training.

Series circuit is a circuit that provides a single path for current flow from the electrical source through all the circuit's components and back to the source (Figure 3-2).

Series circuit laws:

- Current flow is the same at any point in the circuit (Figure 3-3).
- The sum of the individual voltage drops equals the source voltage (Figure 3-4).
- Total circuit resistance is the sum of the individual circuit resistances (Figure 3-5).

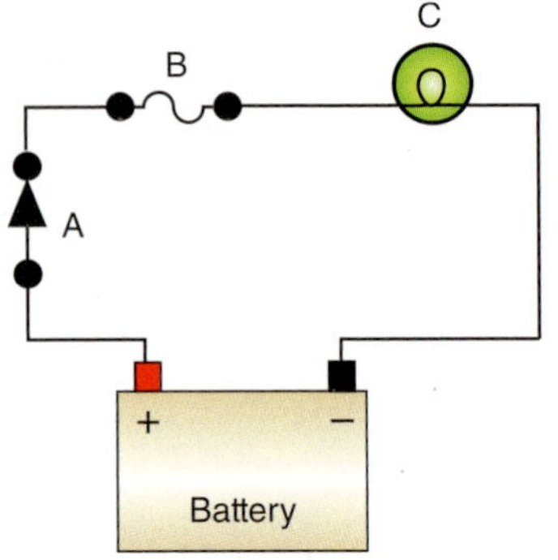

FIGURE 3-2 **A simple series circuit including a switch (A), a fuse (B), and a lamp (C). For all practical purposes, the only load in the circuit is the lamp.**

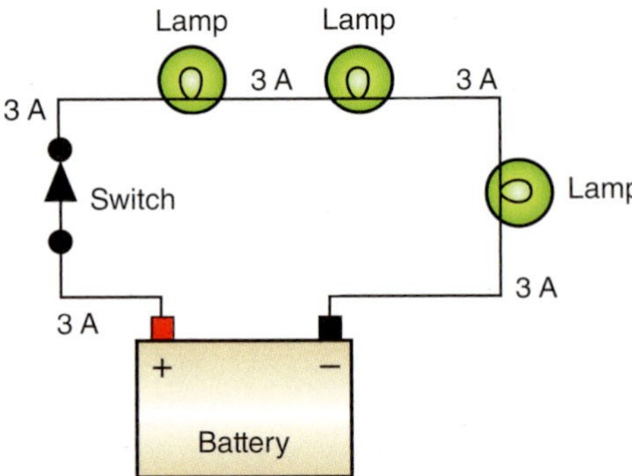

FIGURE 3-3 **Current flow is the same at any point in the circuit.**

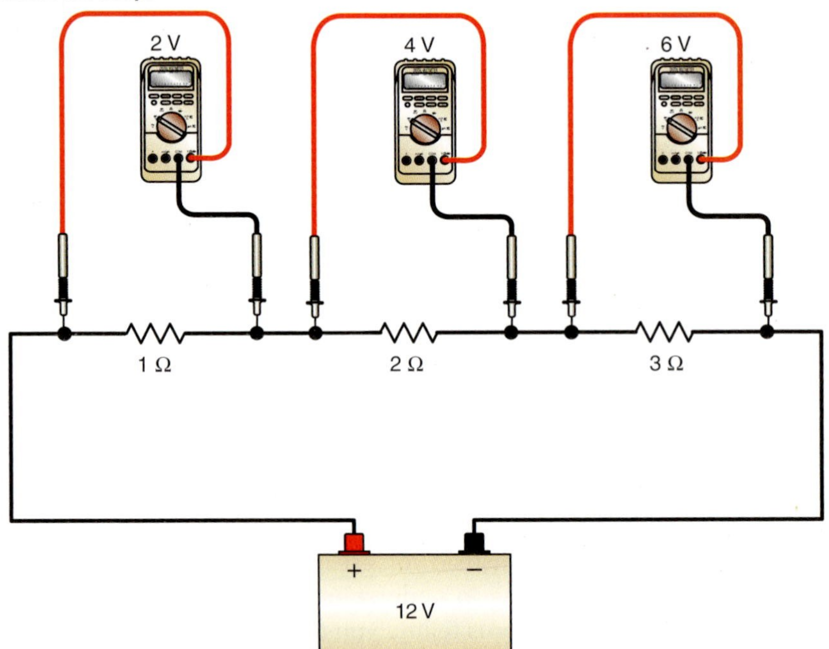

FIGURE 3-4 **The sum of the individual voltage drops equals the source voltage (2V + 4V + 6V = 12V or source voltage).**

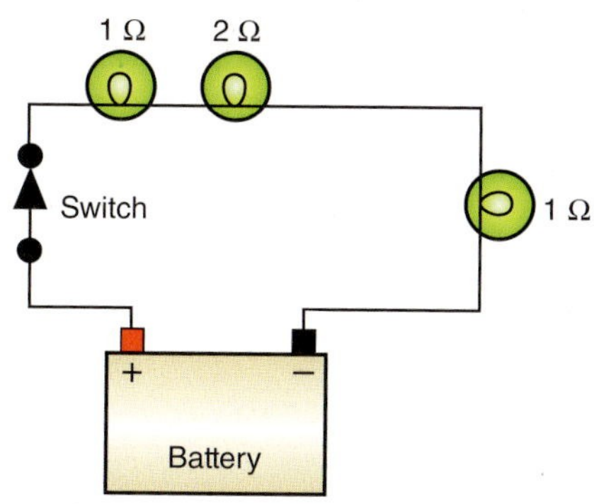

FIGURE 3-5 **Total circuit resistance is the sum of the individual circuit resistances, 1Ω + 2Ω + 1Ω = 4Ω of total circuit resistance.**

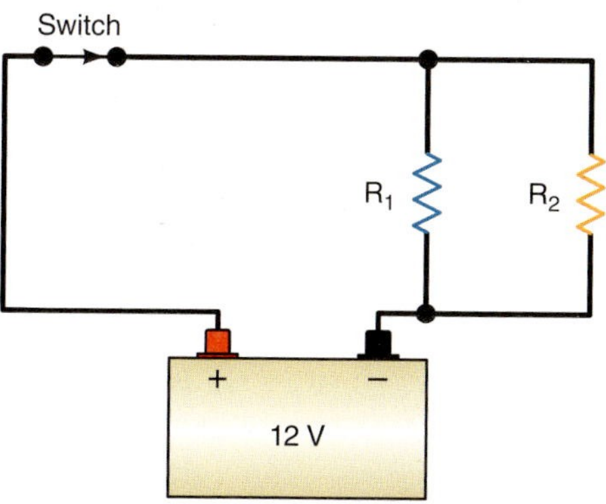

FIGURE 3-6 **There are multiple paths for current to flow in a parallel circuit. If R_1 were to fail, R_2 would still work.**

Parallel circuit is a circuit that provides two or more paths for current flow through all the circuit's components and back to the source (Figure 3-6). In a parallel circuit, each path has separate resistances that operate independently or in conjunction with one another depending on the design of the circuit. Current can flow through more than one branch at a time and voltage is the same across each branch of the circuit. In this type of circuit, the failure of a component in one branch does not affect the operation of the components in the other branches of the circuit.

Parallel circuit laws:

- Voltage is the same across each branch of the parallel circuit (Figure 3-7) and the voltage in each branch is used by the load(s) in that branch. The voltage dropped across each parallel branch will be the same; however, if the branch contains more than one resistor, the voltage drop across each of them will depend on the resistance of each resistor in the branch.

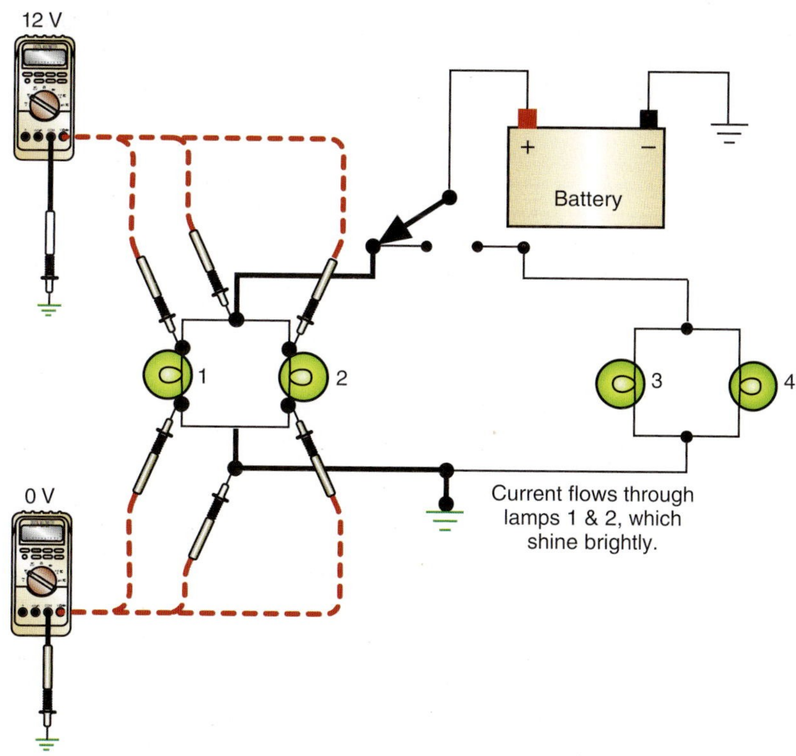

FIGURE 3-7 **Voltage is the same across each branch of the parallel circuit and the voltage in each branch is used by the load(s) in that branch.**

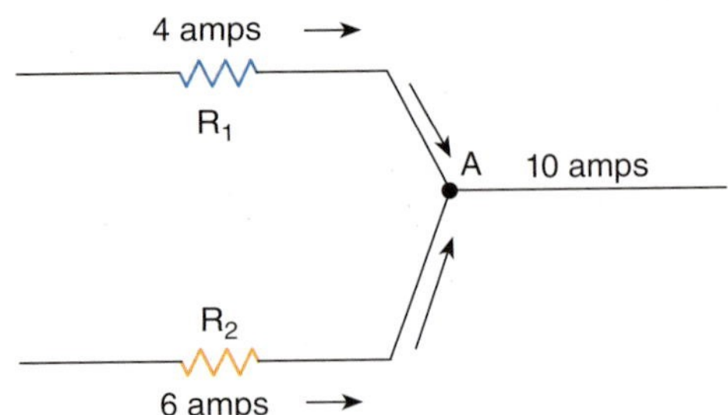

FIGURE 3-8 **Total current in a parallel circuit is equal to the sum of the individual branch circuits.**

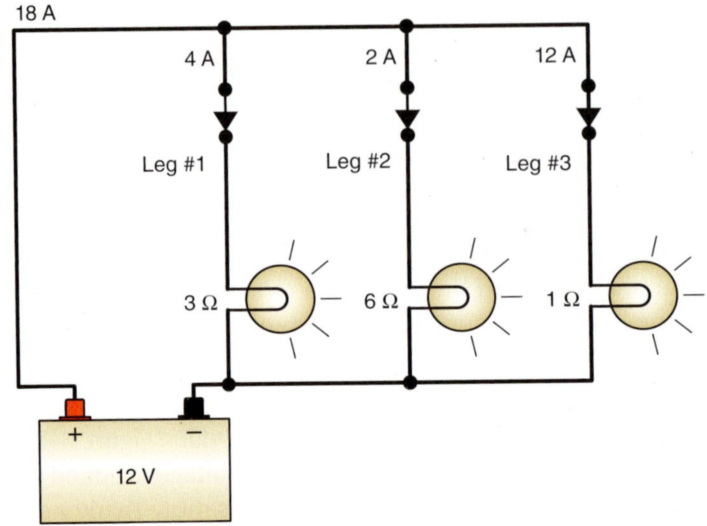

FIGURE 3-9 **Total resistance in a parallel circuit is always less than the smallest resistive branch.**

- Total current in a parallel circuit is equal to the sum of the individual branch currents (Figure 3-8).
- Total resistance in a parallel circuit is always less than the smallest resistive branch (Figure 3-9).

$$R_T = \frac{1}{1/R_1 + 1/R_2 + 1/R_3 ... 1/R_{10}}$$

$$R_T = \frac{1}{1/3\Omega + 1/6\Omega + 1/1\Omega}$$

$$R_T = \frac{1}{0.333 + 0.166 + 1}$$

$$R_T = \frac{1}{1.5}$$

$$R_T = 0.667\Omega$$

The series–parallel circuit has some loads that are in series and some that are in parallel with each other (Figure 3-10).

A voltmeter can be used to check for available voltage at the battery, terminals of any component, or connectors. It can also be used to test voltage drops across electrical circuits, component loads, connectors, and switches. A voltmeter is connected in parallel with the circuit being tested (Figure 3-11). In Figure 3-12, voltage is being tested in a closed 12-volt series circuit with two loads. At test point A, the voltage should be the source voltage of 12 volts. At point B, the 1Ω resistor would have dropped half the voltage (there are two 1Ω loads

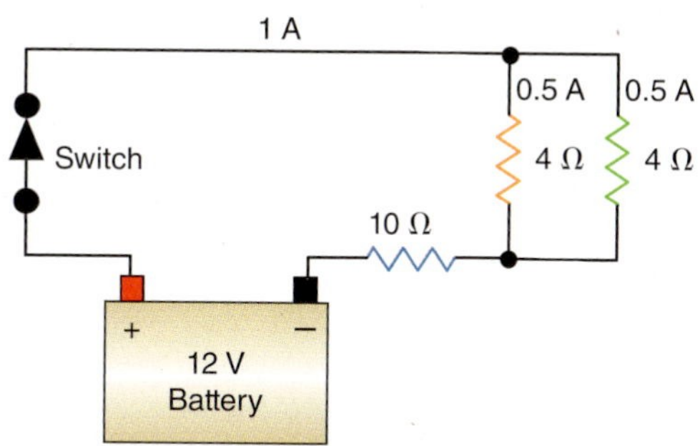

FIGURE 3-10 The series–parallel circuit has some load that are in series and some that are in parallel with each other.

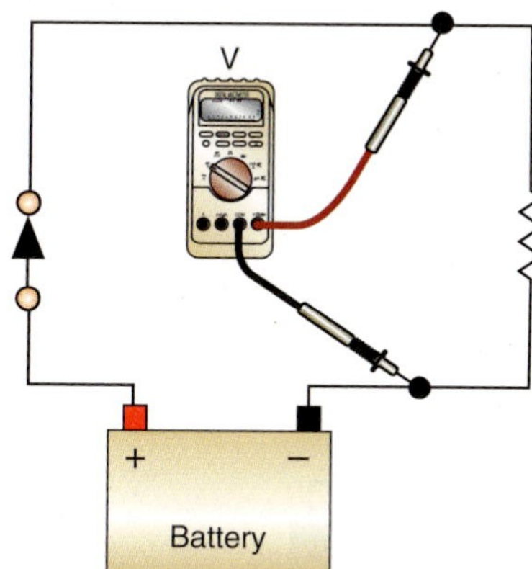

FIGURE 3-11 A voltmeter is connected in parallel with the circuit being tested.

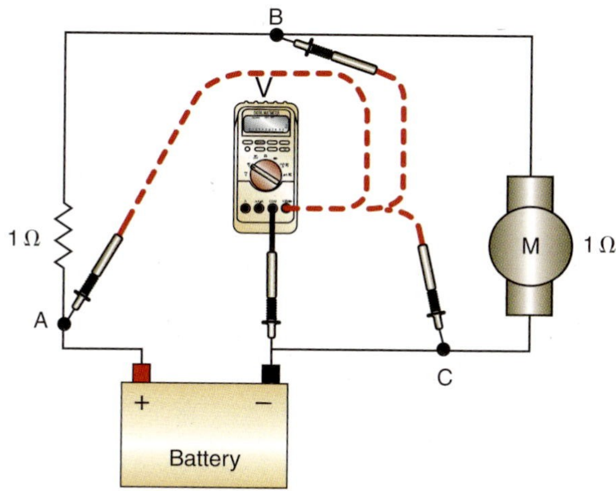

FIGURE 3-12 A voltmeter testing for voltage at various points in the series circuit.

in the circuit) and the meter should read 6 volts. At test point C, all the voltage should have been used up by the two loads in the circuit and the meter reading should be 0 volts. These readings would indicate normal circuit operation.

One of the most useful tests to perform is the voltage drop test. In a circuit, all of the voltage provided by the source power is used (dropped) by the circuit, with nothing left over. The voltage is used by resistance in wiring, connectors, switches, and loads. This loss or use of voltage is called a voltage drop and is the amount of electrical energy converted into another form of energy. Voltage dropped in wiring, connectors, and switches is converted into heat energy. When a circuit or branch of a circuit has only one load (resistance) source, voltage is dropped across that load (Figure 3-13). If there is more than one load in a circuit, each load will use a portion of the voltage. The total of all voltage drops in a circuit should equal source voltage. All of the voltage must be used by the circuit. You should verify that the circuit is turned on and that source voltage is available at the load and that the load drops source voltage. If the component (load) that is in the circuit drops source voltage, then the component is faulty. In Figure 3-14, if both lamp 1 and lamp 2 are the same resistance value (size), they will both share the source voltage equally. Lamp 1 will use 6 volts and lamp 2 will use 6 volts. If there is unwanted resistance in a circuit, the load in the circuit will not drop all source voltage (Figure 3-15). There is some allowable voltage drop by circuit components other than the load, but it is generally limited to a maximum of:

0.2 V (200 mV) for wires and cables

0.3 V (300 mV) for a switch

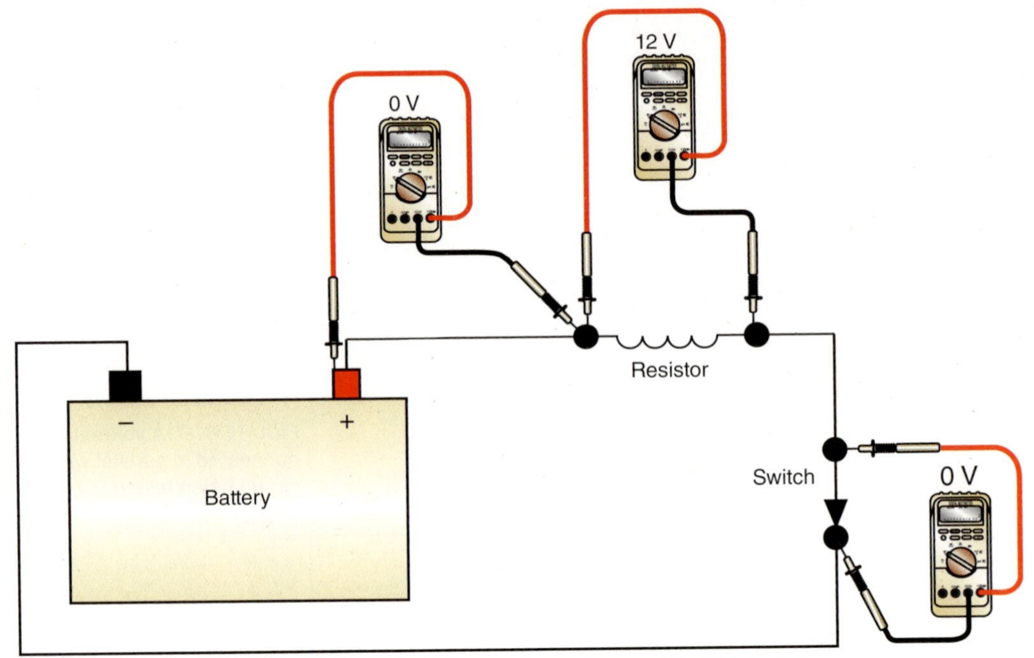

FIGURE 3-13 **Only the load will drop source voltage; the wiring and the switch should not drop significant voltage.**

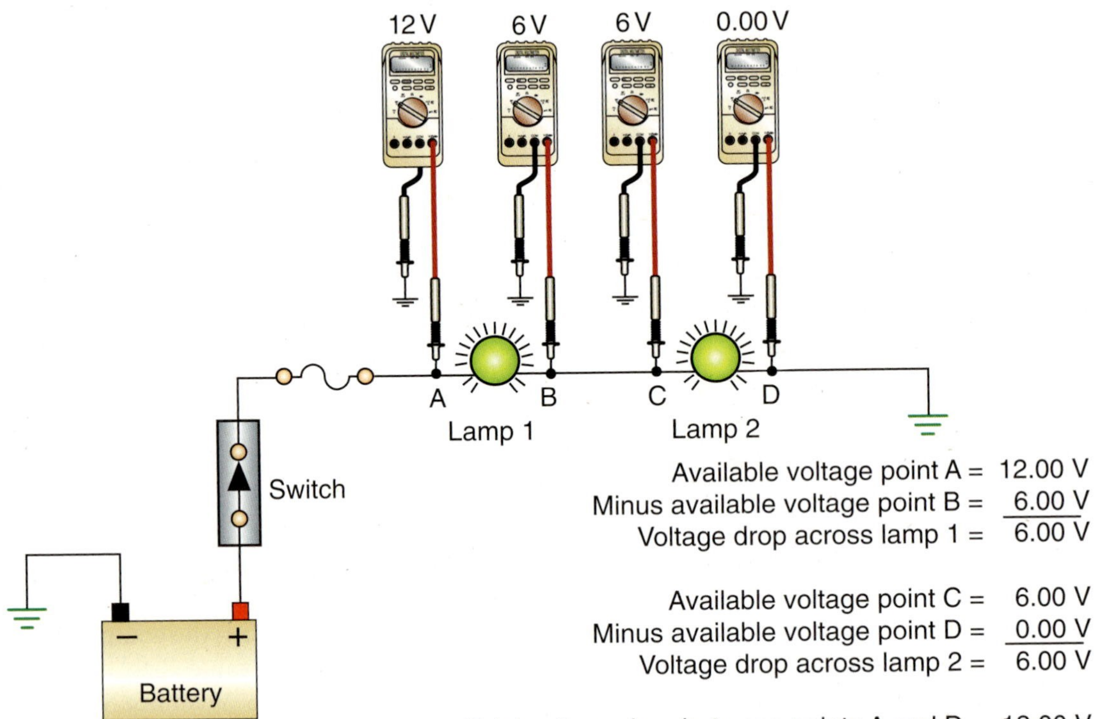

FIGURE 3-14 **There should be source voltage before the first load. If there are two loads in a series circuit, each load will drop a portion of the source voltage. There should be no voltage after the last load.**

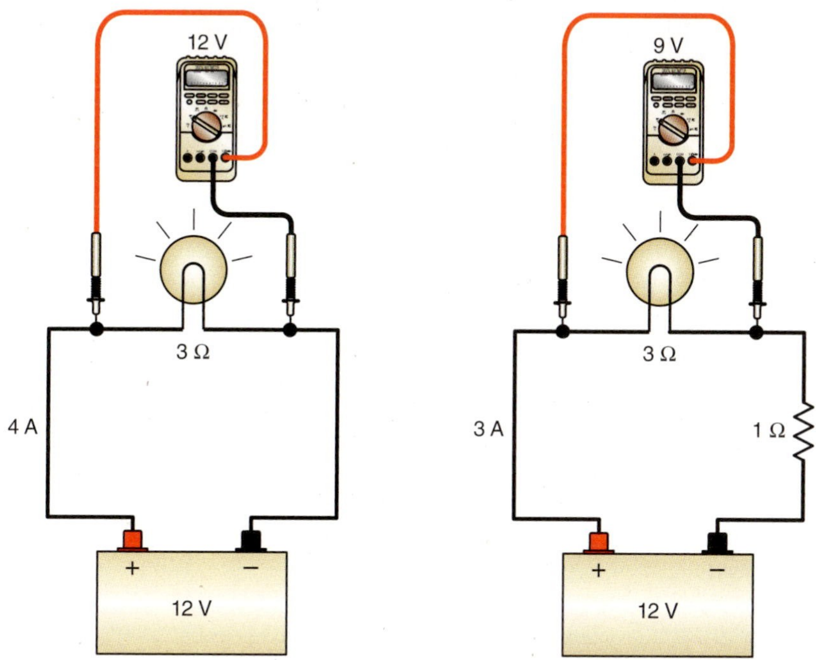

FIGURE 3-15 **Only the load will drop source voltage; the wiring and the switch should not drop significant voltage. But if there is unwanted resistance in a circuit, source voltage will not be available and the bulb would be dimmer than normal or may not light at all.**

***TABLE 3-1*: CIRCUIT TEST CHART**

Type of Defect	Test Unit	Expected Results
Open	Ohmmeter	∞ infinite resistance between conductor ends
	Test light	No light after open
	Voltmeter	Ø volts at end of conductor after the open
Short to Ground	Ohmmeter	Ø resistance to ground
	Test light	Lights if connected across fuse
	Voltmeter	Generally not used to test for ground
Short	Ohmmeter	Lower than specified resistance through load component Ø resistance to adjacent conductor
	Test light	Light will illuminate on both conductors
	Voltmeter	A voltage will be read on both conductors
Excessive Resistance	Ohmmeter	Higher than specified resistance through circuit
	Test light	Light illuminates dimly
	Voltmeter	Voltage will be read when connected in parallel over resistance

0.1 (100 mV) for a ground

0 V for a connection or connectors

Common circuit faults (Table 3-1):

- Short circuits
- Short to ground
- Opens in a circuit

- High resistance (may be caused by corrosion or poor connections)
- Low voltage

When using an ammeter to measure amperage, be sure to first turn power off in the circuit before connecting the multimeter. The circuit must be opened (disconnected) from the load being tested. The multimeter is placed in series in the circuit, recompleting the disconnected circuit (Figure 3-16). Always verify that the multimeter is capable of handling the highest expected amperage in the circuit being tested. What size fuse protects the circuit being tested? Many multimeters are only internally protected to 10 amperes by their internal fuse and many automotive circuits can provide 20 or more amps. In these cases, an inductive amp probe is the best choice of tools to use to avoid multimeter damage (Figure 3-17). The inductive probe eliminates the need to connect the multimeter in series and is a safe noninvasive method of measuring amperage in a circuit.

When using an ohmmeter to measure resistance, the power from the circuit must be removed and the circuit or component should be isolated (Figure 3-18). Ohmmeter leads are placed across or in parallel with the component or circuit being tested.

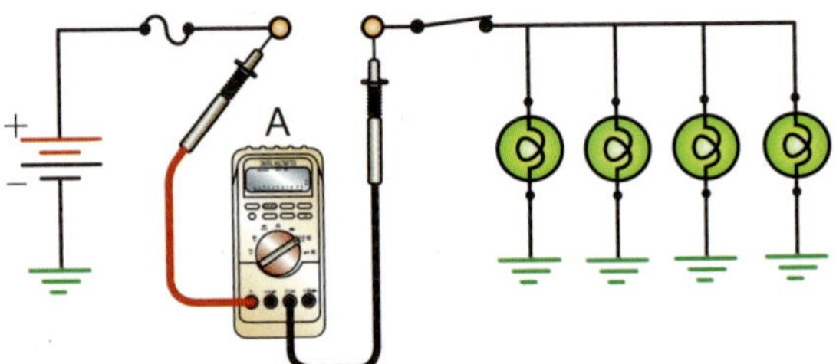

FIGURE 3-16 **When testing amperage, the circuit must be opened (disconnected) from the load being tested and the multimeter is placed in series in the circuit.**

FIGURE 3-17 **A multimeter with an inductive amp probe is a safe, noninvasive method of measuring amperage in a circuit.**

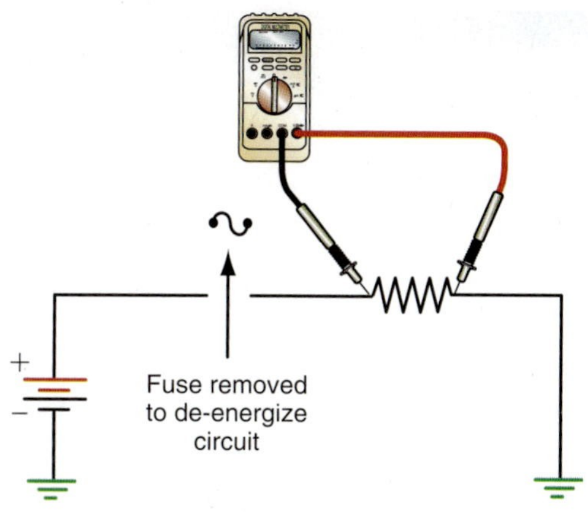

FIGURE 3-18 **When measuring resistance with an ohmmeter, the meter is connected in parallel with power removed from the circuit and the circuit or component isolated.**

FUSES AND CIRCUIT BREAKERS

A **fuse** or **circuit breaker** (Figure 3-19) is used to protect the air-conditioning and heating components and wiring. Usually rated at 20-30 amperes, they should not be replaced with one having a different rating. If a fuse or circuit breaker is rated too low, it will not carry the load and will quickly burn out (blow). If it is rated too high, the device it is intended to protect may be damaged due to excessive current.

A **fuse** is an electrical device used to protect a circuit against accidental overload or unit malfunction.

The fuse or circuit breaker is usually located in the main fuse block. Major circuits are often protected by a fusible link (Figure 3-20) or a maxi-fuse. Fuses may also be located in in-line fuse holders (Figure 3-21).

Circuit breakers are constructed of a bimetallic strip and a set of contacts. Excessive current, caused by a defective component, produces heat that causes the bimetallic strip to bend. When the strip bends, the contacts open and current to the component is interrupted. When there is no current flow, the bimetallic strip cools and the contacts automatically close. This continues until the cause of the problem is corrected.

A **circuit breaker** is a bimetallic electrical device used to protect a circuit against accidental overload or unit malfunction. It automatically resets once it cools down.

FIGURE 3-19 **Fuses and circuit breakers.**

Most factory-installed heater and air-conditioner electrical schematics require two or more pages.

Wiring diagrams are also known as *schematics.*

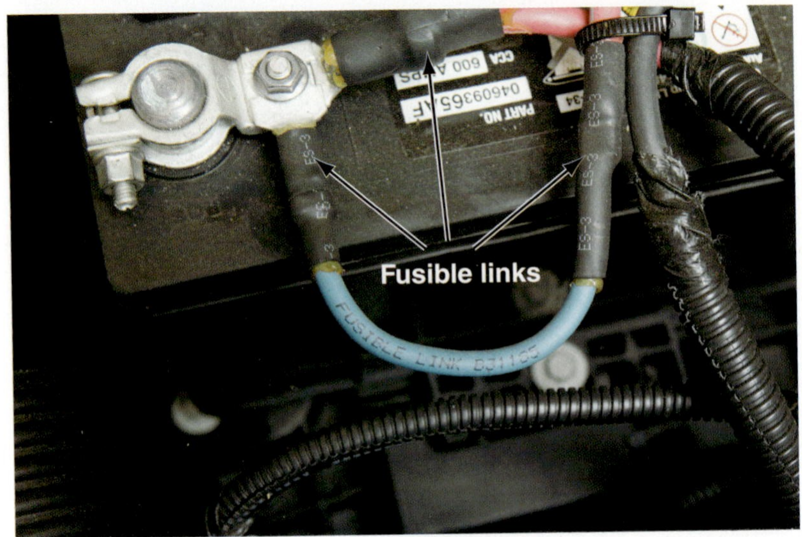

FIGURE 3-20 **A fusible link.**

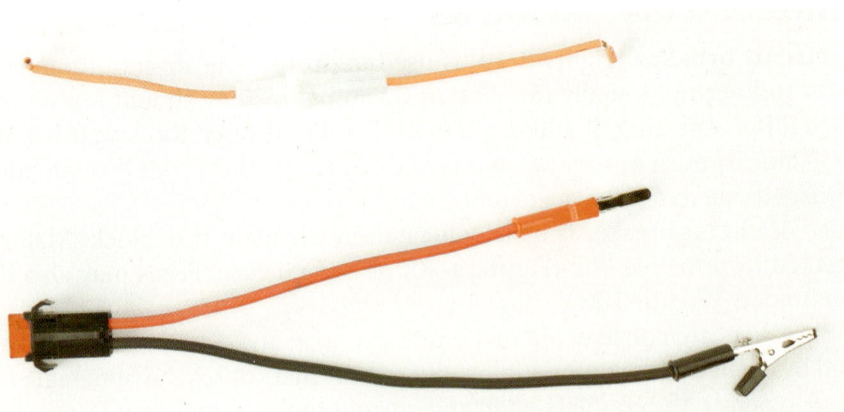

FIGURE 3-21 **In-line fuse holders.**

Circuit Resistance

Resistors are devices constructed to introduce a measured amount of electrical resistance into a circuit. Resistors may be used to limit current flow, and thereby voltage, in a circuit where source current flow and voltage are not required or where too much voltage may cause damage. Resistance in a circuit may also be used to produce heat; an example would be the rear window defroster or light such as bulbs.

There are three common types of resistors used in automotive circuits: fixed, tapped or stepped, and variable. Fixed resistors are designed to have a set value that should not change. These types of resistors are commonly used to control voltage. Tapped or stepped resistors have two or more fixed value resistors with wire taps after each to provide stepped voltage outputs (Figure 3-22). An example of this type of circuit is the basic blower motor speed control circuit, which will be discussed and described in the blower motor section of this chapter.

Variable resistors have an infinite number of resistance values within a range. Two examples of variable resistors are the rheostat and the potentiometer. Rheostats have two wire connections (Figure 3-23), one to the fixed end of the resistor and one to a sliding variable resistance contact or wiper. Moving the sliding contact, wiper, connected to a knob or lever, moves the connection either away from or closer to the fixed end of the tap, causing the

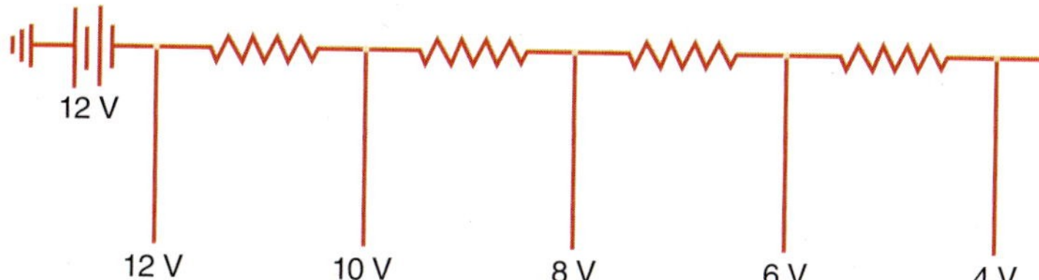

FIGURE 3-22 The use of a stepped resistor assembly is a common application of fixed resistors.

FIGURE 3-23 A rheostat.

FIGURE 3-24 A potentiometer.

resistance to increase or decrease, respectively. A common use for a rheostat is the instrument panel lighting (dimmer) switch, which varies the amount of current to dash light bulbs. Potentiometers are similar to rheostats except that they have three wire connections (Figure 3-24), one at each end of the resistance and one connected to the sliding contact, wiper, with the resistor. Moving the wiper slide contact away from the reference voltage source (+) end of the resistance and toward the output end (−) increases the resistance, which in turn decreases the output voltage on the signal or sliding output side of the circuit. Potentiometers are a common type of input sensor used by vehicle computers to monitor linear movement. Examples are throttle position on the engine throttle body or the mode door position in the climate control duct system of the vehicle.

Another form of variable resistor used is the thermistor (Figure 3-25). A thermistor changes resistance value with a change in temperature. There are two types of thermistors: negative temperature coefficient and positive temperature coefficient thermistors. Thermistors are used to provide compensating voltage in a component or to monitor temperature. As a temperature sender such as a coolant temperature sensor, the thermistor is connected to voltmeter circuit calibrated in degrees. As the temperature of the thermistor increases or decreases the resistance will also change. In a negative temperature coefficient resistor, the resistance goes up as the temperature decreases and in a positive temperature coefficient thermistor the resistance increases with an increase in temperature. These types of sensors are used extensively in the heating and air-conditioning systems used in the automotive industry as well as other areas of the vehicle.

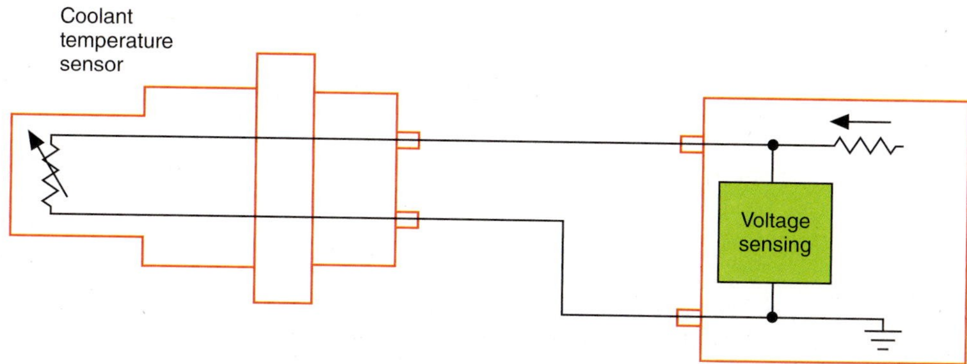

FIGURE 3-25 A thermistor is used to measure temperature. The sensing unit measures the change in resistance and translates this into a temperature valve.

DIODES

A diode is one of the simplest semiconductor devices and functions as an electrical one-way check valve that allows current to flow in one direction only. It is formed by joining P-type (positive) semiconductor material with N-type (negative) semiconductor material. The positive side of the diode is called the anode and the negative side is called the cathode (Figure 3-26) and the point where the two join is called the PN junction. The outer housing of the diode will have a stripe painted around it on one end designating the cathode side.

Diodes may be placed in a circuit in either forward or reverse biased position depending on the function of the diode in the circuit. Reverse biased means that the positive voltage is applied to the negative, cathode side of the diode and negative voltage is applied to the positive, anode side of the diode (Figure 3-27). Diodes are often used in circuits in reverse bias position to function as a clamping or protective device in a circuit. When a coil such as a solenoid or relay is turned off, the magnetic field surrounding the coil collapse and induces a voltage spike in reverse bias into the circuit (Figure 3-28). When these types of circuits are controlled by a microprocessor, a clamping device is needed to protect the solid-state components and diodes wired reverse biased are often chosen. A diode placed in parallel with the coil creates a bypass for the electrons when the circuit is turned off and will redirect this voltage spike back through the coil until all the voltage is used up by the coil.

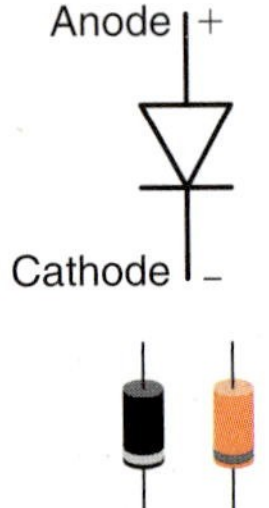

FIGURE 3-26 **The top image is the symbol for a diodes and the lower two images show the location of the painted strip indicating the cathode or negative material side of the diode.**

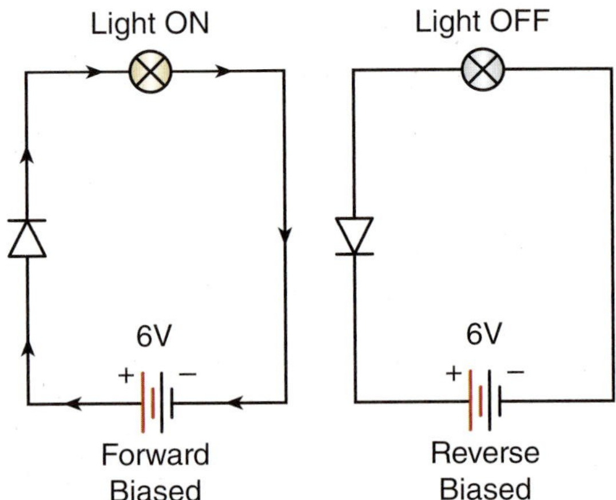

FIGURE 3-27 **A forward-biased diode in the circuit would allow current to flow and the bulb would be on. A reverse-biased diode in the circuit would not allow current to flow and the bulb would be off.**

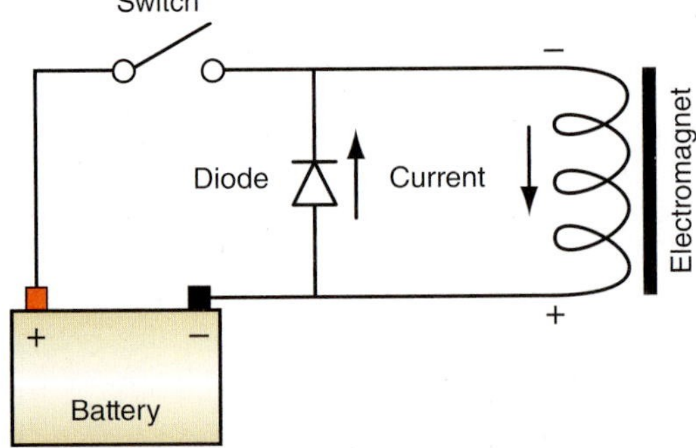

FIGURE 3-28 **When a coil such as a solenoid or relay is turned off, the magnetic field surrounding the coil collapses and induces a voltage spike in reverse bias into the circuit. A diode will redirect this voltage spike back through the coil until all the voltage is used up.**

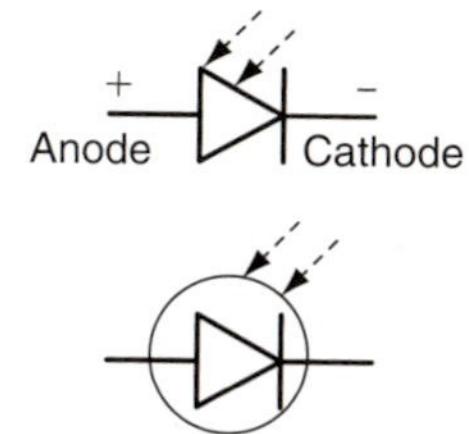

FIGURE 3-29 **The photodiode allows reverse current flow only when a specific amount of light is received.**

Another type of diode is the photodiode (Figure 3-29), which, like a standard diode, only allows current to flow in one direction. However, the direction of current flow is opposite that of a standard diode. Reverse current flow only occurs when the photodiode receives a specific amount of light. Photodiodes operate by the absorption of photons, which are charged particles that carry light and allow a flow of current proportional to the level of light received in the circuit they are connected to. Photodiodes can be used to detect minute amounts of light and can be calibrated for extremely accurate light measurements. As long as the photodiodes' response to light is linear, the output voltage will be proportional to the light level detected by the photodiode.

RELAYS

A relay is an electromagnetic switch that allows a small amount of current to control a large current (Figure 3-30). The low current control circuit is connected to the coil side of the relay. Relays are used where a low current control may be used to activate a high current load. Often, relays are used to control electric motors such as blower motors and air-conditioning compressor clutch coils.

Many relay terminals are identified using the International Standards Organization (ISO) numbering convention (Figure 3-31) for common size and terminal patterns. Even if the circuit you are working on does not use the ISO numbering convention, if you think about all relays based on this convention and terminal identification, the diagnostic process is simplified.

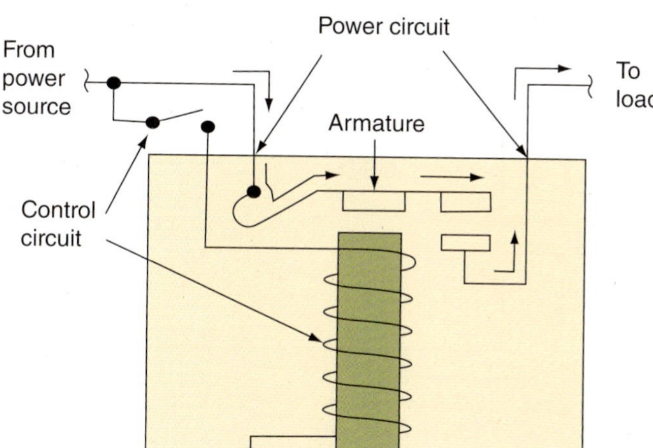

FIGURE 3-30 **A typical relay.**

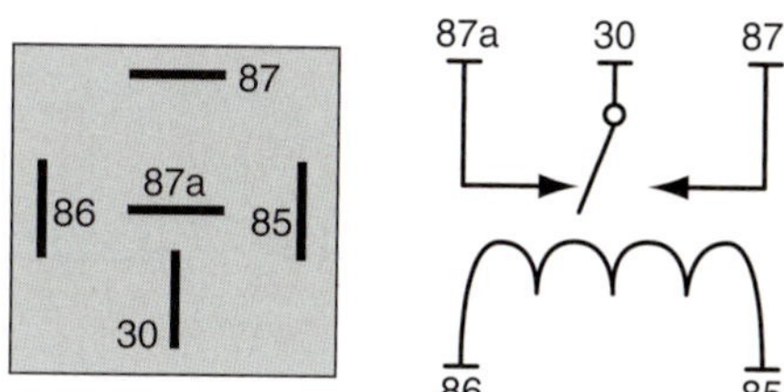

Terminal 86 is the coil high connection on the low current control side
Terminal 85 is the coil low connection on the low current control side
Terminal 30 is the common on the high current side
Terminal 87 is the normally open (NO) contact on the high current side
Terminal 87a is the normally closed (NC) contact on the high current side

FIGURE 3-31 **Relay terminals are identified using the International Standards Organization (ISO) numbering convention.**

The relay low current coil side may be controlled by switching either the power (terminal 86) (Figure 3-32) or ground (terminal 85) side of the circuit (Figure 3-33). The control circuit is wired using a very small diameter wire due to the low current flow required. The coil develops a magnetic field that closes the normally open contact (and opens the normally closed contact if equipped), which in turn energizes the high current load in the circuit. The relay contacts may control either the power (Figure 3-32) or the ground side (Figure 3-33) of the load to energize the circuit. The control circuit amperage is often as low as 0.25 amperes while the high current contact side of the relay may have 25 or more amperes to power the circuit load device.

Relays are often controlled by a microprocessor. When relays are controlled by semiconductors such as transistors, they require some type of voltage suppression device to protect the circuit since solid-state circuits are vulnerable to voltage spikes. Voltage spikes are like bolts of lightning striking the transistors, destroying them. Some computer circuits have voltage suppression built inside the computer, others rely on voltage suppression from within the relay. Diodes, high-ohm resistors, or capacitors can be used for voltage suppression (Figure 3-34). Resistors and diodes are the most common suppression devices used. Generally, a relay is clearly marked if a suppression diode or resistors are used internally in the relay.

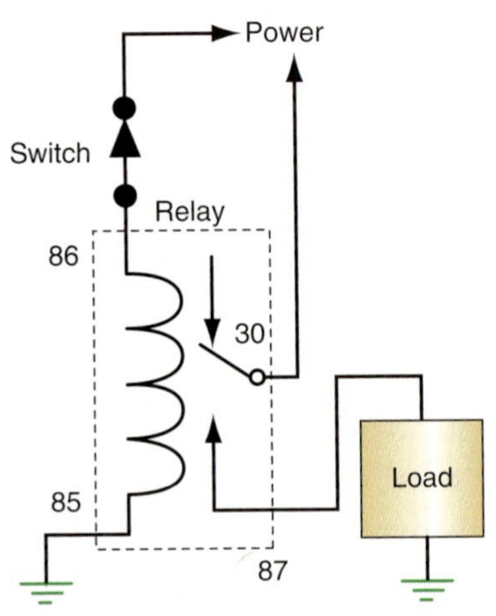

FIGURE 3-32 **The low current coil may be power switched.**

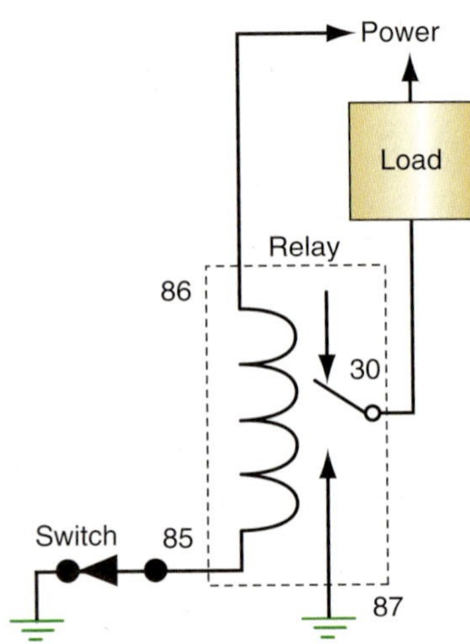

FIGURE 3-33 **The low current coil may be ground switched.**

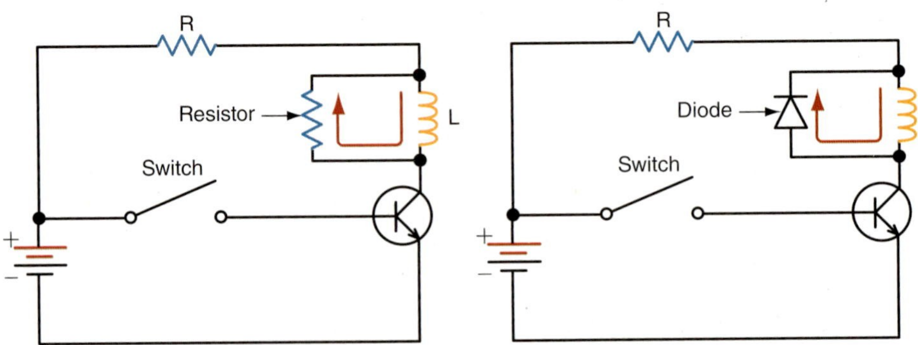

FIGURE 3-34 **Diodes and high ohm resistors can be used for voltage suppression to protect solid-state coil control components in a circuit.**

Relays are rated based on the control voltage for the coil side. In the automotive industry, this means that the typical relay is rated for 12 volts. The contact side of a relay is rated based on the amount of amperes that it can carry. Typical ampere ratings of automotive relays are 20, 25, and 30. It is important that the correct replacement relay is used. Relays may look identical but have different amperage ratings. Always use the relay that is specified for the circuit being repaired.

BLOWER MOTOR

The blower motor fan is an essential part of the heating and air-conditioning system. It is responsible for drawing in air from the cowl or from inside (recirculated) the passenger compartment. The air is then forced through the duct system and directed over the heater core and evaporator and then on to the air distribution network.

The most common blower motor design today is the single-wound brush or brushless motor (Figure 3-35). The motor is usually located in the HVAC housing. A fan switch controls the blower motor fan speed with settings ranging from low to high; some climate control systems in the FULL AUTO mode have the ability to automatically control the full range of blower speeds.

Blower motors may have a single or a double shaft. Some have provisions for flange mounting and may also have provisions for internal cooling. Regardless of style or type, the blower motor drives a squirrel-cage blower to move air across the evaporator and heater core (Figure 3-36).

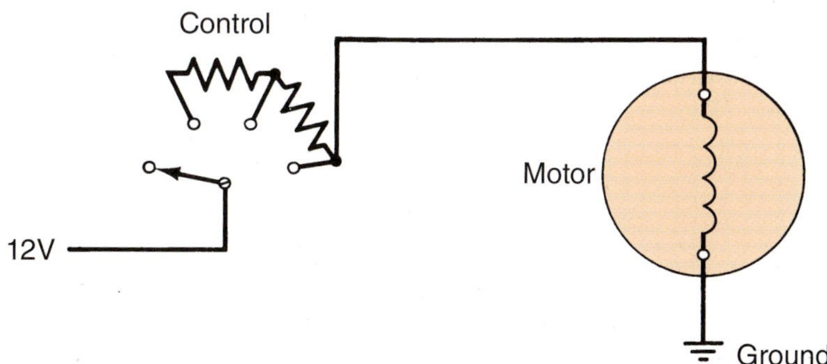

FIGURE 3-35 **A single-wound motor.**

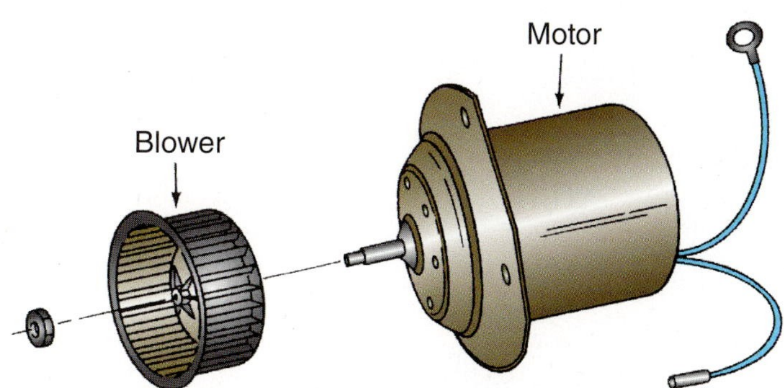

FIGURE 3-36 **A squirrel cage blower with motor.**

The high blower motor speed provides the greatest volume of air across the evaporator core for greatest refrigerant evaporation. Lowering fan speed decreases air volume and allows the slower moving air to remain in contact with the evaporator/heater core for a longer period of time, enabling more heat to be removed/picked up by the air stream. The result is colder/hotter air being discharged at slower fan speeds. Several methods used by manufacturers to regulate blower motor speed will be discussed next.

It should be noted that some replacement motors are reversible, whereas most are not. It is also important to note whether the defective motor turned clockwise or counterclockwise facing the shaft end. The replacement motor selected must turn in the same direction. If the wrong motor is installed, little or no airflow will circulate through the duct system. In addition, some replacement blower motors do not come with a blower cage. In these circumstances, the old cage must be reused. Ensure that the cage fits snugly on the motor shaft and does not slip. Other abbreviations relating to the selection of blower motors include a double shaft (DBL) and threaded shaft (THD) end.

AUTHOR'S NOTE: You may notice a small steel clip on one of the blower motor cage fins; this is a weight to balance the assembly. Do not remove this clip or vibration will result, which could also lead to premature motor bearing failure and customer complaints.

Blower Motor Resistor Block. Historically one of the most popular methods of blower motor speed control on vehicles equipped with manual HVAC control has been through the use of resistors. Wire resistors are used with a single-winding blower motor to regulate motor speed. The blower motor resistor is generally a group of three or four wire resistors integrated into a block assembly (Figure 3-37). Each of the stepped resistors reduces the current flow through the blower motor to change blower motor speed (Figure 3-38).

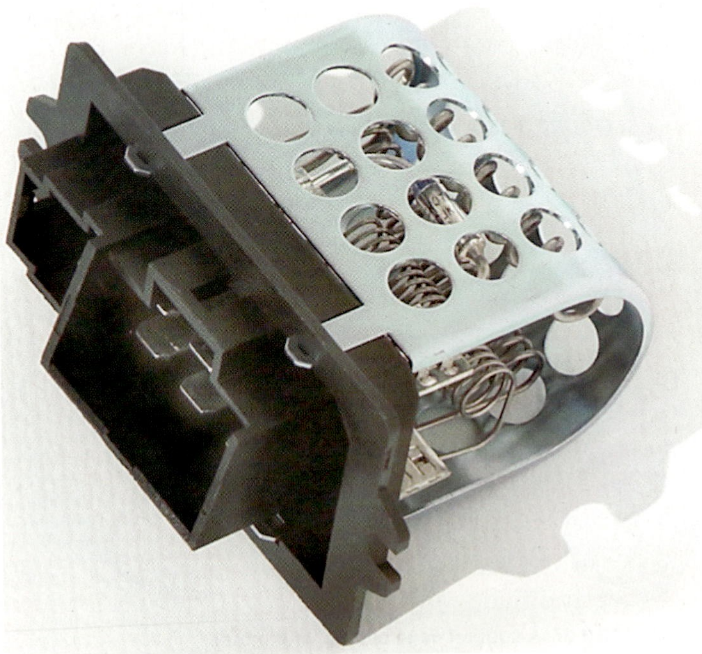

FIGURE 3-37 **The blower motor wire resistor block assembly consists of three to four wire-wound resistors mounted in a frame with integrated connection terminals.**

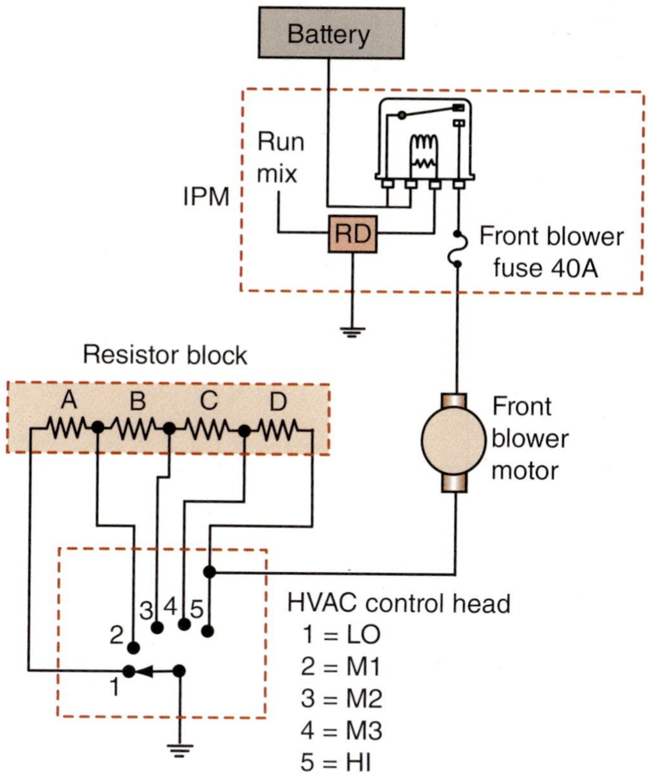

FIGURE 3-38 **Blower motor resistor speed control circuit. The blower motor control switch directs current through one to four resistors (A–D) to regulate motor speed.**

The blower motor switch directs current flow through the correct resistor wire to obtain the selected speed. For low blower speed, the switch is in position 1 and current is directed through all four resistors, A–D, which will drop current in the circuit, limiting motor speed. The resistor assembly may be wired on either the power or ground side of the blower motor. The resistor block is mounted in the HVAC ventilation duct to allow for air stream cooling of the resistors. There are several design variations, including ceramic and credit card type. Regardless of the physical design or the side of the circuit they are wired to, the operation is the same.

AUTHOR'S NOTE: Never remove the resistor block from the air stream and operate the blower motor. The resistor block will heat up rapidly and could be damaged or burn whatever it contacts.

Some blower speed control systems have a high-speed blower control relay (Figure 3-39). When high speed is selected, the resistor block is bypassed and full current is directed through the blower motor. Note that this five-speed blower has two fuses protecting the circuit: one (20A) for the first four speeds and one (30A) for high speed.

Blower Motor Credit Card Resistor. The blower motor credit card resistor works in a similar manner as the wire-wound resistor block, by dropping current flow in the blower motor circuit (Figure 3-40) with the addition of a thermal limiter switch. The thermal limiter switch is a protection device designed to turn off the blower motor at all speeds, except high, if resistor temperatures exceed 363°F (183.89°C). The credit card resistor consists of an integrated

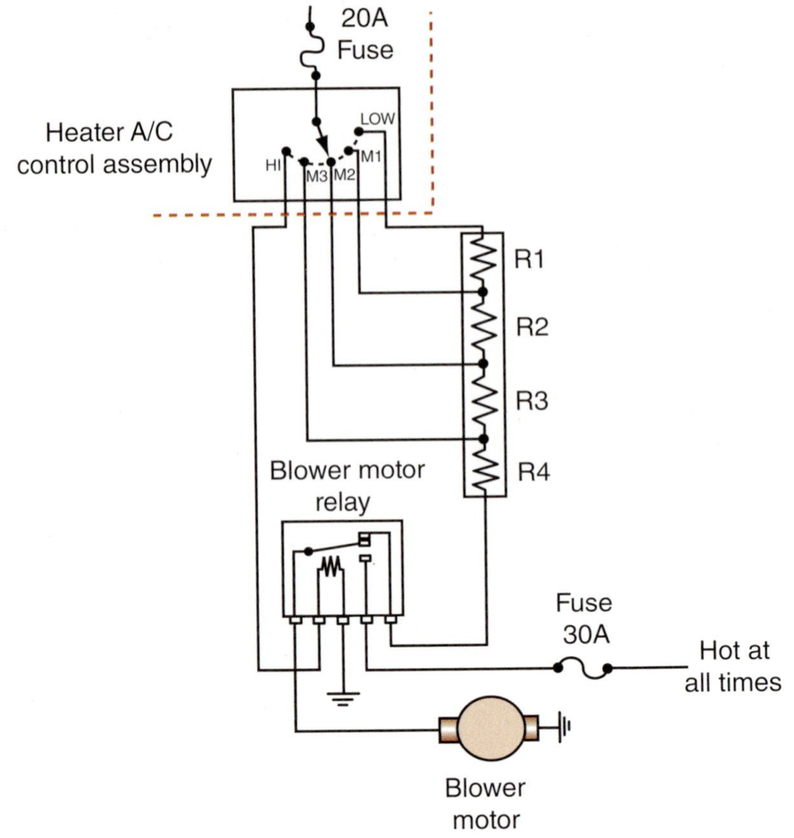

FIGURE 3-39 **High-speed blower motor relay.**

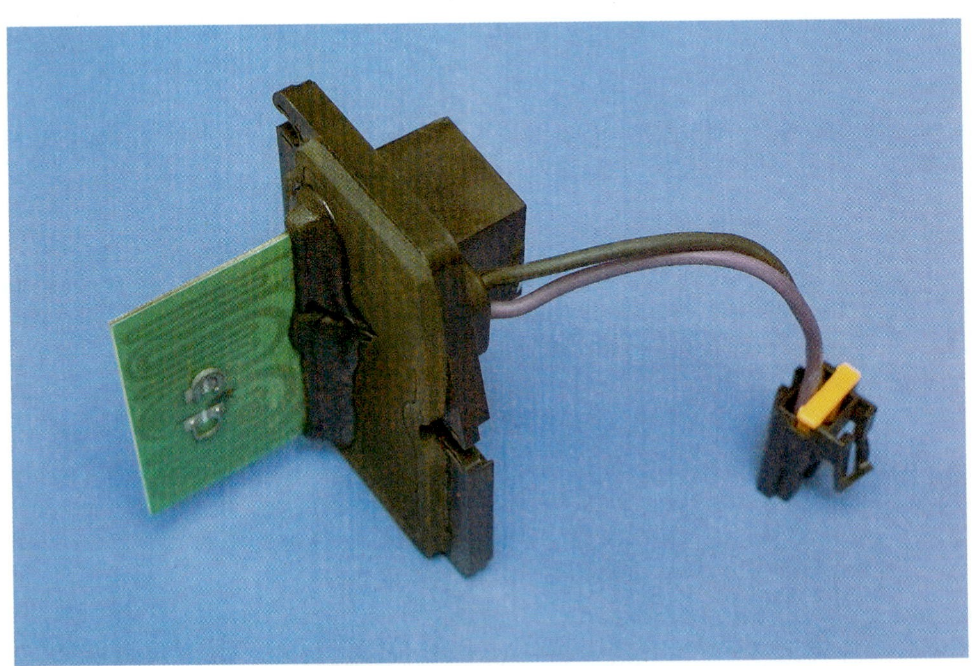

FIGURE 3-40 **The credit card resistor consists of an integrated circuit board mounted to a molded plastic mounting plate and integral connector assembly.**

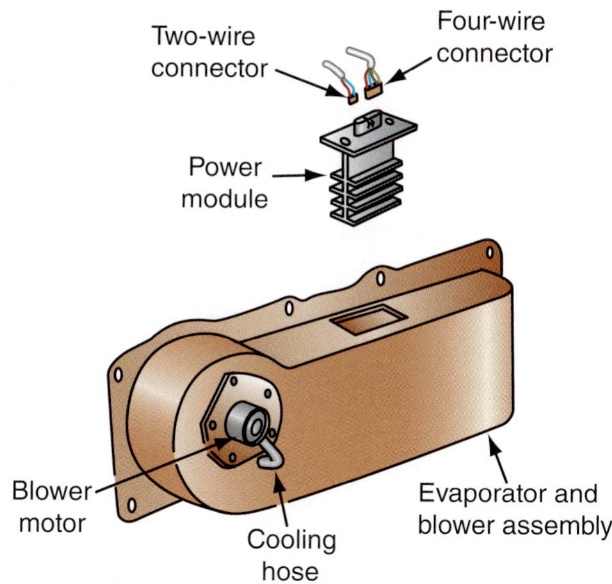

FIGURE 3-41 **A typical power control module located in the HVAC duct system.**

circuit board mounted to a molded plastic mounting plate and integral connector assembly. Like the wire resistor block, it is also mounted in the HVAC ventilation duct to allow for air stream cooling of the resistors.

Blower Motor Power Transistor Module. A power transistor module, also called a power module (Figure 3-41), is a solid-state semiconductor located on the ventilation duct of the HVAC system and cooled by the blower motor fan. The power transistor takes the place of the stepped resistor block and controls blower speeds by varying the amount of resistance in the blower motor's ground circuit.

The climate control module or body control module (BCM) applies a **pulse width–modulated (PWM)** (duty-cycled) voltage (Figure 3-42) to the base of the transistor, which in turn determines the level of conductivity through the power transistor. This type of control provides infinite blower fan speed control within the limits of the motor, between fast and slow, with a smooth transition between speeds. If the amount of ON time is equal to the OFF time in a **cycle**, then the **duty cycle** is 50 percent, and the motor will run at half speed.

Pulse width modulation (PWM) On-off duty cycling of a component. The period of time for each cycle does not change; only the amount of ON time in each cycle changes. The length of time in milliseconds that an actuator is energized.

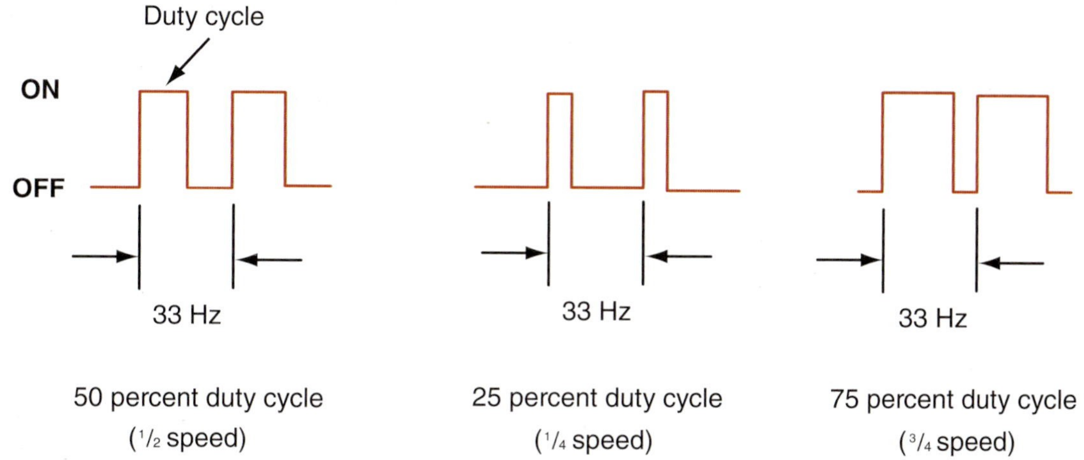

FIGURE 3-42 **Typical duty cycle patterns, as viewed with an oscilloscope, to achieve various blower motor speeds through the use of pulse-width modulation.**

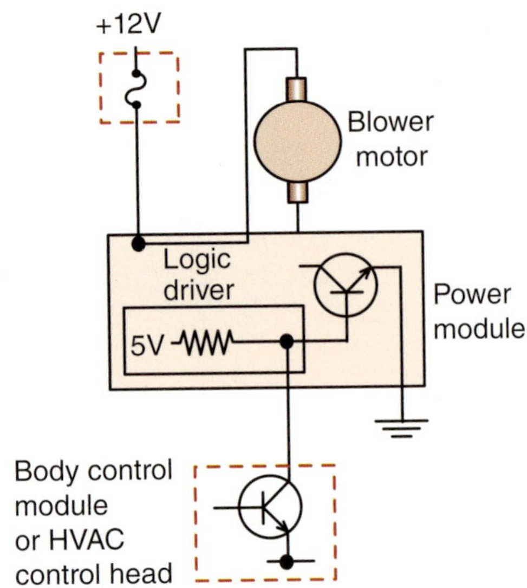

FIGURE 3-43 **The blower motor power transistor is pulse-width modulated by the BCM or climate control module on the ground side of the blower motor with a high- or low-voltage signal.**

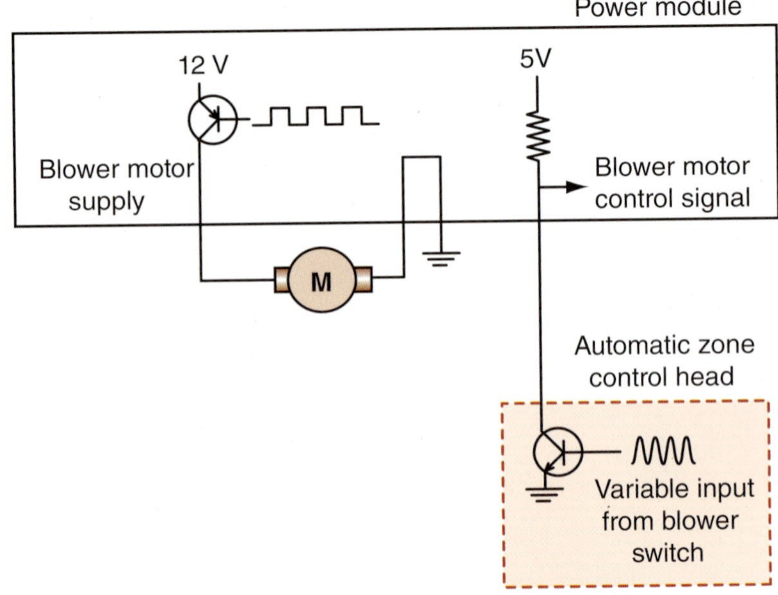

FIGURE 3-44 **The blower motor power transistor is pulse-width modulated by the BCM or climate control module on the power side of the blower motor with a high- or low-voltage signal.**

Regardless of the specific circuit design, the power transistor regulates blower motor current flow to vary fan speed. The BCM or HVAC control module pulse width modulates the signal wire to the power transistor module to provide this speed regulation (Figure 3-43). On some systems, the command signal is pulled high for high blower speeds and on other systems the command signal is set low for high blower speeds (Figure 3-44). Always refer to vehicle-specific service information when diagnosing a fault in the system.

On an automatic climate control system, if the system is set to the FULL AUTO mode, the control module will decide when and at what speed the blower motor should run based on input sensor data and system demands. If the in-car temperature is considerably higher than the selected temperature when using the air conditioning, the control module will send a signal to the power transistor to increase blower motor speed until the temperature inside the vehicle begins to drop. Once the temperature begins to drop, the power control module will reduce the blower speed to a level that will maintain the temperature selected at the control panel.

Electromagnetic Clutch

An electromagnetic clutch (Figure 3-45) is used in automotive air-conditioning systems as a means of engaging the compressor when cooling is desired and disengaging it when cooling is not required. For example, the compressor is disengaged when the air conditioner is not being used or when the desired temperature is reached in the vehicle.

All clutches operate on the same basic principle—that of magnetic attraction. This is accomplished by energizing a stationary field coil. The magnetic attraction of the field coil, in turn, pulls an armature into contact with a rotating member, the pulley.

All automotive air-conditioning systems have an electromagnetic clutch. Not all, however, are used to cycle the compressor for temperature control. For those that do, the electrical circuit to the clutch coil is interrupted when a set of contacts, thermostatically controlled or

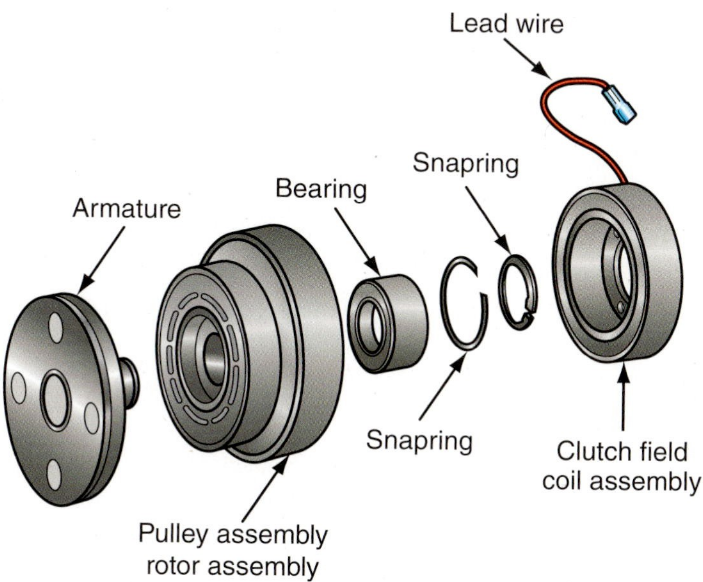

FIGURE 3-45 **Electromagnetic clutch assembly.**

pressure actuated, open as the set temperature or pressure is reached. Those that do not cycle to affect the desired temperature rely on the operation of a variable displacement compressor as a means of temperature control.

Clutch Diode

The clutch coil is an electromagnet with a strong magnetic field when current is applied. This magnetic field is constant as long as power is applied to the coil. When power is removed, the magnetic field collapses and creates high-voltage spikes. These spikes are harmful to delicate electronic circuits of the computer and must be prevented.

A diode placed across the clutch coil (Figure 3-46) provides a path to ground back through the clutch coil until the electrical energy is dissipated, thereby holding the spikes to a safe level. This diode is usually taped inside the clutch coil connector, across the 12-volt lead and ground lead. A diode may be checked with either an analog ohmmeter or DMM set to the diode test mode. Many DMMs have provisions for diode quick-testing procedures; this is the preferred method for testing diodes.

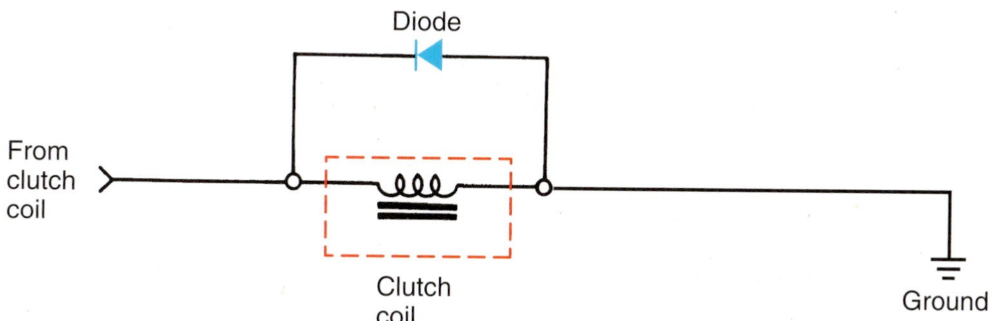

FIGURE 3-46 **A diode is placed across the clutch coil to reduce spikes as the clutch is cycled on.**

TERMS TO KNOW

Amperage (AMP)
Circuit breaker
Diode
Duty cycle
Fuse
Ohms (Ω)
Ohm's law
Pulse-width modulated (PWM)
Voltage

SUMMARY

- Voltage (V) or electromotive force (EMF) is the electrical pressure and is measured in volts (V); it may be either direct current (DC) or alternating current (AC).
- Current is the flow or rate of flow of electrons under pressure in a conductor between two points having a difference in potential and is measured in amperes (A) or amperage.
- Resistance is defined as the opposition to current flow and is measured in ohms (Ω).
- Ohm's law defines the relationship between voltage, current, and resistance. It is the basic electrical law.

REVIEW QUESTIONS

Short-Answer Essays

1. Briefly define the term *voltage.*
2. Briefly define the term *current.*
3. Briefly define the term *resistance.*
4. What are three common types of resistors used in automotive circuits?
5. What can photodiodes be used to detect?
6. Describe a series-parallel circuit.
7. What is a diode?
8. Briefly define Ohm's law.
9. Where are electromagnetic clutches used?
10. Describe parallel circuit principles.

Fill in the Blanks

1. _______________ is the electrical pressure; it may be either direct current (DC) or alternating current (AC).
2. A _______________ is one of the simplest semiconductor devices and functions as an electrical one-way check valve.
3. _______________law defines the relationship between voltage, current, and resistance. It is the basic electrical law.
4. A _______________ changes resistance value with a change in temperature.
5. Many relay terminals are identified using the _______________ numbering convention for common size and terminal patterns.
6. _______________ is the flow or rate of flow of electrons under pressure in a conductor between two points having a difference in potential and is measured in _______________.
7. _______________ are a common type of input sensor used by vehicle computers to monitor linear movement.
8. _______________ have an infinite number of resistance values within a range.
9. _______________ controls the amount of current flow in an electrical circuit.
10. A _______________ can be used to detect minute amounts of light.

Multiple Choice

1. In a parallel circuit:
   A. Total circuit resistance is less than the lowest resistance.
   B. Amperage will decrease as more branches are added.
   C. Total resistance is the sum of all of the resistances in the circuit.
   D. All of the above

2. In a series circuit:
   - A. Total circuit resistance is the sum of the individual circuit resistances.
   - B. The sum of the individual voltage drops equals the source voltage.
   - C. Current flow is the same at any point in the circuit.
   - D. All of the above

3. All of the following are common circuit faults except:
   - A. Short to ground.
   - B. Open in circuits.
   - C. Voltage drop test.
   - D. High resistance.

4. All of the following are true of voltage drops except:
   - A. Voltage drop can be measured with a voltmeter.
   - B. All of the source voltage in a circuit must be dropped.
   - C. Corrosion in a circuit does not cause a voltage drop.
   - D. Voltage drop is the conversion of electrical energy into another form of energy.

5. Ohm's law defines the relationship between all of the following except:
   - A. Voltage
   - B. Corrosion
   - C. Resistance
   - D. Current

6. Resistance is defined as the opposition to current flow and is measured in:
   - A. Volts
   - B. Resistance
   - C. Amperes
   - D. Ohm's

7. Current is defined as the flow of electrons and is measured in:
   - A. Volts
   - B. Resistance
   - C. Amperes
   - D. Ohm's

8. Voltage is defined as electrical pressure and is measured in:
   - A. Volts
   - B. Resistance
   - C. Amperes
   - D. Ohm's

9. A relay is an electromagnetic switch that allows a small amount of current to control:
   - A. A large voltage
   - B. A large current
   - C. A small voltage
   - D. A large resistance

10. A photodiodes can be used to detect minute amounts of:
    - A. Light
    - B. Current
    - C. Voltage
    - D. Resistance

Chapter 4

Engine Cooling and Comfort Heating Systems

Upon Completion and Review of this Chapter, you should be able to:

- Explain the engine cooling system and its components.
- Recognize the various components of the automotive cooling system.
- Identify the different types of radiators.
- Explain the operation and function of the coolant (water) pump.
- Discuss the requirements for a closed cooling system.
- Explain the purpose, advantage, and operation of a thermostat.
- Recognize the safety hazards associated with a cooling system service.
- Explain the operation of various types of cooling fans.

Introduction

Normal operation of the automobile engine produces heat that must be carried away. This excessive engine heat, which is a product of combustion, is transferred to the coolant and then dissipated in the radiator. This is accomplished by two heat transfer principles known as conduction and convention. The cooling system, when operating properly, maintains an operational design temperature for the engine and automatic transmission.

The cooling system functions by circulating a liquid coolant through the engine and the radiator. Engine heat is picked up by the coolant by conduction and is given up to the less hot outside air passing through the radiator by convection. Coolant is also circulated through the heater core, which also uses the convection process to supply heated air to the passenger compartment.

Shop Manual
Chapter 4, page 101

The Cooling System

The purpose of the automotive cooling system is to carry the heat that is generated by the engine during the combustion process away from the engine (Figure 4-1) to maintain a near constant engine operating temperature during varying engine speeds and operating conditions. Due to inefficiencies of the internal combustion engine, as much as 70 percent of the energy from gasoline is converted into heat. The cooling system has a difficult task with internal combustion temperatures, which may exceed 4,500°F (2,482°C). Actually, most of the engine's heat is sent out the exhaust system and is absorbed and dissipated by the cylinder walls, heads, and pistons into the ambient air. The cooling system is designed, therefore, to remove about 35 percent of the total heat produced by the engine.

A defective cooling system may impair air conditioning performance.

Another important function of the cooling system is to allow the engine to reach operating temperatures as quickly as possible. When engines are below operating temperature, exhaust emissions are increased, internal components wear faster, and operation is less efficient.

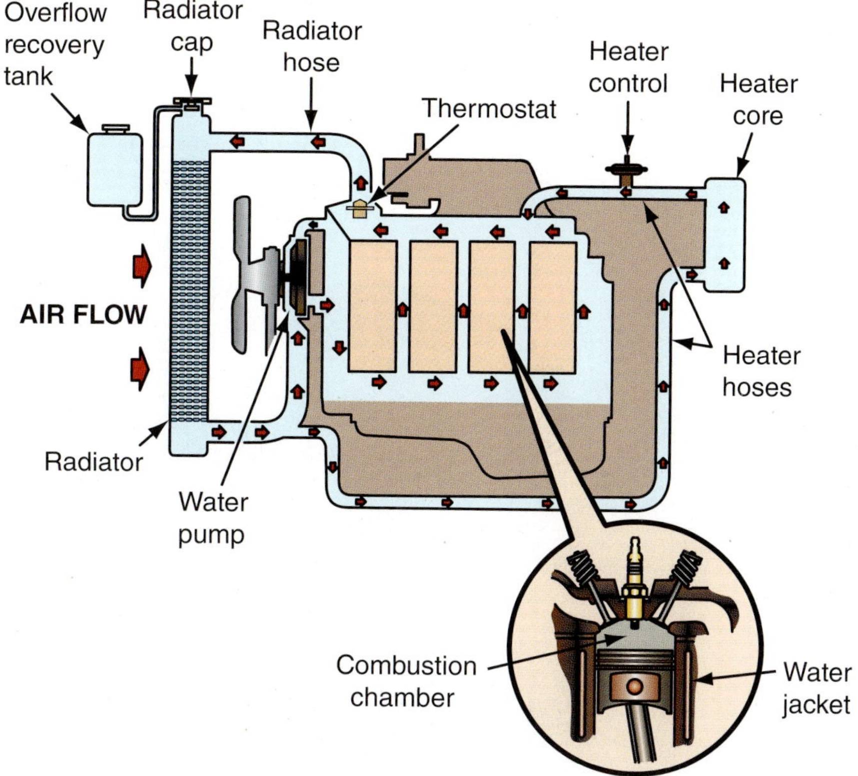

FIGURE 4-1 **A typical engine cooling system and heating core.**

An automobile engine's heat is given up in the radiator by the two heat transfer processes of radiation and convection. The overall surface area of most **radiators** is on the order of about 28 to 35 sq. ft. (2.6 to 3.2 m^2), though their physical size does not imply that they have that much cooling area.

The **radiator** is a coolant-to-air heat exchanger that removes heat from the coolant passing through it.

Heat, as discussed earlier, is always ready to flow from a hot to a less hot object or area. The heat that is picked up by the coolant in the engine block is given off to the less hot air passing over the fins and coils of the radiator. Air movement across the radiator is created in two ways: (1) by the engine fan, known as forced air, and (2) by the forward motion of the car, known as **ram air**.

Ram air is air forced through the radiator by the forward movement of the vehicle.

The high-limit properties of engine lubricating oil necessitate proper heat removal to prevent destroying its formulated lubricating characteristics. On the other hand, removing too much heat lowers the thermal efficiency of the engine. To prevent the removal of too much heat, a condition known as overcooling, a **thermostat** is used in the engine outlet water passage. The thermostat, a temperature-sensitive device, controls the flow of coolant from the engine into the radiator.

Shop Manual
Chapter 4, page 108

In addition to hoses required, the closed cooling system (Figure 4-2) consists of the following:

- Water pump
- Engine water passages
- Cooling fan
- Radiator
- Recovery or expansion tank
- Pressure cap
- Thermostat
- Air baffles and body seals

The **thermostat** is a temperature-sensitive device used to regulate the cooling system operating temperature.

Each of these components is covered individually in this chapter.

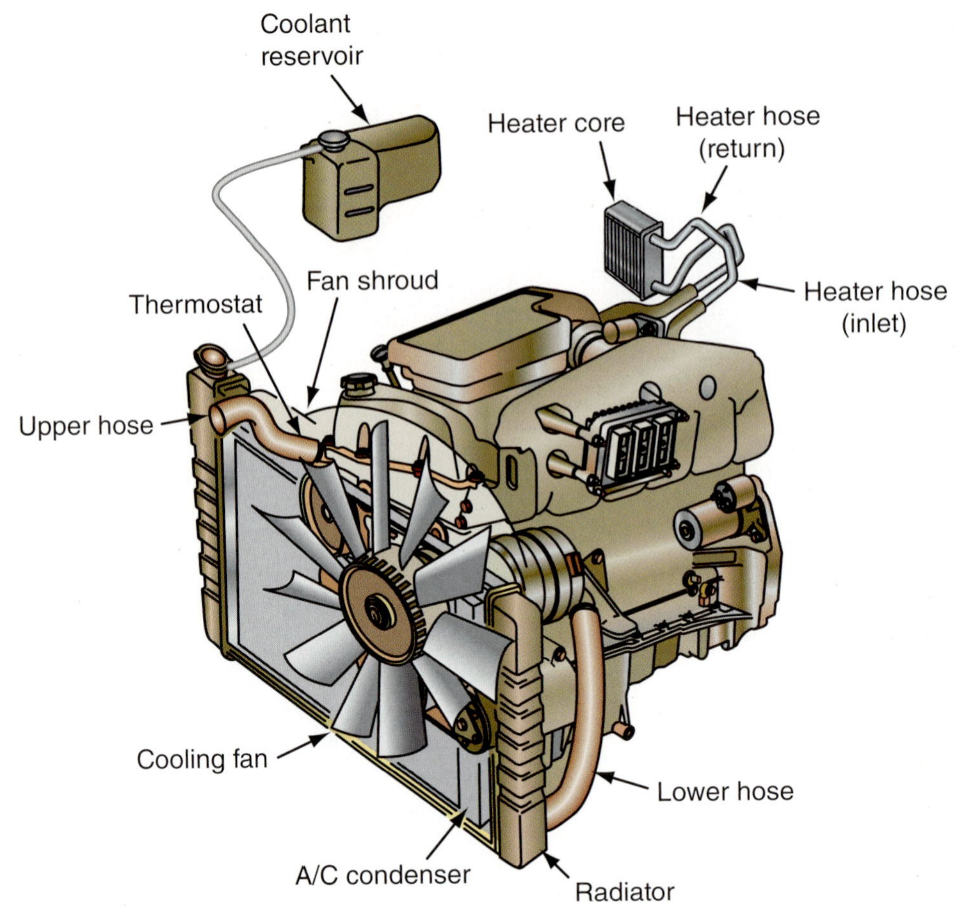

FIGURE 4-2 **A closed cooling system.**

Cooling systems also include a heater core as part of the cooling system circuit. Some car lines may also include a thermostatic vacuum switch (TVS), which is also known as a ported vacuum switch. This vacuum-controlled device advances ignition timing if the engine overheats during prolonged idle periods. Because of modern electronics technology, this component was discontinued on most car lines in the mid to late 1980s, though it was used on some Ford and Jeep car lines through 1991.

If a vehicle was not equipped with a factory-installed air conditioner, the cooling system is most likely not designed to handle the additional heat load. Under normal circumstances, however, the addition of an aftermarket air-conditioning system will cause no problems with the automotive cooling system. This is only true if the cooling system has been well maintained and in good working condition. An aftermarket installation kit often contains a smaller water pump pulley to provide increased coolant flow and a fan system to provide additional airflow.

It is necessary to have a good understanding of the purpose and operation of the cooling system to properly diagnose, troubleshoot, and correct cooling system problems.

Heat Measurement

Today's technicians use both mechanical and electronic thermometers. High-temperature electronic pyrometer probes, which are part of many automotive digital multimeters, are used by touching the surface to be tested or infrared thermometers may be used by simply aiming at the surface or area to be measured (Figure 4-3). The infrared pyrometer is by far the most popular method of determining the temperature of a component. In fact, in some states, they have become a required tool for emission technicians.

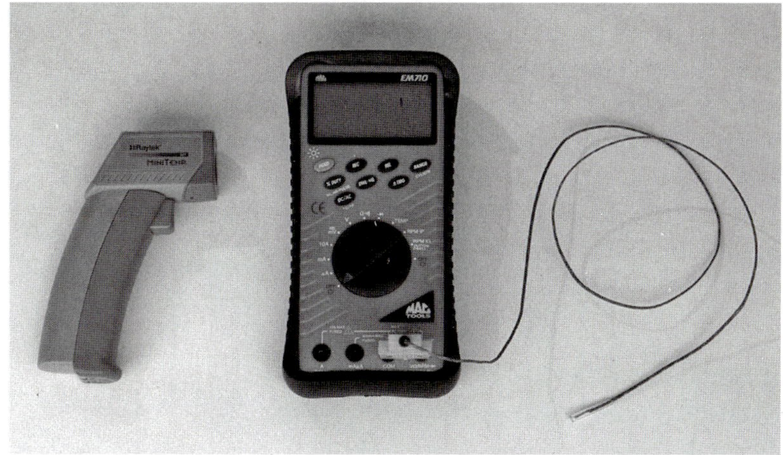

FIGURE 4-3 **Infrared pyrometer and DVOM with temperature probe.**

For safety reasons, temperature measurements are usually taken just after turning the engine off. Beware of moving parts, such as the radiator cooling fan, that can start at any time even when the ignition switch is in the OFF position. Specific methods and procedures for taking temperature measurements are covered throughout this manual, as well as in the Shop Manual where applicable.

Radiator

The radiator is a heat exchanger that consists of a core and two tanks. It is used to remove heat from the coolant passing through it. It performs a critical job, and if the radiator fails to remove excess heat, the engine may overheat and extensive damage, such as blown head gaskets or cracked or warped cylinder heads, may result. Excessive heat could also result in scuffed piston skirts and cylinder walls as well as valve stem and guide damage.

Shop Manual
Chapter 4, page 103

There are two basic radiator design types: (1) cross-flow tube and fin type, and (2) vertical flow tube and fin type (Figure 4-4). In a cross-flow radiator, popular on most late model vehicles, the coolant flows from the inlet tank (generally the upper radiator hose tank) to the outlet tank. In the vertical flow design, coolant flow is from the top tank to the bottom tank. This design is not as popular today. A cross-flow radiator is 40 percent smaller than a downflow radiator and has a flatter design well suited for more compact designs.

The terms "vertical flow" and "down flow" for radiators are used interchangeably for the same design. Radiator tanks are also referred to as "headers."

Radiators are generally constructed of an aluminum (Al) core and high-temperature, nylon-reinforced plastic tanks and are used on most late-model vehicles or with copper (Cu) core with brass tanks on early radiators through the late 1980s and early 1990s. On the aluminum core radiator, the high-temperature plastic tanks are held in place by clinch tabs, which are part of the aluminum header at each end of the core, and a special high-temperature rubber gasket seal between the core header and tank flange edge to prevent leakage (Figure 4-5).

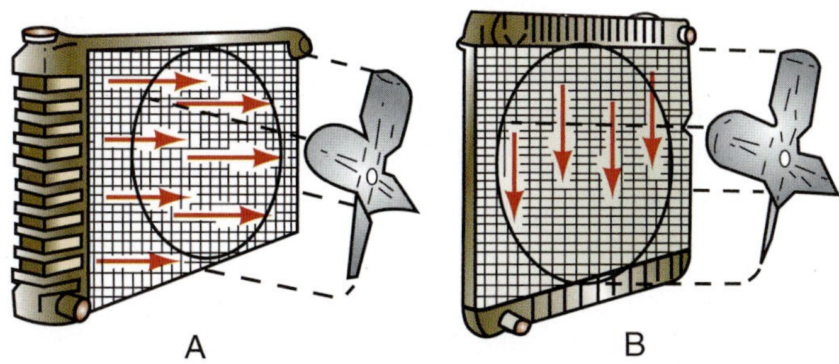

FIGURE 4-4 **Two radiator designs: (A) cross flow and (B) vertical flow.**

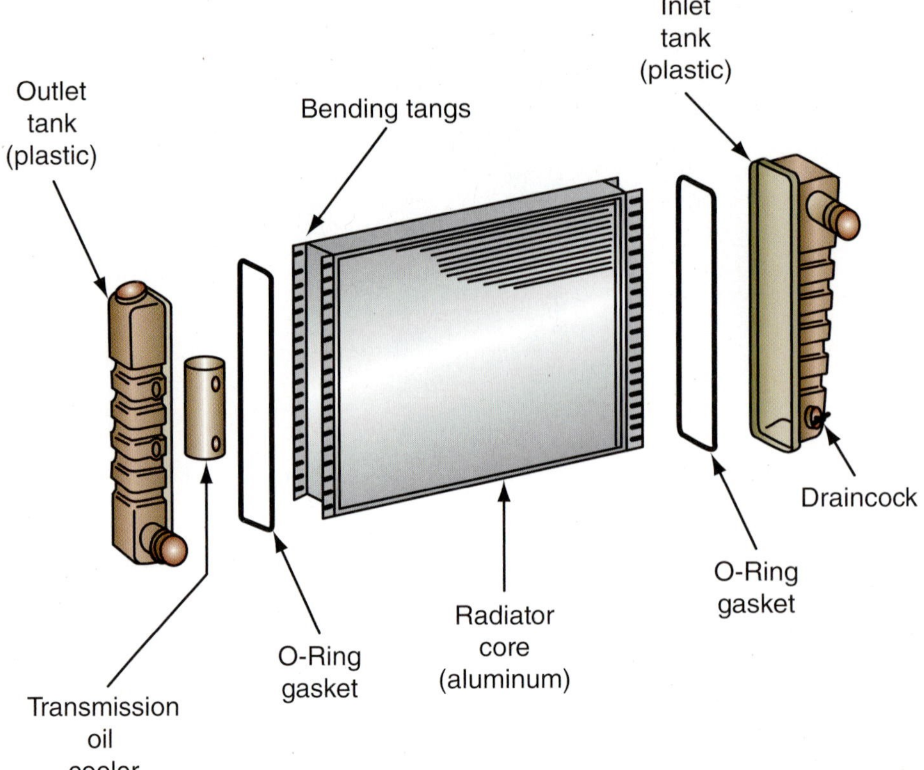

FIGURE 4-5 **A typical aluminum core/plastic header radiator.**

There are, however, occasional variations in how these four materials are combined. Some manufactures have also been using full aluminum radiators on some of their platforms. These radiators offer reduced overall depth and are up to 11 percent lighter, feature a higher burst pressure, and they are fully recyclable, but they are more expensive.

In our discussion, we refer to the first tank as the inlet tank because it receives hot coolant after it passes through the engine and distributes it to the radiator core. The second tank will be referred to as the outlet tank because it collects the less hot coolant after it has passed through the many tubes in the radiator core. At the bottom of one of the tanks, there is a draincock to aid in the removal of coolant.

The inlet tank usually contains a baffle plate to aid in the even distribution of coolant through the passages of the core. The outlet tank may also be considered a storage tank for the coolant after it has given up much of its heat after passing through the core. From the outlet tank, coolant is directed to the water pump inlet via a radiator hose (lower).

The outlet tank often contains an internal coil used as automatic transmission oil cooler (Figure 4-6). Transmission fluid is pumped from the transmission through the coil and back to the transmission. This transmission fluid has no other connection with the engine coolant. Though not common, the transmission oil cooler sometimes develops a leak. Because transmission oil pressure is usually greater than cooling system pressure, the oil will leak into the cooling system, creating an oily, strawberry-colored foam in the cooling system. There could, on the other hand, be a coolant leakage into the transmission depending on pressure differential (Δp). Some vehicles equipped with a trailer-towing package may also have an external transmission cooler.

The cellular core is often called a honeycomb.

Two common designs for cores are the cellular core (A) and the tubular core (B) (Figure 4-7). The tubular core has a series of long, narrow, oblong tubes that connect the inlet and outlet tanks. There are fins around the outside of the tubes to improve heat transfer from the coolant flowing in the tubes. The fins on the core absorb heat from the coolant passing

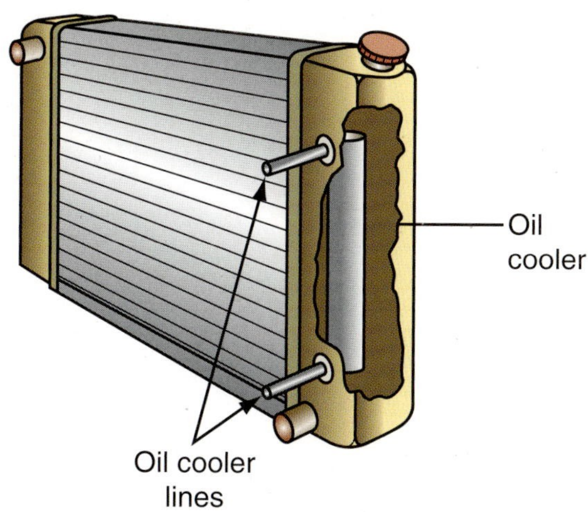

FIGURE 4-6 **Transmission oil cooler details.**

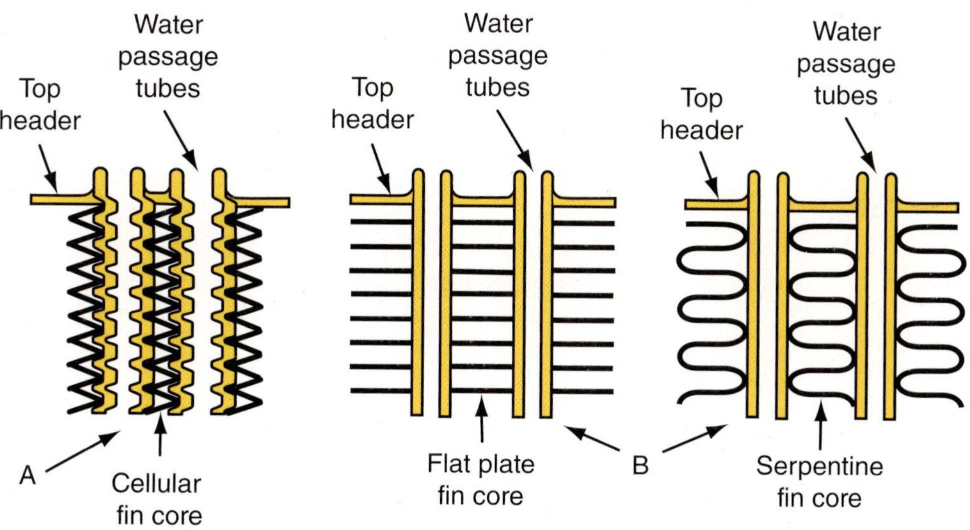

FIGURE 4-7 **Two types of radiator core designs: (A) cellular and (B) tubular.**

through the tubes and release it to the ambient air passing through the core and across the fins and tubes, thus carrying off heat and cooling the coolant (Figure 4-8). A process of soldering together thin, preformed sheets of metal fabricates the cellular core, usually made of aluminum, copper, or brass.

During engine operation, the coolant heats up and expands. As the coolant expands, it is displaced into the **recovery** or **expansion tank**. In addition, as the coolant circulates, air bubbles are allowed to escape. The advantage to allowing air bubbles to escape is that coolant without bubbles absorbs heat much better than if they were allowed to flow through the system.

The **expansion tank** is a pressurized auxiliary tank that is usually connected to the inlet tank on a radiator to provide additional storage space for heated coolant. It is often called a coolant **recovery tank** or an overflow tank when not pressurized and under atmospheric pressure.

Under certain applications, there may be a need to increase the cooling capacity of the system by recommending the upgrade to a heavy-duty or performance radiator. Vehicles that could benefit from an upgrade include those used for towing, carrying heavy loads, off-roading, and other uses that would put increased demand on the cooling system. These radiators will have additional rows of tubes, added thickness, and may have a more efficient design.

The walls of either type of core are not much thicker than the paper you are now reading. Their passages are about the size of a pencil lead. This should give some idea of how fragile radiators are and the care that must be taken to avoid costly damage.

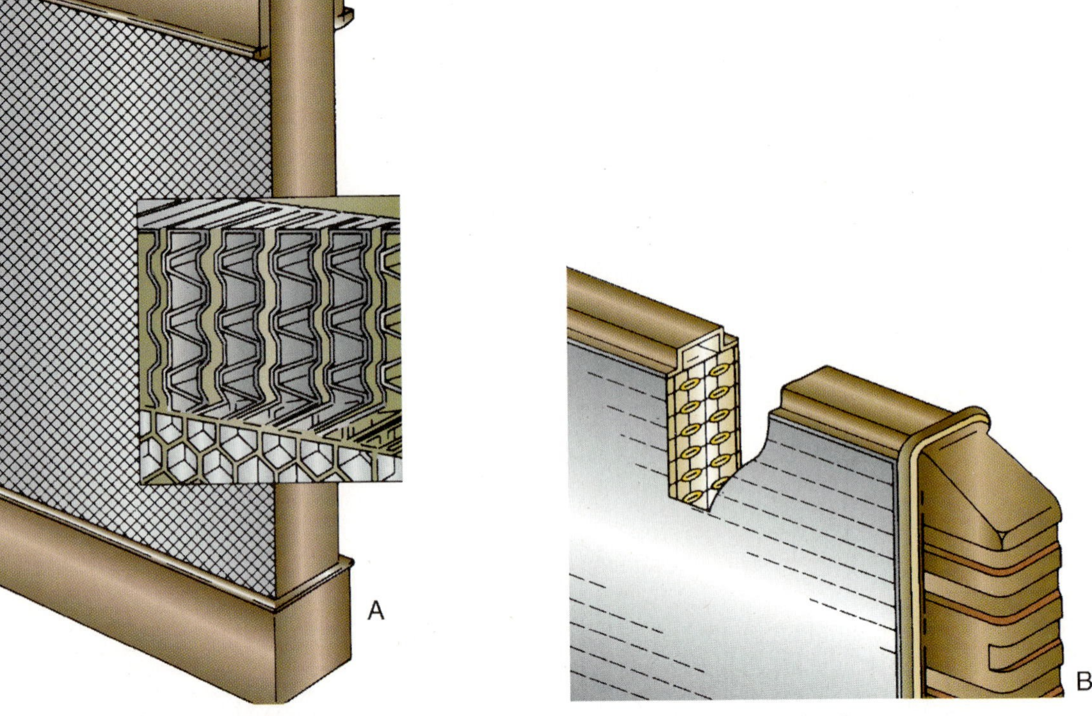

FIGURE 4-8 **Cellular core (A) and tubular core (B) radiator details.**

Electrolysis is the decomposition of an electrolyte (coolant in this case) by the action of an electric current passing through it.

The cooling system should also be checked for signs of **electrolysis**. Electrolysis occurs when an electrical component is not properly grounded and routes itself through the cooling system in search of one. Likely sources are electrical accessories that are bolted to the engine and components in the cooling system, such as the starter motor or engine block to the battery ground connection. The destructive effect of electrolysis may be in the form of recurring pinholes in the coolant tube of the heater and radiator core or where mounting brackets are attached. Increases in the current draw of poorly grounded accessories will increase the destructiveness of electrolysis. Small amounts of voltage may be measurable in a cooling system that has become slightly acidic, and the coolant reacts with the metal in the system but should never exceed a tenth of a volt (0.10V) in engines with aluminum cylinder heads or blocks. Using a digital DC voltmeter may test for this. Connect the negative lead to the battery negative post and place the positive lead into the coolant at the filler neck; make sure not to touch any metal, and note the reading as accessories are turned on, including the starter. If higher voltage is found, determine the source to avoid damage to the cooling system. Cooling system electrolysis is becoming a more frequent problem in today's cars and should not be overlooked.

The primary cause of a radiator's failure is that they develop leaks or become clogged. Whenever a failure occurs, the radiator must be repaired or replaced. It is common practice to perform an off-vehicle radiator leak test by pressurizing the radiator to 20 psig (138 kPa) and submerging it in a test tank. If you are performing a leak test on an aluminum core radiator, however, do not use a tank that has been used for brass/copper radiators. The flux, acids, and caustic cleaner residue in the tank will attack the aluminum and cause early radiator failure. On-vehicle leak testing of an aluminum radiator is performed in the same manner as for a brass/copper radiator cooling system. Due to the high cost of labor, it is generally less expensive to replace a radiator than to have it repaired. Radiator repairs should be attempted only by those with the proper tools, equipment, and knowledge.

Pressure Cap

Concentration

The **pressure cap** both seals and pressurizes the cooling system. One of the functions of the pressure cap is to allow pressure to build up in the cooling system. It is known that water boils at 212°F (100°C) at sea level atmospheric pressure –14.69 psia (101.3 kPa) (Figure 4-9). As pressure builds up in the cooling system, the boiling point of the coolant also goes up. For each pound of pressure (6.9 kPa), the boiling point of the coolant (water) is increased about 3°F (1.7°C). For example, if an 8 psig (55.2 kPa) pressure cap were used, the boiling point of coolant would be raised to 236°F (113.3°C). This allows the cooling system to be safely run at temperatures far above the boiling point of coolant at atmospheric pressure. This enables the engine to reach operating temperature sooner and adds to the overall efficiency of the engine.

The **pressure cap** increases the pressure of the cooling system and allows higher operating temperatures.

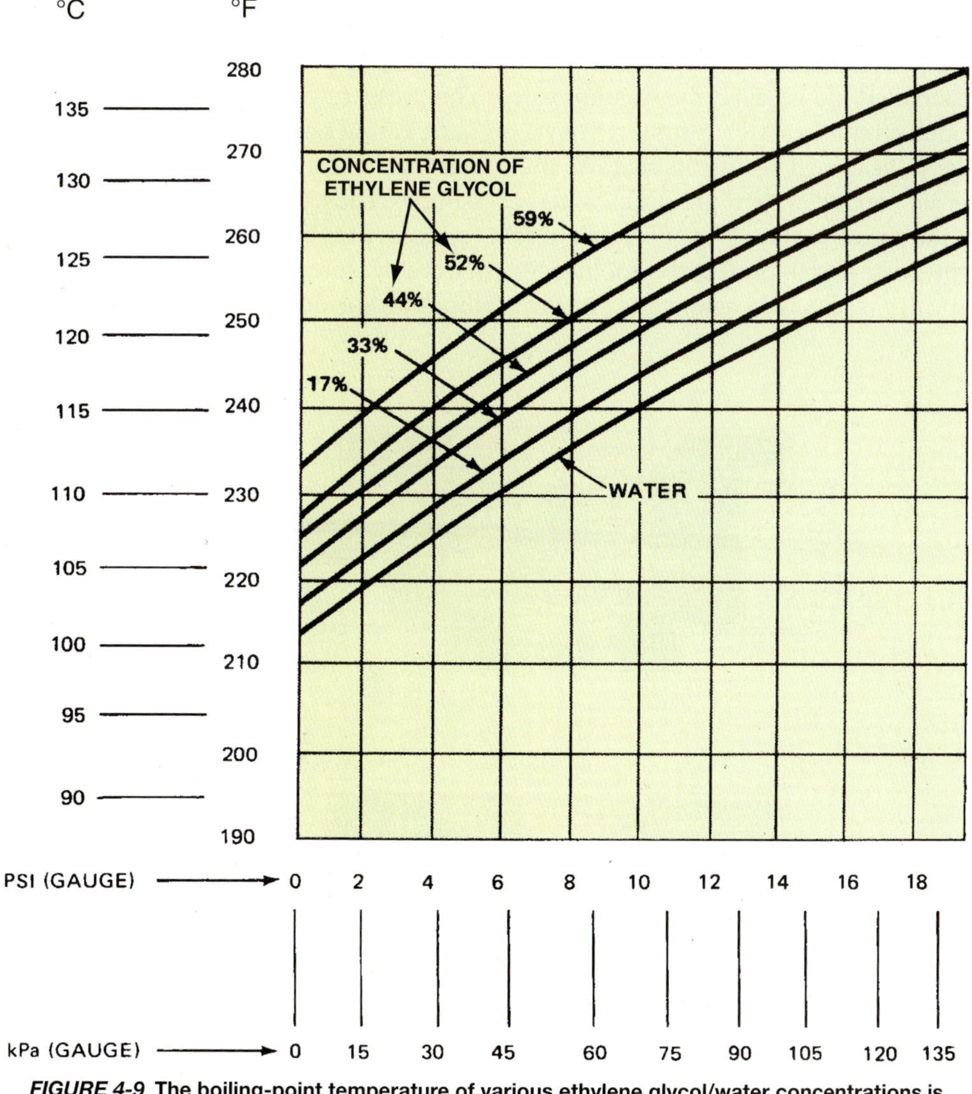

FIGURE 4-9 **The boiling-point temperature of various ethylene glycol/water concentrations is increased by increasing pressure.**

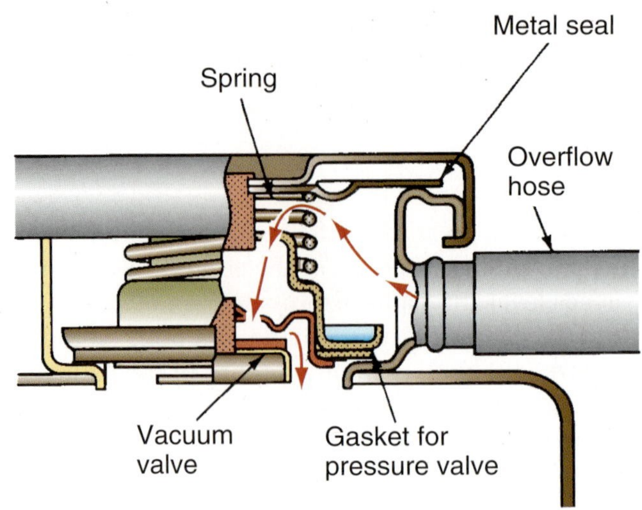

FIGURE 4-10 **Pressure and vacuum valve details of a pressure cap.**

Today's vehicle cooling systems are closed systems, with coolant being displaced to or drawn from a recovery tank (as cooling system pressure increases or decreases) connected to the radiator by a hose connected at the radiator filler neck (Figure 4-10). The pressure cap contains an external seal and two spring-loaded valves; the larger valve is called the pressure valve and the smaller one is called the vacuum valve. The pressure valve contains a spring of a predetermined strength, which is indicated by a pressure rating on the top of the cap. This spring holds the valve closed against its seat. As the coolant heats up, it begins to expand, increasing the internal cooling system pressure. When pressure exceeds the rated value (i.e., 15 psig) on the cap, the pressure valve lifts off its seat to relieve excess pressure in order to maintain correct system pressure and to protect the cooling system from overpressurization (Figure 4-11A). The vacuum valve is also held against its seat by a calibrated spring. When the engine is stopped,

The pressure valve is also referred to as the blow-off valve, whereas the vacuum valve may be referred to as the atmospheric valve. Never exceed the Original Equipment Manufacturer (OEM) specifications for a pressure cap.

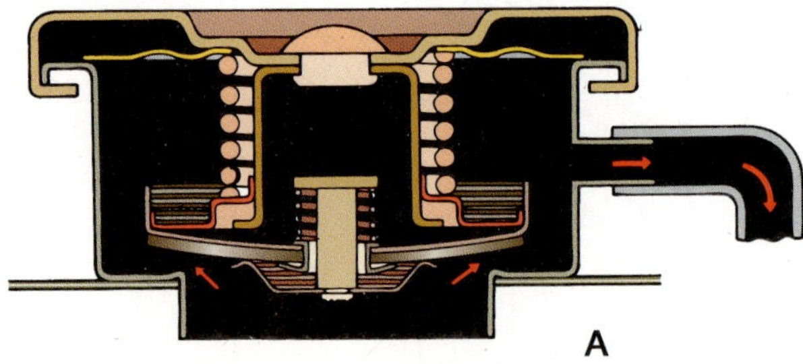

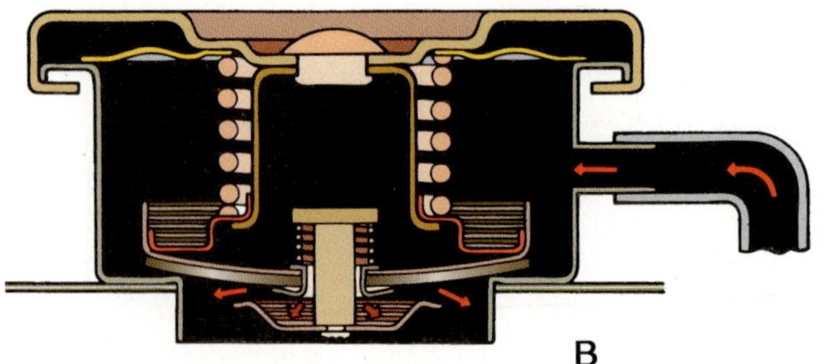

FIGURE 4-11 **Radiator cap details: (A) pressure operation and (B) vacuum operation.**

the cooling systems cool down, the system pressure begins to drop from a positive pressure to a negative pressure, and a vacuum develops inside the cooling system. At a predetermined point, the vacuum valve opens to draw coolant back out of the expansion tank and into the radiator to equalize internal pressure with atmospheric pressure, thereby preventing the radiator and other cooling system components from collapsing (Figure 4-11B).

Due to heat soak, the temperature of the coolant in the engine will increase several degrees a few minutes after the engine is stopped.

A vacuum valve that fails to open to release negative pressure in the cooling system may cause system damage. A collapsed upper radiator hose generally notes this condition as the system cools or during heavy acceleration. Components likely to be damaged by a defective vacuum valve are the radiator and **heater core** tanks, which may collapse, as well as sealing surfaces.

The **heater core** is a heat exchanger used to transfer heat from the engine coolant to the air passing through it. It is used in the comfort heating system to heat the passenger compartment.

Remember that there are three gasket-sealing surfaces that are part of the pressure cap assembly. If the gasket-sealing surface of the pressure valve fails, the system will not become properly pressurized. The outer gasket between the cap top and the radiator neck both seals coolant from leaking out the top of the radiator neck and keeps air from leaking into the system as it cools down. If this seal allows air to leak past during the cool-down process, the coolant in the expansion tank will not be drawn back into the cooling system, and a low coolant level in the cooling system will result. The vacuum valve-sealing surface must also be inspected. Gaskets should be checked for distortion or damage, and the radiator neck-mounting surface must be clean and undamaged. If a good seal is not maintained, overheating and coolant loss will result.

A pressure cap rated at a higher pressure or a lower pressure than the one designed for the system should not be used. A cap with a higher pressure rating could cause the radiator or other cooling system element to rupture or leak. Using a cap of a lower rated value could cause the engine to overheat to a point that could damage internal components. If in doubt about the proper rating for the replacement cap, check the manufacturer's specifications and replace with one of equal design and rating. Radiator caps range from 4 psig (27.6 kPa) to 18 psig (241 kPa). Always remove the radiator cap slowly. Removing the radiator cap on a hot cooling system can cause serious burns as steam and coolant escape. Extreme caution should be used when working on closed systems.

The pressure valve is also referred to as the blow-off valve, whereas the vacuum valve may be referred to as the atmospheric valve. Never exceed the OEM specifications for a pressure cap.

Most cooling system test kits include adapters for both pressurizing the cooling system and pressure testing the pressure cap. The pressure cap should be tested annually or any time service is performed to the cooling system. It is an inexpensive repair that will avoid costly system failures. If the pressure cap fails the test, it should be replaced with one of equal pressure ratings. It should also be noted that special aluminum caps are required on aluminum radiators.

Coolant Recovery System

Coolant recovery systems have been standard equipment on most cars since 1969. Coolant expands by about ten percent of its original volume when it reaches operating temperature. There can be air and vapor present in every cooling system that can cause serious problems if not removed. Coolant without air bubbles trapped in it absorbs heat better than coolant with air bubbles. Air is the leading contributor to the formation of rust and corrosion that causes early cooling system component failure as well as increasing the formation of sludge in the system by breaking down the coolant additives. Air that becomes trapped in the cooling system may create an air lock blocking the flow of coolant. This is especially common on systems that have been recently filled after service. This trapped air and vapor can cause pump cavitation, which inhibits the pump's ability to move the coolant and creates hot spots in the coolant galleys. To eliminate this problem, systems use either a nonpressurized **overflow tank** or a pressurized expansion tank (Figure 4-12). These systems, when properly maintained, prevent air from entering the cooling system and ensure proper coolant capacity. In addition, many systems today also have a manual air bleed valve, which is often placed on top of the thermostat housing (Figure 4-13).

An **overflow tank** (or catch tank) is a nonpressurized coolant recovery tank.

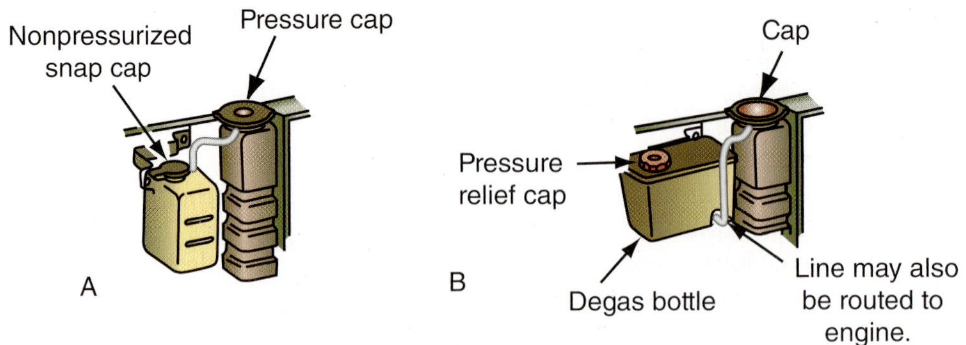

FIGURE 4-12 **Nonpressurized (A) and pressurized (B) coolant recovery tanks.**

FIGURE 4-13 **Manual air bleed on thermostat housing.**

Recovery Tank

A nonpressurized recovery tank, usually of ½ to 1 gal. (1.9 to 3.8 L) capacity, is connected to the pressurized cooling system through the overflow fitting at the filler cap. It is used to capture and store vented coolant and vapor from the radiator as it is being heated to operating temperature. Heating causes expansion, and the excess coolant is expelled through the pressure valve of the radiator cap to the recovery tank (Figure 4-14A). When the coolant cools, it retracts, and the vacuum valve of the radiator cap opens, allowing the same vented coolant to be metered back into the cooling system from the recovery tank (Figure 4-14B). Any vapor is vented to the atmosphere and is not returned to the cooling system as long as the proper level of coolant is maintained in the recovery tank.

The coolant level of the cooling system is easily determined by noting the coolant level in the recovery tank. It is not necessary to remove the radiator cap to check or add coolant. If a low coolant condition is noted, it should be filled to either the hot or cold mark on the side of the recovery tank, depending on engine temperature (Figure 4-15). Unlike the pressurized system, the cap on the recovery tank can be removed at any time for service.

The small hose that connects the recovery tank to the filler neck provisions of the radiator is an important link in the proper performance of the system. Its purpose is to allow coolant to flow back and forth between the recovery tank and the radiator filler neck. If it is clogged or kinked, coolant cannot be transferred from the recovery tank to the radiator. If it is disconnected or leaking, coolant will be lost or ambient air will be drawn into the system. This condition will result in poor performance and early failure of the cooling system.

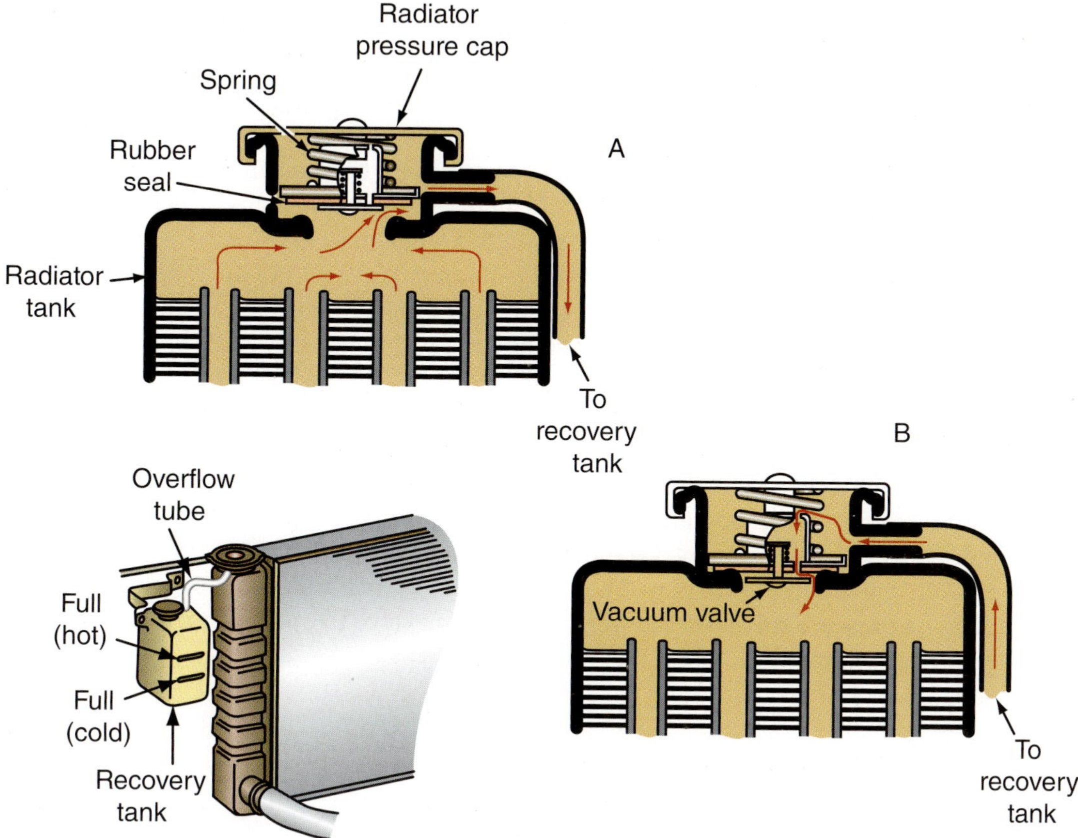

FIGURE 4-14 **Coolant is vented to the recovery tank (A) as internal pressure increases and is returned to the radiator (B) as pressure in the cooling system decreases.**

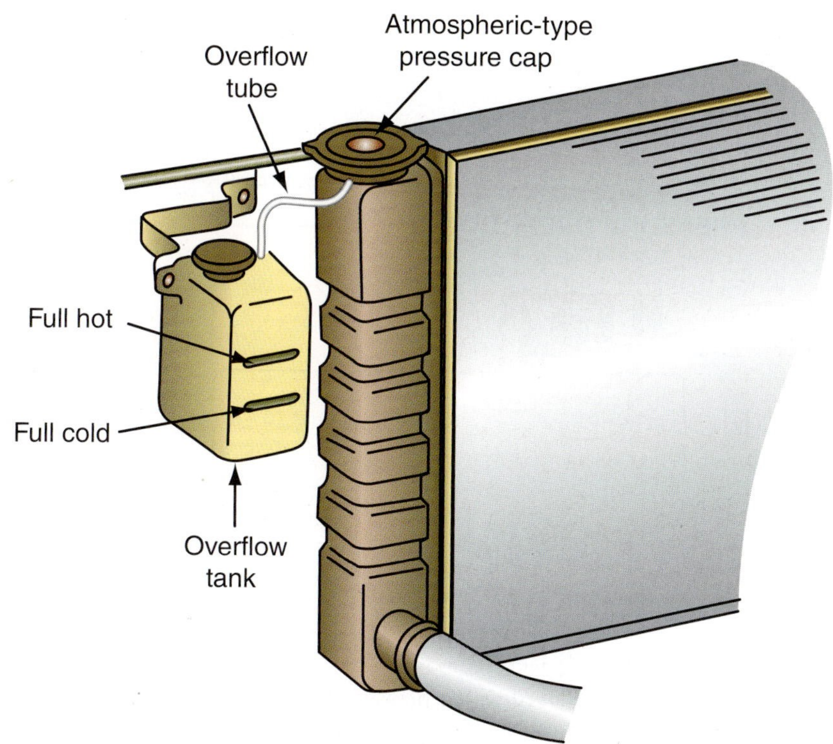

FIGURE 4-15 **A coolant recovery system tank.**

Expansion Tank

Some manufacturers have addressed the problem of trapped air by using a pressurized expansion tank or surge tank placed at the high point in the cooling system to allow air and vapor to rise and be purged out of the cooling system. It is usually mounted on the inner fender (Figure 4-16). It is an integral part of the cooling system and continuously separates and removes air from the system's coolant. When the engine coolant thermostat is open, coolant flows from the top of the radiator outlet tank through a small hose to the expansion tank. The pressurized system recovery tank should have 17–34 oz. (0.5–1.0 L) of air when the coolant is cold, to allow space for coolant expansion. Unlike the nonpressurized recovery tank, the pressurized expansion tank cap must not be removed for service when the coolant is hot, or serious injury could result.

Some expansion tanks on European vehicles use a weighted vacuum relief valve (sometimes referred to as a pressure vent–type cap), which is normally open. The vacuum valve on this style cap hangs freely on the pressure valve and is calibrated with a small weight. Under normal operation, this system operates at atmospheric pressure. If rapid expansion takes place, such as under heavy acceleration, the vacuum valve is closed by the escaping pressure or steam and the pressure valve comes into play. The cap operates in the same way a constant pressure cap operates. As pressure subsides, the weight causes the valve to open, returning the system to atmospheric pressure.

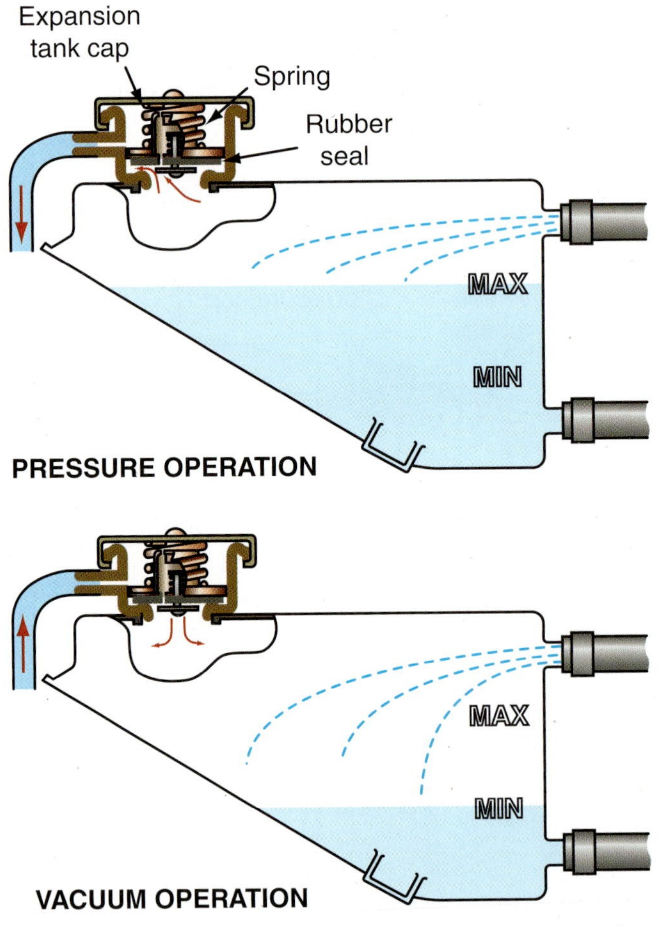

FIGURE 4-16 **Closed system pressurized coolant recovery/expansion tank.**

Engine Block and Cylinder Head Coolant Passages

Water jackets are a collection of passages molded into the engine block and cylinder heads. The passages of the water jackets in the cylinder block and head(s) are designed to control coolant flow and circulation to provide proper cooling around the hot spots of the engine. The water jackets in the cylinder block completely surround the cylinders to dissipate the heat generated during the combustion process. The cylinder head contains water jackets that surround the combustion chamber and contain passages around the valve seats in order to cool them off (Figure 4-17). If the vehicle overheats and the coolant boils, gas pockets may form in the cylinder block or cylinder head, causing hot spots due to the poor heat transfer characteristics of a gas.

In most systems, coolant flows through the water pump to the engine block, around the cylinder wall, and then back to the radiator. The flow then proceeds to the cylinder heads. On the **reverse flow** cooling system, coolant first flows to the hotter cylinder head and around the valves and combustion chamber, where most of the heat is centralized. The coolant then flows to the less hot engine block and cylinder walls. This process of coolant flow results in more even engine temperatures, reduces wear, and increases horsepower and fuel economy.

Reverse flow cooling systems first flow coolant through cylinder head(s), then through the engine block for more even heat transfer.

Expansion Plugs

Expansion plugs are small steel plugs that are pressed into casting holes located in engine blocks and cylinder heads to produce a watertight seal. The casting holes are the result of the casting process. Expansion plugs that fail are a source of external coolant leaks and need to be inspected for signs of leakage, especially if a system is loosing coolant. Their failure is generally due to corrosion from inadequate maintenance of the cooling system, specifically not flushing the coolant at the recommended intervals.

Expansion core plugs are also referred to as freeze plugs. Though they may pop out if the coolant freezes, their purpose is not to protect the engine from damage caused by coolant that freezes.

Coolant Pump

Coolant is circulated through the cooling system by a **centrifugal impeller**–type pump (Figure 4-18), which is usually driven by a belt off the engine crankshaft pulley. This pump may turn as fast as 5,000 revolutions per minute and carry coolant as fast as 10,000 gal./h (631 L/min.). At an average road speed, the coolant may be circulated as much as 160 to 170 gal. (605 to 643 L) per minute.

A **centrifugal impeller**–type pump uses rotational force to pull coolant from the center of the pump to the outside coolant passage connected to the engine.

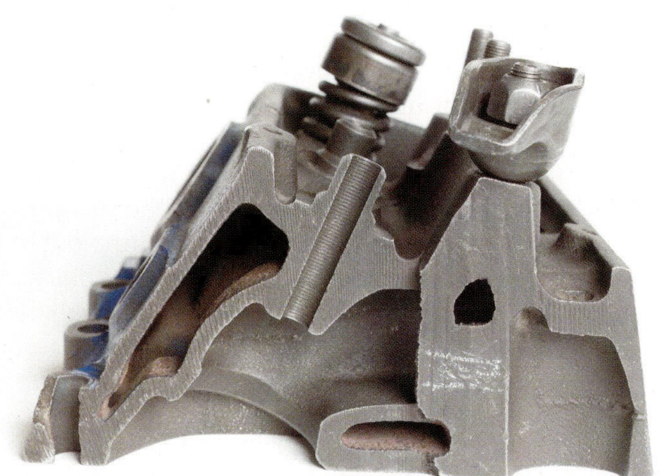

FIGURE 4-17 **Cutaway of cylinder head water jackets.**

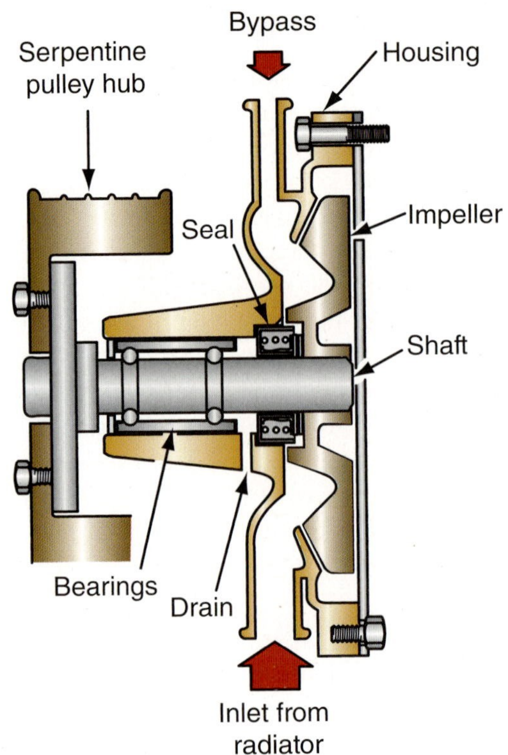

FIGURE 4-18 **Parts of a water pump.**

Coolant pumps are generally referred to as water pumps.

Shop Manual
Chapter 4, page 105

The **bypass** redirects coolant away from the thermostat and back to the water pump to enable coolant circulation in the engine block during the warm-up cycle.

The thermostat outlet hosing is sometimes referred to as the gooseneck.

The pump consists of a housing with an inlet and outlet, an impeller blade, nonserviceable sealed bearing(s), and seals. The impeller is the internal rotating part of the pump that moves coolant through the cooling system. The impeller consists of a flat plate with a series of flat or curved blades or vanes and is mounted to a shaft that passes through the pump casting. This shaft, which is generally stainless steel to prevent rust, is equipped with one or more sealed bearing assemblies and an internal and external seal to support the shaft. As the impeller rotates, coolant is drawn in from the center and is forced outward to the passage entering the engine block by centrifugal force. A belt pulley is mounted on the shaft end opposite the impeller. On vehicles equipped with an engine-mounted cooling fan, the fan is also attached. A defective coolant pump is not generally rebuilt because of the special tools required and is replaced as an assembly.

Some vehicles may have a coolant pump that is driven by the timing belt (Figure 4-19). This coolant pump is generally replaced when the timing belt is serviced as preven-tative maintenance to avoid a future repair because most timing belts are serviced at 90,000 miles and the pump has been in service for millions of rotations (Figure 4-20).

On most engines, the coolant pump inlet is connected to the bottom of the radiator with a rubber hose. This hose is preformed to fit a particular year and model engine, and it generally contains a spiral wire to prevent it from collapsing due to the suction action of the coolant pump impeller when the engine is revved up. The coolant pump outlet is through passages behind the impeller, which pushes the coolant through the engine block. After the coolant has passed through the engine block, it is returned to the radiator through the thermostat housing and upper radiator hose. A centrifugal water pump is a variable-displacement pump. Restricting the flow of coolant does not harm the pump. When the thermostat is closed, restricting coolant flow, coolant circulates through the engine via a **bypass** passage below the thermostat leading from the engine block to the water pump. When the thermostat is open, coolant flow is through the cooling system.

The most frequent cause of coolant pump failure is leaks, which are often the result of bearing failure. Industry studies have linked these leaks and bearing failures to improper

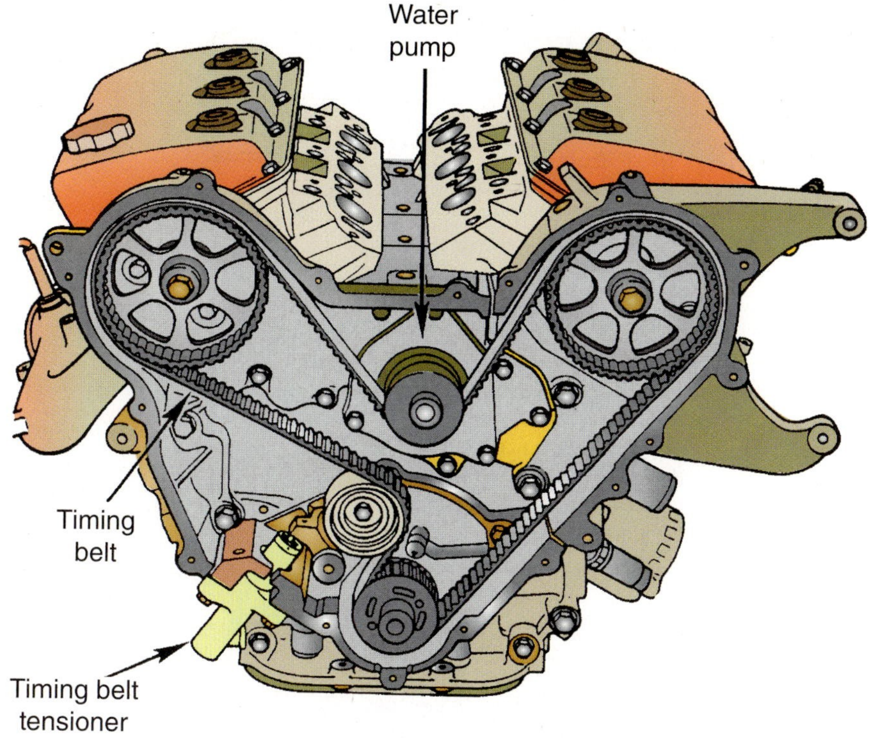

FIGURE 4-19 **A typical timing-belt-driven water pump.**

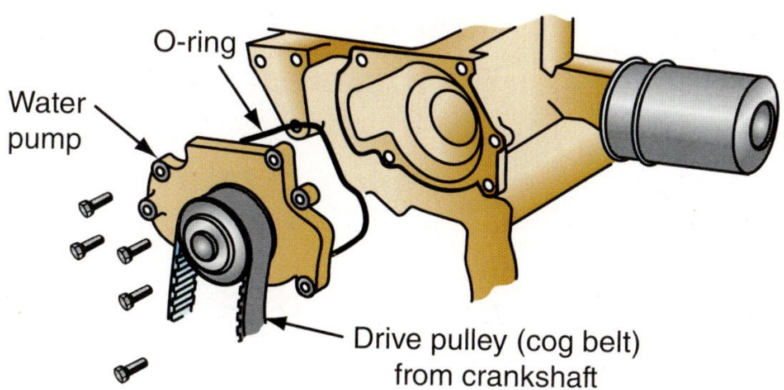

FIGURE 4-20 **Details of a typical timing-belt-driven water pump.**

cooling system maintenance and service as antifreeze additives are depleted or contaminated. We will discuss proper service under the antifreeze section.

Auxiliary Water Pump Hybrid Electric Engine

On hybrid electric vehicles that use a coolant-based heater core, an auxiliary electric water pump may be required when the internal combustion engine is not running, to provide stable heater performance even if the engine is stopped. During normal engine operation, the electric water pump does not operate. Some early designs incorporated a bypass valve that would open when the engine water pump was operating to minimize the resistance to coolant flow (Figure 4-21). Later designs discontinued the bypass valve as new pump designs minimized water flow resistance.

In addition to an electric coolant circulation pump, Toyota hybrid platforms utilize a large vacuum-insulated coolant heat storage tank that recovers and stores hot coolant generated from the engine. This hot coolant is stored at 176°F for up to 3 days and is used to preheat the engine on cold engine start-up to lower engine HC emissions, as well as to supply heat to the

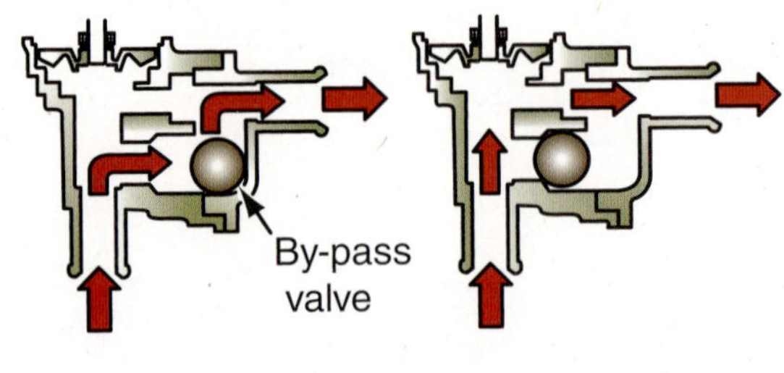

FIGURE 4-21 **A hybrid electric vehicle requires an electric water pump to keep coolant circulating through the heater core when the engine is not running.**

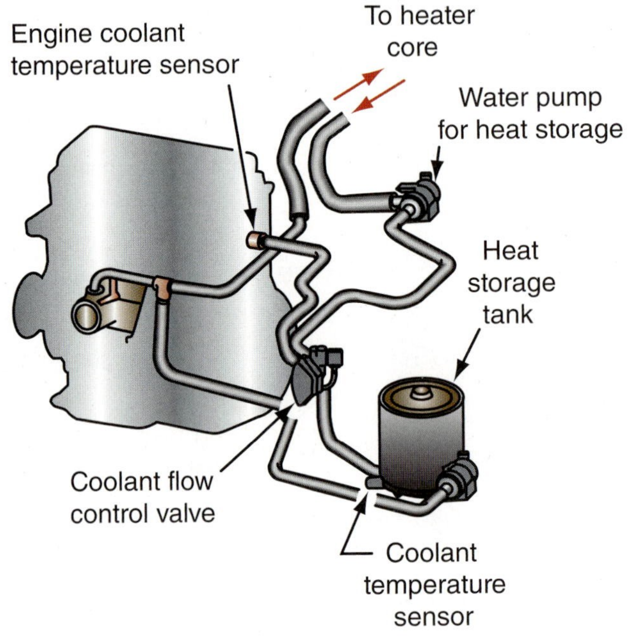

FIGURE 4-22 **A hybrid electric vehicle hot coolant storage tank and water pump.**

heater core (Figure 4-22). When a cold engine is started, the auxiliary electric coolant pump is energized to circulate this hot coolant.

Rotary Water Valve on Hybrid Electric Engine. An electric rotary control valve is used on some hybrid electric platforms to control the flow of hot coolant throughout the cooling system. The control valve can switch between three positions to control the flow of coolant to and from the coolant heat storage tank (Figure 4-23). During the preheat cycle, the water control valve directs the stored heated coolant to the engine cylinder head prior to engine start-up (Figure 4-24). After the engine has started and is in the warm-up cycle, the water control valve directs coolant through the heater core and then back to the engine (Figure 4-25). Once the engine has reached operating temperature, the water control valve allows heated coolant leaving the engine to flow to both the heater core and the coolant heat storage tank so

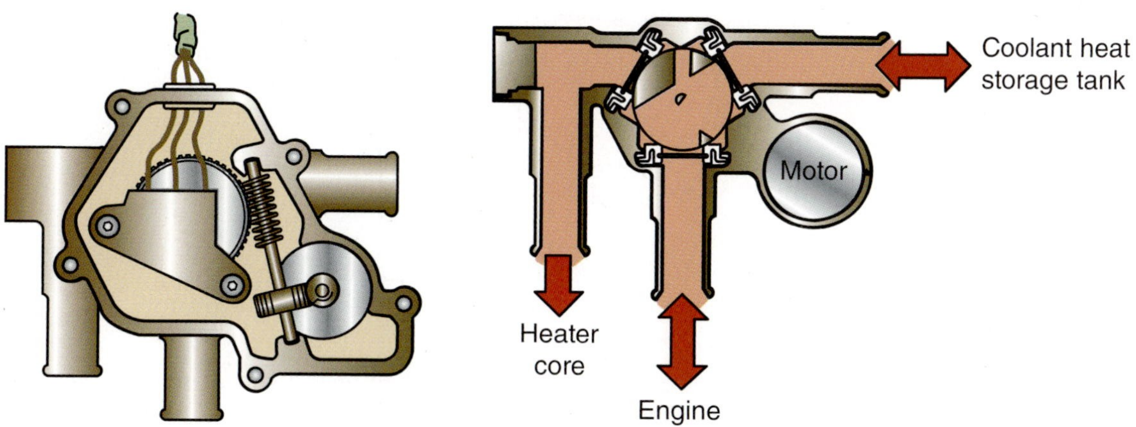

FIGURE 4-23 **A hybrid electric vehicle rotary water control value.**

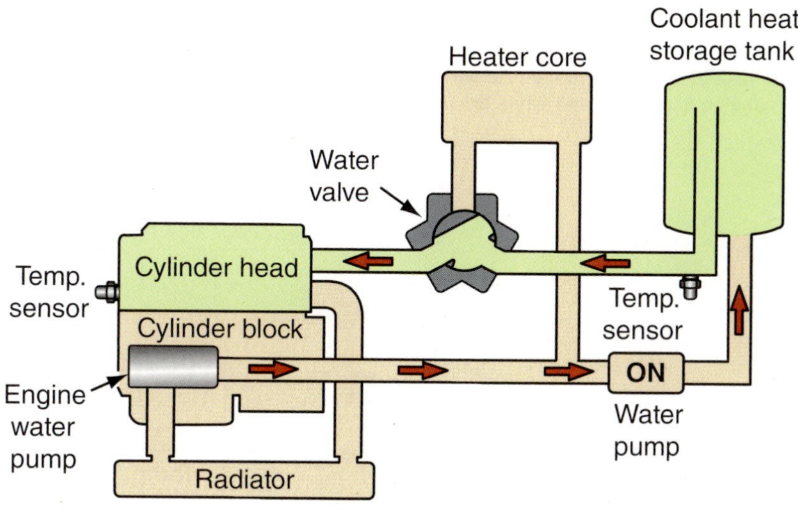

FIGURE 4-24 **A hybrid electric vehicle engine coolant storage system during preheat operation.**

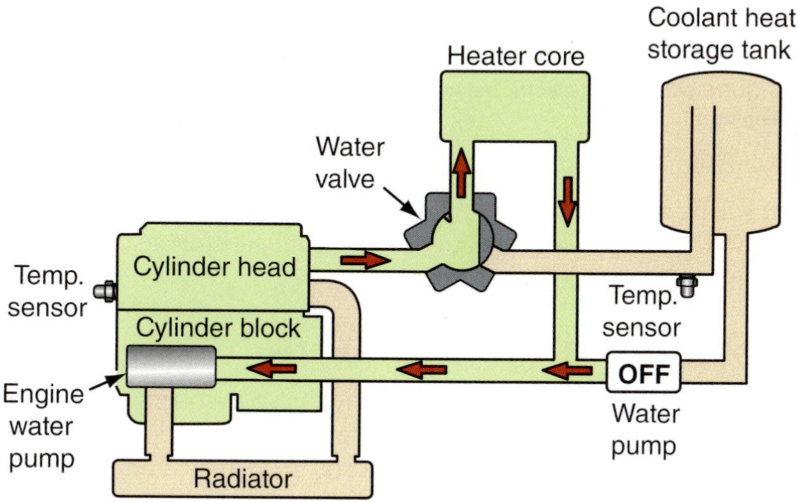

FIGURE 4-25 **Hybrid electric vehicle engine coolant storage system during engine warm-up operation.**

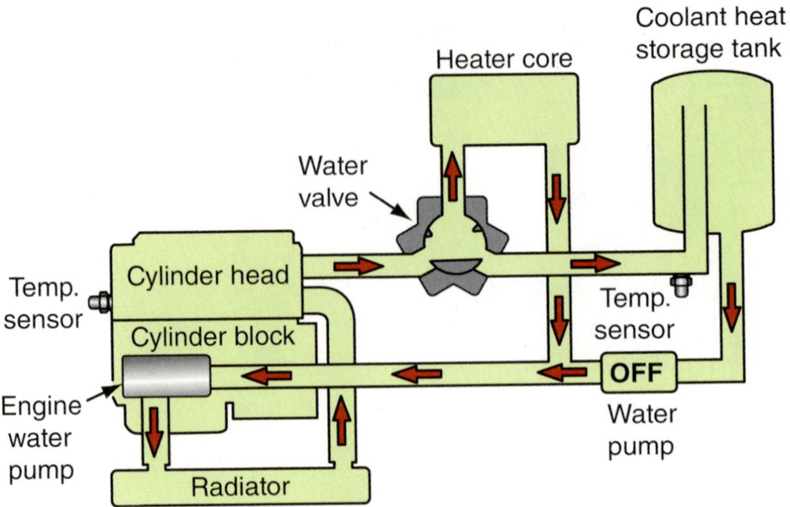

FIGURE 4-26 **Hybrid electric vehicle engine coolant storage system while engine is running and while driving vehicle.**

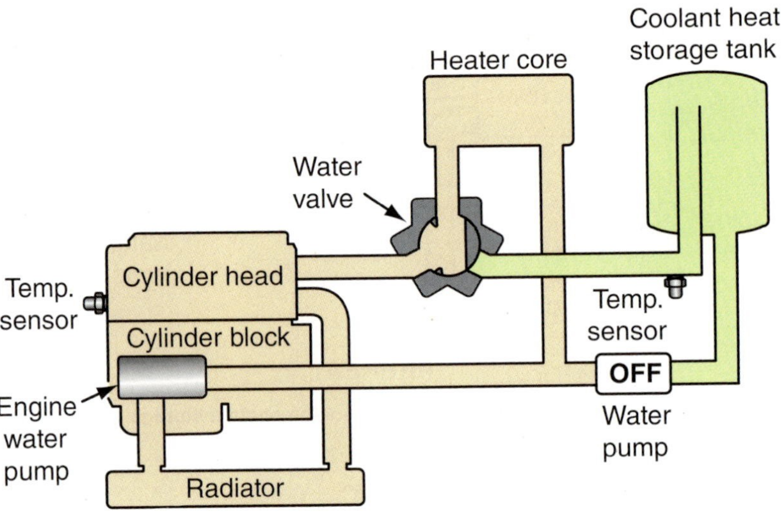

FIGURE 4-27 **Hybrid electric vehicle engine coolant storage system with the engine off.**

that hot coolant will be available during the next preheat cycle (Figure 4-26). After the engine is shut down, the water control valve isolates the coolant heat storage tank so that coolant will not flow to or from it (Figure 4-27).

Fan Shrouds, Air Baffles, and Seals

Air deflectors are commonly damaged by curb impacts. If damaged, they should be repaired or replaced, not discarded.

Airflow can make up to a 30 percent difference in cooling system capacity. There are several air deflectors used in the cooling system to improve overall cooling and air conditioner condenser performance. These include fan shrouds, deflectors, air baffles, and air seals. The radiator fan shroud directs all the air handled by the fan through the radiator, thereby improving the efficiency of the fan. Air deflectors are installed under the vehicle to redirect airflow through the radiator to increase flow. Air baffles are also used to direct airflow through the radiator. Air seals, which are mounted to seal the hood to the body, prevent air from bypassing the radiator and air conditioning condenser. In addition, the air seals help to prevent recirculation of air from under the hood to improve hot weather cooling.

THERMOSTAT

The thermostat is the automatic temperature control component that controls coolant flow in the cooling system and is necessary for efficient engine operation, improving both performance and economy. The primary purpose of the thermostat is to ensure that the minimum operating temperature of the engine is reached as soon as possible. This improves fuel economy and vehicle emissions, as well as preventing the formation of sludge in the engine crankcase. The water pump begins to circulate the coolant the moment the engine is started. The thermostat restricts the circulation of coolant from entering the radiator until the engine has warmed up in order to provide hot coolant to the heater core and improve passenger comfort during the warm-up cycle. This is particularly important for cars driven only a short distance. When the thermostat is closed, a bypass passage in the water pump or coolant passage allows coolant to circulate through the engine. The thermostat also provides a restriction in the cooling system both before and after it has opened. The purpose of this restriction is to provide a pressure difference for the water pump in order to prevent pump cavitation, and it aids in forcing the coolant through the passages in the engine block. Under no circumstances should the thermostat be left out of the system.

The rating of the thermostat is the temperature at which it is designed to begin to open and is usually stamped on the thermostat body. Different engines use thermostats with different temperature ratings, though the most common rating today is 195°F (91°C). Engines are designed to operate at a minimum coolant temperature of 140°F to 195°F (60°C to 91°C). When coolant temperature is below its rated value, the thermostat remains closed. Engine coolant temperature is sensed by a temperature-sensitive element within the thermostat. This causes the normally closed (nc) thermostat to open at a predetermined rating. A typical 195°F (91°C) thermostat will start to open at this rated temperature and will be fully open at 220°F (105°C). This restricts initial circulation to allow for proper engine warm-up. A thermostat's opening and closing are gradual as the temperature of the coolant increases or decreases (Figure 4-28). While the thermostat is closed, coolant flows through the bypass passage in the water jackets. For example, if a thermostat rated at 195°F (91°C) is used, the coolant will not circulate through the cooling system and the radiator until the engine coolant has reached this temperature, and circulation to the radiator will increase as the temperature of the coolant rises. The purpose of a thermostat, then, is to protect against engine **overcooling**.

Overcooling is a condition where the engine never reaches operating temperature due to a thermostat that is opening before the engine reaches operating temperature. Overcooling can also occur if the thermostat has failed in the open position or if it has been removed from the system.

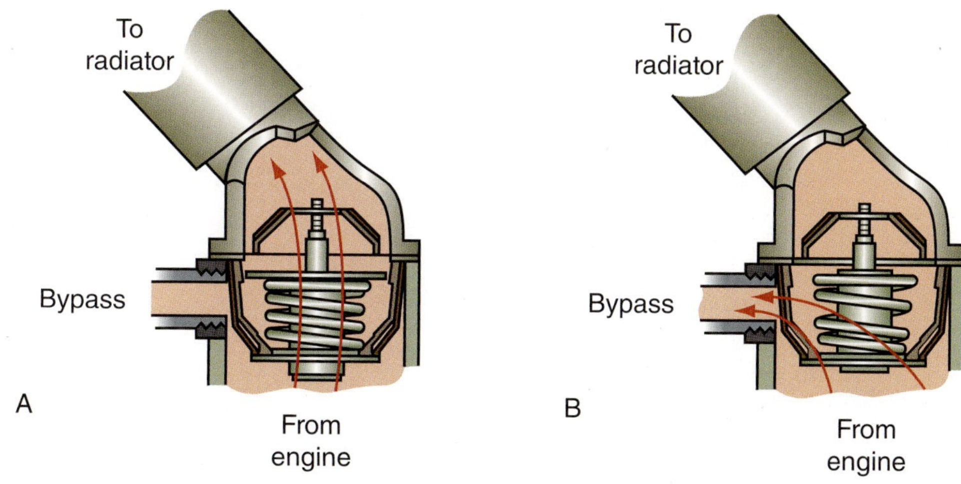

FIGURE 4-28 **A thermostat opening (A) and closing (B).**

The engine should take between 5 and 15 minutes to warm up, depending on ambient air temperature. If the engine requires long warm-ups or if the engine always runs hot, you may need to remove and test the thermostat operation by suspending it in a water bath and heating the water to boiling. You will need to note the temperature at which the thermostat begins to open and the temperature at which the thermostat is fully open and compare this to its rated value.

A thermostat usually starts to open at its rated value and is fully open with a 25°F (14°C) temperature rise.

Unless it is defective, a thermostat will not cause overheating. A thermostat rated at 180°F (82°C) is wide open at 205°F (96°C), and full coolant flow is provided through the cooling system. A thermostat rated at 195°F (91°C) is wide open at 220°F (105°C), and full coolant flow is provided through the cooling system. The 195°F (91°C) thermostat provides neither more nor less coolant flow than the 180°F (82°C) thermostat, but it does change the operating temperature of the engine and the amount of heat the passenger compartment heater can produce. Refer to Chapter 4 of the Shop Manual for additional information on both overcooling and overheating conditions and the manufacturer service information.

The cooling system thermostat (Figure 4-29) is located between the engine and the radiator. It is housed at the outlet of the engine coolant passage under a return hose flange called a thermostat housing and, in most cases, is bolted onto the cylinder head or intake manifold. There is a gasket or O-ring seal between the thermostat housing and mounting surface that must be torqued to specifications to avoid leaks. Thermostats may have either a bleed notch or a jiggle pin that is designed to let trapped air out of the system after refilling in order to eliminate hot spots in the coolant passages during engine warm-up. Many closed cooling systems today require air pockets to be bled out of the system after servicing, using a bleed-off valve for air that may be part of the thermostat housing; see Figure 4-13.

Shop Manual
Chapter 4, page 108

Some manufacturers have chosen to locate the thermostat at the engine inlet. This reduces the risk of thermal shock that may result as cold coolant enters the engine block when the thermostat opens. Inlet thermostats slowly bleed cold coolant into the engine until the entire cooling system comes up to operating temperature.

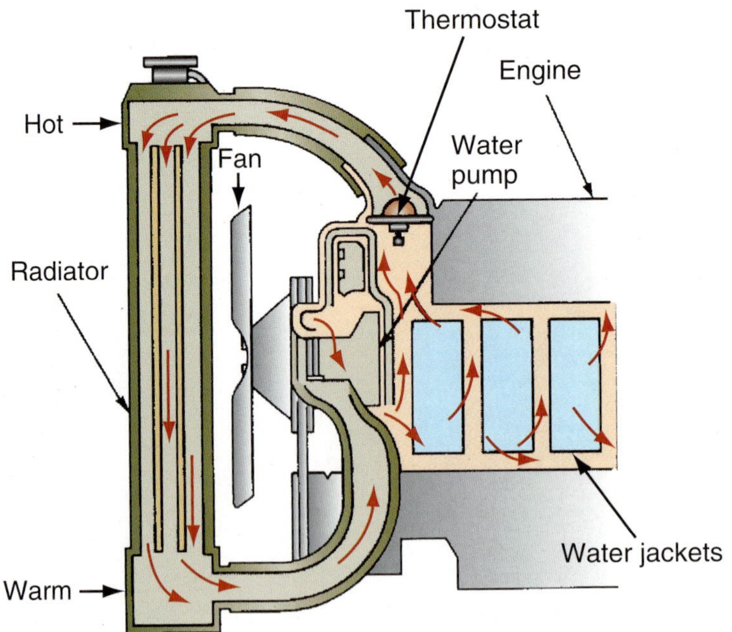

FIGURE 4-29 **Location of thermostat in typical cooling system.**

Six important facts regarding thermostats should be noted.

1. Thermostats are a design component of the engine cooling system and should not be omitted.
2. The design temperature of a thermostat should not be altered.
3. A thermostat will cause engine overheating if stuck closed.
4. A thermostat will cause engine overcooling if stuck open.
5. A vehicle may fail an emissions test due to an improperly functioning thermostat.
6. An improperly functioning thermostat will affect passenger compartment heater efficiency.

Thermostats are a design consideration of the cooling system and should not be permanently removed or changed to one of a different rating.

Solid Expansion Thermostat

The solid expansion thermostat (Figure 4-30) is a heat motor that utilizes a thermally responsive wax pellet sealed in a heat-conducting copper cup containing a flexible rubber diaphragm and piston. This is by far the most popular style thermostat used today. Heat causes expansion of the now liquid wax compound, exerting pressure on the diaphragm and pushing a stainless steel piston plunger up to open a valve. As the element cools, the compound contracts, and a spring is allowed to push the thermostat valve closed.

There are currently three popular variations of the solid expansion type; the balanced sleeve, the reverse poppet, and the three-way thermostat (Figure 4-31). These styles function similarly but with some design differences. The balanced-sleeve thermostat allows pressurized coolant to flow around all of its moving parts, whereas the reverse-flow thermostat opens against the direction of coolant flow and water pump pressure. In this manner, it is able to use water pump pressure to hold it closed when it is cool. The reverse-poppet thermostat is engineered with a self-cleaning, self-aligning stainless steel valve and offers improved coolant flow. The three-way thermostat has a bypass passage located directly below it. When the engine is cold, the thermostat restricts flow to the radiator and allows coolant to flow through the bypass passage. As the thermostat opens, allowing coolant to flow to the radiator, it also closes off the bypass passage.

The solid expansion thermostat is also known as the "pellet" type, so called because of the wax pellet used.

Corrosion and age will cause a calibration change in this type of thermostat. If the thermostat is found to be defective or inoperative, a new one of equal design and rating should be installed.

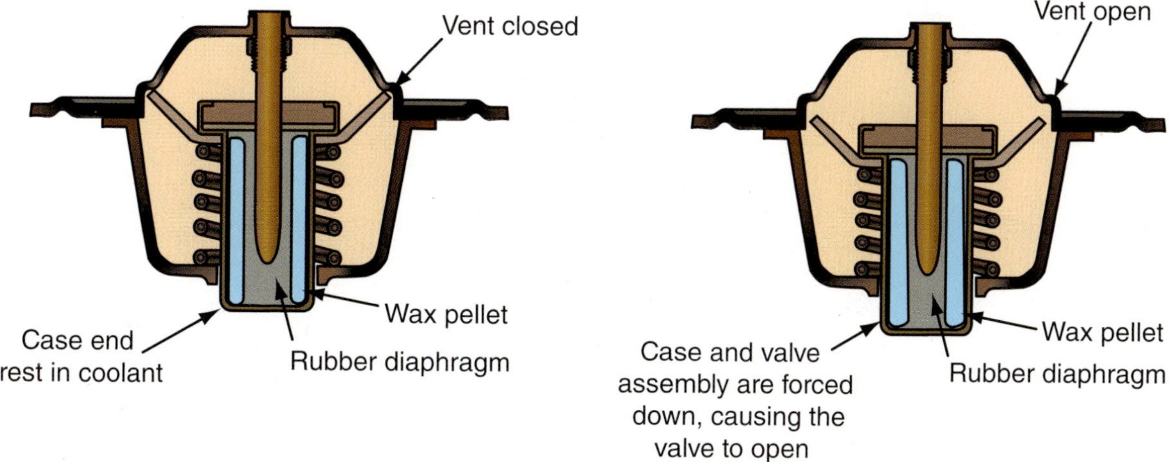

FIGURE 4-30 **A typical solid expansion thermostat, as wax pellet melts it exerts pressure in the capsule pushing value rod out of the pellet.**

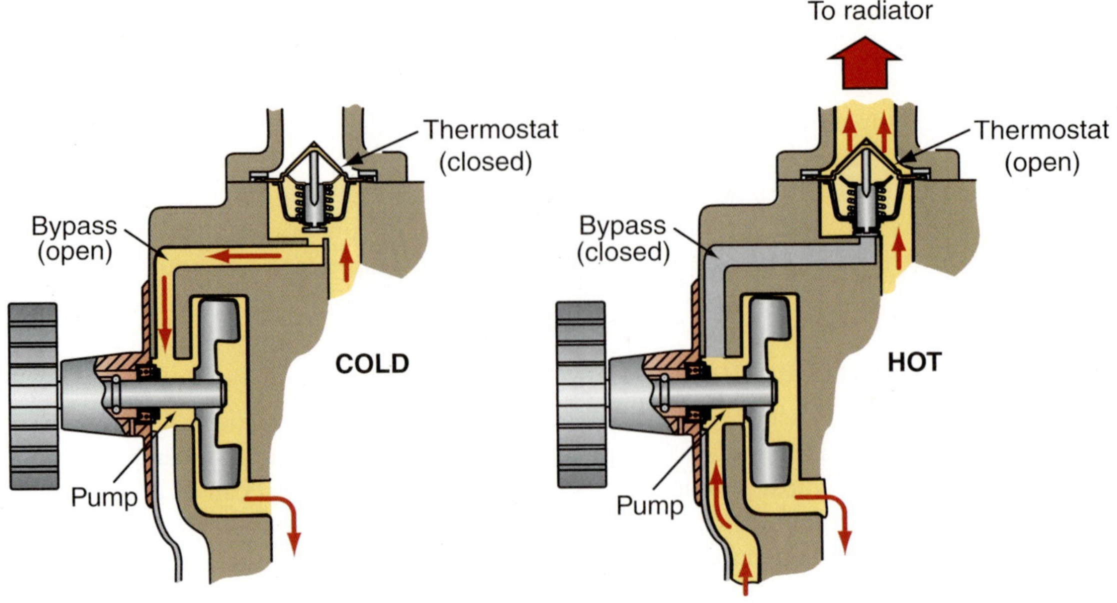

FIGURE 4-31 **Three-way thermostat.**

Electronic Thermostat

The development of the electronically controlled cooling system (ECCS) was to enable the control module to set the operating temperature of the engine to a specified value based on engine load. The optimal operating temperature is determined by a "mapping" program in the power train control module (Figure 4-32). Engine cooling temperature is adapted to the engine's overall performance and load state by heating the thermostat electronically and adjusting the radiator fan speed settings. The advantages of adjusting coolant temperature to current operating demands of the engine are lower fuel consumption, reduced emissions, and longer engine life.

Differences to the conventional cooling system may include thermostat and coolant distributor housing that may be integrated into a single module (Figure 4-33). The thermostat no longer needs to be near or on the cylinder head. The electronic thermostat contains a resistive heating element embedded in the wax pellet (Figure 4-34) that heats the wax pellet to regulate the opening and closing of the thermostat if temperature mapping requires a lower

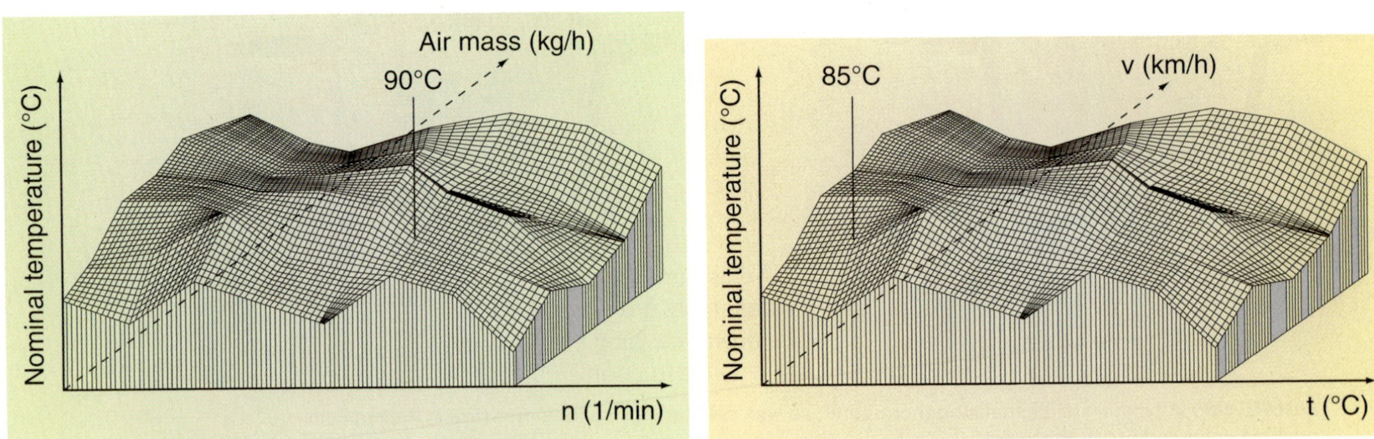

FIGURE 4-32 **An example of a typical electronic thermostat mapping program analysis of optimum engine coolant temperature based on various inputs.**

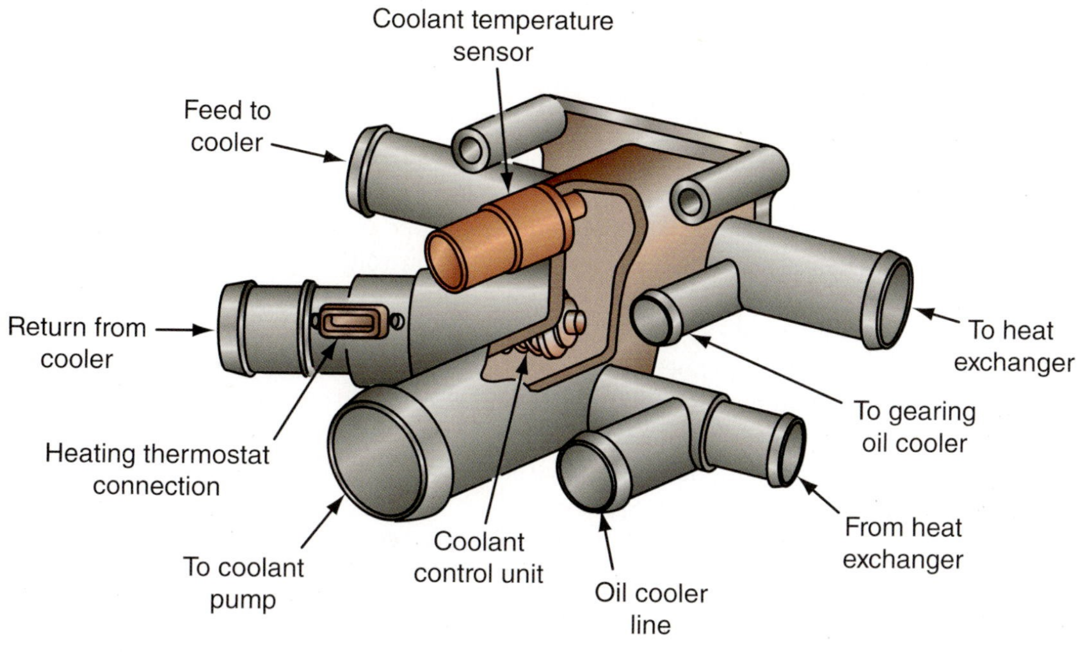

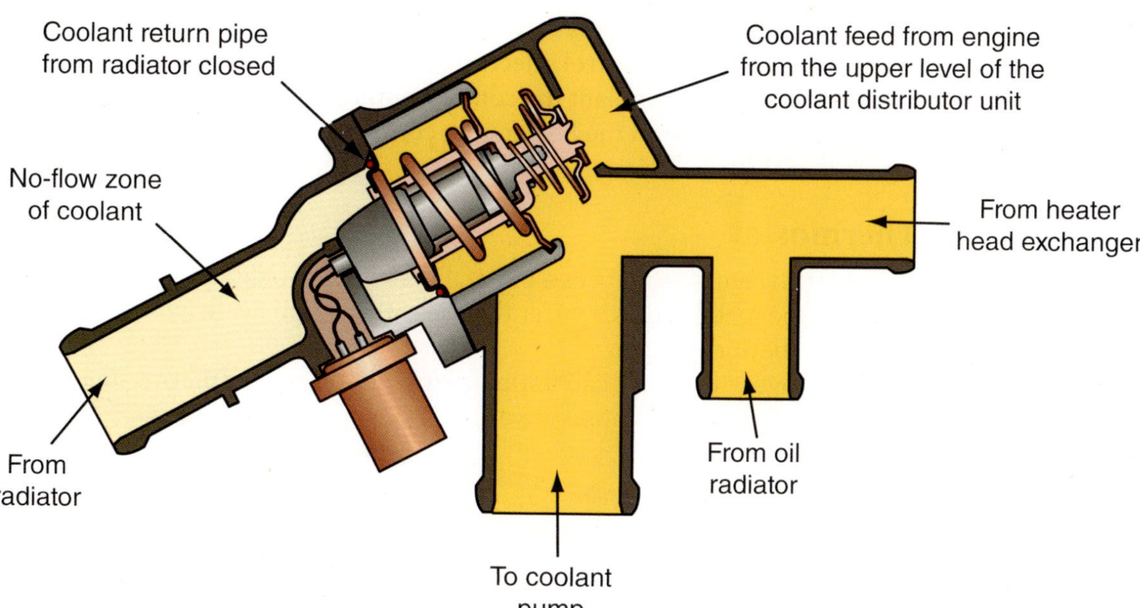

FIGURE 4-33 **A typical electronically controlled cooling system housing containing electronic thermostat.**

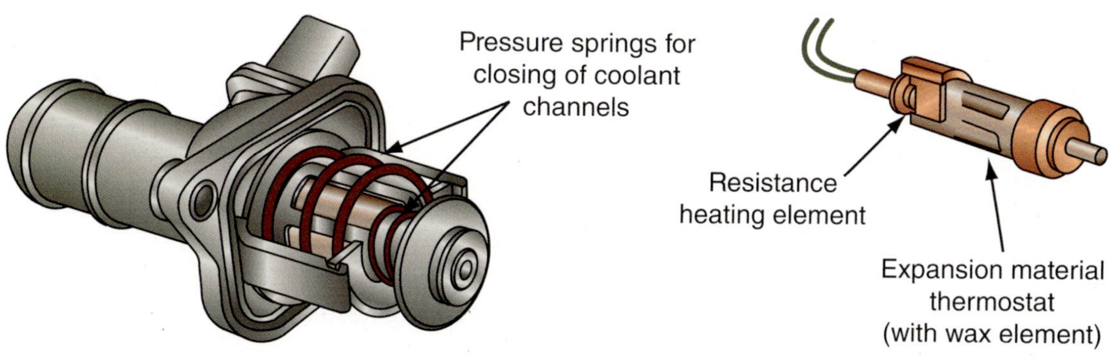

FIGURE 4-34 **A typical electronic thermostat.**

coolant temperature. Or if engine coolant temperature is high, coolant temperature alone can cause the wax pellet to liquefy and expand causing the thermostat to open just as it did on a conventional thermostat but at a much higher temperature than a conventional thermostat. The melting temperature of the wax pellet is 230°F (110°C) compared to a conventional thermostat where the wax pellets began melting at 195°F (90°C). Think of it as a standard thermostat with a heater.

Optimum engine performance is dependent on proper engine temperature for a given engine demand. An example of this in an electronically controlled cooling system is the part-throttle optimum temperature range of 203°F to 230°F (95°C to 110°C) and in the full-throttle optimum temperature range of 185°F to 203°F (85°C to 95°C) depending on load and engine speed as an example. A higher temperature in the part-throttle range improves fuel economy and lowers emissions by improving engine efficiency. A lower coolant temperature at full throttle increases engine power output boosting performance. While maintaining these temperature variations on a conventional cooling system precisely was not possible, it is achievable on an ECCS (Figure 4-35).

One of the effects of this style of engine temperature regulation is a fluctuation of coolant temperature between 230°F (110°C) and 185°F (85°C) when driven between part throttle and full throttle, which in turn would affect heater core temperature. A 45°F (25°C) coolant temperature fluctuation would make the passenger compartment uncomfortable when heating mode is selected and the driver would be constantly readjusting the temperature range selected. To avoid broad temperature fluctuation the electronic control module monitors the passenger compartment temperature control selector. Based on the temperature selected and the actual temperature of the engine coolant the control module will regulate coolant flow to the heater core by regulating flow with the heater core control valve which is either vacuum or electronically activated (Figure 4-36).

Bimetallic Thermostat

The bimetallic thermostat (Figure 4-37) uses a bimetallic strip of two dissimilar metals fused together to form a coil. One metal expands faster than the other when heated, causing the coil to unwind and opening a butterfly valve.

Again, corrosion and age will cause a calibration change in this type of thermostat. It must be replaced if found to be defective or inoperative.

Bellows-Type Thermostat

The bellows-type thermostat (Figure 4-38) is made up of a thin metal bellows assembly filled with a low-boiling point fluid, usually alcohol, and is sealed under a vacuum. They were popular before the advent of the pressurized cooling system. When increased coolant temperature causes the fluid to boil, the bellows expands. This expansion opens the thermostat, allowing engine coolant to circulate through the cooling system. As the volatile fluid cools, the bellows retracts, restricting the flow of engine coolant. The pressurized cooling system made this style obsolete. Pressure in the cooling system would prevent the bellows from opening at the correct temperature.

Thermostat Service

Thermostats generally fail in the closed position.

Thermostats fail and can cause excessive engine wear and waste fuel. If failure occurs while the thermostat is in the closed position, severe engine overheating will result. An extremely hot engine and a cool-to-warm radiator may be noted. If failure occurs while the thermostat is in the open position, a longer than normal warm-up period may be noted by the temperature gauge, or poor passenger compartment heating may be noted. Often a thermostat that is stuck open will go undetected in warmer months, but lack of heater performance reveals this condition during cooler months.

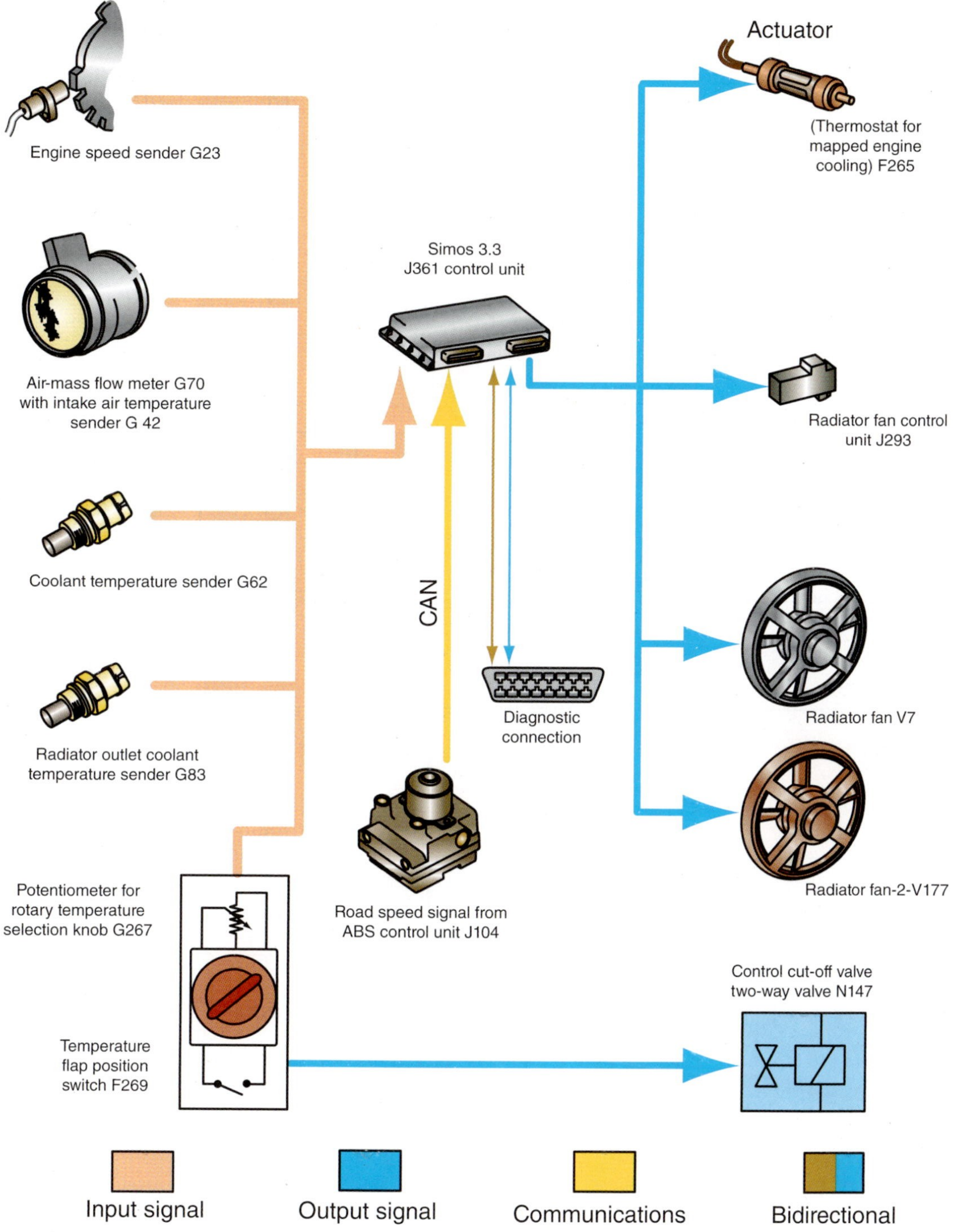

FIGURE 4-35 **An example of a typical electronic thermostat system.**

It should be noted that a defective thermostat could have an adverse effect on the computer's engine control system. A thermostat that is stuck open or opens prematurely may cause the closed-loop status to be delayed, resulting in erratic or fast idle and/or richer than normal fuel conditions which may result in an engine service indicator light illuminating. It is therefore important that the engine coolant thermostat be replaced with the correct temperature range thermostat if it is found to be defective. Thermostats should also be replaced as part of any cooling system repair service.

Shop Manual
Chapter 4, page 109

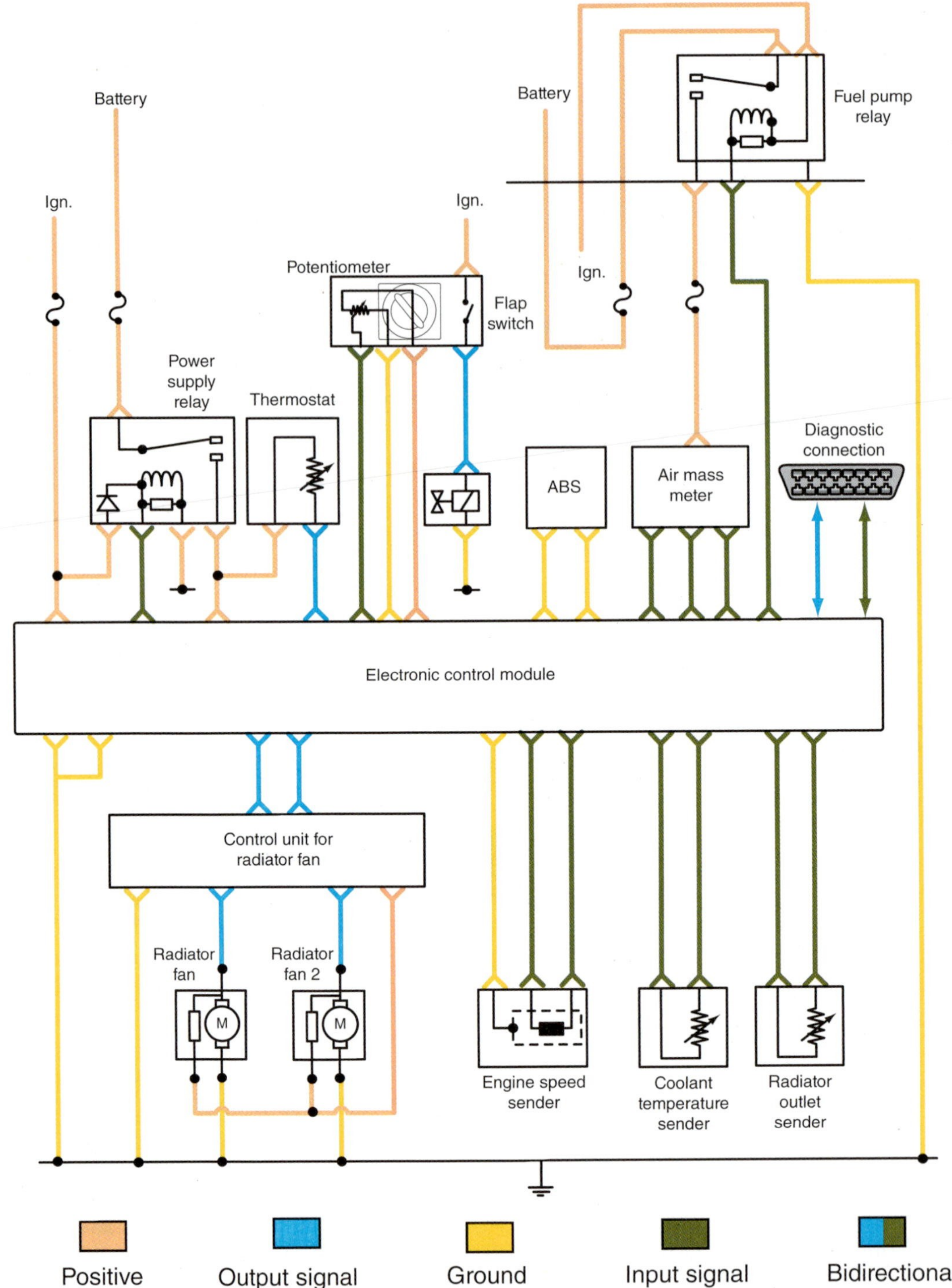

FIGURE 4-36 **An example of a typical electronic thermostat wiring diagram.**

AUTHOR'S NOTE: If a vehicle comes into your shop overheating, do not add coolant or water to the system until it has completely cooled down. The cool-down process is necessary to avoid thermal shock to the engine and cooling system components. Thermal shock could cause components to crack and gaskets to fail, making a bad situation worse.

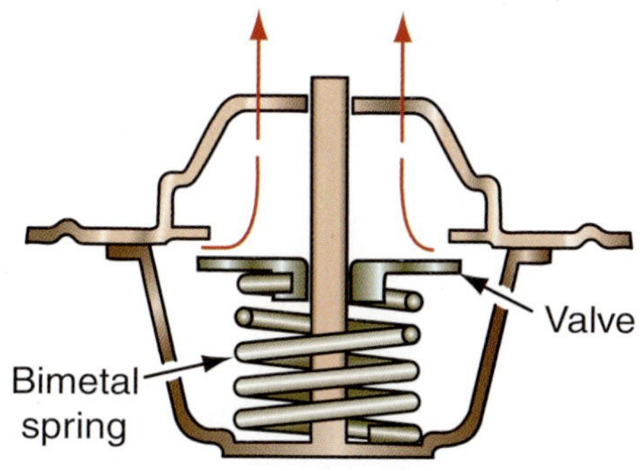

FIGURE 4-37 **A typical bimetal thermostat.**

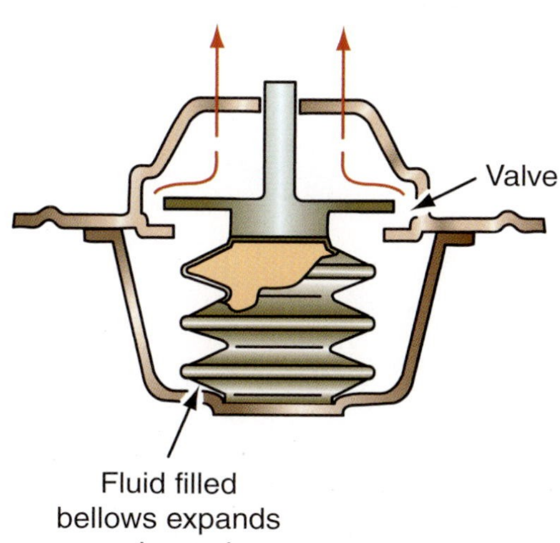

FIGURE 4-38 **A typical bellows-type thermostat.**

PULLEY AND BELT

Two types of belt systems are used to drive the air conditioning compressor and many water pumps, as well as other accessories: the serpentine belt (Figure 4-39) and the V belt (Figure 4-40). The serpentine belt system is found on most late model vehicles. This is often a single-belt system whereby one belt drives all the accessories (Figure 4-41).

Shop Manual
Chapter 4, page 110

Serpentine belts may be constructed of either neoprene or ethylene propylene diene M-class rubber more commonly called EPDM. Today most belts are constructed of EPDM, but vehicles produced prior to 2001 or inexpensive aftermarket replacement belts may still be made of neoprene. Visually it is difficult to tell the two materials apart. Older neoprene belts were designed to last between 50,000–60,000 miles and often showed signs of wear by that time. EPDM belts are designed to last 80,000–100,000 miles and seldom show signs of outward visual wear unless there is a problem. The EPDM belt is more elastic than a standard neoprene belt and resists cracking even at higher mileage. A better indicator of when to replace EPDM belts is rib wear. Belts are designed to have clearance between the rib peaks and the pulley grooves. All belts are exposed to dirt, grit, rocks, road salt, and water. Over time

There are several different sizes of both types of belts.

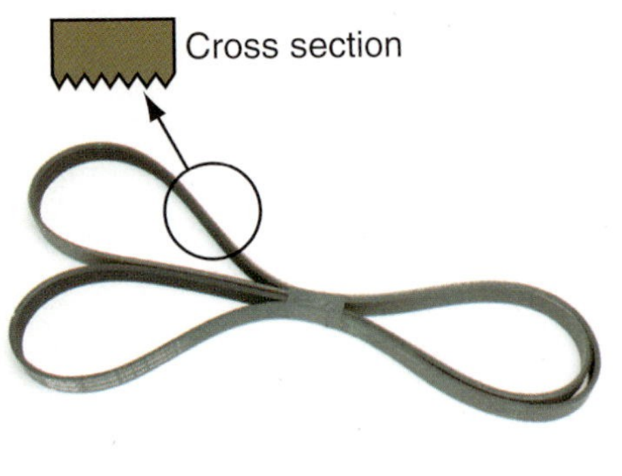

FIGURE 4-39 **A typical serpentine belt.**

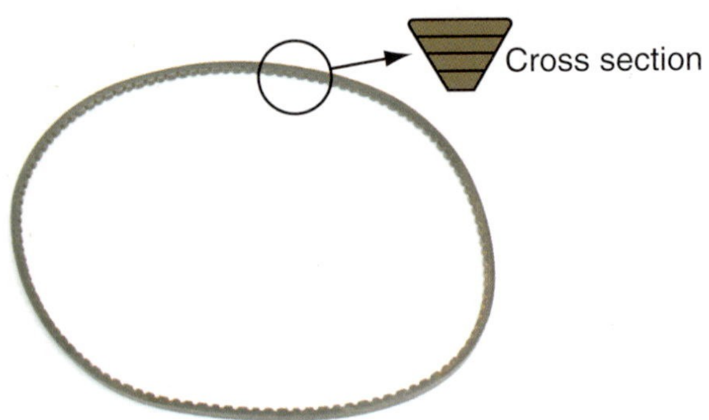

FIGURE 4-40 **A typical V belt.**

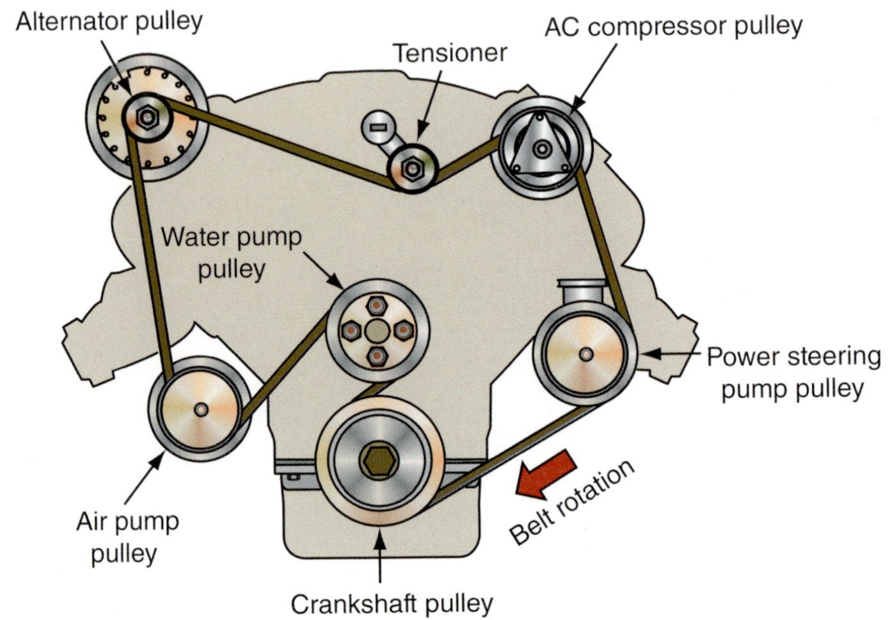

FIGURE 4-41 **A serpentine belt drive system.**

these contaminants cause the EPDM belt to gradually lose material on the ribbing similar to the way a tire wears out causing the belt to ride deeper in the grooves. A 5–10 percent loss of material is enough to cause belt slippage, overheating, and hydroplaning. Hydroplaning occurs when the belt ribs sit deeper in the pulley grooves due to wear and there is not enough room for water to escape. Instead, the water is trapped between the belt ribs and the pulley grooves lifting the belt away from the pulley and causing slippage. A slipping serpentine belt can cause check engine lights associated with misfire codes, air conditioning compressor codes, reduced alternator output, reduced engine cooling, and poor air conditioning performance to name a few. Belts should be checked for wear beginning at 50,000 miles. See Chapter 4 in the Shop Manual for further detail on diagnosing both neoprene and EDPM belts and how to measure for belt grove wear. It is wise to replace the drive belt when any pulley-driven component is replaced.

Another change that has occurred in belts is the use of stretch fit belts that self tension on the drive pulleys. They do not require any mechanical adjustment and are used in limited applications such as a single drive belt for an air conditioning compressor. As the name implies, they contain an elastomeric material and with the use of a special tool they are stretched on to the pulley system, there is no mechanical means to loosen or tighten the belt and as with the standard EPDM belt they are designed to last 80,000–100,000 miles.

The **pitch** and width of belts in the V belt system are important in that they must match those of the drive (engine crank shaft) and driven (engine accessories) pulleys (Figure 4-42). If the compressor and alternator are driven with two V belts, they should be replaced as a pair with a matched set. This is true even though only one of the pair may appear to be damaged.

The **pitch** of a belt is the degree or slope of the V shape of the belt.

Belts should be tensioned in foot-pounds (ft.-lb.) or Newton meters (N · m), according to manufactures' specifications. It is recommended that the belt be retensioned after a "run-in" period of a few hundred miles (kilometers) to ensure proper belt tensioning.

Belts must be replaced with exact duplicates. Their length, width, and groove characteristics are important to ensure proper fit and alignment. Manual belt tension should be set to ½ inch deflection per 12 inches of distance spanned between pulleys. The serpentine system may have a spring-loaded idler pulley used as a belt tensioner. It is therefore not necessary to manually tension this type of belt tension system. If the belt will not remain tight, the tensioner must be replaced.

It should be noted that some coolant pump pulleys turn in the opposite direction from others (Figure 4-43). Therefore, it is possible that the same engine in two different vehicles

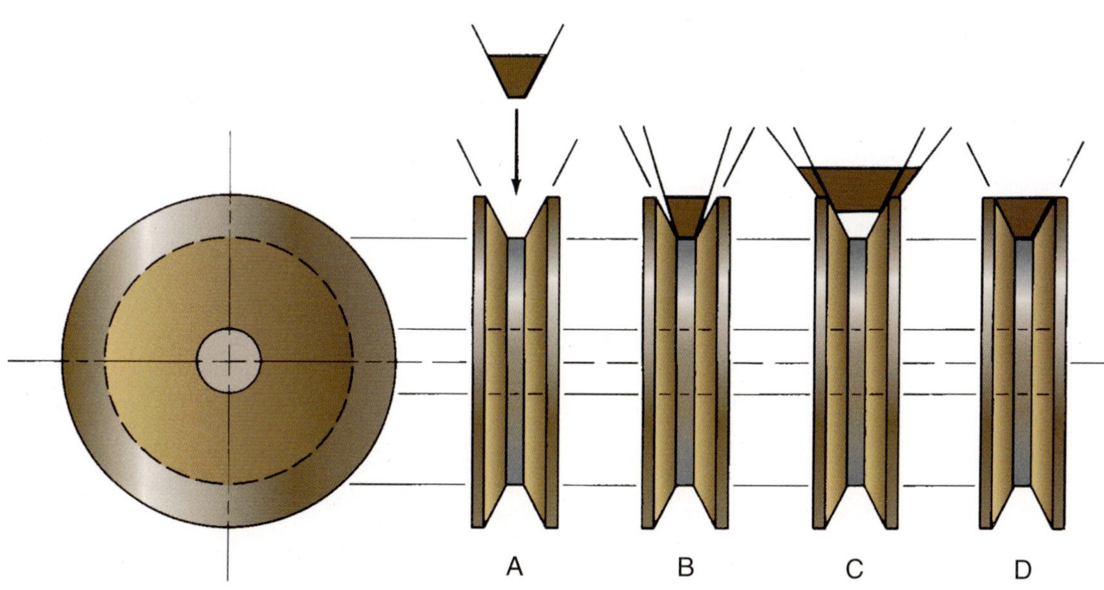

FIGURE 4-42 **The belt should fit the pulley snugly, as shown in A and D. The belt in pulley B is too narrow and has an improper pitch. The belt in pulley C (exaggerated) is too wide and has an improper pitch.**

FIGURE 4-43 **Some coolant pumps turn in opposite directions.**

with either different drive belt styles (V belt versus serpentine belt) or belt routing due to bolt-on belt-driven accessories (one system with A/C and one without) will require two different water pumps and two different mechanical cooling fans. Water pump rotation may be either clockwise or counterclockwise based on drive belt routing.

FANS

The fan is used in the cooling system to increase airflow across the radiator in order to improve the efficiency of the cooling system. Engine coolant fans are essential for idle and low-speed driving to pull sufficient air through the condenser, radiator, and the engine to affect adequate cooling. At road speeds, ram air is sufficient for this purpose. To satisfy the needs for low-speed cooling and to reduce the engine load at high speed, a fan clutch or flexible fan is often used. An electric fan, which replaces the coolant pump-mounted fan, is found on most late-model vehicles.

Engine-Mounted Fan

The engine-driven cooling fan is mounted onto the water pump shaft in front of the pulley (Figure 4-44). Five- or six-blade fans (Figure 4-45) are found on air conditioned vehicles, and four-blade fans are found on vehicles without air conditioning. Fans are made of steel, nylon, fiberglass, or a combination of materials. They are precisely balanced to prevent noise, vibration, coolant pump bearing failure, and seal damage.

Shop Manual
Chapter 4, page 119

It must be noted that engine-driven fans are designed to turn either clockwise or counterclockwise, depending on pulley rotation. For proper airflow, it is important that the replacement fan be suitable for the design. An improper fan will result in little (or no) air circulation and will cause engine overheating.

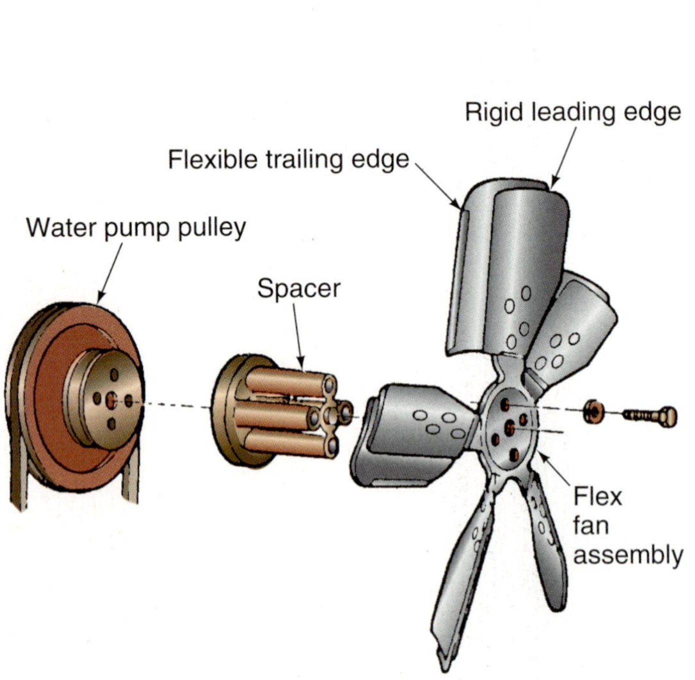

FIGURE 4-44 **A typical engine-mounted cooling fan.**

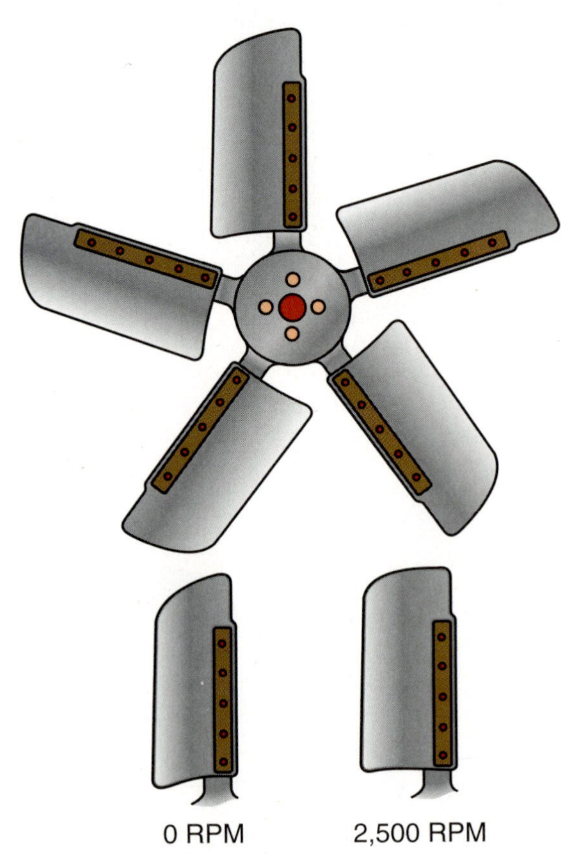

FIGURE 4-45 **As engine speed increases, the pitch of the blade decreases, and this reduces load on the engine.**

Shop Manual
Chapter 4, page 120

Declutching Fan. The declutching fan is used on some rear-wheel-drive vehicles with air conditioning and consists of a fan blade assembly attached to a special clutch. The clutch is attached to the water pump shaft. There are two types of fan clutches: one that is sensitive to engine speed, known as a centrifugal clutch, and one that is sensitive to temperature, often referred to as a thermostatic clutch, which is designed with an internal bimetallic control valve. Either type of fan clutch uses a silicone fluid to engage and disengage the fan blades (Figure 4-46). Either method causes the clutch to be sensitive to engine speed and under-the-hood temperature.

A fan clutch limits the terminal (top) speed of the fan.

The declutching fan is used as a method of solving a problem of increased air needs at low speed and, at the same time, of eliminating air noise problems during high speed. A smaller fan pulley allows for higher fan speed at low engine speed and at the same time provides for increased coolant circulation. The increased air and coolant flow is important when the forward motion of the automobile is not sufficient to produce a strong ram air effect.

The centrifugal clutch is engaged at low engine speed and disengaged at high engine speed. As engine speed is increased, the fluid coupling of the fan clutch increases until the fan reaches its maximum speed. The fan clutch allows the fan to turn at coolant pump speeds up to about 800 rpm. Thereafter, there is slippage that limits the fan speed to between 1,100–1,350 rpm when the engine is cold and between 1,500–1,750 rpm when the engine is hot. The maximum speed of the fan is limited to about 2,000 rpm regardless of engine speed. At maximum speed, the fan will not turn any faster, regardless of how much the engine speed is increased.

The thermostatic clutch has a fluid coupling partially filled with silicone oil. When the temperature is above 160°F (71°C), a bimetal coil spring uncoils or expands. As it expands, it allows additional oil to enter the fluid coupling, causing less slippage and enabling the fan to turn at about the same speed as the coolant pump, up to a maximum of about 2,000 rpm. Minimum fan speed should be between 1,500–2,000 rpm. When the under-hood temperature is below about 160°F (71°C), the opposite will occur; the bimetal coil contracts and fluid is bled off the fluid coupling, causing increased slippage and slowing the fan speed. It now turns at less than coolant-pump speed. Slower fan speeds when additional cooling is not required save fuel, lower engine noise, and increase engine power.

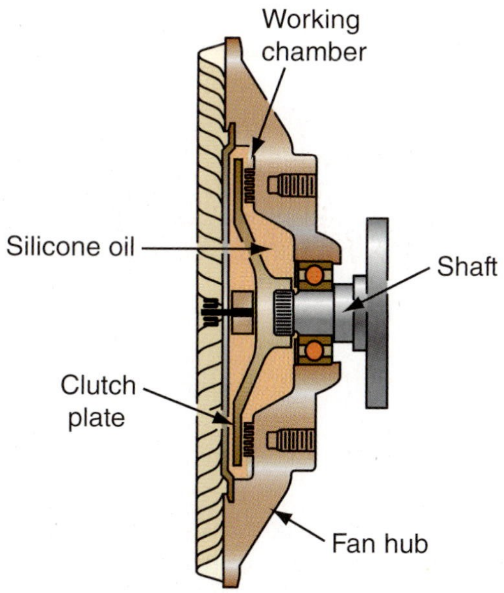

FIGURE 4-46 **Sectional view of a fluid-coupled declutching fan clutch.**

The coolant pump pulley generally turns faster than the crankshaft pulley. The ratio varies from vehicle model to model. For example, if the ratio is determined to be 1.25:1, the coolant pump will turn 1,000 rpm at an engine speed of 800 rpm.

Failure of a fan clutch is usually due to lockup of the clutch or a leak of the silicone fluid, which allows "free-wheeling" of the fan blade. A lockup is generally noted by excessive noise when the engine is revved up and may reduce engine power due to increased drag on the engine. It should be noted, however, that some noise is to be expected during initial cold engine start-up. A fluid leak is most noticeable by a tacky residue substance at the shaft area of the clutch bearing. It is possible for a neglected fan clutch assembly to become detached from the shaft and cause serious damage to the radiator. Also, a defective fan clutch can cause vibrations in the coolant pump shaft, leading to its early failure. The fan clutch is not repairable and, if found to be defective, must be replaced. Many technicians are injured each year by defective fans. Fan blades have been known to come off the hub assembly. Before working under the hood of a vehicle, particularly if looking to locate a noise problem, inspect the fan assembly before starting the engine. Check for loose, bent, or damaged blades.

Electronically Controlled Viscous Cooling Fan. The electronically controlled viscous cooling fan (Figure 4-47) is similar in operation to the standard declutching fan but its operation is controlled by the **power train control module (PCM)**. Several advantages to this system are reduced noise, improved air conditioning performance during idle and city traffic, reduced air conditioning compressor failures because of lower A/C system pressure at low speeds, and improved fuel economy.

The **power train control module (PCM)** is the microprocessor (computer) that monitors input sensors related to engine and transmission operation, interprets this data, and sends commands to output devices.

The electro-viscous (EV) cooling fan design allows for infinitely variable regulation by means of sensors. The regulation process uses data from engine coolant temperature, oil temperature, air charge temperature, engine speed, ignition timing, and air conditioning load. This allows for demand controlled cooling which will lower noise levels, improve coolant temperature for a given load, and reduce fuel consumption.

The silicone oil circulation between the supply chamber and the working chamber affects the fan speed (Figure 4-48). The more silicone oil that is allowed into the working chamber, the higher the fan speed and when the working chamber is empty, the fan is in idle mode (Figure 4-49). There is approximately a five percent slippage rate when fully engaged. But instead of using a bimetallic spring to control oil flow the valve is operated by an electrical solenoid. This solenoid is pulse-width-modulated (PWM) signal from the power train control

FIGURE 4-47 **A typical electro-viscous cooling fan is most easily identified by the addition of an electrical connector.**

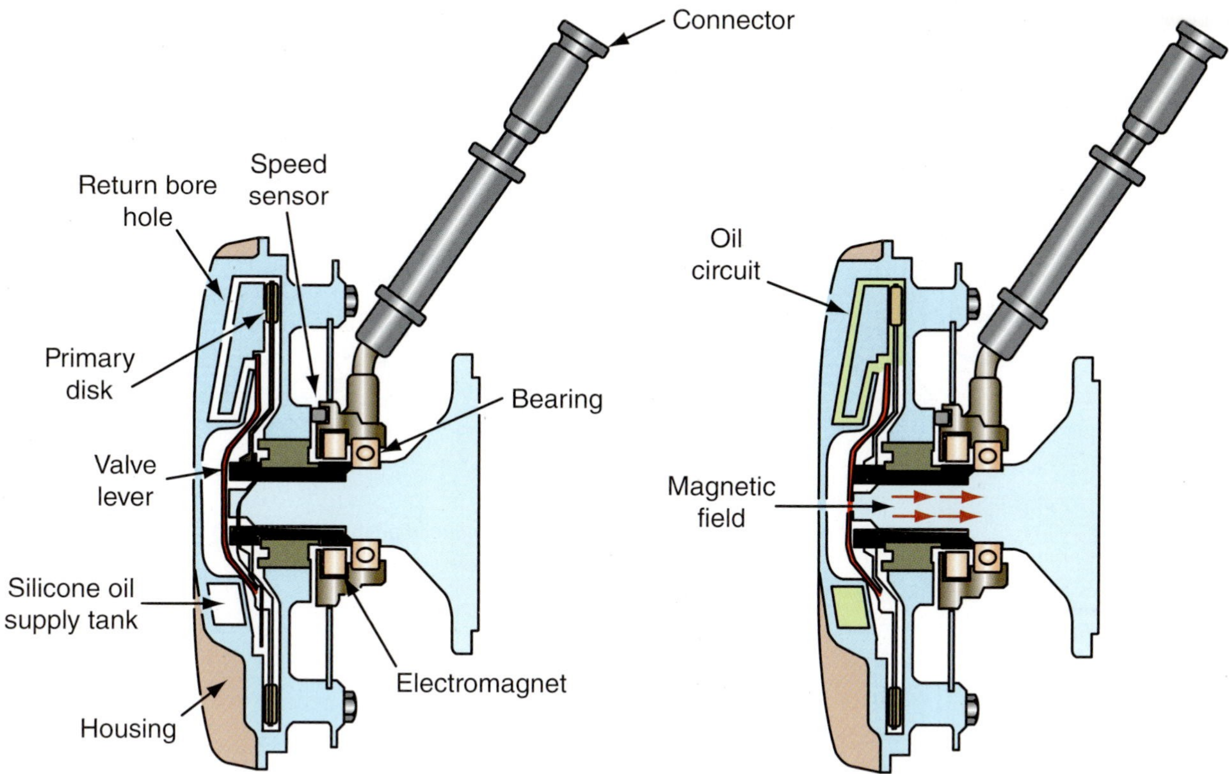

FIGURE 4-48 **A typically electronically controlled cooling system housing containing electronic thermostat.**

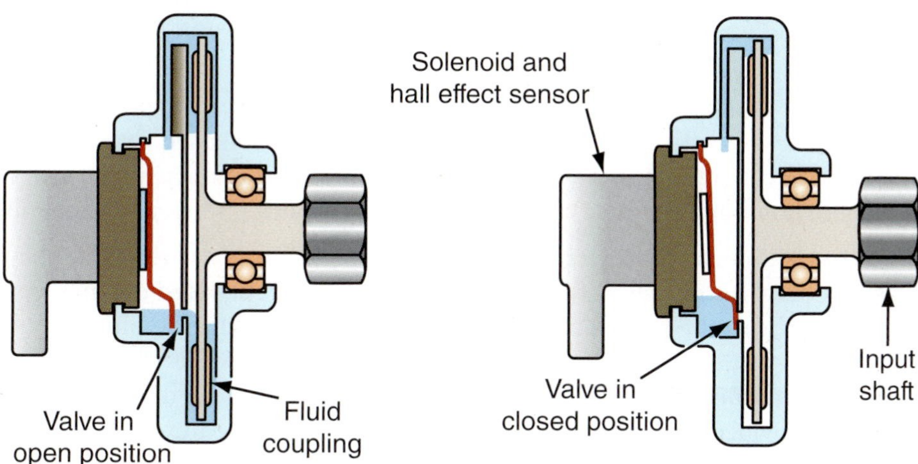

FIGURE 4-49 **Electronic viscous fan clutch operation in the disengaged and engaged mode.**

module (PCM). The EV fan is fully engaged at 100 percent duty cycle and is disengaged at 0 percent duty cycle. To determine duty cycle, the control module uses information from:

- Engine coolant sensor
- Ambient air temperature sensor
- Vehicle speed sensor
- Air conditioning-system pressure sensors
- Electro-viscous fan Hall Effect sensor
- Transmission oil temperature sensor

The Hall Effect sensor provides the control module with fan speed information, every 1 rpm of actual fan speed is represented by a 1 Hz signal from the Hall Effect sensor. The

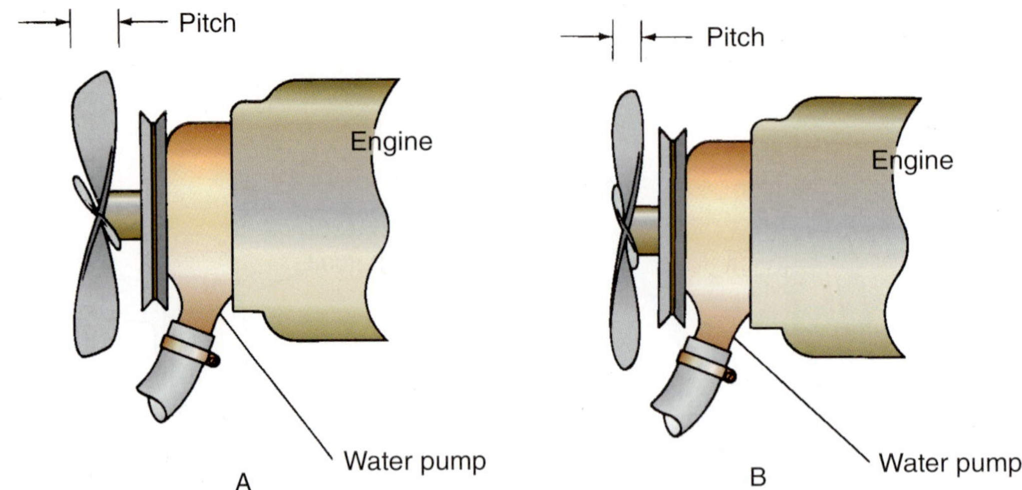

FIGURE 4-50 **Extreme pitch at low speed (A); reduced pitch at high speed (B).**

control module is able to determine if the fan is functioning correctly by comparing the actual fan speed to the desire fan speed. If a fault is determined, a DTC will be set. The 100 percent duty cycle mode is commanded only if:

- The temperature of the engine coolant exceeds 264°F (129°C).
- The temperature of the transmission oil exceeds 304°F (151°C).
- The air conditioning refrigerant pressure exceeds 240 psi (1655 kPa).
- A DTC is set and the control module determines that as a fail-safe the coolant fan should be engaged.

Under all other conditions, the fan duty cycle will be less than 100 percent. The EV fan operation may be tested with an enhanced scan tool which can be used to command the fan on/off in 10 percent increments. The engine speed should be maintained at 2,000 rpm to ensure that there is enough fluid movement to fully engage and disengage the coupling.

Flexible Fans. Flexible, or flex, fans (Figure 4-50) have blades that are made of a material (metal, plastic, fiberglass, or nylon) that will flex, or change pitch, based on engine speed. As engine speed increases, the pitch of the blade decreases. The extreme pitch at low speeds provides maximum airflow to cool the engine and coolant.

At higher engine speeds, the vehicle is moving faster and the need for forced air is provided by ram air provided by the forward motion of the vehicle. The flex blades feather reducing pitch, which in turn saves engine power and reduces the noise level.

Electric Fans

Most late-model vehicles use one or more electric-driven coolant fan motors (Figure 4-51) that are often controlled by the PCM, which have replaced the belt-driven fans. The electric motor and fan assembly are generally mounted to the radiator shroud (Figure 4-52) and are not connected mechanically or physically to the engine coolant pump. Either or both of the following two input methods electronically controls the 12V motor-driven fan:

- Engine coolant temperature switch (thermostat) or sensor
- Air conditioner select switch

There are many variations of electric cooling fan operation. Some provide a cool-down period whereby the fan continues to operate after the engine has been stopped even with the ignition switch in the OFF position. The fan stops only when the engine coolant temperature falls to a predetermined safe value, usually 210°F (99°C). In addition, some air conditioned

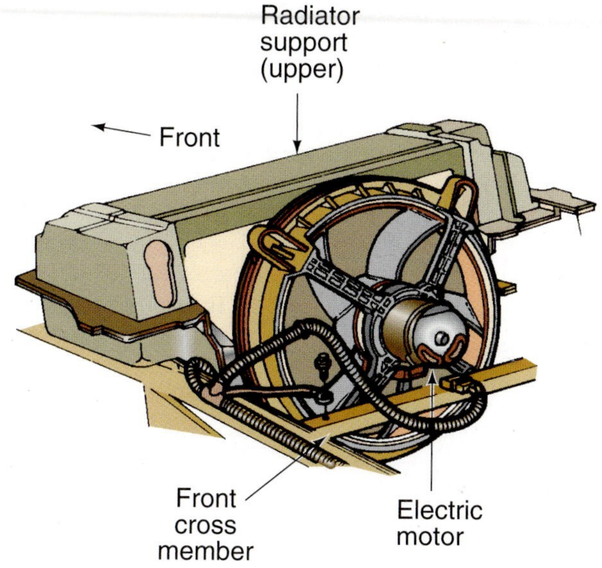

FIGURE 4-51 **An electrically-driven engine cooling fan.**

FIGURE 4-52 **A typical electric cooling fan and radiator shroud assembly.**

vehicles will have two electric cooling fans working independently of each other, depending on temperature conditions.

Because of the many variations of electric fan systems, manufacturers' specifications must be consulted for troubleshooting and repairing any particular year and model vehicle. For example, the computerized engine management system also often plays an important part in controlling the electrically operated engine-cooling fan, and in many systems, the PCM determines when to energize the fan relay control coil. Electric cooling fans may start without warning, even with the ignition in the OFF position. Many technicians are injured each year by defective fans.

There are several methods of controlling fan operation. In the wiring schematic in Figure 4-53, the fan is controlled by both a coolant temperature sensor and the air conditioner selection switch. The cooling fan motor is connected to the 12V battery supply through a normally open set of contacts (points) in the cooling fan relay. A fusible link provides protection for this circuit. During normal operation, with the air conditioner off and the engine coolant below a predetermined temperature of approximately 215°F (102°C), the relay contacts are open and the fan motor does not operate. If the coolant temperature exceeds approximately 230°F (110°C), the engine coolant temperature switch will close (Figure 4-54) to energize the fan relay coil. This action, in turn, will create an electromagnet to close the relay contact, assuming that the ignition switch is in the run position.

Shop Manual
Chapter 4, page 120

The temperature of the coolant in the engine will increase several degrees a few minutes after the engine is stopped.

Dual fan systems often operate independently of each other; either or both may start without warning.

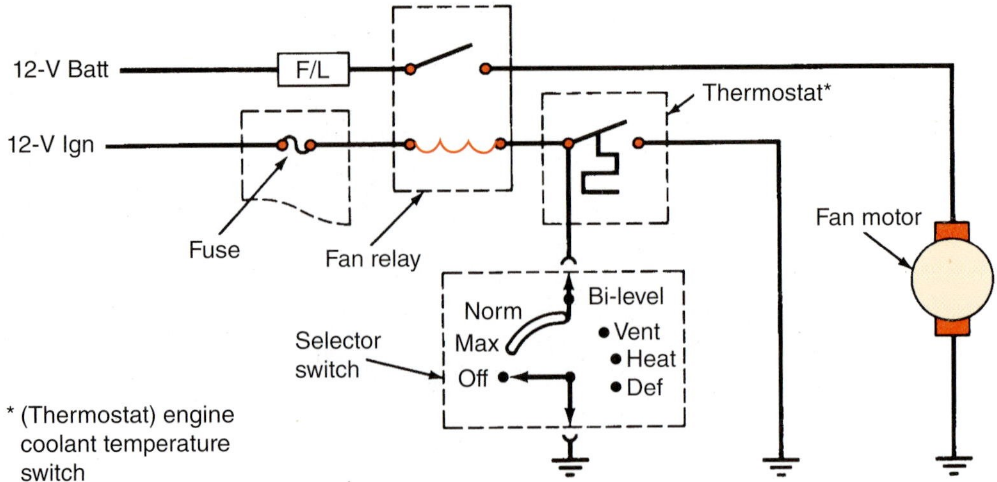

FIGURE 4-53 **A typical cooling fan schematic.**

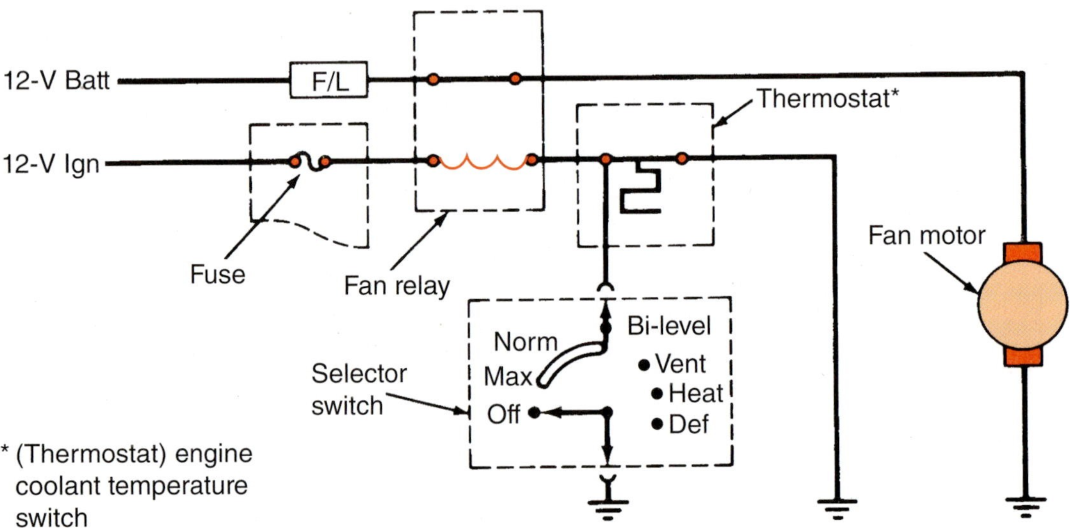

FIGURE 4-54 **The cooling fan switch closes at a predetermined high temperature.**

The 12V supply for the relay coil circuit is independent of the 12V supply for the fan motor circuit. The coil circuit is from the run terminal of the ignition switch, through a fuse in the fuse panel, and to ground through the relay coil control device.

In some systems, if the air conditioner select switch is turned to any cool position (Figure 4-55), regardless of engine coolant temperature, a circuit will be completed through the relay coil to ground (–) through the selection switch. This action closes the relay contacts to provide 12 volts to the fan motor. The fan then operates as long as both the air conditioner and ignition switches are on.

In other systems, the fan does not start when the air conditioner select switch is turned on unless the air-conditioning system high-side pressure is above a predetermined high pressure value. Although, if the air conditioning high-side pressure level is below the predetermined pressure but the engine coolant temperature is above a predetermined level, usually 230°F (110°C), the cooling fan will still come on.

Many fan control systems are ultimately controlled by the PCM, which supplies a ground path to the fan relay coil if it is determined that the fan operation is required. There are several

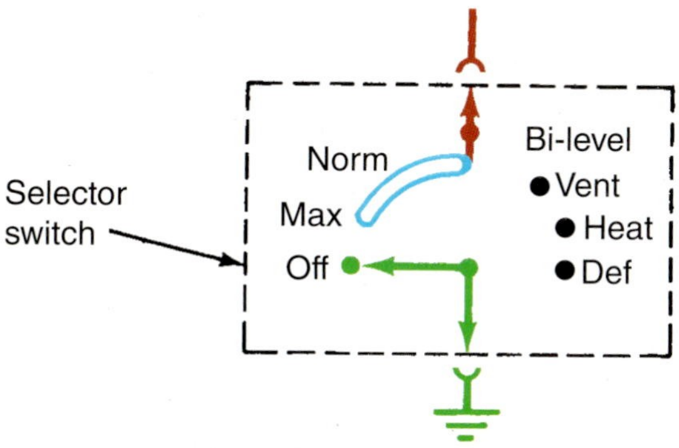

FIGURE 4-55 **The cooling fan will run if the A/C switch is turned to any COOL position.**

inputs that the PCM may look at in order to determine the need to engage the fan. These inputs include, but are not limited to, the following:

- Coolant temperature sensor
- Air conditioning selection switch
- Air conditioning high pressure switch
- Vehicle speed

Some PCM systems also have the ability to vary the fan speed based on cooling requirements. This is usually accomplished through an arrangement of relays arranged in a series/parallel configuration (Figure 4-56), allowing the fan(s) to be operated at low or high speeds.

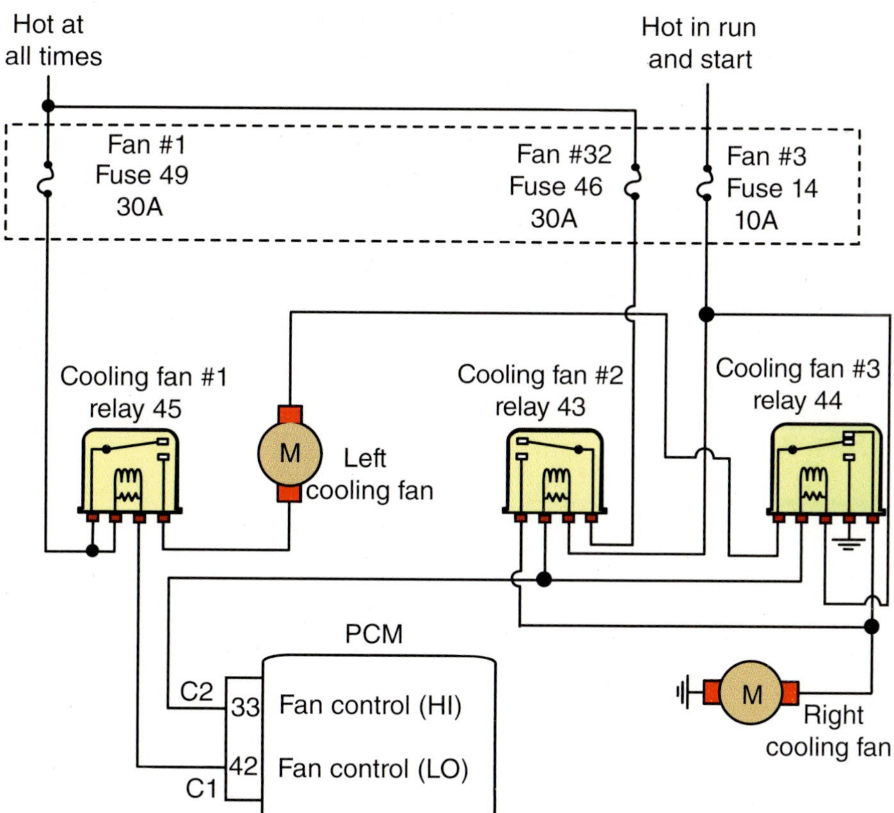

FIGURE 4-56 **Wiring diagram of multi-relay fan control.**

Hoses And Clamps

Shop Manual
Chapter 4, page 123

Radiators usually have two hoses: an inlet (upper) hose and an outlet (lower) hose. Radiator hoses are constructed of an ozone- and oil-resistant reinforced synthetic rubber. They must be the proper length to allow for engine movement on the motor mounts. Preformed hoses (Figure 4-57A) often have a spiral tempered steel wire installed in the lower radiator hose to prevent collapse due to the suction action of the coolant pump impeller. Upper hoses, not subject to this condition, usually do not have this wire. If the original hose has a wire, make sure the replacement hose also has one.

Aside from the radiator hoses, the cooling system also contains heater hoses, a bypass hose in some applications, and may also contain steel piping. These hoses are smaller and somewhat more flexible. For high-heat applications, silicone rubber hoses are available. They are usually identified by a green coloration of the hose and are often found on fleet application vehicles, such as police cruisers.

Universal flexible hoses (Figure 4-57B) have wire inserts available for use when preformed hoses are not available. Because of body and engine parts, hoses must often be critically routed. The universal flexible hoses, however, are not always easily routed. Also, it has been determined that the use of a flexible hose may place an unwanted stress on the radiator connecting flange, resulting in early failure of the radiator due to stress cracks. Preformed hoses designed for the specific vehicle application should always be used whenever possible.

Many types of hose clamp styles are available in several sizes. A popular replacement type is the worm-gear clamp, which has a carbon steel screw and stainless steel band (Figure 4-58). Hose clamp sizes are given by number or letter designations, which are stamped on the side of the clamp. The important consideration is that the clamp is properly positioned and not overtightened, which could cut into the hose. Worm-gear clamp tension should be rechecked periodically to avoid a source of coolant leakage. Many manufacturers have chosen to use constant tension spring clamps (Figure 4-59) because of their ability to maintain consistent

FIGURE 4-57 **Preformed (A) and (B) flexible cooling system hoses.**

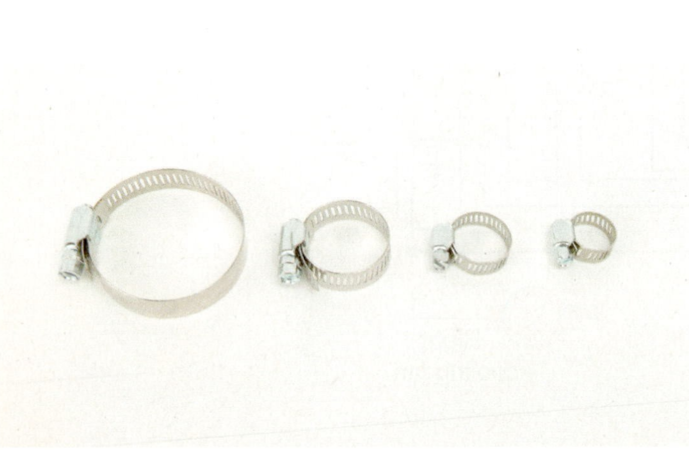

FIGURE 4-58 **A worm-gear clamp.**

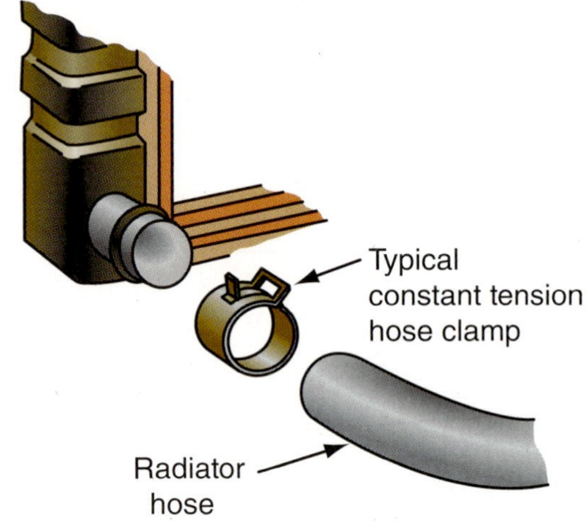

FIGURE 4-59 **A typical constant tension hose clamp.**

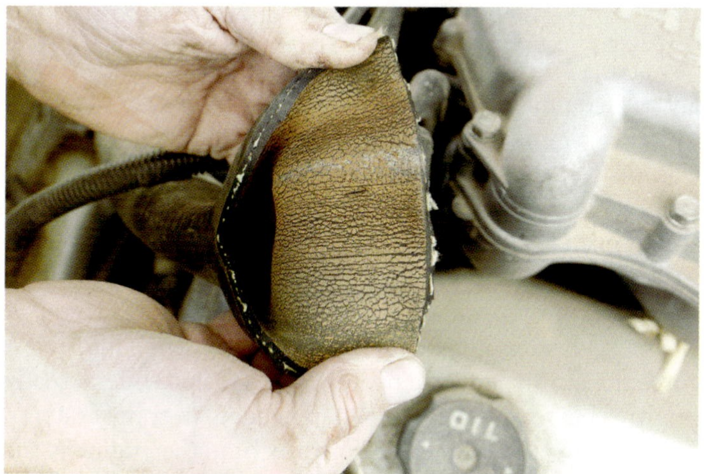

FIGURE 4-60 **A hose that is defective on the inside.**

clamping pressure as components and hoses expand and contract due to thermal cycling, thereby eliminating temperature-related leaks. Some vehicles today use quick-connect couplings on smaller hose connections at the heater core and intake manifold connections. Quick-connect couplings may require special tools to disconnect them, and special care should be exercised when servicing them.

If one hose is found to be defective, all should be replaced.

Engine coolant hoses and clamps should be replaced every four years as a preventative maintenance item. One of the major reasons for vehicles breaking down today is a poor cooling system maintenance service schedule. Avoid damaging components by first loosening the clamp and sliding it out of the way. Then make a small slice lengthwise so the hose may be peeled off.

Hoses should be inspected on a regular basis for defects or contamination. Oil contamination may cause the rubber to soften or swell, whereas heat will cause the hoses to become hard and brittle. Hoses should be checked by squeezing them near the clamp or connection and in the middle, feeling for any differences. Soft spots, cracks, and channels can often be detected in this manner.

Failure of cooling system hoses may also be due to electrochemical degradation (Figure 4-60) or electrolysis. As the coolant ages and the additives break down or are depleted, the coolant, engine, and radiator actually form a galvanic cell, a type of battery. As was noted under the Radiator section, electrolysis occurs when an electrical component is not properly grounded and routes itself through the cooling system in search of a ground.

This electrochemical action causes microscopic cracks on the inside of the hose that allow coolant to reach and weaken the reinforcement material. This action is accelerated by high heat and flexing and continues until the hose develops a leak or ruptures. Damage is generally more severe within an inch or two of the end of the hose where it is attached to a metal component. A sign that the interior of the hose is damaged is indicated by a green residue on the reinforcement fibers at the end of the hose where the coolant was wicked out.

Heater System

The **control valve** is a mechanical valve that regulates coolant flow through the heater core assembly.

The automotive heater system consists of two parts in addition to the hoses and clamps. They are the heater core and the coolant flow **control valve** on some vehicles. The heater housing and duct are part of the passenger compartment's air distribution system.

Heater Core

The automobile heater and heater core are actually part of the engine cooling system, though the heater does not provide the removal of heat from the engine as a normal function. It is meant to provide in-car passenger comfort during the cold winter months. The heater core is mounted in the air distribution duct system and is usually under the dash area of the front passenger side of the vehicle. The heater core (Figure 4-61) resembles a small radiator and also functions as a heat exchanger with the engine coolant flowing from the top of the engine through the heater core and back to the water pump in most designs. Engine heat is picked up by the coolant through the process of conduction and is transferred by convection to the cooler outside air passing through the heater core to the vehicle's interior. An electric blower motor is used to force the air through the heater core. This provides a ready source of heated air to be used to improve passenger comfort when needed. In some systems, the engine coolant is constantly flowing through the heater core any time the engine is running, whereas in other systems, a control valve is used to stop the flow of coolant when heat is not required.

Heater cores may be of tubular or cellular construction similar to the construction of radiators. The tanks on the heater core serve to direct coolant flow through the core. They may be constructed of brass and copper or of plastic and aluminum (Figure 4-62).

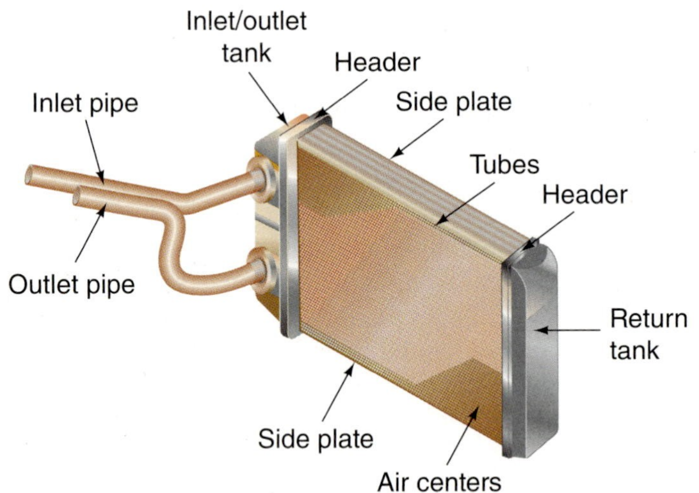

FIGURE 4-61 **A typical heater core with associated elements labeled.**

FIGURE 4-62 **A typical plastic heater core.**

If a vehicle is running hot, a temporary solution that is sometimes effective is to turn the vehicle passenger heater on the maximum hot position to use the heater core as an additional heat exchanger in order to keep a vehicle from overheating.

Leaks are the most common cause of heater core problems. Leaks in a heater core are detected by an obvious loss of engine coolant that is leaking at the firewall or engine compartment ductwork. Other signs that a heater core is leaking may be steam on the windshield when the defroster is on, or the passenger floor carpets could be wet with coolant. A slight leak can also be indicated by a sweet smell coming from the passenger compartment vents. If heater cores are found to be leaking, they are replaced as an assembly and not generally repaired. The replacement heater core should be physically matched to the original to ensure proper fit. In addition, some heater cores are placed higher than the radiator cap in the system. Air can become trapped in these systems, so it is critical to follow manufacturer's recommendations on proper bleeding procedures.

Occasionally, a heater core may become plugged due to poor cooling system maintenance. If this occurs, little or no coolant will flow through the heater core, resulting in no heated air being available for the passenger compartment. A quick method of determining this is by comparing the temperature of the inlet with the outlet hose at the heater core. If the core is blocked or severely restricted, there will be a large temperature difference between the two hoses.

Procedures for replacing the heater core vary with the year, make, and model of the vehicle. It is therefore necessary to consult the manufacturer's repair manual for the proper procedure for replacement.

Control Valve

The heater control valve regulates the flow of coolant through the heater core to control core temperature by opening and closing a passage to increase or decrease flow. The heater control valve may be located in the inlet or outlet line to the heater core. When the control valve is open, a portion of the heated engine coolant circulates through the heater core. This provides a means of providing warm air to the passenger compartment when desired. The heater control valve may be cable operated, vacuum operated, or operated by a bidirectional electric solenoid or motor (Figure 4-63). The control valve, depending on the valve position selected, meters the amount of heated coolant that is allowed to enter the heater core, from full off to full flow. The HVAC control panel temperature selector regulates the operation of most heater control valves, whether actuated by a Bowden cable, vacuum diaphragm, or electrically energized. Some heater core assemblies have a mechanical heater control valve integrated into them to regulate coolant flow through the core.

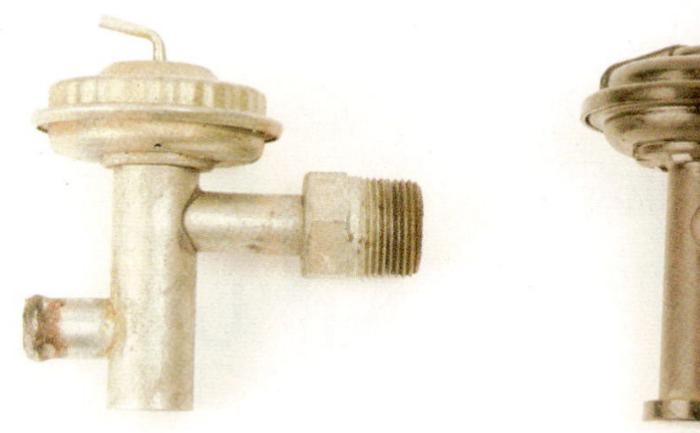

FIGURE 4-63 **Various types of control valves.**

Heater Core and PTC Heater

The engine on a hybrid electric vehicle is relatively small compared to a similar vehicle equipped with only an internal combustion engine. In addition, the hybrid engine is thermally efficient and only runs when needed; therefore, the engine coolant may not always be hot enough to supply adequate heat to the heater core. If the heater core does not become hot, then there will not be enough thermal energy to heat the passenger compartment to a comfortable temperature. In addition to hybrids, positive temperature coefficient (PTC) heaters are also used on some direct injection gasoline and diesel engines due to their low waste heat, which is often insufficient on cold days for fast passenger compartment heating. The PTC heater is also needed during stop-and-go traffic and city driving on many platforms.

To combat this lack of usable thermal energy available consistently in the engine coolant, an electric heating element is required to supply the passenger compartment with heat. Toyota has incorporated two 165 watt 12 volts direct current (vdc) PTC electric heating elements into the conventional heater core (Figure 4-64) to supplement the required thermal energy in order to maintain the passenger compartment at the desired temperature setting when the engine coolant is not warm enough. The PTC thermistor elements are a ceramic honeycomb design that heats the air directly as it passes over them and enters the passenger compartment (Figure 4-65).

The primary source of heat for the heater core is still hot coolant from the engine. When the engine is first started and the cooling system has not come up to temperature, it is advisable to set the heater duct fan to low to maximize heat output from the PTC elements, then increase the fan speed as the engine coolant temperature rises. Lack of sufficient passenger compartment heat during the winter months is a common complaint of hybrid vehicle owners, as well as a decrease in fuel economy during this same time.

ADDITIVES

Using additives is not recommended as a matter of practice.

Many additives, inhibitors, and "remedies" are available for use in the automotive cooling system. These include but are not limited to stop-leak, water pump lubricant, engine flush, and acid neutralizers. Extreme caution should be exercised when using any additive in the cooling

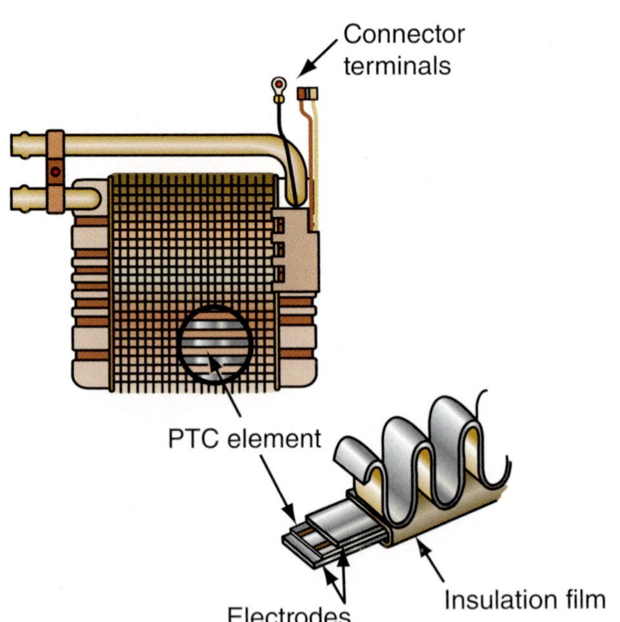

FIGURE 4-64 **A hybrid electric vehicle heater core with PTC heating element.**

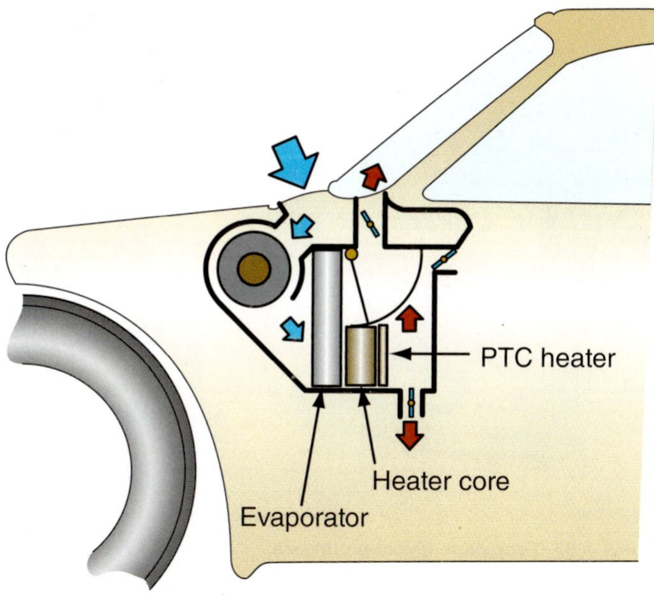

FIGURE 4-65 **A typical location of cooling and heating system elements in the case of duct system.**

system. Read the label directions and precautions to know what the end results of using any additive may be. For example, caustic solutions should not be used in aluminum radiators; alcohol-based "remedies" should not be used in any cooling system.

If a cooling system is maintained in good order by a program of preventative maintenance, additives and inhibitors should not be necessary. Only manufacturer-recommended ethylene glycol-based antifreeze and additives should be added to the cooling system.

Antifreeze is a generic term used to refer to engine coolant mixtures, whether they be ethylene glycol- or propylene glycol-based, used to raise the boiling temperature and lower the freezing temperature of an engine coolant mixture.

ANTIFREEZE

There are four key areas of engine protection. They are freeze protection, heating (boil over) protection, corrosion prevention, and adequate heat transfer.

The coolant in an automotive cooling system is the medium used to transport engine heat to the radiator. A 50/50 mixture of water (H_2O) and ethylene glycol ($C_2H_4[OH]_2$), an **antifreeze**, is the recommended coolant formula generally considered best for the engine. This solution is capable of quickly absorbing and giving up large amounts of heat.

Ethylene glycol, a main ingredient of antifreeze, is poisonous. When ingested, ethylene glycol converts into oxalic acid, (COOH)2, which damages the kidneys and may result in kidney failure and death. Just 2 oz. (59 mL) of ethylene antifreeze can kill a dog; 1 tsp. (4.93 mL) can be lethal to a cat; and 2 tbsp. (29.6 mL) can be hazardous to a child.

Shop Manual
Chapter 4, page 129

Most automobile manufacturers recommend a 50/50 percent mixture of antifreeze and water for adequate year-round cooling system protection. Also, most manufacturers warn against the use of an alcohol-based antifreeze solution or the use of straight water. Instead, they recommend a mixture of **ethylene glycol** (Figure 4-66) or propylene glycol (Figure 4-67) with distilled water (Figure 4-68). Either mixture in the cooling system is sufficient to:

Ethylene glycol ($HOCH_2CH_2OH$) is the base stock used for most automotive antifreezes and is a colorless, viscous liquid in its pure form.

- Lower the freezing temperature point of the coolant
- Raise the boiling temperature point of the coolant
- Help maintain the proper engine temperature
- Provide water pump lubrication
- Inhibit rust and corrosion
- Allow coolant-immersed sensors and switches to operate properly

FIGURE 4-66 **Ethylene glycol antifreeze.**

FIGURE 4-67 **Propylene glycol antifreeze.**

FIGURE 4-68 **Distilled water.**

A BIT OF HISTORY

Prior to 1930, methyl alcohol was the most commonly used engine antifreeze and required constant maintenance to ensure proper freeze protection. Ethylene glycol was first introduced by Prestone in 1927, but it did not become a standard year-round factory fill until the early 1960s.

Propylene glycol ($C_3H_8O_2$) is the base stock used for most automotive antifreezes and is a colorless, viscous liquid in its pure form, has a low toxicity, and is considered an environmentally-friendly alternative to ethylene glycol.

The freezing temperature of water as a coolant, at ambient sea-level atmospheric pressure, is 32°F (0°C) and the boiling point is 212°F (100°C). Water in a radiator with a 15 psi cap will boil at 250°F (120°C). Water should never be used straight as a coolant in an engine because it offers no corrosion protection, lubricating properties, or other necessary additives that are available in antifreezes.

Standard ethylene glycol is generally green or yellow in color but can also be red (Toyota), blue, or pink. The recommended mixture of 50/50 percent ethylene glycol and water (Figure 4-69) has a freeze point of −34°F (−36.7°C) and a boiling point of 265°F (129.4°C) in a radiator with a 15 psi cap. If the percentage of ethylene glycol to distilled water is increased to 70/30 percent, the freeze point is decreased to −84°F (−64.4°C) and the boiling point is increased to 276°F (135.6°C). It is not recommended that the mixture be increased beyond 70 percent ethylene glycol. The ability of the coolant to carry away heat decreases as the percentage of ethylene glycol increases. Straight ethylene glycol only provides freeze protection down to −2°F (−18.9°C) but has a boiling point of 276°F (135.6°C). A straight mixture of ethylene glycol should never be used because ethylene glycol is 15 percent less efficient than water at removing heat and could cause hot spots to develop, resulting in severe engine damage.

The protection level selected should be to the lowest temperature expected In warmer climate zones, such as southern Florida and Southern California, protection is required for the antirust and anticorrosion inhibitors, as well as for anti-boiling protection. In any climate zone, antifreeze is essential for this protection, as well as coolant pump lubrication. The coolant should contain no less than 30 percent antifreeze, which will provide protection to −5°F (−15°C).

A safer alternative to ethylene glycol coolant is antifreeze formulated with **propylene glycol**. Unlike ethylene glycol, propylene glycol is essentially nontoxic and safer for animal life, children, and the environment. Also, propylene glycol is classified "Generally Recognized as Safe" by the U.S. Food and Drug Administration. Low toxicity does not mean that it is safe to drink, but the risk of poisoning is a greatly reduced.

Propylene glycol is available on the aftermarket and not currently used as factory-fill antifreeze. Popular brands are Prestone Low Tox® and Safe Brands Sierra®. Propylene glycol has similar thermal characteristics as ethylene glycol. A 50/50 mixture of propylene glycol and water has a freeze point of −26°F (−32.2°C) and a boiling point of 256°F (124.4°C) in a radiator with a 15 psi cap, whereas a 100 percent mixture has a freeze point of −70°F (−56.7°C) and a boiling point of 370°F (187.8°C) in a radiator with a 15 psi cap.

FIGURE 4-69 **A coolant percentage/protection chart.**

Propylene glycol and ethylene glycol are compatible antifreezes and can be mixed, but mixing the two antifreezes eliminates the low toxicity characteristics of propylene glycol and makes it impossible to test the coolant's strength using a hydrometer. This is due to the differences in the specific gravities of the two coolants.

Antifreezes contain many additives to improve their performance in the cooling system, about 2 to 3 percent of the total volume in pure coolant. Corrosion inhibitors protect metal surfaces from rust, corrosion, and electrolysis. This is particularly important for thin, lightweight aluminum radiators and heater cores. Most ethylene glycol formulated in North America contains inorganic salts of borate, phosphate, and silicate to prevent rust and corrosion of metal components. The additives produce an alkaline mixture with a pH range of 7.5–11.0, depending on the manufacturer. The silicates in the mixture form a protective coating on the surface of metal components and are especially effective at protecting aluminum.

The additive package usually offers enough protection for at least 2 years or 30,000 miles in most vehicles, or longer if mixed with **distilled water**. The ability of a coolant to neutralize acids is referred to as the coolant's reserve alkalinity and varies among manufacturers. Time, heat, dissolved oxygen, and minerals contained in water as well as glycol degradation will eventually

Distilled water is pure water produced by distillation. It is available at grocery stores and pharmacies.

deplete these protective additives, and acids will form. Once the pH level drops to 7.0 or lower, corrosion and electrostatic discharge accelerates, leading to radiator, heater core, water pump, and hose failure. Chemical test strips can be used to check the pH level of the cooling system.

Electrolytic corrosion, as was discussed under Hoses, is of special concern with today's bimetal engines. The different metals contained in radiator cylinder blocks and heads can create an electrochemical action (battery) that promotes corrosion. One metal becomes the anode (aluminum), while the other becomes the cathode (iron), and the coolant is the electrolyte. The higher the percentage of dissolved impurities in the coolant, the greater its ability to conduct electricity, which increases the rate of electrolysis. Aluminum and thin metal components, such as the radiator and heater core, are generally the first to suffer damage.

Other failures can also accelerate the depletion of coolant additives. Air pockets from low coolant levels or improper bleeding during service, exhaust gases leaking into the cooling system due to leaky head gaskets, or cracks in the combustion chamber all lead to rapid coolant failure.

Some European vehicle manufacturers recommend the use of phosphate-free coolants because phosphates can react with calcium and magnesium contained in some water to form sediment and scale, while Asian vehicle manufacturers may recommend using coolants containing phosphates but low or no silicate additives. Regardless of the coolant chosen, under no circumstances should softened water be used in a cooling system. Domestic water softeners use salt (sodium), which is very corrosive to all metals and could lead to serious damage to the cooling system and engine components.

The latest coolant to be developed is the extended-life coolant. General Motors first introduced it in 1995 under the name of Dex-Cool® (Texico/Havoline) and by 1996 was using it in most of their vehicle platforms as the factory fill. It uses a corrosion-inhibiting package referred to as **organic acid technology (OAT)**. It contains organic salts of mono and dicarboxylic acids such as sebasic and octanoic acids plus tolytriazole, and is less alkaline than standard coolants, with a pH of 8.3. Extended-life coolant offers protection for five years or 150,000 miles and is still an ethylene glycol-based coolant. As such, extended-life coolants are compatible with standard coolants but, if combined, will only offer the corrosion protection of the conventional coolant. If the system becomes contaminated or is to be converted to an extended-life coolant, it must be thoroughly drained and flushed to remove all traces of the conventional coolant in order to gain the benefits of the longer-lasting antifreeze formula.

Organic acid technology (OAT) is used in extended-life antifreeze based on carboxylates of organic acids and does not contain silicates, phosphates, borates, amines, or nitrates.

Hybrid organic acid technology (HOAT) G-05 extended-life coolants are ethylene-glycol Glysantin-based formula refrigerants that are low in silicate, have low pH, and are phosphate-free. HOAT coolants use both organic and inorganic carbon-based additives for long life protection.

Both Ford and Chrysler products use **hybrid organic acid technology (HOAT)** extended-life coolants, which are also rated for five years or 100,000 to 150,000 miles. The HOAT coolant is an ethylene-glycol glysantin-based formula that is low silicate, low pH, and phosphate free. HOAT coolant uses both organic and inorganic, carbon-based additives for long life protection. It is commonly marketed as G-05 extended-life coolant and comes in several colors, such as yellow (Ford) and orange (Chrysler). The low-silicate formula is designed to offer aluminum protection and water pump lubrication. Extended-life OAT-based coolant should not be mixed with HOAT-based coolant according to vehicle manufacturers, and Ford has gone as far as to state that orange coolant, meaning Dex-Cool, should not be added to their systems. In addition, German automotive manufacturers require G11 blue, G12 red, and G12 plus purple long-life coolants in their car lines and do not want any blending of coolants to take place. Both G11 and G12 coolants are ethylene glycol-based phosphate, nitrite, and amine free coolants. Beginning in 2010, Ford began a multiyear change over worldwide to an orange OAT based extended life coolant, but does not recommend using this coolant in older models designed for HOAT based coolants. Beginning in 2009, Nissan/Infinity introduced a blue long life coolant. The blue long life coolant is rated for 10 years or 135,000 miles. VW, Audi, BMW, and Volvo are using a hybrid formula based on low silicates and 2-EHA (ethyl hexanoate) organic acid. In addition, Volvo, VW, and Audi do not specify a change interval for their coolant. The industry is a rainbow of colors and it is important to use the engine coolant that is recommended by the manufacturer and flush the cooling system based on the manufacturers guidelines to avoid damage to gaskets, water pumps, and water jackets as well as heat exchangers.

Accidental antifreeze poisoning kills many pets and livestock every year. Beginning in 2010, many states enacted laws that require manufacturers to add a bittering agent (bitter flavor) to antifreeze and engine coolant. The purposes of the laws are to require antifreezes to be produced with an objectionable taste so pets and children will not mistake the poisonous substances for a sweet drink. While newer antifreezes will contain a bitter flavor, older antifreeze will still have a sweet taste and will probably be in vehicles for many years to come. So whether sweet or sour, be sure to clean up spills and dispose of properly, and never leave open pans unattended or keep in unmarked containers.

The percent of antifreeze concentrations in a cooling system mixture can be determined by several methods, which will indicate the freeze point of the mixture and its specific gravity. Common methods for determining the freeze point are the coolant hydrometer, test strips, and the refractometer. A separate hydrometer for propylene glycol and ethylene glycol is required because they have different specific gravities. The most accurate methods for determining the concentration of antifreeze in the mixture are with the use of a refractometer. A refractometer can measure the specific gravity of propylene glycol, ethylene glycol, and battery electrolyte. It does this by analyzing the way the light bends as it passes though the liquid. Ethylene glycol never wears as additives do, however, and none of these methods will indicate the condition of the coolant additives.

Standard propylene glycol and ethylene glycol offer full protection for two years of normal driving. To determine the amount of antifreeze to add to the cooling system after flushing with water, refer to the manufacturer's specifications for total cooling system volume, and divide by two. Example: If the total cooling system volume is 16 quarts (15 liters), you will need to add 8 quarts (7.5 liters) of pure antifreeze, and then finish filling the system with water. Remember that after flushing the system with water, some of the water may still remain in the water jackets.

Although propylene glycol is much less hazardous than ethylene glycol antifreeze, it still should be considered hazardous. At the present time, the Environmental Protection Agency has no restrictions on the disposal of antifreeze unless it is contaminated with lead (Pb). Lead, a byproduct of the material used to solder the seams and joints of the radiator, is considered hazardous in any quantity. State and local governments, however, may have requirements for safe disposal, reclamation, or recycling. To protect animal life and the environment, follow these simple rules:

- Do not mix different types of antifreezes.
- Wipe up and wash away spills.
- Keep stored antifreeze off the floor and away from animals.
- Keep antifreeze in its original container, or use a container that is specifically labeled for the product it contains.
- Store used antifreeze, before recycling or reclamation, in a sealed container that is properly labeled with its contents (in other words, used ethylene glycol).
- Ensure that the vehicle's cooling system has no leaks.

Preventive Maintenance

The cooling system, which is often neglected, is one of the most important systems of the car. If it is kept in good shape and provided with routine preventive maintenance, the cooling system should give years of trouble-free service. The cost of maintenance every 12,000–15,000 miles (19,308–24,135 km) is more than offset by the cost of breakdown and consequent repairs. These repairs, incidentally, often result in expensive engine service.

Mineral-free distilled water is preferred over tap water for maximum cooling system performance and integrity.

Replace belts that are frayed, glazed, or obviously damaged. Replace any hoses that are found to be brittle, soft, or otherwise deteriorated (Figure 4-70).

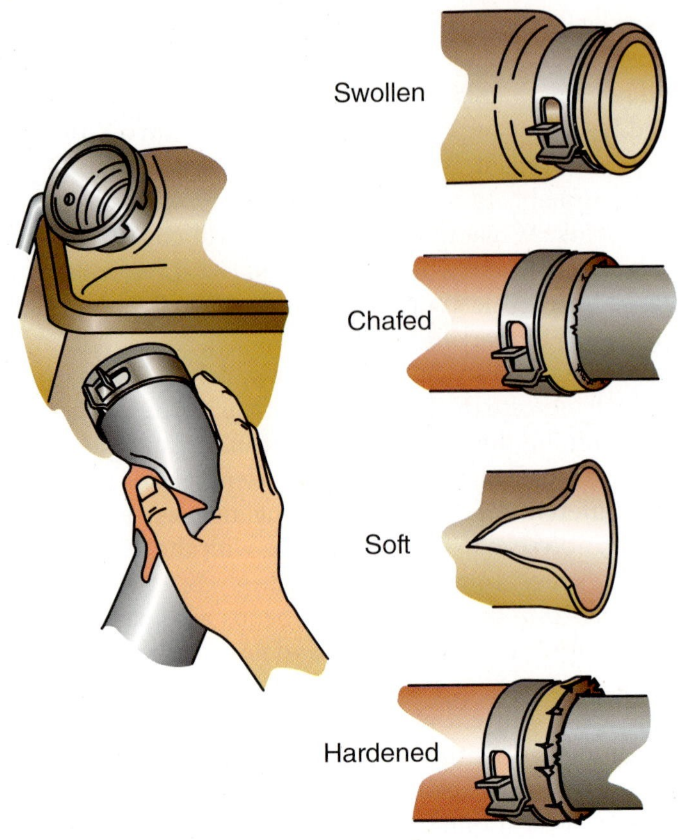

FIGURE 4-70 **Typical hose defects.**

The design consideration of a cooling system is to provide for a minimum sustained road speed operation of 90 mph (144.8 km/h) at an ambient temperature of 125°F (52°C). Another criteria is for 30 minutes of driving in congested stop-and-go traffic in an ambient temperature of 115°F (46°C) without experiencing any overheating problems. These design considerations exceed the conditions that one is likely to encounter in regular day-to-day driving.

If the engine overheats, the problem should be found and corrected. The life of an engine or a transmission that is habitually allowed to overheat is greatly reduced. The high-limit properties of lubricating oil require adequate and proper heat removal to preserve formulated lubricating characteristics.

Terms to Know

Antifreeze
Bypass
Centrifugal impeller
Control valve
Distilled water
Electrolysis
Ethylene glycol
Expansion tank
Heater core
Hybrid organic acid technology (HOAT)
Organic acid technology (OAT)
Overcooling
Overflow tank
Pitch
Power train control module (PCM)
Pressure cap
Propylene glycol
Radiator
Ram air
Recovery tank
Reverse flow
Thermostat

SUMMARY

The preventive maintenance program should include the following procedures:

- Test or replace the thermostat.
- Test or replace the pressure cap.
- Inspect or replace the radiator hose(s).
- Inspect or replace the heater hoses.
- Pressure test the cooling system.
- Test or replace the antifreeze solution.
- Visually inspect the coolant pump, heater, control valve, and belt(s).

REVIEW QUESTIONS

Short-Answer Essays

1. What is the purpose of the automotive cooling system?
2. What are the two types of radiator core that are found in the automotive cooling system?
3. What method(s) is used to increase the boiling point temperature of the coolant in an automotive cooling system?
4. Describe the operation of a typical cooling system thermostat.
5. What are the advantages of engine temperature being adjusted to current operating demands on the electronically controlled cooling system?
6. Describe an advantage of a declutching engine-driven cooling fan.
7. Describe an advantage of an electric-motor-driven cooling fan.
8. Briefly, what is the purpose of the coolant recovery tank?
9. What is a PTC heater and on what vehicles can it be found?
10. What are the advantages of a 50/50 mix of antifreeze and water?

Fill in the Blanks

1. Radiators are constructed of ______________, ______________, and/or plastic.
2. A frequent coolant pump failure is due to ______________ often caused by worn ______________.
3. If the gasket or sealing surfaces of a pressure cap are damaged, the cooling system cannot be ______________.
4. A thermostat failing in the ______________ position will result in engine ______________.
5. Engine-driven fans are balanced to prevent ______________, ______________, coolant pump ______________ failure, and/or seal damage.
6. Many electric cooling fans are ultimately controlled by the ______________ ______________ ______________.
7. Electric cooling fans may ______________ without warning, even with the ignition in the ______________ position.
8. Heater core leaks are detected by a loss of ______________ and a wet ______________ ______________.
9. An antifreeze solution ______________ the freezing temperature and ______________ the boiling temperature of the coolant.
10. A typical cooling system contains a mixture of ______________ antifreeze and ______________ water which offers an excellent balance of both ______________ freeze point and ______________ boiling point protection.

Multiple Choice

1. All of the follow are true about engine thermostats except:
   A. A stuck open thermostat will cause engine overcooling.
   B. A stuck closed thermostat will cause engine overheating.
   C. A faulty thermostat will effect vehicle emissions.
   D. A thermostat may be removed from a system if it is running hot.
2. Engine coolant is designed to provide all of the following except:
   A. Provide water pump lubrication.
   B. Inhibit rust and corrosion.
   C. Lower the boiling point of the solution.
   D. Lower the freezing temperature of the solution.
3. If a thermostat fails in the open position, all of the following will occur, except:
   A. A vehicle may fail emissions test.
   B. A loss of engine coolant.
   C. Poor heater performance.
   D. A longer than normal warm-up period.
4. All of the following statements about engine cooling systems are true EXCEPT:
   A. Extended life coolant and standard life coolant are both ethylene glycol based.
   B. It is ok for a technician to install a thermostat with a low temperature rating than the original thermostat.
   C. Extended life coolant is rated to last in excess of 100k miles.
   D. Electric cooling fans may turn on even when an engine is not running.

5. All of the following are the function of the radiator pressure cap except:
   A. To seal the cooling system.
   B. To lower the freeze point of the coolant.
   C. To raise the boiling point of the coolant.
   D. To pressurize the cooling system.

6. Air in the cooling system can cause all of the following except:
   A. The formation of rust and/or corrosion in the cooling system.
   B. Creating an air lock, blocking coolant flow through the cooling system.
   C. Water pump cavitation inhibiting the pump's ability to circulate coolant.
   D. The coolant hoses bursting due to air pressure.

7. When the cooling system is placed under pressure, for each pound per square inch (psi) of cooling system pressure, the boiling point of the coolant is increased by:
   A. 1°F
   B. 2°F
   C. 3°F
   D. 4°F

8. A common engine thermostat has a temperature rating of:
   A. 185°F
   B. 195°F
   C. 205°F
   D. 225°F

9. Engine coolant loss could be caused by all of the following except:
   A. A leaking transmission cooler line
   B. A leaking engine head gasket
   C. A faulty radiator pressure cap
   D. A leaking transmission oil cooler

10. A vehicle is driven for 15–20 miles and the vehicle's passenger compartment blower motor continues to blow cold air. What is the most likely cause of this condition?
   A. Stuck closed thermostat
   B. Partially restricted heater core
   C. Partially restricted radiator core
   D. Damaged water pump impeller blades

Chapter 5

Air-Conditioning System Operating Principles

Upon Completion And Review Of This Chapter, You Should Be Able To:

- Discuss heat transfer in the refrigerant system.
- Describe the effect humidity has on the air-conditioning system.
- Explain the pressure and vacuum relationship in the air-conditioning system.
- Understand the basic functions of the various air-conditioning components.
- Discuss the role of refrigerant in the system.
- Explain the pressure versus temperature relationship.
- Know the physical state, pressure, and temperature of the refrigerant in different areas of the refrigerant system.

Introduction

Automotive air-conditioning systems are divided into two sections: the high pressure side and the low pressure side. The air-conditioning system is a closed system that circulates a fixed charge of refrigerant through the system over and over again in a looping cycle. The general function of the system is the absorption of heat when the refrigerant boils in the evaporator. It is this process of allowing the correct volume of refrigerant into the evaporator that removes heat and humidity from the passenger compartment. The basic automotive air-conditioning system components consist of these:

- Compressor
- Lines and hoses
- Condenser
- Expansion device
    - Expansion valve
    - Orifice tube
- Evaporator
- Storage vessel
    - Receiver-drier—used with expansion valve
    - Accumulator—used with orifice tube

The two basic system designs are the expansion valve (Figure 5-1) or the orifice tube (Figure 5-2) refrigerant system.

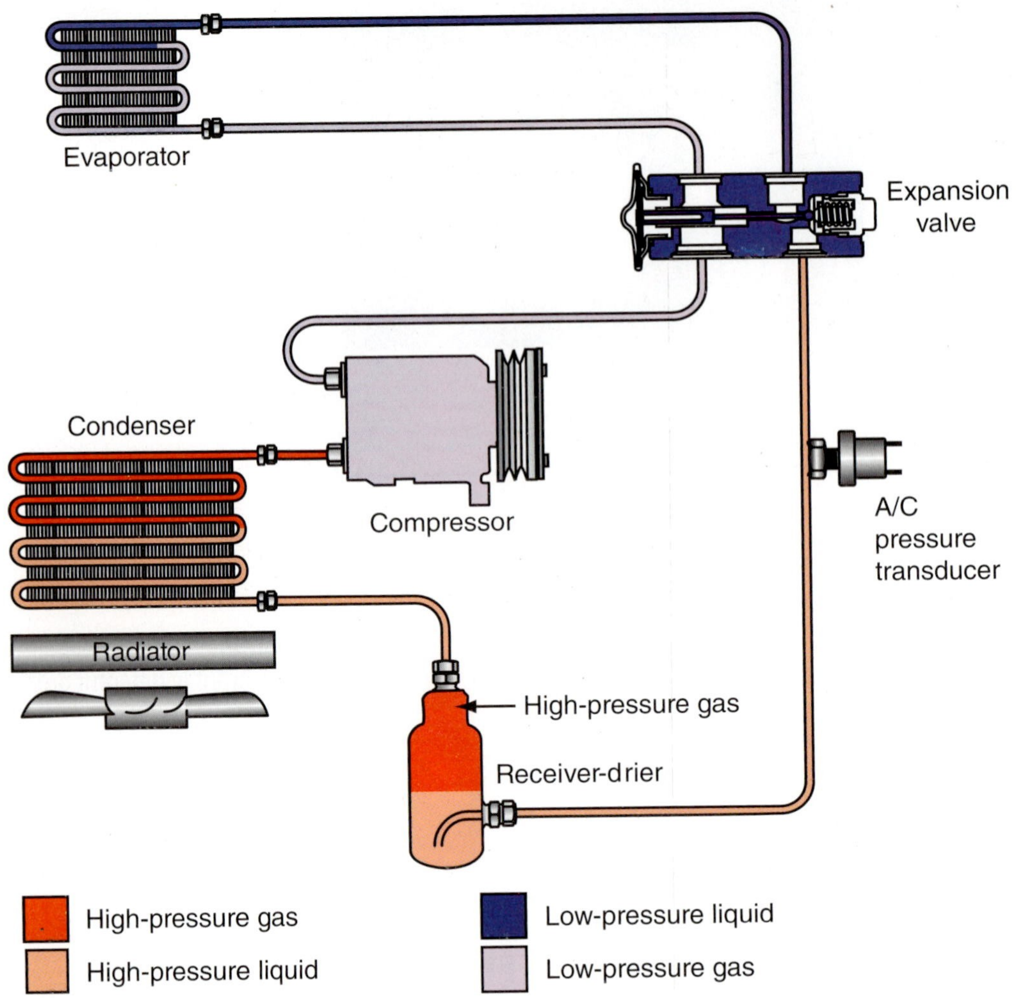

FIGURE 5-1 **Refrigerant cycle and components in the expansion value air-conditioning system.**

Heat Transfer in the Refrigerant System

The basic function of the automotive air conditioner is to remove heat from the passenger compartment and transfer this heat to the outside air. The system also dehumidifies the air before it enters the passenger cabin.

To perform these functions, the air-conditioning system uses two heat exchangers. One of these heat exchangers is the evaporator core, which is located in the passenger compartment duct work. Its role is to remove heat from the incoming or recirculated air as it flows across the cooling fins of the core and to transfer this heat energy into the refrigerant system (Figure 5-3). The cooled air is then directed into the passenger compartment. The other heat exchanger in the system is the condenser, which is located in front of the engine radiator and just behind the body grille assembly. Its role is to remove the heat that was absorbed by the evaporator from the refrigerant system and transfer (radiate) this heat back into the atmosphere (Figure 5-4).

The following is a summary of heat transfer as it relates to the vehicle refrigerant system:

- Convection—the heat content of the outside air is transferred to the evaporator core as the air is blown across the core fins.
- Conduction—the heat is absorbed by the cooler refrigerant gas flowing inside the coils of the evaporator core.

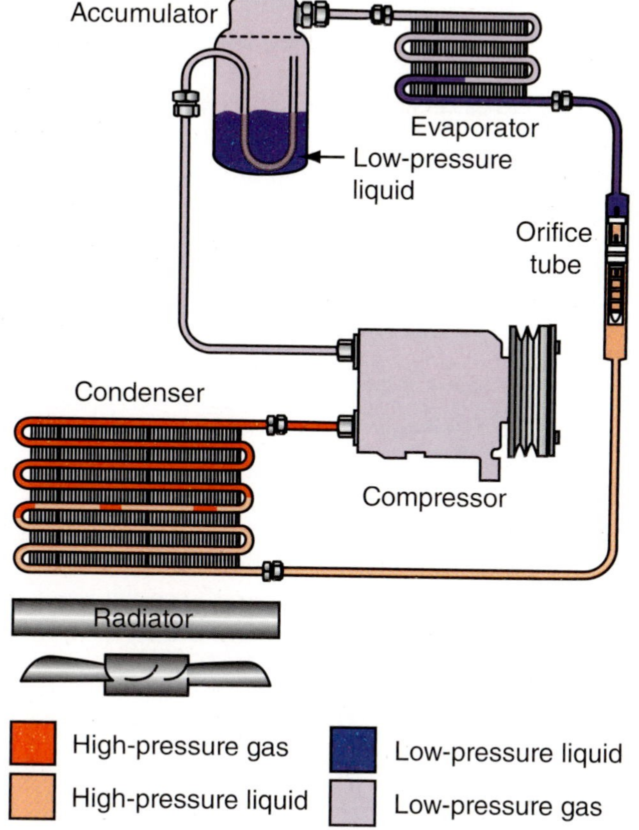

FIGURE 5-2 **Refrigerant cycle and components in the orifice tube air-conditioning system.**

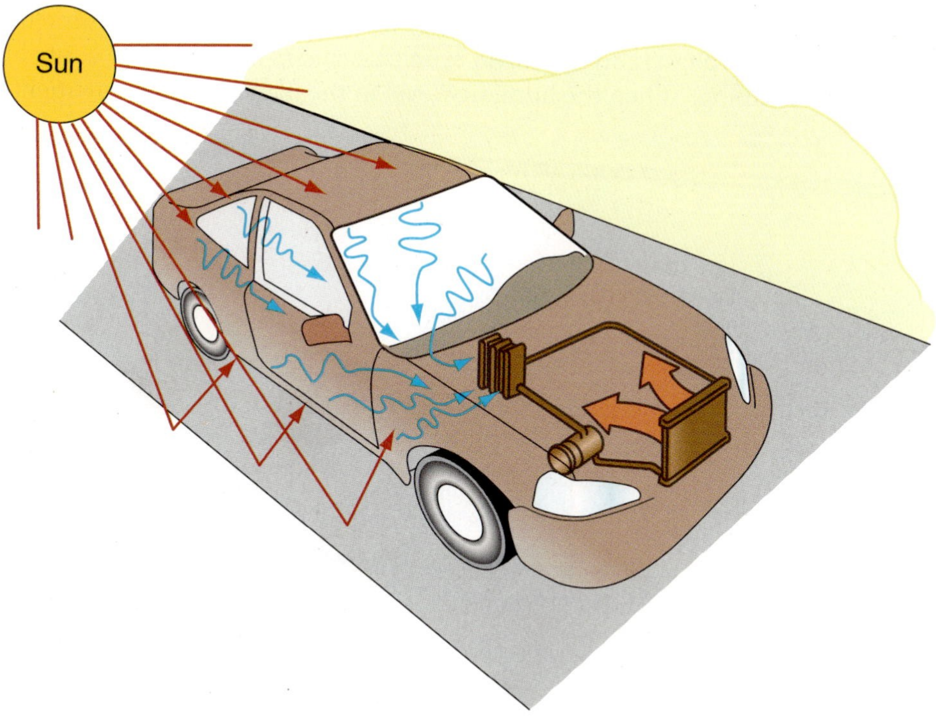

FIGURE 5-3 **Heat transfer in the refrigerant system.**

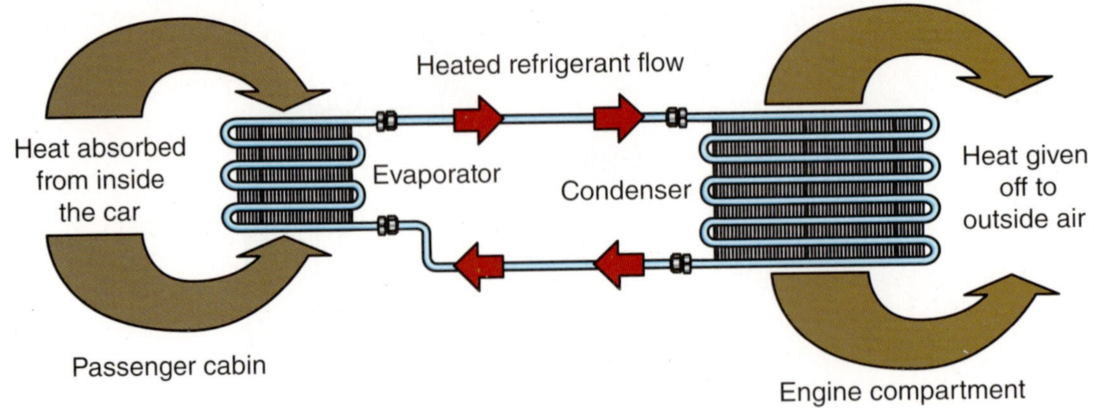

FIGURE 5-4 **Heat movement through the refrigerant system.**

- Convection—the heat absorbed by the refrigerant is circulated to the condenser.
- Conduction—the cooler condenser fins absorb the heat from the refrigerant.
- Radiation—the condenser fins then give off heat to the cooler air.
- Convection—the air then moves the heat away from the condenser.

Air Conditioning and Humidity

To maximize the effectiveness of the evaporator core's ability to remove heat from the passenger compartment, the moisture content of the air must be reduced. The moisture content in air (humidity) contains heat; thereby increased humidity increases the heat load on the refrigerant system (Figure 5-5). To improve the efficiency and the operation of the refrigerant system and provide maximum cooling of the passenger compartment on hot humid days, the air-conditioning system must remove both heat and moisture from the air. Thus maximum refrigerant performance and passenger compartment cooling is achieved by selecting the air Recirculation mode (air inside the passenger compartment is recirculated through the evaporator core) on the climate control panel instead of the Fresh air mode (outside air).

Humidity in the passenger compartment can be increased by many factors including but not limited to hot muggy days, rainy days, and even passengers in the vehicle contributing to humidity load by breathing. When the humidity level in the passenger compartment rises, it

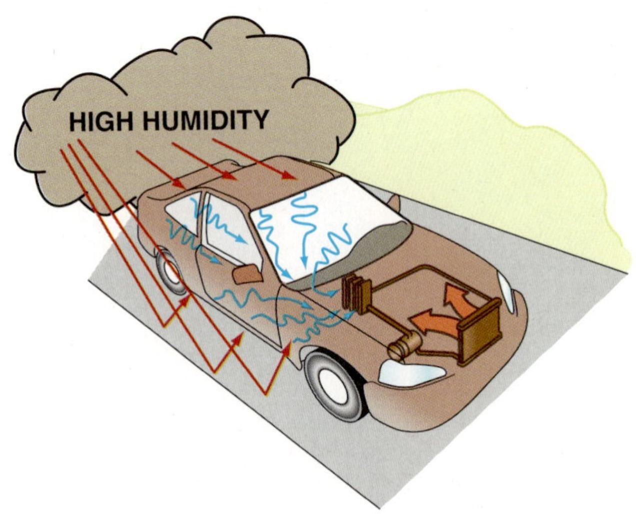

FIGURE 5-5 **Humidity increases the heat load on the refrigerant system.**

can create a safety concern beyond passenger comfort if the moisture condenses on the inside of the windows, obscuring the driver's vision. This condensation on the inside of the windows occurs when the water vapor suspended in the air contacts a surface (window) that is cooler than the air temperature and the moisture condenses into water droplets on the cold surface. This same process can be used to lower the humidity level of the air in the passenger compartment. As air is passed over the cold evaporator core fins, the air gives up its heat, and the moisture contained in the air condenses against the surface of the evaporator (Figure 5-6). If the air Recirculation mode is selected, the air in the passenger compartment is looped from the passenger compartment back through the evaporator core, providing maximum dehumidification of the passenger compartment air.

If we analyze the process a little further and think about the principle of latent heat and the change of state as a vapor becomes a liquid, we will recall that a large amount of thermal energy (Btu's) must be exchanged for this to occur. This means that the evaporator must absorb hundreds of Btu's in heat, removing energy as humid air is passed over it. If humid outside air is continually drawn over the evaporator, thermal energy is used to condense the moisture out of the air stream during the dehumidification process, meaning that less energy will be available to lower air temperature further and thus will not be available to lower passenger compartment temperature. In other words, the energy used to condense out moisture and dehumidify the air will overstress the system, and the air-conditioning system will not be able to cool the air as well as it could if the air were dryer (less humid). During air-conditioning system performance testing and analysis, it is advisable to run the air conditioner with the Recirculation mode selected and the vehicle windows closed in order to achieve maximum system efficiency. In addition, if this procedure were used on each vehicle tested, results would be more consistent regardless of humidity levels on a particular day.

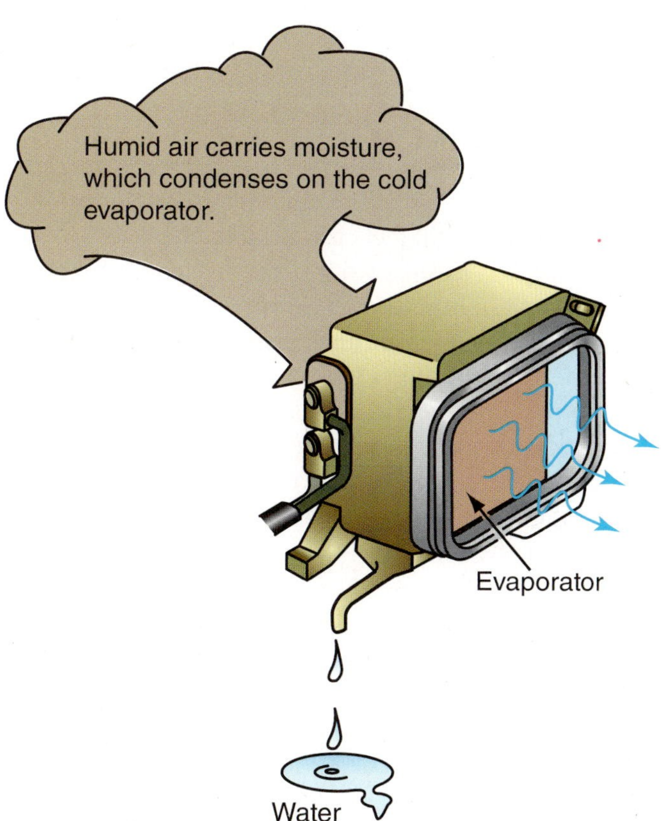

FIGURE 5-6 **The dehumidification process produced by the refrigerant system evaporator.**

AUTHOR'S NOTE: Whether your home has a window air conditioner or a central air-conditioning system, both use the benefit in performance achieved by recirculating the air inside the room back through the evaporator to lower room humidity and thus increase the efficiency of the system by cooling ever dryer air.

The atmospheric relative humidity has a dramatic effect on the effectiveness of the air-conditioning system. The vehicle air-conditioning system will be able to lower the passenger compartment temperature substantially further on a 90-degree day at 30 percent relative humidity than it can on a 90-degree day at 80 percent relative humidity. The higher the relative humidity level of the air, the less energy will be available for cooling the air stream. It should be stated that by itself the effect of removing humidity from the air will have a positive effect on passenger comfort as high humidity levels prevent the body from dissipating heat easily on its own. So, lowering humidity alone will increase personal comfort on a hot day as the body's evaporational cooling will improve.

The dew point temperature of the air is another indicator of the humidity level. The higher the dew point temperature, the more moisture the air contains by volume, because warm air is less dense and can hold more water vapor. Dew point is a better comfort indicator than relative humidity, the higher the dew point temperature the more uncomfortable humans are. At a dew point temperature 50°F (10°C) if the ambient air temperature is 60°F (16°C) and the relative humidity is 100 percent we still feel comfortable because our body is able to give up heat because the air temperature is far below our body surface temperature. Dew points above 65°F (18°C) make it feel sticky and humid outside while dew points less than 65°F are comfortable with respect to the stickiness of the air. The higher the dew point temperature is, the more moisture that is in the air. The higher the dew point is above 65°F (18°C), the stickier it will feel outside. If the dew point temperature is 75°F (24°C) or above the air really feels sticky and humid. The dew point temperature is the temperature at which the humidity level becomes 100 percent; that is to say, it is the temperature at which air has become saturated and cannot retain any additional water vapor. Dew point is a better comfort indicator than relative humidity, the higher the dew point temperature the more uncomfortable humans are. At a dew point temperature 50°F (10°C) if the ambient air temperature is 60°F (16°C) and the relative humidity is 100 percent we still feel comfortable because our body is able to give up heat because the air temperature is far below our body surface temperature. Dew points above 65°F (18°C) make it feel sticky and humid outside while dew points less than 65°F are comfortable with respect to the stickiness of the air. The higher the dew point temperature is, the more moisture that is in the air. The higher the dew point is above 65°F (18°C), the stickier it will feel outside. If the dew point temperature is 75°F (24°C) or above the air really feels sticky and humid. It is this dew point temperature, when water vapor cannot remain in the vapor state and condensation begins to form on any surface, that is cooler than the air temperature. This occurs at the refrigerant system's evaporator core, where the air temperature is lowered to below the dew point temperature of the air flowing across it, and water vapor condenses on the evaporator.

THE RELATIONSHIP OF PRESSURE AND VACUUM IN THE AIR-CONDITIONING SYSTEM

Pressure must be understood in order to properly diagnose an air-conditioning system. We are exposed to air pressure every day in the form of atmospheric pressure. This is the pressure created by the atmosphere that surrounds the earth and extends nearly 600 miles above the earth's surface, and which is held in place by the gravitational forces generated by the earth's rotation and magnetic field (Figure 5-7).

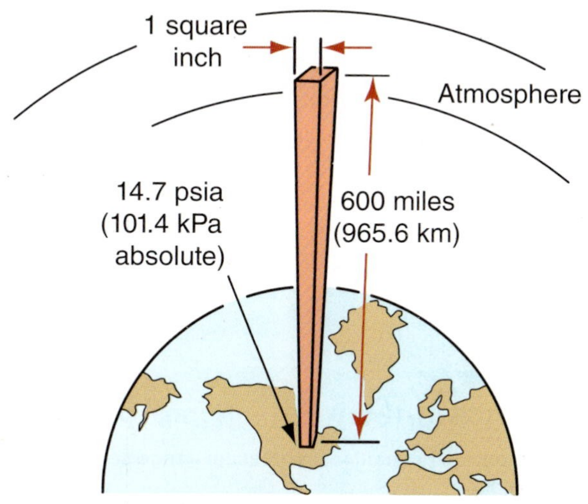

FIGURE 5-7 **Atmospheric pressure at sea level is 14.7 psia (101.4 kPa Absolute).**

Atmospheric pressure at sea level is 14.7 pounds per square inch absolute (psia) (101.4 kPa absolute).

Most of us only notice a difference in pressure when it is either more or less than the normal level (14.7 psia). In the automotive field, we usually use gauges that are calibrated to 0 pounds per square inch at atmospheric pressure, called 0 psig. A pressure gauge reading of 30 psi means 30 psig—pounds per square inch gauge. On a set of absolute gauges, the gauge would read 14.7 psia at sea level and the 30 psig reading previously stated would read 44.7 psia on a set of absolute gauges (Figure 5-8). Pounds per square inch absolute is the amount of pressure measured above absolute zero. As stated above, this is the gauge pressure plus 14.7 psi. Automotive air-conditioning pressures listed and gauge sets used for diagnosis and repair are normally in psig unless otherwise stated (Figure 5-9).

We also deal with pressures below atmospheric pressure (0 psig) when servicing automotive air-conditioning systems. Any pressure below atmospheric pressure is referred to

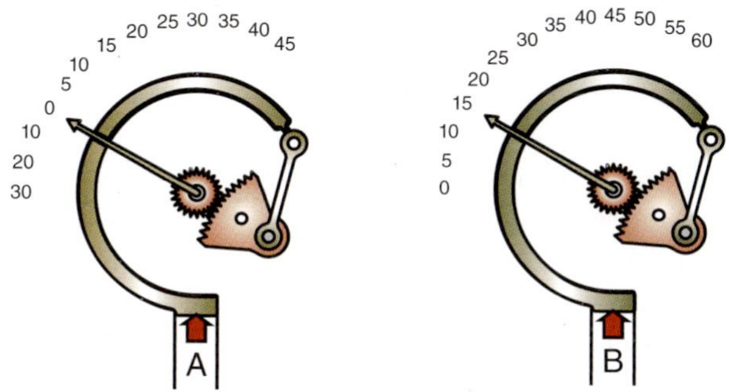

FIGURE 5-8 **The two gauges illustrated above show a comparison of the low-side gauge scales at atmospheric pressure sea level. Gauge A is scaled in pounds per square inch gauge (psig). Gauge B is scaled in pounds per square inch absolute (psia). Atmospheric pressure at sea level is 14.7 psia (101.4 kPa absolute).**

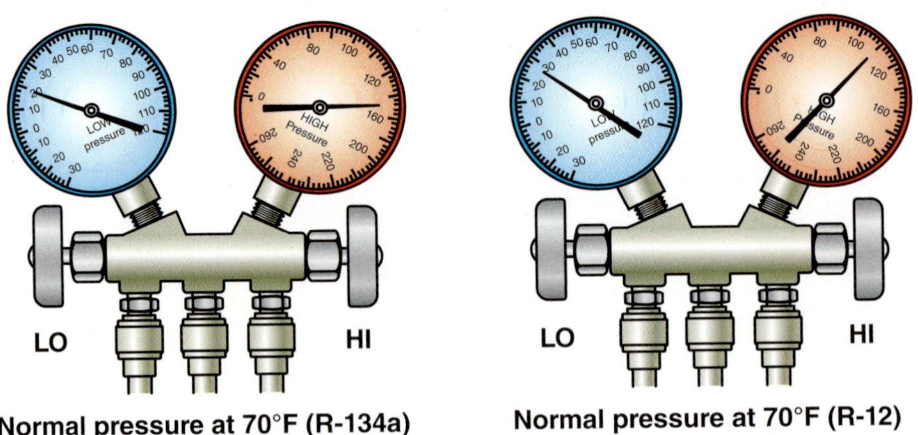

FIGURE 5-9 **Automotive air conditioning pressure gauge sets used for diagnosis and repair are normally in psig.**

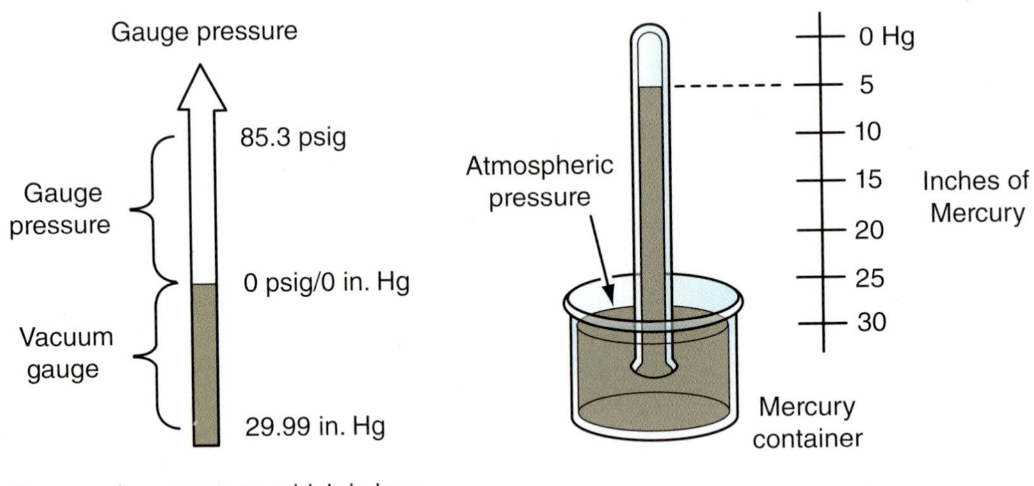

FIGURE 5-10 **Vacuum is any pressure that is less than atmospheric pressure with 29 in. Hg being a perfect vacuum.**

as a vacuum or partial vacuum. A standard psig pressure gauge is only capable of reading a value if the pressure is greater than atmospheric pressure (0 psig). To read a pressure value less than atmospheric pressure, a vacuum gauge is required. A vacuum is generally calibrated in inches or centimeters of mercury (Hg). A complete vacuum (devoid of all air) is 29 inches Hg and is only achievable at sea level. A partial vacuum is any vacuum between 0 and 29 inches Hg (Figure 5-10). The low-side gauge (blue gauge) on a compound set of air-conditioning gauges reads both positive pressures (0–120 psig) and negative pressures (vacuum) levels from 0–30 in. Hg.

Air Conditioning 101

To understand how an air-conditioning system functions, you must first know all the components that make up that system. You must also learn what role each of these components plays in the system and how each component interrelates with one another.

During the discussion on air-conditioning components that follows in this chapter, it may be helpful to refer to Figure 5-25 and Figure 5-26 to understand how the basic components relate to one another.

The remainder of this chapter will discuss the various components in the air-conditioning system and the role that each plays in the overall system, as well as the physical state of the refrigerant in various areas of the system. Knowing basic component location and terminology is required of all automotive technicians. Both customers and your peers in the trade expect you to know and use basic trade terminology and have an in-depth knowledge of the systems you are working on. This is all part of being a professional technician. In addition, it will be all but impossible to diagnose and repair a system if you do not grasp the basic operating principles and relationships of the various system components.

COMPRESSOR

The refrigerant compressor (Figure 5-11) is the heart of the refrigerant system. Its purpose is to compress and pump refrigerant through the system. This specially designed pump raises the pressure of the refrigerant from approximately 20 to 30 psi to approximately 180 to 220 psi. As you may recall from Chapter 2, according to the laws of physics, when a gas is compressed, its pressure and temperature are increased proportionally. By increasing the refrigerant's pressure, we also increase the temperature at which it will condense.

Compressor designs may use one or more **reciprocating pistons**, rotary vanes, or scroll-type design. They may have a fixed displacement or a variable displacement. There are many compressor designs and styles in use today, and these will be addressed in more detail in Chapter 8. Though many compressors may look alike, they are not interchangeable. Refer to manufacturer specifications for the correct compressor for each application.

Reciprocating piston(s) move(s) up and down or back and forth in a linear motion.

The compressor is one of the points in the air-conditioning system where there is a separation between high and low pressure. The low side is also referred to as the suction side of the system and connects the compressor inlet to the evaporator side of the system. The high side of the system is also referred to as the discharge side and connects the compressor outlet to the condenser inlet.

The refrigerant leaves the evaporator as a low-pressure gas (vapor) with as much heat as it can transport for its pressure. It goes into the compressor on the low-pressure (suction) side. There it is compressed by compressor action into a high-pressure vapor and is then pumped out of the smaller, outlet (discharge) side of the compressor.

The compressor increases refrigerant vapor pressure.

FIGURE 5-11 **Typical compressors.**

The compressor pumps the low-pressure refrigerant vapor out of the evaporator by suction, raises its pressure, and then pumps it, under high pressure, into the condenser.

Quick-Connect Valve

The service valve used for R-134a systems is a positive-coupled, quick-connect type. There is a Schrader valve similar to the R-12 Schrader access fitting, with the exception that it is recessed further into the fitting (Figure 5-12). The high-side service port, at 16 mm, is larger than the low-side service port, which is 13 mm. Service hoses for automotive air-conditioning system use must have unique fittings that attach to the service ports (Figure 5-13) and refrigerant cylinder. This combination, required by the Environmental Protection Agency (EPA), prevents reversing the hoses as well as introducing the wrong refrigerant.

Service Valves

Some compressors may be equipped with service valves, though most have Schrader-type service valves (Figure 5-14) located on suction and discharge lines.

On some systems, the suction and discharge service valves are located on the compressor cylinder heads. The suction and discharge lines or hoses are connected to the compressor at the service valves. If the service valve is not found on the compressor, the suction service valve will be found somewhere between the evaporator outlet and compressor inlet. The discharge service valve will be found somewhere between the compressor outlet and condenser inlet.

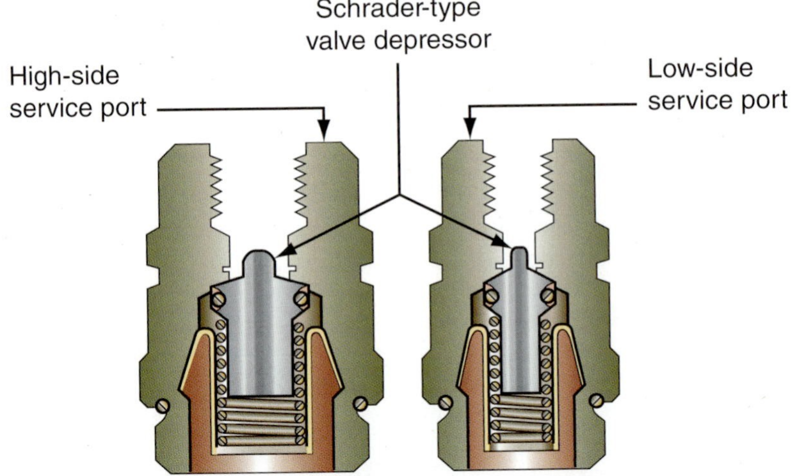

FIGURE 5-12 **R-134a (HFC-134a) service port details.**

FIGURE 5-13 **Typical R-134a service port adapters.**

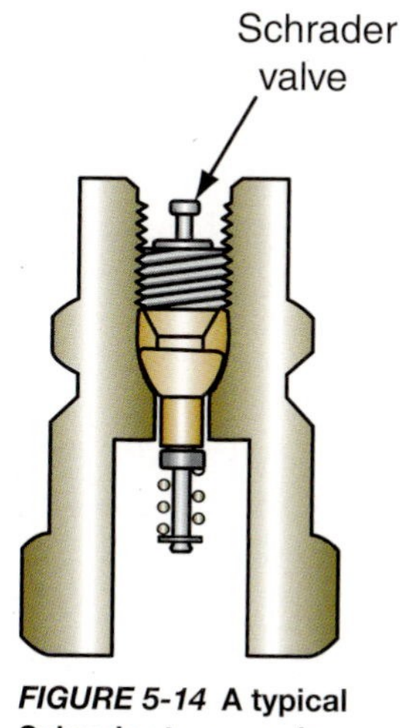

FIGURE 5-14 **A typical Schrader-type service valve for R-12 systems.**

Hand Shutoff Valve

The hand-type shut-off service valve located on the compressor, once a very popular type, is likely encountered today only on older R-12 mobile air-conditioning system applications. It may also be found on some off-road air-conditioning systems using the two-cylinder Tecumseh and York compressors. Though no longer found frequently, it is important that the technician be familiar with its operation. Unlike the now familiar Schrader-type valve, the shut-off service valve has, for all practical purposes, three positions: front seated, back seated, and midpositioned.

Shop Manual
Chapter 5, page 166

Front Seated. When the valve stem is turned clockwise (cw) all the way in, it is said to be front seated (Figure 5-15A). This is not a normal operating position for either hand valve. The compressor should not be operated with the discharge service valve front seated. To do so will result in serious compressor damage due to excessive high pressure. It may possibly burst, causing personal injury. Do not operate the compressor with service valve(s) in the front-seated position.

Back Seated. When the service valve is turned counterclockwise (ccw) all the way out, it is said to be in the back-seated position. In this position (Figure 5-15B), the gauge port is closed, and the line port and compressor circuit are open. This is the normal operating position for both service valves.

Midpositioned. When the service valve is in midposition (Figure 5-15C), it is said to be cracked. In this position, the line port, gauge port, and compressor are in the circuit. This is not the normal operating position, but it is the position for service when the manifold and

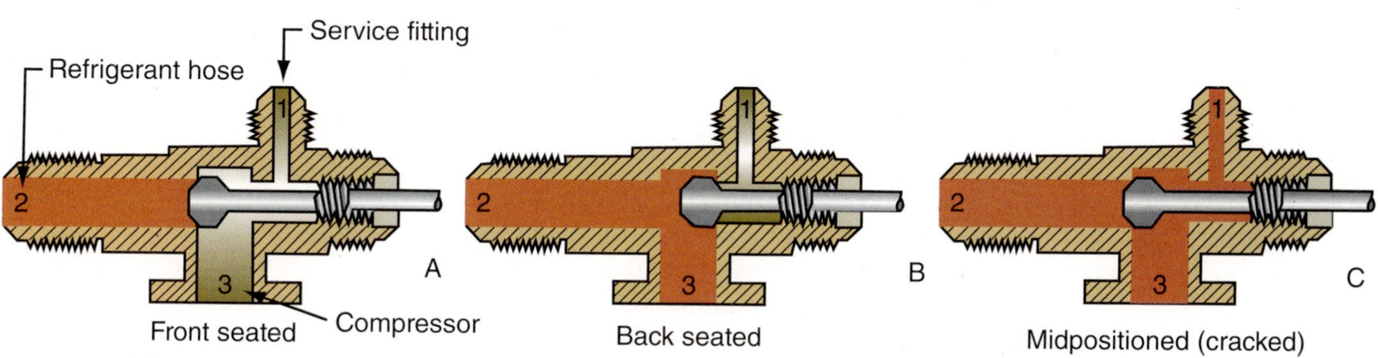

FIGURE 5-15 **(A) Service value in the front-seated position; (B) back-seated position; and (C) midposition (cracked).**

gauge set is installed. To midposition a service valve, turn the stem two turns clockwise off its back-seated position. This is sufficient to allow system pressure to the gauge port. Do not front seat, then midposition.

It should be noted that the hand shut-off type service valve is not to be midpositioned or front seated unless the manifold and gauge set is attached. The compressor should not be operated with the high-side service valve front seated.

Schrader-Type Valve

The Schrader-type service valve is self-opening when a manifold hose is attached. It operates in much the same manner as a tire valve. Because of its design, the Schrader-type service valve may only be cracked, or back seated.

Condenser

The **condenser** is a heat exchanger located in front of the vehicle radiator. It is the component of the refrigerant system in which refrigerant is changed from a gas to a liquid by the removal of heat.

Superheated is the process of adding heat intensity to a liquid above its boiling point without vaporization or heating a gas above its saturation point so that a drop in temperature will not cause reconversion to liquid.

The **condenser** is a heat exchanger for the superheated refrigerant in the system. Refrigerant containing heat is compressed by the compressor and, in both a high-pressure and high-temperature (**superheated**) gaseous state, flows to the condenser. This super-heated high-pressure vapor enters the top of the condenser. As it passes through the condenser coils, the outside air passing over the coils and fins picks up heat from the refrigerant. This occurs because the outside air at this point has less heat than the refrigerant in the coil. As the heat leaves the refrigerant, the refrigerant condenses, changing from a high-pressure vapor to a high-pressure liquid, which exits at the bottom of the condenser. This pressured liquid refrigerant is a liquid at a temperature lower than the minimum temperature (saturation temperature) at a given pressure required to keep it from boiling (changing from a liquid to a gas). Subcooling occurs in the condenser when additional heat is removed below the condensation temperature. The difference between the saturation temperature and the actual liquid refrigerant temperature is the amount of subcooling. Subcooling increases the efficiency of the refrigerant system by increasing the amount of heat removed per pound of refrigerant circulated. Stated another way, less refrigerant is pumped through the system to maintain the desired refrigerant temperature which in turn reduces the amount of time a compressor must run. Subcooling prevents the liquid refrigerant from changing to a vapor before it reaches the evaporator. In a refrigerant system, inadequate subcooling will prevent the proper metering of refrigerant into the evaporator due to flash gassing (change of state to a vapor), resulting in poor system performance.

The condenser is that part of the air conditioner that removes heat from the refrigerant and dissipates this heat to the outside air. The engine cooling system fan pulls air through the condenser, which is located in front of the radiator. Air passes through the condenser, then through the radiator. Ambient air is also forced through the condenser and radiator by the forward movement of the vehicle. This is known as ram air.

Receiver-Drier Accumulator

The **receiver-drier** and suction-line **accumulator** are tank-type devices that have nearly the same external appearance. The functions of the two devices are somewhat different, however.

The function of the receiver-drier is to store a liquid refrigerant reserve to ensure a constant liquid supply to the expansion valve. A strainer and a drying agent (called a desic-cant) are in the receiver-drier to remove moisture and clean the refrigerant.

The function of the accumulator is to catch and trap liquid refrigerant from the evaporator to protect the compressor. The accumulator also contains a strainer and desiccant for refrigerant cleaning and purification.

Receiver-Drier

From the condenser, the high-pressure liquid refrigerant enters the receiver-drier (Figure 5-16), where it is stored until it is needed by the expansion valve. The receiver-drier performs three functions in the automobile air-conditioning system:

- The receiver section is the storage tank for excess (reserve) liquid refrigerant that is necessary for proper operation of the air-conditioning system.
- The drier section collects small droplets of moisture that may have entered the system at the time of installation or repair.
- The pickup tube ensures a vapor-free stream of liquid to the expansion valve.

Accumulator

The suction-line accumulator (Figure 5-17) is located at the outlet of the evaporator and before the inlet of the compressor. The purpose of the accumulator is to trap excess liquid refrigerant, preventing it from entering the compressor. Liquid refrigerant in the compressor could cause serious damage. The accumulator is a storage container for refrigerant vapor and any small amount of liquid refrigerant that did not reach the vapor point as it passed through the evaporator as well as lubrication oil that may be atomized and traveling with the refrigerant.

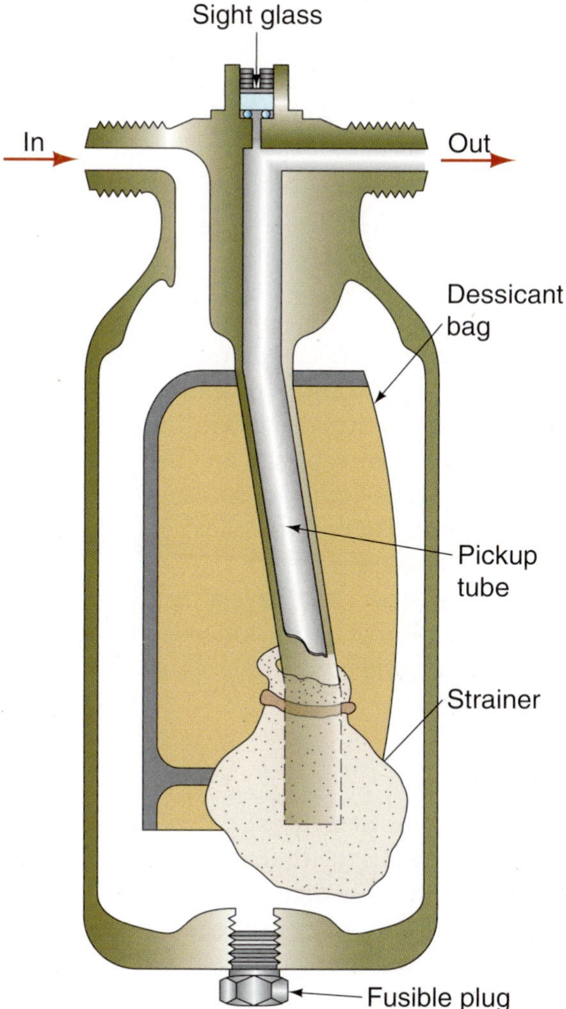

FIGURE 5-16 **Cutaway of a typical receiver-drier.**

The **receiver-drier** is a storage container on the high pressure side of the system between the condenser and the thermostatic expansion valve. It is used to separate out refrigerant vapor from liquid refrigerant, allowing only liquefied refrigerant to travel onto the thermostatic expansion valve, and it contains a drying agentinadesiccant bag inside the container.

The **accumulator** is a storage container on the low-pressure side of the system between the evaporator and the compressor. It is used to separate out refrigerant liquid from vaporized refrigerant, allowing only vaporized refrigerant to travel onto the compressor assembly. It contains a drying agent in a desiccant bag inside the container.

The receiver-drier or accumulator contains the drying agent known as desiccant.

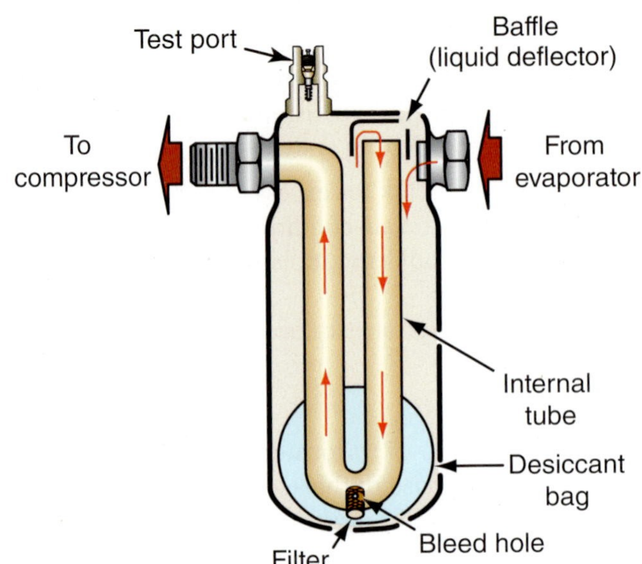

FIGURE 5-17 **Cutaway of a typical accumulator.**

The system will have either an accumulator or a receiver-drier, not both.

The **thermostatic expansion valve** is the component in the refrigerant system that regulates the rate of flow of refrigerant into the evaporator core through the use of a variable valve opening as governed by a remote bulb-sensing evaporator temperature. It is located just before the evaporator and after the receiver-dryer. It is one of the points in the air-conditioning system where there is a separation between a high and low pressure. A system that uses a thermostatic expansion valve also uses a receiver-dryer assembly.

Any liquid sinks to the bottom of the accumulator. As the vaporized refrigerant is drawn out of the accumulator through the internal tube, a small amount of liquid, if present, is drawn into the tube by venturi action through the bleed hole in the bottom loop. This bleed hole keeps the accumulator from becoming flooded with liquid refrigerant and refrigerant oil and allows a small volume of liquid refrigerant and lubricating oil to pass on to the compressor. In this manor, a compressor hydrostatic lock condition is avoided. Hydrostatic lock occurs if liquid refrigerant in excess of the compressors compressed cylinder volume were to be drawn in on the intake stroke, compressor seizure and damage would result. It is important to never overfill a refrigerant system with lubricating oil or refrigerant.

In addition to storing refrigerant, the accumulator contains a desiccant bag which contains a drying agent. This drying agent captures any moisture that could have entered the system during original assembly or service procedures. Moisture is one of the worst enemies of a refrigerant system and can cause both physical damage to components and system performance complaints.

The accumulator is used in systems that have a fixed orifice tube as a metering device. Those systems that have a thermostatic expansion valve have liquid-line receivers and not suction accumulators. The system will have either an accumulator or a receiver-drier, not both, depending on system design.

METERING DEVICES

At the present time, there are two types of metering devices used in automotive air-conditioning systems. The most widely used device is the **thermostatic expansion valve**, more commonly called an expansion valve and often abbreviated TXV.

Another device, originally introduced by General Motors and now found on many car lines, is the expansion tube, more commonly referred to as an orifice tube.

- Systems that have an expansion valve will have a receiver-drier in the liquid line before the device.
- Systems that have an orifice tube will have an accumulator in the suction line after the device.

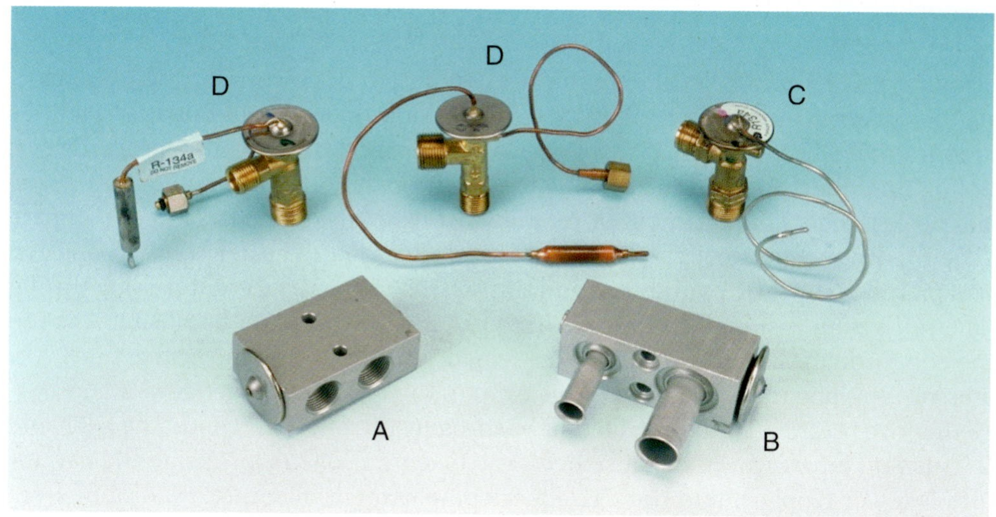

FIGURE 5-18 **Thermostatic expansion values: (A) H-valve; (B) block type; (C) internally equalized; (D) externally equalized.**

Expansion Valves

The expansion valve (Figure 5-18) controls the amount of refrigerant entering the evaporator under ever-changing heat load conditions. Heat load conditions of the car's interior depend on many factors, such as the number of occupants and heat gain from the sun through windows and the car's body, as well as heat gain from the engine compartment and exhaust system.

Expansion Tubes

The **expansion tube**, better known as a fixed orifice tube (FOT), is a nonadjustable device that has a fixed orifice metering element and a fine-mesh strainer. Unlike the expansion valve, the FOT (Figure 5-19) has no remote bulb, no moving parts, and does not vary the amount of refrigerant entering the evaporator in the same manner. The FOT meters the proper amount of refrigerant into the evaporator based on a pressure differential (high side to low side). A pressure differential is known as delta P (Δp).

The **expansion tube** is the component in the refrigerant system that regulates the rate of flow of refrigerant into the evaporator core. It is often referred to as the FOT and is a fixed metering device equipped with a filter screen. It is located between the condenser and the evaporator core and is one of the points in the air-conditioning system where there is a separation between high and low pressure. A system that uses a fixed orifice tube also uses an accumulator assembly placed between the evaporator and the compressor.

AUTHOR'S NOTE: The easiest way to determine whether you have a fixed orifice tube system with an accumulator or an expansion valve system with a receiver-drier is to pay attention to line diameter. Accumulators are located in the larger diameter suction line and receiver-driers are located in the smaller diameter liquid line.

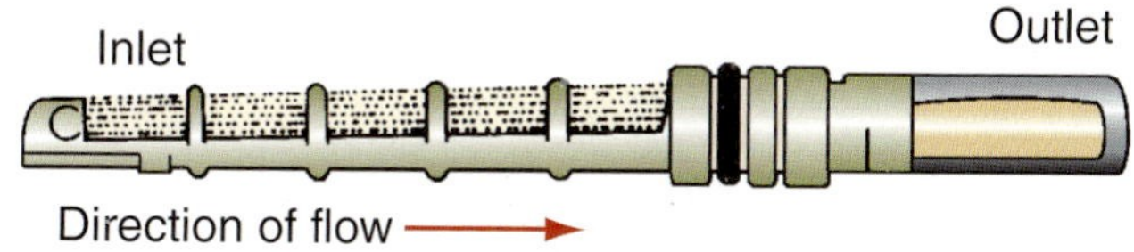

FIGURE 5-19 **A typical fixed orifice.**

EVAPORATOR

The evaporator is physically located in the air distribution duct work for the passenger compartment comfort heating and cooling system and looks like a small radiator (Figure 5-20). The evaporator is a heat exchanger that removes heat from the air flowing across the evaporator cooling fins. The source of the air blown over the evaporator fins and coils may be from outside the passenger compartment or recirculated from inside the passenger compartment when the MAX mode or Recirculation mode is selected on the heater control panel. Heat in the air is picked up by the fins and coils and transferred to the refrigerant passing through the coil. The refrigerant inside the **evaporator** coil is at a low pressure because it was metered into the coil through the small orifice of the expansion valve or by the orifice tube. Also, the compressor is pulling refrigerant out of the evaporator by a suction action.

As the low-pressure liquid refrigerant absorbs heat from the evaporator coils and fins, it boils, turning into a vapor. Because heat was taken out of the air inside the car, its temperature is lower (cooler), and the passenger compartment becomes conditioned or more comfortable. As air continues to recirculate over the evaporator coil, more heat is removed and the air continues to cool. Actual temperature control is by the action of a thermostat or a low-pressure control.

Humidity is an important factor in the quality and temperature of the air delivered to the interior of the car. The service technician must understand the effect that relative humidity (RH) has on the performance of the system. Relative humidity is the term that is used to denote the amount of moisture in the air. For example, a relative humidity of 80 percent means that the air contains 80 percent of the moisture that it can contain at a given temperature.

When the relative humidity is high, the evaporator has a double function. It must lower the air temperature as well as the temperature of the moisture carried in the air. The process of condensing the moisture in the air transfers a great amount of heat energy in the

Expansion tubes are often referred to as fixed orifice tubes.

The **evaporator** is a heat exchanger that removes heat from the air flowing across the evaporator cooling fins and into the passenger compartment.

The evaporator core contains no moving parts to wear out.

FIGURE 5-20 **A typical evaporator core.**

evaporator. Consequently, the amount of heat that can be absorbed from the air in the evaporator is greatly reduced.

The evaporator capacity required to reduce the amount of moisture in the air is not wasted, however. Lowering the moisture content in the air in the vehicle adds to the comfort of the passengers. The average person is comfortable at a temperature of 78°F to 80°F at a relative humidity (RH) of 45 to 50 percent.

Hoses and Lines

Hoses and lines carry refrigerant, as a liquid or vapor, from one component to another in the system. Hoses are constructed of a special synthetic reinforced rubber. Because of the properties of refrigerants and the high pressure of the system, only this type of hose should be used.

Rust in the system is undesirable and will cause early component failure.

Lines may be constructed of aluminum (Al) or steel tubing. Any good grade of aluminum tubing may be used, provided it is rated at a working pressure of 400 psig (2,760 kPa) or higher. If steel is used, it must be clean and dry.

Just like the other components of the system, each hose or line is referred to by name. Follow the system layout for identification of the following components (Figure 5-21).

Suction Line

The suction line should be cool to the touch.

The **suction line** is also referred to as the low-pressure line or the low-pressure vapor line. It connects the evaporator outlet to the compressor inlet. This line, which usually has the largest diameter in the system, carries low-pressure refrigerant vapor from the evaporator to the compressor. The suction line is cool to the touch.

The **suction line** is the line connecting the evaporator outlet to the compressor inlet.

Discharge Line

Also referred to as the high-pressure discharge line, the **discharge line** connects the compressor outlet to the condenser inlet. This line carries high-pressure refrigerant vapor.

In a properly operating system, this line is hot. It may be very hot in an improperly operating system, so taking care to avoid burns is important in many cases.

The **discharge line** is the line connecting the compressor outlet to the condenser inlet.

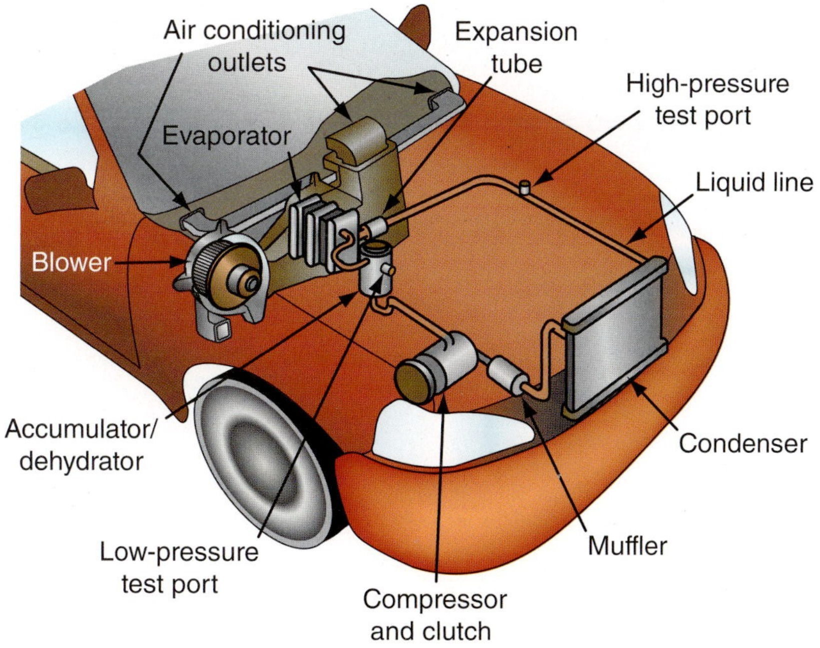

***FIGURE 5-21** A typical system layout.*

Liquid Line

The **liquid line** connects the condenser to the receiver-dryer inlet and the receiver-dryer outlet with the expansion valve inlet.

The **liquid line** is also referred to as the high-pressure liquid line. It connects the condenser outlet to the receiver-drier inlet. It also connects the receiver-drier outlet to the evaporator metering device inlet. This line, which is usually warm, may be hot under certain conditions. This line carries high-pressure liquid/vapor from the condenser to the receiver-drier and high-pressure liquid from the receiver-drier to the metering device.

REFRIGERANT

Refrigerant is the term used when referring to the fluid that is used in an automotive air-conditioning system. By definition, refrigerant is "a gas used in mechanical refrigeration systems." Actually, there are many types of refrigerant in use today, depending on application (Figure 5-22). Refrigerant is any fluid or vapor that is used to transfer heat from one area or space to another. One may not think of water (H_2O) as a refrigerant, but, when used to remove engine heat from a vehicle, it is a refrigerant. As a matter of fact, water is assigned a refrigerant number: R-718. This refrigerant evaporates (boils) and condenses at 212°F (100°C) at sea level atmospheric pressure (14.696 psia or 101.3 kPa absolute). Refrigerant-12, which has been used in automotive air-conditioning systems for many years, was also used on other applications such as domestic refrigeration. Refrigerant-12, more commonly known as R-12 or CFC-12, has the highest human safety factor of any refrigerant available that is capable of withstanding high pressures and temperatures without deteriorating or decomposing. The boiling point of R-12 at sea level atmospheric pressure is −21.67°F(−29.8°C). Therefore, it must be kept contained to prevent it from immediately boiling away.

The basic chemical, a fluorinated hydrocarbon known as carbon tetrachloride (CCl_4) was selected. It met the requirements most closely with only a few minor changes. Carbon tetrachloride (CCl_4) consists of one atom of carbon (C) and four atoms of chlorine (Cl). To change carbon tetrachloride (CCl_4) into a suitable refrigerant, two of the chlorine (Cl) atoms were removed, and two atoms of fluorine (F) were introduced in their place.

FIGURE 5-22 **Two types of refrigerant are used in the automotive air-conditioning system: (A) R-12 and (B) R-l34a.**

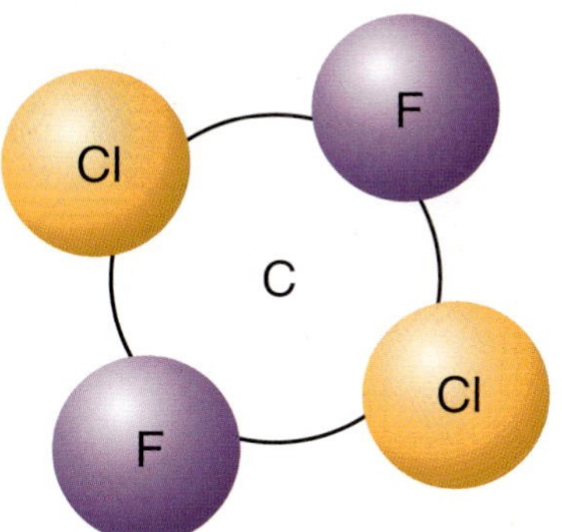

FIGURE 5-23 **The chemical structure of R-12 (CFC-12).**

Carbon tetrachloride (CCl_4), a cleaning agent, is not considered safe for personal use.

The new compound, known as dichlorodifluoromethane, is R-12. R-12 has many applications in various types of domestic and commercial refrigeration and air-conditioning systems as well as automotive air conditioners. The chemical symbol for R-12 is CCl_2F_2. This means that one molecule of this refrigerant contains one atom of carbon, two atoms of chlorine, and two atoms of fluorine (Figure 5-23).

Until the late 1980s, R-12 was considered ideal for automotive use because of its relatively low operating pressures. Its stability at high and low operating temperatures is also desirable. It does not react with most metals such as iron (Fe), aluminum (Al), or copper (Cu). Liquid R-12, however, may cause discoloration of chrome (Cr) and stainless steel (SS) if large quantities are allowed to strike these surfaces.

Hoses with liners are called barrier hoses.

R-12 is soluble in mineral oil and does not react with rubber. Some synthetic rubber compositions, however, may deteriorate if used as refrigerant hose. Synthetic rubber hose, such as Buna N, designated for refrigeration service is to be used. In applications since the late 1980s, refrigeration hoses have been lined with nylon, nytril, or polyamide veneer to provide a better barrier against leaks.

R-12 is odorless in concentrations of 20 percent or less. In greater concentrations, it can be detected by the faint odor of its original compound, carbon tetrachloride (CCl_4).

R-12 does not affect the taste, odor, or color of water or food. It was believed that it was not harmful to animal or plant life. Recent discoveries, however, dispel this belief.

Unfortunately, it has been determined that R-12 is, by far, the leading single cause of ozone depletion. The United States and 22 other countries signed an agreement in 1987 known as the Montreal Protocol. At that meeting, it was agreed that the production of chlorofluorocarbon (CFC) refrigerants would be phased out in a timely manner. The automotive air-conditioning industry was the first to be regulated because it was found that the automotive industry was the greatest offender. It was determined that 30 percent of all R-12 released to the atmosphere was from mobile air-conditioning systems.

Protocol: The plan of a scientific experiment or treatment.

The production of R-12 ended in the United States on December 31, 1995. Importing virgin R-12 into the United States from other countries is illegal, with the exception of limited quantities that are used for such medical purposes as metered-dose inhalers. It is, however, legal to import used and recovered R-12 under close scrutiny of the federal government. Those who wish to export R-12 to the United States must first petition the EPA with specific and verifiable information about its source. The industry, then, must now rely on surplus and recycled R-12 or equipment conversion to another type refrigerant, such as R-134a.

An alternate refrigerant has been developed to take the place of R-12. Tetrafluoro-ethane, referred to as R-134a, has many of the same characteristics of R-12 but poses no threat to the ozone. It does not contain ozone-depleting chlorine. Its chemical formula is CF_3CFH_2 (Figure 5-24). Though chemically it is referred to as HFC-134a, it is generally called R-l34a in the industry.

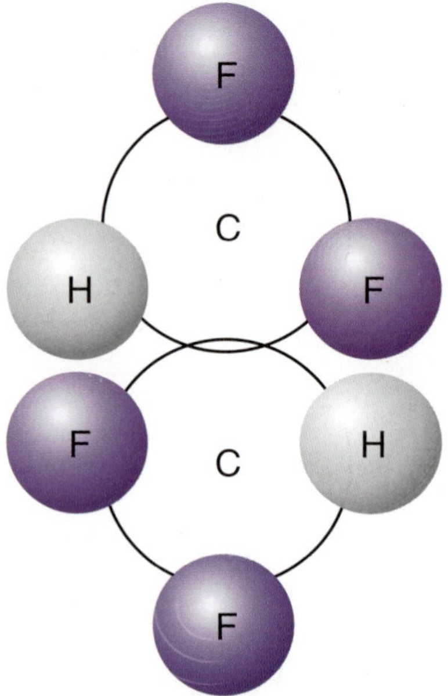

FIGURE 5-24 **The chemical structure of R-134a (HFC-134a).**

The substance that flows through the refrigeration system is referred to as the refrigerant. It circulates through the system to provide a cooling effect by absorbing heat from the evaporator. Liquid refrigerant flows into the evaporator and absorbs the heat transferred to it from the air flowing across the outside of the evaporator. This heat causes the refrigerant to vaporize as heat is drawn away from the evaporator and air stream. Refrigerant in the liquid form is clear and translucent. The refrigerant used in all new vehicles is R-134a (HFC-134a).

Use the proper refrigerant.

Several refrigerants have been approved by the EPA for use in automotive air-conditioning systems. Only two types, however, are approved by the automotive industry for use. The use of a refrigerant not approved by industry may void manufacturer's warranties on the system as well as on replacement components.

The two industry approved refrigerants are:

1. R-12 (CFC-12)
2. R-134a (HFC-134a)

Other refrigerants approved by the EPA for automotive use include:

- FRIGC FR-12
- Freeze-12
- Free Zone (also, RB-276)
- GHG-HP
- GHG-X4 (also, Autofrost and Chill-It)
- GHG-X5
- Hot Shot (also, Kar Kool)
- Ikon-12
- R-406A (also, GHG)

Many substances could be used as refrigerants but may have undesirable properties that make them unsuitable in a mobile air-conditioning system. The following is a list of the most important properties of a refrigerant:

1. A refrigerant must not be explosive or flammable.
2. A refrigerant must not be hazardous and a leak should be easily detectable.

3. The refrigerant must be highly stable and allow for repeated use without decomposing or changing its properties.
4. The refrigerant must not cause damage to parts or materials used in the compressor or other components.
5. The refrigerant must vaporize easily in the evaporator.
6. The larger the latent heat value at the vaporization point of the refrigerant, the smaller the volume of refrigerant that will be required for circulation and the smaller the total size of the refrigeration system.
7. The critical temperature of the refrigerant must be higher than the condensation temperature of the system.
8. Evaporator pressure must be higher than atmospheric pressure.

The characteristics and properties of refrigerant R-134a made it a very good choice for use in automotive air-conditioning systems. It is nonexplosive, noncorrosive, nonflammable, nonpoisonous, odorless, and harmless to food and clothing.

Never use air in combination with refrigerants. Both HCFC-22 and R-134a fall into a category of refrigerants known as combustible. If exposed to oxygen, they will both burn when under pressure or when exposed to high temperatures, but they will not ignite in air at atmospheric pressure and temperature.

De Minimis Release

The practice of purging small amounts of refrigerant in the course of repair and service is known as a *de minimis* release. The word *minimis* is not to be confused with the Latin word *minimus*, which means "smallest." *De minimis* is taken from the Latin phrase "De *minimis non curat lex*," which means "The law does not concern itself with trifles."

Refrigerant R-134a (HFC-134a)

Refrigerant 134a (HFC-134a) is at present the automotive industry's refrigerant of choice and replaced R-12 in automotive systems in the early 1990s. By 1995, all new cars manufactured contained R-134a. When servicing a refrigerant system, the most important characteristic of R-134a is its relationship between temperature and pressure. If the pressure of R-134a is high, the temperature will also be high. If the pressure is low, the temperature will also be low. At atmospheric pressure, R-134a boils at −16.4°F (−26.8°C) and water boils at 212°F (100°C). If we place R-134a under 10 psig of pressure, it will boil at −12.8°F (−10.6°C) and water boils at 250°F (121°C) (Figure 5-25). If R-134a is released into room temperature air at atmospheric pressure, it would instantly vaporize into a gas as it absorbs heat from the air. R-134a will also condense back to a liquid state if it is placed under pressure and heat content is removed.

The graphs in Figure 5-26 represent the relationship of pressure and temperature of R-134a. The curve in the graph shows the change-of-state point of R-134a between a liquid and a gas under various pressures and temperatures. The upper portion of the graph is R-134a in the vapor state and the lower portion of the graph is R-134a in the liquid state. In example 1, we can see that refrigerant in the vapor (gaseous) state can make a change of state to a liquid by increasing the pressure on the refrigerant without changing its temperature. In example 2, we can see that refrigerant in the vapor (gaseous) state can make a change of state to a liquid by decreasing the temperature of the refrigerant without changing its pressure. The refrigerant system utilizes both of these principles at the condenser, where pressure has been increased (by the compressor) and temperature has been decreased (air flow across the condenser) to allow the refrigerant to change back to a liquid.

In example 3, we can see that refrigerant in the liquid state can make a change of state to a gas (vapor) by decreasing the pressure on the refrigerant without changing its temperature.

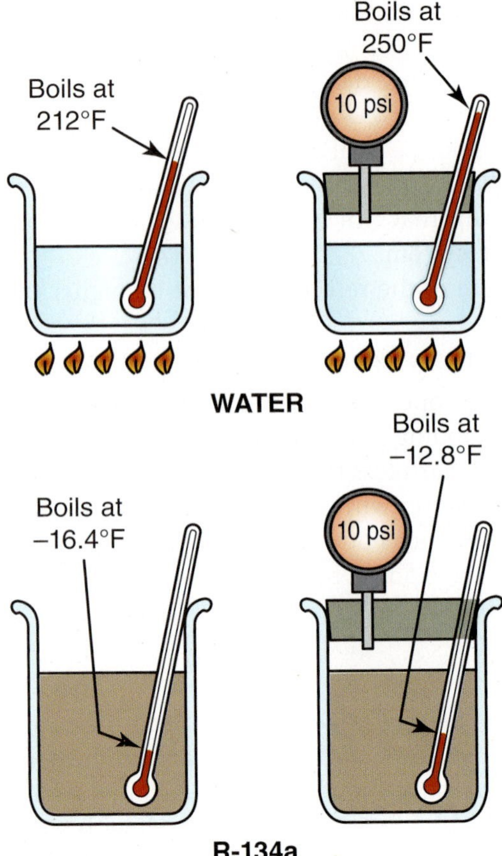

FIGURE 5-25 **Boiling point of refrigerant R-134a compared to water at both atmospheric pressure and at 10 psig.**

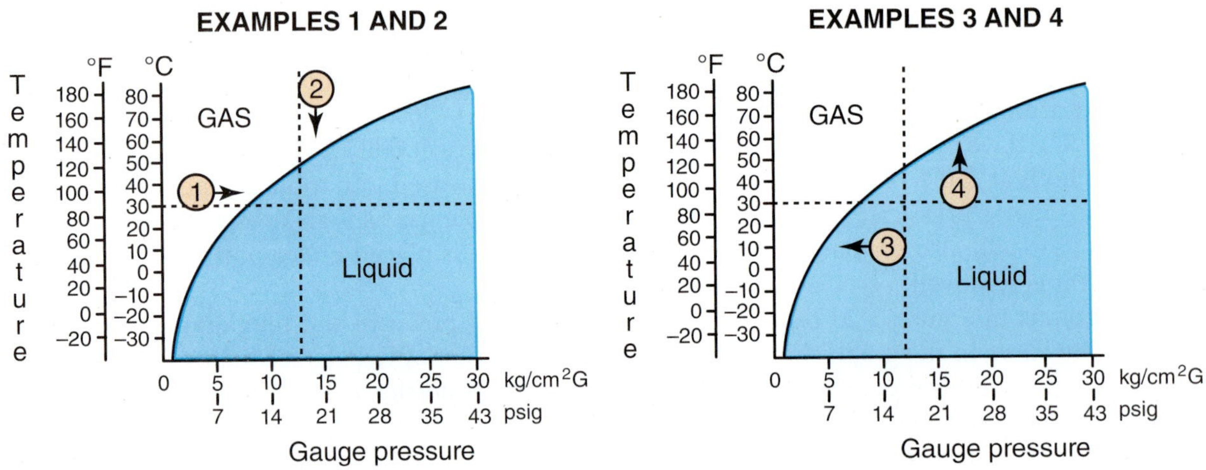

FIGURE 5-26 **Refrigerant saturation curve. In Examples 1 and 3, if temperature is held constant, refrigerant may be a gas or a liquid, depending on pressure. In Examples 2 and 4, if pressure is held constant, refrigerant may be a gas or a liquid, depending on temperature.**

In example 4, we can see that refrigerant in the liquid state can make a change of state to a gas (vapor) by increasing the temperature of the refrigerant without changing its pressure.

The basic conditions of R-134a based on the temperature-pressure fundamental at various locations in the refrigerant system is the foundation on which system operation and diagnosis is based.

REFRIGERANT HFO-1234YF (R-1234YF)

R-1234yf is classified as HFO-1234yf. HFO stands for hydrofluoro olefin and has a chemical structure of CF3CF = CH22,3,3,3-tetrafluoropropene. From a performance standpoint, R-1234yf is almost identical to R-134a (Figure 5-27), with similar system sizes and operating pressures, it is compatible with current R-134a components, and it has similar temperature–pressure relationship tables to R134a. Honeywell-DuPont states that R-1234yf is a potential direct substitute for R-134a. Cross contamination with R-134a is not a problem from a system durability standpoint, though performance would be affected. The desiccant that is currently recommended is XH7 and XH9, which is the current desiccant in use with R-134a, in the same amount (size) that is currently used.

Some are calling R-1234yf a drop-in refrigerant, as has been indicated by Honeywell-DuPont, though the EPA has indicated it will not accept it as such in existing R-134a systems or vice versa. As such, technicians will have very little difficulty diagnosing or working on this new refrigerant system as working pressures and diagnosis are virtually identical. It will be very similar to the transition from R-12 to R-134a as far as serviceability and diagnosis are concerned and in many respects simpler since there will be no retrofitting allowed and R134a and R1234yf have almost identical pressure–temperature relationships. Shop owners may not feel the same way because R-1234yf, like all refrigerants, will require the purchase of a dedicated recovery/recycling/recharging machine (J2927) equipped with a new refrigerant analyzer (J2912) and new leak detection equipment (J2913). Honeywell-DuPont states that current R-134a leak detection equipment will work on R-1234yf systems as long as it was manufactured to meet J2913 requirements. Verify that your leak analyzer meets the new SAE J2913 standard, as older equipment may have been manufactured before this requirement. Like all SNAP refrigerants, R-1234yf will have a unique service connection fitting to avoid cross contamination. One concern the service industry has is since R-1234yf and R-134a are so similar from a functionality standpoint, we may find an R-1234yf system filled with the lower-cost R-134a after system repairs. This is one reason that the SAE J2927 standard required that a refrigerant identifier be integrated into recovery/recycling/recharging equipment. Unlike the changeover from R-12 to R-134a, where R-134a was much less expensive than R-12, it is

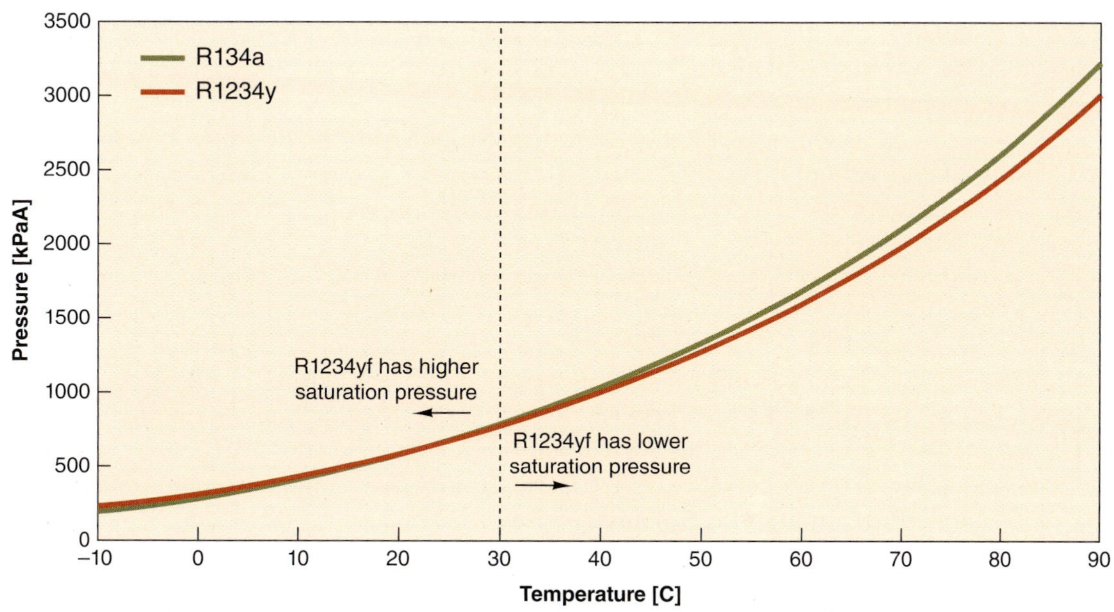

FIGURE 5-27 **R1234yf refrigerant performance compared to R134A refrigerant. Both are very similar at working pressures and temperatures.**

Acute Toxicity Exposure Limit	
Refrigerant	**ATEL (ppm)**
R-12	18,000
R-134a	50,000
R-744	50,000
R-1234yf	101,000

FIGURE 5-28 **This chart is an estimate of the maximum exposure limit for a short time period of less than 30 minutes with no adverse health effects. This is meant as an example only and should not be used as an actual guide for computing exposure time.**

expected that R-1234yf is considerably more expensive than R-134a. Some have called for a tax on R-134a that will raise its price to be more in line with R1234yf, but this is not a popular idea and is unlikely to gain political support.

As a refrigerant, both the capacity and the coefficient of performance (COP) of R-1234fy systems are within 5 percent of R-134a systems. In addition, R-1234yf has a lower compression ratio and lower discharge temperatures, 54°F (12°C) lower at peak conditions. Also, further improvements are expected in systems designed to use R-1234yf, such as improved TXV optimization of superheat and lower AP suction line performance. Furthermore, R-1234yf has a lower acute toxicity exposure limit (ATEL) as compared to either R-12 or R-134a (Figure 5-28).

There are 18 new standards and revisions published by SAE related to both R-134a and HFC-1234fy refrigerants for use in mobile air-conditioning systems (MAC):

- J639 – Safety Standards for Motor Vehicle Refrigerant Vapor Compression Systems (revised 2/2011).
- J2064 – R134a Refrigerant Automotive Air-Conditioned Hose (revised 2/2011).
- J2099 – Standard of Purity for Recycled R-134a (HFC-134a) and R-1234yf (HFO-1234yf) for Use in Mobile Air-Conditioning Systems (revised 2/2011).
- J2297 – Ultraviolet Leak Detection: Stability and Compatibility Criteria of Fluorescent Refrigerant Leak Detection Dyes for Mobile R-134a and R-1234yf (HFO-1234yf) Air-Conditioning Systems (revised 2/2011).
- J2670 – Stability and Compatibility Criteria for Additives and Flushing Materials Intended for Aftermarket Use in R-134a (HFC-134a) and R-1234yf (HFO-1234yf) Vehicle Air-Conditioning Systems (revised 2/2011).
- J2762 – Method for Removal of Refrigerant from Mobile Air-Conditioning System to Quantify Charge Amount (revised 2/2011).
- J2772 – Measurement of Passenger Compartment Refrigerant Concentrations Under System Refrigerant Leakage Conditions (revised 2/2011).
- J2773 – Standard for Refrigerant Risk Analysis for Mobile Air-Conditioning Systems (revised 2/2011).
- J2842 – R-1234yf and R744 Design Criteria and Certification for OEM Mobile Air-Conditioning Evaporator and Service Replacements (revised 2/2011).
- J2843 – R-1234yf (HFO-1234yf) Recovery/Recycling/Recharging Equipment for Flammable Refrigerants for Mobile Air-Conditioning Systems (revised 2/2011).
- J2844 – R-1234yf (HFO-1234yf) New Refrigerant Purity and Container Requirements for Use in Mobile Air-Conditioning Systems (revised 2/2011).
- J2844 – R-1234yf New Refrigerant Purity and Container Requirements Used in MVAC Systems

- J2845 – R-1234yf (HFO-1234yf) and R-744 Technician Training for Service and Containment of Refrigerants Used in Mobile A/C Systems (revised 2/2011).
- J2851 – R-1234yf (HFO-1234yf) Refrigerant Recovery Equipment for Mobile Automotive Air-Conditioning Systems (revised 2/2011).
- J2888 – HFO-1234yf Service Hose, Fittings, and Couplers for Mobile Refrigerant Systems Service Equipment (revised 2/2011).
- J2911 – Procedure for Certification that Requirements for Mobile Air-Conditioning System Components, Service Equipment, and Service Technician Training Meet SAE J Standards (revised 2/2011).
- J2912 – Performance Requirements for R-134a and R-1234yf Refrigerant Diagnostic Identifiers for Use with Mobile Air-Conditioning Systems (revised 2/2011).
- J2913 – R-1234yf (HFO-1234yf) Refrigerant Electronic Leak Detectors, Minimum Performance Criteria (revised 2/2011).
- J2927 – R-1234yf Refrigerant Identifier Installed in Recovery and Recycling Equipment for Use with Mobile A/C Systems (revised 2/2011). For larger shops with more than one R/R/R machine, a portable refrigerant analyzer that meets J2912 may be connected via a USB cable to the R/R/R unit, thereby reducing investment costs.

R-1234yf Refrigerant System Design

Fundamentally the R-1234yf system is almost identical to R-134a system design and operation. The temperature–pressure characteristics of R-1234yf are shown in Figure 5-29 and are very similar to that of R-134a.

R-1234yf Fahrenheit Pressure/Temperature Chart

PSIG	°F	PSIG	°F	PSIG	°F
9	0	77	72	200	132
11	4	80	74	206	134
14	8	83	76	212	136
16	12	86	78	218	138
19	16	89	80	223	140
19	17	92	82	229	142
20	18	96	84	236	144
23	22	99	86	242	146
24	23	102	88	248	148
25	24	106	90	255	150
28	28	110	92	261	152
31	32	113	94	268	154
33	34	117	96	275	156
35	36	121	98	282	158
37	38	125	100	289	160
38	40	129	102	296	162
40	42	133	104	304	164
42	44	137	106	311	166
44	46	142	108	319	168
47	48	146	110	326	170
49	50	148	111	334	172
51	52	150	112	342	174
53	54	155	114	351	176
56	56	160	116	359	178
58	58	164	118	368	180
61	60	169	120	376	182
63	62	174	122	385	184
66	64	179	124	394	186
68	66	184	126	403	188
71	68	190	128	413	190
74	70	195	130		

Evaporator

Condenser

R-1234yf Celsius Pressure/Temperature Chart

kPa	°C	kPa	°C	kPa	°C
62	-18	544	23	1363	55
75	-16	562	24	1398	56
90	-14	581	25	1432	57
105	-12	601	26	1468	58
120	-10	620	27	1504	59
137	-8	641	28	1541	60
155	-6	661	29	1578	61
174	-4	682	30	1616	62
194	-2	704	31	1654	63
214	0	726	32	1694	64
225	1	748	33	1733	65
236	2	771	34	1774	66
248	3	794	35	1815	67
260	4	818	36	1857	68
272	5	842	37	1900	69
284	6	866	38	1943	70
297	7	891	39	1987	71
309	8	917	40	2032	72
323	9	943	41	2078	73
336	10	970	42	2124	74
350	11	997	43	2171	75
364	12	1024	44	2219	76
379	13	1052	45	2267	77
394	14	1081	46	2317	78
409	15	1110	47	2367	79
425	16	1140	48	2418	80
440	17	1170	49	2523	82
457	18	1201	50	2631	84
473	19	1232	51	2743	86
490	20	1264	52	2859	88
508	21	1297	53	2979	90
526	22	1330	54		

Evaporator

Condenser

FIGURE 5-29 **Temperature/pressure chart for R-1234yf refrigerant.**

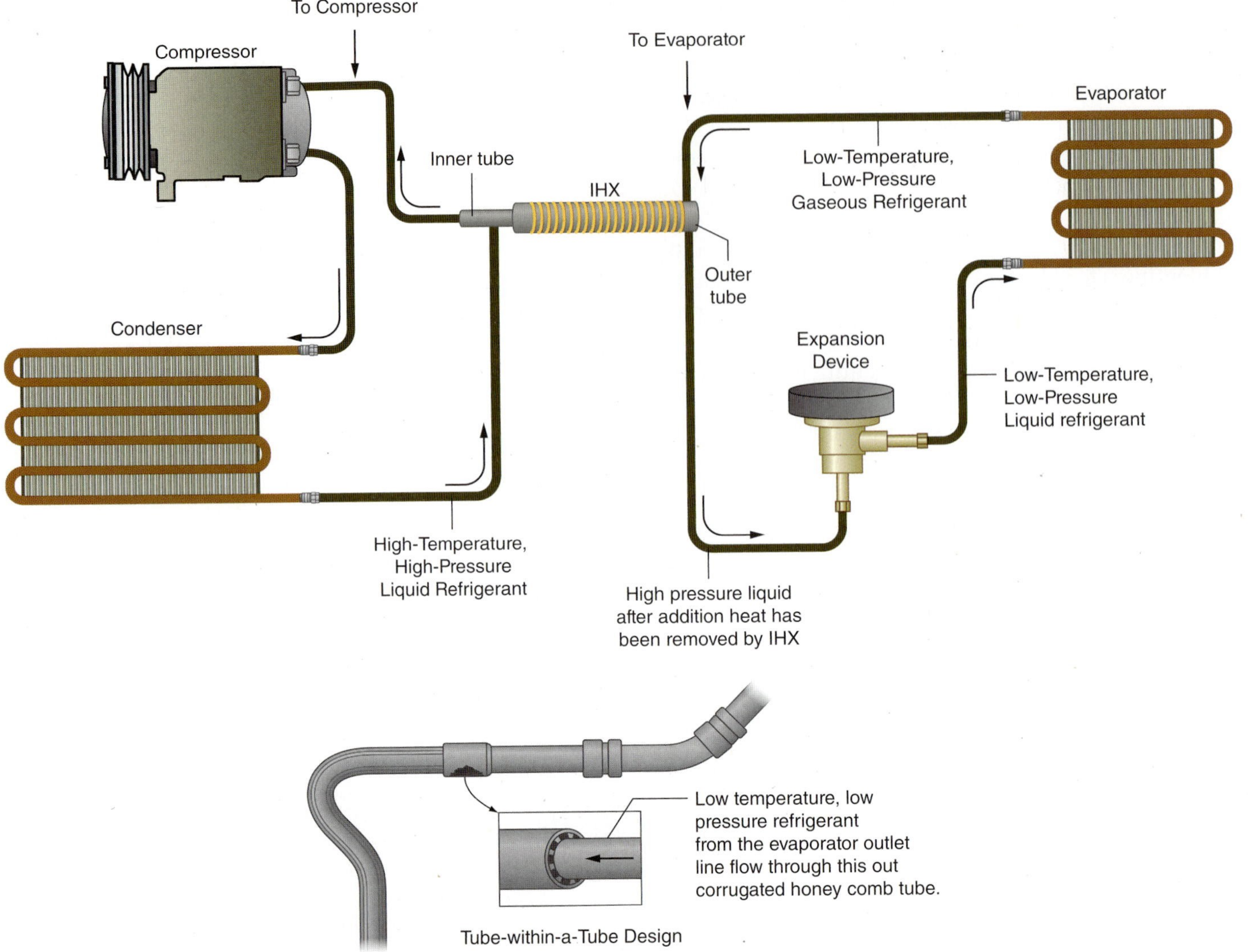

FIGURE 5-30 **R1234yf system performance is improved with the addition of an internal heat exchanger (IHX) between the high side liquid line and the low side vapor return line. Otherwise the system is fundamental the same as an R134A system. In fact, some manufacturers are using an IHX to improve performance on smaller R134A systems making an R134A and an R1234yf system virtually identical.**

One difference is the use of an internal heat exchanger (Figure 5-30) to improve system performance to meet that of an R-134a system.

Pressure versus Temperature Relationship

The temperature at which refrigerant vaporizes or condenses is the saturation temperature. The saturation temperature of a gas or liquid increases or decreases based on the pressure applied to the refrigerant. The refrigerant in the air-conditioning system cannot stay a gas at temperatures below a corresponding pressure. Nor can a refrigerant stay a liquid at temperatures above a corresponding pressure. The refrigerant cycle of repeated heat absorption at the evaporator and heat transfer at the condenser cannot happen without pressurizing and raising the temperature of the refrigerant (Figure 5-31).

In order for the refrigerant in the evaporator to boil at the correct temperature, the pressure of the refrigerant must be lowered. If the pressure is lowered the proper amount, the temperature will eventually be less than that of the surrounding air. When this pressure is reached, the refrigerant will absorb the heat from the air, and a change of state from a liquid

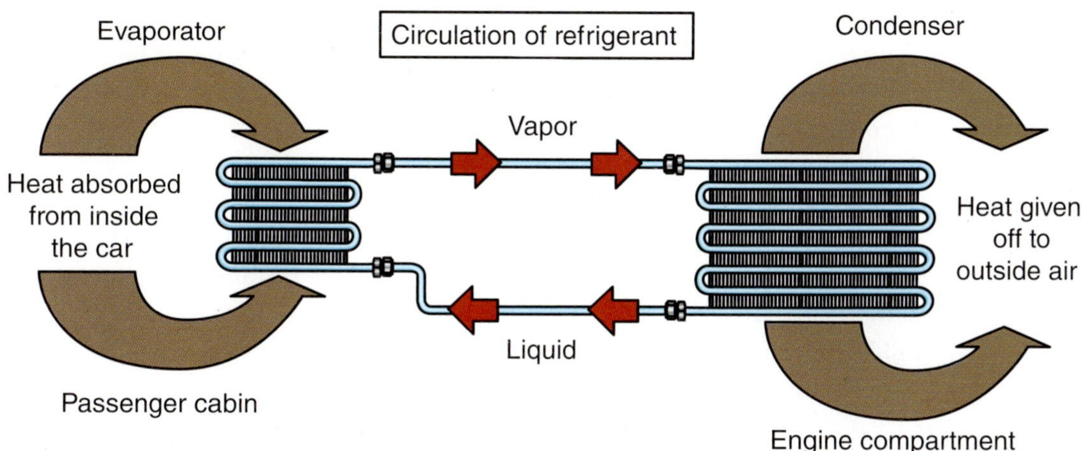

FIGURE 5-31 **Refrigerant cycle of heat exchange.**

to a vapor (vaporization process) will occur. This process will transfer a large quantity of heat from the air to the refrigerant. The net effect is that the air temperature will be lowered to enable cooling of the passenger compartment.

Now that the refrigerant has absorbed a large quantity of heat, it must release this heat before the next cycle through the evaporator core. For the refrigerant to release this heat, the pressure must be increased, which will also raise its temperature. The compressor will raise the refrigerant pressure and heat concentration. This increase in pressure and temperature is required so that the refrigerant can give off its heat at the condenser and a change of state can occur from a vapor to a liquid (condensation process). The cycle can now be repeated to cool more air entering the passenger compartment.

Refrigerant R-12 (CFC-12)

A refrigerant, known as R-12 or CFC-12, was used in automotive air-conditioning systems through the early 1990s. Because of environmental concerns, its production and use has been phased out. Certain system changes, however, have to be made in order to use the new refrigerant. There is no drop-in refrigerant available that is approved for automotive use.

Other Refrigerants

In 1994, the EPA established the Significant New Alternatives Policy (SNAP) Program to review alternatives to ozone-depleting substances. Under authority of the 1990 Clean Air Act, the EPA also examines potential substitute refrigerants as to their flammability, effects on global warming, and toxicity. As of this writing, the agency has determined that ten "new" refrigerants, including R-134a, are acceptable for use as an R-12 replacement in motor vehicle air-conditioning systems. They are all, however, "acceptable subject to use conditions." All alternate refrigerants except R-134a are "blends," which means that they contain more than one component in their composition.

Do not contaminate recovery system equipment or cylinders.

"Acceptable subject to use conditions" indicates that the EPA believes these refrigerants, when used in accordance with the use conditions, to be safer for human health and for the environment than the R-12 they are meant to replace. This designation, however, is not intended to imply that the refrigerant will work as satisfactorily as R-12 in any specific system. Also, it is not intended to imply that the refrigerant is perfectly safe regardless of how it may be used.

The EPA does not test refrigerants and therefore does not specifically approve or endorse any one refrigerant over any others. The agency reviews all of the information about a refrigerant submitted by its manufacturer and independent testing laboratories. The EPA does not determine what effect, if any, a "new" refrigerant may have on vehicle warranty.

Some refrigerant manufacturers use the term *drop-in* to imply that their refrigerant will perform identically to R-12 and that no modification is required for its use. The term also implies that the alternate refrigerant can be used alone or mixed with R-12. The EPA believes the term *drop-in* confuses and obscures at least two important regulatory points:

1. Charging one refrigerant into a system before extracting the old refrigerant is a violation of the SNAP use conditions and is therefore illegal.
2. Certain components may be required by law, such as hoses and compressor shutoff switches. If these components are not present, they must be installed. Five blends, for example, contain HCFC-22 and require barrier hoses.

It may also be noted that system performance is affected by such variables as outside temperature, relative humidity, and driving conditions. Therefore, it is not possible to ensure equal performance of any refrigerant under all of these conditions.

Contaminated is a term generally used when referring to a refrigerant cylinder or system that fails a purity test and is known to contain foreign substances such as other incompatible or hazardous refrigerants.

The service facility must have service and recovery equipment specifically designed for each type of refrigerant that is to be serviced. This means that at least two systems are required: one for R-12, and one for R-134a. A third set is required if **contaminated** systems are to be serviced, and a fourth set is required if a blend refrigerant is to be used.

Each new alternate refrigerant must be used with a unique set of fittings attached on the service ports, all recovery and recycling equipment, on can taps and other charging equipment, and on all refrigerant containers. A unique label must be affixed over the original label to identify the type of refrigerant as well as lubricant used in the air-conditioning system.

Handling Refrigerant

All refrigerants must be properly stored, handled, and used.

Liquid refrigerant can cause blindness if sprayed into the eyes. Also, if liquid refrigerant comes into contact with the skin, frostbite may result.

A refrigerant container should never be exposed to heat above 125°F (51.7°C). This means that it should not be allowed to come into contact with an open flame or any type of heating device, and it should not be stored in direct sunlight. The increase in refrigerant pressure inside the container, known as *hydrostatic pressure*, as a result of excessive heat can become great enough to cause the container to explode.

If refrigerant is allowed to come into contact with an open flame or heated metal, a poisonous gas is created. Anyone breathing this gas may become ill. Remember—refrigerant is not a toy. Refrigerant should be handled only by a properly trained and experienced automotive service technician.

The term *Freon* is frequently used when referring to refrigerant. Freon and Freon-12 are registered trademarks of E. I. DuPont de Nemours and Company. These terms, then, should be used only when referring lo refrigerant manufactured by this company. A new term, *SUVA*, is used by DuPont to identify a new ozone-friendly group of refrigerants that includes R-134a (HFC-134a). Both refrigerants, as well as others, are also produced by several other manufacturers and are packaged under various tradenames.

Either refrigerant is available in sizes from "pound" cans to 1-ton (907-kg) cylinders. Actually, the "pound" can of R-12 contains 12 oz. (340 g). The R-134a can also contains 12 oz. (340 g) (Figure 5-32).

After November 15, 1992, it became unlawful to sell or distribute to the general public R-12 in containers of less than 20 lb. (9 kg). Proper certification is now required for the purchase of refrigerant, in any quantity, to ensure that those dispensing refrigerants are knowledgeable in their profession and may be held accountable for their actions. Some states also require special licensing of the service facility in addition to the federal certification requirements. Also, to legally service automotive air-conditioning systems, the service facility must have proper, adequate, and EPA-approved refrigerant service equipment for each type refrigerant they wish to service.

FIGURE 5-32 **Typical "pound" cans of R-12 (CFC-12) and R-134a (HFC-134a).**

The industry has adopted a standard color code to identify refrigerant containers. An R-12 container is white, and an HCFC-22 container is green. Light blue identifies an R-134a container, which must not be confused with R-114, which is packaged in a dark blue container. This color code, however does not apply to "pound" cans.

It cannot be overemphasized that R-12, a chlorofluorocarbon refrigerant; HCFC-22, a hydrochlorofluorocarbon; and R-l34a, a hydrofluorocarbon refrigerant, are not compatible with each other. They must not be mixed under any circumstances or in any other manner substituted one for the other. Mixing refrigerants, even in small quantities, will result in exceptionally high pressures that may cause serious damage to system components, such as the evaporator and hoses. An improper refrigerant may also cause damage to the system due to the incompatibility of the lubricant and desiccant The appropriate equipment—such as manifold and gauge set, recovery system, and charging station—must be used for each refrigerant.

Air-Conditioning Circuit

To better understand the function of an automotive air-conditioning system, it is helpful to know the physical state of the refrigerant in the various sections of the system. Actually, there are only six such states to be considered:

1. Low-pressure vapor (A)
2. Low-pressure liquid (B)
3. Low-pressure vapor and liquid (C)
4. High-pressure vapor (D)
5. High-pressure liquid (E)
6. High-pressure liquid and vapor (F)

Following is a brief overview of each of these states. For component location, refer to callouts (A through F) in Figure 5-33 for an expansion valve system or Figure 5-34 for an orifice tube system.

Low-Pressure Vapor

The refrigerant is a low-pressure vapor in the section of the system from the evaporator outlet to the compressor inlet (A). This includes any devices found in the suction line, such as a suction-line drier, muffler, or accumulator.

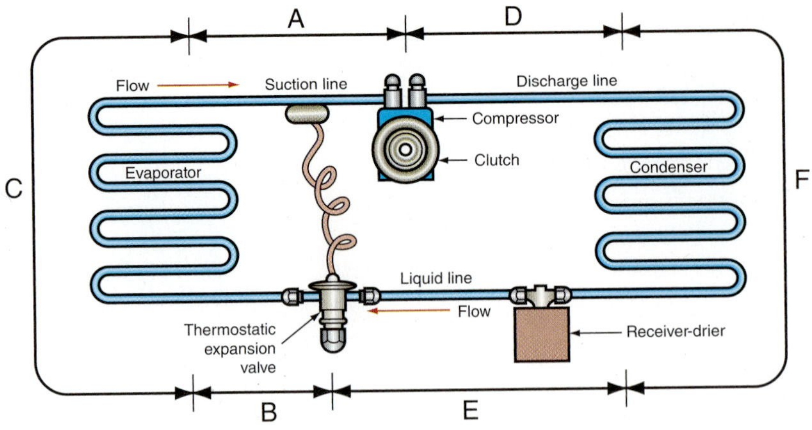

FIGURE 5-33 **Thermostatic expansion valve system.**

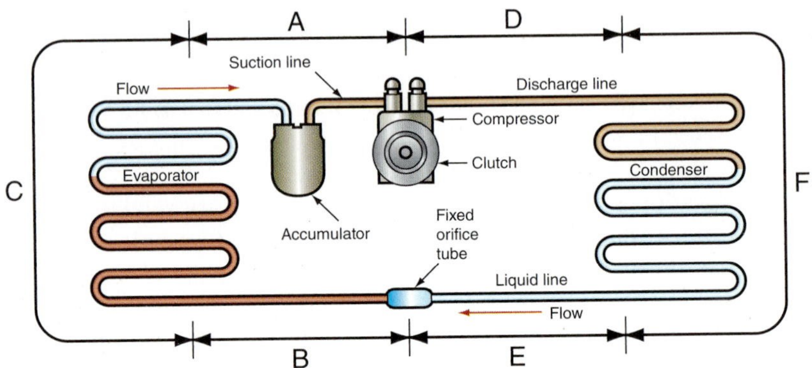

FIGURE 5-34 **Orifice tube system.**

Low-Pressure Liquid

Immediately after the metering device, the entrance to the evaporator (B) is the only part of the system that may contain low-pressure liquid. Even this section contains vapor, called flash gas, having just passed through the metering device.

Low-Pressure Vapor and Liquid

In the evaporator (C), low-pressure liquid refrigerant boils as it picks up heat and is changed to low-pressure vapor.

High-Pressure Vapor

The refrigerant is at high pressure in a vapor state in the line from the compressor outlet to the condenser inlet (D). This includes any devices that may be in the discharge line, such as a muffler.

High-Pressure Liquid

The high-pressure liquid refrigerant section extends from the condenser outlet to the metering device inlet (E). This includes any devices in the liquid line, such as receiver, drier, and sight glass.

High-Pressure Liquid and Vapor

In giving up its heat, the high-pressure refrigerant vapor is changed to liquid in the condenser (F).

System Review

Heat is picked up inside the automobile driver/passenger compartment from the air passing through the coils and fins of the evaporator. This heat is picked up by the liquid refrigerant as it evaporates. The heat-laden refrigerant vapor is then pumped by the compressor into the condenser on the outside of the automobile, usually located in front of the radiator. In the condenser, the refrigerant's heat is given up to the less hot air passing across the coils and fins as it condenses back to a liquid.

The transfer of heat in an air-conditioning system is accomplished by two pressure and two temperature systems: a low-pressure system of 21–35 psig (145–241 kPa) with a low temperature of 24°F–40°F (−4°C to 4°C), and a high-pressure system of 180–220 psig (1,241–1,517 kPa) with a high temperature of 123°F–137°F (51°C–58°C). Any time there is a pressure change, there is a temperature change. During this pressure-temperature change with R-134a and R-12, there is also a change of state (Figure 5-31). In the low-pressure side, the change is from a liquid to a vapor; in the high-pressure side, the change is from a vapor to a liquid.

It is important that you understand this pressure–temperature relationship as you use the manifold and gauge set as a diagnostic tool. The manifold and gauge set is used as a diagnostic tool to determine many system problems that relate to abnormal gauge pressures.

If you will recall from Chapter 2, superheat is the added heat intensity given to a gas after the complete evaporation of a liquid. In a refrigerant system, if all the liquid refrigerant in the evaporator core at a given point has gone through a change of state from a liquid to a gas as it picked up heat it may only be 75 percent of the way through the core. These refrigerant molecules still have 25 percent of the evaporator core left to travel through. As the refrigerant gas continues to travel through the evaporator, this gas will pick up additional heat from the core as more heat is given up by the air passing over its surface. Even though the refrigerant is at the same pressure, it will become hotter than the pressure/temperature chart in Figure 5-35 may indicate it should be. This increase in heat above the normal pressure/temperature relationship is called superheat. This phenomenon only occurs when there are no liquid refrigerant molecules nearby. Refrigerant systems are designed to maintain approximately 10°F (12°C) of superheat in the refrigerant leaving the evaporator so that the gas returning to the compressor is several degrees away from the condensation point of the refrigerant. This is to avoid the risk of liquid refrigerant entering the compressor. The compressor is designed to be a vapor pump and would be damaged if it had to compress liquid refrigerant.

The pressure of the refrigerant is increased by the compressor. The compressor pumps low-pressure refrigerant vapor from the evaporator to the condenser at a high pressure. The pressure of the refrigerant is decreased by the metering device (expansion valve or orifice tube) at the inlet of the evaporator.

The flow of refrigerant is regulated into the evaporator by a metering device such as a TXV or an FOT. Just before entering the metering device, the refrigerant is a high-pressure liquid. Refrigerant is metered into the evaporator through a small orifice, changing it to a low-pressure liquid.

From what we have outlined, it may be concluded that the compressor is the dividing line, low- to high-side, and the metering device is the dividing line, high- to low-side. Whenever necessary, refer to the basic principles previously discussed.

Temperature °F	Pressure psig	Temperature °F	Pressure psig
-5	4.1	39.0	34.1
0	6.5	40.0	35.0
5.0	9.1	45.0	40.0
10.0	12.0	50.0	45.4
15.0	15.1	55.0	51.2
20.0	18.4	60.0	57.4
21.0	19.1	65.0	64.0
22.0	19.9	70.0	71.1
23.0	20.6	75.0	78.6
24.0	21.4	80.0	86.7
25.0	22.1	85.0	95.2
26.0	22.9	90.0	104.3
27.0	23.7	95.0	113.9
28.0	24.5	100.0	124.1
29.0	25.3	105.0	134.9
30.0	25.3	110.0	146.3
31.0	27.0	115.0	158.4
32.0	27.8	120.0	171.1
33.0	28.7	125.0	184.5
34.0	29.5	130.0	198.7
35.0	30.4	135.0	213.5
36.0	31.3	140.0	229.2
37.0	32.2	145.0	245.6
38.0	33.2	150.0	262.8

A: English Temp/Pres Chart for HFC-134a (R-134a)

Temp. °F	Press. psig	Temp. °F	Press. psig	Temp. °F	Press. psig	Temp. °F	Press. psig	Temp. °F	Press. psig
0	9.1	35	32.5	60	57.7	85	91.7	110	136.0
2	10.1	36	33.4	61	58.9	86	93.2	111	138.0
4	11.2	37	34.3	62	60.0	87	94.8	112	140.1
6	12.3	38	35.1	63	61.3	88	96.4	113	142.1
8	13.4	39	36.0	64	62.5	89	98.0	114	144.2
10	14.6	40	36.9	65	63.7	90	99.6	115	146.3
12	15.8	41	37.9	66	64.9	91	101.3	116	148.4
14	17.1	42	38.8	67	66.2	92	103.0	117	151.2
16	18.3	43	39.7	68	67.5	93	104.6	118	152.7
18	19.7	44	40.7	69	68.8	94	106.3	119	154.9
20	21.0	45	41.7	70	70.1	95	108.1	120	157.1
21	21.7	46	42.6	71	71.4	96	109.8	121	159.3
22	22.4	47	43.6	72	72.8	97	111.5	122	161.5
23	23.1	48	44.6	73	74.2	98	113.3	123	163.8
24	23.8	49	45.6	74	75.5	99	115.1	124	166.1
25	24.6	50	46.6	75	76.9	100	116.9	125	168.4
26	25.3	51	47.8	76	78.3	101	118.8	126	170.7
27	26.1	52	48.7	77	79.2	102	120.6	127	173.1
28	26.8	53	49.8	78	81.1	103	122.4	128	175.4
29	27.6	54	50.9	79	82.5	104	124.3	129	177.8
30	28.4	55	52.0	80	84.0	105	126.2	130	182.2
31	29.2	56	53.1	81	85.5	106	128.1	131	182.6
32	30.0	57	55.4	82	87.0	107	130.0	132	185.1
33	30.9	58	56.6	83	88.5	108	132.1	133	187.6
34	31.7	59	57.1	84	90.1	109	135.1	134	190.1

C: English Temp/Pres Fen CFC-12 (R-12)

Temperature °C	Pressure kPa	Temperature °C	Pressure kPa
-15.0	63	5.0	247
-12.5	83	7.5	280
-10.0	103	10.0	313
-7.5	122	12.5	345
-5.0	142	15.0	381
-4.5	147	17.5	422
-4.0	152	20.0	465
-3.5	157	22.5	510
-3.0	162	25.0	560
-2.5	167	27.5	616
-2.0	172	30.0	670
-1.5	177	32.5	726
-1.0	182	35.0	785
-0.5	187	37.5	849
0.0	192	40.0	916
0.5	198	42.5	990
1.0	203	45.0	1066
1.5	209	47.5	1146
2.0	214	50.0	1230
2.5	220	52.5	1315
3.0	225	55.0	1385
3.5	231	57.5	1480
4.0	236	60.0	1580
4.5	242	65.0	1795

B: Metric Temp/Pres Chart for HFC-134a (R-134a)

EVAPORATOR TEMPERATURE °C	EVAPORATOR PRESSURE GAUGE READING KILOPASCAL (GAUGE)	(ABSOLUTE)	AMBIENT TEMPERATURE °C	HIGH PRESSURE GAUGE READING KILOPASCAL (GAUGE)
-16	73.4	174.7	16	737.7
-15	81.0	182.3	17	759.8
-14	87.8	189.1	18	784.6
-13	94.8	196.1	19	810.2
-12	100.6	201.9	20	841.2
-11	108.9	210.2	21	868.7
-10	117.9	219.2	22	901.8
- 9	124.5	225.8	23	932.2
- 8	133.9	235.2	24	970.8
- 7	140.3	241.6	25	1 020.5
- 6	149.6	250.9	26	1 075.6
- 5	159.2	260.5	27	1 111.5
- 4	167.4	268.7	28	1 143.2
- 3	183.2	268.7	29	1 174.9
- 2	186.9	288.2	30	1 206.6
- 1	195.8	288.2	31	1 241.1
0	206.8	308.1	32	1 267.3
1	218.5	319.8	33	1 294.8
2	227.8	329.1	34	1 319.7
3	238.7	340.0	35	1 344.5
4	249.4	350.7	36	1 413.5
5	261.3	362.6	37	1 468.6
6	273.7	375.0	38	1 527.9
7	287.5	388.8	39	1 577.5
8	296.6	397.9	40	1 627.2
9	303.3	404.6	42	1 737.5
10	321.5	422.8	45	1 854.7

D: Metric Temp/Pres Chart for CFC-12 (R-12)

FIGURE 5-35 **Typical temperature-pressure charts.**

SUMMARY

- The automotive air-conditioning system is a combination of mechanical systems circulating a chemical medium (refrigerant). The refrigerant absorbs heat from the air entering or recirculating through the passenger compartment and transfers this heat to the outside air.
- The conditioned air in the passenger compartment contains less moisture, which adds to the cooling sensation felt by the passengers. In principle, the air-conditioning system does not cool; it removes heat from the air entering the passenger compartment.
- The air-conditioning compressor is the heart of the heating and cooling system. The compressor is one of the points in the air-conditioning system where there is a separation between high and low pressure. Its purpose is to pull refrigerant into the compressor through the suction line and compress and pump refrigerant out the discharge (pressure) line.
- The condenser is a heat exchanger for the superheated refrigerant in the system. It removes the heat energy from the refrigerant that was gained in the evaporator.
- The evaporator is a heat exchanger that removes heat from the air flowing across the evaporator cooling fins in the passenger compartment duct system to cool the passenger compartment.
- The accumulator and receiver-drier are storage and distribution components used to clean and dry the refrigerant.
- Refrigerant lines are of barrier design for R134a refrigerant and are used to transport refrigerant through the system.
- The air-conditioning system is designed to maintain in-car temperature and humidity at a predetermined level.

TERMS TO KNOW

Accumulator
Air conditioning
Condenser
Contaminated
Discharge line
Evaporator
Expansion tube
Liquid line
Receiver-drier
Reciprocating piston(s)
Suction line
Superheated
Thermostatic expansion valve

REVIEW QUESTIONS

Short-Answer Essays

1. Why is the air-conditioning system's efficiency improved when the Recirculation mode is selected?
2. How is liquid refrigerant used to lower passenger compartment temperature?
3. What can affect the saturation temperature of gas or liquid refrigerant?
4. Explain the term dew point temperature.
5. What is the basic function of the automotive air-conditioning system?
6. What is the difference between the accumulator and the receiver-drier?
7. Where is the receiver-drier located, and what state is the refrigerant in that flows through it?
8. Where is the accumulator located, and what state is the refrigerant in that flows through it?
9. What is the purpose of a compressor?
10. How does the expansion tube differ from the expansion valve?

Fill in the Blanks

1. Humidity in the passenger compartment can be increased by many factors, including but not limited to, ______________, ______________, and even ______________ in the vehicle contributing to humidity load by ______________.
2. When the pressure of R-134a is high, the temperature will be ______________. When the pressure is low, the temperature will be ______________.
3. The temperature at which refrigerant vaporizes or condenses is ______________ the temperature.
4. An air-conditioning system that uses a fixed orifice tube has a(n) ______________ in the suction line.
5. An air-conditioning system that uses a thermostatic expansion valve has a(n) ______________ in the liquid line.
6. The ______________ is a heat exchanger for the superheated refrigerant in the system.
7. The ______________ is one of the points in the air-conditioning system where there is a separation between high and low pressure.

8. There is a direct relationship between temperature and _______________ in an automotive air-conditioning system.

9. The refrigerant changes from a _______________ to a _______________ in the condenser.

10. Two types of metering devices are the _______________ valve and the _______________ _______________ tube.

Multiple Choice

1. All of the following are forms of heat transfer that take place as it relates to the vehicle air-conditioning system except:
   A. Convection
   B. Conduction
   C. Radiation
   D. Perspiration

2. All of the following are important properties that an automotive refrigerant must exhibit except:
   A. The refrigerant must be highly stable and allow for repeated use without decomposing or changing its properties.
   B. A refrigerant must be explosive or flammable.
   C. The critical temperature of the refrigerant must be higher than the condensation temperature of the system.
   D. Evaporator pressure must be higher than atmospheric pressure.

3. R-134a, when used as a refrigerant, exhibits a predict able relationship between pressure and temperature, and a stable change of state point between a liquid and a gas under various pressures and temperatures. All of the following are true of the temperature pressure relationship of R-134a except:
   A. Refrigerant in the vapor (gaseous) state can make a change of state to a liquid by increasing the pressure on the refrigerant without changing its temperature.
   B. Refrigerant in the vapor (gaseous) state can make a change of state to a liquid by decreasing the temperature of the refrigerant without changing its pressure.
   C. Refrigerant in the liquid state can make a change of state to a gas (vapor) by increasing the pressure on the refrigerant without changing its temperature.
   D. Refrigerant in the liquid state can make a change of state to a gas (vapor) by increasing the temperature of the refrigerant without changing its pressure.

4. In a normally operating system, the refrigerant is in different physical states (vapor or liquid) at the various sections of the system. All of the following are physical states the refrigerant would be found in at various specific locations except:
   A. A low-pressure vapor at the suction line
   B. A low-pressure vapor and liquid at the evaporator
   C. A high-pressure vapor at the discharge line
   D. A high-pressure liquid at the condenser

5. Under normal operating conditions all of the following occurs inside the evaporator, except:
   A. The refrigerant absorbs heat.
   B. Heat entering the refrigerant causes it to change state.
   C. The refrigerant removes heat from the outside air drawn across evaporator core.
   D. The refrigerant changes from a vapor to a liquid.

6. Technician A says the condenser is a heat exchanger for the superheated refrigerant in the system. Technician B says the accumulator is a heat exchanger that removes heat from the air flowing across the evaporator cooling fins.
   Who is correct?
   A. A only
   B. B only
   C. Both A and B
   D. Neither A nor B

7. Technician A says the compressor is one of the points in the air-conditioning system where there is a separation between a high and a low pressure. Technician B says the metering device is one of the points in the air-conditioning system where there is a separation between a high and a low pressure. Who is correct?
   A. A only
   B. B only
   C. Both A and B
   D. Neither A nor B

8. The metering device changes high-pressure liquid to
   A. A low-pressure
   B. A high-pressure
   C. A low-pressure liquid
   D. Any of the above, vapor depending on the outside air vapor temperature

9. *Technician* A says the fixed orifice tube can vary the amount of refrigerant allowed to flow to the evaporator.
   Technician B says the thermostatic expansion valve has a fixed opening and cannot vary the amount of refrigerant allowed to flow to the evaporator.
   Who is correct?
   A. A only
   B. B only
   C. Both A and B
   D. Neither A nor B

10. *Technician* A says that the line or hose that connects the compressor outlet to the condenser inlet is called the discharge line.
    Technician B says that the line or hose that connects the evaporator outlet to the compressor inlet of a TXV system is called a suction line.
    Who is correct?
    A. A only
    B. B only
    C. Both A and B
    D. Neither A nor B

Chapter 6

REFRIGERANT SYSTEM COMPONENTS

UPON COMPLETION AND REVIEW OF THIS CHAPTER, YOU SHOULD BE ABLE TO:

- Explain the purpose and operation of an automotive air conditioner compressor.
- Identify and compare the state of the refrigerant in each section of the automotive air-conditioning system.
- Discuss the **change of state** of the refrigerant:
  a. In the evaporator
  b. In the condenser
- Explain the purpose of:
  a. The receiver-drier
  b. The accumulator
- Compare the function of the thermostatic expansion valve (TXV) to the fixed orifice tube (FOT).

INTRODUCTION

In the following discussion, the purpose and function of each component of the basic automotive air-conditioning system are discussed. The sizes of the system refrigerant hoses are given, as are the states of the refrigerant in each of them. The state of the refrigerant in each of the components is also discussed (Figure 6-1). Although hose sizes may vary slightly from vehicle to vehicle, the state of the refrigerant at various points in all systems is basically the same.

For the purpose of this discussion of the air-conditioning cycle, we will start with the compressor. It is the compressor's function to circulate the refrigerant throughout the system.

THE REFRIGERATION CYCLE

What we learned in Chapter 2 was that to cool down an object, heat must be removed or given off. In an automotive refrigerant system, we use a pressurized refrigeration system. The refrigerant in the automotive refrigerant system circulates under pressure in a sealed closed loop circuit continually changing from a liquid to a gas (vapor) and from a gas back to a liquid while the system is operating. The entire refrigerant cycle exhibits several processes as the refrigerant changes state, from liquid to vapor and from vapor to liquid.

For the following discussion refer to Figure 6-2. Refrigerant is compressed in the vapor state at the compressor, entering as a low pressure vapor and leaving as a high pressure vapor. The compressor raises both the pressure and the temperature of the refrigerant. From there it moves on to the condenser where the pressurized heat (Btu) concentrated refrigerant gives up its heat energy to the surrounding air being drawn across its core, making a change of state as it condenses from a high pressure vapor into a high pressure liquid. Next, the pressure of the refrigerant drops as it passes through the metering device. The refrigerant enters the

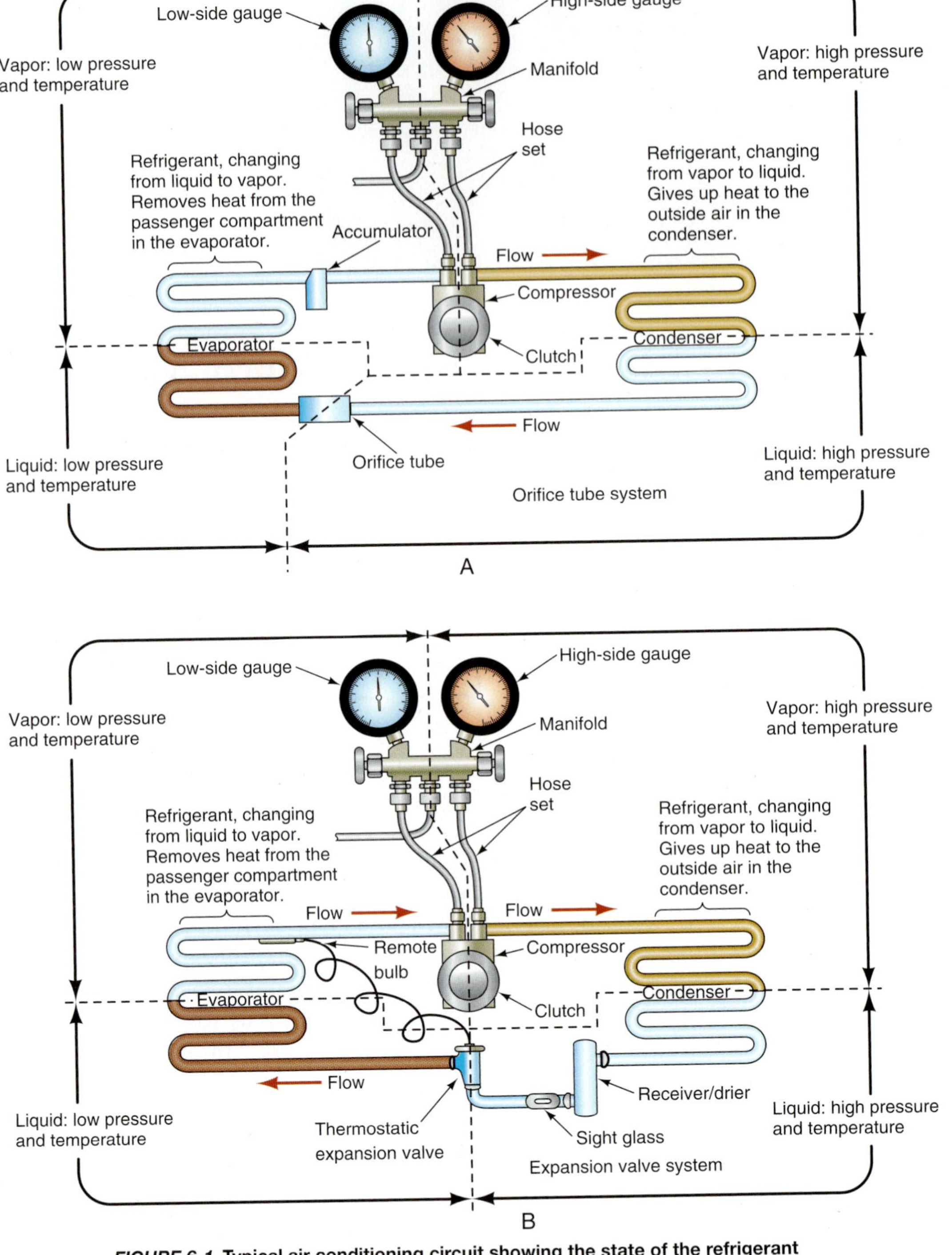

FIGURE 6-1 **Typical air conditioning circuit showing the state of the refrigerant in each section: (A) orifice tube system and (B) expansion valve system.**

metering device as a high pressure liquid and leaves as a low pressure liquid. It is second of the two dividing lines between the high pressure side and low pressure sides of the refrigerant system, the first was the compressor. The low pressure liquid refrigerant enters the evaporator core where heat is absorbed from the air traveling across its surface causing another change of state as the low pressure liquid refrigerant boils as it absorbs heat from the air blowing across it and changes from a low pressure liquid into a low pressure vapor. Cool air is not produced; heat is removed from the air flowing into the vehicles duct system and finally to the passenger

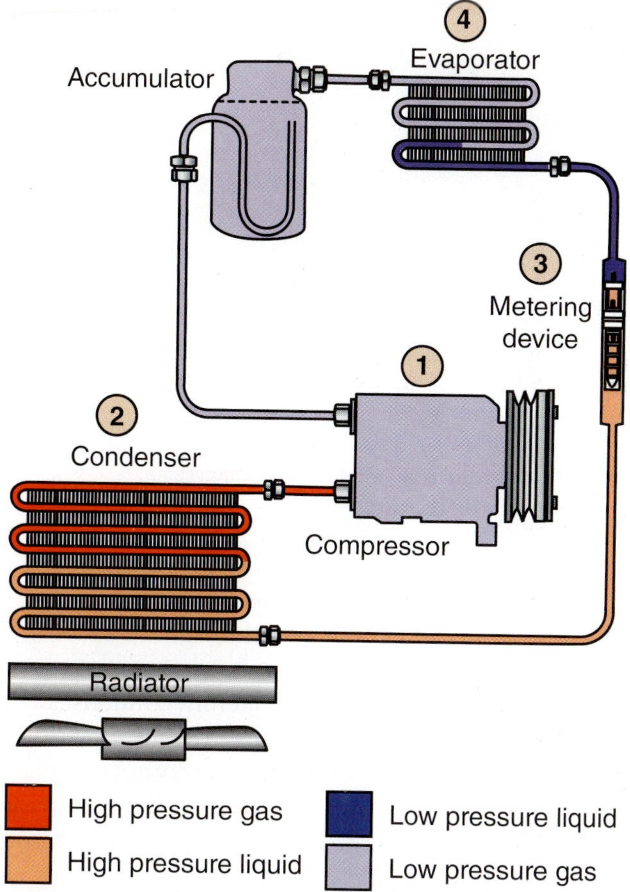

FIGURE 6-2 **The compressor (1) is one of the dividing line between the low pressure and high pressure side of the system, the refrigerant is a low pressure vapor on the inlet side and a high pressure vapor on the discharge side. The condenser (2) is a heat exchange where a change of state takes place from a high pressure vapor to a high pressure liquid. The metering device (3) is the other dividing line between the high pressure and low pressure side of the system, the refrigerant is a high pressure liquid on the inlet side and a low pressure liquid on the outlet side. The evaporator (4) is a heat exchanger where a change of state takes place from a low pressure liquid to a low pressure vapor.**

compartment. So, the refrigerant system picks up heat at the evaporator core that it removed from the air entering the passenger compartment and gives up this heat at the condenser to the outside air passing across its surface transferring this heat back into the environment. This process is repeated continuously as the system is operated. This is the basic air-conditioning circuit from which all of the other automotive refrigerant circuits are patterned. A good understanding of the basic circuit makes an understanding of other circuits much easier.

To reiterate what was stated previously, the compressor draws in (suction) vaporized refrigerant from the evaporator and increases the gas pressure thereby separating the low side of the system from the high side. The metering device used by automotive refrigerant systems is either a thermostatic expansion valve (TXV) or a fixed orifice tube (FOT) and is the device that controls the refrigerant flow into the evaporator and as such separates the high side from the low side of the system. This information is also important when trying to locate where the manufacturer located both the high side and the low side service ports when trying to attach a manifold and gauge set.

A BIT OF HISTORY

Recovery of R-134a became mandatory on November 15, 1995, and the recycling of R-134a became mandatory on January 29, 1998.

FIGURE 6-3 **A typical air conditioning compressor.**

Compressor

There are three basic types of compressors: reciprocating, rotary, and scroll.

Reciprocating motion is to move to and fro, fore and aft, or up and down.

Electromagnetic is a temporary magnet created by passing electrical current through a coil of wire. A clutch coil is a good example of an electromagnet.

The compressor (Figure 6-3) in an air-conditioning system is a pump especially designed to raise the pressure of the refrigerant and circulate it through the system. According to the laws of physics, when the pressure of a gas or vapor is increased, its temperature is also increased. When pressure and temperature are increased, refrigerant condenses more rapidly in the next component, the condenser.

With some exceptions, automotive air-conditioning compressors are of the same design, **reciprocating** piston. This means that the pistons move in a linear motion, back and forth or up and down. The only exceptions are the rotary vane (RV) and scroll compressors, which are both increasing in popularity.

The automotive air-conditioning system uses a fixed- or variable-displacement compressor to move the refrigerant and to compress low-pressure, low-temperature refrigerant vapor from the evaporator into a high-pressure, high-temperature vapor to the condenser.

A label is generally found on the compressor to identify the type of refrigerant for which it is designed. This is a requirement to comply with the rules of the Environmental Protection Agency (EPA).

Compressors are belt driven from the engine crankshaft through an **electromagnetic clutch** pulley (Figure 6-4). When not energized, the compressor clutch pulley rotates freely without turning the compressor shaft. When voltage is applied, the electromagnetic clutch coil is energized, and the pulley engages with a clutch plate, often referred to as an armature, mounted on the compressor shaft. The magnetic field locks the clutch plate and pulley together as one unit to drive the compressor shaft.

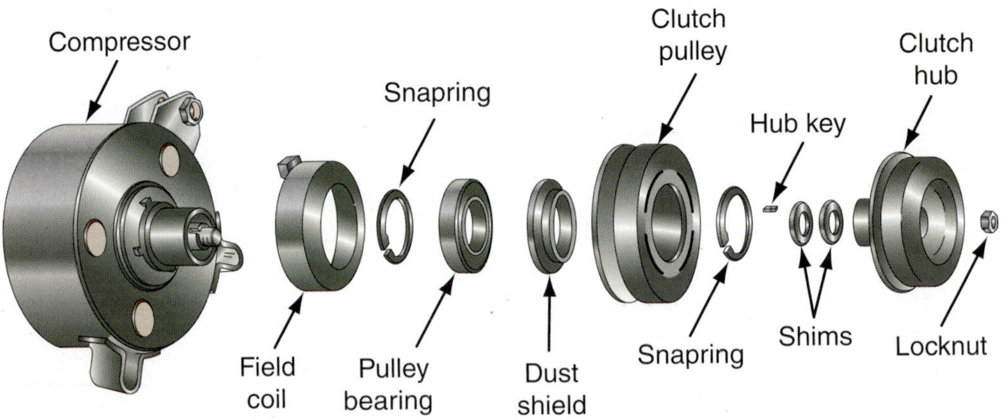

FIGURE 6-4 **A typical compressor showing clutch details.**

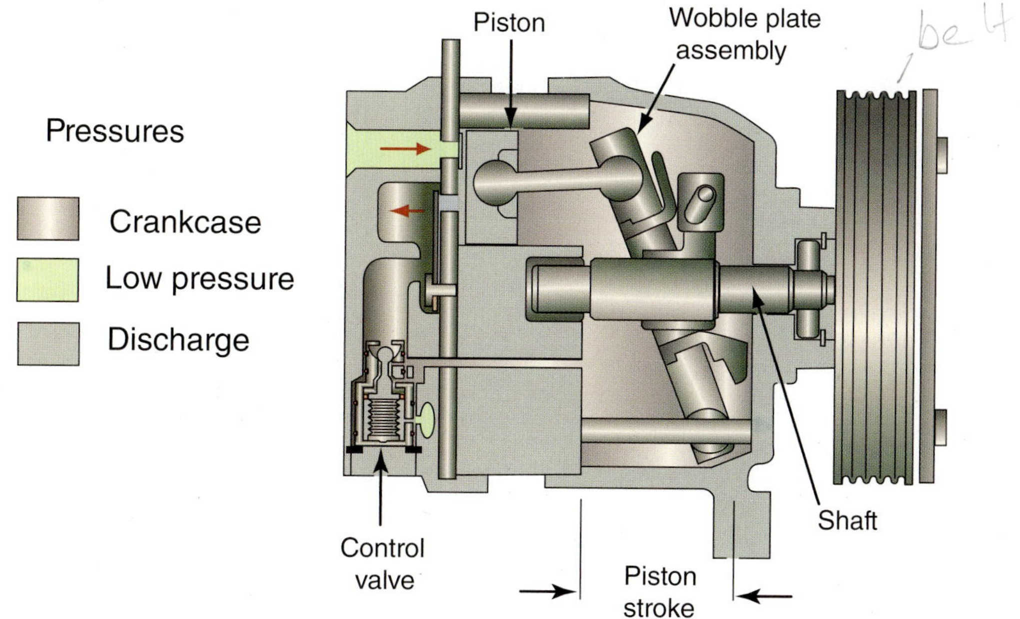

FIGURE 6-5 **Cutaway of a typical swash plate compressor.**

Some compressors cannot be repaired and, if defective or damaged, must be replaced. Many rebuilt replacement compressors contain lubricant that must be drained and replaced with the proper type and amount. Most new compressors, on the other hand, are supplied without lubricant. Fill the rebuilt or new compressor with the same amount and type lubricant as removed from the defective compressor or as recommended in the manufacturer's service manual.

Piston motion is caused by action of a crankshaft or a swash plate, often referred to as a wobble plate. Some swash plate compressors have double-ended pistons, such as General Motors DA-6, whereas others have single-ended pistons, such as Sanden's SD-5 (Figure 6-5). A more detailed explanation of compressors is given in Chapter 9 of this Manual, with trouble-shooting and repair procedures given in Chapter 9 of the Shop Manual. The following brief description is of the operation of the pistons and valves of a typical reciprocating compressor.

The intake stroke of the compressor is also called the suction stroke.

The compression stroke of the compressor is also called the discharge stroke.

Operation

Each piston in a compressor has a set of **reed valves** and valve plates—one suction and one discharge valve. Assume a simple two-cylinder compressor for the following description of operation (Figure 6-6). While one piston is on the suction (intake) **stroke**, the other piston is on the discharge (**compression**) stroke. The piston draws refrigerant into the compressor through the suction valve and forces it out of the compressor through the discharge valve. When the piston is on the suction (or intake) stroke, the discharge valve is held closed by the higher pressure above it. At the same time, the suction valve is opened to allow low-pressure refrigerant vapor to enter. When the piston is on the compression (or discharge) stroke, refrigerant vapor is forced through the discharge valve; the suction valve is held closed by this same pressure.

Reed valves are the leaves of steel located on the valve plate of a compressor that allow refrigerant to enter or leave the compressor.

Stroke is the distance a piston travels from its lowest point to its highest point.

Compression is the act of reducing volume by pressure.

The compressor separates the low side from the high side of the system (Figure 6-7). Refrigerant entering the compressor is a low-pressure, slightly superheated, vapor. When the refrigerant leaves the compressor, it is a high-pressure, highly superheated, vapor.

The compressor in some air-conditioning systems has service valves that are used to access the air-conditioning system. The manifold and gauge set is connected into the system at the service valve ports. All service procedures—such as recovering, evacuating, and charging the system—are performed with the use of a manifold and gauge set and the proper recovery and charging equipment.

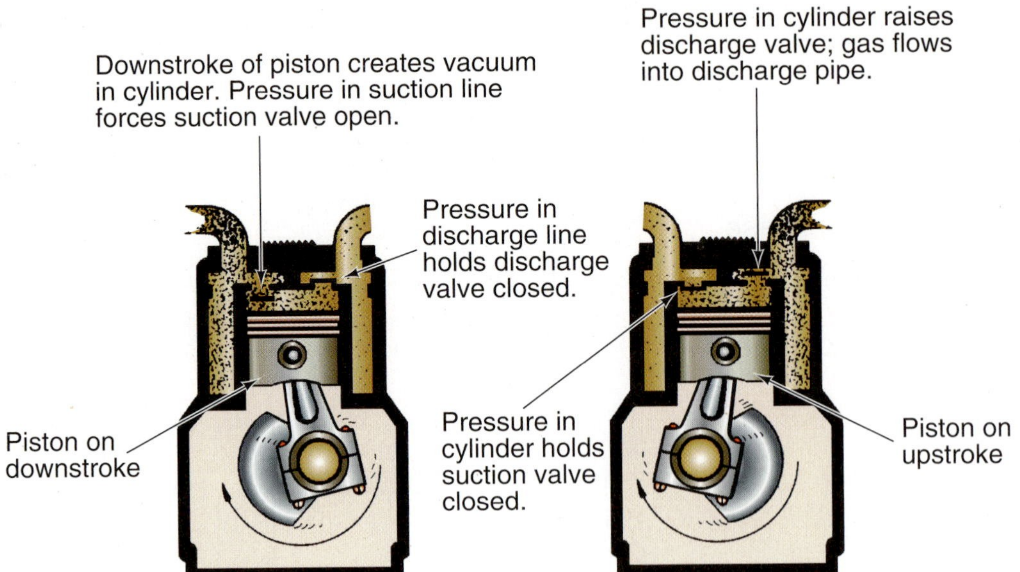

FIGURE 6-6 **Typical operation of a single-cylinder compressor.**

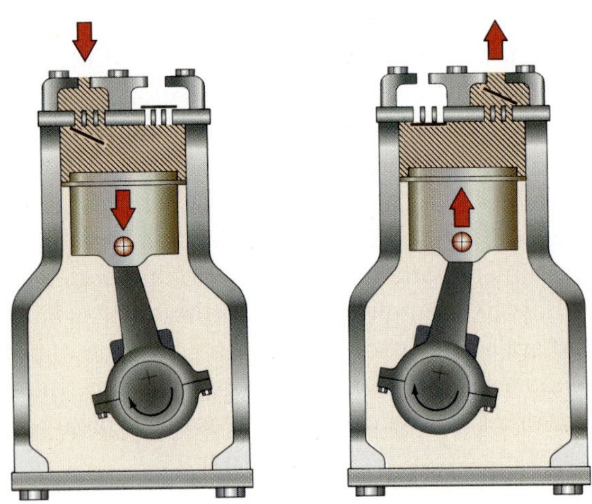

FIGURE 6-7 **Refrigerant entering the compressor is low-pressure vapor and is high-pressure vapor when leaving.**

Refrigeration lubricant, most often referred to as oil, is stored in the compressor and is essential in keeping the internal parts of the compressor lubricated. A small amount of this lubricant circulates with the refrigerant through the system. The velocity of the refrigerant through the tubes and hoses, however, allows this lubricant to return to the compressor. Chapter 9 will provide greater detail on the various oils used in refrigerant systems.

Hoses And Lines

Refrigerant fluid and vapor lines may be made of aluminum or steel. Hoses are usually made of synthetic rubber covered with nylon braid for strength and have an inner lining of nylon to ensure integrity and to form a barrier wall to prevent refrigerant leakage. This hose design is classified as a barrier hose and is found on all R-134a refrigerant systems. Older R-12 systems used hoses with inner liners typically made of Buna "N," a synthetic rubber. Buna "N," which is not affected by R-12, is not acceptable for R-134a systems. Barrier hoses with a nylon lining are compatible with both R-134a and R12 systems.

There are several ways to identify whether a hose or pipe is on the high-pressure side or low-pressure side of the system:

- The low-side pressure lines are cold to the touch when the system is operating normally and are larger in diameter than those on the high-pressure side.
- The high-side pressure lines are hot to the touch when the system is operating normally and are smaller in diameter than those on the low-pressure side.

Special consideration must be given for hoses and other components used in an R-134a air-conditioning system. Many materials that were compatible for an R-12 system, such as nitrile or epichlorohydrin, cannot be used for R-134a service. For example, O-rings used with fittings in an R-12 system, such as nitrile, and those used in an R-134a system, such as neoprene, are not interchangeable.

Most early R-12 (CFC-12) hoses are not compatible with R-134a (HFC-134a) refrigerant.

Standard hose sizes are given a number designation, such as #6, #8, #10, and #12. Size #6 is usually used for the liquid line, #8 or #10 as the hot gas discharge line, and #10 or #12 as the suction line. Figure 6-8 gives the inside diameter (ID) and outside diameter (OD) of two types of hoses used in automotive air-conditioning service.

Most early R-12 hoses were not barrier hoses, which have a nylon liner inside the hose designed to stop the leakage of the smaller particles of R-134a refrigerants. Though not of a barrier design, the old hoses used in R-12 systems are oil soaked on the inside with the mineral oil lubricant used in these systems. The idea is that R-134a is not compatible with mineral oil and will not go through it. Although true in most cases, the constant refrigerant pressure eventually opens a path and allows the refrigerant to escape to the atmosphere. Most original equipment manufacturer (OEM) retrofit procedures, as far back as 1984, do not require replacing hoses. Many feel, however, that if a retrofit is to be done properly, all nonbarrier hoses should be replaced with barrier hoses (Figure 6-9). The EPA does not require replacement of hoses or seals during the retrofit of a vehicle from R-12 to R-134a. Always refer to manufacturer recommendations. Figure 6-10 is a comparison table of R-134a and R-12 refrigerants. As the table indicates, barrier hoses are an option for the R-12 system. The best practice is to always follow the manufacturer's procedures and recommendations when retrofitting an air-conditioning system.

Shop Manual
Chapter 6, page 188

When refrigerant changes from vapor to liquid, it gives up heat.

When refrigerant changes from liquid to vapor, it takes on heat.

Do not mix refrigerants.

The hose type is generally distinguished by the crimp style used on the fitting (Figure 6-11). Either the finger style crimp, used with a nonbarrier hose having a barb fitting, or the bubble style beadlock crimp, used specifically with a nylon barrier hose, will generally be found. Also, fittings with worm gear hose clamps are used with nonbarrier-type hoses. A worm gear hose clamp should never be used with nylon-barrier hose fittings.

Depending on design and application, several types of fittings are used to connect the hoses to the various components of an air-conditioning system. The line connections on

Hose Size	Inside Diameter (ID)		Outside Diameter (OD)			
			Rubber Hose		Nylon Hose	
	English	Metric	English	Metric	English	Metric
#6	5/16 in.	7.94 mm	3/4 in.	19.05 mm	15/32 in.[1]	11.9 mm[3]
#8	13/32 in.	10.32 mm	59/64 in.[1]	23.42 mm[3]	35/64 in.[1]	13.89 mm[3]
#10	1/2 in.	12.7 mm	1-1/32 in.[2]	25.8 mm[4]	11/16 in.[1]	17.46 mm[3]
#12	5/8 in.	15.87 mm	1-5/32 in.[2]	29.37 mm[4]	NA	NA

[1] ± 1/64 inch
[2] ± 1/32 inch
[3] ± 0.4 mm
[4] ± 0.8 mm

FIGURE 6-8 **Inside and outside diameter of hoses used for air conditioning service.**

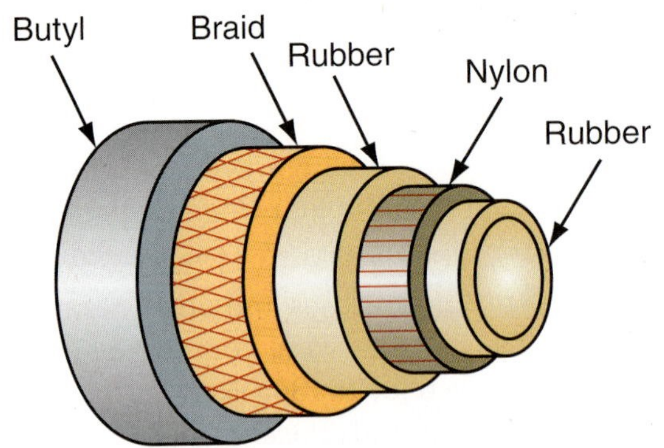

FIGURE 6-9 Barrier hose details.

	R-134a System	R-12 System
Chemical Name	Tetrafluoroethane	Dichlorodifluoromethane
Refrigerant container identification	Labeled R-134a sky-blue container	Labeled R-12 white container
Refrigerant container fitting	½" x 16 ACME	7/16" x 20, ¼" Flare
Boiling point at sea level	–15.07°F (–26.15°C)	–21.62°F (–29.79°C)
Desiccant	XH7, XH9	XH5, XH7, XH9
Hose construction	Barrier nylon liner required	Barrier nylon liner optional
Valve core	M6 thread, O-ring seal	TV thread, Teflon seal
Refrigerant oil	Polyalkylene glycol (PAG) or polyol ester (Ester)	Mineral based
Refrigerant oil hygroscopicity	2.3%–5.6% by weight	0.005% by weight
Condenser	Improved heat transfer design	Standard

FIGURE 6-10 R-134a/R-12 comparison table.

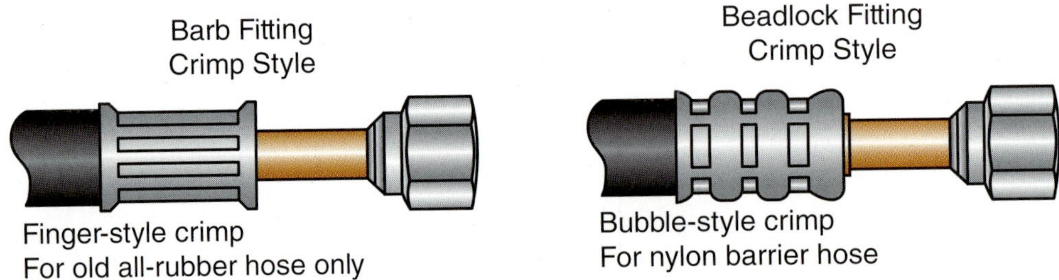

FIGURE 6-11 Hose crimp styles.

anR134a system have moved the O-ring to the midpoint (captured) of the connection from the base of the connection that was used on R-12 systems (Figure 6-12). There are many types and styles of line connections, including block joint fittings (Figure 6-13) and spring lock fittings (Figure 6-14).

Specially designed O-rings are used in conjunction with line connections to provide a seal between joints. If an air-conditioning system component or line is removed for service, the O-ring seals must be replaced (Figure 6-15). A small amount of clean refrigerant oil should be applied to the new O-ring (Figure 6-16) before installation.

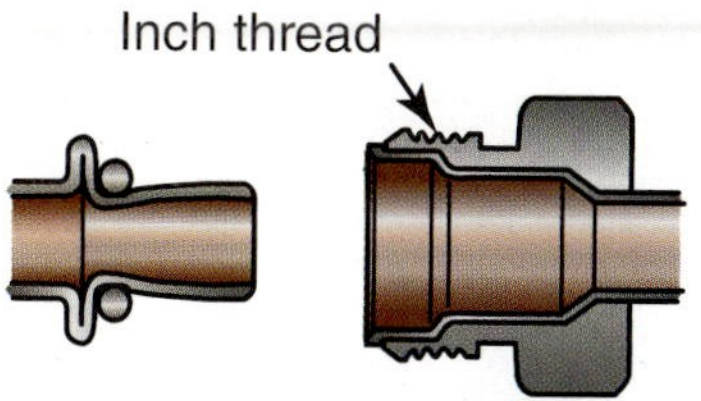

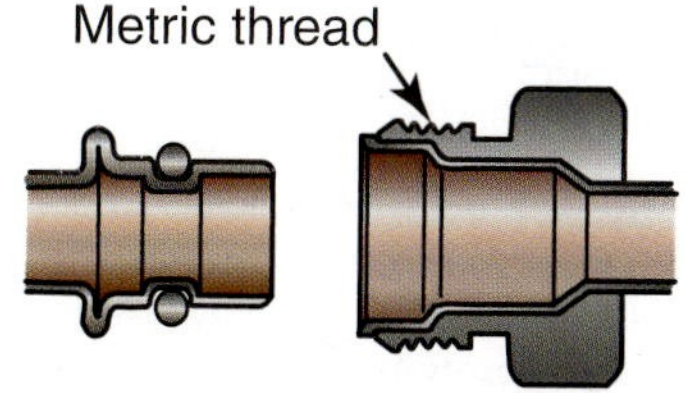

FIGURE 6-12 **Threaded nut-type pipe connections.**

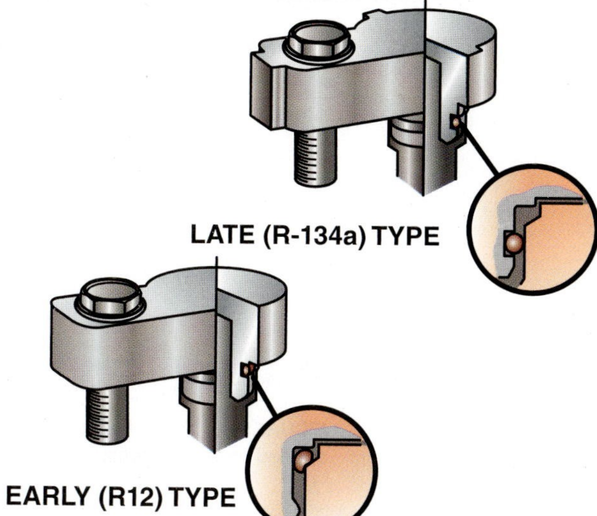

FIGURE 6-13 **Block fitting-type pipe connections.**

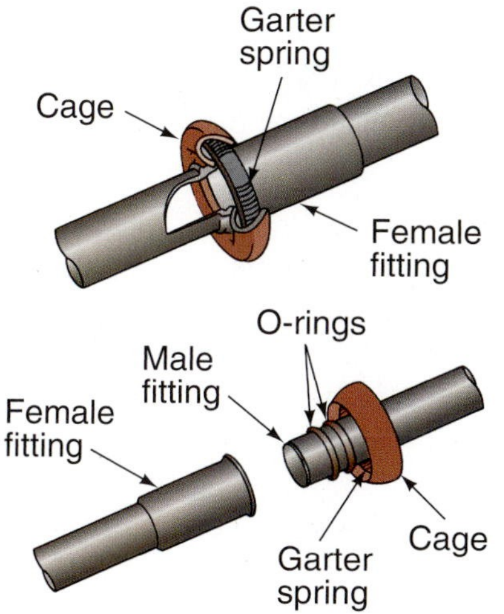

FIGURE 6-14 **Spring lock fitting used on some refrigerant system lines.**

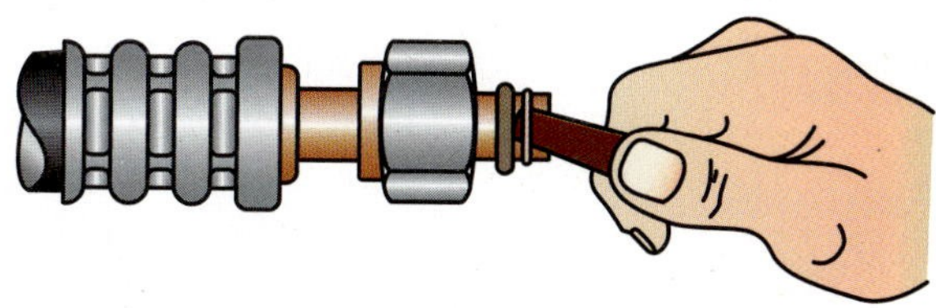

FIGURE 6-15 **New O-rings should always be installed once they are disturbed.**

AUTHOR'S NOTE: Only clean refrigerant oil should be used to lubricate refrigerant seals or connections. Never use petroleum-based oils, grease, or silicone as a lubricant.

Some air-conditioning systems require a muffler on the compressor discharge line to reduce noise and vibration caused by compressor pulses. The muffler contains several chambers or baffles that redirect refrigerant flow, generating a cancelling frequency to reduce noise (Figure 6-17).

DISCHARGE LINE

The hose leaving the compressor contains high-pressure refrigerant vapor. This hose, which is made of synthetic rubber, generally has a nylon **barrier** lining. It typically has a $^{13}/_{32}$ in (10.3 mm) inside diameter and often has extended preformed metal (steel or aluminum) ends

Barrier is a term given to something that stands in the way, separates, keeps apart, or restricts—an obstruction. Barrier hoses are used on R134a systems due to their smaller molecular size than R12 refrigerant.

FIGURE 6-16 **A small amount of oil should be applied to new O-rings.**

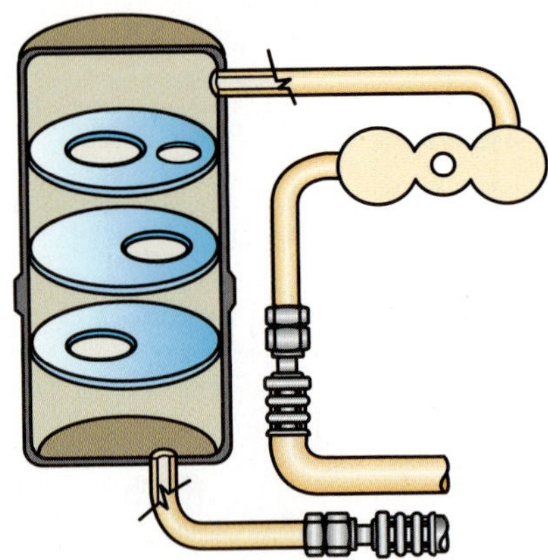

FIGURE 6-17 **Refrigerant system high-side discharge muffler used to dampen noised and vibration generated by compressor pump pulsations.**

with fittings. It is known as the hot gas discharge line and connects the outlet of the compressor to the inlet of the condenser.

Under normal operating conditions, this line is very warm. During certain system malfunctions, however, it is very hot. Because of the refrigerant temperature and pressure in this line, it is generally the most susceptible to leaks.

Ideally, the refrigerant will be all liquid when it leaves the condenser.

Condenser

The condenser, which is located in front of the engine cooling radiator, is a heat exchanger made up of cooling fins and tubes that carry refrigerant. The condenser provides a rapid transfer of heat from the refrigerant passing through the tubes to the air passing through the fins and across the tubes. Part of a preventive maintenance service is to ensure that the condenser is clean and free of all debris. If found to be bent, a fin comb may be used to straighten the condenser fins.

Heat-laden refrigerant in the vapor state liquifies or condenses in the condenser. To do so, the refrigerant must give up its heat. As cooler air passing over the condenser carries its heat away, the vapor condenses. Heat that is removed from the refrigerant in the condenser as it changes from a vapor to a liquid, is the same heat that was absorbed in the evaporator as it changed from a liquid to a vapor.

The refrigerant from the compressor is almost 100 percent vapor as it enters the condenser. On certain occasions, a very small amount of vapor may condense in the hot gas discharge line. The amount is so small, however, that it is not considered in the overall operation of the system.

The refrigerant is not always 100 percent liquid when it leaves the condenser, however. Only a certain amount of heat can be dissipated by the condenser at any given time. A small percentage of refrigerant, then, may leave the condenser in the vapor state. This condition does not affect overall system performance because the next component is a long liquid line or a receiver-drier.

The refrigerant in the condenser is a combination of liquid and vapor under high pressure. To avoid personal injury, extreme care must be exercised when servicing the condenser.

The inlet of the condenser must be at the top so the refrigerant vapor, as it condenses, will collect at the outlet at the bottom of the condenser. To ensure that all of the refrigerant vapor has condensed to a liquid when leaving the condenser, some systems are equipped with a small, second (auxiliary) condenser. This auxiliary condenser, called a subcondenser, provides the additional heat transfer surface required in some air-conditioning systems for the refrigerant to condense to a liquid.

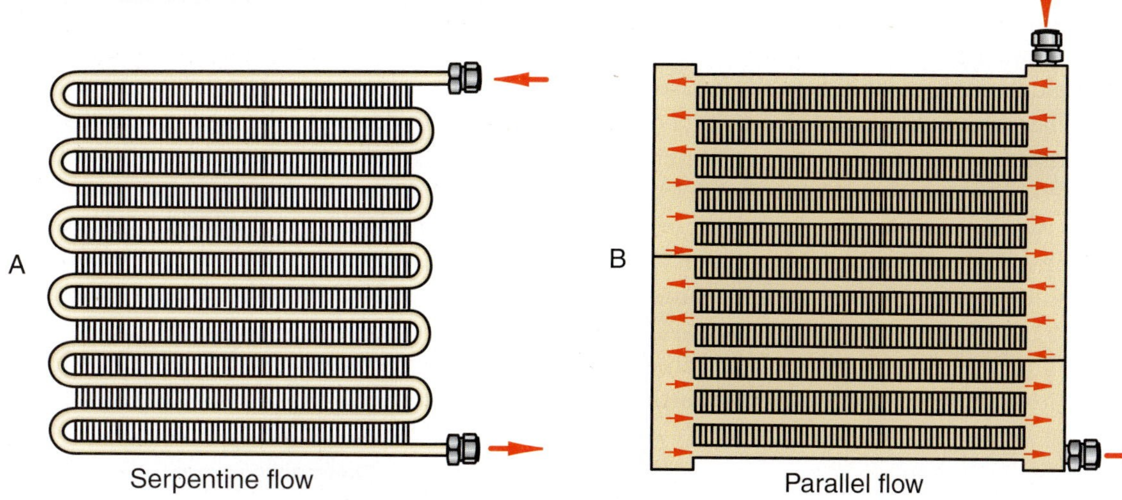

FIGURE 6-18 **Two common flow paths through air conditioning condensers are (A) serpentine flow, and (B) parallel flow.**

The outlet of a condenser in a TXV system is connected to the receiver-drier, then to the metering device, by the liquid line. In a FOT system, the outlet of the condenser is connected by a liquid line to the metering device.

The condenser for an R-134a system has a larger capacity and is designed for increased heat transfer compared to the standard R-12 system. This is necessary due to the differences in heat transfer characteristic of R-134a. The volume of gas entering the condenser is about 1,000 times the volume of the condensed liquid leaving the condenser. The efficiency of the condenser affects the overall performance of the refrigerant system.

There are several condenser designs and flow paths in use today. The two flow paths for refrigerant through the condenser's tubing are either a serpentine path flowing back and forth on older designs or a multipass flat tube parallel cross flow (Figure 6-18). The condenser may be a tube and fin with older designs using 3/8 in. tubing and newer designs using a 6 mm tube. They have similar heat transfer characteristics of the serpentine design and 15 percent better heat transfer than older 3/8 in. designs. Condenser tubes may also be extruded aluminum tubing with honeycomb serpentine passages for increased surface area and airflow for improved heat transfer (Figure 6-19). This design is physically smaller for the same level of heat transfer, making them popular for compact car designs. All aluminum parallel flow condensers with

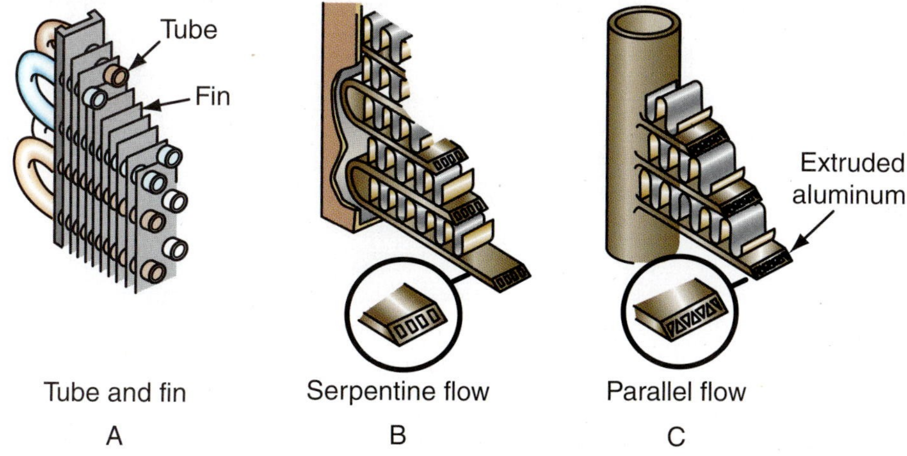

FIGURE 6-19 **Condensers may be (A) tube and fin, extruder aluminum; (B) serpentine; or (C) parallel flow.**

extruded aluminum tubing with honeycomb are currently popular with original equipment manufacturers (OEM) where limited space and airflow is a concern. The multipass extruded aluminum designs are virtually impossible to flush after a catastrophic compressor failure and should be replaced if a restriction is suspected.

The subcooling or supercooling condenser is similar in design to the multipass extruded aluminum designs but have even smaller passages and a modulator assembly with an integrated receiver dryer (Figure 6-20). On the modulated flow design the condenser is divided into two sections for condensing and supercooling. The modulator separates the refrigerant in the middle of the cycle after the first pass that is still in the gaseous state and recirculates it with liquid refrigerant to cool it again, which enables almost 100% liquid refrigerant too pass on to the metering device (Figure 6-21).

The modulator performs the same function as the conventional receiver-drier with the major difference being the design of the condenser (Figure 6-22). After the refrigerant's first

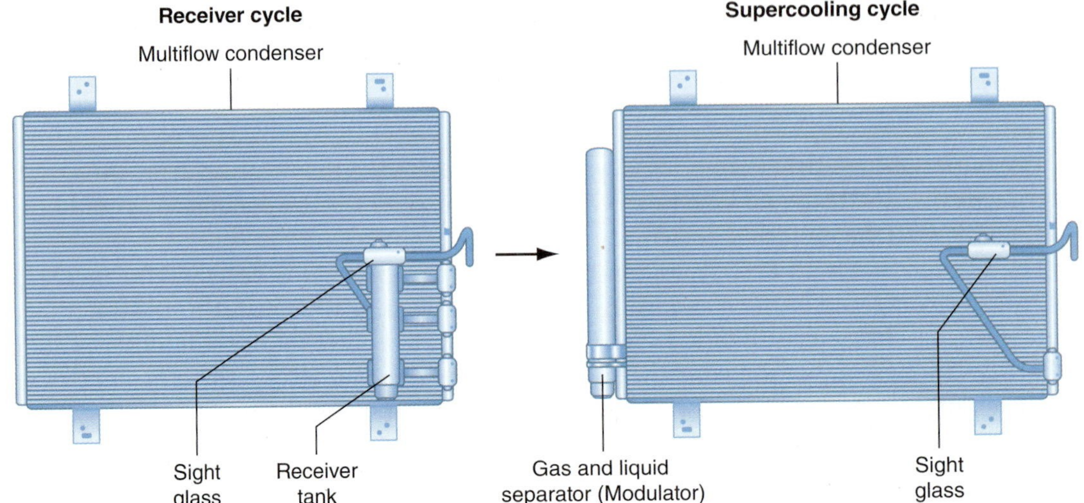

FIGURE 6-20 **On a super- or sub-cooling condenser there is a modulator assembly that contains the desiccant so there is no need for a separate receiver dryer that is required on a conventional system.**

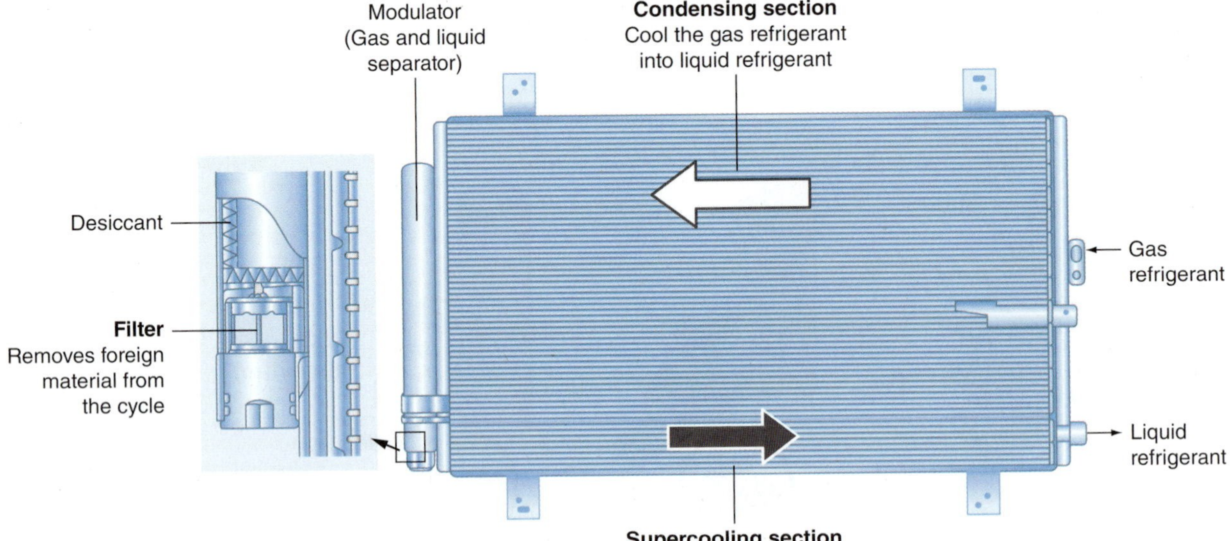

FIGURE 6-21 **On the modulated flow design the liquid refrigerant and any gas still remaining is cooled again after passing through the modulator which enables almost 100% liquid refrigerant too pass on to the metering device.**

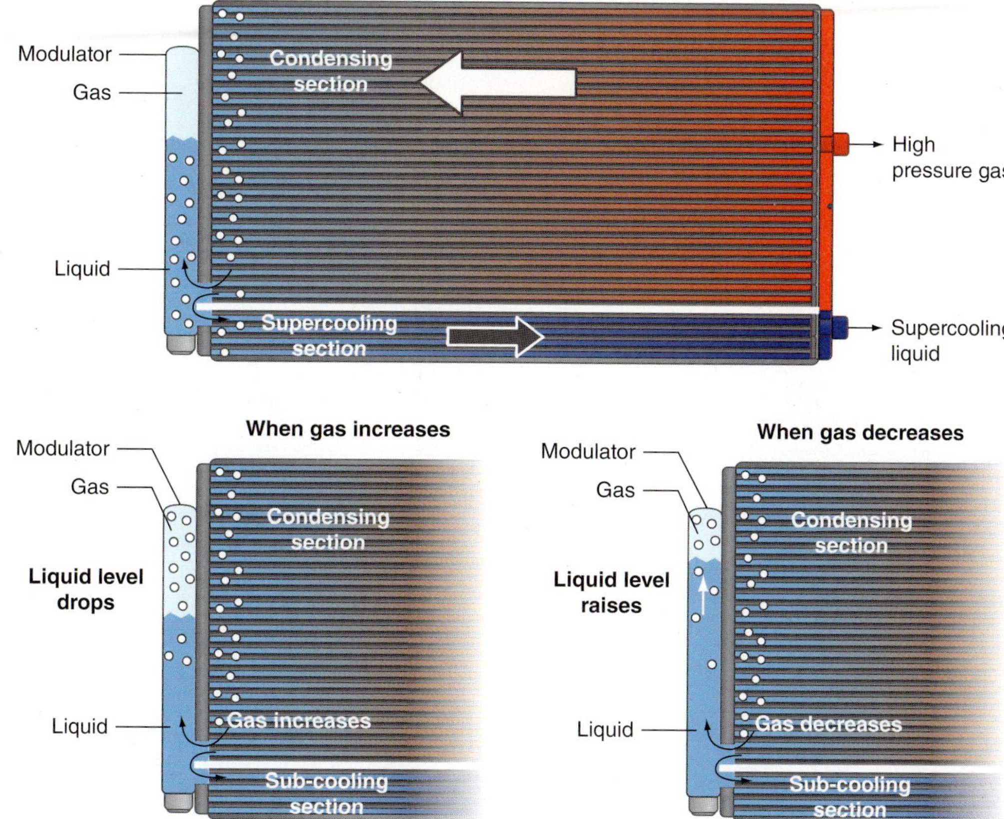

FIGURE 6-22 **On the modulated flow design the liquid refrigerant and any gas still remaining is cooled again after passing through the modulator which enables almost 100% liquid refrigerant too pass on to the metering device.**

pass through the condenser, only liquid refrigerant is allowed to pass through the lower supercooling section of the condenser. The liquid level in the modulator is maintained at a balancing point with liquid and gaseous particles. If the amount of incoming gas increases under high-load operating conditions, the liquid level is pushed downward. This process allows additional liquid refrigerant to be supplied to the metering device. When system load and demand is low, the liquid level inside the modulator will increase as incoming gas flow decreases and the gaseous refrigerant condenses. The excess liquid is stored in the modulator just as it would be in a receiver-drier until system heat load and demand increases. There may be a sight glass on the liquid line leaving the condenser, but bubbles in the sight glass are only an indicator of low refrigerant level and should not be used as a charge indicator. The bubble-free point in a super cooling condenser is 50–100 grams of refrigerant less than the optimal charge level (Figure 6-23). Proper charge level is approximately 100–150 grams of refrigerant above the point at which bubbles in the sight glass disappear. Cooling efficiency will be insufficient if gas charging is stopped at the bubble-free point and will leave the system undercharged. Overcharging will also cause insufficient cooling. It is important to follow manufacturer charging-level recommendations and not charge any system based on sight glass bubbles.

Some modulators have a serviceable desiccant bag while others integrate a nonserviceable desiccant. If a condenser with a serviceable desiccant is to be flushed the removable service plug and desiccant must first be removed prior to beginning the flushing procedure. Once the flushing is complete, a new desiccant bag should be installed. Those condensers that integrate a nonserviceable receiver-drier cannot be flushed and must be replaced if a failure in either is suspected. If a catastrophic compressor failure occurs, these condensers must be replaced as an assembly when the new compressor is installed since debris and contaminated oil will be trapped in the condenser and desiccant.

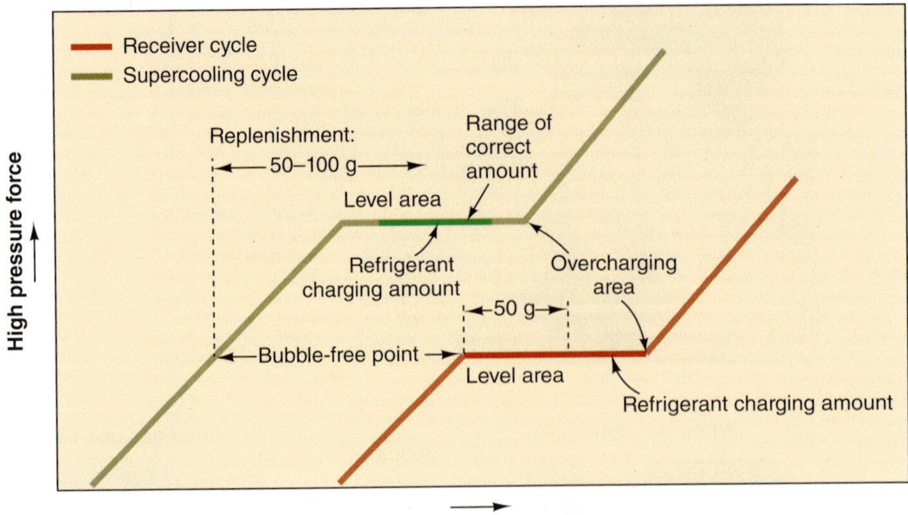

FIGURE 6-23 **On subcooling condenser designs the proper refrigerant charge level is 100–150 grams more than on a traditional receiver drier designed system.**

Condenser problems usually result from external clogging, damage, or from leaks. External clogging is caused by dirt, bugs, leaves, or other foreign debris that collects onthe condenser fins and restricts airflow. This lessens the condenser's ability to transfer heat, resulting in poor cooling of the car interior.

The condenser and the radiator must be kept clean for best performance. To clean the condenser, use a soft bristle brush (such as a hair brush) and a strong stream of water. Take care not to bend the fins, which would also restrict the flow of air.

Clean the radiator/condenser assembly with clean water directed from the back (engine side) to the front. If air is used, use only low-pressure air to prevent damage to the delicate and fragile fins of the radiator. Do not use a steam cleaner to remove debris from the condenser. To do so may cause an increase in air-conditioning system pressure.

Because temperature and pressure are high in the condenser, a leak is not always successfully repaired. It is generally recommended that the condenser be repaired by a professional or be replaced if it is found to be leaking.

Receiver-Drier

Shop Manual
Chapter 6, page 204

The receiver-drier is often called a "receiver" or "drier."

The receiver-drier, often referred to simply as a drier, is located in the high-pressure side of the air-conditioning system between the condenser outlet and the metering device inlet. Construction of the drier is such that refrigerant vapor and liquid are separated to ensure that 100 percent liquid is available at the metering device, the TXV.

The receiver-drier (Figure 6-24) is used in systems that have a TXV as a metering device. The receiver-drier has five basic functions: it acts as a reservoir to store liquid refrigerant from the condenser and ensures a vapor-free liquid column to the TXV under a wide range of heat loads. The liquid refrigerant will sink to the bottom while any vapor present will rise to the top of the container. The receiver absorbs moisture and acts as a filter to remove solid contamination. In addition, the receiver-drier dampens compressor pulses, ensuring a steady flow of refrigerant.

The receiver-drier is a cylindrical container with both the inlet and outlet tubes at the top of the assembly (Figure 6-25). Some models have a sight glass at the top, but this feature has been phased out. The outlet pickup tube extends almost to the bottom of the container to ensure that only liquid refrigerant will flow out of the receiver-drier.

FIGURE 6-24 **Typical receiver-drier assembly.**

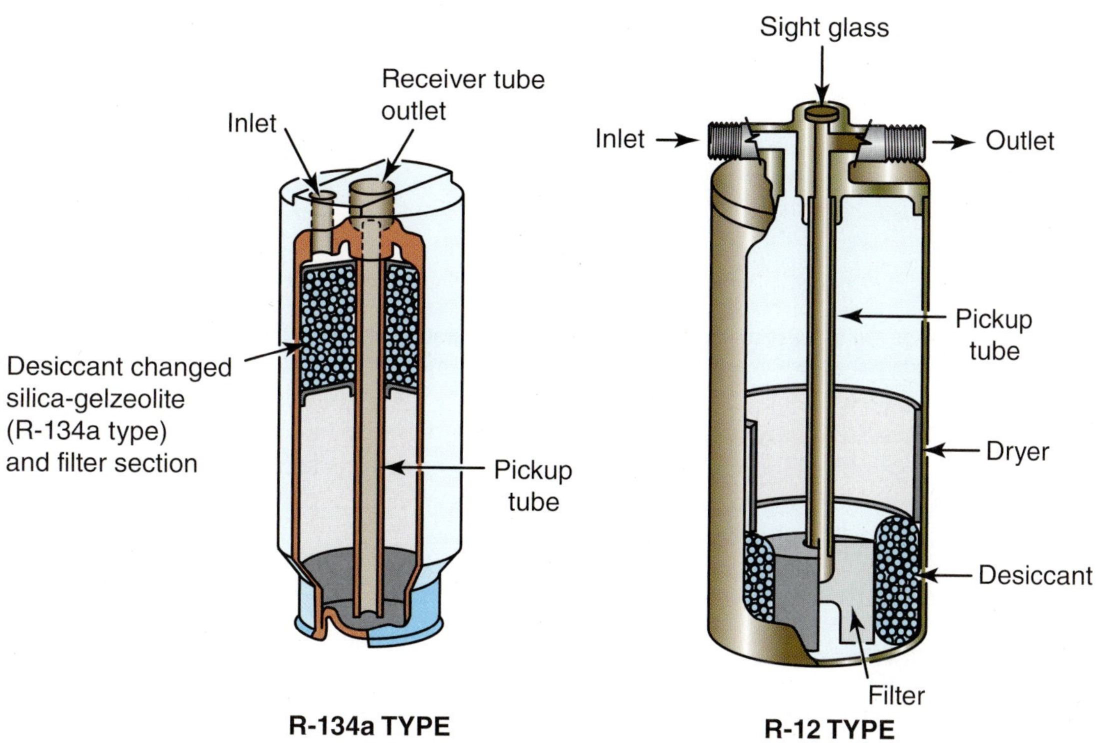

FIGURE 6-25 **Both R-134a and older R-12 receiver-driers are similar in design and appearance.**

The amount of refrigerant demanded by the evaporator varies depending on heat load and compressor output. When the heat load is low, such as on cooler days, the thermostatic expansion valve is only partially open, resulting in low refrigerant flow to the evaporator, and the level of liquid refrigerant in the receiver-drier will be high. But on days when the heat load is high, such as on hot humid midsummer days, the system demand for refrigerant is high because the thermostatic expansion valve is open, resulting in higher refrigerant flow to the evaporator, and the level of liquid refrigerant in the receiver-drier will be low. The receiver-drier acts as an accumulator of refrigerant. It must be able to store enough liquid refrigerant for both cool (low-heat load), and hot, humid (high-heat load) days. In addition, the receiver-drier allows for the expansion and contraction of refrigerant when the system is not in operation but under-hood temperatures and varying ambient air temperatures are high.

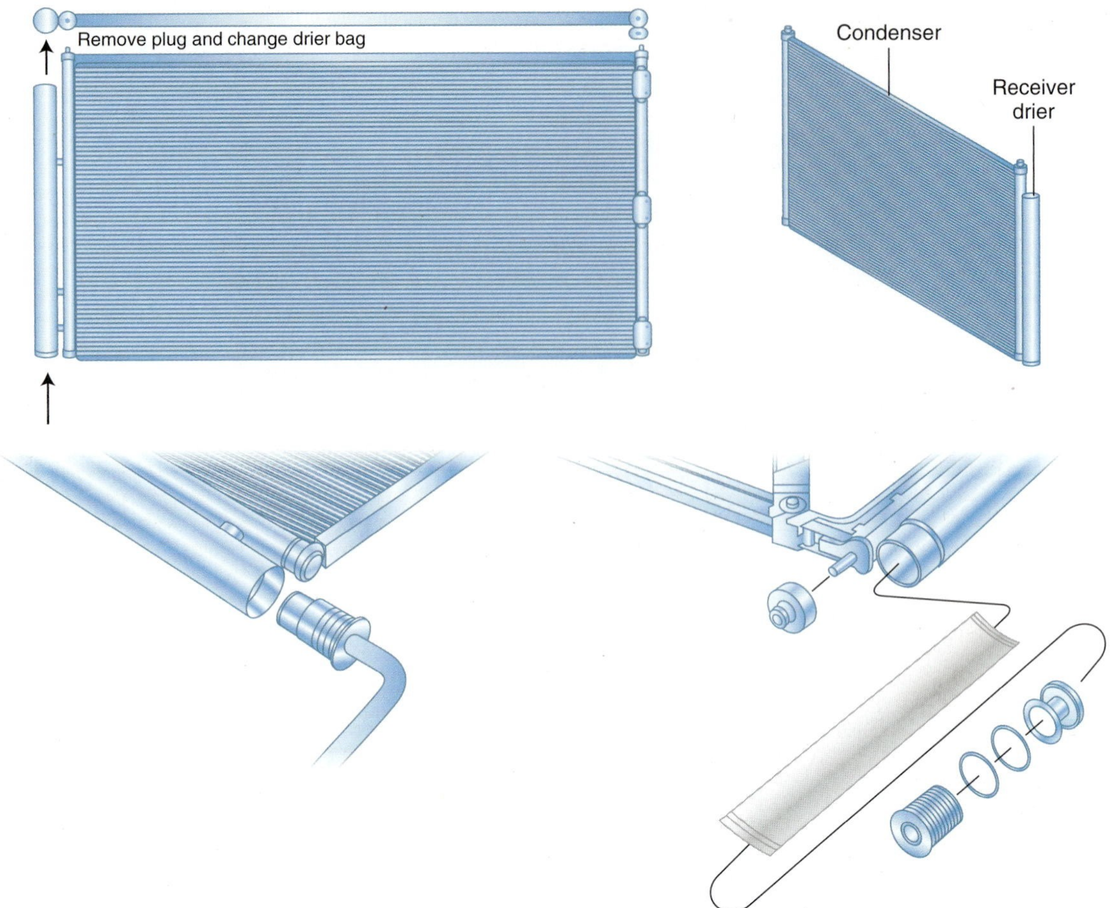

FIGURE 6-26 **On some systems the receiver drier is integrated into the outlet liquid tank of the condenser assembly and may also contain a serviceable desiccant bag.**

On some smaller systems today, the receiver-drier is integrated into the liquid line outlet of the condenser assembly (Figure 6-26). When the receiver-drier is located in the liquid outlet tank of the condenser, the desiccant bag may also be serviceable separately. If it is serviceable, there will be a removable plug on either end of the drier.

Receiver Section

The receiver section of the receiver-drier is a tank-like storage compartment. This section holds the proper amount of reserve refrigerant required to ensure proper performance of the air-conditioning system under variable operating conditions. The receiver also ensures that a steady flow of liquid refrigerant can be supplied to the thermostatic expansion valve.

Drier Section

Desiccant is a drying agent used to remove excess moisture in refrigeration systems. The desiccant is located in a bag in the receiver-drier or accumulator.

The drier section of the receiver-drier is generally nothing more than a fabric bag filled with **desiccant**, which is a chemical drying agent that can absorb and hold a small quantity of moisture to prevent it from circulating through the system. The desiccant used in an R-12 receiver-drier may not be compatible with R-134a refrigerant. The desiccants are classified XH5 (silica-gel), XH7 (molecular sieve), and XH9 (zeolite). Only XH7 and XH9 are acceptable for use on R-134a systems. Therefore, to ensure that the desiccant will be compatible with the refrigerant and refrigeration lubricant in the system, use only replacement components that are designated for a particular application. The receiver-drier should be changed any time a major component of the air-conditioning system has been replaced. Most vendors will not honor warranty claims on a new or rebuilt replacement compressor if the receiver-drier is not replaced at the time the system is serviced.

Screen/Strainer

A screen or strainer is included inside the receiver-drier. This is intended to prevent the circulation of any debris throughout the system that may have entered during careless service procedures. This screen cannot be serviced as a component; the receiver-drier must be replaced as an assembly if the screen is found to be restricted.

A condenser-mounted receiver-drier should be mounted as level as possible, usually by adjusting its bracket or the condenser mounting brackets. The vertical-type drier should be mounted in a position as vertical as possible, with no more than a 15-degree slant off vertical.

Receiver-driers are available in a variety of sizes and have different fittings for different applications. Universal-type receiver-driers are available and may be used if exact replacement units are not available. Driers are also available without the receiver. This type drier is usually used in series with a receiver-drier in a "problem" air-conditioning system when a great deal of debris is encountered. It should be noted that receiver-driers and driers are not omnidirectional. They are designed for refrigerant to flow in one direction only. Most driers are marked IN and OUT or have an arrow (→) to indicate direction of refrigerant flow. Remember, the refrigerant flow is away from the condenser and toward the evaporator.

Liquid Line

High-pressure liquid refrigerant moves from the condenser or receiver-drier through a hose or tube called the liquid line to the evaporator metering device. The liquid line, which is usually made of aluminum, is generally ¼ in. to $\frac{5}{16}$ in. (6.3–7.9 mm) inside diameter. In some installations, such as a dual evaporator system, the inside diameter of the liquid line may be as large as ⅜ in. (9.5 mm). This line is sized so the refrigerant flow is not restricted yet maintains the constant pressure that is required to ensure proper metering of the refrigerant into the evaporator.

The liquid line may also be made of copper, steel, or a combination of rubber or nylon and copper, steel, or aluminum. As its name implies, the state of refrigerant in the liquid line is liquid under high pressure.

The liquid line is usually identified as the smallest line in the system.

The two types of TXV are internal equalized and external equalized. They are not interchangeable.

Thermostatic Expansion Valve

The TXV (Figure 6-27), located at the inlet side of the evaporator, is the metering device for the system. The TXV separates the high side of the system from the low side of the system. A small variable **orifice** in the valve allows only a small amount of liquid refrigerant to enter the evaporator. The amount of refrigerant passing through the valve is governed by the

Orifice is a small hole of calibrated dimensions for metering fluid or gas in exact proportions.

FIGURE 6-27 **Typical thermostatic expansion values: (A) internal equalized and (B) external equalized.**

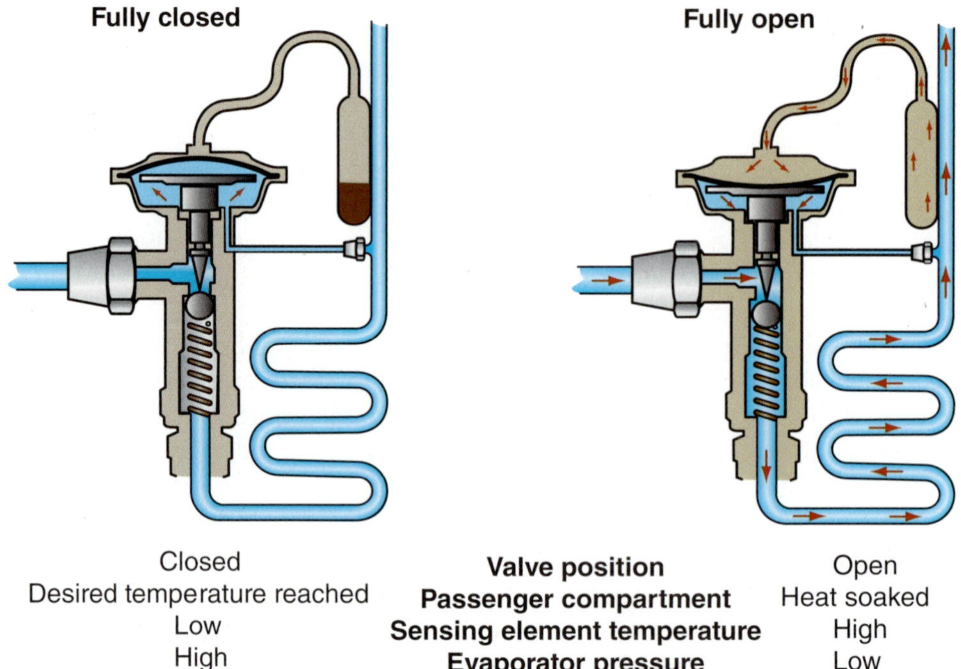

FIGURE 6-28 **Expansion valves provide a variable restriction to refrigerant flow.**

evaporator temperature. A tapered pin is raised or lowered in an orifice to change the size of the opening up to 0.008 in. (0.2 mm) diameter when the valve is wide open (Figure 6-28).

The thermostatic expansion valve is designed to provide two basic functions:

- Based on various heat loads and compressor output, the expansion valve meters the flow of refrigerant into the evaporator.
- Refrigerant pressure is reduced as high-pressure liquid entering the valve passes through a restriction and suddenly expands and vaporizes into a low-pressure mist.

Refrigerant, as it passes through the thermostatic expansion valve and immediately after it, is 100 percent liquid. A very small amount of liquid refrigerant, known as flash gas, vaporizes immediately after passing through the valve due to the severe pressure drop. All of the liquid refrigerant soon changes state; as the pressure drops, the liquid refrigerant begins to boil. All liquid should boil off before reaching the outlet of the evaporator. As it boils, it must absorb heat from the air passing over the coils and fins of the evaporator. The air, then, feels cool; heat is being removed from the air; cold air is not being created.

Shop Manual
Chapter 6, page 196

At the point of total evaporation, the refrigerant is said to be saturated. The saturated vapor continues to pick up heat in the evaporator and in the suction line until it reaches the compressor. The refrigerant is then said to be superheated.

There are three distinct physical types of thermostatic expansion valves. They are the external equalizing type, the internal equalizing type, and the box or H-Block internal equalizing type (Figure 6-29).

The thermostatic expansion valves used in automotive air-conditioning systems are designed for a specific use and are manufactured as precision components. They are calibrated to provide the correct amount of refrigerant and superheat in the evaporator. No attempt should be made by the inexperienced to disassemble, repair, or adjust the expansion valve. It is possible, however, to clean or replace the inlet screen (strainer) should it become clogged.

The thermostatic expansion valve acts as a variable metering orifice that regulates the flow of refrigerant by sensing the system's temperature and pressure. The expansion valve has a sensing element called a remote bulb attached to the evaporator outlet line (suction line) by a capillary tube (Figure 6-30). This bulb, which is attached to the evaporator tailpipe, senses outlet temperature.

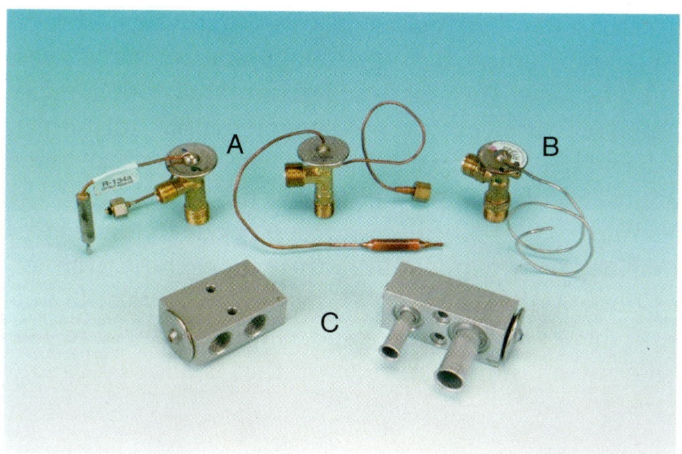

FIGURE 6-29 **Three distinct types of thermostatic expansion values, (A) the external equalizing type, (B) the internal equalizing type, and (C) the Box or H-Block internal equalizing type.**

FIGURE 6-30 **The (H-block) thermostatic expansion valve showing the remote bulb attached to the outlet of the evaporator.**

The expansion valve is regulated by the temperature-sensing bulb that is tightly clamped to the evaporator outlet tube. In operation, the sensing bulb senses the temperature of the refrigerant as it leaves the evaporator core. Expansion valves are either externally or internally pressure regulated (equalized) through an equalization tube to provide a variable restriction to refrigerant flow. The expansion valve allows the air-conditioning system to maintain peak performance regardless of the passenger compartment temperature change that affects the thermal load on the evaporator.

The amount of refrigerant allowed to flow is determined by the movement of the internal valve. The openings and closings of the valve are regulated by the difference between the following:

- The internal return spring pressure returning the valve to its seat (closed).
- The refrigerant pressure at the inlet of the evaporator on internally equalized expansion valves or the refrigerant pressure at the outlet of the evaporator on externally equalized expansion valves.
- The temperature-sensing bulb gas pressure. The temperature-sensing bulb pressure is relative to the evaporator outlet temperature.

If the refrigerant outlet line is warm due to high thermal load (high passenger-compartment or intake air temperature), the vapor temperature of the refrigerant leaving the evaporator outlet will also be high. The charge of refrigerant (or other volatile liquid) in the sensing bulb expands (high pressure), putting pressure on the valve diaphragm through a small capillary tube. The diaphragm then forces a needle valve off its seat, and the valve opens to allow more refrigerant to enter the evaporator (Figure 6-31).

As the evaporator outlet tube becomes cooler due to low thermal load (cool passenger-compartment or intake air temperature), the vapor temperature of the refrigerant leaving the evaporator outlet will also be low. The charge of volatile liquid in the sensing bulb will contract and there will be less pressure on the diaphragm. The needle valve will then close, decreasing the amount of refrigerant that is allowed to enter the evaporator core (Figure 6-32).

AUTHOR'S NOTE: A special insulating cover wraps around the sensing bulb on the evaporator outlet pipe to insulate it from external heat sources. It is critical that the bulb make good contact with the outlet pipe and that the insulating cover be properly placed around it after replacement of the expansion valve.

As a precision metering device, the thermostatic expansion valve also senses system pressure. The valve either has an external or internal pressure equalization tube that allows the pressure of the vaporized refrigerant (evaporator outlet) to oppose the gas pressure from the temperature-sensing bulb. This pressure is on the opposite side of the expansion valve's diaphragm from that of the sensing bulb. Increases or decreases in refrigerant velocity and pressure are affected by compressor speed.

The main difference between the external or internal pressure-equalized valve is the side the evaporator pressure is received from. The internally equalized expansion valve uses low-side pressure from the inlet side of the evaporator to apply opposing pressure to the

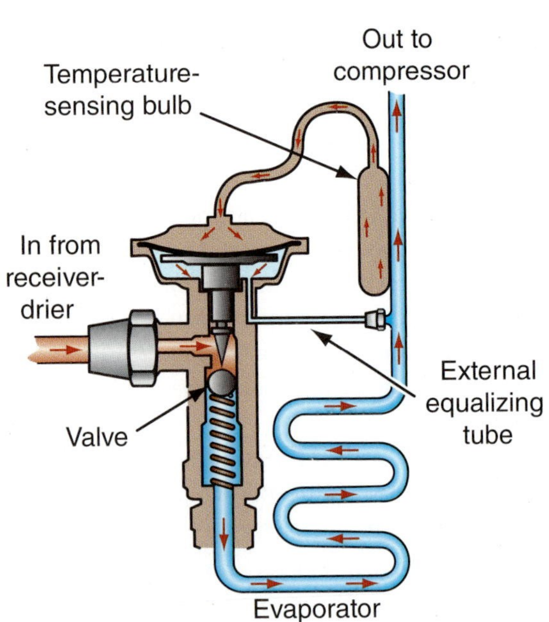

FIGURE 6-31 **Expansion valve open when heat load is high.**

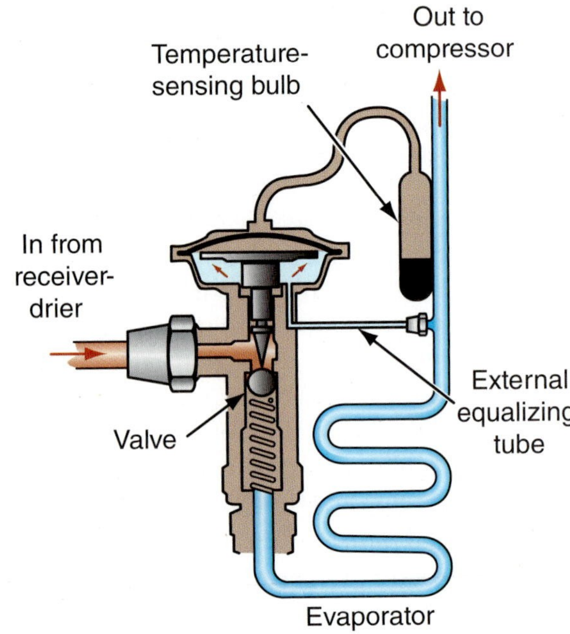

FIGURE 6-32 **Expansion valve closed when heat load is low.**

diaphragm. The externally equalized expansion valve uses low-side pressure from the outlet side of the evaporator to apply opposing pressure to the diaphragm.

Under normal conditions, a certain length of the evaporator outlet contains superheated refrigerant vapor (Figure 6-33). Superheat is the difference between the inlet and outlet temperatures of the evaporator core. The outlet temperature should be higher than the inlet temperature. If the evaporator is filled with refrigerant, the pressure at the evaporator outlet (suction) line increases and the evaporator temperature will decrease; the amount of superheated refrigerant will also decrease. The sensing bulb will react to the low temperature by contracting the gas (lowering pressure in bulb) and relieving pressure on the diaphragm. While this is occurring, the high pressure in the evaporator or outlet line will increase pressure on the opposite side of the diaphragm through the equalization tube (passage), accelerating the closing of the valve, and limiting additional refrigerant from entering the evaporator.

If there is less than the ideal amount of refrigerant in the evaporator, the refrigerant will vaporize faster and cause a greater portion (length) of the evaporator outlet to be superheated. This in turn will cause the temperature of the outlet to increase and the pressure to decrease. The sensing bulb will react to the high temperature by expanding the gas (increasing pressure in the bulb) and relieving pressure on the diaphragm. The low pressure in the evaporator or outlet line will decrease pressure on the opposite side of the diaphragm through the equalization tube (passage), accelerating the opening of the valve, allowing additional refrigerant to enter the evaporator.

The pressure-sensing connections (equalization tube) to the thermostatic expansion valve diaphragm have a dampening effect to keep the expansion valve from opening and closing erratically. This allows a quicker response time by opening and closing the valve rapidly in response to changes in heat load and compressor speed. Using a manifold gauge set and watching the low-side pressure, it is possible to see some of the minute, rapid, and continuous changes in pressure that take place. After the air-conditioning system has been turned on and allowed to stabilize, the low-side gauge needle may fluctuate smoothly, 3–4 psig, as the expansion valve makes minute adjustments to maintain passenger-compartment temperature. If the expansion valve is found to be defective, the sensing bulb may have lost its charge or the internal parts may be seized due to corrosion or foreign matter. The sensing bulb cannot be recharged. If seized, the valve may be in the fully open or fully closed position. If either is the case, the valve must be replaced as an assembly.

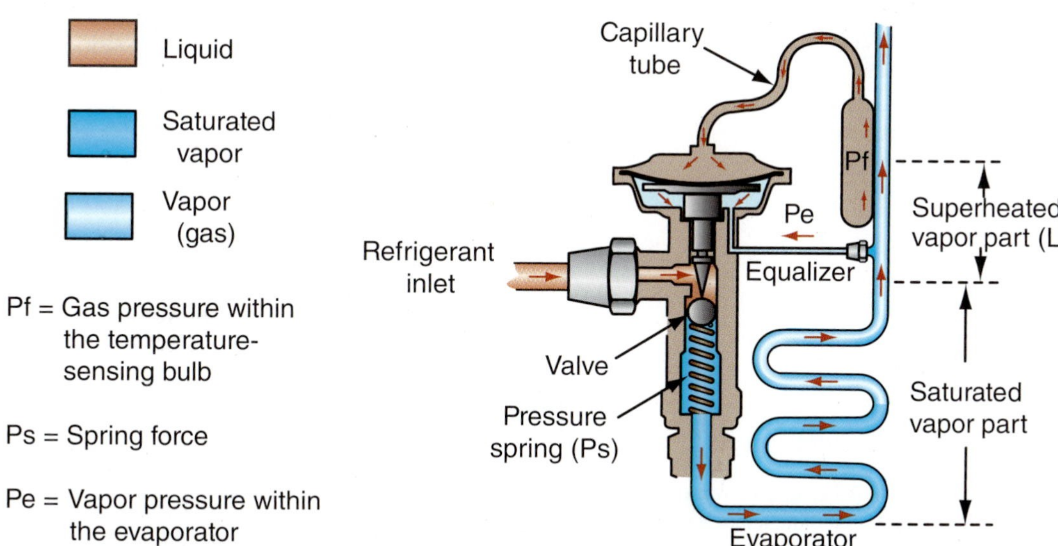

FIGURE 6-33 **An increase/decrease in the area of superheat at the evaporator outlet will cause the expansion valve to open or close.**

H-valve is an expansion valve with all parts contained within.

The H-valve is internally equalized.

Block valve: Another term for H-valve.

H-Valve

The **H-valve** (Figure 6-34), often called a block valve, is used on many car lines. The most common block-type expansion valve is internally regulated (Figure 6-35). The refrigerant enters the valve through the high-pressure liquid line and passes through a variable restriction that regulates the pressure to the evaporator. As the refrigerant leaves the evaporator, it travels back through the H-valve's upper outlet port and passes over the temperature-sensing sleeve (internal sensing bulb) contained in the passage, transferring some heat to the refrigerant contained in the power dome diaphragm. This causes the refrigerant contained in the power dome to expand and contract based on the temperature of the refrigerant leaving the evaporator. The expansion and contraction exerts pressure on the sensing cavity diaphragm, causing the valve pin to move up and down and in turn regulates the flow of refrigerant through the evaporator core, thus regulating core temperature. Its purpose, like the standard TXV, is to sense suction line refrigerant temperature. Operation and function of the H-valve, or block valve, are essentially the same as for the TXV.

The equalizing passage, whether it is internal or external, is a direct passage to the low side of the system to the opposite side of the power dome diaphragm that the sensing bulb connects to. The equalizing pressure ensures smooth, consistent opening and closing of the

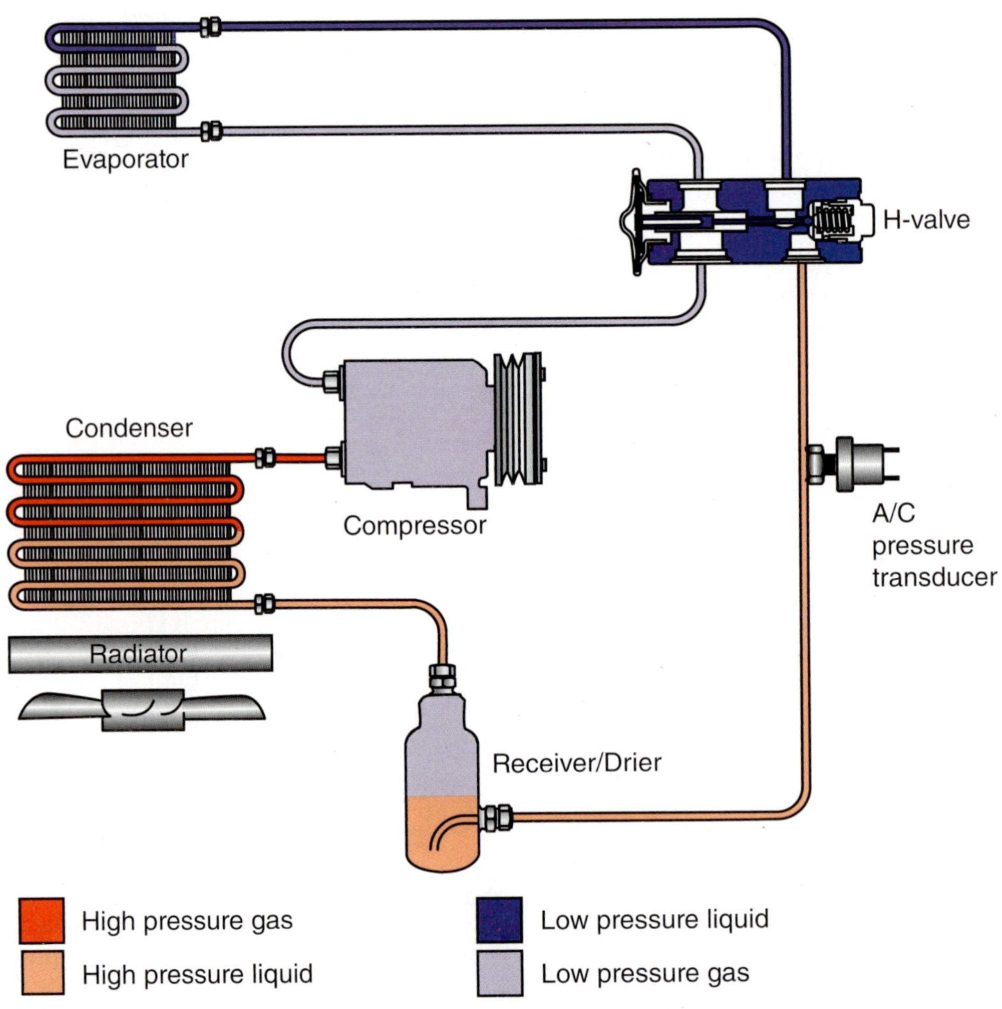

FIGURE 6-34 **An H-valve, also known as a block valve or just H-block, is located between both the evaporator inlet an outline lines. It has the unique role of have high pressure liquid (1), low pressure liquid (2), and low pressure vaporized (3) refrigerant at various locations flowing either into or out of it.**

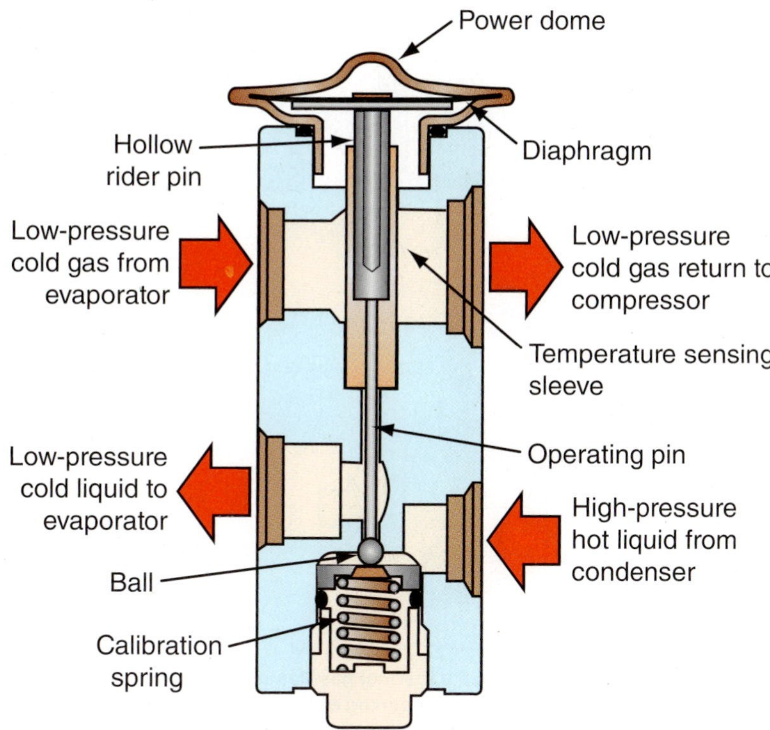

FIGURE 6-35 **Details of H-block style expansion valve.**

expansion valve. It allows for fine adjustments, thus reducing broad temperature fluctuations of the evaporator core, and more consistent temperature levels are maintained within an acceptable operating range.

A similar design used on some vehicles is a new generation expansion valve (Figure 6-36), which does not have an external power dome like a standard H-valve but instead has an internal thermal head. The internal thermal head controls a globe valve at the H-valve inlet

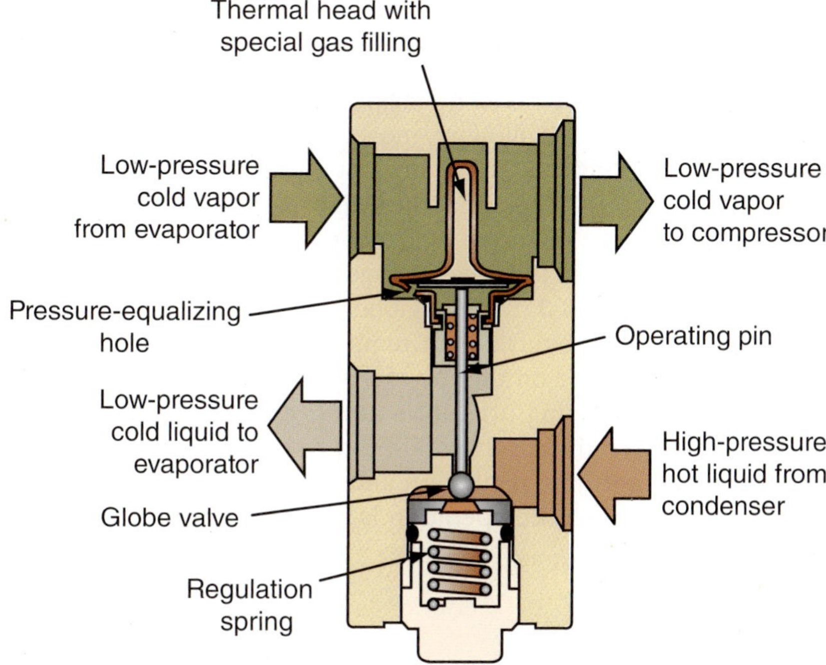

FIGURE 6-36 **New generation H-valve has an internally mounted thermal head and control diaphragm located in the outlet passage of the evaporator discharge instead of an external power dome.**

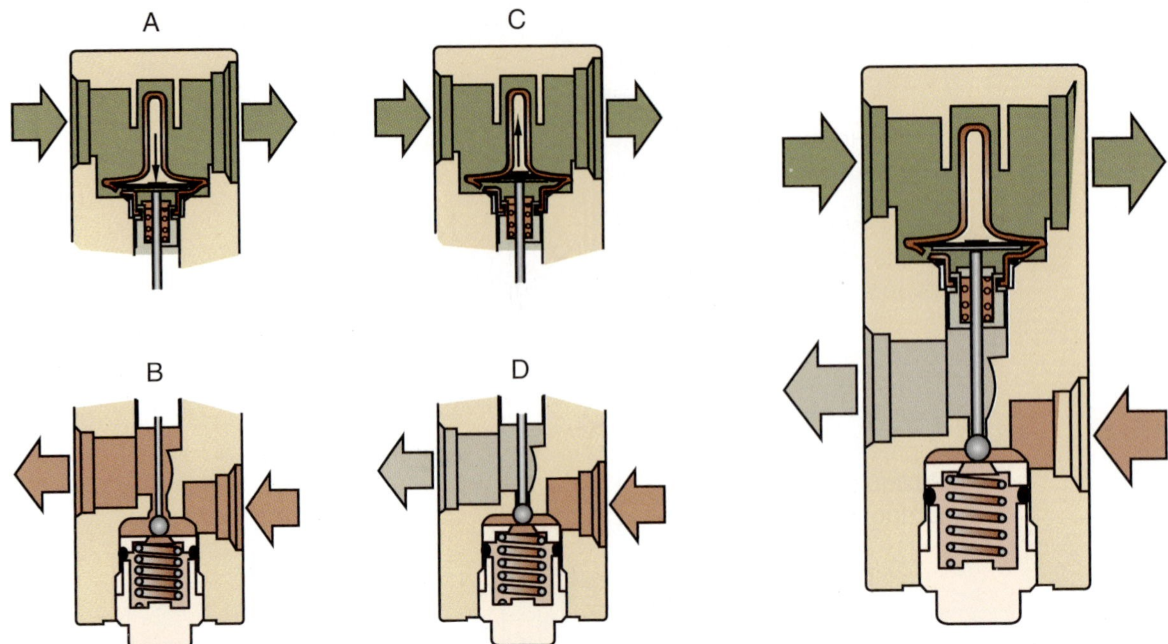

FIGURE 6-37 **New generation expansion valves have a thermal gas dome that is in direct contact with the evaporator outlet refrigerant (A) which allows precise control of the opening of the globe valve (B) to allow more refrigerant into the evaporator when heat loads are high. When the evaporator refrigerant outlet temperature drops and heat load is low the pressure in the thermal head decreases (C) and the globe valve closes (D) decreasing the amount of refrigerant entering the evaporator. This design allows for faster response time to changing evaporator heat load conditions.**

which regulates refrigerant flow. The thermal head is a capsule filled with a special gas on one side of a membrane (diaphragm) and the other side is connected to the evaporator outlet through pressure equalizing passages. The globe valve is actuated by a push rod attached to the membrane. The operation of the new generation H-valve is similar to the other styles of thermal expansion valve that have been described. An increase in cooling load will cause an increase in the evaporator outlet temperature as the refrigerant leaving contains more heat. This in turn will cause the special gas contained within the thermal head to expand, which will increase the chamber pressure and push against the diaphragm (Figure 6-37A). The pressure increase on the diaphragm will cause the globe valve to open the inlet passage to the evaporator allowing more refrigerant to flow into the evaporator (Figure 6-37B), thereby lowering the temperature of the evaporator. When the heat load is low and the temperature of the refrigerant at the evaporator outlet drops the pressure in the thermal head will also decrease (Figure 6-37C). The pressure decrease on the diaphragm will cause the globe valve to close the inlet passage to the evaporator reducing refrigerant flow into the evaporator (Figure 6-37D). The globe valve opening ratio is dependent on the evaporator refrigerant outlet line refrigerant temperature. The expansion valve regulates the evaporator temperature based on the thermal heat load it detects. Though this design is essentially the same in operating principles as all other expansion valves, having the thermal head placed directly in the refrigerant outlet passage allows for precise control of expansion valve refrigerant flow regulation and precise evaporator temperature based on thermal load. So even though the operation is similar, its response time to changes in thermal load is enhanced.

The Orifice Tube

The orifice tube (Figure 6-38) is a calibrated restrictor used as a means of metering liquid refrigerant into the evaporator. Its purpose is to meter high-pressure liquid refrigerant into the evaporator as a low-pressure liquid. The orifice tube establishes a pressure differential at the

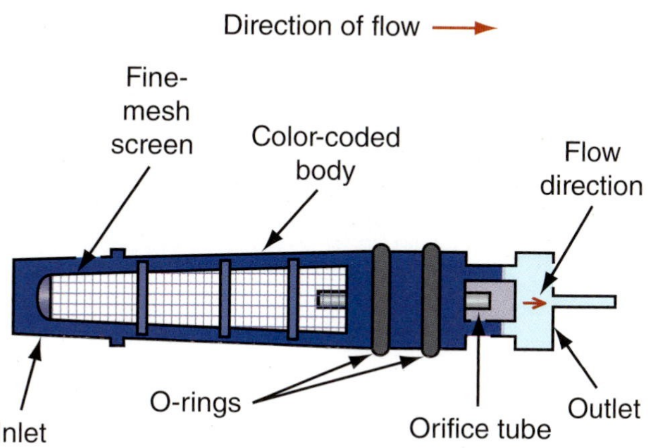

FIGURE 6-38 **A typical fixed orifice tube (FOT).**

restriction, with the high-pressure liquid line and condenser on one side and the low-pressure liquid line and evaporator on the other side.

The amount of refrigerant entering the evaporator with an orifice tube system is dependent on the size of the orifice, subcooling of the refrigerant, and the pressure difference (Ap) between the inlet and outlet of the orifice device. It is frequently referred to as a FOT because of its fixed orifice and tubular shape. The orifice tube is available in sizes ranging from 0.047 in. (1.19 mm) to 0.072 in. (1.83 mm), depending on application, and are generally color coded (Figure 6-39).

Fine-mesh filter screens protect the inlet and outlet of the orifice tube. If foreign matter blocks or partially blocks the orifice, the air-conditioning system will not function to full efficiency. If the blockage is severe enough, the system may not function at all.

The fixed orifice tube, with few exceptions, is located in a cavity in the liquid line or at the inlet connection of the evaporator and is easily accessible. Exceptions are that some vehicles have an inaccessible FOT located inside the liquid line. If found to be defective, the

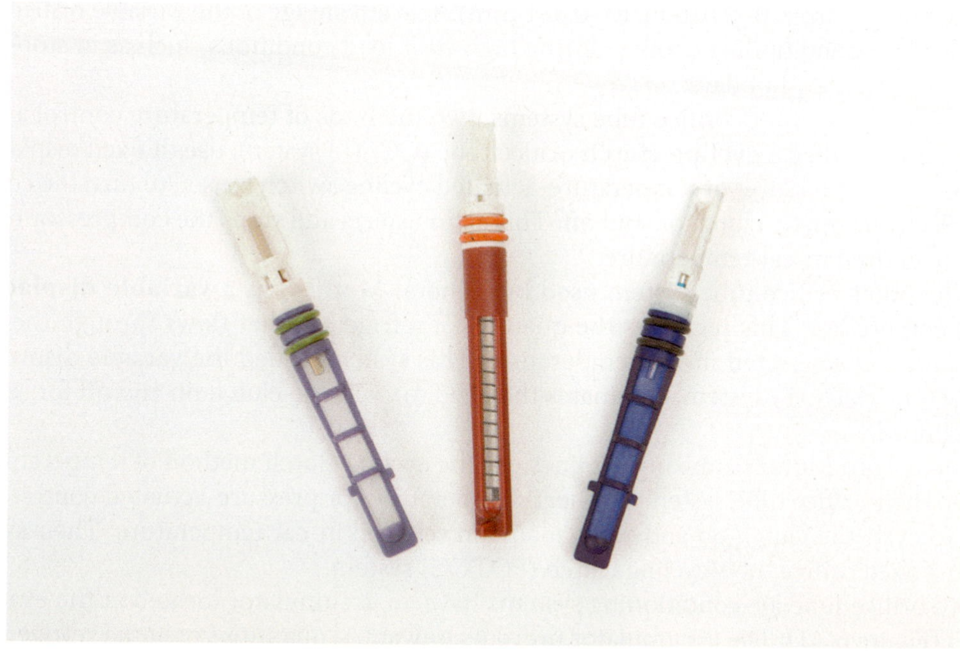
FIGURE 6-39 **Fixed orifice tubes are available in a variety of sizes and are generally color coded.**

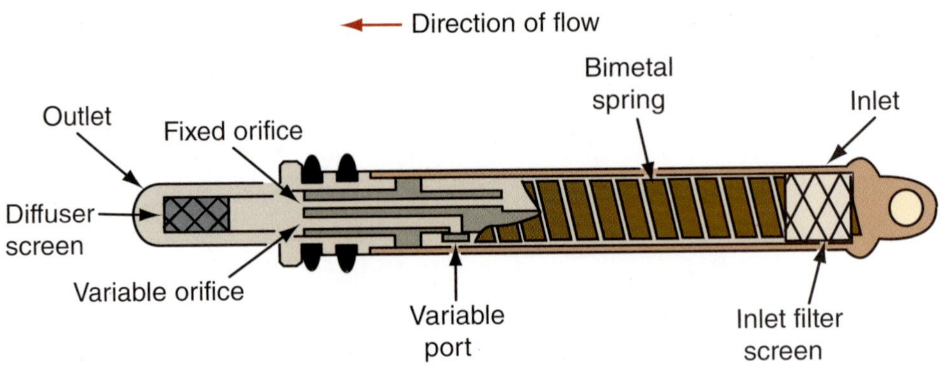

FIGURE 6-40 **Details of a variable orifice tube.**

liquid line either has to be replaced, or a repair kit must be used to replace a section of the liquid line containing the FOT. Procedures for replacing the FOT are found in Chapter 6 of the Shop Manual.

Some "aftermarket engineers" replace the standard original equipment manufacturers (OEM) fixed orifice tube with an aftermarket Smart variable orifice valve (VOV). The Smart VOV utilizes system pressure to move a metering piston relative to a fixed opening in the sleeve. This is claimed to compensate for reduced compressor output at idle speeds and increase the cooling performance. The Smart VOV manufacturer claims that it is a "drop in" replacement for ineffective OEM orifice tubes and that it can offer a "dramatic improvement on factory R-134a systems." Before making any changes to an automotive air-conditioning system, it is strongly suggested that manufacturer's recommendations be followed.

Shop Manual
Chapter 6, page 198

The factory-installed air-conditioning system in some Jeep models, beginning in 1999, are equipped with a variable orifice tube. The Jeep design, however, is slightly different from after-market VOVs. There are two parallel paths for refrigerant to flow through the variable orifice valve (Figure 6-40). One is a fixed orifice opening, and the other is a variable orifice opening. As the temperature of the refrigerant flowing though the VOV changes, a bimetal coil opens or closes the variable port. High temperatures cause the port to close. The opening through the fixed orifice tube is normally 0.047 in. (1.1938 mm), and the variable orifice tube opening ranges from 0–0.015 in. (0–0.381 mm). The advantage of the variable orifice tube is improved air-conditioning cooling during high-heat load conditions, such as in stop-and-go traffic or extremely hot days.

Cycling clutch systems turn the compressor clutch on and off to control evaporator temperature.

The orifice tube, if clogged, may be cleaned.

Variable displacement (VD) changes the displacement of the compressor by changing the stoke of the piston(s).

On General Motors orifice tube systems, two methods of temperature control are used. One method, called a **cycling clutch** orifice tube (CCOT) system, uses a fixed displacement compressor. A pressure- or temperature-actuated cycling switch is used to turn the compressor's electromagnetic clutch on and off. This action starts and stops the compressor to maintain the desired in-car temperature.

The other orifice tube system used by General Motors has a **variable displacement (VD)** compressor. This regulates the quantity of refrigerant that flows through the system to maintain the selected in-car temperature. This system, called the variable displacement orifice tube (VDOT) system, eliminates the need to cycle the clutch on and off for temperature control.

Many Ford Motor Company car lines use the cycling clutch method of temperature control on their orifice tube systems. Either a temperature- or pressure-actuated control may be used to cycle the clutch on and off to maintain selected in-car temperature. Their system is called a fixed orifice tube/cycling clutch (FOTCC) system.

All orifice tube air-conditioning systems have an accumulator located at the evaporator outlet (Figure 6-41). The accumulator prevents unwanted quantities of liquid refrigerant and oil from returning to the compressor at any one time.

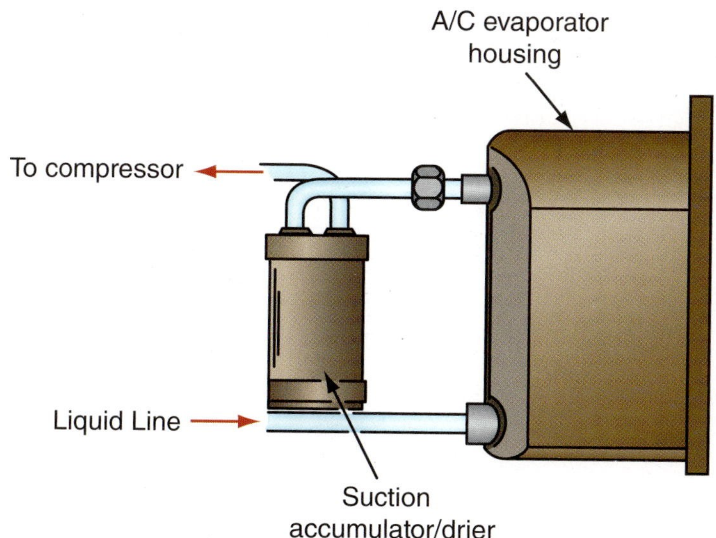

FIGURE 6-41 **A typical accumulator, located at the outlet of the evaporator.**

Orifice Tube Failure

The primary cause of failure of the FOT is clogging of the metering element orifice and strainer screen. This is often caused by failure of the desiccant inside the accumulator. Cleaning a clogged FOT seldom provides satisfactory results. The time and expense of having to do the repairs a second time far outweigh the cost of a new FOT. For this reason, if the FOT is found to be clogged, it should be replaced. Also, if the FOT is clogged, the accumulator should be replaced as well.

Evaporator

The evaporator's purpose is to cool and dehumidify the incoming air when the air-conditioning system is operating. The evaporator (Figure 6-42) is that part of the air-conditioning system where the refrigerant vaporizes as it picks up heat. The expansion device (orifice tube or thermal expansion valve) allows low-pressure, low-temperature, atomized liquid refrigerant into the evaporator. As this cold refrigerant flows through the evaporator core, the heat from the warmer air passing over the evaporator fins will transfer its thermal energy (heat) into the cooler refrigerant, lowering the air's temperature and causing the refrigerant to vaporize (boil) in the evaporator. The refrigerant has now received enough heat to change from a low-pressure, low-temperature liquid into a low-pressure, low-temperature gas. The expansion device and the compressor work in conjunction to meter the exact amount of refrigerant into the evaporator to provide maximum efficiency. This process will ensure that all the liquid refrigerant will be in the vapor state by the time it leaves the evaporator outlet.

The evaporator core is housed inside an insulated box section of the passenger compartment air duct system called the evaporator housing. The case is designed to direct all airflow through the core and reduce radiant heat loads on the evaporator. The common designs of evaporators are the laminated (drawn cup) type (Figure 6-43) and the serpentine fin type (Figure 6-44). The laminated design has more refrigerant passage area and allows for a U-turn pattern of flow in the evaporator from front to rear and left to right, which improves cooling capacity. Important considerations in the design of evaporators are

- Size and length of the tubing
- Number and size of the fins
- Number of return bends
- Amount of air passing through and past the fins
- The **heat load** (NOTE: Heat load refers to the amount of heat, in Btu's, to be removed.)

Heat load is the load imposed on an air conditioner due to ambient temperature, humidity, and other factors that may produce unwanted heat.

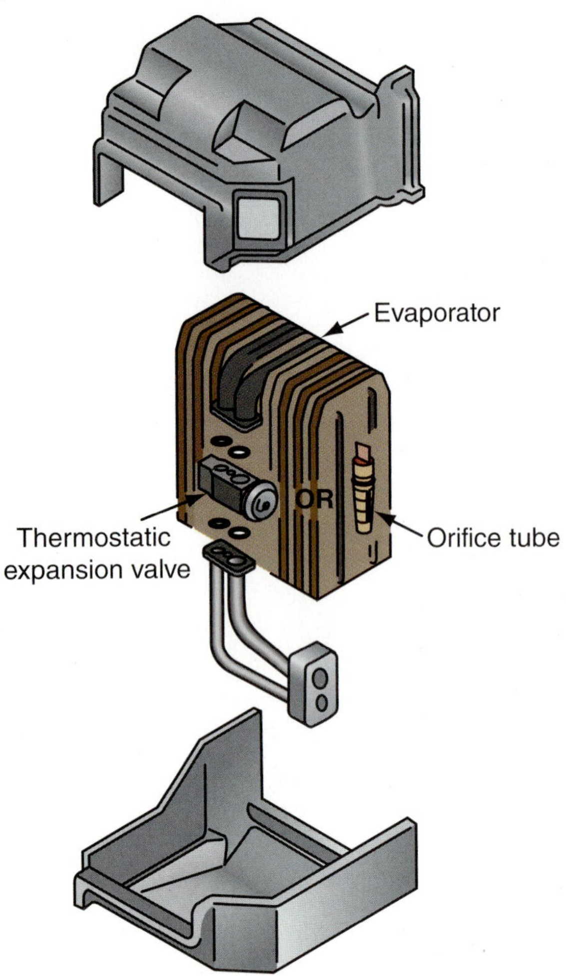

FIGURE 6-42 A typical evaporator core.

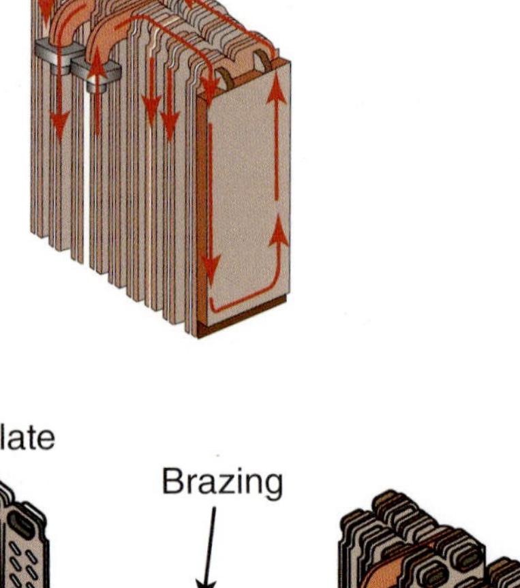

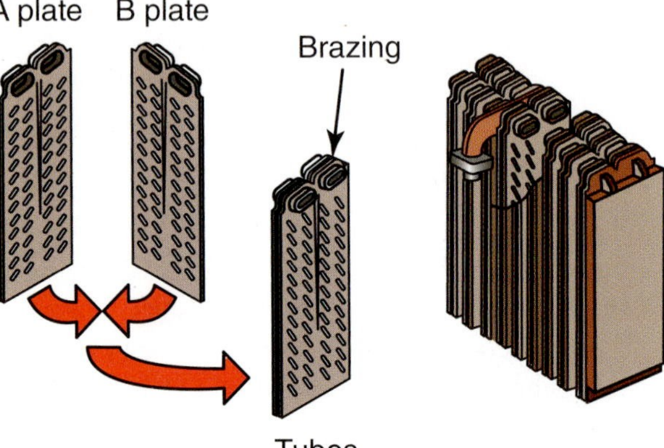

FIGURE 6-43 Laminated core evaporator assembly construction.

Flooded refers to a condition whereby too much refrigerant is metered into the evaporator.

Starved refers to a condition whereby too little refrigerant is metered into the evaporator.

Automotive air-conditioning compressors are not designed to "pump" liquid.

Refrigerant, as it leaves the evaporator, should be a low-pressure, slightly superheated vapor. If too much refrigerant is metered into the evaporator, it is said to be **flooded**. As a result, a flooded evaporator will not cool well because the pressure of the refrigerant in the evaporator is high, and it does not boil away as rapidly. When the evaporator is full of liquid refrigerant, there is no room for expansion. In this case, the refrigerant cannot vaporize properly, which is necessary if the refrigerant is to take on heat. A flooded evaporator also allows an excess of liquid refrigerant to leave the evaporator. The result is that serious damage can be done to the compressor. An accumulator is included in a FOT system to prevent liquid slugging of the compressor. There is no superheat if the evaporator is flooded.

If too little refrigerant is metered into the evaporator, the system is said to be **starved**. Again, the unit does not cool because the refrigerant boils off too rapidly, long before it passes through the evaporator. Under this condition, the superheat is very high.

Under ideal conditions, the refrigerant should boil off about two-thirds to three-quarters of the way through the evaporator. At this point, the refrigerant is said to be saturated. It has picked up all of the latent heat required to change from a liquid to a vapor without undergoing a temperature change.

From this point, the vaporized refrigerant will pick up additional heat before leaving the evaporator. The refrigerant will also pick up under-hood heat in the suction line before reaching the compressor. This superheat is sensible heat that is added to a vapor, raising its temperature without increasing its pressure. The ideal superheat for an air-conditioning system is between 10°F and 20°F (5.6°C–11.1°C). In humid regions of the

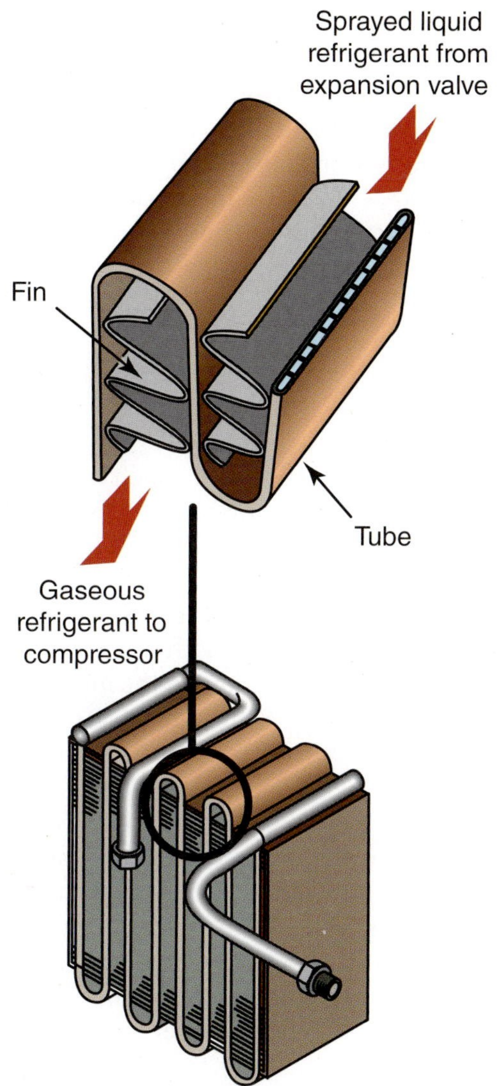

FIGURE 6-44 **Construction of the Serpentine fin evaporator core assembly.**

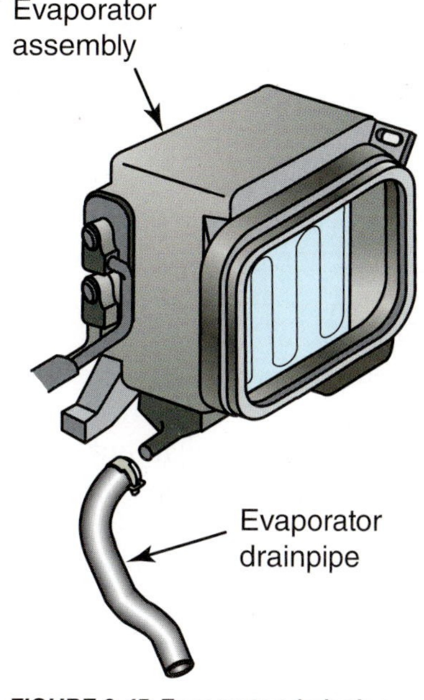

FIGURE 6-45 **Evaporator drainpipe located at the bottom of the case assembly.**

country, it is the addition of superheat that often causes the suction line to sweat and, in some cases, ice over.

During normal operation, the air blowing across the evaporator contains some moisture. This moisture is removed from the air and collects on the surface of the evaporator. The moisture will drain to the bottom of the evaporator housing in the HVAC housing and then out a drainpipe at the bottom of the case assembly (Figure 6-45). It is normal to see a puddle of water forming under a vehicle while the air conditioning is operating; this indicates that the drain vent is not blocked. A blocked vent can lead to moisture building up in the case, causing bacterial growth and odor as well as water dripping into the passenger compartment.

Latent heat cannot be measured with a thermometer.

The dehumidification process adds to passenger comfort. In addition, the air-conditioning system is used to control the fogging of the vehicle's interior windows. In fact,many vehicles' climate control systems will engage the air-conditioning system anytime the defrost mode is selected at any temperature range to aid in clearing the windshield.

There are three problems that could occur with the evaporator, resulting in poor cooling:

- Leaks
- Dirty cooling fins
- Blocked or kinked refrigerant passages

AUTHOR'S NOTE: While on a summer vacation with my family, my daughter noticed a large puddle of water forming on the passenger side floor. Shortly after she noticed the puddle, water mist began to blow out of the dash vents. After arriving at our destination and recovering from the laughter that this incident had generated, I crawled under the vehicle and, with a piece of coat hanger, carefully cleaned out the case vent tube drain. No less than a gallon of water drained from the HVAC case, and another two gallons of water had to be wet/dry vacuumed off the passenger compartment floor. This was one vacation that was not soon forgotten.

Evaporators on R-1234yf systems must meet more stringent SAE J2842, which imposes severe durability testing due to the fact that the refrigerant is mildly flammable. If an evaporator is removed on an R-1234yf system, it must be replaced. No used or repaired evaporators should be used on an R-1234yf system.

In 2010, Toyota introduced a new design evaporator for their Prius platform. The design is essentially two evaporators face-butted together into one assembly with a secondary restriction device called an injector built-in. After the refrigerant passes though the metering device it has two parallel paths into the evaporator dual core (Figure 6-46). The core has an injector, which is a specially shaped tube that produces a pumping action from the pressurized refrigerant flow, creating a venturi effect. The majority of the refrigerant, called the drive flow, flows through the injector while a smaller volume of refrigerant, called the suction flow, flows into a capillary tube and continues through the downwind evaporator where it vaporizes, reducing the temperature of the air passing over the core. At the same time, a venturi effect occurs as the refrigerant passes through the tapered nozzle of the injector, drawing refrigerant out of the downwind evaporator core. This is referred to as the jet pump effect. The drive and suction flow refrigerant mix before they pass into a wider section of pipe in the injector assembly called the diffuser (Figure 6-47). Since the diffuser is larger in diameter,

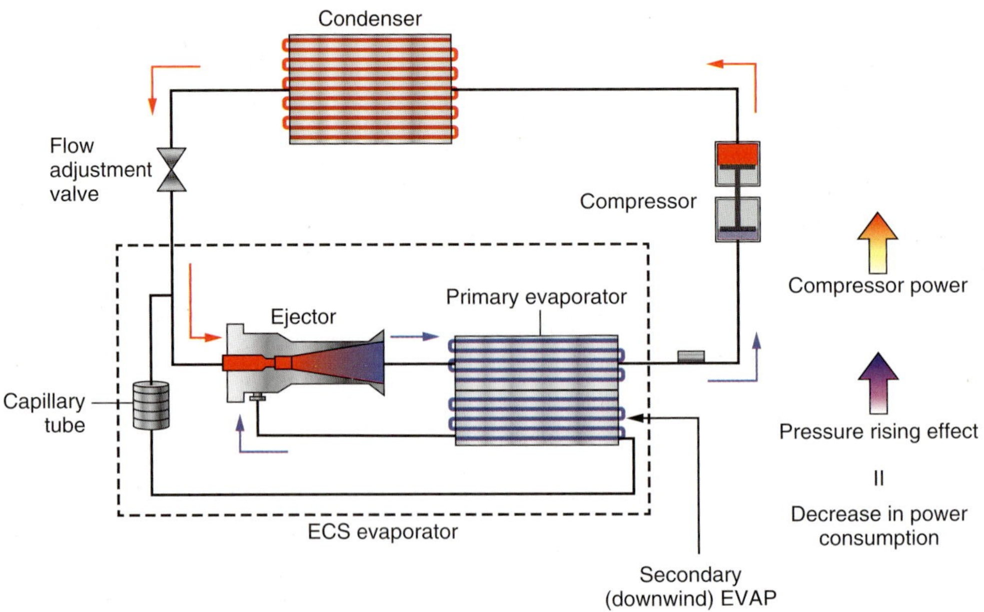

FIGURE 6-46 **Toyota introduced an injector cycle evaporator core on the 2010 Prius.**

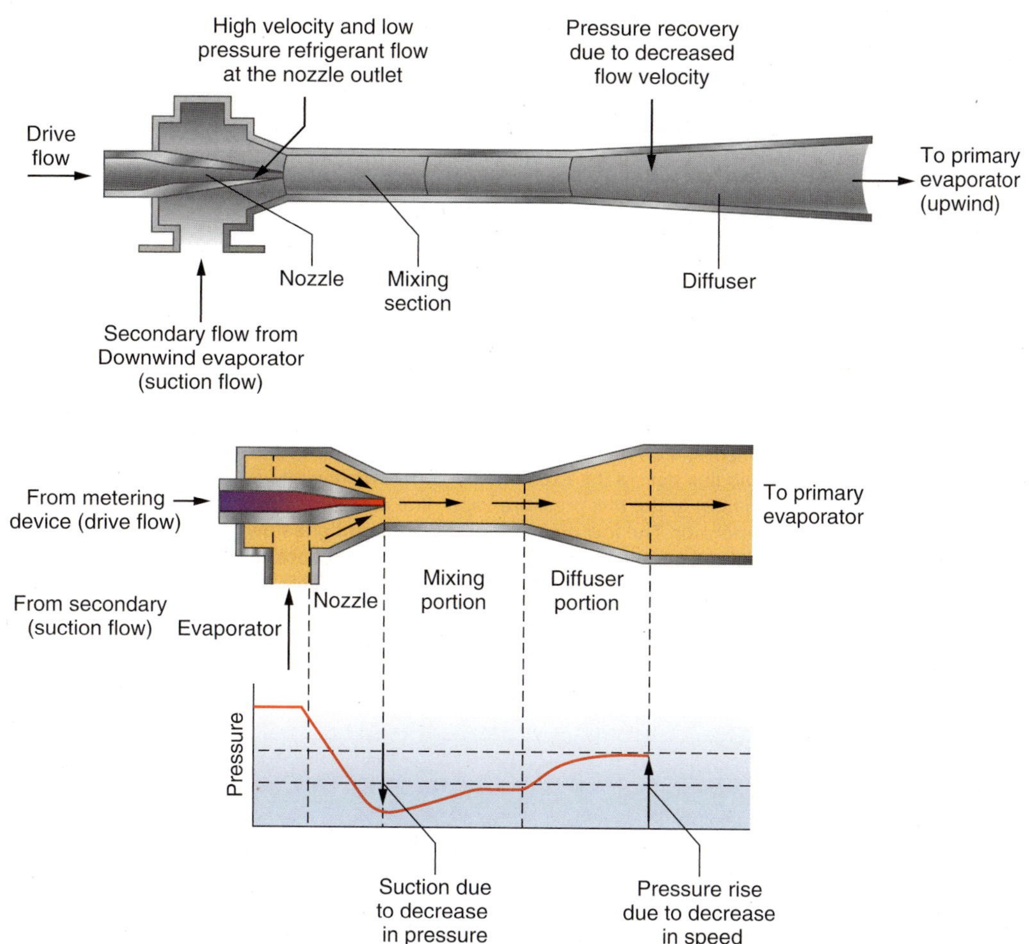

FIGURE 6-47 **Evaporator injection assembly located internally in the dual core evaporator.**

the mixed refrigerant slows down, raising the pressure in this section of the injector. This pressure increase translates into compressor work recovery, meaning the compressor does not need to work as hard to achieve the same cooling effect. This higher-pressure refrigerant passes into the upwind evaporator, removing additional heat from the air that previously passed over the downwind evaporator, improving cooling system performance. In effect the upwind evaporator is cooled by the downwind evaporator. The system is also referred to as a two-temperature evaporator.

INTERNAL HEAT EXCHANGER (IHX)

The internal heat exchanger is generally used on R-1234yf systems but can also be used on R-134a systems to improve their efficiency. The internal heat exchanger (IHX) (Figure 6-48) replaces the suction and liquid refrigerant lines with a coaxial tube that allows heat to be transferred from hot high-pressure liquid line and the cold vapor-filled suction line. The internal heat exchanger is located before the metering device (Figure 6-49). By increasing the amount of thermal energy that can be transferred away from the high-pressure liquid line before the refrigerant passes through the metering device, the evaporator core cooling capacity is increased without increasing system size. This increased cooling capacity in a smaller system reduces energy consumption used by the air-conditioning system, which improves fuel economy, lowers carbon emissions, and increases the air-conditioning system's overall efficiency.

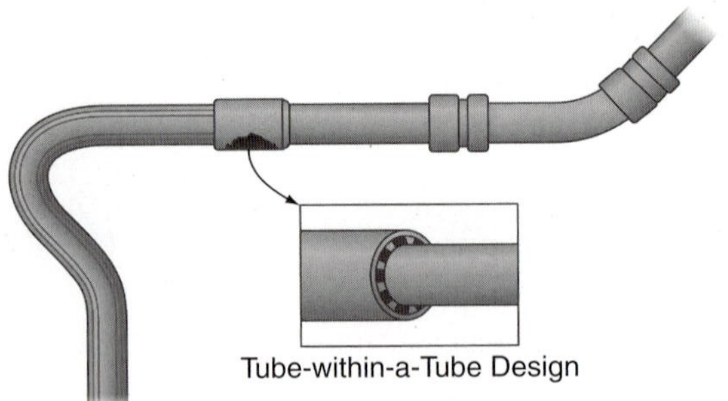

FIGURE 6-48 **An internal heat exchanger (IHX) between the high side liquid line and the low side vapor return line is used to improve system performance by removing thermal heat energy from the high pressure liquid line before the expansion device and moving this heat to the low pressure vapor line after the evaporator. The low pressure cold refrigerant gas flows around the hot high pressure gas line in effect cooling the refrigerant before it passes into the evaporator allowing the evaporator to cool more effectively.**

A/C system with an Internal Heat Exchanger (IHX):

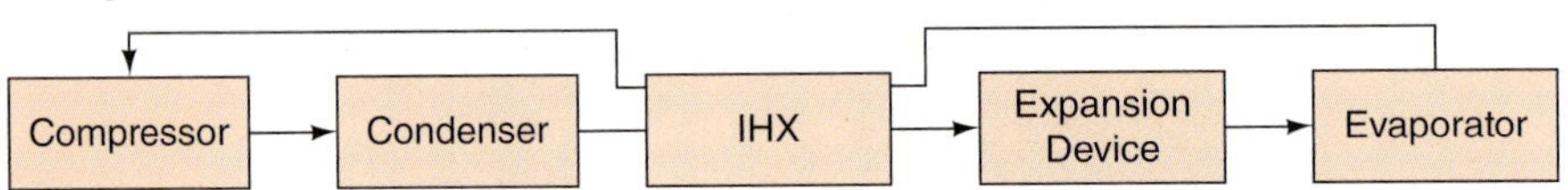

FIGURE 6-49 **The internal heat exchanger (IHX) is located before the metering device.**

Shop Manual
Chapter 6, page 203

The accumulator may be considered a liquid trap.

Accumulator

The accumulator, a tank-like vessel, is located at the outlet of the evaporator (Figure 6-50). It is an essential part of an orifice tube air-conditioning system. The orifice tube, under certain conditions, may meter more liquid refrigerant into the evaporator than can be evaporated. If it were not for the accumulator, excess liquid refrigerant leaving the evaporator would enter the compressor, causing damage.

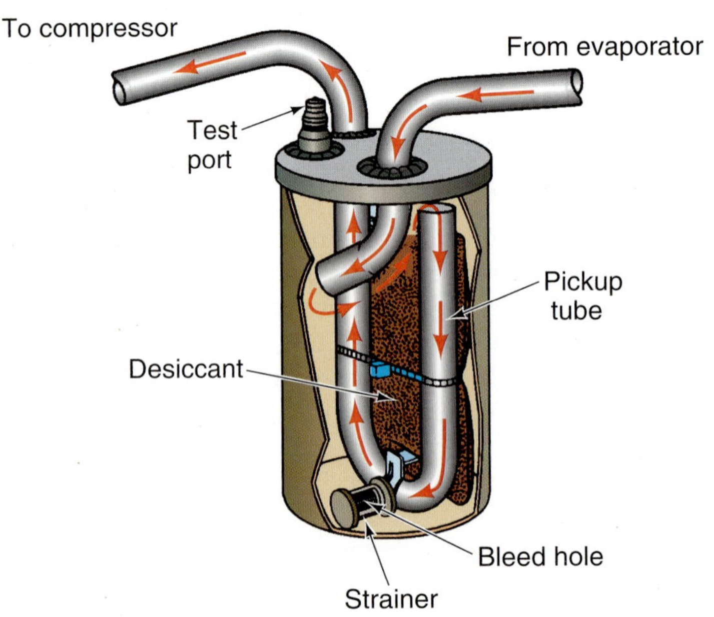

FIGURE 6-50 **Construction details of the accumulator.**

To prevent this problem, all refrigerant and oil leaving the evaporator enters the accumulator, where the liquid (heavier than vapor) falls to the bottom of the tank. A U-shaped pickup tube ensures that only refrigerant vapor enters and leaves the accumulator to the compressor inlet, while trapping the liquid refrigerant and oil. A metered oil bleed hole (orifice) at the bottom of the U-bend meters a small amount of liquid (when present) into the suction line. This orificeis calibrated to ensure that liquid is metered in an amount that will vaporize before it reaches the compressor. The orifice also allows small quantities of refrigerant oil to return to the compressor.

Another important function of the accumulator is that it contains the desiccant, a chemical drying agent. The desiccant attracts, absorbs, and holds moisture that may have entered the system due to improper or inadequate service procedures. The desiccant is not serviced as a component of the accumulator. If desiccant replacement is indicated, the accumulator must be replaced as an assembly. The desiccant used in R-12 accumulators may not be compatible in an R-134a system. The desiccants are classified XH5, XH7, and XH9. Only XH7 and XH9 are acceptable for use on R-134a systems. To be sure of system compatibility, use only replacement components specifically designated for a particular application.

A fine-mesh screen is placed in the accumulator to catch and prevent the circulation of any debris that may be in the system. This screen cannot be serviced; if it is clogged, the entire unit must be replaced as an assembly.

The accumulator should be replaced any time a major component of the air-conditioning system is replaced or repaired. Most vendors will not honor the warranty on new or rebuilt compressors unless the accumulator is replaced at the time of service.

AUTHOR'S NOTE: In general, students have difficulty in the beginning distinguishing what type of system they are dealing with, whether orifice tube or expansion valve. First, find the lines entering and leaving the evaporator. Follow the outlet line, which leads to the compressor. If there is a metal canister attached to this line you have found the accumulator and you are dealing with an orifice tube system. If no canister is found, you are dealing with an expansion valve system.

WHAT TYPE SYSTEM

There are basically two methods of temperature control for automotive air-conditioning systems:

- Cycling clutch
- Noncycling clutch

Cycling Clutch

The cycling clutch system relies on two methods for temperature control:

1. Temperature cycling switch. The temperature cycling switch is a temperature-sensitive switch that cycles the compressor clutch on and off at **predetermined** temperature levels.
2. Pressure cycling switch. The pressure cycling switch, as its name implies, is sensitive to system pressure and turns the compressor clutch on and off at predetermined pressure levels.

Predetermined is a set of fixed values or parameters that have been programmed or otherwise fixed into an operating system.

Noncycling Clutch

The noncycling clutch system relies on a variable displacement (VD) compressor to control the in-car temperature. The amount of refrigerant permitted to flow through the system is controlled by the compressor's ability to alter the stroke of the pistons as required by varying system conditions.

The only purpose of the clutch in a noncycling system, then, is to disengage the compressor when the air conditioner is not in use and to engage the compressor when the driver calls for cooling.

Terms to Know

Barrier
Change of state
Compression
Cycling clutch
Desiccant
Electromagnetic
Flooded
Heat load
H-valve
Orifice
Predetermined
Reciprocating
Reed valve
Starved
Stroke
Variable displacement (VD)

SUMMARY

- The compressor is the prime mover of the refrigerant. Its purpose is to pump refrigerant throughout the air-conditioning system.
- Refrigerant changes state: vapor to liquid in the condenser, and liquid to vapor in the evaporator.
- The receiver-drier ensures a gas-free liquid supply to the metering device.
- The accumulator ensures that no liquid refrigerant is returned to the compressor.
- A thermostatic expansion valve (TXV), H-valve, or orifice tube (OT) is essential to meter the proper amount of refrigerant into the evaporator. These restrictive devices establish a pressure differential between the high side and the low side in the system.

REVIEW QUESTIONS

Short-Answer Essays

1. How does compressor action increase the condensation rate of refrigerant?
2. Briefly, how does a compressor "pump" refrigerant?
3. Why is it important that the inlet of the condenser be at the top?
4. What are two purposes of the receiver-drier?
5. Explain the term *flash gas* and what causes it.
6. What two factors determine how much refrigerant enters the evaporator in a fixed orifice tube system?
7. Briefly describe the state of the refrigerant as it leaves the evaporator in a properly operating system.
8. What is the primary purpose of an accumulator?
9. Briefly describe the refrigeration cycle.
10. What are the two basic functions of the thermostatic expansion valve?

Fill-in-the-Blanks

1. The compressor's function is to ______________ the refrigerant throughout the system.
2. Refrigerant ______________ is stored in the ______________ and is essential to keeping the internal parts of the compressor lubricated.
3. Heat-laden refrigerant gives up its heat in the ______________ as it changes from a ______________ to a ______________.
4. High-pressure liquid refrigerant moves through a hose called a ______________.
5. As cold refrigerant flows through the evaporator core, the heat from the warmer air passing over the evaporator fins will transfer its ______________ into the cooler refrigerant ______________ the air's temperature and causing the refrigerant to ______________ in the evaporator.
6. All orifice tube systems have a(n) ______________ at the ______________ of the evaporator.
7. The evaporator is said to be ______________ if too little refrigerant is metered into it and ______________ if too much refrigerant is metered into it.
8. The ______________, a drying agent, is found in the ______________ or ______________.
9. The compressor separates the ______________ side of the system from the ______________ side of the system; the metering device separates the ______________ side from the ______________ side of the system.
10. The air-conditioning system low- and high-pressure pipes and hoses are different in diameter. To identify the pipes, the high-side line is ______________ in diameter than the low-side line.

Multiple Choice

Refer to Figure 5-51 to answer questions 1 through 7.

1. *Technician A* says that the component identified by A is a receiver-drier.

   Technician B says that the component identified by D is a condenser.

   Who is correct?

   A. A only (if the illustration depicts a thermostatic expansion valve [TXV] system)

   B. B only

   C. Both A and B

   D. Neither A nor B

2. *Technician A* says that the component identified by B is an evaporator.

   Technician B says that the component identified by A is a condenser.

   Who is correct?

   A. A only C. Both A and B

   B. B only D. Neither A nor B

3. What component is shown as C?

   Technician A says that it is an accumulator.

   Technician B says that it is a receiver-drier.

   Who is correct?

   A. A only C. Both A and B

   B. B only D. Neither A nor B

4. What is the state of the refrigerant in line 4?

   Technician A says that it is high-pressure vapor.

   Technician B says that it is high-pressure liquid.

   Who is correct?

   A. A only C. Both A and B

   B. B only D. Neither A nor B

5. What is the state of the refrigerant in line 5?

   Technician A says that it is low pressure.

   Technician B says that it is a vapor.

   Who is correct?

   A. A only C. Both A and B

   B. B only D. Neither A nor B

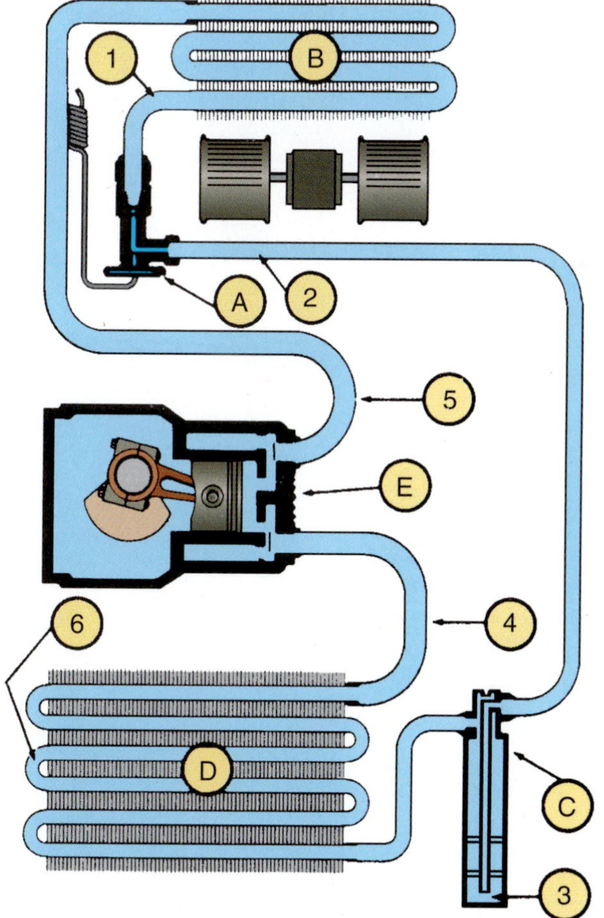

FIGURE 6-51 **The automotive air-conditioning system.**

6. Which of the following statements identifies the correct state and pressure of the refrigerant throughout the system?
   A. The refrigerant is low pressure at points 1, 2, and 3 in the system and high pressure at points 4, 5, and 6 in the system.
   B. The refrigerant is low pressure at points 1 and 5 in the system and high pressure at points 2, 3, 4, and 6 in the system.
   C. The refrigerant is high pressure at points 1, 2, and 3 in the system and low pressure at points 4, 5, and 6 in the system.
   D. The pressure varies from high pressure to low pressure throughout the system, depending on heat loads.

7. What is the state of the refrigerant as it immediately-enters the evaporator, line 1?

   Technician A says it is all liquid with some flash gas.

   Technician B says it is all vapor.

   Who is correct?

   A. A only C. Both A and B
   B. B only D. Neither A nor B

8. What is the purpose of the desiccant?

   Technician A says it is to clean the refrigerant.

   Technician B says it is to dry the refrigerant.

   Who is correct?

   A. A only C. Both A and B
   B. B only D. Neither A nor B

9. One of the functions of the air-conditioning compressor is to increase the pressure of the refrigerant. Another function of the compressor is to increase:
   A. The heat (temperature) of the refrigerant
   B. The volume of the refrigerant
   C. The vaporization point of the refrigerant
   D. Pump the liquid refrigerant

10. *Technician A* says that the ideal superheat for an automotive air-conditioning system is 10–20°F.

    Technician B says that flash gas is a contributing factor of superheat.

    Who is correct?

    A. A only C. Both A and B
    B. B only D. Neither A nor B

Chapter 7

REFRIGERANT SYSTEM SERVICING AND TESTING

UPON COMPLETION AND REVIEW OF THIS CHAPTER, YOU SHOULD BE ABLE TO:

- Describe a refrigerant system performance test.
- Explain how moisture collects in an air-conditioning system.
- Explain the importance of a moisture-free system.
- Discuss noncondensable gas contamination.
- Describe the methods used to remove moisture from a system.
- Describe the leak test procedures for an automotive air-conditioning system using soap trace solutions, halide leak detectors, halogen leak detectors, and dye solutions.
- Discuss the other types of leak detector devices available to the automotive technician.
- Discuss the acceptable methods for charging a system with refrigerant.

INTRODUCTION

Testing and servicing an automotive air-conditioning system is a skill that is generally developed with practice and experience. This chapter gives a basic fundamental understanding of these procedures, including leak testing, **moisture** removal, refrigerant recovery, charging the system, and diagnostics.

Moisture is defined as droplets of water in the air: humidity, dampness, or wetness.

PERFORMANCE TESTING

An air-conditioning system performance test is an initial test that determines whether the refrigerant system is operating as designed. As a service technician, you should always follow the manufacturer's service and diagnostic information. There are two basic types of performance tests that may be done to aid in analyzing system operation: the load test and the unloaded test.

Many service manuals specify a loaded test in which a heat load is applied to the evaporator in the form of outside ambient air temperature and humidity. The test should be performed in a shaded area, not in direct sunlight. The front windows are left open and the hood is opened. The vent fan is turned on high and the air-conditioning system is turned on. The ambient air humidity is measured. Next, the engine is run at 1,500 rpm and the system is allowed to stabilize for 10 minutes. The air temperature leaving the center vents is then measured.

Shop Manual
Chapter 7, page 238

The unloaded test has you select the Recirculation mode, which dries the air as it is recirculating during system stabilization. The test should be performed in a shaded area, not in direct sunlight. Close all windows and doors. Open the vehicle hood. The vent fan is turned on high and the air-conditioning system is turned on and set to Recirculation mode.

Next, the engine is run at 1,000 rpm and the system is allowed to stabilize for 15 minutes. The air temperature leaving the center vents is then measured. The unloaded test strips the air in the passenger compartment of heat and humidity as it is recirculated through the evaporator, allowing for the lowest possible vent temperatures to be achieved. The temperature–pressure performance charts in Figure 7-1 for the unloaded performance test require that the engine speed be 1,000 rpm (±50 rpm) to be accurate.

The key difference between the loaded and unloaded performance tests is how the tests deal with humidity and ambient heat load. Because the loaded performance test allows outside ambient air to pass over the evaporator, the test yields different results, depending on ambient temperatures and humidity levels. The unloaded performance test's use of recirculated dry air allows for consistent achievement of the lowest possible vent temperatures and provides a consistent performance comparison of similar vehicles, regardless of ambient air temperatures or humidity levels on a particular day.

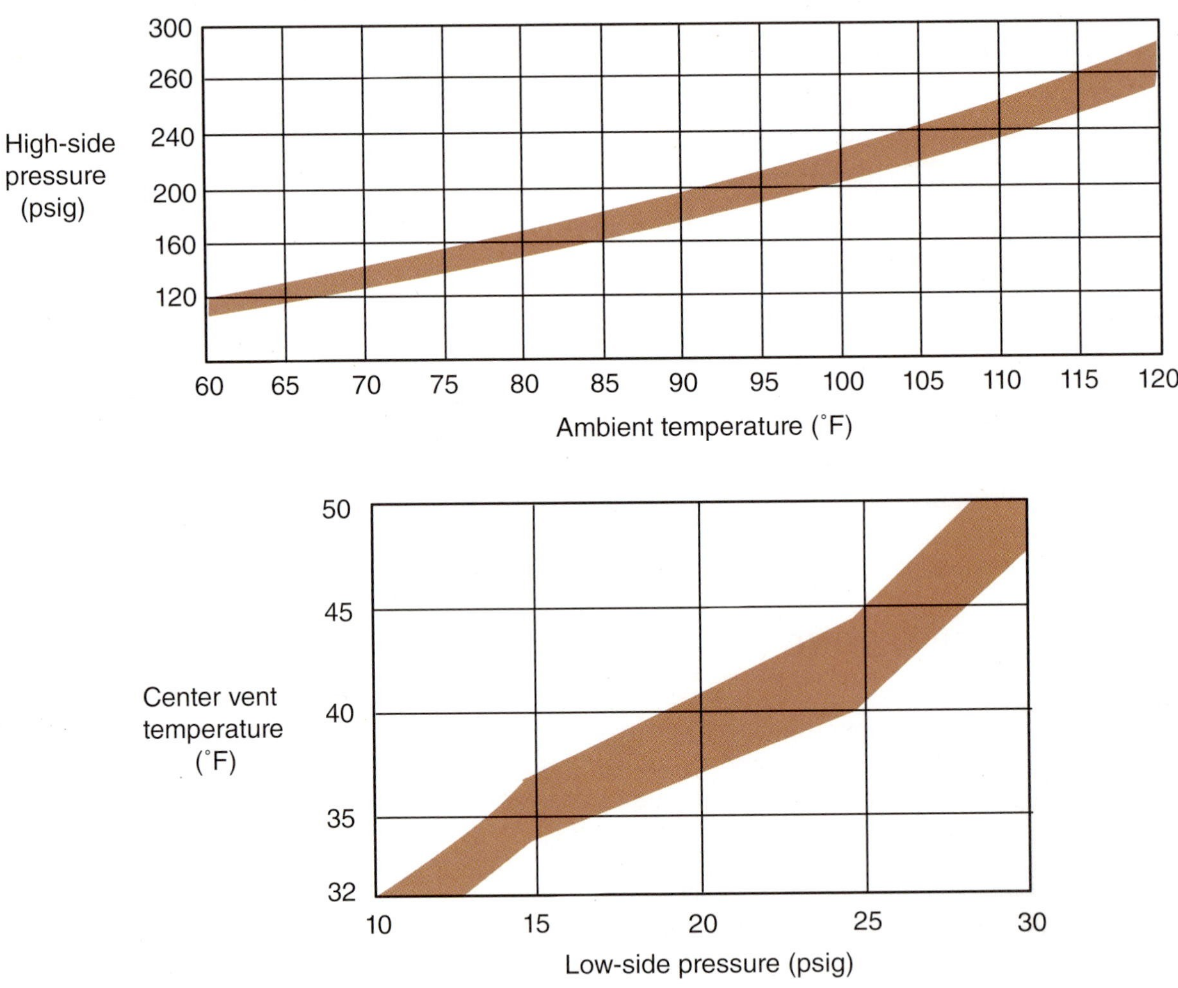

Ambient temperature ________ ˚F
Low-side pressure (1000 rpm) ________ (psig) **Rear A/C OFF**
High-side pressure (1000 rpm) ________ (psig) **Rear A/C OFF**
Low-side pressure (1000 rpm) ________ (psig) **Rear A/C ON**
High-side pressure (1000 rpm) ________ (psig) **Rear A/C ON**

Center vent temperature ________ - ˚F
Rear vent temperature ________ ˚F
Low-side pressure (3000 rpm) ____ ˚F
Left corner vent temperature ____ ˚F
Right corner vent temperature ____ ˚F

FIGURE 7-1 **R-134a air-conditioning system unloaded performance test data chart requires that the vehicles windows and doors are closed and that the recirculation vent control mode be selected for accurate test results.**

Refrigerant Analyzer

Before attempting to recover refrigerant or charge an automotive air-conditioning system, one should use a refrigerant analyzer, also called a refrigerant identifier, to test samples taken from the air-conditioning system or storage container to determine their purity. Recovered refrigerant should contain less than two percent impurities.

It is important to follow the instructions that are included with the analyzer to obtain the test sample. The analyzer, such as the one shown in Figure 7-2, will display the following:

Shop Manual
Chapter 7, page 236

- R-134a percentage, a fault is triggered if the refrigerant purity is less than 98 percent or better by weight.
- R-12 percentage, a fault is triggered if the refrigerant is less than 98 percent pure.
- R-22 percentage.
- HC, if the sample contains hydrocarbons (flammable material) some units will sound a audible warning signal.
- Air Contamination also referred to as noncondensable gas contamination.
- J2099–Standard of Purity for Recycled R-134a (HFC-134a) and R-1234yf (HFO-1234yf) for Use in Mobile Air-Conditioning Systems (revised 2/2011).

For recycled R-134a refrigerant that has been directly removed and intended to be returned, the system must meet:

- Moisture 50 parts per million (PPM) by weight; R-134a and its oil have a much higher affinity for water and are harder to keep dry
- Refrigerant oil: 500 ppm by weight
- Noncondensable gases (air) 150 PPM by weight

- J2844–Standard of Purity for Recycled R-1234yf
- J2912 R-1234yf (HFO-1234yf) Refrigerant Identification Equipment standard

Contaminated or **cross-contaminated** refrigerant can cause system and equipment damage as well as present diagnostic problems if you are not aware of the contamination. Cross-contamination can cause reduced system performance, lubrication problems, and chemical breakdown within the system, which may contribute to an acidic condition within the system, leading to component failure.

Contaminated is a term generally used when referring to a refrigerant cylinder or system that fails a purity test and is known to contain foreign substances, such as other ncompatible or hazardous refrigerants.

Cross-contaminated refrigerant is a refrigerant that contains some refrigerant other than the type designated on the system service label.

In the event that the recovery/recycling equipment is contaminated with cross-contaminated refrigerant, the unit will have to be cleaned out; the recovery tank will have to be sent out to a refrigerant reclamation station; and the filters and dehydrators will have to be replaced. The Environmental Protection Agency (EPA) prohibits the venting of any automotive refrigerant

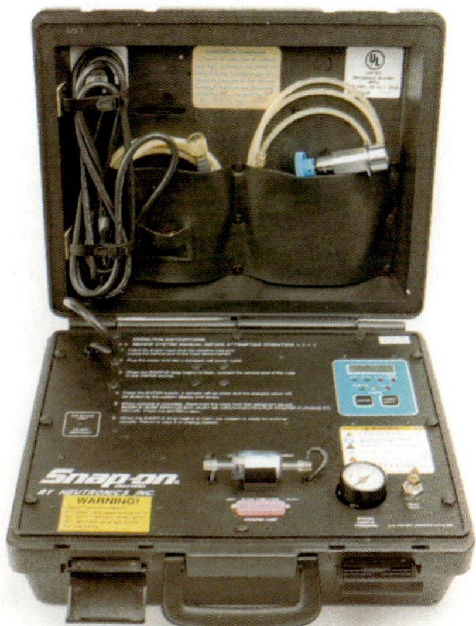

FIGURE 7-2 **A typical refrigerant analyzer.**

into the atmosphere, even if the refrigerant is not identifiable by the service facility. The EPA maintains a list of "Refrigerant Reclaimers" that can be ordered at no charge from the EPA Stratospheric Ozone Protection Hotline at 1-800-296-1996 or through the EPA Web site at www.epa.gov/ozone/title6/609. Besides the cost of servicing the unit, there is also the lost time involved while waiting for the recovery tank to be returned after reclamation or the purchase of a spare recovery tank. The bottom line is, if you suspect system contamination and your facility plans on servicing contaminated systems, you must have a dedicated recovery-only unit for recovering contaminated refrigerant to be sent out for reclamation later. Remember to never fill the DOT-approved recovery tank beyond 60 percent of its gross weight capacity. Also, consult local and state regulations on the storage of hazardous or combustible materials.

Statistics on refrigerant contamination indicate that:

- 63 percent of the systems that are contaminated involve R-12 systems that are contaminated with R-134a. This is generally due to the backyard mechanic who can purchase 1-pound cans of R-134a in a retrofit kit; is not required to obtain a Section 609 license to purchase R-134a; and has no way of removing the old refrigerant.
- 18 percent of the contaminated R-12 systems are contaminated with R-22.
- 17 percent of the contaminated R-12 systems are contaminated with three or more refrigerants, including hydrocarbons.
- 2 percent of the contaminated R-134a systems are contaminated with three or more refrigerants, including hydrocarbons.
- There are 50 percent more contaminated R-134a systems seen each year. It was once thought that contamination was likely to occur only in the older R-12 systems, but recent evidence indicates that the occurrence of R-134a system contamination is occurring at an alarming rate. This is due in part to the fact that the **Significant New Alternative Policy (SNAP)** program rule does not apply to R-134a replacement refrigerants.

SNAP stands for **Significant New Alternative Policy** program and sets Environmental Protection Agency (EPA) guidelines for the use of alternative refrigerants and recovery/recycling equipment.

When purchasing refrigerant identification equipment, ensure that it meets the standards set by SAE J1771. Manufacturers are required to label their equipment, stating the unit's level of accuracy.

The Society of Automotive Engineers (SAE) has set purity standards for recycled R-12 and R-134a refrigerants so that they provide proper system performance and longevity. The purity standard for R-134a is J2099 and specifies a limit in parts per million (ppm) for moisture (15 ppm by weight), noncondensable gases (air) (330 ppm by weight), and refrigerant oil (4,000 ppm by weight). The purity standard for R-12 is J1991 and specifies a limit in parts per million (ppm) for moisture (15 ppm by weight), noncondensable gases (air) (150 ppm by weight), and refrigerant oil (500 ppm by weight).

Sealant Contamination

Shop Manual
Chapter 7, page 241

Refrigerant sealants for small refrigeration system leaks have become a popular remedy when compared to the cost and labor involved with some component replacement. The sealant is added to the system through the high or low side of the system, depending on manufacturer recommendations. The sealant travels through the system with the refrigerant. There are two types available, seal swellers and stop leak. Seal swellers do just what the name implies; they contain chemicals that soften and swell O-rings and seals. Stop-leak sealers work by sealant chemical interaction with moisture. When the sealant begins to leak out of the system, it reacts with the moisture in the air and forms an epoxy, thus sealing the leak. Unfortunately, some sealant contamination has been identified as a major cause of system performance problems and recovery/recycling equipment failures. Damage caused by the use of sealants is not covered by manufacturer warranties. Refrigerant identifiers do not detect the presence of sealants in the refrigerant system. You must use a sealant identifier product specifically designed for this purpose (Figure 7-3). One company, Neutronics Inc., produces one model called QuickDetect™ Sealant Detector, which quickly identifies the presence of stop-leak

FIGURE 7-3 **A simple-to-se Neutronics Inc. refrigerant system sealant identifier will determine whether an air-conditioning system has been contaminated by sealant.**

sealants. Once you have determined that sealant is not present in the system, the refrigerant can be safely recovered. In addition to damage to equipment, if the refrigerant system contains excess amounts of moisture, the sealant can become activated inside the system, causing restriction in refrigerant system components including but not limited to the expansion valve, FOT, evaporator, and condenser. Systems should be checked for the presence of refrigerant sealants prior to attaching any other air-conditioning equipment, including refrigerant identification equipment.

Noncondensable Gas

Shop Manual
Chapter 7, page 265

A noncondensable gas cannot condense into a liquid. Air in the refrigerant system is considered a noncondensable gas because at no time are refrigerant system pressures great enough or temperatures low enough to cause air (78 percent nitrogen, 21 percent oxygen, and 1 percent other inert gases) to condense into a liquid. Noncondensable gas takes up space in the refrigerant system condenser, which in turn effectively reduces the condenser surface area. This in turn will increase system operating pressure on the high side of a thermostatic expansion value (TXV) system and possibly both high- and low-side pressures on an FOT system. A system is considered contaminated if it contains 150 ppm by weight for R-134a or 330 ppm by weight for R-12.

Refrigerant system air contamination may be caused by:

- Refrigerant system low-side leak
- Improper system evacuation or insufficient evacuation time
- Weak evacuation pump

Anytime that a system is open for repair service, there is the possibility that air may enter the system. Also, air will enter a closed system if the ambient air pressure is greater than the system refrigerant pressure. For example, assume that an air-conditioning system is low on refrigerant due to a leak on the low side of the system. If the low-pressure cutoff switch is defective or there is a problem that permits the low-side pressure to fall below zero gauge (ambient) pressure, air may enter the system at the same location that the refrigerant leaks out of the low side. To remove air from an air-conditioning system, it must be leak free and thoroughly evacuated with a quality vacuum pump. To remove air from recovered refrigerant, follow the equipment manufacturer's instructions included with the recovery/recycle equipment.

If the refrigerant system is contaminated with a noncondensable gas, poor refrigerant system cooling and performance will result. An air-contaminated system will cause static (system off and stabilized) system pressure to be higher than that of one containing a pure charge of refrigerant (R134a or R-12). Air contamination will cause the operating pressure on the high side of the refrigerant system to be higher than normal, and the high-side gauge may fluctuate or flutter slowly. This is caused by the fact that air takes up space in the condenser and will occupy even more space when heated. A recovered tank of refrigerant may be checked for excess air contamination by comparing tank pressure to a standard temperature-pressure chart (Figure 7-4). Prior to testing the recovery tank, it must be out of direct sunlight and at a stable temperature above 65°F (18.3°C) for at least 12 hours. If the recovery tank is suspected

STANDARD TEMPERATURE-PRESSURE CHART FOR R-134a

°F	PSI	°F	PSI	°F	PSI	°F	PSI	°F	PSI
65	69	77	86	89	107	101	131	113	158
66	70	78	88	90	109	102	133	114	160
67	71	79	90	91	111	103	135	115	163
68	73	80	91	92	113	104	137	116	165
69	74	81	93	93	115	105	139	117	168
70	76	82	95	94	117	106	142	118	171
71	77	83	96	95	118	107	144	119	173
72	79	84	98	96	120	108	146	120	176
73	80	85	100	97	122	109	149		
74	82	86	102	98	125	110	151		
75	83	87	103	99	127	111	153		
76	85	88	105	100	129	112	156		

METRIC TEMPERATURE-PRESSURE CHART FOR R-134a

°C	kPa	°C	kPa	°C	kPa
18	476	29	676	40	945
19	483	30	703	41	979
20	503	31	724	42	1007
21	524	32	752	43	1027
22	545	33	765	44	1055
23	552	34	793	45	1089
24	572	35	814	46	1124
25	593	36	841	47	1158
26	621	37	876	48	1179
27	642	38	889	49	1214
28	655	39	917		

STANDARD TEMPERATURE-PRESSURE CHART FOR R-12

°F	PSI	°F	PSI	°F	PSI	°F	PSI	°F	PSI
65	74	75	87	85	102	95	118	105	136
66	75	76	88	86	103	96	120	106	138
67	76	77	90	87	105	97	122	107	140
68	78	78	92	88	107	98	124	108	142
69	79	79	94	89	108	99	125	109	144
70	80	80	96	90	110	100	127	110	146
71	82	81	98	91	111	101	129	111	148
72	83	82	99	92	113	102	130	112	150
73	84	83	100	93	115	103	132	113	152
74	86	84	101	94	116	104	134	114	154

METRIC TEMPERATURE-PRESSURE CHART FOR R-12

°C	kg/cm²	°C	kg/cm²	°C	kg/cm²
18	5.2	28	7.0	38	9.0
19	5.3	29	7.1	39	9.2
20	5.5	30	7.2	40	9.4
21	5.6	31	7.5	41	9.6
22	5.8	32	7.7	42	9.9
23	6.0	33	7.9	43	10.0
24	6.1	34	8.1	44	10.4
25	6.3	35	8.3	45	10.7
26	6.6	36	8.5	46	10.9
27	6.8	37	8.7	47	11.0

FIGURE 7-4 **Standard temperature-pressure charts for both R-134a and R-12 can be used to determine whether a refrigerant system is contaminated.**

to be contaminated, a manifold gauge may be connected to the tank; the pressure in the tank will be higher for a given temperature than the chart pressure listed for that temperature. For example, if the tank pressure is 100 psig at 80°F, the tank is contaminated. A refrigerant identifier may also be used to detect the presence of noncondensable gases.

One common mistake that causes air to be left in the refrigerant system after a repair is improper evacuation. The most common mistake is not opening both service port fittings during the evacuation process. To ensure proper system evacuation, both service valves must be open so that both the high and low side of the refrigerant system is put under a vacuum at the same time.

Moisture and Moisture Removal

Moisture is a small quantity of diffused or condensed liquid. Actually, any substance that is not "dry" may be considered "moist." Moisture in an air-conditioning system is one of its greatest perils. Even a slight amount of free moisture, i.e., a water (H_2O) vapor, in an air-conditioning system can play havoc. When it is heated and mixed with refrigerant and oil, it can cause sludge and can form harmful acids that erode system components.

Atmospheric air contains moisture (humidity), and this moisture can enter the refrigerant system during service. The higher the humidity level, the greater the moisture content for any given quantity of air. Moisture removal, however, is not as simple as air removal. To remove all moisture from a system, one must use a vacuum pump capable of achieving and maintaining a deep vacuum for several hours. Any time the refrigerant system is opened for service, extreme care must be taken to limit the system's exposure to atmospheric air. Systems should never be left empty and open for longer than is necessary to complete a repair.

In a deep vacuum, moisture will boil off and be carried out of the system as a vapor. The length of time required depends on three major factors: the amount of moisture in the system; the ambient temperature; and the efficiency of the vacuum pump, which relates to the amount of vacuum that will be applied to the air-conditioning system. Assuming sea-level atmospheric pressure, an air-conditioning system evacuation should be conducted when the ambient temperature is 60°F (15.6°C) or above. Moisture will boil at this temperature in a vacuum of 29.4 in. Hg (2.3 kPa absolute) or better. A temperature-vacuum chart is given in Figure 7-5 for moisture removal at various vacuum levels.

The average service facility may not possess or maintain equipment capable of achieving a sufficient vacuum for complete moisture removal. Because of this, if a high level of moisture is suspected in an automotive air-conditioning system, the general remedy is to replace the accumulator or receiver-drier that contains a desiccant and then evacuate the system to remove excess moisture. Periodic maintenance of the vacuum pump is essential to ensure maximum performance. The vacuum pump lubricant should be changed at frequent intervals recommended by the manufacturer.

New refrigerant is considered to be moisture free. The moisture content of new refrigerant should not exceed 10 ppm. This information is given on the label of the refrigerant container. The SAE purity standards for recycled refrigerant set the maximum allowable moisture content levels to ensure that the refrigerant provides proper system performance and longevity. Standard J2099 states that the maximum allowable of moisture content for R134a is 15 ppm by weight, and standard J1991 states that the maximum allowable moisture content for R12 is 15 ppm by weight.

The term *virgin refrigerant* is used to identify new refrigerant as opposed to recovered refrigerant.

Every time a component is removed from the system for repair or replacement, air is inadvertently introduced into the system. As a result, there is always the danger of moisture entering the system. Refrigerant and refrigeration oil, particularly R-134a and its oil, **absorb** moisture readily when exposed to air. To keep the system as moisture-free as possible, all

To **absorb** is to take in or suck up; to become a part of.

PRESSURE/BOILING POINT RELATIONSHIPS			
Boiling Point of Water (°F)	**Inches of Vacuum**	**Microns**	**Pressure (PSI)**
212	0.00	759,968	14.696
205	4.92	535.000	12.279
194	9.23	525.526	10.162
176	15.94	355,092	6.866
158	20.72	233.660	4.519
140	24.04	149,352	2.888
122	26.28	92.456	1.788
104	27.75	55,118	1.066
86	28.67	31,750	.614
80	28.92	25.400	.491
76	29.02	22,860	.442
72	29.12	20,320	.393
69	29.22	17,780	.344
64	29.32	15,240	.295
59	29.42	12,700	.246
53	29.52	10, 160	.196
45	29.62	7.620	.147
32	29.74	4,572	.088
21	29.82	2,540	.049
6	29.87	1,270	.0245
−24	29.91	254	.0049
−35	29.915	127	.00245
−60	29.919	25.4	.00049
−70	29.9195	12.7	.00024
−90	29.9199	2.54	.000049

FIGURE 7-5 **The table illustrates the boiling point of water (H_2O) under a vacuum. Use the table to determine the level of evacuation needed based on ambient air temperature.**

automotive air-conditioning systems have an accumulator or a receiver-drier that contains desiccant, a drying agent (Figure 7-6). Desiccants are chemicals that are capable of absorbing and holding moisture.

The oil used with R-134a(HFC-134a) is very hygroscopic.

Any moisture introduced into the system in excess of the amount that the desiccant can handle is free in the system. Even one drop of free moisture (H_2O) cannot be controlled and may cause irreparable damage to the internal parts of the air conditioner.

Moisture in concentrations greater than 20 ppm may cause serious damage. For an idea of how small an amount 20 ppm is, consider one small drop of water in a system having a capacity of 6 lb. (0.94 mL) of refrigerant. That one small drop amounts to 20 ppm, or twice the amount that is desired.

Hydrochloric acid (HCl) is a corrosive acid produced when water (H_2O) and R-12 are mixed, as occurs within an automotive air-conditioning system.

Refrigerants react chemically with water (H_2O) to form **hydrochloric acid (HCl)**. Heat, which is generated in the system, is an aid in the acid-forming process. The greater the concentration of moisture in the system, the more concentrated are the corrosive acids that are formed.

Hydrochloric acid (HCl) corrodes all the metallic parts of the system, particularly those made of steel. Iron (Fe), copper (Cu), and aluminum (Al) parts are damaged by the acid as well. The corrosive process also creates oxides that are released into the refrigerant as particles of

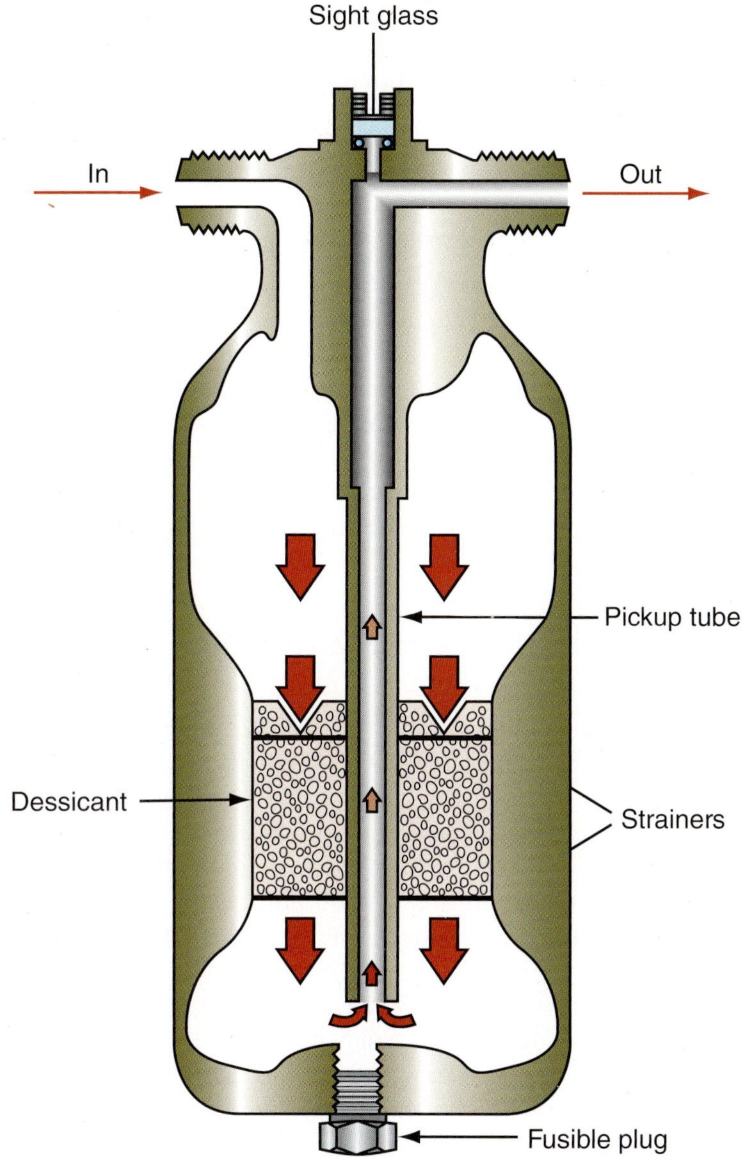

FIGURE 7-6 **Cutaway of receiver-drier showing construction details.**

metal to form a sludge. Further damage is caused when oxides plug the fine-mesh screens in the metering device, compressor inlet, and the drier or accumulator.

There are commercially available "remedies" that claim to prevent moisture freeze-up problems. There is also a notion that system freeze-up can be avoided by adding 0.07 oz. (0.002 mL) of alcohol per pound (0.03 mL) of refrigerant. However, the addition of alcohol to the system may cause even greater damage. The drier seeks out alcohol even more than moisture and, in doing so, releases all of its moisture to the system. This can cause severe damage to the system components. Once a system is saturated with moisture, irreparable damage is done to the inside of the system. If the moisture condition is neglected long enough, pinholes caused by corrosion appear in the evaporator and condenser coils and in any metal tubing used in the system. Any affected parts must be replaced.

The average automotive air-conditioning system contains 3 lb. (1.42 mL) or less of refrigerant.

Additives, which are marketed under various trade names, are available that make the claim of increasing the cooling effect and/or stopping leaks. Most additives have a negative effect on the performance of an air-conditioning system, however, and are not recommended.

An additive may cause more problems than it will solve.

Pump down is the process of evacuation whereby a liquid is changed to a vapor by lowering pressure exerted on the liquid, thus lowering its boiling point.

Whenever there is evidence of moisture in the system, a thorough system cleanout is recommended. Such a cleanout should be followed by the installation of a new receiver-drier or accumulator and a complete system **pump-down** evacuation using a vacuum pump, charging station, or recovery system.

Prevention

The automotive air-conditioning service technician can prevent unwanted debris and moisture from entering a system by following a few basic rules:

- Always install the receiver-drier or accumulator last.
- Always immediately cap the open ends of hoses and fittings.
- Do not work around water, outside in the rain, or in very humid locations.
- Do not allow new refrigerant or refrigeration oil to become contaminated.
- Keep the refrigeration oil container capped when it is not being used.
- Develop clean habits: Do not allow dirt to enter the system; keep all service tools clean; and never charge a unit with refrigerant without first ensuring that all air and moisture have been removed.

Do not remove the protective caps from the accumulator or receiver-drier until it is to be installed.

Evacuating the Refrigeration System (Moisture Removal)

As discussed earlier, many problems can arise due to moisture in an automotive air-conditioning system. After any repairs have been made to the system, it must be "pumped down" (evacuated) to remove any moisture that may have entered during the process.

Shop Manual
Chapter 7, page 250

Moisture removal from a system can cause serious problems for the service technician who is not properly equipped. A vacuum pump is an essential tool for air-conditioning service. Other methods may be used, but the vacuum pump is by far the most efficient means of moisture removal. Typical vacuum pumps suitable for automotive service are shown (Figure 7-7).

Pressure below atmospheric is referred to as a vacuum.

A pressure below 0 lb. (0 psig) or 0 kiloPascal (0 kPa) gauge pressure is referred to in terms of inches of mercury (in. Hg) on the English scale or kiloPascals absolute (kPa absolute) on the metric scale. Moisture is removed from the air-conditioning system by creating a vacuum. In a vacuum, the moisture within the system boils. The action of the vacuum

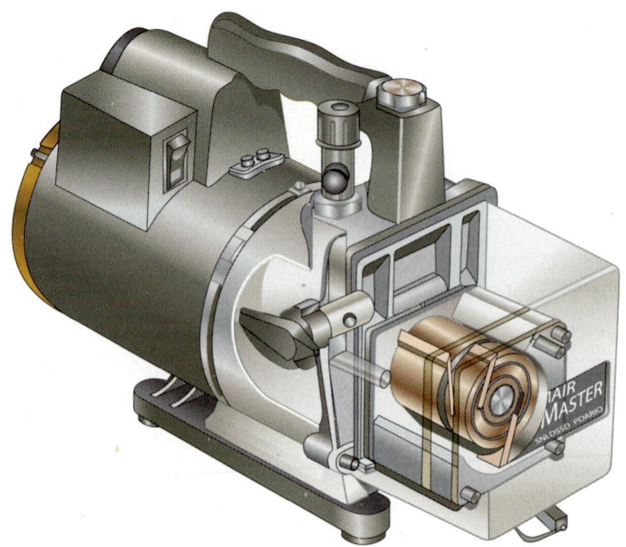

FIGURE 7-7 **A typical two-stage vacuum pump.**

pump then pulls the moisture in the form of a vapor from the system. When the pressure is increased on the discharge side of the pump, the vapor again liquifies. This process usually occurs inside the pump.

A minimum of 30 minutes is required to ensure moisture removal with a compressor speed of about 1,500 revolutions per minute (rpm or r/min.). The compressor is lubricated by refrigeration oil in its pump. Some of the oil is picked up in the refrigerant vapor. If the compressor runs dry of oil when it is operated as a vacuum pump, it may be seriously damaged.

Refrigeration oil is hygroscopic and will absorb moisture.

A good vacuum pump is capable of pumping a vacuum pressure of 29.76 in. Hg (0.81 kPa absolute) or more. At this pressure, water boils at 40°F (4.44°C). In other words, if the ambient temperature is 40°F (4.44°C) or higher, moisture will boil out of the system.

At 0 in. Hg (101.3 kPa absolute) at sea level, water boils at 212°F (100°C). To find the boiling point of water in a vacuum (absolute pressure), use the table in Figure 7-5. Note that the boiling point is lowered only 112°F (62.2°C) to 100°F (37.7°C) as the pressure is decreased from 0 in. Hg (101.3 kPa absolute) at sea level to 28 in. Hg (0.98 kPa absolute). However, the boiling point drops by 120°F (66.6°C) as the pressure decreases from 28 in. Hg (0.98 kPa absolute) to 29.91 in. Hg (0.30 kPa absolute).

The degree of vacuum and the amount of time the system is under a vacuum determines the amount of moisture removed. When evacuating a refrigerant system, it is important to apply the proper vacuum level for the ambient temperature that exists. To determine the minimum vacuum levels required to remove moisture from a refrigerant system refer to the chart in Figure 7-8. The deeper the vacuum or the longer the time, the more moisture is removed. The removal of moisture from a system can be compared to the boiling away (vaporization) of water in an open saucepan on a hot burner. It is not sufficient to cause the water to boil; time must be allowed for it to boil away.

Contaminated vacuum pump oil can reduce the efficiency of the vacuum pump by 10 percent or more.

The recommended minimum pumping time is 30 minutes below 29 in. Hg. If time allows, however, a four-hour pump down achieves much better results. Vacuum pump manufacturers' specifications and recommendations should be followed for proper maintenance. For example, changing oil on a regular basis is essential to ensure maximum efficiency.

If a system is suspected of being extremely contaminated with water that has caused expansion valve icing or if the evaporator exhibited a whistling noise caused by ice restriction, the evacuation process may be made more efficient by increasing the temperature of the components to which the moisture is suspected to have migrated (in other words, expansion valve, evaporator core). Start the vehicle and allow it to reach operating temperature. Once at operating temperature, turn the blower motor on high, select the Recirculation mode, set the temperature control to hot, and close all windows and doors. Begin the evacuation process. This procedure will increase the evaporator temperature to above 150°F and greatly reduce the evacuation time.

The most accurate method for measuring the effectiveness of system evacuation is with the use of a thermistor vacuum gauge or micron meter. A thermistor vacuum gauge measures 29.00 in. Hg −29.9199 (the last inch of vacuum) with extreme accuracy in very small increments called microns. An inch of vacuum contains 25,400 microns. A refrigerant system is considered under a deep vacuum when it is evacuated down to 700–300 microns (Figure 7-4). When all the water is removed from the system, you will achieve the lowest evacuation vacuum possible. The ultimate goal is to achieve the deepest possible vacuum to ensure that 100 percent of the water is removed. The refrigerant system should hold a minimum of a 700-micron vacuum for three minutes after turning off the vacuum pump and isolating the system if all the moisture and air has been removed. If the gauge pressure rises above 700 microns, the evacuation process must be repeated.

The most efficient evacuation pumps are dual-stage 6 or 8 cubic feet per minute (cfm) vacuum pumps, which will produce vacuums as low as 20 microns. The larger the vacuum

BOILING TEMPERATURES OF WATER AT CONVERTED PRESSURES			
Temperature in (°F)	Inches in Vacuum	Microns	Pounds Sq. in (Pressure)
212°	000	759,968	14,696
205°	5.00	535,000	12,279
194°	9.81	525,526	10,162
176°	16.02	355,092	6,866
158°	20.80	233,680	4,519
140°	24.12	149,352	2,888
122°	26.36	92,456	1,788
104°	27.83	55,118	1,066
86°	28.75	31,750	.614
80°	29.00	25,400	.491
76°	29.10	22,860	.442
72°	29.20	20,320	.393
69°	29.30	17,780	.344
64°	29.40	15,240	.295
59°	29.50	12,700	.246
53°	29.60	10,160	.196
45°	29.70	7,620	.147
32°	29.82	4,572	.088
21°	29.90	2,540	.049
6°	29.95	1,270	.0245
24°	29.99	254	.0049
35°	29.99.5	127	.00245

FIGURE 7-8 **Refer to chart to determine the minimum vacuum levels required to remove moisture from a refrigerant system for a given temperature.**

pump, the shorter the evacuation time. As a rule, you cannot buy too large a vacuum pump. This is one area where you should not cut costs.

Moisture Removal at High Altitudes

The information given for moisture removal by a vacuum pump applies to normal atmospheric pressures at sea level, 14.696 (14.7) psig (101.3 kPa absolute). At higher altitudes, the boiling point must be reduced to a point below the ambient temperature. Moisture (H_2O) boils at a lower temperature at higher altitudes. However, it must be pointed out that vacuum pump efficiency is reduced at higher altitudes.

For example, the altitude of Denver, Colorado, is 5,280 ft. (1,609.3 m) above sea level. Water (H_2O) will boil at 206.2°F (96.78°C) at this altitude. The maximum efficiency of a vacuum pump, however, is reduced at this altitude. A vacuum pump that can pump 29.92 in. Hg (0.27 kPa absolute) at sea level will pump only 25.44 in. Hg (15.44 kPa absolute) at this altitude. Note in Figure 7-10 that water (H2O) boils at about 130°F (54.4°C) at this pressure.

The English formula for determining vacuum pump efficiency at a given atmospheric pressure is:

$$AP_L / AP_S \times PRE = APE$$

where AP_L = atmospheric pressure in your location

AP_S = atmospheric pressure at sea level

PRE = pump rated efficiency

APE = actual pump efficiency

Assume that a vacuum pump has a rated efficiency of 29.92 in. Hg at sea level (0.27 kPa absolute) and that the atmospheric pressure at Denver is 12.5 psia (86.18 kPa absolute). To determine the actual efficiency at this location, the formula is:

$$\frac{12.5}{14.7} \times 29.92 = 25.44 \text{ in. Hg}$$

The metric formula for determining vacuum pump efficiency at a given atmospheric pressure is:

$$\text{Atmospheric Pressure at Sea level} - \text{Atmospheric Pressure in Your Location} + \text{Original Efficiency} = \text{Actual Efficiency}$$

Assuming the same conditions previously mentioned, the formula is applied in the following manner:

$$101.32 - 86.18 + 0.27 = 15.41 \text{ kPa absolute}$$

In this example, the ambient temperature must be raised above 130°F (54.44°C) if the vacuum pump is to be efficient for moisture removal. To increase the ambient temperature under the hood, the automobile engine can be operated with the air conditioner turned off. The compressor, condenser, and some of the hoses may be heated sufficiently; however, some other parts, such as the evaporator and the receiver-drier, will not be greatly affected. Do not overheat components containing refrigerant. Hydrostatic pressure can build up rapidly and rupture the component.

Evacuating an automotive air-conditioning system when the ambient temperature is below, say, 60°F (16°C) is generally very inefficient. To remove moisture at this temperature, the vacuum pump must pull a minimum of 29.4 in. Hg (2.03 kPa absolute). Unless well maintained, many shop vacuum pumps will not reach the level required for adequate moisture removal at low temperatures.

Another method of moisture removal is the *sweep* or *triple evacuation* method. Although this method cannot remove all the moisture, it should be sufficient to reduce the moisture to a safe level if the system is otherwise sound and a new drier is installed.

Mixed Refrigerant Types

It is not uncommon to find that R-134a has been added to an R-12 system or vice versa. Sometimes it is found that a blend refrigerant has been added to an R-12 or R-134a system. If this is the case, the refrigerant must be recovered, and the system should be evacuated and charged with the proper refrigerant. It is not possible to separate the refrigerants with currently available equipment. Recovery should be made using equipment dedicated to contaminated refrigerant and stored in a dedicated cylinder also designated for contaminated refrigerant.

In addition to the flammable refrigerants, identified below as hazardous, the following refrigerants are NOT approved for automotive use under the SNAP program rule of the

EPA. R-176 contains R-12 and, therefore, is inappropriate as an R-12 substitute, and R-405A contains perfluorocarbons, which have been implicated in global warming.

Hazardous Refrigerant

A refrigerant identifier must be used to ensure the purity of the refrigerant before service is performed on the vehicle. Hazardous refrigerants are generally those that contain an excessive amount of a flammable substance and are therefore not approved by the SNAP program. If a system is known or suspected to contain a flammable substance, it must be recovered in a manner consistent with the specific instructions provided with the particular recovery equipment used.

At the present time, flammable blend refrigerants that are NOT approved for automotive (or any other) use include OZ-12®, HC-12a®, and Duracool 12a®.

Leak Detectors

A cold leak is generally the most difficult to locate.

There are two general types of automotive air-conditioning system leaks: cold leaks and hot leaks. A cold leak occurs when the system is not at its operating temperature and pressure, such as when it is parked overnight. A hot leak occurs at periods of high pressure, such as when the vehicle is moving slowly in heavy traffic.

The methods of detecting leaks in an air-conditioning system range from using an inexpensive soap solution to using an expensive self-contained electronic instrument.

Shop Manual
Chapter 7, page 245

Leak Detection Using a Soap Solution

A soap solution is an efficient method of locating small pinhole-sized leaks. Leaks often occur in areas of limited access where it is impractical or unhandy to use a halide or **Halogen** leak detector.

A sudsing liquid detergent may also be used undiluted.

Halogen refers to any of the five chemical elements-astatine (At), bromine (Br), chlorine (Cl), fluorine (F), and iodine (I)-that may be found in some refrigerants.

If a commercial soap leak detector solution is not available, mix one-half cup of soap powder with water to a thick consistency. The mixture should be just light enough to produce suds when applied with a small brush. When the mixture is applied to the area of a suspected leak, bubbles will reveal the leak (Figure 7-9).

FIGURE 7-9 **Bubbles reveal the point of the leak.**

Leak Detection Using Visible Dye

Shop Manual
Chapter 7, page 247

To locate a difficult leak, it may be desirable to **inject** a dye solution into the system. The dye approved for automotive air-conditioning system use is available in either yellow or red. When it is used properly, it will not affect the overall performance of the system. After injection, the automobile is driven for a few days. The leak can then be visibly detected by the dye trace at the point of the leak.

CFCs are no longer manufactured, as of 1995.

Once a dye is introduced into the system, it will remain there until the system is completely cleaned out, the oil changed, and the accumulator or receiver-drier replaced. Even then, a slight residue may be noticed in the sight glass.

To **inject** is to insert by force or pressure.

One manufacturer, E. I. DuPont de Nemours and Company, produced Refrigerant-12 with a red dye. Called Dytel®, this refrigerant is charged into the system in the same manner as any other refrigerant. It is soon to become scarce, however, as supplies dwindle due to the CFC phaseout.

It is important to note that most leak detector dye solutions are compatible only with certain systems. For example, a dye solution intended for an R-12 system is not satisfactory for use in an R-134a system. Conversely, R-134a dye should not be introduced into an R-12 system. When retrofitting a system that contained dye, it is important that all residual dye be flushed out of the system.

Before using a dye solution, be sure to determine that it is approved for such use and will not affect air-conditioning system performance or any vehicle warranty conditions. This assurance generally should be given in writing by the product manufacturer.

Fluorescent Leak Detectors

There are several manufacturers of fluorescent leak detectors. A metered amount of ultraviolet-sensitive dye is injected into the system (Figure 7-10). The air conditioner is operated for a few minutes to allow time for the dye to circulate. An ultraviolet lamp is then used to pinpoint the leak. When the ultraviolet light beams come in contact with the dye that has leaked out of the system, the dye will give off a fluorescent glow. It is advisable to first wash the engine compartment before installing dye into the system. In addition, it is easier to detect the dye if ship light levels are low. Though it is not inexpensive, the ultraviolet method of leak detection is most effective for locating small, difficult-to-find leaks.

One-half oz. (0.015 mL) per year is equal to 1 lb. (0.472 mL) in 32 years.

Some automobile manufacturers add fluorescent dye, called scanner solution, to factory-installed air conditioners. The refrigerant identification label under the hood will identify the presence of a leak-detecting agent installed in the system. Ford Motor Company started this practice in 1996. Many technicians add scanner solution to systems being serviced for future

FIGURE 7-10 **A typical fluorescent leak detector.**

troubleshooting. As with any additive, it is important to use the proper scanner solution to ensure system compatibility. More critical than type of refrigerant is the type of lubricant in the system. The scanner solution has either mineral, alkyl benzine, PAG, or polyol ester base stock to match the system lubricant.

Using the improper scanner solution can contaminate an otherwise healthy system. Just 0.3 oz. (0.89 mL) is sufficient to treat a system with a refrigerant capacity up to 2.9 lb. (1.21 L). The average system, however, requires 0.5 oz. (14.79 mL) of scanner solution. This is sufficient for a system with a capacity up to 4.9 lb. (2.33 L) of refrigerant.

Ultraviolet Leak Detection: Stability and Compatibility Criteria of Fluorescent Refrigerant Leak Detection Dyes for Mobile R-134a and R-1234yf (HFO-1234yf) Air-Conditioning Systems must meet J2297 standards (revised 2/2011).

Because the scanner solution is soluble in the lubricant, the refrigerant is recyclable and is accepted by most manufacturers and reclaimers. If retrofitting a system, be sure that all trace of the scanner solution is removed with the lubricant before introducing a new refrigerant.

Electronic (Halogen) Leak Detectors

The electronic (halogen) leak detector is the most sensitive of all leak detection devices. The initial purchase price of a halogen leak detector, however, exceeds the cost of most fluorescent leak detectors. In addition, this more sophisticated device requires routine maintenance in order to maintain accuracy.

Electronic leak detectors must be capable of detecting a refrigerant loss rate of 0.15 oz (4g) per year. This value corresponds to one part of refrigerant in 10,000 parts of air or 100 parts per million (ppm).

Electronic leak detectors are either corded or cordless (Figure 7-11). The corded leak detector operates on 120 volts (V), 60 hertz (Hz). The cordless leak detector is portable and operates from a rechargeable battery. Both units are simple to operate and easy to maintain. When the halogen leak detector comes in contact with refrigerant vapor, the audible click noise emitted from the device will become more rapid. If the unit is also equipped with an LED light bar, additional lights will illuminate. The halogen leak detector also allows the technician to diagnose the leak the same day the vehicle is in for the repair. This saves the customer both time and money and avoids the aggravation involved to both you and the customer of having to make a return visit.

Shop Manual
Chapter 7, page 247

A popular portable halogen leak detector is the model 5750 manufactured by TIF Industries (Figure 7-12). This instrument, which is powered by two "C" cell alkaline batteries,

FIGURE 7-11 **Two types of electronic leak detector: (A) cordless and (B) corded.**

FIGURE 7-12 **A portable electronic leak detector that may be used for R-12 or R-134a.**

requires no warm-up period and may be used to detect CFCs, HCFCs, or HFCs (R-12, R-22, or R-134a). In addition, this instrument calibrates itself automatically while in use to ignore ambient concentrations of gas and pinpoint leaks much more easily.

An important consideration in the selection of an electronic (halogen) leak detector is that it complies with current SAE standards and EPA regulations. Beginning 2008 SAE J2791 standard for R134a leak detecting equipment went into effect. This SAE standard was created to improve leak detection equipment sensitivity requirements to a sensitivity level of 0.15 oz/year (4g/Yr). It was created in response to the need to reduce refrigerant HFC-134a emissions and to establish minimum performance criteria for handheld electronic leak detection equipment. Leak detectors that meet this standard will have at least three sensitivity scales that can be selected manually: (A) 0.15 oz/year (4g/Yr), (B) 0.25 oz/year (7g/Yr), and (C) 0.5 oz/year (14g/Yr). It must be calibrated to detect a refrigerant leak with two seconds from a distance of 3/8 in (9.5 mm) moving at a rate of 3 in. (75 mm) per second and must be able to self clear itself with two seconds once moved away from the leak. In addition, the leak detector must not false trigger in the presence of engine oil or transmission oil. This is a great improvement over earlier refrigerant leak detection equipment.

Equipment produced prior to 2008 was required to meet SAE standard J1627 which stated that equipment had to be capable of detecting a leak at the rate of 0.5 oz/year (14g/Yr) or less per year. Make sure you are using the correct equipment for the vehicle you are working on. Please note that some older electronic leak detection equipment manufactured under the SAE J1627 regulation was build to a sensitivity level that already meets the SAE J2791 standard.

A BIT OF HISTORY

The halide leak detector (Figure 7-13) was once one of the most popular tools for locating R-12 refrigerant leaks. Today, though, the halide leak detector is considered an obsolete and potentially unsafe tool for refrigerant detection by the air-conditioning industry.

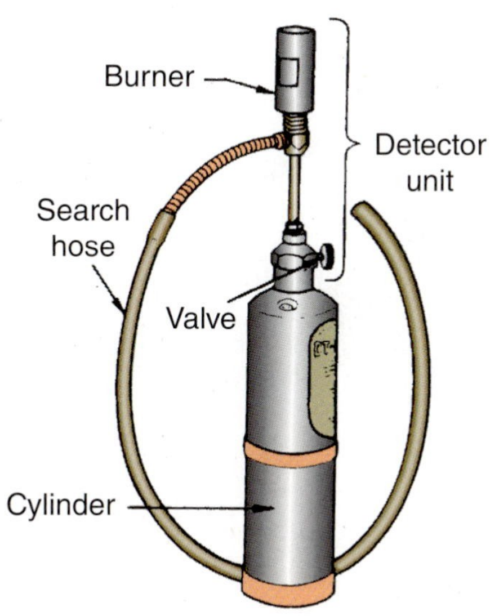

FIGURE 7-13 **A typical halide leak detector for CFC-12 (R-12).**

Refer to individual equipment specifications data sheets for older equipment to see if equipment updating is required.

With the advent of R-1234yf, a new standard for electronic leak detection equipment was developed. The new standard is SAE J2913 and went into effect in 2011. Leak detectors meeting this standard are able to differentiate between a 4 gram per year (high sensitivity), 7 gram per year (medium sensitivity), and 14 gram per year (low sensitivity) (0.141, 0.247, and 0.5 oz). Combination electronic leak detectors are available that meet both J2913 and J2791 and can be used to detect either R-134a or R-1234yf.

Other Types

Any leak, regardless of how slight, emits an inaudible noise.

Several other types of leak detectors, such as ultrasonic units, are available. Space does not permit the description of each type in this text. The technician should contact local refrigeration suppliers for additional information. This will allow a comparison of the different makes and models before purchase. When making a selection, do not overlook the commercial refrigeration parts supply houses. They often have a greater variety to choose from than the automotive parts supply houses.

Regardless of what type of leak detection device you are using, you need to take a systematic approach to detecting leaks. This means you should test the system when it is cold, having sat for several hours or overnight, and again when it is at operating temperature. You need to begin by looking at the most likely areas for a leak to occur. These areas include connection points of lines and components, as well as areas that contain gaskets and seals (Figure 7-14).

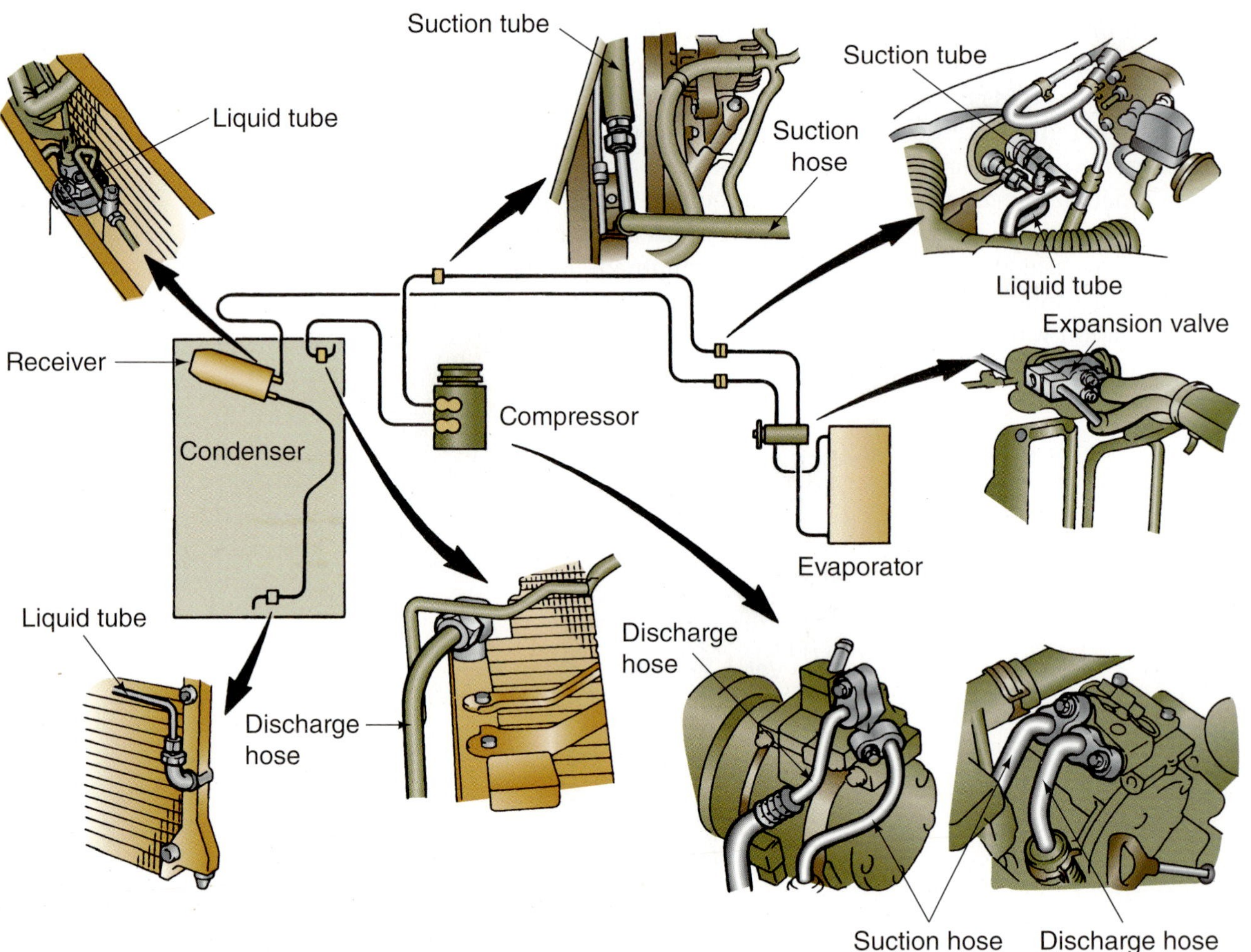

FIGURE 7-14 **First check connection points of fittings for signs of refrigerant leakage.**

Recovery/Recycling/Recharging Systems

There are many manufacturers of refrigerant recovery/recycling/recharging (R/R/R) systems, each of which produces several models. When considering recovery equipment, there are three terms to become familiar with: recover, reclaim, and recycle.

Recover

To **recover** refrigerant is to remove it in any condition and store it in an external container without necessarily processing it further. Under certain conditions, this refrigerant is returned to the system from which it was removed. It may also be sold to a reclamation center where it is processed to new product specifications.

To **recover** is to remove refrigerant in any condition from a system and transfer it to an external storage container without necessarily testing or processing it in anyway.

Reclaim

To **reclaim** refrigerant is to remove it from a system and reprocess it to new product specifications (ARI 700-88 standards). Analytical testing is required. This is an off-site procedure by a laboratory equipped to run such tests.

To **reclaim** is to process used refrigerant to new product specifications by means that may Include distillation. This process requires that a chemical analysis of the refrigerant be performed to determine that appropriate product specifications are met. This term implies the use of equipment for processes and procedures usually available only at a reprocessing facility.

Recycle

To **recycle** refrigerant is to remove it from a system and reduce contaminants by oil separation and filter drying to remove moisture, acid, and particulate. This is an on-site procedure, and analytical testing is not required.

One of the major points of the Clean Air Act was to ensure recovery and recycling of refrigerants, specifically R-12, instead of allowing it to be vented into the atmosphere. A service facility that services mobile air-conditioning systems must have the proper recovery/recycling equipment under the EPA Clean Air Act, or it will be charged with "Intent to Vent." The SAE, in conjunction with the EPA, has established guidelines for the recovery and recycling of both R-134a and R-12 refrigerants.

- Effective January 1, 1992, no service facility could service mobile air-conditioning systems unless it acquired approved recovery/recycling equipment and trained and certified service personnel performing said services.
- Recovery of R-134a became mandatory in November 1995.
- Recycling of R-134a became mandatory on January 29, 1998.

Beginning in 1992, R-134a was phased into new vehicle production air-conditioning systems. With the 1994 model year, all vehicles sold in the United States contained the ozone friendly refrigerant R-134a.

- Must have dedicated recovery/recycling equipment for each alternative refrigerant serviced. Refrigerants may not be mixed, and the term *drop-in refrigerant* used by some alternative refrigerants should not imply refrigerants can be mixed. The EPA does not allow the mixing of any refrigerant.
  - Contaminated refrigerant must also be recovered into a dedicated recovery unit for future redemption.
- Beginning on June 1, 1998, refrigerant blends may be recycled, provided that recycling equipment meets Underwriter's Laboratories (UL) standard and refrigerant is returned to the vehicle from which it was removed.
- Beginning December 2007 SAE J2788 standard for R134a recovery/recharge equipment went into effect. This standard requires that any new equipment produced must do a better job at evacuating and recovering refrigerant from the system being serviced. This new equipment must be capable of measuring and displaying the amount of refrigerant recovered to an accuracy of +/−1 oz. In addition, this standard also requires that the recharge accuracy level must be +/−0.5 oz. This regulation is especially important since many systems today are much smaller and contain much less refrigerant when fully charged. It is not unusual to see refrigerant system volumes of less than one pound.

To **recycle** is to clean refrigerant for reuse by oil separation and pass it through other devices such as filter-driers to reduce moisture, acidity, and particulate matter. Recycling applies to procedures usually accomplished in the repair shop or at a local service facility.

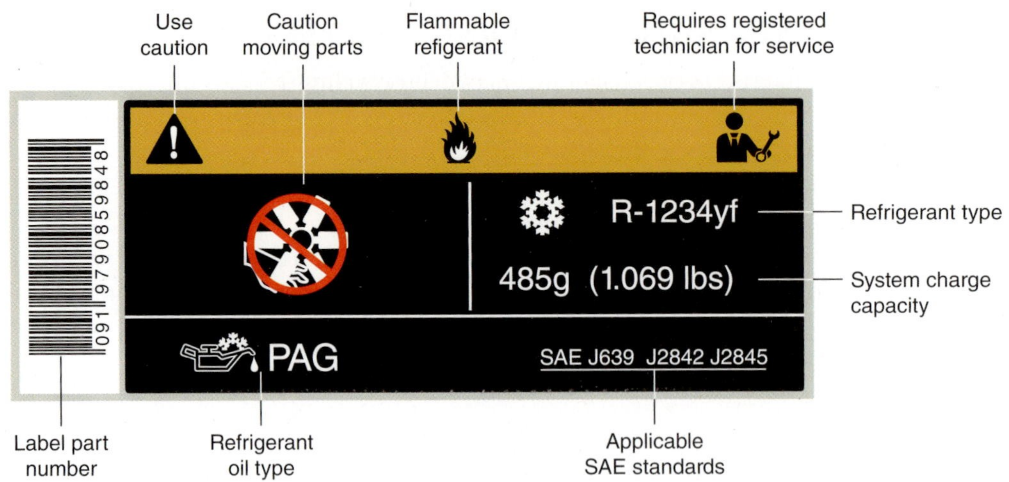

FIGURE 7-15 **The R-1234yf decal.**

- SAE J2788H is specified for recovery equipment intended for use on hybrid vehicles that use high-voltage (HV) electric air-conditioning compressors, the "H" suffix denotes hybrid. It is imperative that a high-voltage-driven compressor oil is not cross contaminated during service since the HV compressor uses a nonconductive dielectric POE oil compared to a PAG oil that is conductive in a conventional belt-driven R-134a system.

WARNING: R-1234yf is a mildy flammable refrigerant, never smoke or expose refrigerant to an open flame, hot surface, high-energy ignition source, sparks, or secondary ignition components. This warning is also on the underhood label as is indicated by the flammability symbol (Figure 7-15).

Recovered refrigerant should not be used in another application unless it is first reclaimed.

R-1234yf refrigerant recovery/recycling/recharge (R/R/R) equipment must meet SAE J2843, charging accuracy with +/− ½ oz, automatic oil drain, automatic air purge, and integrated internal J2927 refrigerant identifier (Figure 7-16) or external J2912 refrigerant identifier plugged in. The standard requires that the vehicle's air-conditioning system be analyzed for purity prior to refrigerant recovery or transfer. The equipment must have either a built-in identifier or be inoperable if one is not connected, generally through a USB connector. An acceptable reading that must be received prior to recovery/recycling is 98 percent pure or greater. If an unacceptable reading is detected, such as contamination by another gas (R-134a), the refrigerant is considered contaminated and the equipment will not allow recovery to proceed. If a system is contaminated, a separate recovery-only machine must be used to recover and store the contaminated refrigerant for either reclamation or disposal at an EPA-approved facility.

Before the recovery/recycling/recharge equipment can recharge a vehicle's refrigerant system, it must first pass an automated precharge leak test to detect the possibility of a gross system leak greater than 0.3 g/s before the system is charged. During the pressurized portion of the test, the technician must turn the vehicle's HVAC system blower motor to high, turn off the A/C, and set air distribution to floor discharge. Next, the technician must use a J2913-compliant leak detector set to low sensitivity (14 g/year) into the center duct of the floor discharge. Once the technician has set up the vehicle and leak detection equipment, the R/R/R machine will install 15 percent of the programmed charge into the vehicle's refrigerant system. The technician must then monitor the leak detector for the next 5 minutes or until a leak is detected, whichever comes first. After 5 minutes, the sytem will ask a series of questions:

1. Was the leak test performed? Y/N
2. If yes, Was a leak found? Y/N. If the technician answers yes then the machine will only allow recovery and evacuation.

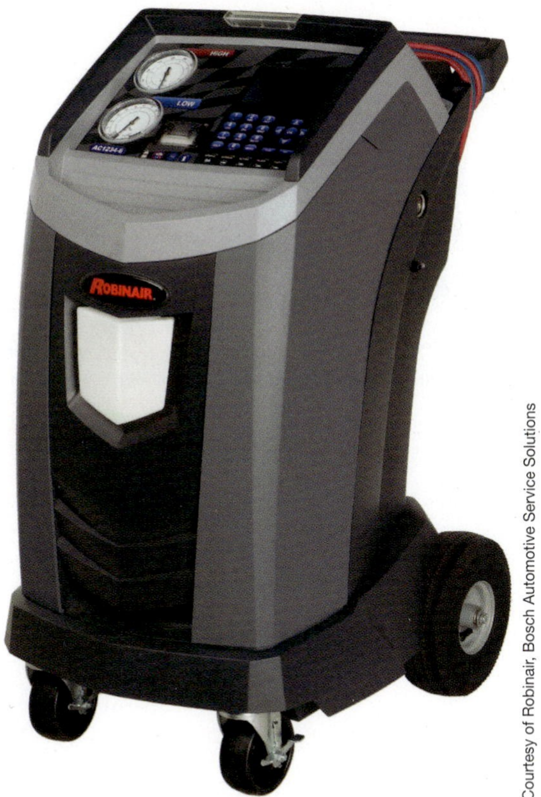

FIGURE 7-16 **The R-1234yf recover/recycle/recharge equipment must meet SAE J2843 and looks very similar to a conventional R-134a machine but has an integrated refrigerant identifier that meets SAE J2927.**

3. If no, Is there an auxiliary evaporator? If no, then the machine will allow the recharge procedure to continue and will complete the system recharge. If the technician answered yes, then the machine will instruct the technician to perform a rear evaporator leak test as was performed for the front evaporator.
4. Next, was an auxiliary evaporator leak test performed? If yes, was a leak found? Again, if a leak was noted then only recovery and evacuation will be allowed. If no, then the machine will complete the system recharge.

The J2843 equipment is a little more time-consuming to use but it is intended to be as environmentally conscious as possible and force technicians to follow proper protocol.

The make and model of the equipment selected for recovering and recycling refrigerants should be based on the needs of the service facility. For example, if the air-conditioning service is occasional, a simple recovery unit like the one shown in Figure 7-17 should suffice. This inexpensive unit can recover about 0.5 lb. (0.236 mL) per minute. A separate recovery unit, pictured in Figure 7-18, frees up the service technician for other service work while a system is being evacuated. This unit costs about twice as much as the first unit and removes about 0.78 lb. (0.368 mL) per minute. Neither system, however, can be used for recycling.

Most service centers impose a recovery fee in addition to their regular charges.

Many recovery systems are designed to be used for CFC and HCFC recovery only. Some recovery systems may be used for both CFC and HFC refrigerant. These systems have special provisions to prevent mixing the refrigerants. The automotive service technician is primarily concerned with two refrigerants: R-12, a CFC, and R-134a, an HFC. For full service, if a CFC/HFC combination system is not available, the service facility must have two recovery systems: one dedicated for R-12 service (Figure 7-19) and another dedicated for R-134a service (Figure 7-20). This may seem to be an expensive investment, but considering the high cost of refrigerants, an early payback may be realized.

Shop Manual
Chapter 7, page 254

FIGURE 7-17 A mechanical refrigerant pump used to recover refrigerant.

FIGURE 7-18 A typical recovery unit.

FIGURE 7-19 An R-12 recovery unit suitable for automotive service.

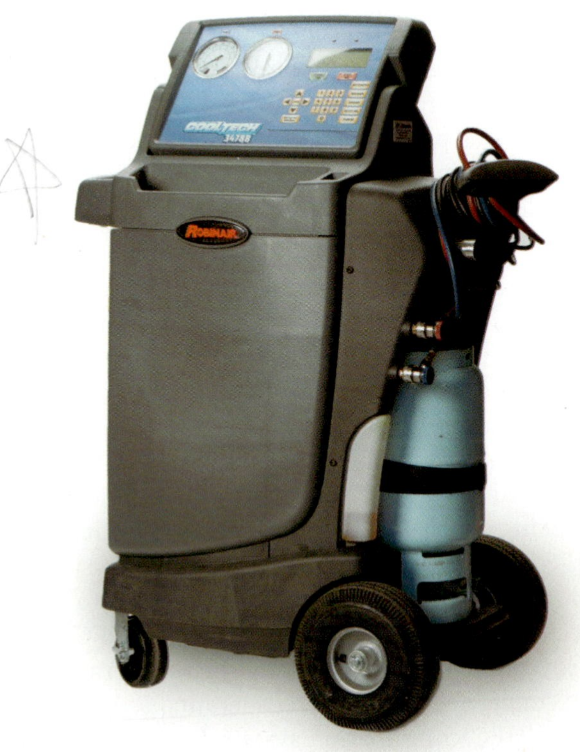

FIGURE 7-20 An R-134a Recovery/Recycle/Recharge unit that meets suitable for servicing today's refrigerant systems.

Most systems manufactured for automotive air-conditioning system service are used for recovering, evacuating, recycling, and recharging. Current recover–recycle–recharge systems automatically control the functions of the equipment. This eliminates the need for personal attention and the requirement for a separate vacuum pump and charging station. Current equipment also has automatic air purge capabilities to detect and vent air from the storage cylinder.

A single-pass recycling machine cleans and filters refrigerant as it is being recovered. A multipass recycling machine recovers the refrigerant in one operation and recycles it through multiple filters, driers, and separators in another operation.

It is important that the manufacturer's maintenance and operational instructions be followed for optimum equipment performance and service. Improper start-up procedures, for example, may induce unwanted air into the system.

Record Keeping Requirements

The EPA has established that service shops must maintain records of the name and address of any facility to which refrigerant is sent. In addition, if refrigerant is recovered and sent to a reclamation facility, the name and address of that facility must be kept on file. Service shops are also required to maintain records (on-site) showing that all service technicians are properly certified.

Service shops must certify to the EPA that they have purchased or acquired and are properly using approved refrigerant recovery equipment for both R-12 and R-134a. This is accomplished by sending a form to the EPA certifying that the shop owns the equipment. This only has to be done once, and it is not required if additional equipment is purchased. Shops must also certify that each person using the equipment has been properly trained and that technicians who repair or service R-12 and R-134a have been certified by an EPA-approved organization. Additional information regarding EPA regulations and how they relate to the mobile air-conditioning industry may be found at the EPA Web site, www.epa.gov/ozone/title6/609.

PDA Diagnostics

A personal digital assistant (PDA) has become as common as the cell phone, and today PDAs are also part of the automotive diagnostic industry. There are programs and interfaces that allow you to use them as onboard diagnostic scan tools, digital storage oscilloscopes, and as air-conditioning system diagnostic and testing tools.

The PDA tool is not intended to replace the other air-conditioning service equipment that you have but is just one more tool to aid you in quickly and accurately diagnosing and repairing air-conditioning problems. One advantage of these tools is that they take a systematic approach to testing and analyzing an air-conditioning system. A systematic approach generally only comes with years of experience, and even then no two technicians approach a problem in the same way.

One of these systems is produced by Neutronics and is called the Master A/C System Technician (Figure 7-21). This is an all-in-one tool for identifying refrigerant and checking the system pressures. It also provides a PDA interface with updatable cartridges for step-by-step diagnostic procedures.

AUTHOR'S NOTE: Being a service technician today involves more than just staying up to date on current service trends and technology. It also involves knowing what changes are occurring at both the state and federal levels and how these changes affect your ability to perform certain services on vehicles. To stay in tune with changes in our industry, you must read trade journals and join trade organizations. Two good organizations to look into are Society of Automotive Engineers (SAE), www.sae.org and Mobile Air Conditioning Society (MACS), www.macsw.org.

Higher than normal high-side pressure does not necessarily indicate an overcharge of refrigerant.

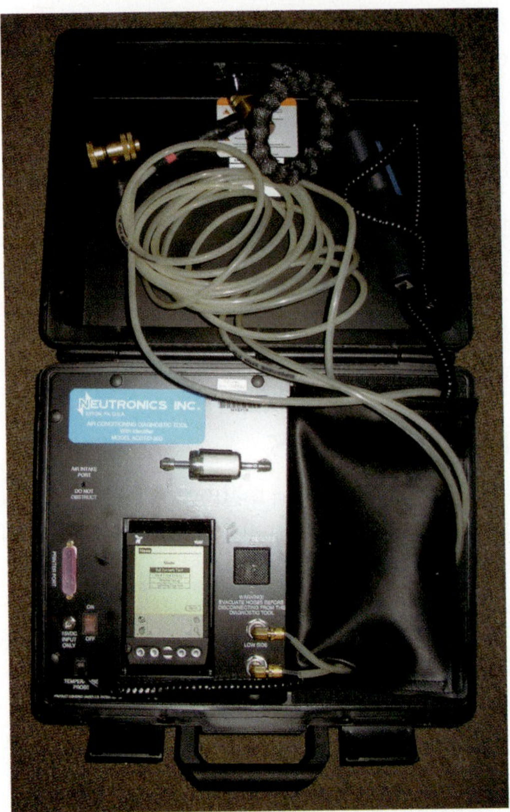

FIGURE 7-21 **A simple-to-use Neutronics Inc. PDA-based Master A/C System Technician is a diagnostic tool designed to analyze refrigerant system pressures and temperatures to determine the most likely cause of system malfunctions.**

Charging the System with Refrigerant

Shop Manual
Chapter 7, page 264

Charging the air-conditioning system with the correct type and quantity of refrigerant is, perhaps, the most important service procedure that the technician will perform. Proper operation and durability of the system may be directly linked to this procedure. An improperly charged system not only results in less than maximum performance but also leads to inaccurate diagnosis that may result in unnecessary repairs. Before charging an air-conditioning system, determine the type of refrigerant.

An undercharged system will:

- Result in inadequate cooling under high load conditions.
- Cause the compressor to cycle rapidly due to the action of the clutch cycling pressure switch (if equipped).
- Aid in early compressor failure.

An overcharged system will result in:

- Higher than normal high-side pressures.
- Reduced cooling capacity under any load condition.
- Improperly operating pressure controls.
- Early compressor failure.

Even a minor error in refrigerant charge level (weight) can affect air-conditioning system performance. This is particularly true of small capacity systems, those under 1.5 lbs. of refrigerant. Air-conditioning systems today are smaller than ever before and more efficient and still offer outstanding performance. In addition, systems manufactured today are much less susceptible to leaks and can go five years or longer on the original factory charge.

To address the issue of small capacity refrigerant systems, the EPA adopted SAE standard J2788 effective December 2007, covering the accuracy level of future refrigerant recovery, recycling, and recharging equipment. Standard J2788 supersedes the previous standard J2210. Refrigerant recovery, recycling, and recharging equipment must now be certified J2788 compliant. This equipment must be capable of measuring and displaying the amount of refrigerant recovered to an accuracy level of ±1 oz. When recharging a refrigerant system with an SAE J2788-compliant machine, it must charge a system with ±0.5 oz. of accuracy. It should be noted that refrigerant service equipment built prior to the SAE J2788 standard taking effect may have displays down to a tenth of a pound, but this does not mean that they are accurate to a tenth of a pound. Always consult the specifications information for the model and manufacturer of the equipment you are using, especially if it was manufactured prior to 2008.

There are several generally accepted methods of charging an automotive air-conditioning system. Not all methods, however, are recommended for all systems. The manufacturer's specifications and procedures should be followed to avoid problems. The air-conditioning system may be charged by weight, chart or graph, low-side and high-side pressure, and superheat. The correct charge is critical for an R-134a system. Many recommend initially undercharging by 5 to 10 percent, testing system performance, then adding refrigerant if necessary. A better choice would be to phase out older equipment in the shop build to meet the older SAE J2210 standard and replace it with recovery/recycling/recharging equipment that meets or exceeds the current SAE J2788 standard for accuracy. Even on a very small 0.8 lbs. system J2788 equipment would charge the system within +/− 3.9% of the recommended amount of refrigerant and a 2 lbs. system would be charged within +/− 1.5% of the recommended amount of refrigerant.

Diagnosis

Air-conditioning problems may often be diagnosed quickly simply by checking the function of the components of the system. The following should be on the basic checklist for quick diagnosis:

- Belt tension
- Clutch operation
- Radiator/condenser fan operation
- Evaporator blower operation
- Proper airflow from registers
- Observe sight glass, R-12 systems
- Refrigerant charge
- Suction line
- Liquid line
- Service valves/ports
- Lines, hoses, and connections
- Inspect for debris in front of condenser

Belt Tension

Is the belt tight? A loose belt will slip under the heavy heat loads of the air-conditioning system. Also, if it is a serpentine belt (Figure 7-22), another defective accessory component, such as the alternator, may cause the belt to slip. This is also a good opportunity to check the belt for wear, cracks, and glazing, a sign of early failure.

Clutch

With the engine running and the air-conditioning controls set for maximum cooling, make sure that the clutch is fully engaged. If it does not engage or if it slips, check for low voltage at the clutch coil. Other problems that could affect clutch operation are:

- Outside ambient air temperature that is too cold
- Open high-pressure switch

Determine whether the clutch is engaged but slipping, or whether the belt is slipping on the clutch. Both appear nearly the same.

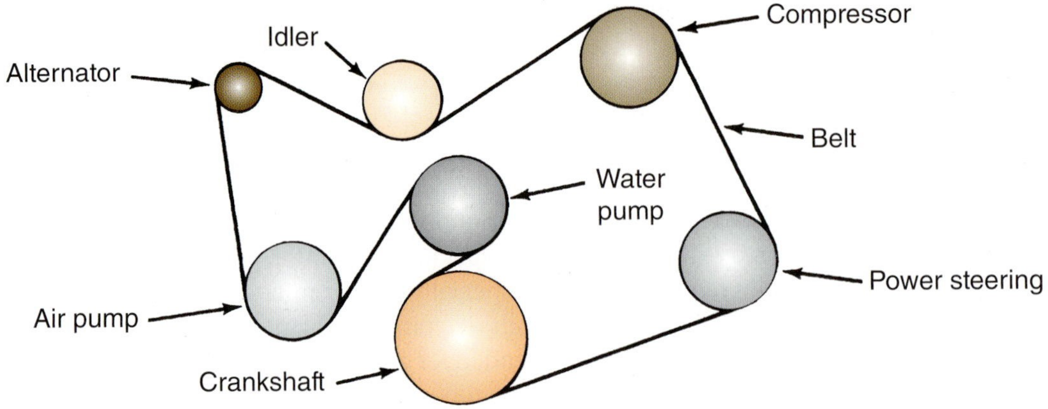

FIGURE 7-22 **Routing of a typical serpentine belt.**

- Open low-pressure switch
- Low refrigerant charge
- Excessive clutch air gap
- Thermostat that is out of adjustment

Also, note the condition of the clutch mating surfaces. If they are covered with grease or oil, a defective compressor shaft seal may be indicated.

Radiator-Condenser Fan

On car lines equipped with water-pump-mounted direct-drive fans, the fan should turn when the engine is running. If it does not, the belt may be slipping due to a defective (seized) water pump. Whatever the reason, the problem must be corrected.

On some car lines equipped with an electric fan (Figure 7-23), the fan will operate any time the compressor clutch is engaged. On others, the fan only operates during high-temperature conditions. If the fan does not operate as required, check for:

- A blown fuse or open circuit breaker
- A defective high-pressure switch

FIGURE 7-23 **A typical electrical cooling fan.**

- An inoperative relay (some models)
- A defective wiring or connector

An inoperative fan may result in serious problems associated with high head pressure at slow and idle engine speeds, such as compressor lockup (seizure) or hose rupture.

Blower Operation

The blower motor should operate any time the air-conditioning system controls are in the ON position. If it does not, the cause of the problem must be located and corrected. If it operates, make sure that it operates in all speeds. Most have at least an HI-MED-LO speed, whereas some have two or more MED speeds. A typical blower and motor assembly is shown in Figure 7-24.

A more detailed description of blower motor operation and diagnosis will be covered in Chapter 10.

Airflow

Check to be sure that the air is flowing from the proper registers (Figure 7-25) in all modes. Refer to Chapter 9 of this manual for a description of the proper airflow for the different modes of operation.

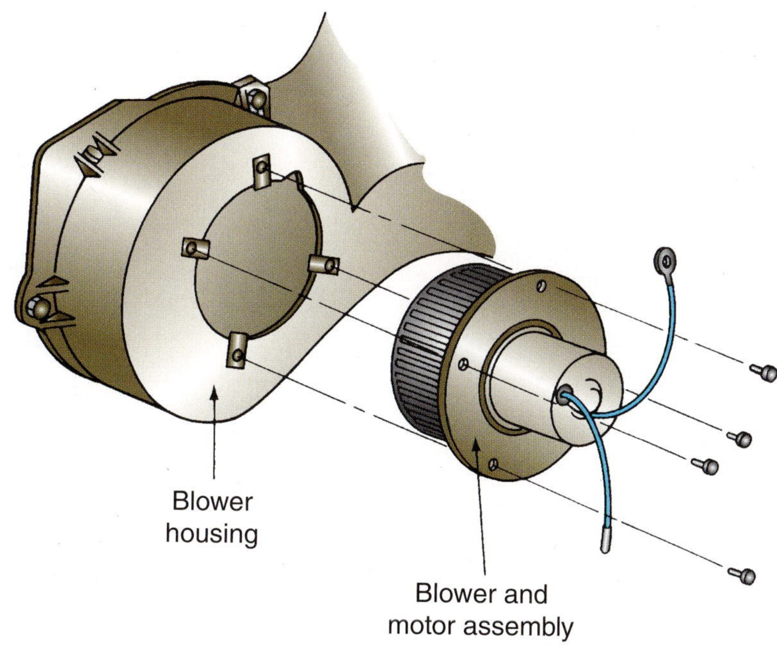

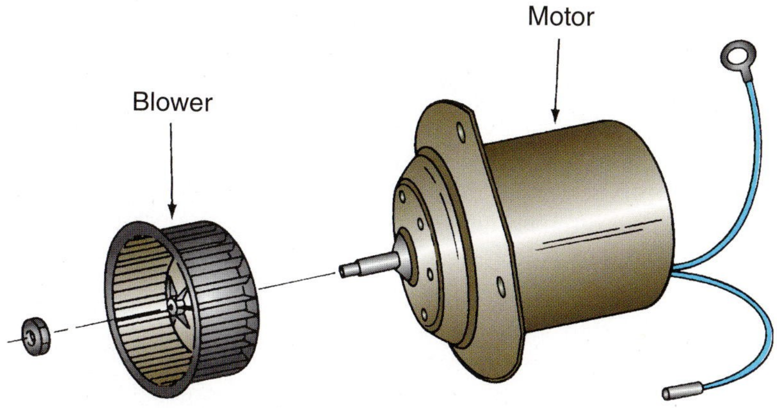

FIGURE 7-24 **A typical blower motor assembly.**

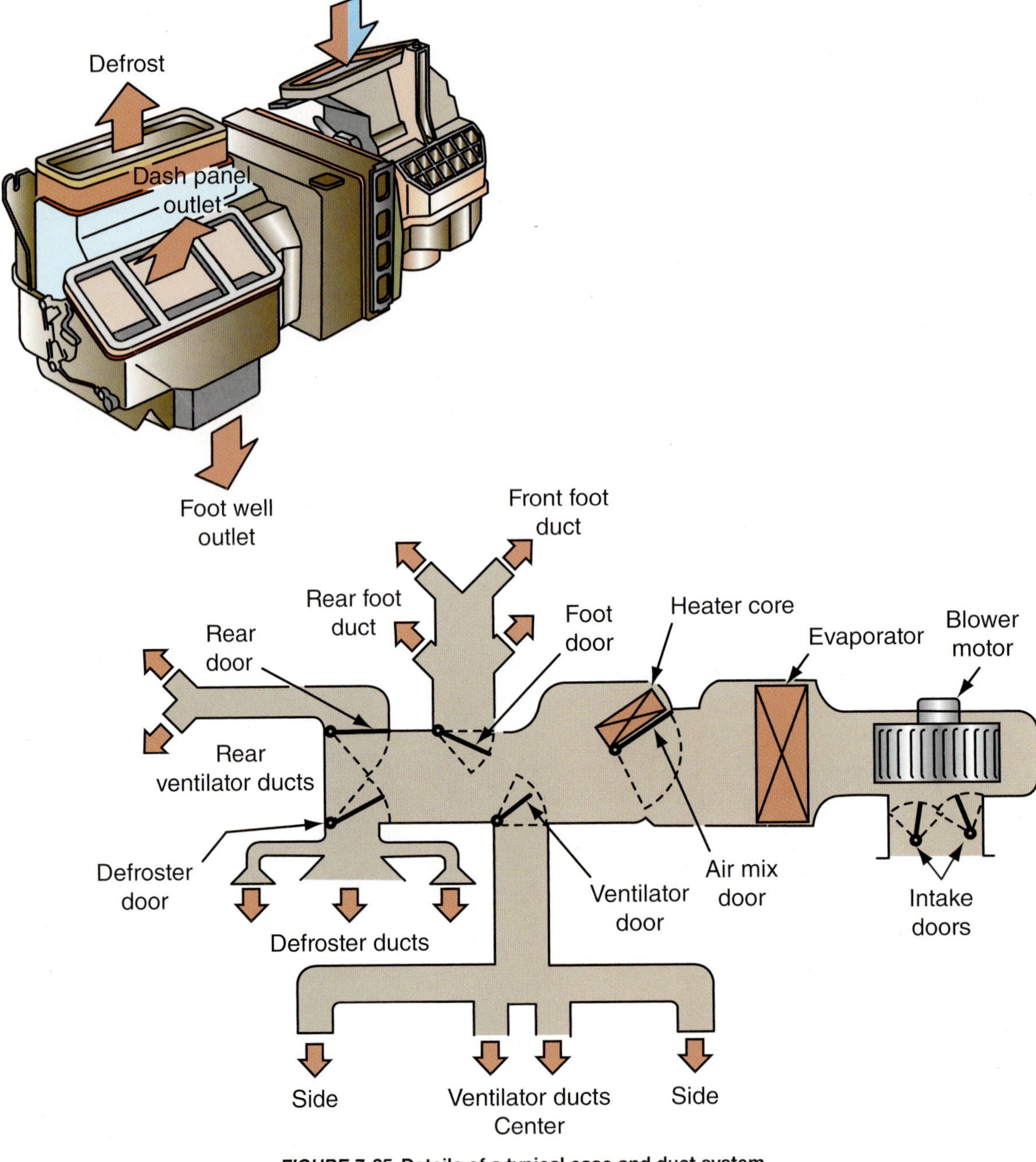

FIGURE 7-25 **Details of a typical case and duct system.**

The sight glass, if present, may be a part of the receiver-drier or it may be found in the liquid line.

Sight Glass

A sight glass is a window into the high-side liquid line of the refrigerant system. It is generally located on top of the receiver-drier or in the high-side liquid line. The sight glass should be clear when the compressor clutch is engaged and the system is operating properly. This check may be made on many, but not all, R-12 systems. Seldom is a sight glass (Figure 7-26) included in an R-134a system because the properties of the oil will erroneously indicate a low charge. Continuous bubbles in the sight glass generally indicate air trapped in the system. Continuous foam indicates that the charge is low, and oil streaks indicate no liquid refrigerant in the system. When the compressor clutch cycles off and on, it is normal to observe bubbles for a short period of time, particularly when the compressor clutch cycles off.

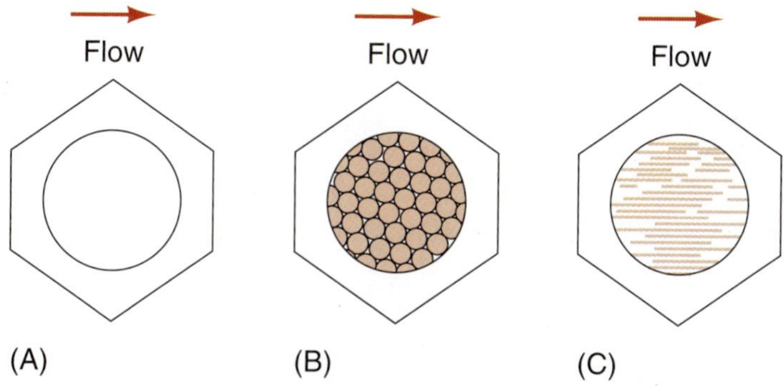

FIGURE 7-26 **Conditions found in a sight glass: (A) clear, (B) foamy or bubbly, and (C) cloudy.**

System Leaks

Visually inspect all hoses, tubing, components, fittings, and service ports for refrigerant leaks. Generally, but not always, a leak will be marked by a trace of refrigeration oil.

Suction Line

The suction line is the hose or line that connects the evaporator outlet to the compressor inlet. It should feel cool to the touch when the air-conditioning system is in operation. If it does not feel cool, install the manifold and gauge set to perform a performance test.

A second suction line may be found between the evaporator outlet and the accumulator inlet.

Liquid Line

The liquid line connects the condenser or receiver-drier outlet to the evaporator or metering device inlet. It should be warm or even hot to the touch. If it is not warm, install a manifold and gauge set and perform a performance test. It must be noted, however, that on some car line models the metering device is at the outlet of the condenser. The orifice tube is located at the condenser outlet to reduce or eliminate a hissing noise problem caused by refrigerant passing through it. The line between the metering device outlet and the evaporator inlet, in this case, will be cool.

A second liquid line may be found between the metering device outlet and the evaporator inlet.

Service Ports

Service valve ports (Figure 7-27) are often a problem area for leaking. They are often neglected when leak testing because hoses are generally connected to them. The hose(s) should be disconnected to ensure that no refrigerant is leaking past a defective valve seat. The Schrader-type valve may be replaced if it is found to be leaking. Note that there is a distinct difference in the service port used for R-12 as compared with R-134a. In an R-12 system, the low- and high-side ports will be the same size or the high-side port will be smaller. The opposite is true for an R-134a system, which has a larger high-side service port.

Learn to recognize the difference in the two types of fittings.

Hoses and Fittings

Throughout the years of automotive air-conditioning system service, refrigerant hoses have been the greatest cause of refrigerant leakage problems. Since the introduction and requirement of barrier-type refrigerant hoses, however, the problem has been greatly reduced. The problem of leaks still exists at the fittings (Figure 7-28). This is particularly true for hoses equipped with spring lock couplers. When replacing fittings, gaskets, or O-rings, be sure to use the proper component. For example, an O-ring may have a round, oval, or square profile. Also, there may be one or two O-rings used on a particular fitting. It must also be noted that some O-rings are refrigerant specific: an R-12 O-ring may not be used on an R-134a system and vice versa. There are O-rings available, however, that are compatible with both refrigerants.

FIGURE 7-27 **A typical R-134a service valve port cap.**

FIGURE 7-28 **A leak at a line fitting detected with the use of a soap solution.**

Refrigeration Oil

The moving parts of a compressor must be lubricated to prevent damage during operation. Oil is used on these moving parts and on the seals and gaskets throughout the system as well. In addition, oil is picked up by the refrigerant, which circulates through the system. This refrigerant and oil combination also helps to maintain the thermostatic expansion valve in a proper operating condition.

The oil that must be used in an automobile air-conditioning system is a nonfoaming sulfur-free grade specifically formulated for use in certain types of air-conditioning systems. This special oil is known as refrigeration oil, and it is available in several grades and types

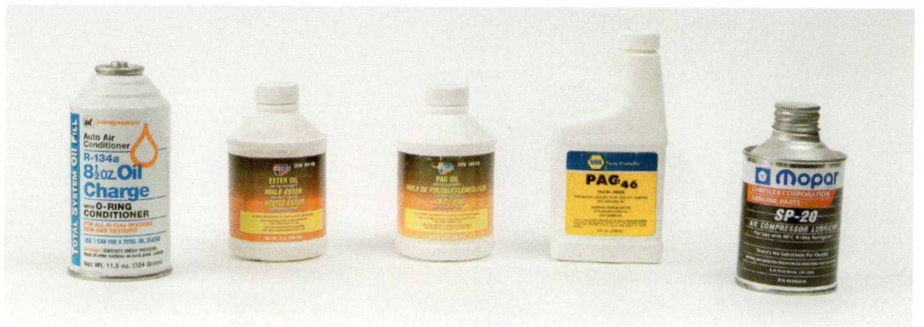

FIGURE 7-29 **Several types and grades of oil are used in automotive air-conditioning systems.**

(Figure 7-29). The grade and type to be used are determined by the compressor manufacturer and the type of refrigerant in the system. To replace oil that may be lost due to a refrigerant leak, pressurized oil is available in disposable cans. Generally, this container contains two fluid oz. (59 mL) of oil and a like measure of refrigerant. The refrigerant provides the necessary pressure required to force the oil into the system.

Mineral oil, used in R-12 air-conditioning systems, is clear to light yellow in color. Some synthetic oils used in R-134a air-conditioning systems may be blue or some other color. An impurity in refrigeration oil can cause a color change ranging from brown to black. Mineral oil is practically odorless, and a strong odor indicates that the oil is impure. Some synthetic oils, on the other hand, have a pungent odor though not impure. In either case, to ensure optimum system protection and performance, impure oil must be removed and replaced with clean, fresh oil. The receiver-drier or suction accumulator should also be replaced and a good pump down (system evacuation) performed before the air-conditioning system is recharged.

Lubricants for HFC Refrigerants

Some of the substitute automotive refrigerants are compatible with conventional mineral oil or alkyl benzene lubricants. However, R-134a refrigerant is not miscible with conventional mineral oil or alkyl benzene lubricants. Lack of miscibility can lead to system operational problems due to insufficient lubrication. When the two fluids are miscible, the lubricant is carried back to the compressor. When not miscible, however, the lubricant can accumulate in the various components of the air-conditioning system.

Miscible refers to the mixing ability and compatibility of two products.

Lubricant that accumulates in the condenser can reduce heat transfer and restrict the flow of liquid refrigerant. This condition can cause vapor pockets to form in the liquid stream as it flows through the metering device into the evaporator. Lubricant that accumulates in the evaporator will reduce heat transfer and restrict refrigerant vapor flow. Poor lubrication return to the compressor often leads to excessive wear of the compressor due to lubricant starvation.

Always select the correct lubricant and amount for the system being serviced. The amount of refrigerant oil contained in a refrigerant system varies according to the design and size of the system. As a rough guideline the distribution of oil throughout the refrigerant system is 50 percent in the compressor, 10 percent in the condenser, 10 percent in the fluid container (receiver dryer/accumulator), 20 percent in the evaporator, and 10 percent in the suction hose (Figure 7-30). When a refrigerant component is replaced always refer to the service information for the recommended amount of oil that should be added to the system.

AUTHOR'S NOTE: In addition to remembering to add the recommended amount of oil to the refrigerant system after component replacement, do not forget to pay attention to the amount of oil removed from the system during the refrigerant recovery process that also has to be added back to the system.

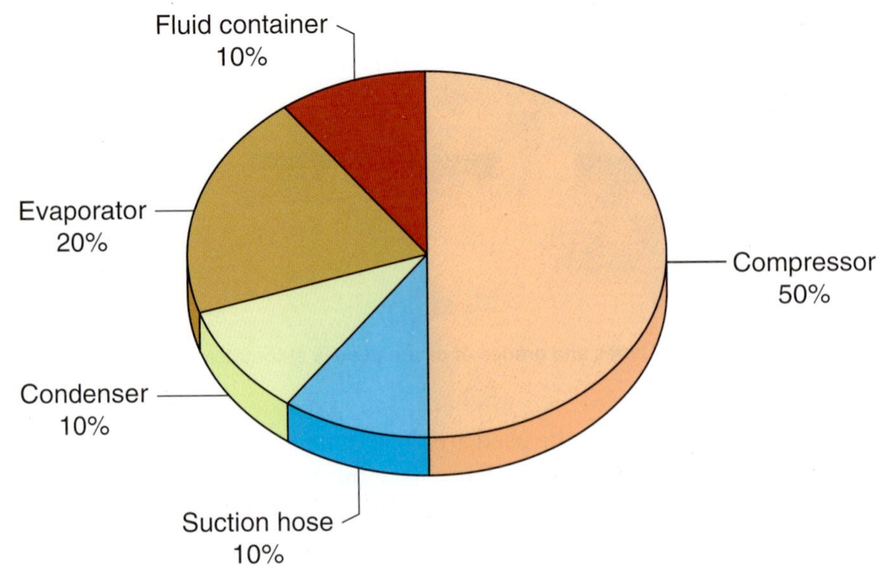

FIGURE 7-30 **The pie chart roughly represents the distribution of oil throughout the refrigerant system.**

When servicing the hybrid electric vehicle refrigerant system do not assume that it takes the same refrigerant oil as the nonhybrid vehicle refrigerant system. When servicing a hybrid vehicles refrigerant system, it is imperative that the correct refrigerant oil be used. The hybrid electric vehicle uses insulated refrigerant oil designed to minimize the conductivity of electricity through the compressor case in the event of a circuit failure. Many Toyota hybrid electric vehicles call for ND-11 refrigerant oil. In addition, it is even more critical than ever to properly evacuate the refrigerant system after service to the required vacuum levels for proper moisture removal. Always refer to vehicle manufacturers recommendations for correct refrigerant oil.

Synthetic Lubricants. Two different types of synthetic lubricant—polyalkylene glycols (PAG) and neopentyl polyol esters (POE)—are available for use with HFC refrigerants. Though first used in automotive air-conditioning in 1992, PAG is not a new lubricant. It has been used for years in compressors for natural gas production and compressors handling other "difficult" gases.

Polyalkylene Glycol (PAG). PAGs are extremely hygroscopic and can absorb several thousand parts per million (ppm) of water when exposed to moist air—100 times more than mineral oil, which generally contains less than 100 ppm of water. High water content causes corrosion and copper plating problems in some refrigeration systems when PAG lubricants are used. PAG is also sensitive to chlorine-containing contaminants such as residual R-12 remaining in a system that has been converted.

PAG, however, was selected in 1992 as the lubricant of choice for automotive air-conditioning systems in which R-134a is used as the refrigerant. This selection was based on previous success and due to deadline time constraints to phase out R-12. PAG lubricant is available in low- and high-viscosity grades. Always follow the manufacturer's specifications when adding or changing compressor lubricant.

R-1234yf refrigerant systems use specially formulated PAG oil and previous PAG oils formulated for R-134a systems should not be used.

Polyol Ester (POE). Neopentyl POEs are a group of organic esters that have an application in systems having a wide operating temperature range where good lubricating characteristics are desired. These lubricants also possess good miscibility with HFCs.

Various POE compositions enhance miscibility with HFCs and their resistance to copper plating in refrigeration systems. They also have excellent thermal stability, low volatility, low deposit-forming tendencies, high flash points, and high auto-ignition temperatures.

On the downside, POEs are also hygroscopic, susceptible to **hydrolysis**, and incompatible with certain elastomers. These esters can absorb several thousand parts per million of water; however, they are less absorbent than most PAGs. Therefore, like PAGs, POEs must be handled so excess moisture does not enter the refrigeration system. POE should only be used in automotive air-conditioning systems that specifically require this type lubricant. A few vehicle manufacturers, for example, recommend POE lubricant be used in specific R-12 air-conditioning systems that are being retrofitted to R-134a. Some hybrid electric vehicles specify POE (ester oil) in their refrigerant systems. These POE oils have a lower conductivity of electricity than PAG oils and offer proper insulation between the compressor housing and the high-voltage compressor drive motor.

Hydrolysis is a chemical reaction in which a substance reacts with water to form one or more other substances, such as hydrochloric acid, in an air-conditioning system.

Over time, an air-conditioning system may be exposed to moisture contamination during service. Although PAG oil is more hygroscopic (absorbs water) than ester oil, moisture remains suspended in PAG oil. This further emphasizes the need to evacuate the air-conditioning system beyond the minimum of 30 minutes. One disadvantage of ester oil is that under certain conditions (excess moisture contamination), moisture can drop out of ester oil to form small droplets of water that can cause corrosion and possible freeze-up in the expansion device. Remember, less than one drop exceeds the allowable moisture content in most refrigerant systems.

Safety

Personal protection equipment—such as rubber- or PVC-coated gloves or barrier creams and safety goggles—should be worn when handling lubricants. Prolonged skin contact or any eye contact can cause irritations and discomfort, such as stinging and burning. One should avoid breathing the vapors produced by these lubricants, and they should only be used in a well-ventilated area. Keep them in tightly sealed containers to prevent moisture contamination by humidity and to ensure that their vapors do not escape.

The Classification of Refrigeration Oil

The classification of refrigeration oil is based on three factors:

- Viscosity
- Compatibility with refrigerants
- Pour point

Viscosity. The viscosity rating for a fluid is based on the time, in seconds, required for a measured quantity of the fluid to pass through a calibrated orifice when the temperature of the fluid is 100°F (37.8°C). The resistance to flow of any fluid is judged by its viscosity rating. The higher the viscosity number, the thicker the fluid.

Compatibility. Refrigeration oil must be compatible with the refrigerant with which it is to be used. It should be noted that refrigeration oil now used in an R-12 system is a mineral oil designated YN-9. This oil cannot be used in an R-134a system. Conversely, a poly-alkylene glycol oil, designated YN-12, for an R-134a system with a reciprocating compressor, may not be used in an R-12 system. A second polyalkylene glycol oil is used in systems with a rotary compressor. There may be as many as three polyalkylene oils, each formulated for a particular application. Incompatible oil mixtures may cause serious damage to the air-conditioning system.

The Saybolt Universal Viscosity (SUV) is defined as the time, in seconds, required for 60 cubic centimeters (cm) of oil at 100°F (37.8°C) to flow through a standard Saybolt orifice.

Refrigeration oil, to be compatible with the refrigerant used in the system, must be capable of existing (remaining an oil) when mixed with the refrigerant. In other words, the oil is not changed or separated by chemical interaction with the refrigerant.

The compatibility of a refrigeration oil with a refrigerant is determined by a test called a floc test F. This test is performed by placing a mixture containing 90 percent oil and 10 percent refrigerant in a sealed glass tube. The mixture is then slowly cooled until a waxy substance appears. The temperature at which the substance forms is recorded as the floc point.

Do not mix oils.

Compatible: Capable of forming a mixture that does not separate and is not altered by chemical interaction.

Pour Point. The temperature at which an oil will just flow is its pour point. This temperature is recorded in degrees Fahrenheit. The pour point is a standard of the American Society for Testing Materials (ASTM).

The pour point temperature for oil and lubricant used in high-temperature refrigeration systems, such as automotive air-conditioning, is between −40°F (−40°C) and −10°F (−23.3°C), depending on grade and type.

The following is a list of manufacturers and the popular compressors used with the recommended type of refrigerant oil. Always refer to specific vehicle service information for proper lubricant selection and quantity.

OEM	Compressor Manufacturer	OEM-Recommended Oil	ISO PAG Viscosity
Acura	Nippondenso	ND8	46
Alfa Romeo	Nippondenso	ND8	46
	Sanden	SP10	46
Audi	Nippondenso	ND8	46
	Sanden	SP10	46
	Zexel (Diesel ZXL100 Kiki, Tama)		46
BMW	Nippondenso	ND8	46
	Seiko-Seiki	SP20	100
Chrysler/Jeep	Mitsubishi	PAG-56	46
	Nippondenso	ND8	46
	Sanden	SP20	100
	Sanden	SP10, SP15	46
Ford	Ford	YN-12a, YN-12b	46
	Harrison	UCON-488	133
	Nippondenso	ND8	46
	Panasonic		46
	Sanden	SP20	100
	Sanden	SP10	46
General Motors	Harrison	UCON-488	133
	Nippondenso	ND8	46
	Nippondenso	ND9	150
	Sanden	SP20	100
Honda	Nippondenso	ND8	46
	Sanden	SP10	46
Hyundai	Ford	YN-12a, YN-12b	46
	Nippondenso	ND8	46
	Sanden	SP20	46
	Sanden	SP10	100
Infiniti	Calsonic	PAG S (ZXL 100)	46
	Zexel	PAG S	46
	Zexel	PAG E	100
Isuzu	Zexel	ZXL200	100
Jaguar	Harrison	UCON-488	133
	Nippondenso	ND8	46
	Sanden	Ester	
Land Rover	Sanden	Ester	
Mazda	Nippondenso	ND8	46
	Nippondenso	ND9	150

OEM	Compressor Manufacturer	OEM-Recommended Oil	ISO PAG Viscosity
	Panasonic	DS-83P	46
	Sanden	SP20	100
	Zexel	ZXL100PG	46
Mercedes-Benz	Harrison	UCON-488	133
	Nippondenso	ND8	46
	Sanden	SP10	46
	York	Ester	
Mitsubishi	Mitsubishi	PAG-56	46
	Nippondenso	ND8	46
Nissan	Atsugi	Type R	100
	Calsonic	PAG S	46
Ford	FS10, FX15, VF2	PD46	46
Peugeot	Sanden	SP20	100
	Sanden	SP10	46
Porsche	Nippondenso	ND8	46
Rover	Nippondenso	ND8	46
	Sanden	SP20	100
	Sanden	SP10	46
Saab	Nippondenso	ND8	46
	Sanden	Ester	
	Seiko-Seikl	SP20	100
Saturn	Zexel	ZXL200PG	100
Subaru	Calsonic	PAG R	100
	Sanden	SP20	100
	Zexel	ZXL200PG	100
	Zexel	ZXL100	46
Toyota	Nippondenso	ND8	46
	Nippondenso	ND9	150
Volkswagen	Sanden	SP10	46
	Sanden	SP20	100
	Zexel	ZXL100	46
Volvo	Harrison	UCON-488	133
	Sanden	Ester	
	York	Ester	
	Zexel	Ester	

Servicing Tips

The oil level of the compressor should be checked each time the air conditioner is "opened" for service. Always check the manufacturer's recommendations before adding oil to the air-conditioning system.

When the oil is not being used, the container must remain capped. Always be sure that the cap is in place and tightly secured. Refrigeration oil is very hygroscopic; it absorbs moisture. Moisture is very damaging to the air-conditioning system. It should be noted that polyalkylene glycol oil is 100 times more hygroscopic than mineral oil. It is also ten times more expensive. For these reasons, it is suggested that refrigeration oil be purchased in small containers (Figure 7-31). Refrigeration oil is packaged in 1 qt. (0.946 L), 1 gal. (3.785 L), 5 gal. (18.925 L), and larger containers.

Hygroscopic: Readily taking up and retaining moisture.

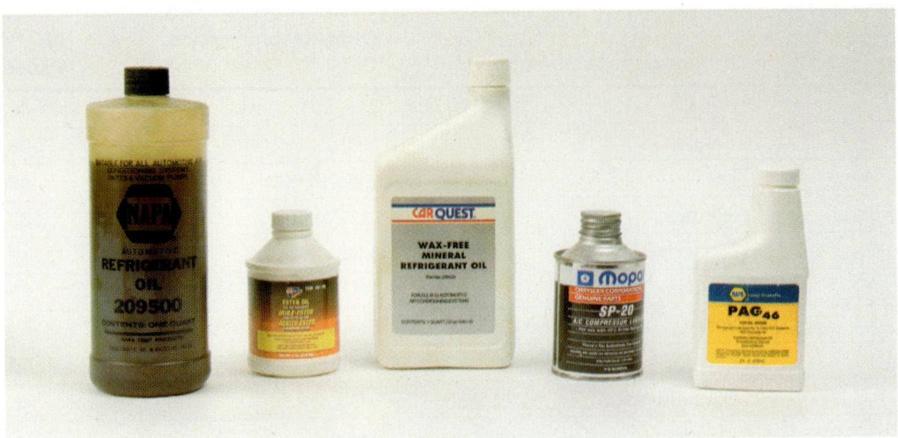

FIGURE 7-31 **Refrigeration oil is packaged in several container sizes.**

Do not refill disposable cylinders.

In conclusion, the properties of a good refrigeration oil are low wax content, good thermal and chemical stability, low viscosity, and a low pour point. A few simple rules are listed here for handling refrigeration oil:

DO

- Use only approved refrigeration oil.
- Be sure that the cap is tight on the container when not in use.
- Replace oil if there is any doubt of its condition.
- Avoid contaminating the oil.
- Dispose of used oil in a proper manner.
- Always use safety goggles and gloves when handling lubricants.
- Avoid breathing the fumes of lubricants.

DO NOT

- Transfer oil from one container to another.
- Return used oil to the container.
- Leave the oil container uncapped,
- Use a grade or type of oil other than that recommended for the air conditioner being serviced.
- Dispose of used oil in an improper manner.
- Overfill an air-conditioning system with oil.

AUTHOR'S NOTE: Compressor input shaft seals are a source of refrigerant leakage causing a low system charge level. It is a common practice for technicians to slip a note card in the air gap between the compressor clutch hub and pulley to see whether oil is detected. Oil on the card indicates that the shaft seal is leaking.

TERMS TO KNOW

Absorb
Contaminated
Cross-contaminated
Halogen
Hydrochloric acid (HCl)
Inject
Moisture
Pump down
Reclaim
Recover
Recycle
Significant New Alternative Policy (SNAP)

SUMMARY

- The least expensive method of leak testing is using a commercially available soap solution.
- Moisture collects in an air-conditioning system during service procedures and improper or careless service.
- Moisture mixed with refrigerant and oil forms corrosive acids and sludge in the system.

- Refrigerant must be recovered, not vented, as prescribed by the EPA.
- Manufacturers' procedures and specifications should be followed when charging an air-conditioning system with oil or refrigerant.
- Many system malfunctions can be diagnosed by visual inspection.

REVIEW QUESTIONS

Short-Answer Essays

1. Explain how moisture enters the system.
2. Describe one type of commonly used leak detector.
3. Is it important to maintain a moisture-free system? Why?
4. Explain the key difference between the load and unloaded air-conditioning system performance test?
5. How is the temperature-pressure chart used for system diagnosis?
6. What is the acceptable method of adding dye to the system?
7. Briefly describe how a halogen leak detector will react when it comes in contact with raw refrigerant.
8. What is a noncondensable gas and how does it affect air-conditioning system operation?
9. What does SAE standard J2788 apply to and what are the details of this standard?
10. Describe the process of removing moisture by vacuum.

Fill in the Blanks

1. A ______________ type of leak warrants the use of a dye trace solution.
2. An air-conditioning system ______________ is an initial test that determines whether the refrigerant system is operating as designed.
3. The maximum moisture content allowable in new refrigerant is ______________.
4. When evacuating a refrigerant system it should be placed under a ______________ for a minimum of ______________ minutes to ensure moisture removal from the system.
5. A ______________ acid is formed by the chemical combination of refrigerant and moisture.
6. The symbol ______________ (English) is used to denote a vacuum pressure.
7. The sensitivity of the electronic leak detector is ______________ per year.
8. ______________ or ______________ refrigerant can cause system and equipment damage and present problems if you are not aware of its presence.
9. In Denver, Colorado, water (H_2O) boils at a temperature of ______________ .
10. Air in the refrigerant system is considered a ______________ because at no time are refrigerant system pressures great enough or temperatures low enough to cause air to condense into a liquid.

Multiple Choice

1. Vacuum pump efficiency is being discussed.
   Technician A says that vacuum pump efficiency is greatest at sea level.
   Technician B says that altitude has little or no effect on vacuum pump efficiency.
   Who is correct?
   A. A only
   B. B only
   C. Both A and B
   D. Neither A nor B
2. Fluorescent leak detector scanner solution is being discussed:
   Technician A says once introduced into the system, it cannot be removed.
   Technician B says a ultraviolet light is required for detecting a leak.
   Who is correct?
   A. A only
   B. B only
   C. Both A and B
   D. Neither A nor B
3. *Technician A* says that a sight glass is often found in an R-12 system.
   Technician B says that a sight glass is seldom found in an R-134a system.
   Who is correct?
   A. A only
   B. B only
   C. Both A and B
   D. Neither A nor B
4. *Technician A* says that service ports may be a source of leaks.
   Technician B says that hose connections are sometimes a source of leaks. Who is correct?
   A. A only
   B. B only
   C. Both A and B
   D. Neither A nor B

5. *Technician A* says that an overcharged system will result in lower than normal low-side pressures.

   Technician B says that an overcharged system will result in a slight increase in cooling capacity, but only under low load conditions.

   Who is correct?

   A. A only
   B. B only
   C. Both A and B
   D. Neither A nor B

6. Refrigerant system air contamination may be caused by:

   A. Refrigerant system low-side leak.
   B. Improper system evacuation or insufficient evacuation time.
   C. Weak evacuation pump.
   D. All of the above.

7. An R134a recovery tank pressure is determined to be 115 psig at 85°F. What does this temperature-pressure relationship indicate?

   A. The recovery tank is contaminated with moisture.
   B. The recovery tank is contaminated with air.
   C. The recovery tank is overcharged.
   D. The recovery tank is at normal operating pressure.

8. Moisture removal from a refrigerant system is being discussed. The length of system evacuation time required depends on all of the major factors listed below except:

   A. The amount of moisture in the system.
   B. The ambient temperature.
   C. The efficiency of the vacuum pump.
   D. The amount of oil in the system.

9. The most accurate method for measuring the effectiveness of system evacuation is with the use of a:

   A. Kelvinometer.
   B. Micrometer.
   C. Thermistor vacuum gauge.
   D. Mercury (Hg) vacuum gauge.

10. To address the issue of small capacity refrigerant systems, the EPA adopted a new SAE standard effective December 2007, covering the accuracy level of future refrigerant recovery, recycling, and recharging equipment. The SAE number for this new standard is:

    A. J1850
    B. 2197
    C. J2210
    D. J2788

Chapter 8

DIAGNOSIS OF THE REFRIGERATION SYSTEM

UPON COMPLETION AND REVIEW OF THIS CHAPTER, YOU SHOULD BE ABLE TO:

- Diagnose six system malfunctions by gauge readings.
- Identify the low and high side of the air-conditioning system.
- Read and understand temperature-pressure charts.
- Discuss temperature-pressure relationships.
- Identify differences between R-12 (CFC-12) and R-134a (HFC-134a) systems.
- Identify differences between thermostatic expansion valve (TXV) and fixed orifice tube (FOT) systems.
- Understand the proper handling of refrigerants.
- Understand the proper handling of refrigeration oil.

INTRODUCTION

Study the air-conditioning system diagram (Figure 8-1), and note the dividing line between the high side and low side of the system and the manifold and gauge set connected into the system.

As illustrated, any condition (translated to pressure) that occurs in the low side of the system will be indicated on the low-side (compound) gauge. At the same time, any condition in the high side of the system will be indicated on the high-side (pressure) gauge.

There is a direct relationship between the pressure and temperature of refrigerant. For any given pressure, there is a corresponding temperature. For example, if the pressure of the low side of the system is 35 psig (241 kPa), the temperature of the evaporating refrigerant will be 38°F (3.3°C) for Refrigerant-12 and 40°F (4.4°C) for Refrigerant-134a. This is the temperature of the refrigerant, not the temperature of the air passing through the evaporator. Actual air temperature will be several degrees warmer. For convenience of illustration, a typical temperature-pressure chart is given in Figure 8-2.

SYSTEM DIAGNOSIS

Knowing the temperature of the ambient air entering the condenser, the normal high-side pressure can be determined by referring to Figure 8-3. This assumes that an unloaded refrigerant system performance test, described in Chapter 6, is being performed. For example, if the system is operating properly and the ambient air temperature is 85°F (29.4°C), the proper high-side pressure should be approximately 180 psig (1241 kPa) for an R-134a system. Allowances must be made in all readings to provide for slight errors in thermometers and gauges, as well as relative humidity.

The diagnostic process starts with a systematic approach to the customer complaint, followed by a detailed rational analysis to assure that the fault is corrected (Figure 8-4). One

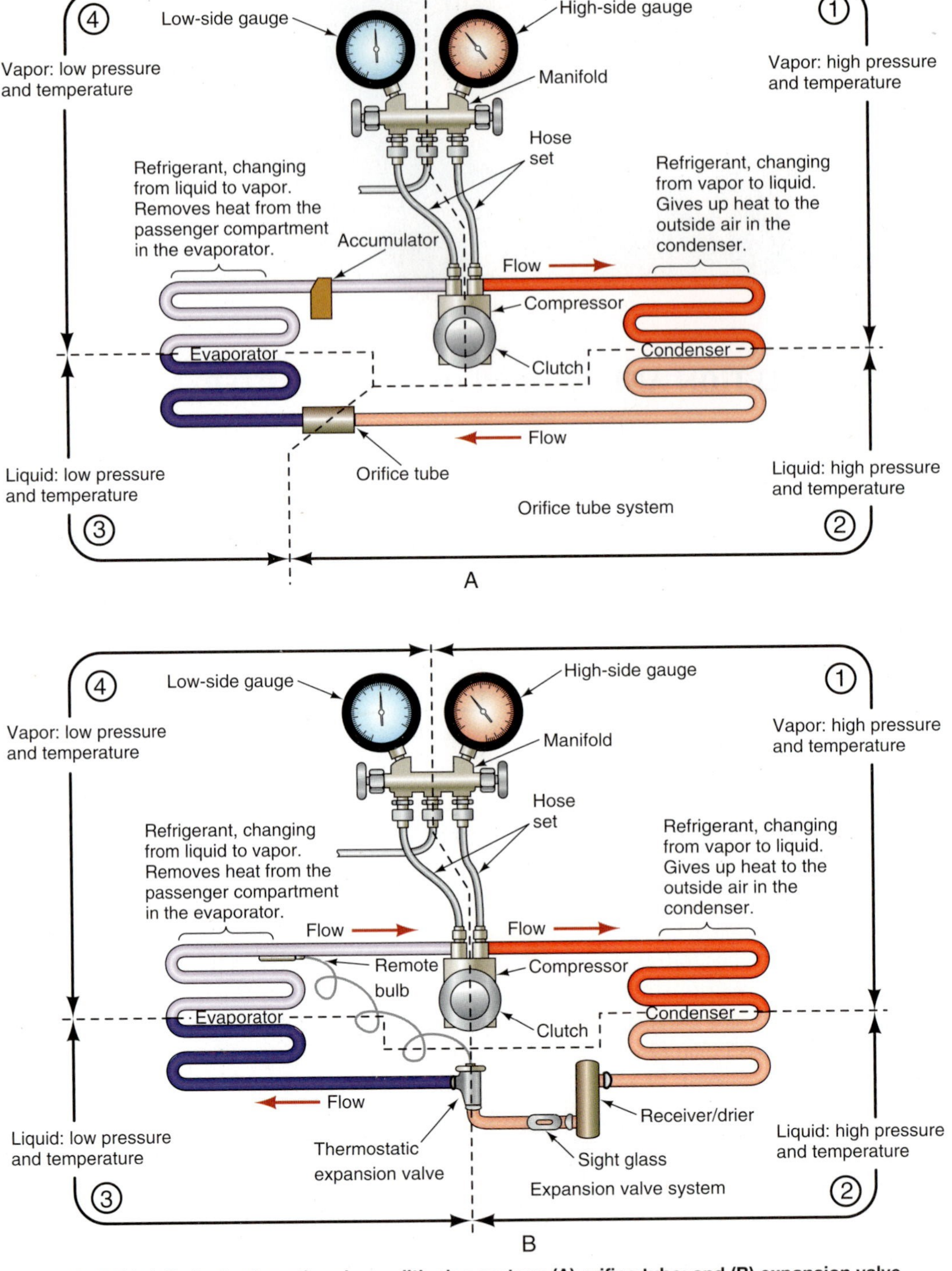

FIGURE 8-1 **Typical automotive air-conditioning system: (A) orifice tube; and (B) expansion valve.**

There are no valid temperature–pressure relationships for a malfunctioning system, such as an undercharge of refrigerant.

of the first steps is to establish a baseline before you perform any repairs. The initial air-conditioning performance test can serve as this baseline because it records system pressures, ambient air temperature, vent temperature, and relative humidity levels. In addition to being able to verify a repair with this baseline, you will be in a position to give a customer some relative facts after the repair is completed. An example would be telling a customer: "Your air-conditioning system had a vent temperature of 70°F (21°C) in the shop, where the temperature was 85°F (29.4°C). Your system refrigerant pressure was determined to be low. A leak at the condenser inlet line was located. After the leak was repaired, the system was recharged with the proper amount of refrigerant. The outlet temperature from your car's dash ventilation ducts is now down to 38°F (3.3°C)."

TEMPERATURE-PRESSURE CHART					
TEMPERATURE		CFC-12 PRESSURE		HFC-134A PRESSURE	
°F	°C	PSIG	KPA	PSIG	KPA
−30	−23.3	*5.5	37.9	*9.7	66.9
−25	−31.7	*2.3	15.9	*6.8	46.9
−20	−28.9	0.6	4.1	*3.6	24.8
−15	−26.1	2.4	16.5	*0.2	1.4
−10	−23.3	4.5	31.0	2.0	13.8
−5	−20.6	6.7	46.2	4.2	29.0
0	−17.8	9.2	63.4	6.5	44.8
5	−15.0	11.8	81.4	9.1	62.7
10	−12.2	14.6	100.7	11.9	82.1
15	−9.4	17.7	122.0	15.3	105.5
20	−6.7	21.0	144.8	18.4	126.9
25	−3.9	24.6	169.6	22.0	151.7
30	−1.1	28.5	196.5	26.1	180.0
35	1.7	32.6	224.8	30.4	209.6
40	4.4	37.0	255.1	35.1	242.0
45	7.2	41.7	287.5	40.1	276.5
50	10.0	46.7	322.0	45.5	313.7

TEMPERATURE-PRESSURE CHART					
TEMPERATURE		CFC-12 PRESSURE		HFC-134A PRESSURE	
°F	°C	PSIG	KPA	PSIG	KPA
55	12.8	52.1	359.2	51.3	353.7
60	15.6	57.7	397.8	57.3	395.1
65	18.3	63.8	439.9	64.1	442.0
70	21.1	70.2	484.0	71.2	490.9
75	23.9	77.0	530.9	78.7	542.6
80	26.7	84.2	580.6	86.8	598.5
85	29.4	91.8	633.0	95.3	657.1
90	32.2	99.8	688.1	104.4	719.8
95	35.0	108.3	746.7	114.0	186.0
100	37.8	117.2	808.1	124.2	856.4
105	40.6	126.6	872.9	135.0	930.8
110	43.3	136.4	940.5	146.4	1,009.4
115	46.1	146.8	1,012.2	157.5	1,086.0
120	48.9	157.7	1,087.3	171.2	1,180.4
125	51.7	169.1	1,166.0	184.6	1,272.8
130	54.4	180.0	1,241.1	198.7	1,370.0

FIGURE 8-2 **A typical temperature-pressure chart for R-12 and R-134a in English and metric values.**

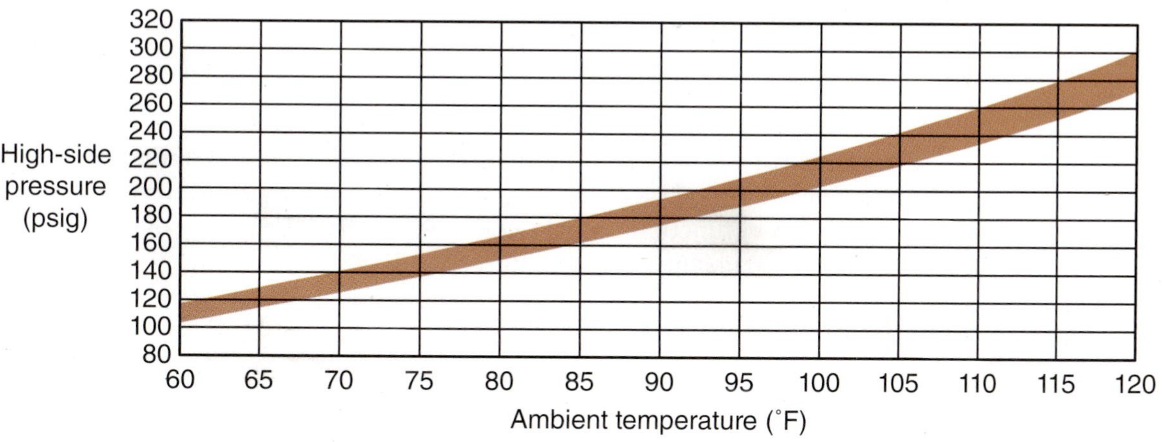

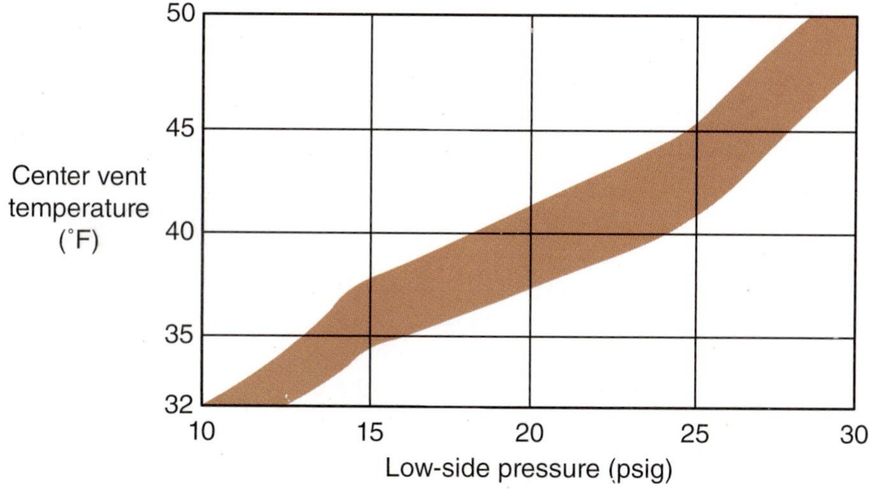

FIGURE 8-3 **This unloaded performance test data chart for an R-134a systems high- and low- side refrigerant pressures is an example of the type of chart provided by manufacturers for refrigerant system analysis.**

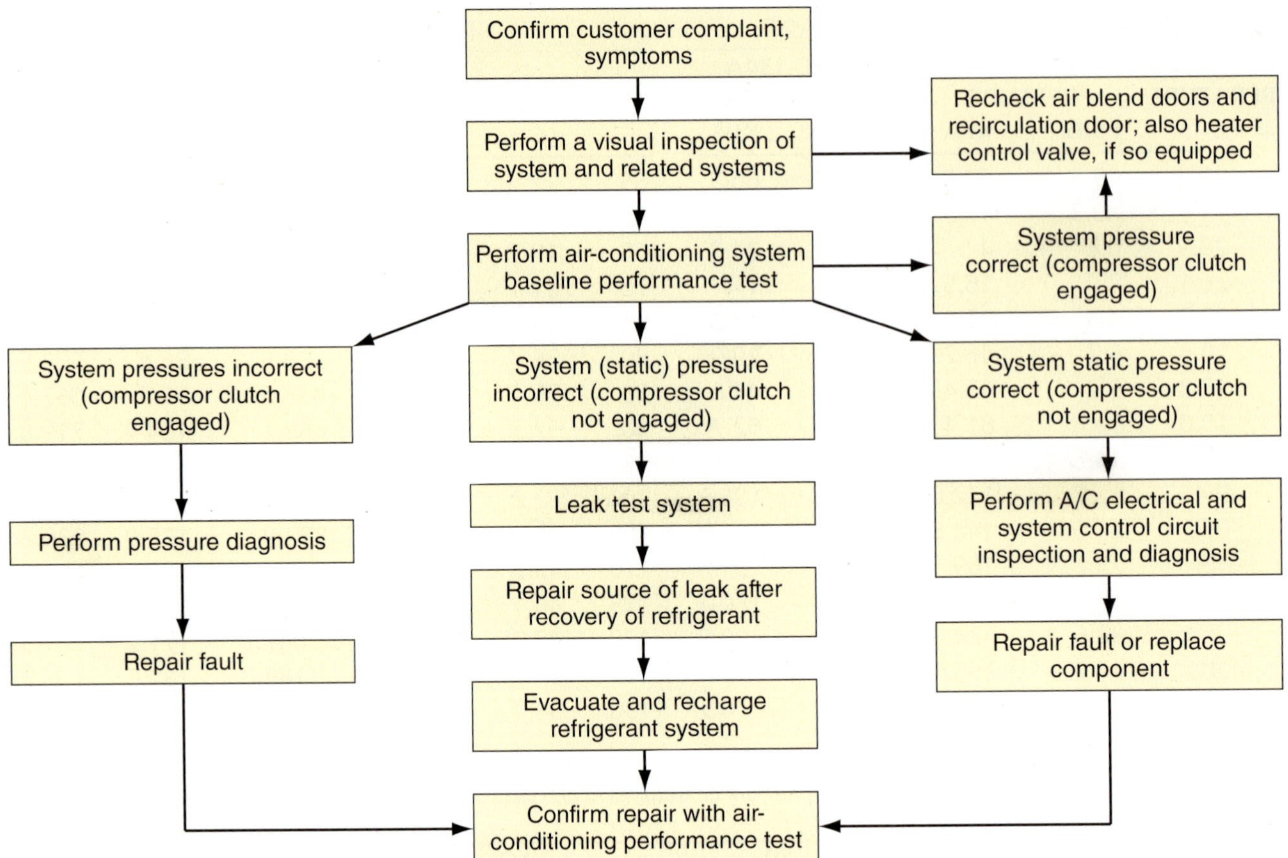

FIGURE 8-4 **Starting air-conditioning system diagnosis by following a logical sequence such as this will improve diagnostic results.**

Malfunction refers to a component's failing to work as designed.

There are seven basic conditions for the automotive air-conditioning system: one condition indicates normal operation, and six conditions indicate a system **malfunction**. Following is a brief description of system function or malfunction for each of the seven conditions. We will assume that the outside ambient air temperature is 90°F (32°C).

Temperature and Pressure Relationships of R-12 (CFC-12)

R-12 (CFC-12) was a desirable refrigerant for automotive use because the temperature on the Fahrenheit scale and English system pressure values in the 20–70 psig range are very close to the corresponding temperatures of 20°F–70°F.

There is only a slight variation between the temperature and pressure values of the refrigerant in the 20–70 psig range (Figure 8-5). In this range, the assumption for R-12 is made that for each pound of pressure recorded, the temperature is the same. For example, for a pressure of 23.8 psig, the corresponding temperature is 24°F (Figure 8-6). This value, however, is the temperature of the evaporating refrigerant, not the temperature of the outside surface of the evaporator coil or the air passing over it. Unfortunately, this close correlation does not exist in the metric system (Figure 8-7).

Evaporator coil temperature must be maintained above 32°F (0°C).

The objective in automotive air-conditioning is to allow the evaporator to reach its coldest point without icing. Because ice forms at 32°F (0°C), the fins and cooling coils of the evaporator must not be allowed to reach a colder temperature. Because of the temperature rise through the walls of the cooling fins and coils, the temperature of the refrigerant may be several degrees cooler than that of the air passing through the evaporator.

For example, a pressure gauge reading of 28 psig (193.06 kPa) in an R-12 system means that the evaporating temperature of the refrigerant is about 30°F (−1.1°C). Because of the

Temp. °F	Press. psig	Temp. °F	Press. psig	Temp. °F	Press. psig	Temp. °F	Press. psig	Temp. °F	Press. psig
0	9.1	35	32.5	60	57.7	85	91.7	110	136.0
2	10.1	36	33.4	61	58.9	86	93.2	111	138.0
4	11.2	37	34.3	62	60.0	87	94.8	112	140.1
6	12.3	38	35.1	63	61.3	88	96.4	113	142.1
8	13.4	39	36.0	64	62.5	89	98.0	114	144.2
10	14.6	40	36.9	65	63.7	90	99.6	115	146.3
12	15.8	41	37.9	66	64.9	91	101.3	116	148.4
14	17.1	42	38.8	67	66.2	92	103.0	117	151.2
16	18.3	43	39.7	68	67.5	93	104.6	118	152.7
18	19.7	44	40.7	69	68.8	94	106.3	119	154.9
20	21.0	45	41.7	70	70.1	95	108.1	120	157.1
21	21.7	46	42.6	71	71.4	96	109.8	121	159.3
22	22.4	47	43.6	72	72.8	97	111.5	122	161.5
23	23.1	48	44.6	73	74.2	98	113.3	123	163.8
24	23.8	49	45.6	74	75.5	99	115.1	124	166.1
25	24.6	50	46.6	75	76.9	100	116.9	125	168.4
26	25.3	51	47.8	76	78.3	101	118.8	126	170.7
27	26.1	52	48.7	77	79.2	102	120.6	127	173.1
28	26.8	53	49.8	78	81.1	103	122.4	128	175.4
29	27.6	54	50.9	79	82.5	104	124.3	129	177.8
30	28.4	55	52.0	80	84.0	105	126.2	130	182.2
31	29.2	56	53.1	81	85.5	106	128.1	131	182.6
32	30.0	57	55.4	82	87.0	107	130.0	132	185.1
33	30.9	58	56.6	83	88.5	108	132.1	133	187.6
34	31.7	59	57.1	84	90.1	109	135.1	134	190.1

FIGURE 8-5 **English temperature–pressure chart for R-12 (CFC-12).**

ENGLISH TEMPERATURE-PRESSURE CHART					
Low-Side Pressure psi		Temperature °F	High-Side Pressure psi		Temperature °F
Absolute	Gauge		Absolute	Gauge	
25.9	11.2	4	120	105	60
27	12.3	6	124	109	62
28.1	13.4	8	128	113	64
29.3	14.6	10	132	117	66
30.5	15.8	12	137	122	68
31.8	17.1	14	141	126	70
33	18.3	16	147	132	72
34.4	19.7	18	152	137	74
35.7	21	20	159	144	76
37.1	22.4	22	167	152	78
38.5	23.8	24	175	160	80
40	25.3	26	180	165	82
41.5	26.8	28	185	170	84
43.1	28.4	30	190	175	86
44.7	30	32	195	180	88
46.4	31.7	34	200	185	90
47.8	33.1	36	204	189	92
49.8	35.1	38	208	193	94
51.6	36.9	40	215	200	96
53.5	38.8	42	225	210	98
55.4	40.7	44	235	220	100
57.3	42.6	46	243	228	102
59.3	44.6	48	251	236	104

FIGURE 8-6 **A pressure of 23.8 psig corresponds to a temperature of 24°F.**

EVAPORATOR TEMPERATURE °C	EVAPORATOR PRESSURE GAUGE READING KILOPASCAL (GAUGE)	(ABSOLUTE)	AMBIENT TEMPERATURE °C	HIGH PRESSURE GAUGE READING KILOPASCAL (GAUGE)
-16	73.4	174.7	16	737.7
-15	81.0	182.3	17	759.8
-14	87.8	189.1	18	784.6
-13	94.8	196.1	19	810.2
-12	100.6	201.9	20	841.2
-11	108.9	210.2	21	868.7
-10	117.9	219.2	22	901.8
- 9	124.5	225.8	23	932.2
- 8	133.9	235.2	24	970.8
- 7	140.3	241.6	25	1 020.5
- 6	149.6	250.9	26	1 075.6
- 5	159.2	260.5	27	1 111.5
- 4	167.4	268.7	28	1 143.2
- 3	183.2	268.7	29	1 174.9
- 2	186.9	288.2	30	1 206.6
- 1	195.8	288.2	31	1 241.1
0	206.8	308.1	32	1 267.3
1	218.5	319.8	33	1 294.8
2	227.8	329.1	34	1 319.7
3	238.7	340.0	35	1 344.5
4	249.4	350.7	36	1 413.5
5	261.3	362.6	37	1 468.6
6	273.7	375.0	38	1 527.9
7	287.5	388.8	39	1 577.5
8	296.6	397.9	40	1 627.2
9	303.3	404.6	42	1 737.5
10	321.5	422.8	45	1 854.7

FIGURE 8-7 **Metric temperature–pressure chart for R-12 (CFC-12).**

temperature rise through the fins and coils, the air passing over the coil is more on the order of about 34°F or 35°F (1.1°C or 1.7°C).

Temperature and Pressure Relationships of R-134a (HFC-134a)

The performance characteristics of R-134a and R-12 refrigerants are very similar. R-134a refrigerant has a lower normal low-side pressure than R-12 and a higher normal high-side pressure than R-12 refrigerant. The power consumption of R-134a is slightly higher, but the refrigerant capacity is 3 to 5 percent less than an R-12 system. The vapor pressure of both systems is essentially equal under normal operating temperatures.

Like R-12, R-134a has its own unique temperature and pressure relationship, as shown in Figure 8-8. The metric equivalent is given in Figure 8-9. Note that the evaporating temperature and pressure of R-134a is reasonably close in the 10 − 40°F (−12.2 − +4.4°C) range. In fact, the temperature and pressure is nearly the same at 15°F (−9.4°C).

A pressure gauge reading of 26 psig (179 kPa) means that the evaporating temperature of the refrigerant in the evaporator is about 30°F (−1.1°C). This may be favorably compared with the temperature and pressure relationship of R-12.

Temperature and Pressure Relationships of R-1234yf (HFO-1234yf)

Current research has shown that R-1234yf and R-134a refrigerants have similar performance characteristics, in fact even closer than the similarity in pressure seen between R-134a and R-12. Refrigerant R-1234yf has a slightly higher normal low-side pressure than R-134a and a slightly higher normal high-side pressure than R-134a refrigerant, but they are so close that

Temperature °F	Pressure psig	Temperature °F	Pressure psig
-5	4.1	39.0	34.1
0	6.5	40.0	35.0
5.0	9.1	45.0	40.0
10.0	12.0	50.0	45.4
15.0	15.1	55.0	51.2
20.0	18.4	60.0	57.4
21.0	19.1	65.0	64.0
22.0	19.9	70.0	71.1
23.0	20.6	75.0	78.6
24.0	21.4	80.0	86.7
25.0	22.1	85.0	95.2
26.0	22.9	90.0	104.3
27.0	23.7	95.0	113.9
28.0	24.5	100.0	124.1
29.0	25.3	105.0	134.9
30.0	25.3	110.0	146.3
31.0	27.0	115.0	158.4
32.0	27.8	120.0	171.1
33.0	28.7	125.0	184.5
34.0	29.5	130.0	198.7
35.0	30.4	135.0	213.5
36.0	31.3	140.0	229.2
37.0	32.2	145.0	245.6
38.0	33.2	150.0	262.8

FIGURE 8-8 **English temperature–pressure chart for R-134a (HFC-134a).**

Temperature °C	Pressure kPa	Temperature °C	Pressure kPa
-15.0	63	5.0	247
-12.5	83	7.5	280
-10.0	103	10.0	313
-7.5	122	12.5	345
-5.0	142	15.0	381
-4.5	147	17.5	422
-4.0	152	20.0	465
-3.5	157	22.5	510
-3.0	162	25.0	560
-2.5	167	27.5	616
-2.0	172	30.0	670
-1.5	177	32.5	726
-1.0	182	35.0	785
-0.5	187	37.5	849
0.0	192	40.0	916
0.5	198	42.5	990
1.0	203	45.0	1066
1.5	209	47.5	1146
2.0	214	50.0	1230
2.5	220	52.5	1315
3.0	225	55.0	1385
3.5	231	57.5	1480
4.0	236	60.0	1580
4.5	242	65.0	1795

FIGURE 8-9 **Metric temperature–pressure chart for R-134a.**

most technicians will probably not notice the difference. The vapor pressure of both systems is essentially equal under normal operating temperatures. Technicians that are used to working on R-134a systems will find that there will be very little difference in working on and diagnosing an R-1234yf system other than the service warning that it is a mildly flammable refrigerant.

The temperature–pressure relationship of R-1234yf is shown in Figure 8-10. A pressure gauge reading of 28 psig (193 kPa) means that the evaporating temperature of the refrigerant in the evaporator is about 28°F (−2.2°C). This is very similar to the R-134a pressure–temperature relationship.

Condition One: Normal Operation (Figure 8-11)

- Low-side gauge: Normal pressure
  - R-134a (HFC-134a): 30–31 psig (207–214 kPa)
  - R-12 (CFC-12): 32–33 psig (221–228 kPa)
- High-side gauge: Normal pressure
  - R-134a (HFC-134a): 204–210 psig (1,407–1,448 kPa)
  - R-12 (CFC-12): 185–190 psig (1,276–1,310 kPa)

Results from a performance test should be compared to an air-conditioning performance table for the type of refrigerant in the vehicle's air-conditioning system. Figure 8-12 shows the normal operating pressure ranges for R134a under various ambient air temperature and humidity levels, and the performance table in Figure 8-13 shows the normal operating pressure ranges for R1234yf under various ambient air temperature and humidity levels.

When a refrigerant system is operating as designed, the normal flow of refrigerant in a thermostatic expansion valve (TXV) system is as shown in Figure 8-14; Figure 8-15 shows the normal flow of refrigerant in a fixed orifice tube (FOT) system.

Assuming a cycling-clutch system, the desired "**average**" temperature of the evaporator should be about 35°F (1.7°C). To achieve a "theoretical average" temperature, the thermostat should cycle the compressor clutch OFF at about 27°F (−2.8°C) and back ON

Average is a single value that represents the median.

R-1234yf Fahrenheit Pressure/Temperature Chart

PSIG	°F	PSIG	°F	PSIG	°F
9	0	77	72	200	132
11	4	80	74	206	134
14	8	83	76	212	136
16	12	86	78	218	138
19	16	89	80	223	140
19	17	92	82	229	142
20	18	96	84	236	144
23	22	99	86	242	146
24	23	102	88	248	148
25	24	106	90	255	150
28	28	110	92	261	152
31	32	113	94	268	154
33	34	117	96	275	156
35	36	121	98	282	158
37	38	125	100	289	160
38	40	129	102	296	162
40	42	133	104	304	164
42	44	137	106	311	166
44	46	142	108	319	168
47	48	146	110	326	170
49	50	148	111	334	172
51	52	150	112	342	174
53	54	155	114	351	176
56	56	160	116	359	178
58	58	164	118	368	180
61	60	169	120	376	182
63	62	174	122	385	184
66	64	179	124	394	186
68	66	184	126	403	188
71	68	190	128	413	190
74	70	195	130		

Evaporator (17 °F–50 °F); Condenser (132 °F–180 °F)

R-1234yf Celsius Pressure/Temperature Chart

kPa	°C	kPa	°C	kPa	°C
62	-18	544	23	1363	55
75	-16	562	24	1398	56
90	-14	581	25	1432	57
105	-12	601	26	1468	58
120	-10	620	27	1504	59
137	-8	641	28	1541	60
155	-6	661	29	1578	61
174	-4	682	30	1616	62
194	-2	704	31	1654	63
214	0	726	32	1694	64
225	1	748	33	1733	65
236	2	771	34	1774	66
248	3	794	35	1815	67
260	4	818	36	1857	68
272	5	842	37	1900	69
284	6	866	38	1943	70
297	7	891	39	1987	71
309	8	917	40	2032	72
323	9	943	41	2078	73
336	10	970	42	2124	74
350	11	997	43	2171	75
364	12	1024	44	2219	76
379	13	1052	45	2267	77
394	14	1081	46	2317	78
409	15	1110	47	2367	79
425	16	1140	48	2418	80
440	17	1170	49	2523	82
457	18	1201	50	2631	84
473	19	1232	51	2743	86
490	20	1264	52	2859	88
508	21	1297	53	2979	90
526	22	1330	54		

Evaporator (-10 °C–11 °C); Condenser (51 °C–82 °C)

FIGURE 8-10 **R-1234yf pressure/temperature chart.**

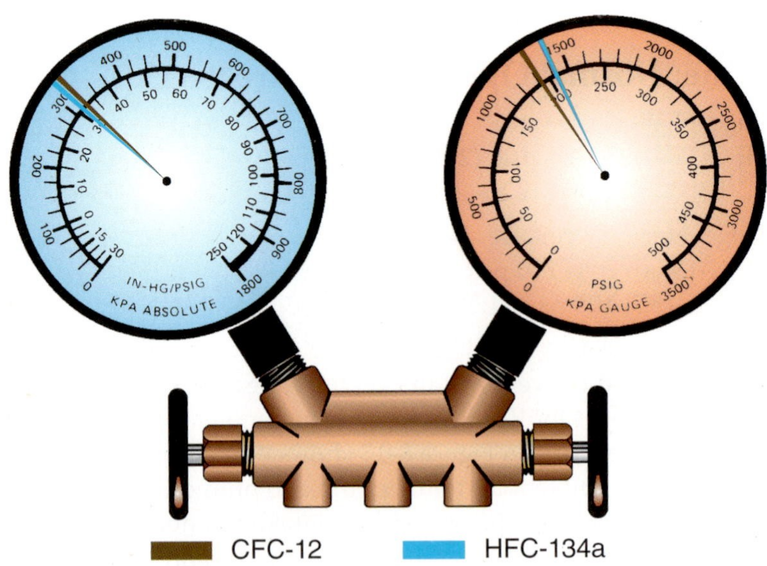

FIGURE 8-11 **Manifold gauge reading 1; system normal.**

Ambient Air Temperature	Relative Humidity	Refrigerant Service Port Pressure: Low Side	Refrigerant Service Port Pressure: High Side	Maximum Center Discharge Air Temperature
55°F – 65°F (13 – 18°C)	0–100%	22–29 psi (151–199 kPa)	129–168 psi (888–1,157 kPa)	43°F (6°C)
66°F – 75°F (19 – 24°C)	Below 40%	22–28 psi (151–192 kPa)	149–215 psi (1,026–1,481 kPa)	43°F (6°C)
	Above 40%	22–34 psi (151–234 kPa)	152–210 psi (1,047- 1,446 kPa)	46°F (7°C)
76°F – 85°F (25 – 29°C)	Below 35%	22–32 psi (151–220 kPa)	179–220 psi (1,233–1,515 kPa)	48°F (8°C)
	35–50%	22–33 psi (151–227 kPa)	179–225 psi (1,233–1,550 kPa)	50°F (13°C)
	Above 50%	24–37 psi (165–254 kPa)	179–212 psi (1,233–1,460 kPa)	55°F (13°C)
86°F – 95°F (30 – 35°C)	Below 30%	24–36 psi (165–248 kPa)	202–241 psi (1,391–1,660 kPa)	55°F (13°C)
	30–50%	25–38 psi (172–261 kPa)	202–238 psi (1,391–1,639 kPa)	65°F (18°C)
	Above 50%	28–40 psi (192–275 kPa)	200–235 psi (1,378–1,619 kPa)	66°F (19°C)
96°F – 105°F (36 – 41°C)	Below 20%	28–40 psi (192–275 kPa)	231–270 psi (1,591–1,860 kPa)	64°F (17°C)
	20–40%	29–42 psi (199–289 kPa)	231–267 psi (1,591–1839 kPa)	66°F (19°C)
	Above 40%	31–43 psi (213–296 kPa)	228–270 psi (1,570–1,860 kPa)	70°F (21°C)

FIGURE 8-12 **An example of an air-conditioning performance table of the normal operating ranges for R134a under various ambient temperature and humidity levels.**

Ambient Temperature	Relative Humidity	A/C Performance Table: Low Side Service Port Pressure	A/C Performance Table: High Side Service Port Pressure	Maximum Left Center Discharge Air Temperature
13 – 18°C (55 – 65°F)	0–100%	199–261 kPa (29–38 psi)	1233–1481 kPa (179–215 psi)	10°C (50°F)
19 – 24°C (66 – 75°F)	Less than 35%	241–316 kPa (35–46 psi)	1419–1639 kPa (206–238 psi)	12°C (52°F)
	Greater than 40%	227–310 kPa (33–45 psi)	1336–1612 kPa (194–234 psi)	13°C (55°F)
25 – 29°C (76 – 85°F)	Less than 35%	241–316 kPa (35–46 psi)	1419–1639 kPa (206–238 psi)	13°C (55°F)
	35–60%	254–323 kPa (37–47 psi)	1460–1667 kPa (212–242 psi)	14°C (57°F)
	Greater than 60%	254–344 kPa (37–50 psi)	1481–1729 kPa (215–251 psi)	17°C (61°F)
30 – 35°C (86 – 95°F)	Less than 30%	261–337 kPa (38–49 psi)	1522–1750 kPa (221–254 psi)	15°C (59°F)
	30–50%	268–358 kPa (39–52 psi)	1550–1798 kPa (225–261 psi)	17°C (61°F)
	Greater than 50%	282–385 kPa (41–56 psi)	1591–1860 kPa (231–270 psi)	18°C (64°F)
36 – 41°C (96 – 105°F)	Less than 20%	275–351 kPa (40–51 psi)	1632–1853 kPa (237–269 psi)	17°C (61°F)
	20–40%	289–378 kPa (42–55 psi)	1646–1901 kPa (239–276 psi)	18°C (64°F)
	Greater than 40%	310–399 kPa (45–58 psi)	1688–1949 kPa (245–283 psi)	20°C (68°F)
42 – 46°C (106 – 115°F)	Less than 20%	296–372 kPa (43–54 psi)	1743–1949 kPa (253–283 psi)	18°C (64°F)
	Greater than 20%	310–399 kPa (45–58 psi)	1770–1998 kPa (257–290 psi)	20°C (68°F)
47 – 49°C (116 – 120°F)	Below 30%	330–406 kPa (48–59 psi)	1867–2080 kPa (271–302 psi)	2°C (70°F)

FIGURE 8-13 **An example of an air conditioning performance table of normal operating ranges for R-1234yf under various ambient temperatures and humidity levels.**

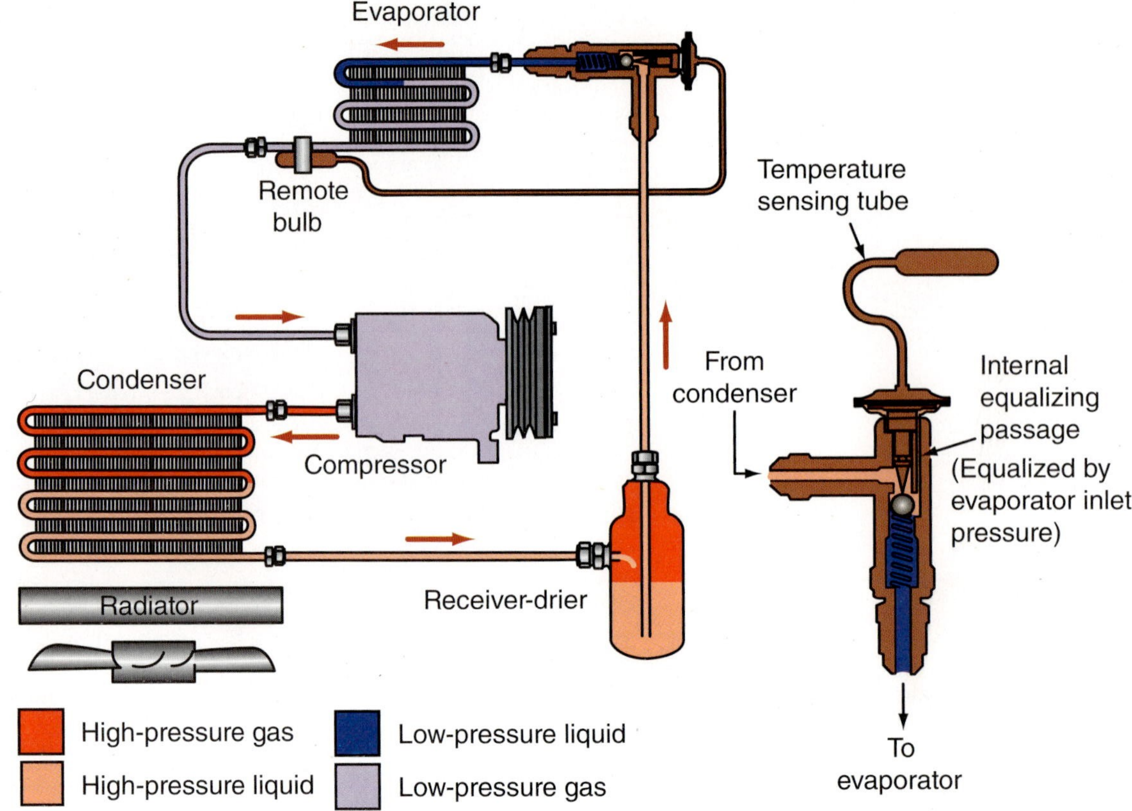

FIGURE 8-14 **The normal state of refrigerant in a TXV system.**

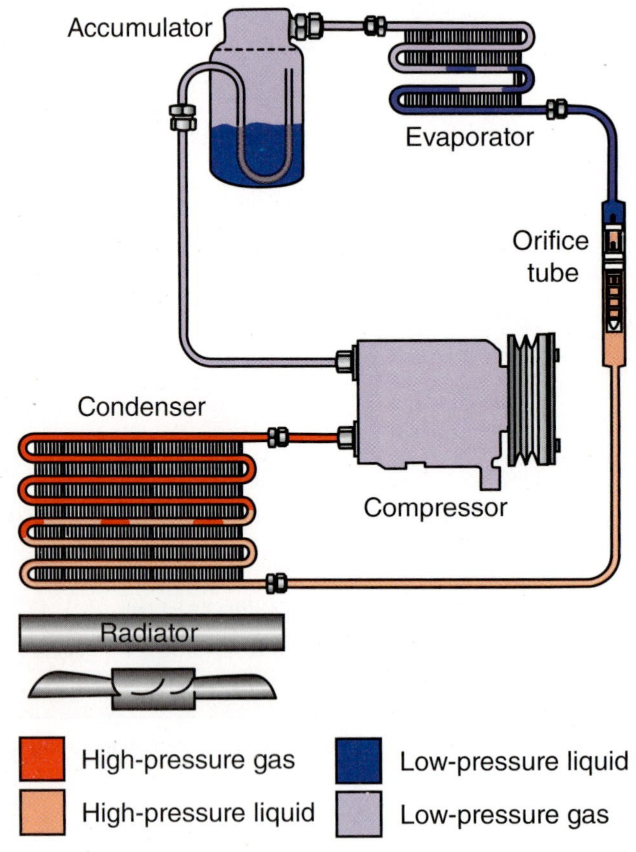

FIGURE 8-15 **The normal state of refrigerant in a FOT system.**

at about 39°F (3.9°C). According to the R-12 temperature-pressure chart, the gauge reading should be a low of 26 psig (179 kPa) and a high of 36 psig (248 kPa). For R-134a systems, the off cycle should be at 24 psig (165 kPa) and the on cycle at 37 psig (255 kPa).

It should be noted that the theoretical average is seldom accomplished in actual operation. Therefore, it is suggested that manufacturers' specifications be consulted for the operating range of any particular vehicle.

Actually, the concern is with air temperature, not with refrigerant temperature. A low of 14–15 psig (96.5–103.4 kPa) and a high of 40–50 psig (275.8–344.7 kPa), then, is a more realistic indication for the low-side gauge.

The high-side gauge should indicate pressure shown in the temperature-pressure chart for any given ambient temperature, plus or minus a few psig (kPa).

> Moisture in the system causes harmful acids.

Condition Two: Insufficient Cooling (Figure 8-16)

- Low-side gauge: Low pressure
  - R-134a (HFC-134a): 0–12 psig (83 kPa)
  - R-12 (CFC-12): 0–15 psig (103 kPa)
- High-side gauge: Normal to slightly low pressure
  - R-134a (HFC-134a): 208 psig (1,434 kPa)
  - R-12 (CFC-12): 190 psig (1,310 kPa)

There are three major possible causes for this condition. They are as follows:

1. A thermostat that is improperly adjusted (temperature), out of adjustment (mechanical), or defective
2. A **restriction** in the low side of the system
3. Moisture in the system

> A **restriction** is a blockage in the air-conditioning system caused by a pinched line, foreign matter, or moisture freeze-up.

The TXV **control thermostat** may be defective or improperly adjusted. It must be adjusted so its electrical contacts will open at the desired low temperature to allow the clutch to cycle off. The differential must be adjusted so the clutch will cycle back on after a predetermined temperature rise.

> A **control thermostat** is a temperature-actuated electrical switch used to cycle the compressor clutch on and off, thereby controlling the air-conditioning system temperature.

A defective thermostatic expansion valve that limits refrigerant flow will cause evaporator refrigerant starvation (Figure 8-17), resulting in insufficient cooling. Similarly, a restricted FOT that limits refrigerant flow will also cause evaporator refrigerant starvation (Figure 8-18).

Another indication of a defective thermostat is that the evaporator coil may be icing over. Ice on the coil blocks the flow of air passing through it. If the system works fine for a while

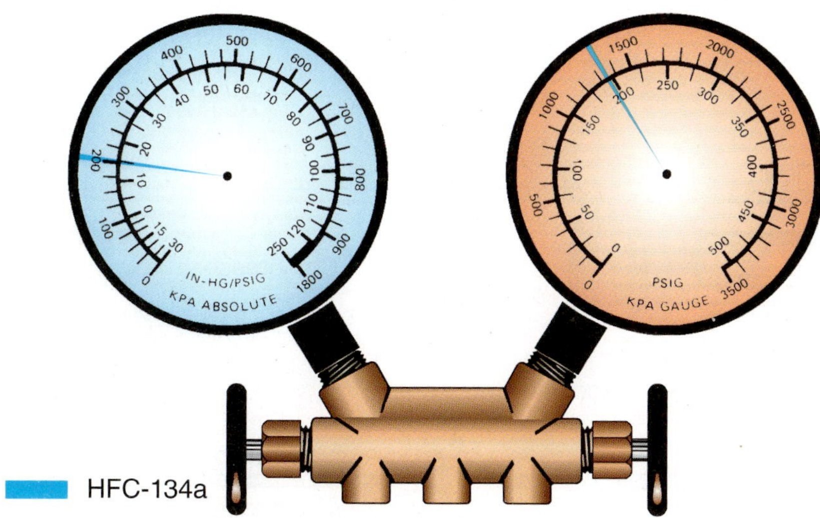

FIGURE 8-16 **Manifold gauge reading indicating insufficient cooling due to improperly adjusted temperature control, restriction of the low side, or moisture in the system.**

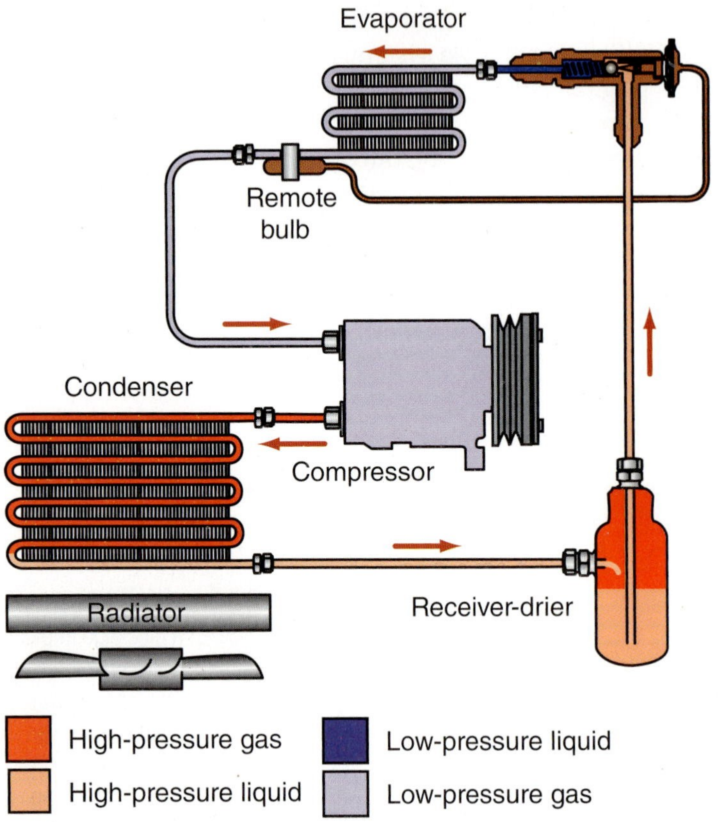

Evaporator
Remote
bulb
Condenser
Compressor
Radiator
Receiver-drier
High-pressure gas
High-pressure liquid
Low-pressure liquid
Low-pressure gas

FIGURE 8-17 **Defective thermostatic expansion valve, which limits refrigerant flow, will cause evaporator refrigerant starvation.**

Accumulator
Evaporator
Orifice
tube
Condenser
Compressor
Radiator
High-pressure gas
High-pressure liquid
Low-pressure liquid
Low-pressure gas

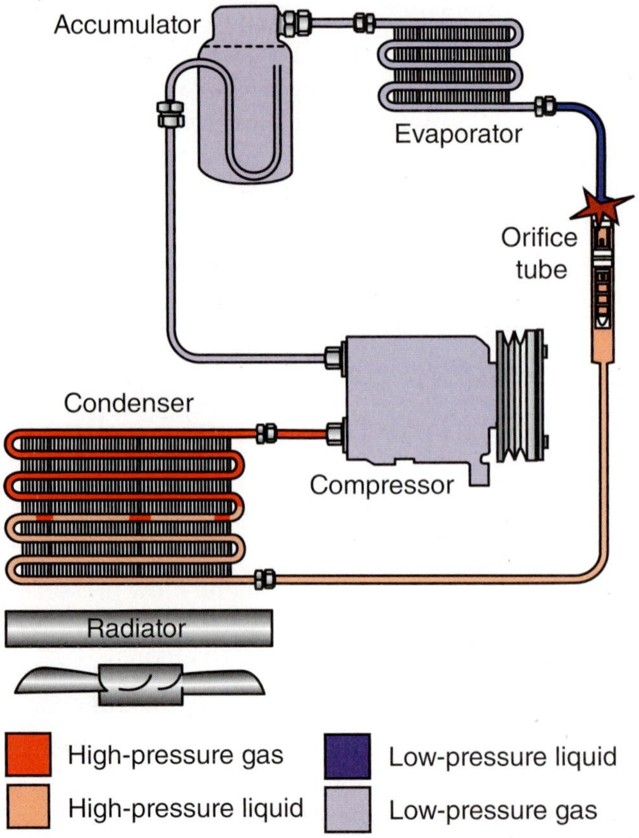

FIGURE 8-18 **Restricted FOT, which limits refrigerant flow, will cause evaporator refrigerant starvation.**

A **metering device** is a component that regulates the proper amount of refrigerant in the evaporator. The two common types for automotive applications are the TXV and the FOT.

and then becomes warmer, there may be moisture in the system. If reduced airflow is also noted, the evaporator freeze-up is indicated. If evaporator freeze-up is suspected, check for a frozen evaporator line or the compressor clutch not cycling. Causes for this condition are:

1. A defective cycling clutch switch that is stuck in the ON position.
2. An improperly positioned evaporator core sensor (fin sensor) that is not touching the evaporator core or not inserted into the fins.

There may be a restriction in the low side of the system (Figure 8-19) between the **metering device** outlet and the compressor inlet. The screen at the inlet of the metering device may be clogged. If there is excess moisture in the system, it will collect and freeze in the screen at the inlet of the metering device. Refer to Figure 8-1a; restriction will be located in quadrant 3 or 4.

B
Suction line
Flow
Discharge line
Compressor
Evaporator
Condenser
Clutch
Accumulator
Fixed
orifice
tube
A
Liquid line
Flow

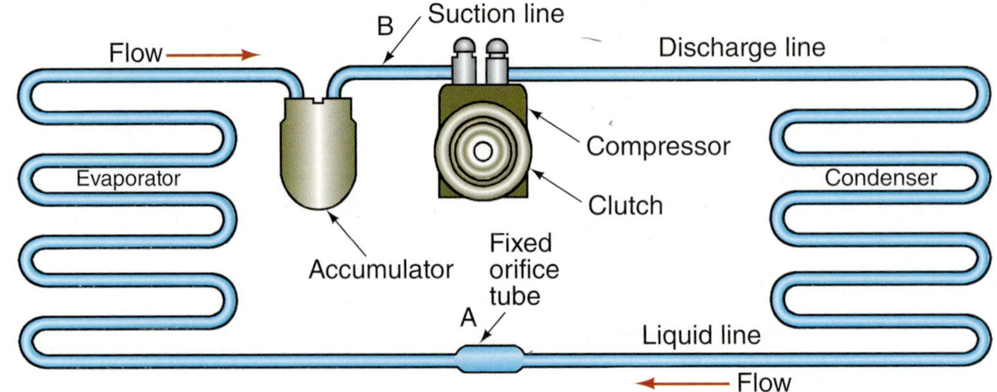

FIGURE 8-19 **Look for a restriction in the low side of the system between the metering device outlet (A) and the compressor inlet (B).**

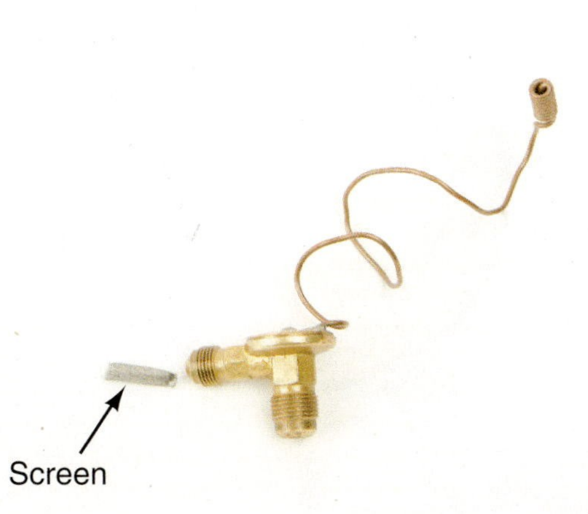

FIGURE 8-20 **Inlet screen of thermostatic expansion valve.**

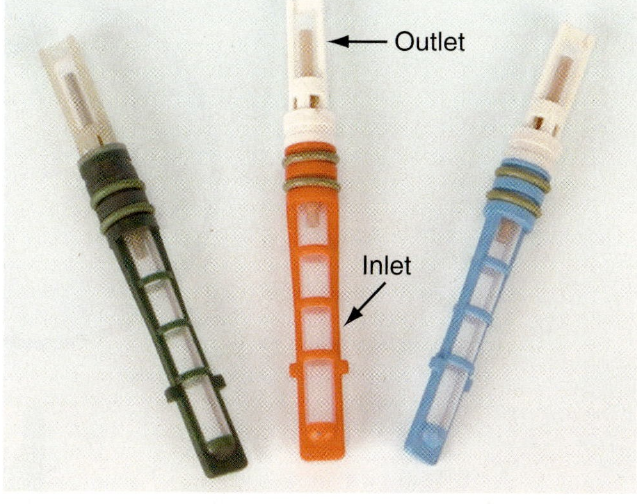

FIGURE 8-21 **Orifice tube showing inlet screen.**

This condition may be checked by carefully feeling the receiver-drier or condenser outlet, along the liquid line, and finally the metering device inlet. All should be warm. If any part is cool, a restriction is indicated at that point. Only the inlet of the metering device should be warm; the outlet should be cool. In fact, the outlet of the metering device is perhaps the coolest part of the system under normal operating conditions.

A restriction in the system is generally "marked" by a temperature difference.

If the inlet screen of the expansion valve (Figure 8-20) is found to be the problem, it may be cleaned. After cleaning the screen, the receiver-drier must be replaced. Also, the receiver-drier must be replaced if the restriction proves to be at its outlet. This is true even though the liquid line and expansion valve inlet screen may be clean. If the screen on an orifice tube (Figure 8-21) is found to be the problem, the orifice tube and the accumulator should be replaced. A temperature change at the outlet of the accumulator is expected and does not indicate a problem.

Another problem that can cause the symptoms of Condition Two is moisture in the system. If there is moisture in the system that is not absorbed by the desiccant in the receiver-drier or accumulator, it may freeze at the metering device inlet. The inlet may then become very cold, the same symptom indicated by a clogged inlet screen. To determine whether moisture is the problem, turn the air conditioner off for 10 to 15 minutes, then turn it back on. If the gauge reading immediately goes to an abnormal condition, the screen is probably clogged. If the gauge reading is normal for a few minutes, then goes to abnormal, there is probably excess moisture in the system. This condition is corrected by replacing the receiver-drier or accumulator.

Before condemning a component, though, make sure the system is fully charged with refrigerant. A slightly low system charge will cause rapid cycling of the air-conditioning compressor clutch when the engine speed is raised. On noncycling-clutch systems (variable displacement compressors), the low side will be low (15–30 psig for R-134a) and the high-side will also be low (110–150 psig for R-134a). You may also notice foamy bubbles in the sight glass, if so equipped, and the evaporator outlet line will be warm.

Condition Three: Insufficient Cooling or No Cooling (Figure 8-22)

- Low-side gauge: Very low pressure to low pressure
  - R-134a (HFC-134a): 15 psig (103 kPa)
  - R-12 (CFC-12): 18 psig (124 kPa)
- High-side gauge: Low pressure
  - R-134a (HFC-134a): 139–144 psig (958–993 kPa)
  - R-12 (CFC-12): 130–135 psig (896 kPa)

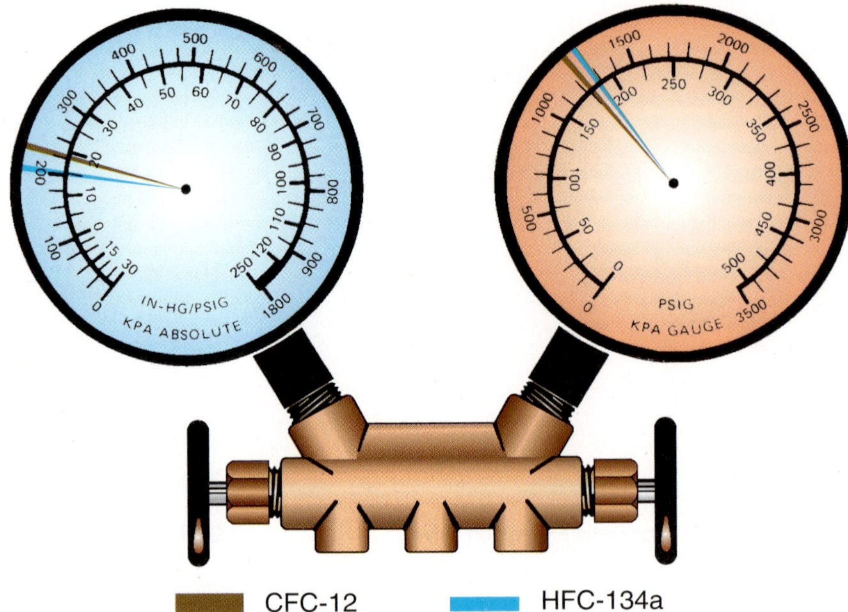

FIGURE 8-22 **Manifold gauge reading indicating insufficient or no cooling due to undercharge of refrigerant, clogged metering device inlet screen, defective metering device, or excessive moisture in the system.**

If the low-side gauge pressure is moderately low, the most probable cause is an undercharge of refrigerant. If the low-side gauge pressure is very low, possibly in a vacuum, there are four other possible causes, all relating to the metering device:

If there is a loss of refrigerant, there is a leak in the system.

1. Clogged inlet screen
2. Defective valve or tube
3. Moisture in the system
4. High-side restriction

Shop Manual
Chapter 8, page 294

Loss of refrigerant resulting in an undercharge is usually caused by a leak. This condition may also be noted by bubbles in the sight glass, if so equipped. To correct this condition, the cause of the leak must be located and repaired. The system must then be properly evacuated and charged with refrigerant. The effects of an undercharged thermal expansion valve refrigerant system may be seen in Figure 8-23. Similar refrigerant undercharge effects in an orifice tube system may be seen in Figure 8-24. Symptoms of an undercharged refrigerant system may include poor cooling, compressor cycling rapidly, and warm evaporator outlet line.

The screen in the metering device may be clogged or there may be moisture in the system, as outlined in Condition Two.

Shop Manual
Chapter 8, page 304

An expansion valve may be defective and completely closed. The most probable cause for this condition is that the remote sensing bulb may have lost its charge of volatile gas. If this is the cause, the valve will not regulate (open), and it must be replaced.

If the problem still cannot be found, look for a restriction in the high side before the metering device. Refer to Figure 8-1. The restriction will be located in quadrant 2 or 3. It should be noted that the gauge reading may be high if the restriction is found shortly after the low-side service fitting.

Condition Four: Insufficient Cooling or No Cooling (Figure 8-25)

- Low-side gauge: Low pressure
  - R-134a (HFC-134a): 20 psig (138 kPa)
  - R-12 (CFC-12): 22 psig (152 kPa)

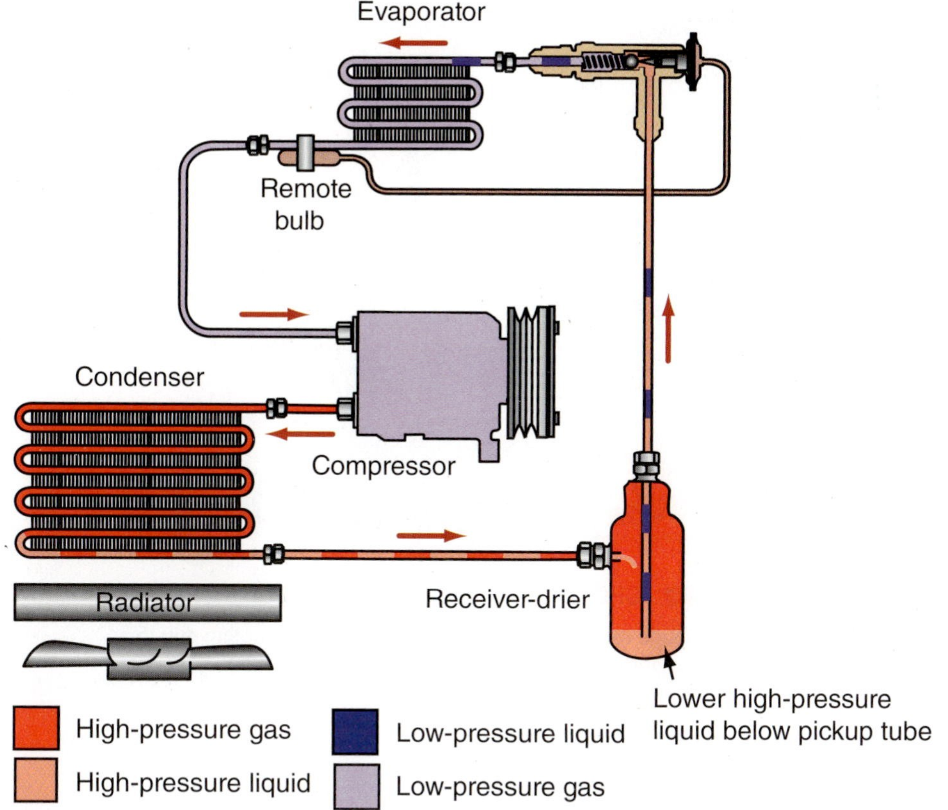

FIGURE 8-23 **Effects of refrigerant undercharge on a thermal expansion valve refrigerant system.**

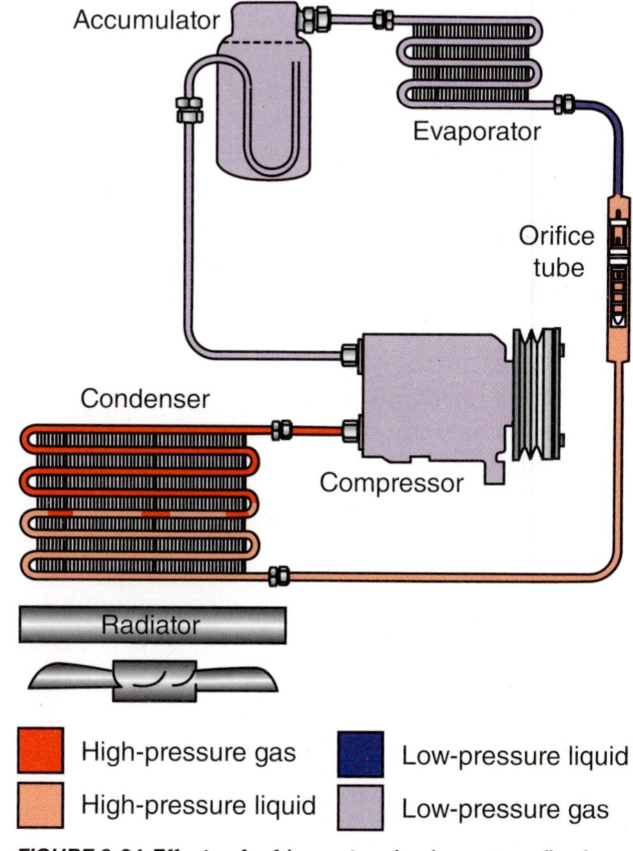

FIGURE 8-24 **Effects of refrigerant undercharge on a fixed orifice tube refrigerant system.**

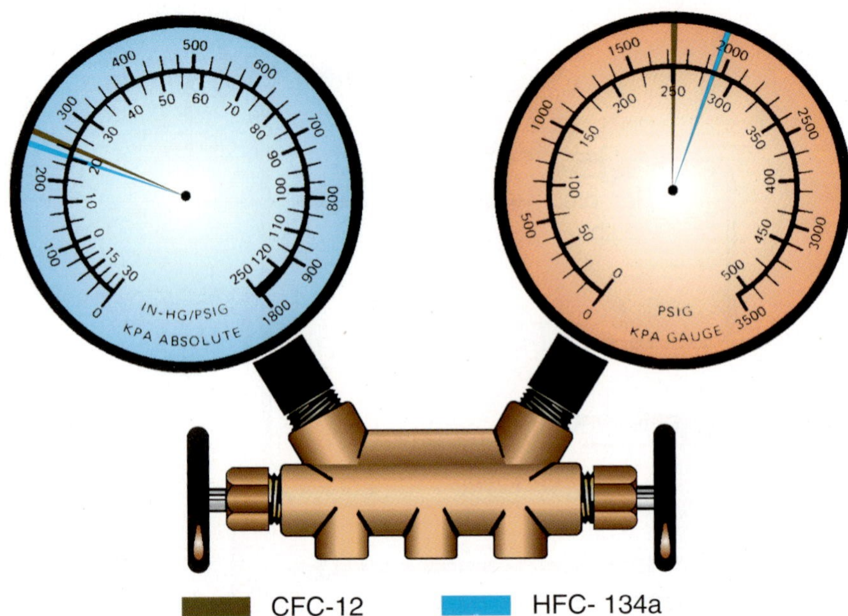

FIGURE 8-25 **Manifold gauge reading indicating insufficient or no cooling due to a restriction in the high side of the system, such as a bent or kinked tube.**

- High-side gauge: High to extremely high pressure
  - R-134a (HFC-134a): 281 psig (1,937 kPa)
  - R-12 (CFC-12): 250 psig (1,724 kPa)

The most probable cause of this condition is a restriction in the high side of the system. The restriction may be anywhere from the compressor outlet to the receiver-drier or fixed orifice tube inlet. The closer to the compressor, the higher the high-side gauge pressure will be. Refer to Figure 8-1. The restriction will be located in quadrant 1 or 2.

A moderately high high-side pressure may indicate a clogged receiver-drier (Figure 8-26) or liquid line. An extremely high pressure may indicate a restriction, such as a bent tube in the condenser closer to the compressor. The probable location of high-side restrictions is shown in Figure 8-27.

In any event, the restriction must be located and corrected. Often, a marked temperature change will be noted at the point of restriction. The upstream side of the restriction will be very hot while the downstream side will be cooler. High-side restrictions can cause extremely high temperatures. Be careful to avoid personal injury.

If the high-side pressure is abnormally high and the high-side lines and hoses are vibrating or pulsating, the refrigerant system may be overcharged with refrigerant oil. This will also be accompanied by high duct temperatures and poor cooling performance. This condition is difficult both to diagnose and to repair. In excess, oil in the refrigerant system acts as a heat insulator and takes up space. Too much refrigerant oil in the system will cause the compressor to work harder to displace it and in the process increase system temperatures and pressures. Typically the high-side gauge will read a higher pressure than a static temperature-pressure chart would indicate. In a system that contains an excess amount of oil, the outside temperature of the high-side line will be cooler than the static temperature-pressure chart would indicate, but the pressure will be higher than the measured temperature would indicate on the static temperature-pressure chart.

To repair a system that is overcharged with oil, the refrigerant must first be recovered; then the compressor must be removed and drained, and the evaporator, condenser, lines, and hoses should be drained or flushed. In addition the receiver-drier/accumulator must be replaced. After reassembly, the system must be evacuated, filled with the proper type and amount of oil, and recharged with the proper amount of refrigerant.

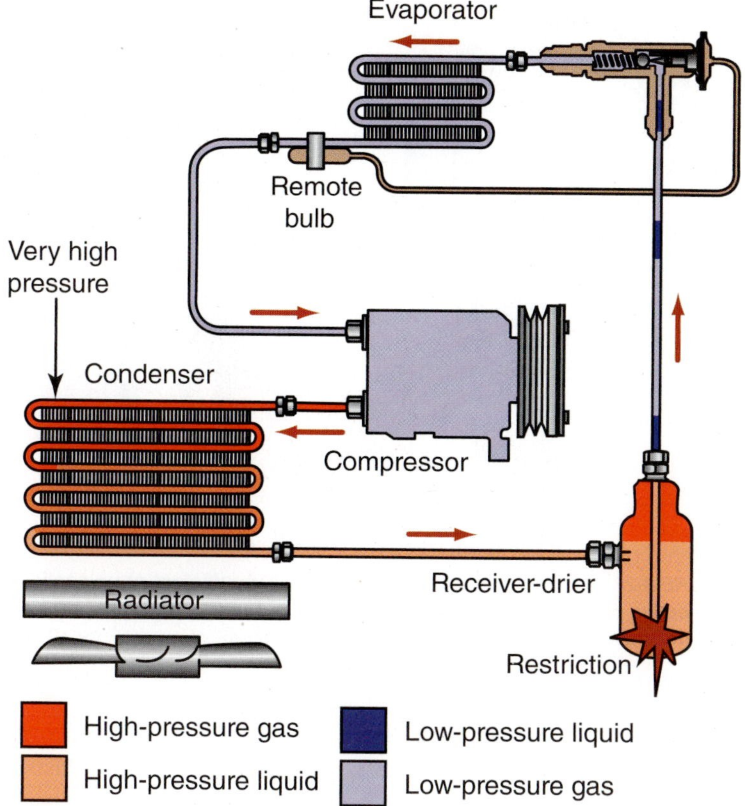

FIGURE 8-26 **Effects of partially restricted receiver-drier on a thermal expansion valve refrigerant system.**

FIGURE 8-27 **A damaged return bend at the condenser inlet will result in very high high-side pressure.**

Condition Five: Insufficient Cooling or No Cooling (Figure 8-28)

- Low-side gauge: High pressure
  - R-134a (HFC-134a): 43 psig (296 kPa)
  - R-12 (CFC-12): 44 psig (303 kPa)
- High-side gauge: Low pressure
  - R-134a (HFC-134a): 150 psig (1,034 kPa)
  - R-12 (CFC-12): 140 psig (965 kPa)

A mechanical malfunction can cause an electrical malfunction.

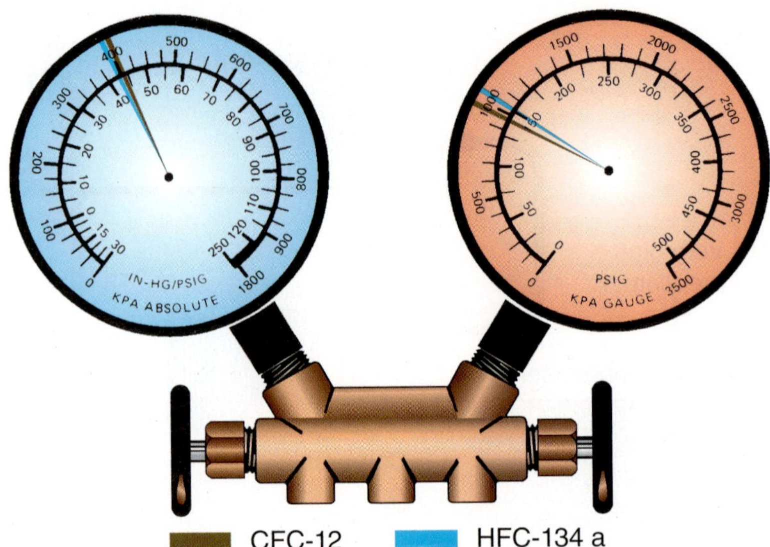

FIGURE 8-28 **Manifold gauge reading indicating insufficient or no cooling due to a defective clutch coil or temperature control. The condition may also be caused by a defective clutch or compressor.**

This problem may be caused by either an electrical or a mechanical condition. It may be caused electrically by a defective clutch coil or a defective thermostat. Also, inspect for these defective components:

1. Cycling switch
2. Pressure switch(es)
3. Ambient air temperature switch
4. Evaporator temperature sensor (fin sensor)

Mechanically, this condition may be caused by either of the following two problems:

1. A defective clutch
2. A defective compressor
    a. Valve plate(s)
    b. Head gasket(s)
    c. Broken piston ring

The symptoms for no or poor compressor action for a TXV system (Figure 8-29) and an FOT system (Figure 8-30) are similar and include poor to no cooling, warm evaporator outlet, and all lines warm to touch. In addition, the low- and high-side pressures will equalize quickly after the air-conditioning system is turned off.

To determine whether the problem is due to electrical or mechanical defects, if the air conditioner is operational, visually inspect the clutch center bolt to determine whether the compressor crankshaft is turning properly. If it is turning properly, the problem is probably a defective compressor or valve plate assembly. If the compressor is turning erratically, disconnect the clutch wire and connect it to a digital multimeter (Figure 8-31). If there are at least 10.8 volts present required for proper clutch operation, the problem may be a defective clutch coil or clutch assembly. First, however, check to ensure that the clutch coil is properly grounded.

Shop Manual
Chapter 8, page 295

If the multimeter does not indicate at least 10.8 volts, the probable cause is a defective relay, electrical control device, or loose wire.

Determine whether the system is TXV or FOT equipped before diagnostics.

Also, listen for compressor noise, which is an indication of a defective compressor. If the problem is determined to be the compressor, the valve plate or gaskets may be defective. In either case, it will be necessary to remove the compressor head and valve plate assembly to determine and repair the cause or to replace the compressor.

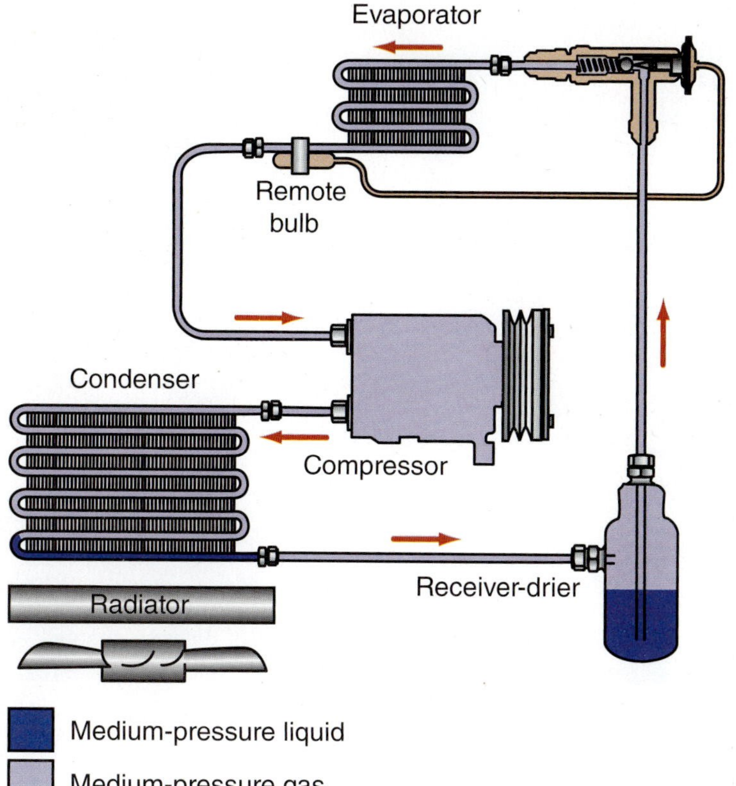

FIGURE 8-29 **No or poor compressor action on a thermal expansion valve refrigerant system. Symptoms include poor or no cooling, warm evaporator outlet, and all lines warm to touch.**

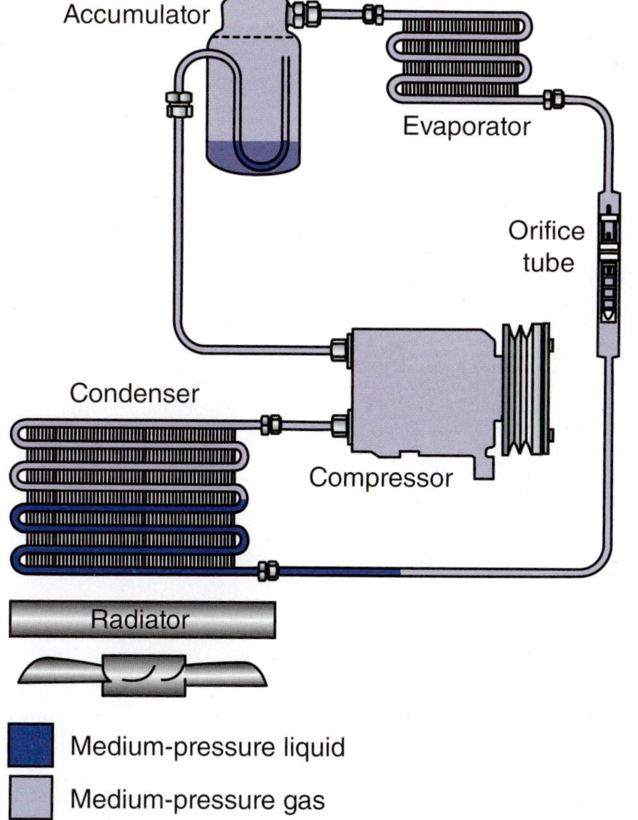

FIGURE 8-30 **No or poor compressor action on a fixed orifice tube refrigerant system. Symptoms include poor to no cooling, warm evaporator outlet, and all lines warm to touch.**

FIGURE 8-31 **Checking for voltage at the clutch coil connector.**

Condition Six: Insufficient Cooling (Figure 8-32)

- Low-side gauge: High pressure
  - R-134a (HFC-134a): 38 psig (262 kPa)
  - R-12 (CFC-12): 40 psig (276 kPa)
- High-side gauge: Normal pressure
  - R-134a (HFC-134a): 184 psig (1,269 kPa)
  - R-12 (CFC-12): 170 psig (1,172 kPa)

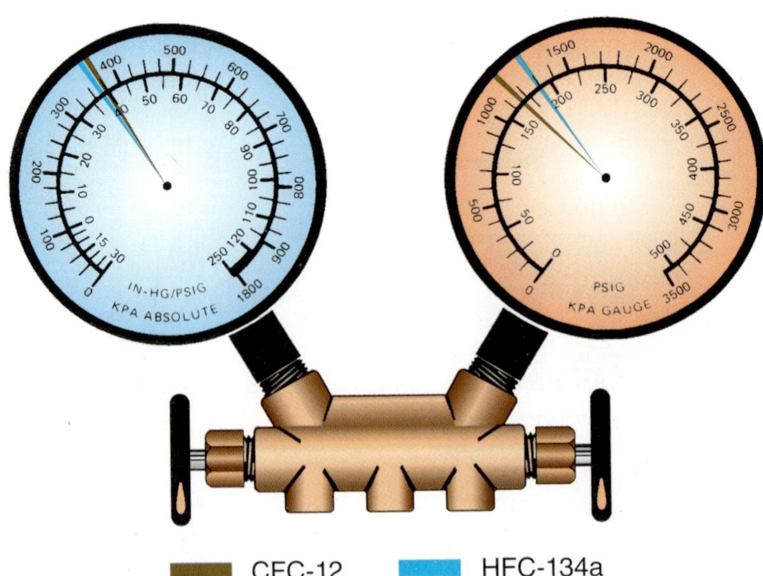

FIGURE 8-32 **Manifold gauge reading indicating insufficient cooling due to a defective thermostatic value or mispositioned remote bulb.**

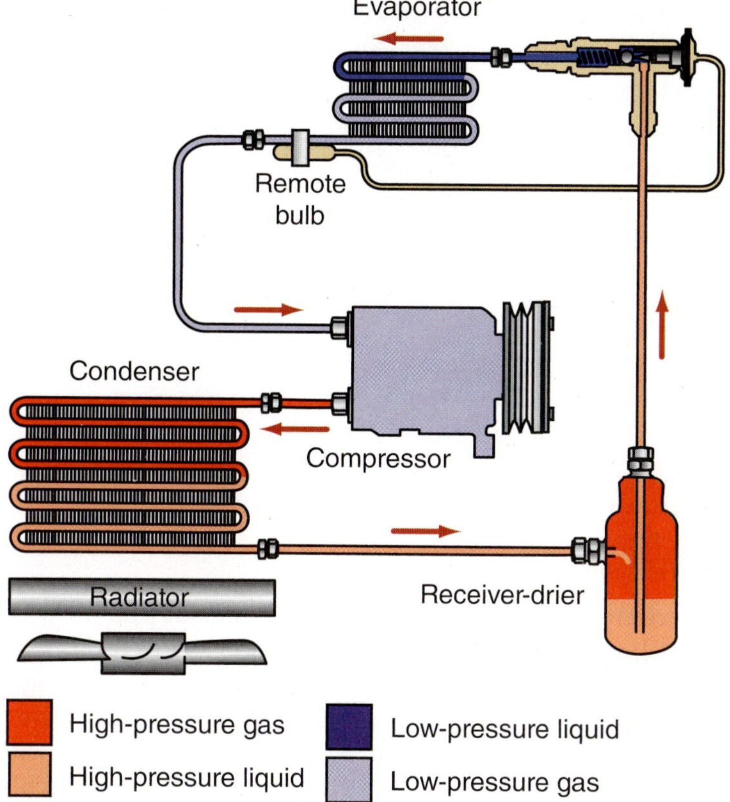

FIGURE 8-33 **A TXV that is stuck in the open position or is slow to respond to temperature changes will result in poor system performance. Symptoms include poor to fair cooling and evaporator outlet warm to the touch.**

This condition is found only in systems equipped with a TXV. The condition, then, is caused by a defective expansion valve. Unlike expansion valve problems of Conditions Two and Three, however, this indication is that the expansion valve is stuck in the open position or is not closing because the **remote bulb** is not making proper contact with the evaporator outlet tube or has lost its charge (Figure 8-33). A similar condition could occur on a FOT system if the orifice tube was missing (removed) from the system.

A **remote bulb** is a sensing device connected to the expansion valve by a capillary tube. This device senses the evaporator outlet temperature and transmits pressure to the expansion valve control diaphragm for proper operation.

First, make sure that the remote bulb and the evaporator outlet tube are clean and that the two mating surfaces make good mechanical contact with each other. A small piece of cork (no-drip) tape wrapped around the remote bulb and the outlet tube helps to ensure good "sensing" conditions. This tape also acts as an insulator for the remote bulb, preventing it from sensing and being influenced by ambient air.

If the remote bulb is securely fastened to the outlet tube and the condition is not corrected, the expansion valve is probably defective and must be replaced.

One method for determining whether the internally regulated thermostatic expansion valve is functioning correctly is to cool it externally with low pressure CO_2. With the use of a low pressure CO_2 regulator, allow the gas from a discharge line to be bled directly over the expansion valve. This will chill the valve, causing both a pressure and an evaporator temperature change.

If the TXV proves to be functioning properly and the problem still exists, check the heater control valve and the blend air door. A defective heater control valve may let heated coolant flow through the heater core, thereby creating an environment that promotes a higher than normal low-side pressure. Though not as likely, the blend air door, if mispositioned, may create a similar environment.

If air or excessive refrigerant is in the system, recover, evacuate, and recharge the system.

Shop Manual
Chapter 8, page 296

Condition Seven: Insufficient or No Cooling (Figure 8-34)

- Low-side gauge: High pressure
  - R-134a (HFC-134a): 37 psig (255 kPa)
  - R-12 (CFC-12): 42 psig (290 kPa)
- High-side gauge: High to extremely high pressure
  - R-134a (HFC-134a): 263 psig (1,813 kPa)
  - R-12 (CFC-12): 235 psig (1,620 kPa)

There are several possible causes for this condition:

1. Air in the system
2. An overcharge (excess) of refrigerant
3. An overcharge (excess) of oil
4. Condenser air passages (fins) clogged
5. Defective cooling fan(s)
6. An overheating engine
7. Incorrect refrigerant
8. Contaminated refrigerant

The symptoms of a system overcharged with refrigerant include poor to fair cooling, evaporator outlet warm to cool, and poor to no cooling during stop and go traffic, but highway driving may be okay. The compressor clutch may frequently cycle on/off as the high-pressure switch detects high high-side pressure. Too much air in the system may also cause these symptoms. In addition, when the air-conditioning system is turned off on a system with air contamination, the pressure will drop 30 psig quickly but then falls gradually until high- and low-side pressures equalize. The effects of an overcharged thermal expansion valve refrigerant system may be seen in Figure 8-35. Similar effects of a refrigerant overcharge occur in an orifice tube system and may be seen in Figure 8-36.

When cleaning air-conditioning components in a charged system, use cool or warm water only.

A visual inspection of under-the-hood conditions should determine whether the problem of Condition Seven is due to clogged condenser air passages or an overheating engine. If the problem is determined to be an excess of refrigerant, oil, or air in the system, it is most difficult to determine which of these is the cause.

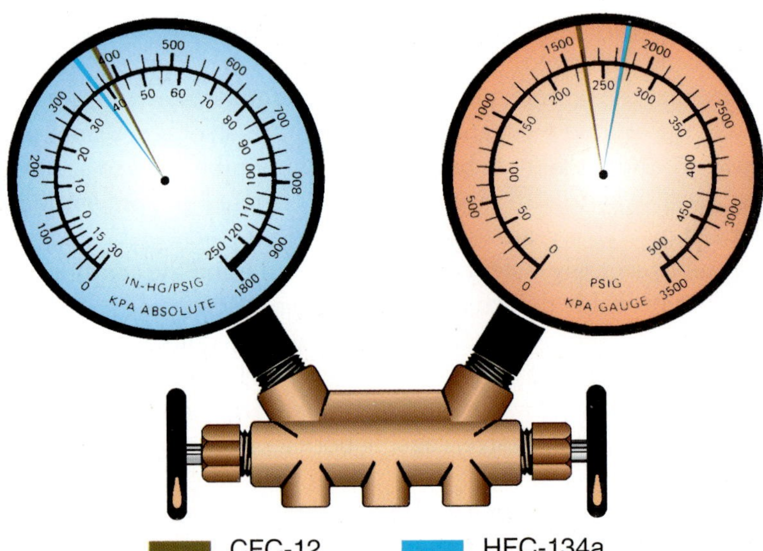

FIGURE 8-34 **Manifold gauge reading indicating insufficient or no cooling due to air in the system, overcharge of lubricant or refrigerant, clogged condenser, defective cooling fan(s), or an overheating engine.**

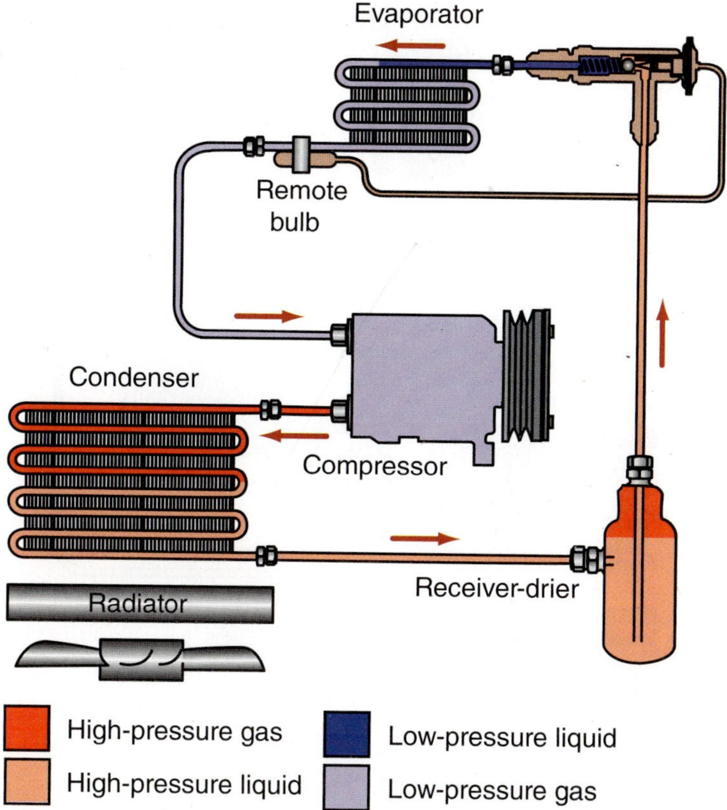

FIGURE 8-35 **Effects of refrigerant overcharge on a thermal expansion valve refrigerant system include high low-side and extremely high high-side pressures resulting in poor to no cooling and evaporator outlet warm to the touch.**

If the condenser air passages are clogged, heat cannot be carried away. This will result in moderately high pressures and insufficient cooling. Condenser clogging is generally caused by dirt, leaves, bugs, or other foreign material lodged in the fins.

The condenser may be cleaned with a strong stream of detergent and water (H_2O) such as may be found at a do-it-yourself car wash. Whenever possible, if space between the radiator and condenser permits, clean the condenser in the opposite direction of the airflow. Take care not to damage the delicate tubes and fins of the radiator. Use warm, not hot, water.

A defective or inoperative coolant fan can give the same symptoms as a blocked condenser in shop conditions. At road speeds, however, ram air may suffice for heat removal.

High **head pressure** may also be caused by a kinked hose or a restriction. The high-side restriction may be anywhere between the compressor outlet and receiver-drier or metering device inlet. See also Condition Four.

Head pressure is the pressure of the refrigerant from the compressor discharge port to the metering device.

An overheating engine causes an additional heat load (ambient conditions), which in turn will cause high head pressure conditions. Overheating engines may be caused by:

- Loss of coolant
- Slipping belts
- Improper engine timing
- A defective water pump
- A defective thermostat or radiator cap

Engines and engine service are covered in Delmar Learning's *Today's Technician* series *Automotive Engine Repair & Rebuilding,* 5th edition, and *Automotive Engine Performance,* 6th edition (Figure 8-37).

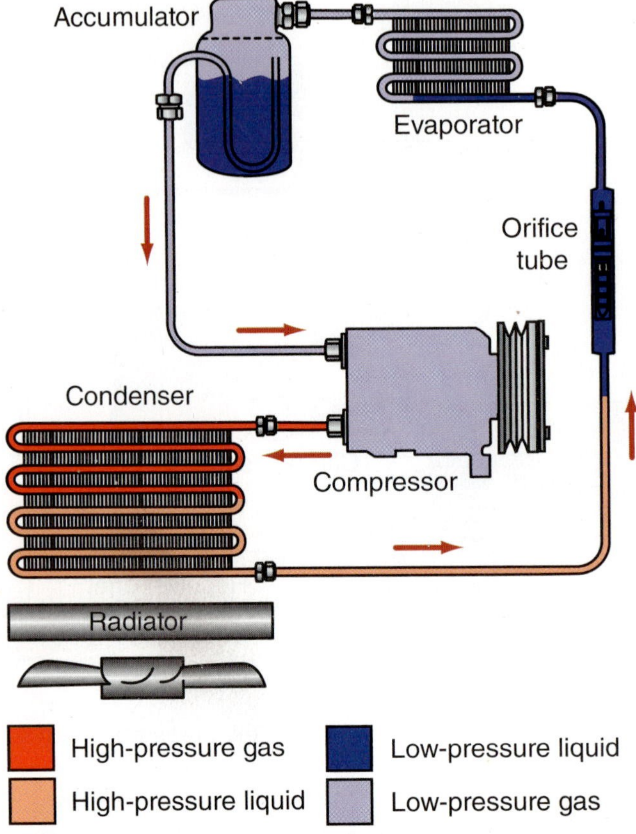

FIGURE 8-36 **Effects of refrigerant overcharge on a fixed orifice tube refrigerant system include high low-side and extremely high high-side pressures, resulting in poor to no cooling and evaporator outlet warm to the touch.**

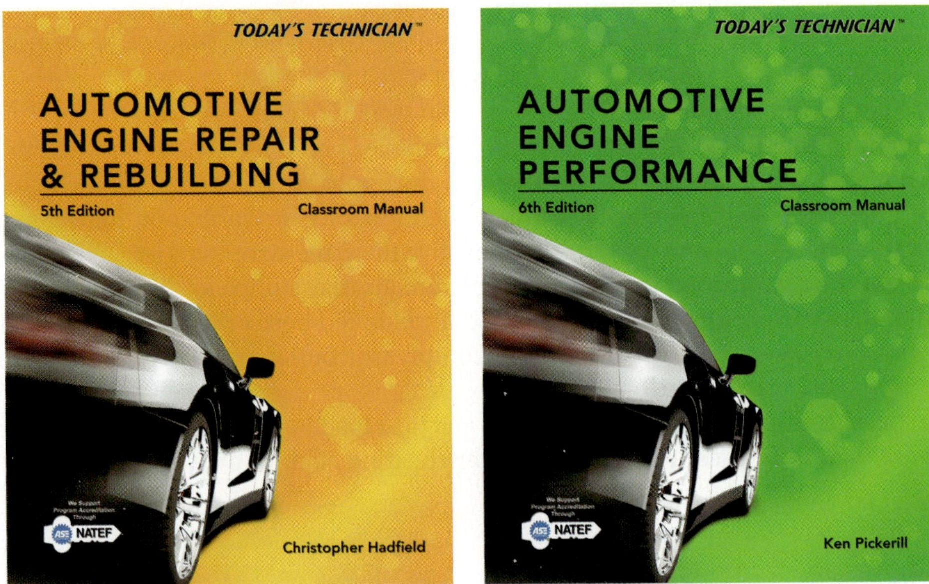

FIGURE 8-37 **Engine service and engine performance are covered in the *Today's Technician* series.**

This condition may be caused by air in the system, which can result, for example, through a low-side leak. If the low side goes into a vacuum while running, ambient air will be drawn into the system. Many systems have a low-pressure switch to prevent the system operation in a vacuum. Improper evacuation or failure to evacuate the system before charging with refrigerant can also result in air contamination, as can using improperly recycled refrigerant.

This condition may also be caused by an overcharge of refrigerant. Excess refrigerant may be bled off using the standard practice procedures for recovering refrigerant. Because it is almost impossible to determine whether the cause is air or excess refrigerant, it is advisable to recover, evacuate, and recharge the system.

Shop Manual
Chapter 8, page 296

Another cause for this condition is excessive oil in the compressor. If no oil has been added, however, this is not likely to be the problem.

Moisture Contamination

If a refrigerant system is contaminated with moisture, the pressure in the system will swing between a vacuum to normal on the low-side gauge and between low to normal on the high-side gauge. The air-conditioning system may operate normally at first, but as the system runs, a no-cooling condition may exist. This condition is often described as intermittent air-conditioning operation by the driver. The moisture in the system freezes in the TXV or FOT and causes a temporary blockage. After the blockage occurs, the evaporator warms up and the ice melts, returning the system to normal until the process repeats itself. If moisture in the refrigerant system is suspected, the old refrigerant must be recovered and the receiver-drier/accumulator must be replaced. The system must then be evacuated to 700 microns or less to ensure that the moisture has been removed from the system before recharging it with a new refrigerant.

Restriction Diagnosis

Diagnosing the effects of restrictions in a refrigerant system based on gauge readings is sometimes difficult and misleading because of the location of the refrigerant system service ports. It is important to keep in mind where the service ports are in relation to the compressor discharge and suction ports, and the normal restriction created by the TXV or FOT.

High-Side Restrictions

A restriction on the high side of the system will have different pressure effects depending on where the restriction is. The closer the restriction is to the compressor, the higher the pressure will be. The high-side gauge, however, will not indicate the high-pressure condition if the restriction is before the high-side service port (Figure 8-38); in fact, the gauge reading will be lower than normal. By also looking at the low-side gauge reading, clues can be gained. The low-side gauge may indicate normal or slightly low pressure, or a vacuum, depending on how severe the high-side restriction is. If the high-side restriction is after the service valve, the high-side gauge reading will be higher than normal (Figure 8-39). This is often the case with a restriction at the TXV or FOT.

Additional information will be gained by measuring component temperatures with a thermocouple. The line temperature will be higher than normal before the restriction and will be lower than normal after the restriction. An increase in pressure raises the temperature before the restriction, which then acts like an orifice, allowing for a pressure differential as refrigerant flows through; this causes some boil- off to occur, lowering the temperature.

Low-Side Restrictions

A restriction on the low side of the refrigerant system is not as likely because of the larger diameter of the lines and hoses. But it is still possible for restrictions to occur in the evaporator core if the system is contaminated, and if the system has an accumulator, the desiccant may have failed. If a low-side restriction occurs, the low-side gauge reading will be low or even a vacuum, depending on where the service fitting is located in relation to the restriction (Figure 8-40). If the compressor is not able to circulate the refrigerant because of the low- side restriction, the high-side gauge reading will also be low. Finding the restriction by looking

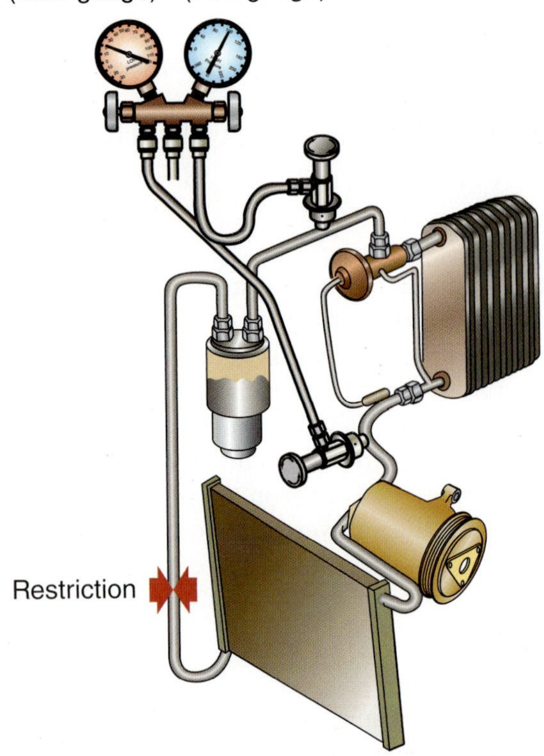

FIGURE 8-38 **The high-side pressure gauge reading may be lower than normal if the high-side restriction is before the high-side service port.**

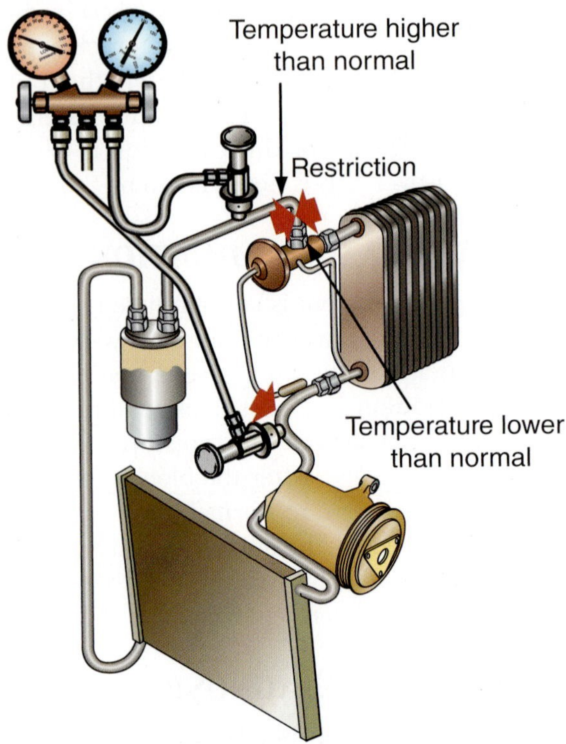

FIGURE 8-39 **The high-side pressure gauge reading will be higher than normal if the high-side restriction is after the high-side service port. A temperature differential will be detectable before and after the restriction.**

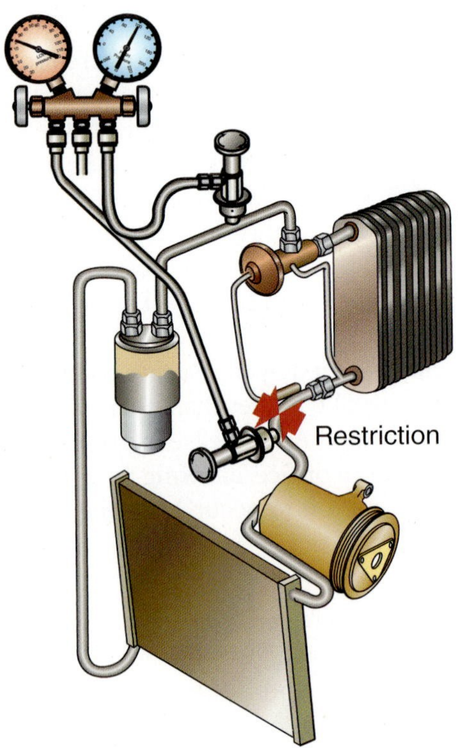

FIGURE 8-40 **The low-side pressure gauge reading will be lower than normal or possibly even a vacuum.**

for temperature differentials before and after the low-side restriction is still possible, but the temperature differential on either side will not be as great, and some temperature difference between the evaporator inlet and outlet is expected.

Preventive Maintenance

Preventive maintenance (PM) pays off in the long run. Whenever servicing an automotive air-conditioning system, potential problems can sometimes be discovered before they occur. A thorough visual check of the mechanical and electrical system is well worth the time invested.

Mechanical

Check the air-conditioning system for damaged hoses (Figure 8-41) and connections that may be caused by rubbing or chafing. Slight oil staining may indicate a refrigerant leak. With the engine off, inspect the belt(s) for glazing and cracking. Heavy glazing may indicate a slipping belt. Some glazing, however, is acceptable. Inspect the hoses and hose connectors. Soft or brittle hoses are an indication that they are deteriorating and should be replaced. Do not neglect the heater hoses. A slight leak in a heater hose is often overlooked.

Electrical

Check for loose connections and frayed (Figure 8-42) or broken wires. If the problem is a "blown" fuse or circuit breaker, it is possible that a bare wire is intermittently "shorting" to ground due to vibration.

The blower motor can sometimes give an indication of other problems. If, for example, the blower speed increases when the engine is revved up, the battery may be undercharged; a defective voltage regulator may cause overcharging; or the battery ground cable may be corroded or poorly connected. A slipping compressor clutch may be an indication of improper adjustment or low voltage supplied to the coil.

A fully charged automotive battery has a surface charge of 12.8–13.2 volts.

The fuse, circuit breaker, and fusible links are provided for protection against an electrical fire. Never bypass a circuit protection device simply because it "blows" frequently. If it blows, there is a reason. To prevent further damage, locate the problem and correct it.

Never take chances—if in doubt, disconnect the battery cable. Always disconnect the battery ground (—) cable, not the positive (+) cable (Figure 8-43). If the wrench grounds out while disconnecting the ground cable, no harm is done. If, however, the wrench grounds out while disconnecting the positive cable (with the ground cable attached), the electrical system is "shorted" and may be damaged.

FIGURE 8-41 **Check for defective hoses.**

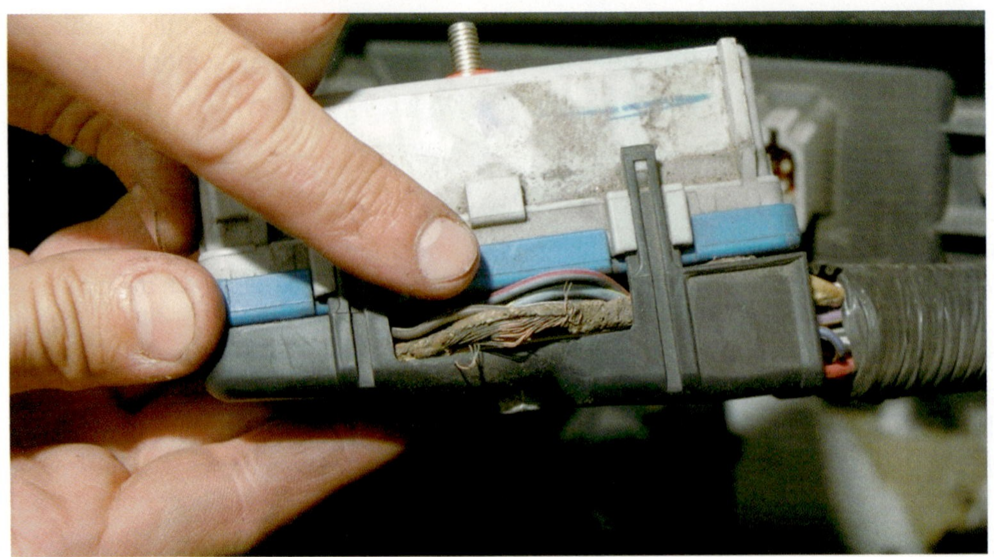

FIGURE 8-42 **Check for frayed or broken wires.**

FIGURE 8-43 **Always disconnect the battery ground (−) cable.**

A fully charged 12V automotive battery is capable of delivering very high current. In a matter of seconds, a shorted electrical system can cause extensive damage to the automobile's wiring harness.

From the information given related to system conditions, it should now be obvious that the manifold and gauge set is a very important tool in air-conditioning service. It is not only used to service the system, but also as a diagnostic tool.

Advanced Diagnostic Tools

As was noted in Chapter 7, software-based system diagnostics have become integrated into the automotive air-conditioning service industry. One of these tools, the Neutronics Inc. Master A/C System Technician (Figure 8-44), attaches to the high-side and low-side pressure service fittings and uses a temperature sensor to check ambient air temperature and component temperature. The technician follows the diagnostic steps listed and inputs requested information, such as component temperatures. Using the FlexTemp's™ temperature probe to check component temperature differences, the software uses the Delta T method, low-and high-side pressures, as well as look-up tables to calculate charge levels on FOT systems. On

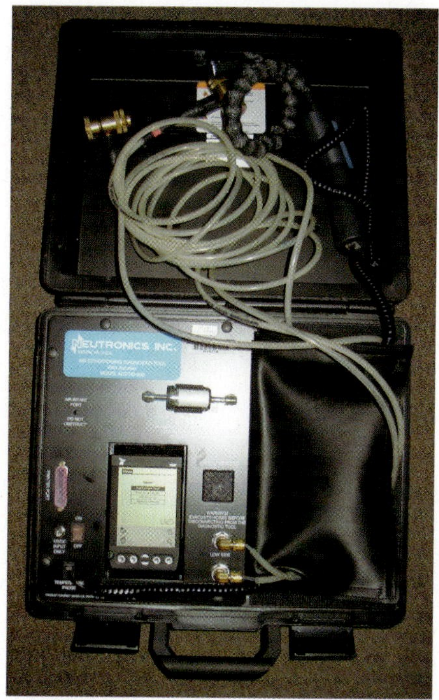

FIGURE 8-44 **Air-conditioning system electronic analyzer used to diagnose system performance.**

thermostatic expansion valve systems, it bases its calculations on the temperature of refrigerant leaving the condenser and high-side line pressure to determine system charge levels.

Temperature and pressure comparisons are a standard practice to determine superheat, subcooling, and system change levels and performance. On an FOT system, the outlet line from the evaporator should be 3°F–10°F (2°C–6°C) colder than the evaporator inlet line temperature on a properly charged system. If the evaporator outlet line is warmer than the inlet line, the system is low on refrigerant. If the evaporator outlet line is colder than 10°F (2°C–6°C), the system may be overcharged. If you do not own a sophisticated diagnostic tool, you can use two contact thermometers and a manifold gauge set to perform a Delta T system charge level test. Run the engine at 1,000 rpm and, if equipped with a cycling-clutch switch, jump the connector so the compressor will run continuously. Check both the evaporator inlet and outlet line temperatures simultaneously. Compare your results to a Delta T chart for the manufacturer and type of vehicle you are working on.

By checking both superheat and subcooling, you can gain information above and beyond what your pressure gauges are telling you. When you adjust the system charge based on superheat, you are charging the system based on the amount of air passing over the evaporator. You should not, however, base your refrigerant charge levels on superheat results for thermal expansion valve systems. TXV systems control superheat by automatically adjusting to evaporator refrigerant temperature. Superheat will help you to determine whether the TXV is working, though.

A method for determining the superheat capacity of the refrigerant is to:

1. Check the low-side pressure reading of the system and convert it to a temperature based on the type of refrigerant in the system. An example of a pressure-to-temperature comparison chart was given in Figure 8-2.
2. Place the blower fan on high speed setting.
3. Next, check the temperature of the low-side (suction) line about 6 in. before the compressor.
4. Now calculate the difference between the low-side line temperature and the saturation temperature based on the chart (low-side line temperature minus saturation temperature).

5. If the ambient air temperature is 75°F–85°F (23.89°C–29.44°C), the superheat should be 12°F–15°F (−11.11°C to −9.44°C). If the ambient air temperature is above 85°F (29.44°C), the superheat should be 8°F–12°F (−13.33°C to −11.11°C).

   Example for an R-134a system:

   Low-side line temperature is 45°F (7.22°C)
   Low-side line pressure is 30.4 psig, saturation temperature will be 35°F (1.6°C) 45°F (7.22°C) − 35°F (1.67°C) = 10°F (12.22°C) superheat

6. If superheat is low, then the evaporator may be flooded, and if the superheat is high, then the evaporator may be starved for refrigerant. Do not adjust charge levels until subcooling has been tested.

Subcooling

7. To determine subcooling, check the high-side pressure reading of the system and convert it to a temperature based on the type of refrigerant in the system (saturation temperature).
8. Next, check the high-side liquid-line temperature as close to the evaporator as possible, but before the metering device (in other words, FOT or TXV).
9. Now calculate the difference between the high-side liquid-line temperature and the saturation temperature based on the chart (saturation temperature minus high-side temperature). The subcooling temperature should be 12°F–15°F (−11.11°C to −9.44°C). It should be noted that the temperature of the liquid line at the condenser outlet and the temperature of the liquid line at the metering device inlet should be within 2°F (−16.67°C) of each other. If not, there could be a restriction in the liquid line.

   Example for an R-134a system:

   High-side liquid line temperature is 195°F (90.55°C)

   High-side liquid line pressure is 198.7 psig; saturation temperature will be 180°F (82.22°C) 180°F (82.22°C) − 195°F (90.55°C) = 15°F (9.44°C) subcooling

Using the information gained from both the superheat and the subcooling test, we will have some idea of how the system is operating.

1. If both the superheat and subcooling are low, the TXV is stuck open or the wrong orifice tube is installed (too large).
2. If both the superheat and subcooling are high, look for a restriction at the metering device, evaporator, or refrigerant line.
3. If the superheat is low and subcooling is high, the system may be overcharged.
4. If the superheat is high and subcooling is low, the system may be undercharged.

Another method for determining system performance and charge levels is to compare the on and off times of the compressor clutch. The charts in Figure 8-45 represent typical on and off times of the compressor clutch on a cycling-clutch system compared to ambient air temperature. The charts in Figure 8-46 depict the typical duct discharge temperature and the expected high- and low-side system pressure for a typical R-134a system at various ambient temperatures. The system should operate within the shaded areas. More detailed information is given in Chapter 8 of the Shop Manual.

Leaks

Service valves and protective caps (Figure 8-47) are among the most common causes of refrigerant leaks. The high- and low-pressure service (Schrader) valves are not perfect seals and may leak a small amount of R-134a. The level of leakage may not exceed 0.25 ounces per year. The service valve cap and O-ring are the primary seal, and the Schrader valve is considered the secondary seal. It is imperative that the service valve cap seal be in good condition. When servicing a refrigerant system, the initial untorquing of the service cap may release a short wisp

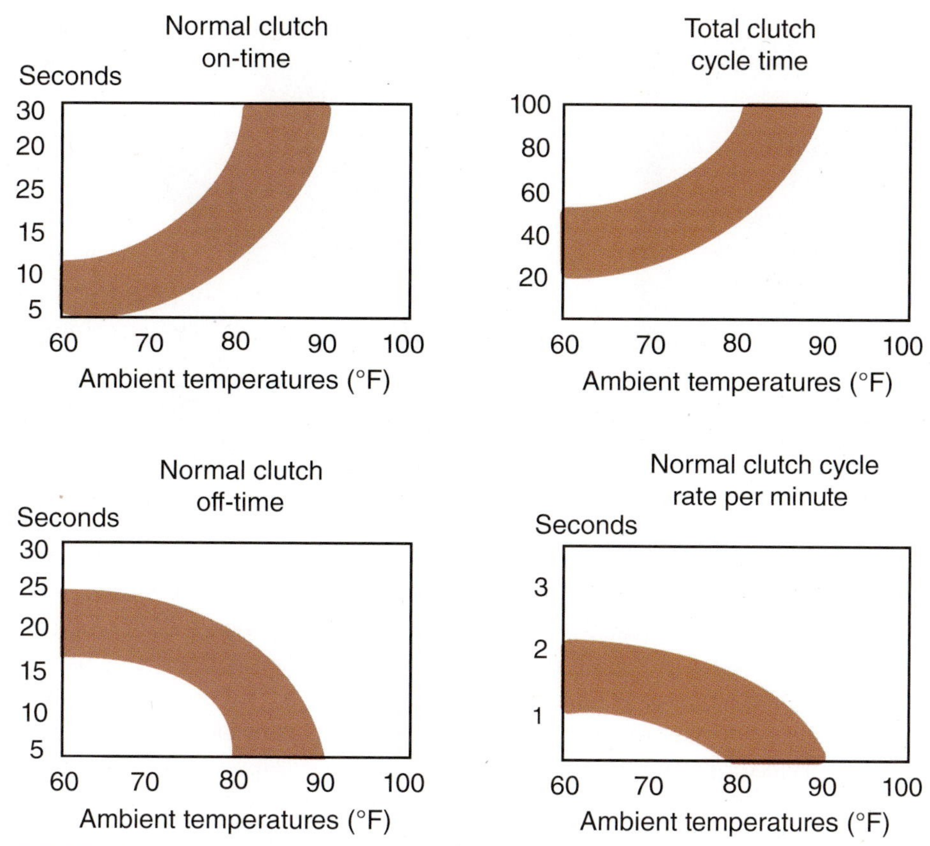

FIGURE 8-45 **Typical diagnostic and testing chart for cycling clutch systems.**

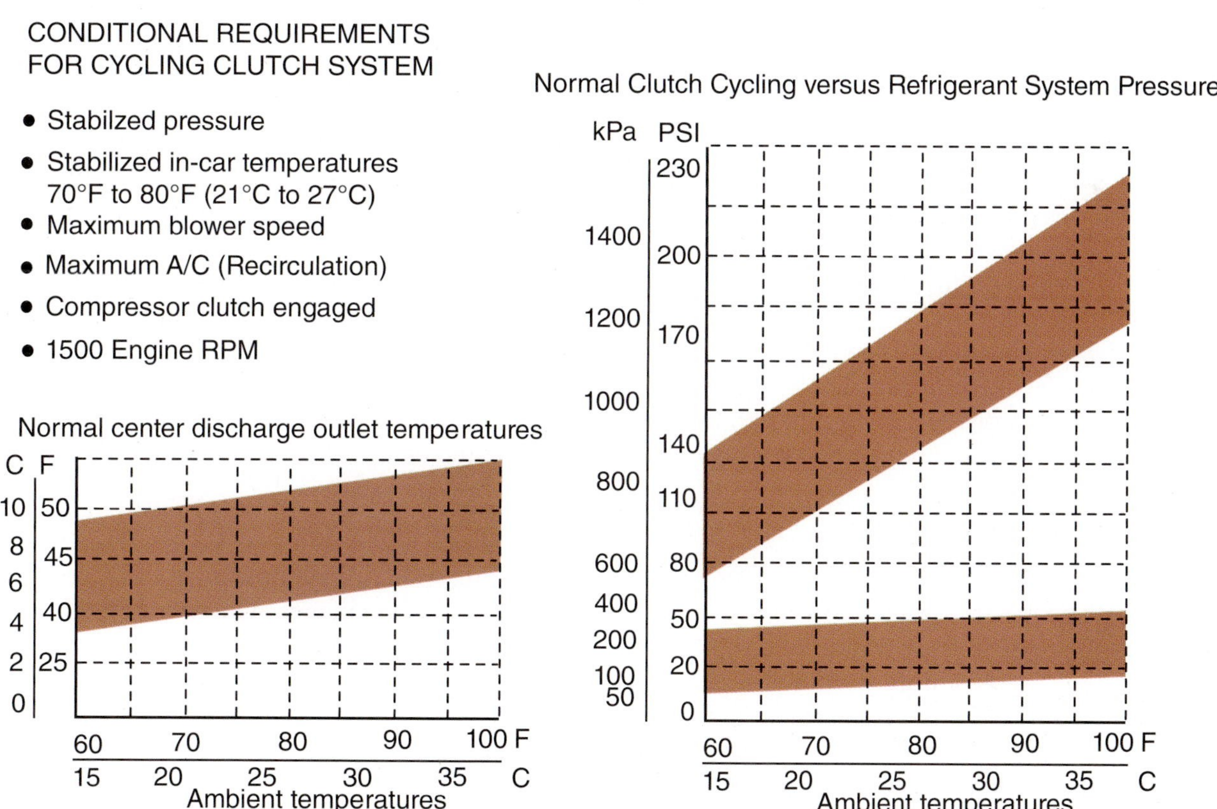

FIGURE 8-46 **Typical duct discharge temperature and expected high- and low-side system pressures of a typical R-134a system.**

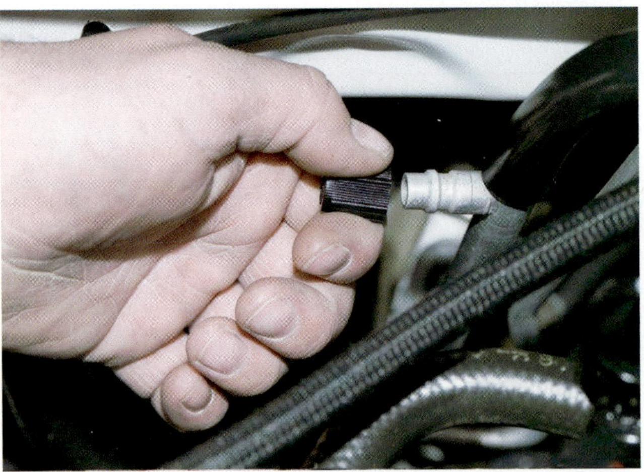

FIGURE 8-47 **Service valve with protective cap removed.**

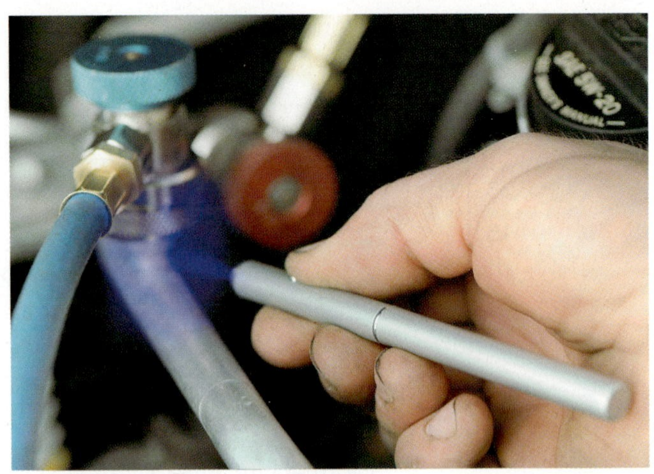

FIGURE 8-48 **Service valves with hoses connected are often neglected when leak testing.**

of refrigerant. On systems with fluorescent leak dye, installed dye may be present in the service fitting. Because the caps are the final seal, this is normal as long as it is not excessive. Always verify with an electronic leak analyzer after blowing out the service fitting with low pressure shop air. As much as a pound (0.45 kg) of refrigerant per year can escape from the service valve if the cap is missing or the O-ring is defective. Leak testing the service valve is often neglected because the service hoses are generally connected to them (Figure 8-48). Service valves should be leak tested with the caps and the service hoses removed. System integrity should not depend on the sealing power of a protective cap. The primary purpose of the cap is to keep debris out of the service valve. If found to be leaking, the service valve should be repaired or replaced as applicable. Often, a new Schrader assembly is sufficient to stop most leaks.

An insufficient refrigerant charge, for any reason, will cause oil to become trapped in the evaporator. Oil also leaks out with refrigerant at the point of a leak. Any oil loss due to any reason can result in compressor seizure.

The compressor circulates a small amount of oil through the system with the refrigerant. Oil pumped out of the compressor in small quantities is mixed with the refrigerant in the condenser. This oil enters the evaporator with the refrigerant, and if the evaporator is properly flooded with refrigerant, passes to the compressor through the low-pressure line. Some of the oil passes to the compressor in small droplets. Most of the oil, however, is swept along the walls of the refrigerant lines by the velocity of the refrigerant vapor. This oil is returned to the compressor as a mist. If the evaporator is starved of refrigerant, oil will not return to the compressor in sufficient quantity to keep it properly lubricated. The major cause of premature compressor failure is a lack of lubricant. The tendency of a customer to have refrigerant added to the system "every few months or so" is a sure sign that the compressor is doomed. If the system is leaking refrigerant, it is a good bet that it is also leaking lubricant, and compressor failure is sure to follow.

High Pressure

As refrigerant pressure increases in an air-conditioning system, its temperature also increases. The resulting high temperature quickly accelerates the failure of a contaminated system. An increase in temperature of only 15°F (8°C) doubles the chemical reaction rate in the system. High temperature starts a chain of harmful reactions even in a clean system. Contamination, resulting from high temperature, may cause seizure of the compressor bearings.

The top of the condenser is often the highest point of the system; air, lighter than refrigerant, seeks the highest point.

High heat may also cause the refrigerant in the system to decompose or break down. High heat can cause synthetic rubber parts to become brittle and susceptible to cracking and breaking.

As temperature and pressure in an air-conditioning system increase, stress and strain on compressor discharge valves increase. If this condition is not corrected, the discharge reeds in the valve plates may fail.

High pressure and the accompanying high temperature can be caused by air in the air-conditioning system. Air can enter the system through careless or incomplete service procedures. Systems that have been opened to the atmosphere during service procedures must be properly evacuated. If the system is not properly evacuated, the results, most surely, will be an air-contaminated system.

A system with air contamination does not operate at full efficiency. Air in the air-conditioning system can cause oil to oxidize. Oxidized oil forms gums and varnishes that coat the inside walls of the tubes, reducing the efficiency of the heat transfer process. Still more damaging, air usually carries moisture into the system in the form of humidity.

When an air-conditioning system is operated with a low-side pressure below atmospheric pressure (14.696 psig at sea level), air will be drawn into the system through the leak. This occurs when a noncycling system is low on refrigerant; the low side often operates below atmospheric pressure in a vacuum. If a system contaminated in this way is recharged without proper evacuation procedures, high temperature and pressure conditions will result. Air, a noncondensable gas, has a tendency to collect in the condenser during the off cycle.

CONNECTIONS

If there is an O-ring compression fitting, avoid overtightening. Overtightening can cause O-ring damage, resulting in a leak or early failure. Before assembly, inspect the fitting for burrs, which may cut the O-ring. It is important that the proper O-ring be used for the type of refrigerant and the type of fitting. It is also important to follow the manufacturer's recommendations for selecting O-rings.

When a connection is made with an O-ring compression fitting, place the gaskets or O-ring over the tube before inserting it into the connection (Figure 8-49). Use a torque and backing wrench to ensure a proper connection. Again, follow the manufacturer's specifications for proper torque requirements.

RESTRICTIONS

Most restrictions are caused by dirt, foreign matter, or corrosion. Corrosion is generally due to excess moisture in the system. Contaminants can lodge in filters and screens and can block the flow of refrigerant through the system. Filters are found in the receiver-drier and suction line accumulator, usually as a means to hold the desiccant in place. Screens are generally found:

- At the metering device inlet
- In the receiver-drier or accumulator
- At the compressor inlet

A restriction in the system can cause a "starved" evaporator. This can result in reduced cooling, poor oil return, and eventually, if not corrected, compressor seizure.

A contaminated system may have both a strainer and a drier.

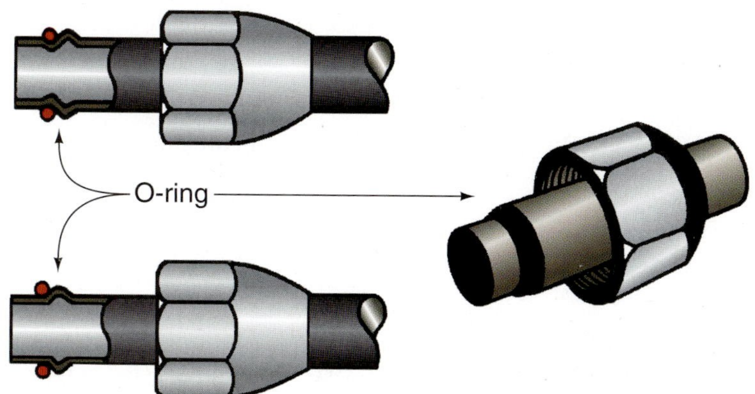

FIGURE 8-49 **Place the O-ring over the fitting before inserting it into the connection.**

FIGURE 8-50 **A supplemental liquid line filter/drier.**

Supplemental aftermarket liquid-line filters are available for installating in air-conditioning systems that have been contaminated (Figure 8-50). The filter should be installed in the system:

1. After repeated metering device plugging
2. When a seized compressor has been replaced

The liquid-line filter contains a screen and a filter pad. It does not contain a desiccant. The fine-mesh screen catches larger particles and holds the filter in place. The filter catches smaller particles and filters the refrigerant oil.

The filter is installed in the liquid line between the condenser outlet and the evaporator inlet. Filters are available with or without an expansion tube orifice. The filter without an orifice is generally preferred. This type can be installed anywhere in the liquid line, preferably close to the metering device. A filter with an orifice is required when the installation is to be made in the low-pressure side of the system beyond the original expansion tube location. This installation, which is usually found on General Motors vehicles, requires that the original expansion tube be removed from the system.

Contamination

Contamination by foreign matter has many sources, including:

- Failed desiccant
- Preservative oils
- Lint
- Soldering or brazing fluxes
- Loose corrosion flakes

Any of these materials in the air-conditioning system can cause:

- Compressor bearings to seize
- Metering device failure
- Corrosion of metal parts
- Decomposition of refrigerant
- Breakdown of the oil

Corrosion and the by-products of corrosion can clog metering device screens, ruin compressor bearings, and accelerate the failure of compressor discharge valves. Moisture is the primary cause of corrosion in the air-conditioning system.

In fact, the greatest enemy of an air-conditioning system is moisture. When combined with the metals found in the system, moisture causes the formation of iron hydroxide and aluminum hydroxide. When combined with refrigerant, moisture can form three acids:

1. Carbonic (H_2CO_3)
2. Hydrochloric (HCl)
3. Hydrofluoric (HF)

Carbonic acid is a weak solution generally found in solutions of carbon dioxide in water.

Avoid contact with hydrochloric and hydrofluoric acids; both are very poisonous.

Moisture also causes metering devices to freeze up. As the operating temperature of the evaporator is reduced to the freezing point, moisture collects in the metering device orifice and freezes. This, in turn, restricts the flow of refrigerant into the evaporator. The result is an erratic or poor cooling condition of the evaporator.

High temperature and foreign matter are responsible for many refrigerant system difficulties. In most cases, it is the presence of moisture that accelerates these conditions. The acids that result from the combination of high pressure, moisture, and refrigerant cause damaging corrosion.

SUMMARY

- There are but six basic abnormal conditions for air-conditioning system diagnosis.
- The evaporator coil temperature is kept above 32°F (0°C) to prevent freeze-up.
- A misadjusted or defective thermostat may cause evaporator freeze-up.
- An undercharge of refrigerant is an indication of a leak in the system.
- A restriction in the high side of the system will cause high high-side pressure.
- A defective thermostatic expansion valve (TXV) can cause the same symptoms as a low-side restriction.
- Air in the system can cause the same symptoms as an overcharge of refrigerant or oil. An overheated engine can cause high high-side pressure.
- The proper cure for an undercharged system is to first repair the leak, then recharge the system.
- Observe all safety practices when handling refrigerants.

TERMS TO KNOW

Average
Control thermostat
Head pressure
Malfunction
Metering device
Remote bulb
Restriction

REVIEW QUESTIONS

Short-Answer Essays

1. What effect would a low-side restriction have on a system and how would it be detected?
2. How can moisture in the system cause a problem?
3. How does one determine whether a problem is due to electrical or mechanical failure?
4. How does one clean a dirty condenser?
5. How is a baseline used in a refrigerant system diagnosis test?
6. Why must the temperature of an evaporator be kept above 32°F (0°C)?
7. What is the symptom of a refrigerant system that is overcharged with refrigerant?
8. What does current research indicate about R-1234yf and R-134a refrigerants, performance characteristics?
9. What effect would an overcharge with refrigerant oil have on a system and how would it be detected?
10. What effect would a high-side restriction have on a system and how would it be detected?

Fill in the Blanks

1. ______________ on the evaporator coil ______________ the flow of air through it, resulting in poor or ______________ cooling.
2. Very low low-side pressure may be caused by a ______________ screen in the metering device ______________.
3. A high low-side pressure is an indication that the thermostatic expansion valve (TXV) ______________ restricting the flow of ______________.
4. High high-side pressure may be caused by excessive ______________, oil, or ______________ in the system.
5. ______________ is the primary cause of corrosion in the air-conditioning system.
6. Most restrictions are caused by ______________, or ______________.
7. By checking both ______________ and ______________, you can gain information above and beyond what your pressure gauges are telling you.
8. If a low-side restriction occurs, the low-side gauge reading will be ______________ or even a ______________.
9. If a refrigerant system is contaminated with moisture, the pressure in the system will swing between a ______________ to ______________ on the low-side gauge and between ______________ to on the high-side gauge.
10. The ______________ and ______________ ______________ is an important tool for air-conditioning service.

Multiple Choice

1. *Technician A* says that a gauge reading of 21 psig (145 kPa) corresponds to an R-12 evaporator temperature of about 20°F (−6.7°C).

   Technician B says that a gauge reading of 21 psig (145 kPa) corresponds to an R-134a evaporator temperature of about 23.5°F (−4.7°C).

   Who is correct?

   A. A only
   B. B only
   C. Both A and B
   D. Neither A nor B

2. *Technician A* says that if R-134a evaporating temperature is 30°F (−1.1°C), the gauge pressure should be about 26 psig (179 kPa).

   Technician B says that a low-side gauge pressure of 26 psig (179 kPa) indicates a refrigerant evaporating temperature of 27°F (−2.8°C) in an R-12 system. Who is correct?

   A. A only
   B. B only
   C. Both A and B
   D. Neither A nor B

3. During a refrigerant system performance test the following results are recorded: outside ambient temperature of 90°F, high-side gauge reading of 145 psi, and low pressure gauge reading of 14 psi. What is indicated by these results?

   A. Faulty refrigerant compressor
   B. Normal refrigerant system operation
   C. Low refrigerant levels
   D. A restriction on the high side of the refrigerant system

4. A customer complains of poor air-conditioning cool ing performance. During a system performance test, the low-side gauge reading is 50 psig; the high-side gauge reading is 310 psig; and the ambient air temperature is 81°F (27.22°C). What is the least likely cause of the above results?

   A. Restricted airflow through condenser
   B. Refrigerant overcharge
   C. Refrigerant undercharge
   D. Overcharge of refrigerant oil

5. Temperature and pressure comparisons are a standard practice to determine all of the following, except:

   A. System change levels
   B. Superheat
   C. Subheat
   D. Subcooling

6. During a refrigerant system performance test, the high-side pressure instantly goes over 370 psig when the compressor is engaged. Which of the following is the most likely cause?

   A. An overcharged system
   B. An undercharged system
   C. Moisture in the system
   D. Expansion valve stuck open

7. An R-134a thermal expansion valve system contains the proper refrigerant charge. During a system performance test, the low-side gauge reading is 40 psig and the high-side gauge reading is 140 psig with an ambient air temperature of 87°F (30.55°C). Which of the following is the most likely cause of these results?

   A. An overcharged system
   B. A restricted evaporator
   C. A restricted condenser
   D. An expansion valve stuck open

8. During a system performance test, the low-side gauge is reading a vacuum. Which of the following is the most likely cause?

   A. An overcharged system
   B. Incorrectly connecting the gauge hoses to the wrong service ports
   C. An expansion valve stuck open
   D. A restricted expansion valve

9. During a system performance test, the low-side gauge reading and the high-side gauge reading are about the same. Which of the following is the most likely cause?
   A. An overcharged system
   B. Moisture in the refrigerant system
   C. A malfunctioning compressor
   D. A restricted expansion valve

10. An R-134a thermal expansion valve system contains the proper refrigerant charge. During a system performance test, the low-side gauge reading and the high-side gauge reading are in the high range and the system has poor cooling performance. Which of the following is the most likely cause?
    A. A restricted airflow across the evaporator
    B. A restricted airflow across the condenser
    C. An expansion valve stuck open
    D. A malfunctioning compressor

Chapter 9

COMPRESSORS AND CLUTCHES

UPON COMPLETION AND REVIEW OF THIS CHAPTER, YOU SHOULD BE ABLE TO:

- State the purpose and describe the function and operation of a magnetic clutch in an air-conditioning system.
- Compare fixed and variable displacement compressors.
- Discuss electric motor–driven compressors.
- Discuss and explain the operating principles of a reciprocating compressor.
- Discuss and explain the operating principles of a scroll compressor.
- Discuss and explain the operating principles of a rotary compressor.

Kilometer is the metric conversion for the English mile, and liter (litre) is the metric conversion for the English quart or gallon.

Auxiliary components are those such as the rear evaporator in a dual air-conditioning system, which is often referred to as an "auxiliary evaporator."

The low-pressure condition exists in an air-conditioning system from the metering device outlet to the compressor inlet.

Low pressure is a relative term used to describe normal pressure in the low side of an air-conditioning system.

INTRODUCTION

There are many different types, makes, and models of compressors used for automotive air-conditioning applications.

A prime consideration for new compressor design is to help to reduce overall vehicle weight. Overall vehicle weight is decreased by reducing the weight of individual components. A reduction in overall (gross) vehicle weight provides greater economy or more miles per gallon (kilometers per liter) of fuel.

The compressor, as well as other **auxiliary** components, must also be designed to be efficient and durable to withstand long hours of heavy use.

Chapter 5 of this manual generally covered the refrigeration system, and Chapter 6 considered other basic components of the system. If necessary, refer to these chapters to review the compressor's role in a vehicle air-conditioning system.

FUNCTION

The compressor in an automotive air-conditioning system serves two important functions. It creates a low-pressure condition within the system, and it compresses refrigerant vapor from a low pressure to a high pressure, thereby increasing its temperature. It is important that these two functions be accomplished at the same time.

Low-Pressure Condition

One function, creating a **low-pressure** condition at the compressor inlet, aids in the removal of heat-laden refrigerant vapor from the evaporator. This low-pressure condition is essential to allow the refrigerant metering device to admit the proper amount of liquid refrigerant into the evaporator (Figure 9-1).

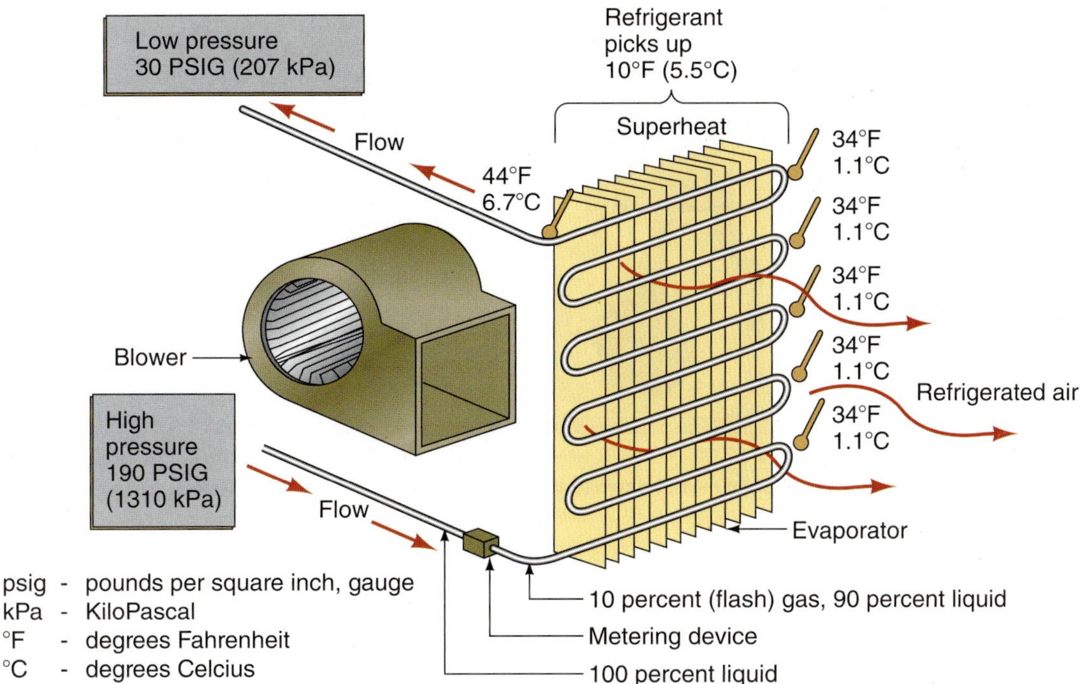

FIGURE 9-1 **The intake stroke of the compressor creates a low-pressure condition to draw refrigerant into the evaporator.**

High pressure is a relative term used to describe refrigerant pressure in the high side of an air-conditioning system.

The high-pressure condition exists from the compressor outlet to the metering device inlet.

Compress Refrigerant

The second function of the compressor is to compress the low-pressure refrigerant vapor into a **high-pressure** refrigerant vapor. This increased pressure raises the heat content of the refrigerant. High pressure with high heat content is essential if the refrigerant is to condense, giving up its heat, in the condenser. It is in the condenser that the refrigerant vapor is changed to a liquid. While it is a slightly lower temperature, it is still at a high pressure until it again reaches the metering device, generally at the inlet of the evaporator.

Failure of either function of the compressor will result in a loss or reduction of the circulation of refrigerant within an air-conditioning system. Without proper refrigerant circulation in the system, the air conditioner will not function properly or may not function at all.

Design

Several types of compressors are used in automotive air-conditioning systems. Regardless of the type, however, with few exceptions, most compressors are basically of the reciprocating piston design. Reciprocating means that the piston moves up and down, to and fro, or back and forth (Figures 9-2 and 9-3).

Multicylinder compressors have a set of valves for each piston. A set consists of one suction and one discharge valve. The suction and discharge valves operate conversely of each other. Discharge pressure (piston on the upstroke) forces the **suction valve** closed and the **discharge valve** open.

In a two-cylinder compressor, for example, when piston one is on the upstroke, the other piston is on the downstroke. Piston two is then forcing the suction valve open while the high pressure behind the discharge valve is holding it (discharge valve) closed.

The **suction valve** is on the low-side suction port (intake) of the refrigerant compressor. It is a mechanical one way check valve that only allows refrigerant to flow into the compressor.

The **discharge valve** is on the high-side discharge port (pressure) of the refrigerant compressor. It is a mechanical one-way check valve that only allows refrigerant to flow out of the compressor.

A **crankshaft** is the part of a reciprocating compressor to which the wobble plate or connecting rods are attached. It is splined to the clutch plate and receives reciprocating power from the drive pulley assembly when the clutch is engaged.

An **axial plate** is the part of an automotive air-conditioning compressor piston assembly that rotates as a part of the drive shaft.

Swash plate is another term used for a wobble plate.

A **wobble plate** is a type of offset concentric plate attached at an angle. It is found on some compressor crankshafts and is used to move the pistons up and down as the shaft is turned. Another term used for a wobble plate is a swash plate.

Rotary compressors use vanes attached to a rotor assembly and are driven by the input shaft to compress refrigerant.

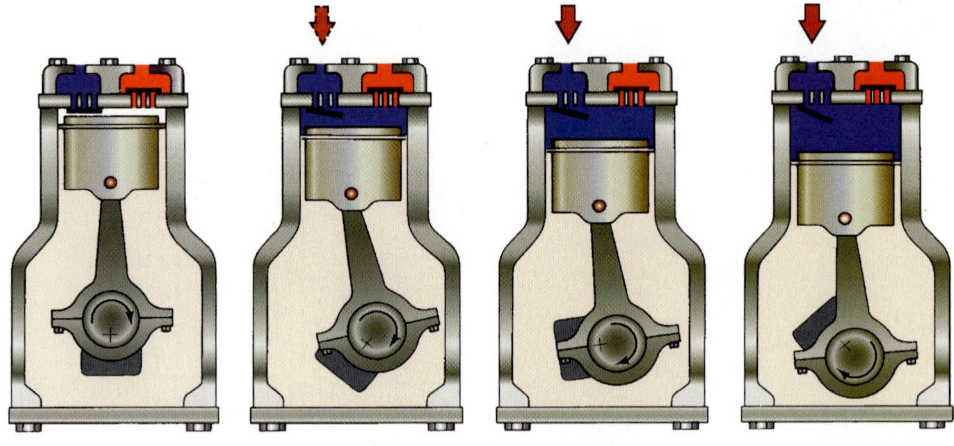

FIGURE 9-2 **The piston(s) moves down (top to bottom) during the suction (intake) stroke.**

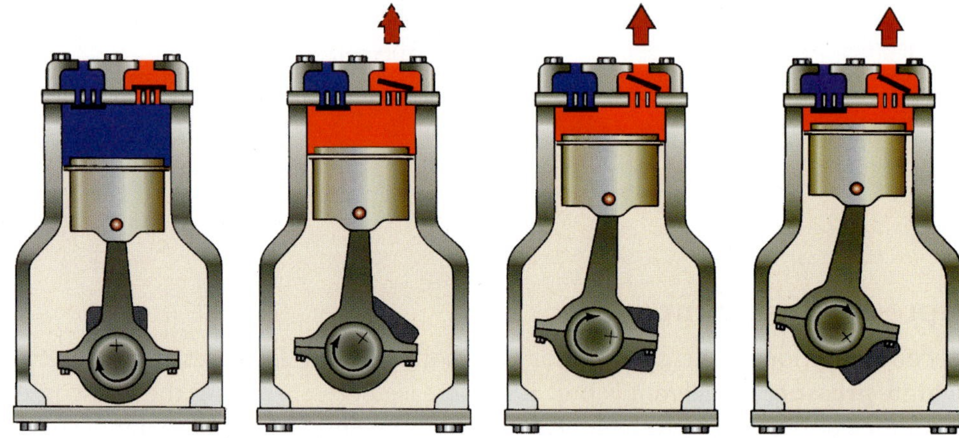

FIGURE 9-3 **The piston(s) moves up (bottom to top) during the compression (discharge) stroke.**

Two basic methods of driving the piston of a reciprocating compressor are by **crankshaft** or **axial plate**. The axial plate is often called a **swash plate** or **wobble plate**.

The exceptions, **rotary** and **scroll** compressors, found on a limited number of car lines beginning in the early 1990s, are discussed later in this chapter.

Crankshaft

Driving the piston of a reciprocating compressor by crankshaft (Figure 9-4) is an operation that is very similar to that of an automobile engine. The main difference is that a compressor crankshaft drives the piston, whereas in an engine, the piston drives the crankshaft. The compressor crankshaft is driven directly or indirectly off the engine crankshaft by means of pulleys and belts (Figures 9-5, 9-6, 9-7, and 9-8).

Axial Plate

The other method of driving the piston of a reciprocating compressor is by an axial plate pressed on the main shaft, providing a reciprocating motion of the piston (Figure 9-9). The axial plate is driven directly or indirectly by the main shaft off the engine crankshaft by means of pulleys and belts.

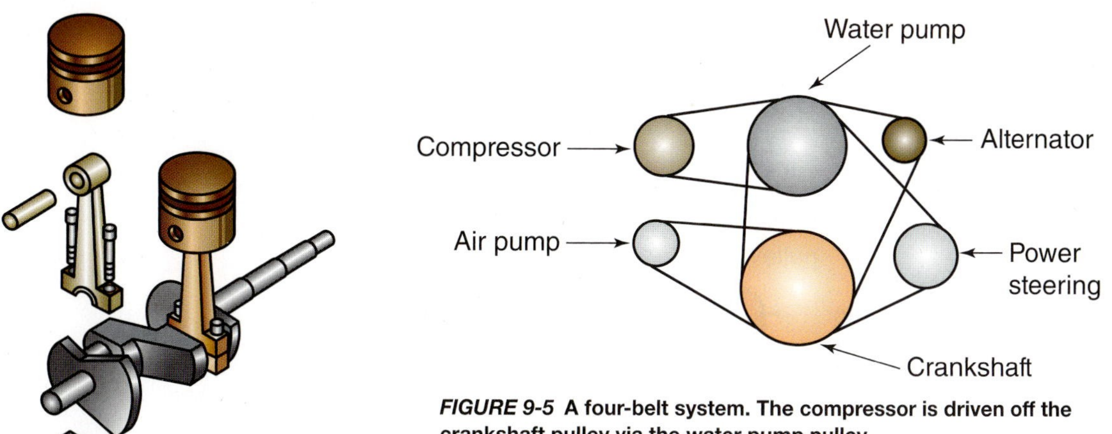

FIGURE 9-4 **Details of a piston driven by action of the crankshaft in a compressor.**

FIGURE 9-5 **A four-belt system. The compressor is driven off the crankshaft pulley via the water pump pulley.**

A **scroll** compressor is a spiral corkscrew design compressor used in limited applications to produce a more continuous, steady supply of refrigerant pressure.

Turning the crankshaft causes a to-fro, fore-aft, or up-down action of the piston(s).

Turning the axial plate will create the same conditions as turning a crankshaft.

Other terms used for V-rib or serpentine belt are poly-rib and microgroove.

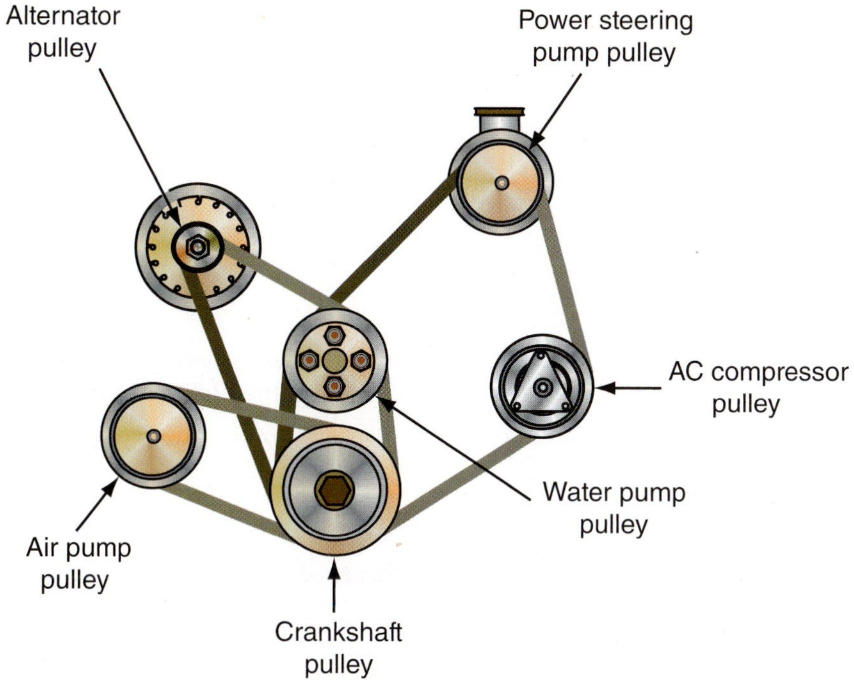

Alternator pulley
Power steering pump pulley
AC compressor pulley
Water pump pulley
Air pump pulley
Crankshaft pulley

FIGURE 9-6 **A three-belt system. The compressor is driven off the crankshaft pulley with the alternator used for belt tensioning.**

Shop Manual
Chapter 9, page 335

Clutch

All belt-driven automotive air-conditioning compressors have an electromagnetic clutch attached to the crankshaft or main shaft (Figure 9-10). The clutch provides a means of turning the compressor on and off. An idler pulley or tensioner is provided to adjust belt tension. Most compressors are driven off the crankshaft by a single belt, along with such other accessories as the power steering pump, alternator, and water pump. This system is known as a **serpentine** drive. This serpentine belt is tensioned by a spring-loaded or manually adjustable idler pulley assembly, which generally rides on the back (flat) side of the belt. Refer to Figure 9-8 shown earlier.

A **serpentine** belt is a flat or V-grooved multiribbed design that winds through all of the engine accessories to drive them off the engine crankshaft pulley.

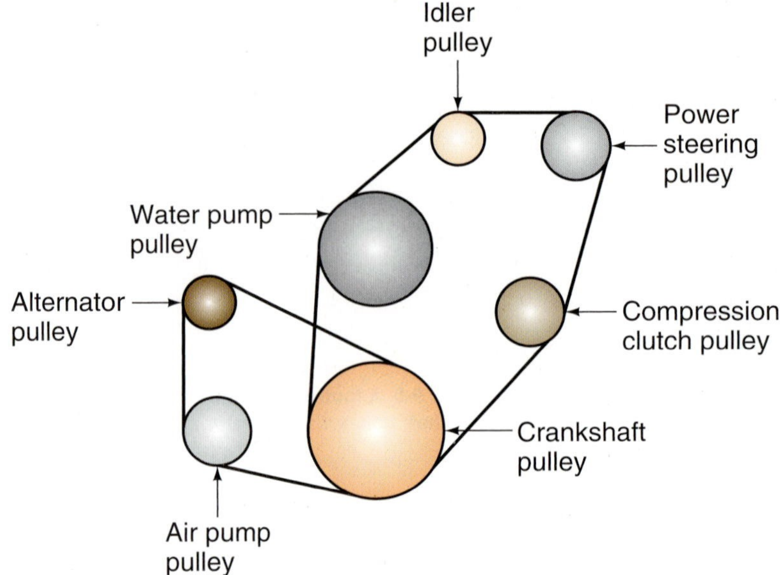

FIGURE 9-7 **A two-belt serpentine drive system. The compressor is tensioned by a manually adjusted idler pulley.**

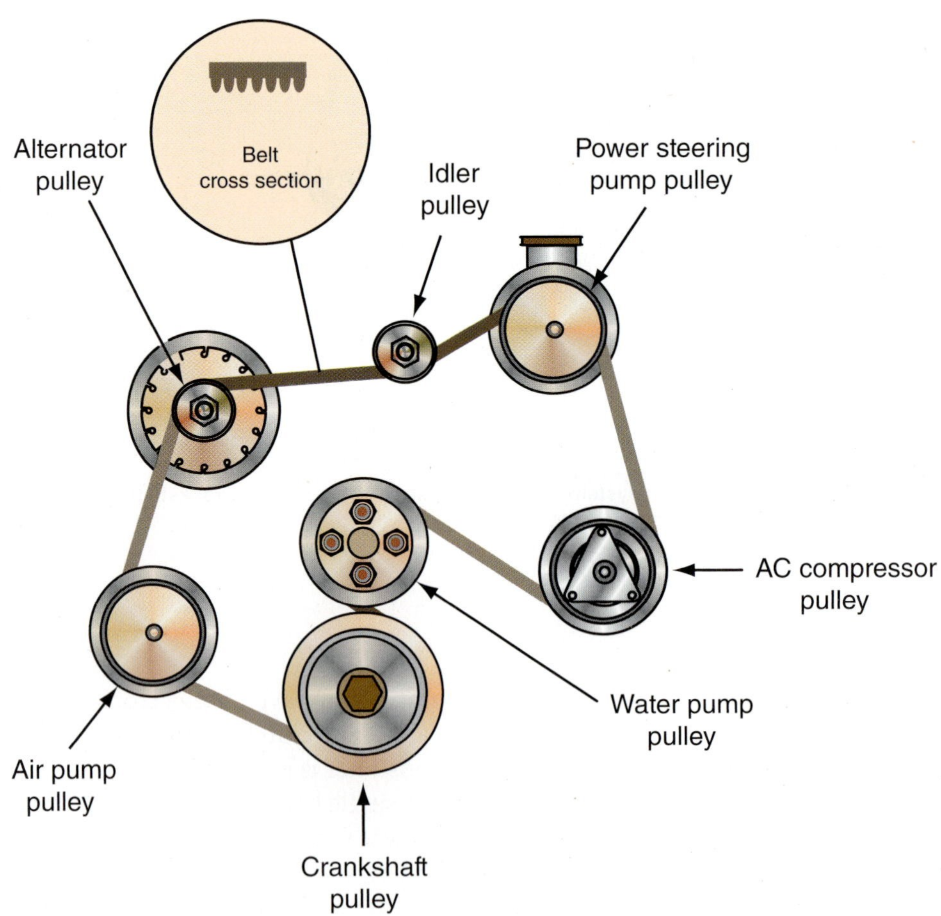

FIGURE 9-8 **A single-belt drive system. The belt is tensioned by a spring-loaded idler pulley.**

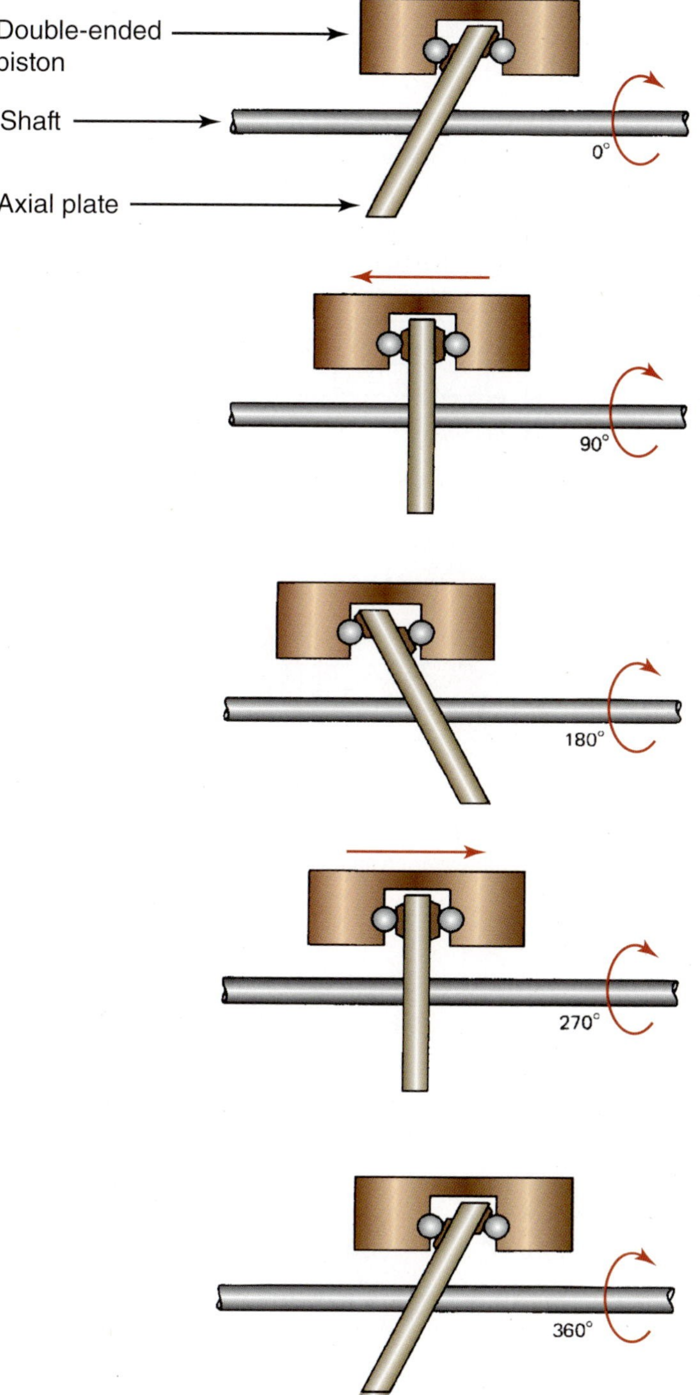

FIGURE 9-9 **The piston(s) is moved back and forth or to and fro by an axial plate.**

The three main parts of the compressor clutch assembly are the coil magnet, the pulley and idler bearing, and the clutch plate (shoe) (Figure 9-11). The clutch drive plate is splined to the input shaft of the compressor.

The compressor clutch is a large electromagnet that, when energized, draws the clutch plate into the clutch pulley. The magnetic field holds the clutch plate tightly against the clutch pulley as long as current is supplied to it. This in turn engages the drive pulley to the compressor input shaft, causing the shaft to spin. When the compressor clutch is not engaged,

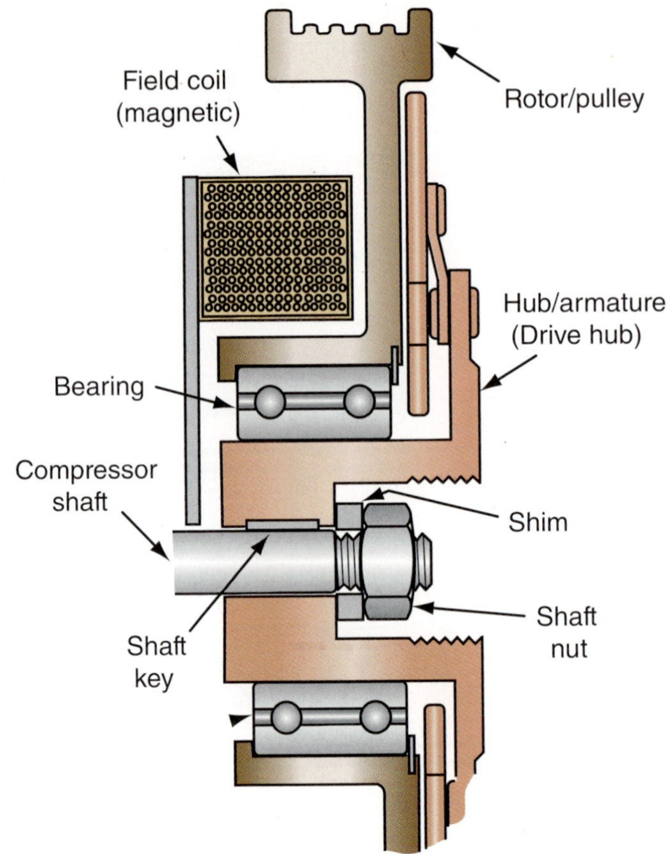

FIGURE 9-10 **An electromagnetic clutch provides a means of turning the compressor on and off.**

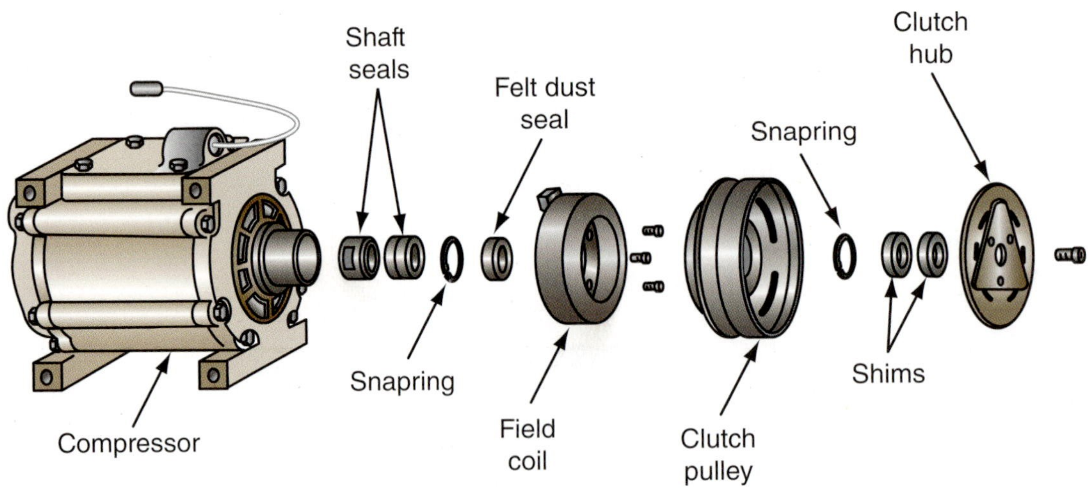

FIGURE 9-11 **The three mains parts of the compressor clutch assembly are the field coil, clutch pulley, and the clutch hub.**

the input shaft of the compressor does not spin and the drive pulley freewheels on a sealed bearing assembly. If a noise is heard when the clutch is not engaged, it is generally an indication of a faulty bearing, which in many cases can be serviced without the replacement of the entire compressor assembly. If, however, the noise is only heard when the compressor clutch is engaged, this may indicate that there is an internal problem with the air-conditioning compressor.

The operation of the air-conditioning clutch today is controlled by the heater control head and often uses a solid-state control module. The power train control module (PCM) is also integrated into the system.

The PCM controls compressor clutch engagement when:

- The low-pressure switch senses pressure below 25 psig.
- The high-pressure switch senses pressure above 450 psig.
- Coolant temperature is above 230°F (110°C).
- Engagement is delayed for 5–10 seconds when the engine is first started.
- Engine speed is below 400 rpm.
- The throttle is opened above 80 percent.

When the air-conditioning switch on the climate control head assembly is first turned to the on position, the PCM runs a logic loop to check the operation of its sensors to make sure they are within operational parameters prior to allowing the compressor clutch to engage (Figure 9-12). The PCM will also look at the air-conditioning system high- and low-pressure transducers to verify that the system contains the proper refrigerant charge before allowing the compressor to engage. The idle speed will also be raised to compensate for the added load on the engine.

Most air-conditioning compressor clutch circuits also use a clamping diode placed across the clutch coil to prevent unwanted electrical spikes as the clutch is disengaged, which could damage control modules and relay contacts. The air-conditioning clutch can generate a voltage

A BIT OF HISTORY

Early automotive air-conditioning systems did not have a convenient driver-operated means of engaging and disengaging the compressor. Most compressors were belt- and pulley-driven off the crankshaft or accessory device. If one wished to disable the compressor for the winter months, the belt was removed. It was then replaced for the warm months. In-car temperature was controlled by means of a hot gas bypass valve. The compressor operated any time the engine was running and the hot gas bypass valve simply routed the unwanted gas from the compressor discharge back to the suction side of the system. This inefficient method of temperature control has not been used for automotive service since the mid-1960s.

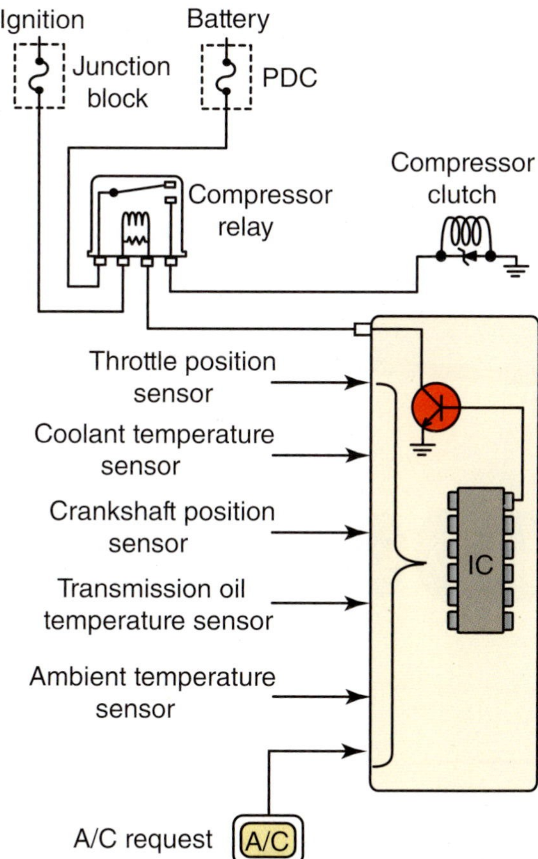

FIGURE 9-12 When an air-conditioner clutch request is made, the PCM will process the request after input sensor data and refrigerant system pressure are analyzed.

spike of over 200 volts, and the diode provides a path back through the coil assembly for this unwanted voltage. The diode is wired in parallel with the air-conditioning compressor clutch and may be part of the field coil assembly, or it may be located in the engine wiring harness near the clutch. Other systems use a bidirectional **zener clamping diode** that turns on at voltages above the 60-volt level. Use a digital volt ohmmeter (DVOM) set to diode testing to inspect the integrity of the diode. In addition, clutch coil resistance and amperage draw should be measured and compared to the manufacturer's specifications.

A **zener clamping diode** is a semiconductor rectifier diode one-way voltage gate that allows current to flow when voltage levels increase above its threshold voltage.

When servicing compressor clutches, be sure to properly set the air gap of the clutch plate to the pulley hub to the manufacturer's specifications (generally 0.020 in. [0.50 mm]), using a nonmagnetic feeler gauge. If the air gap is set too loose, the clutch may not engage or may slip. If the air gap is set too tight, the clutch may drag, causing noise and leading to overheating of the clutch coil.

Air-conditioning compressor clutches can fail for many reasons. They may slip if improperly adjusted or if the proper current is not supplied. Coil assembly may develop a short or an open. The bearing may fail; a noise may develop; or the clutch may drag due to weak return springs, to name a few failures.

The hot gas is no longer used as a method of temperature control in automotive air-conditioning systems.

Types of Compressors

According to a leading compressor rebuilder, there are currently over 160 makes and models of remanufactured compressors readily available for use in mobile air-conditioning systems. The various types include reciprocating piston (Figure 9-13), scroll (Figure 9-14), rotary vane (Figure 9-15), and scotch yoke (Figure 9-16).

The **suction** side is another term used to describe the low side of the refrigerant system.

Reciprocating (Piston-Type) Compressors

Reciprocating, piston-type mobile air-conditioning compressors, depending on their design, may have one, two, four, five, six, seven, or ten pistons (cylinders). Tecumseh manufactured a single-cylinder compressor for use with aftermarket air-conditioning systems in compact imports.

A two-cylinder V-type compressor was manufactured by Chrysler Air-Temp but was discontinued due to its heavy weight. Two-cylinder, in-line reciprocating compressors manufactured by Nippondenso, Tecumseh, and York may be found on some early model vehicles, as well as some heavy duty and off-road equipment.

Chrysler Air-Temp manufactured the only V-type compressor for automotive use. It is now discontinued.

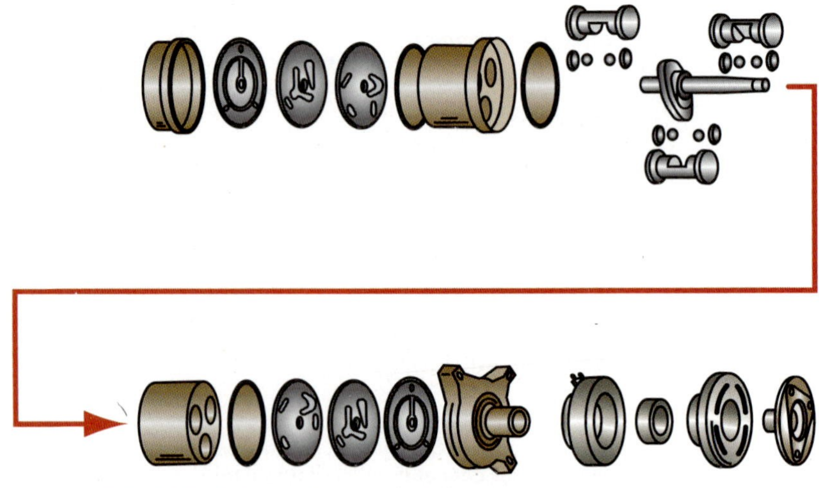

FIGURE 9-13 **A typical reciprocating piston compressor details.**

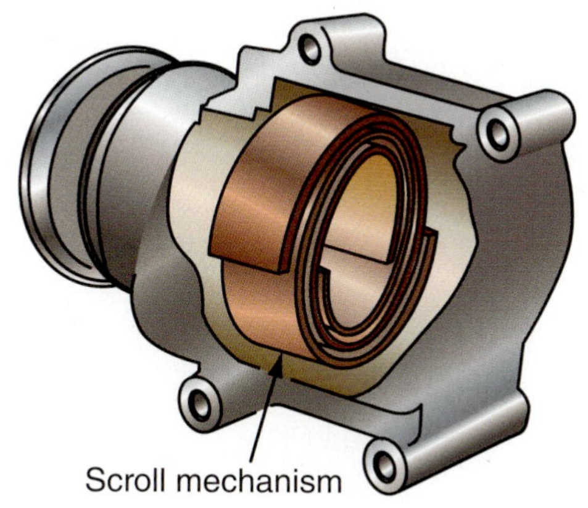

FIGURE 9-14 **Scroll compressor details.**

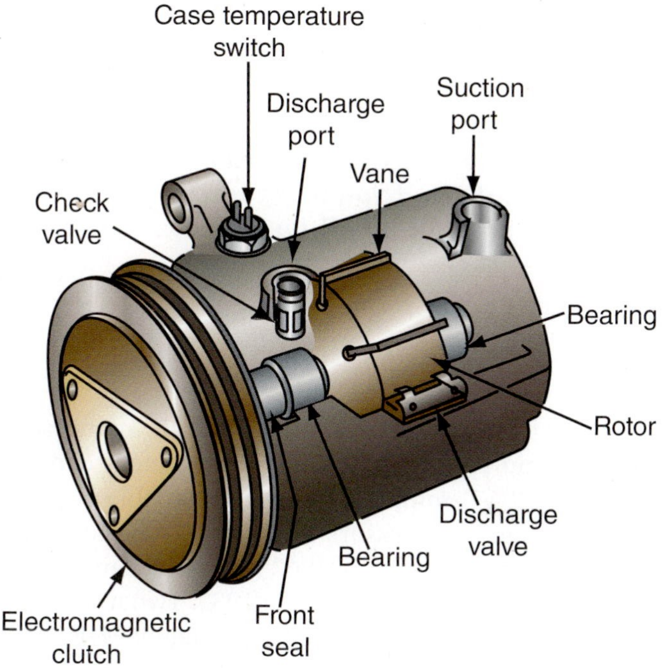

FIGURE 9-15 **A typical rotary vane compressor details.**

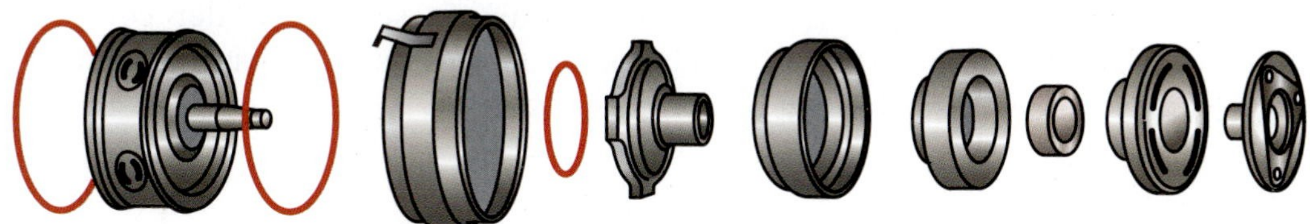

FIGURE 9-16 **A typical scotch yoke compressor details.**

A four-cylinder, radial-design scotch yoke reciprocal compressor, manufactured by Harrison (Frigidaire) as their model R-4, is available in either standard or lightweight versions. A similar compressor, model HR-980 by Tecumseh, was produced through the late 1980s. A version of the radial scotch yoke reciprocating compressor design was produced by Keihin in Japan for use on some Honda automobiles.

FIGURE 9-17 **The Calsonic V-5 compressor.**

AUTHOR'S NOTE: In order to understand what can physically go wrong in an air-conditioning system, you must know how an air-conditioning compressor is built and how it functions. Even if you never intend to overhaul the assembly, diagnosing a system failure will be all but impossible without this level of understanding.

Shop Manual
Chapter 9, page 364

The **intake** stroke of the compressor creates a negative pressure that draws refrigerant in from the low side of the system.

The **compression** stroke of the compressor creates a positive pressure greater than that contained on the high-side hot gas line leaving the compressor, and it thus forces refrigerant under pressure through the system when the exhaust valve is opened.

Sanden (Sankyo), Harrison (Frigidaire), and Calsonic manufacture a five-cylinder compressor. The Sanden compressor is a positive displacement compressor; the Harrison V-5 and Calsonic V-5 (Figure 9-17) compressors are of a variable displacement design.

Six six-cylinder axial design compressors are currently available. The Harrison model A-6 was manufactured for over 20 years, from 1962 through the mid-1980s. This compressor was superseded by a lighter version, model DA-6, in 1982. Two more changes soon followed: the "Harrison Redesigned" HR-6 and the "High Efficiency" HR6HE version. Ford and Chrysler also have models of a six-cylinder compressor similar to one developed originally by Nippondenso.

A six-cylinder variable displacement compressor by Calsonic, model V-6, is very similar in appearance to Calsonic's model V-5 compressor.

Honda Air Device Systems (HADS) manufactures a seven-cylinder compressor (Figure 9-18) for use with refrigerant R-134a air-conditioning systems. Harrison manufactures a seven-cylinder variable displacement compressor (Figure 9-19).

Action

Low-pressure refrigerant vapor is compressed to high-pressure refrigerant vapor by action of the pistons and valve plates. For each piston, there is one intake (suction) valve and one outlet (discharge) valve mounted on a valve plate. For simplicity of understanding, a single-cylinder (piston) compressor is discussed.

By action of the crankshaft, the piston travels from the top of its stroke to the bottom of its stroke during the first one-half revolution. On the second one-half revolution, the piston travels from the bottom of its stroke to the top of its stroke. The first action, top to bottom, is called the **intake** or suction stroke; the second action, bottom to top, is called the **compression** or discharge stroke.

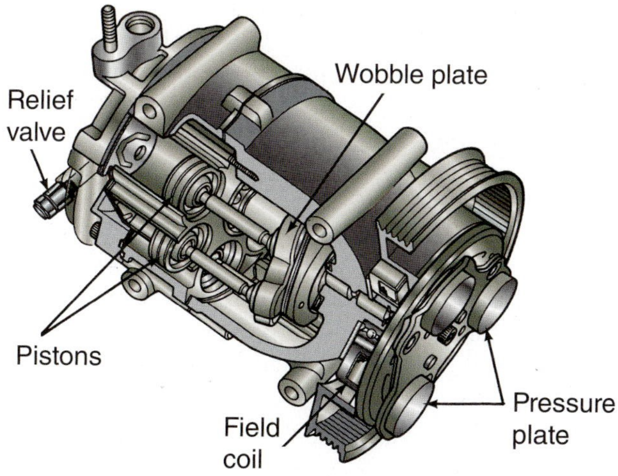

FIGURE 9-18 **Honda Air Device Systems (HADS) compressor details.**

FIGURE 9-19 **Harrison's seven-cylinder variable displacement compressor.**

The piston is fitted with a piston ring to provide a seal between the piston and the cylinder wall. This seal helps to provide a negative (low) pressure on the down or intake stroke, and a positive (high) pressure on the up or **exhaust** stroke.

Exhaust is the final stage that occurs as the piston is moving up on the compression stroke and pressures exceed the preset calibration on the exhaust valve, forcing refrigerant in the high side of the system.

The Intake Stroke. During the intake stroke, a low-pressure area is created atop the piston and below the intake (suction) and exhaust (discharge) valves. The higher pressure atop the intake valve, from the evaporator, allows this valve to open, admitting low-pressure heat-laden refrigerant vapor into the compressor cylinder chamber. The discharge valve is held closed during this time period. The much higher pressure atop this valve, as opposed to the low pressure below it, prevents it from opening during the intake stroke.

The Discharge Stroke. During the compression stroke, a high-pressure area is created atop the piston and below the intake and exhaust valves. This pressure becomes much greater than that above the intake valve and closes that valve. At the same time, the pressure is somewhat greater than that above the exhaust valve. The pressure difference is great enough to cause the exhaust valve to open. This allows the compressed refrigerant vapor to be discharged from the compressor.

Compression action is repeated over 6,000 times each minute at road speeds in an R-4 compressor.

Continuous Action. This piston action is repeated rapidly—once for each revolution of the crankshaft; perhaps 600 times each minute at curb idle. At over-the-road speeds, the action may be repeated 1,500 or more times each minute for each cylinder of the compressor.

Rotary Vane Compressors

The rotary vane compressor, by design, provides the greatest cooling capacity per pound of compressor weight. It has no pistons and only one valve: a discharge valve. The discharge valve actually serves as a check valve to prevent high-pressure refrigerant vapor from entering the compressor through the discharge provisions during the off cycle, or when the compressor is not operating. The function of the rotary vane compressor is the same as that of the piston- or reciprocating-type compressor. Its operation, however, is entirely different.

The concept and use of rotary-type compressors for refrigeration service is not new. Two basic types of rotary vane compressors have been available for nonautomotive refrigeration use for many years: rotating vane and stationary vane. York was first to introduce this compressor to the automotive marketplace in the early 1980s. Only about 50,000 rotary vane compressors were manufactured by York before being discontinued, however. With some exceptions, a rotary vane compressor was also found on some Geo Prism and Toyota Corolla and Tercel car lines as early as 1989. In 1993, Panasonic manufactured a rotary compressor that was introduced on some Ford car lines. The Zexel rotary compressor was also used on Nissan's Altima.

Shop Manual
Chapter 9, page 361

Operation of Rotary Compressors

Follow the illustration shown (Figure 9-20) for a brief description of the operation of a rotary vane compressor.

- The compressor shaft turns a rotor assembly that has vanes that extend to the wall of the cylinder block.

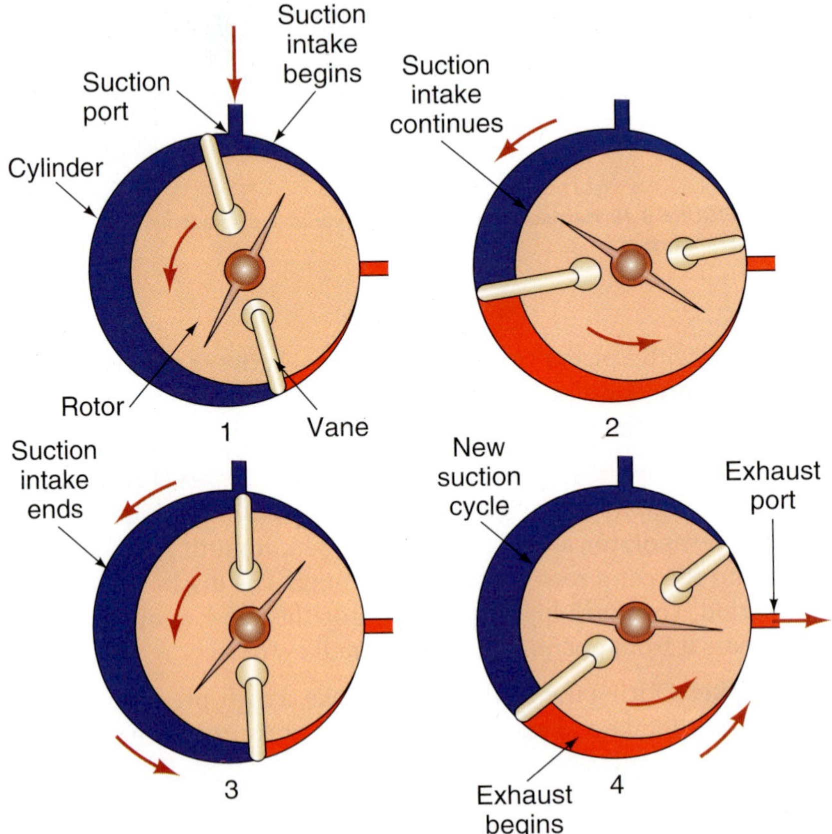

FIGURE 9-20 **The operational sequence of a rotary vane compressor.**

- This forms a compression chamber, or several chambers if there is more than one vane.
- The rotating vanes then draw in refrigerant vapor through the suction ports.
- Compression of the refrigerant starts after the vanes have crossed the suction ports, increasing the refrigerant pressure and temperature.
- The hot vapor is then forced out through the discharge valves to the condenser.

A rotary vane compressor has fewer moving parts than does a reciprocating compressor.

Scroll Compressors

A scroll compressor has only one moving part, the scroll.

The scroll compressor was first introduced by Copeland Corporation for use in residential heat pump systems.

Although the scroll compressor was first patented in 1909, it did not meet practical application until it was introduced by Copeland Corporation in 1988 for use in home air conditioners and heat pumps. Sanden introduced the scroll compressor to the automotive marketplace in 1993. Its unique design is considered by many to be a major technological breakthrough in compressor design and its use has expanded.

A scroll compressor is composed of two scrolls: one is a fixed scroll and the other is a moveable scroll attached to the rotating input shaft. When combined, they create a spiral configuration with the moveable scroll orbiting eccentrically without rotating inside the fixed scroll, thereby trapping and pumping refrigerant (Figure 9-21). The compressor housing forms the fixed scroll, and the compressor input shaft attaches to the orbiting scroll (Figure 9-22). The design of the two scroll halves creates a varying volume of space between them that allows for the suction and compression of the refrigerant.

A scroll compressor has only one cylinder with a compression stroke (output) for every 360 degrees of rotation. If you look closely at Figure 9-23, you will notice that a new intake chamber forms after each 360 degrees of rotation. In other words, for every 360 degrees of rotation, a new compression chamber is formed and a discharge pulse occurs. Thus a scroll compressor is a one-cylinder compressor.

Follow the illustration shown (Figure 9-23) for a brief description of the operation of a scroll compressor.

- Compression in the scroll compressor is achieved by the interaction of a orbiting scroll and a stationary scroll.
- Refrigerant vapor enters the compressor suction port and an outer opening of one of the orbiting scrolls.

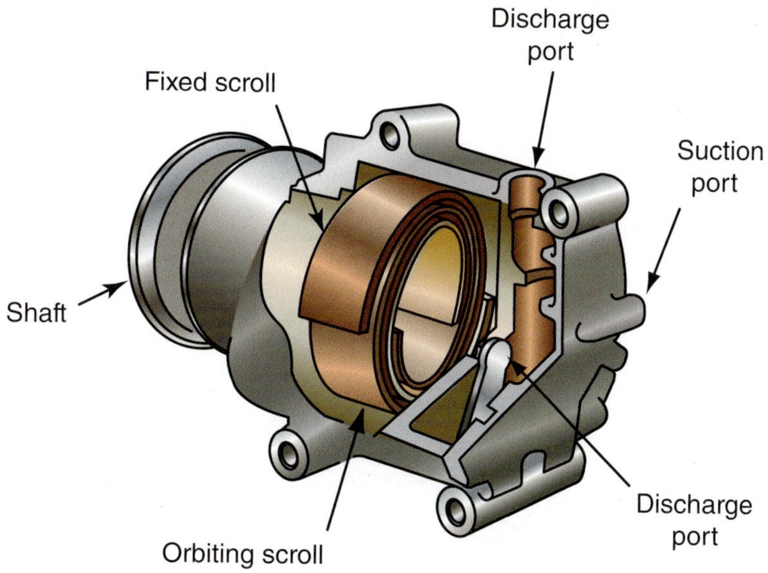

FIGURE 9-21 **A scroll design refrigerant compressor.**

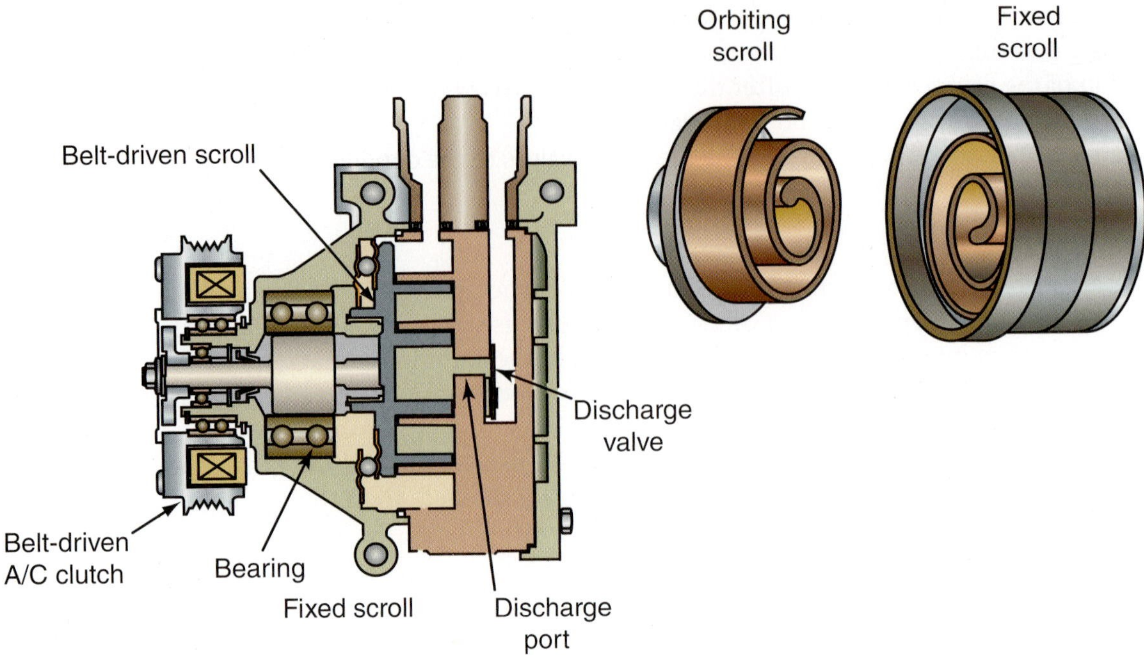

FIGURE 9-22 **The compressor housing forms the fixed scroll, and the compressor input shaft attaches to the rotating scroll.**

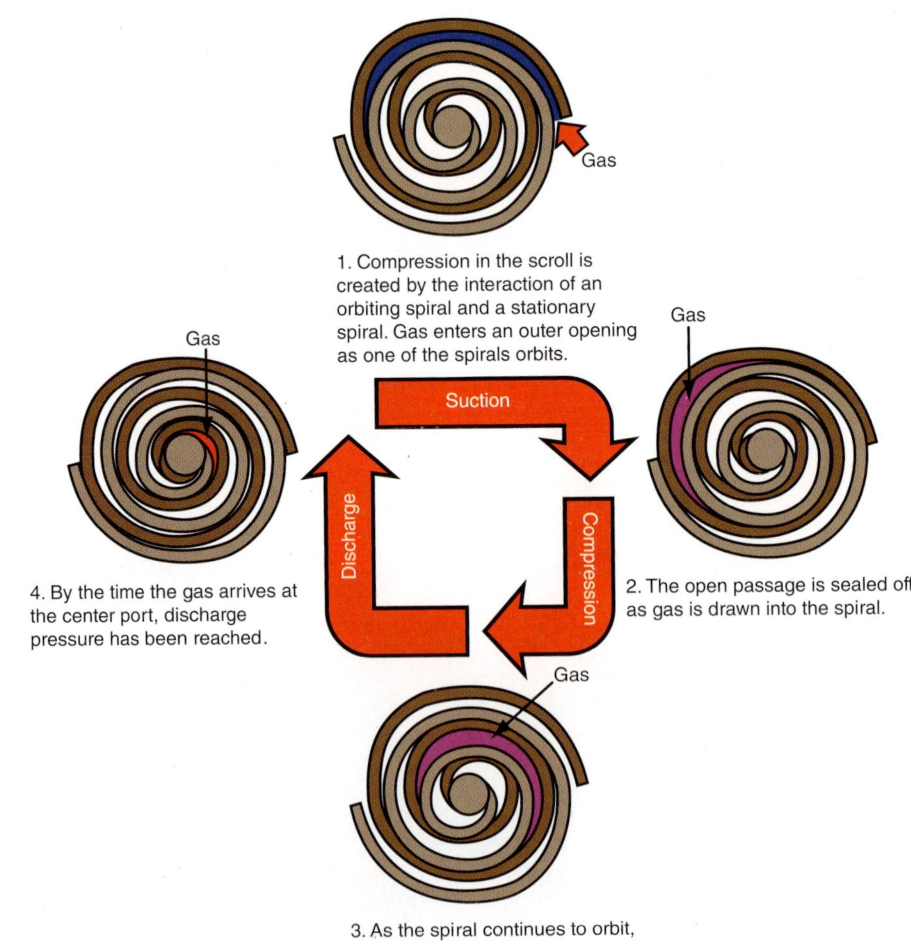

FIGURE 9-23 **The operational sequence of a scroll compressor.**

- This open passage allows refrigerant vapor to be drawn into the passage of the scroll, which is then sealed off.
- As the scroll continues to orbit, the passage becomes smaller and the refrigerant vapor is compressed.
- As the refrigerant pressure increases, the discharge valve is pushed open when the pressure in the scroll passage rises above both the valve spring tension and the discharge line pressure (pressure differential is greater in the compressor than in the high-side port), and the refrigerant flows into the discharge port and the high-side discharge line.
- Refrigerant gas is discharged from the compressor once every shaft rotation. So, if the shaft is rotating at 2,000 rpm, there are 2,000 pump pulses every minute, generating a virtually continuous flow of refrigerant through the system.
- As the refrigerant vapor is discharged from the compressor discharge port, its temperature and pressure have been increased.

This brief explanation is of just one vapor passage of the scroll. During actual operation, all vapor passages of the scroll are in various stages of compression at the same time. This provides a nearly continuous suction and discharge pressure at all times.

Scotch Yoke Compressors

In a scotch yoke compressor (Figure 9-24), opposed pistons are pressed into opposite ends of a yoke riding upon a slider block located on the shaft eccentric. Rotation of the shaft moves the yoke, with attached pistons, in a reciprocating motion. Counterweights are used to balance the rotating assembly. A suction reed valve is located at the top of each piston, and a discharge valve plate is located at the top of each cylinder. Like all reciprocating compressors, low-pressure refrigerant is drawn into the cylinder through the suction valve on the intake stroke and is forced out through the discharge valve on the exhaust stroke at a high pressure.

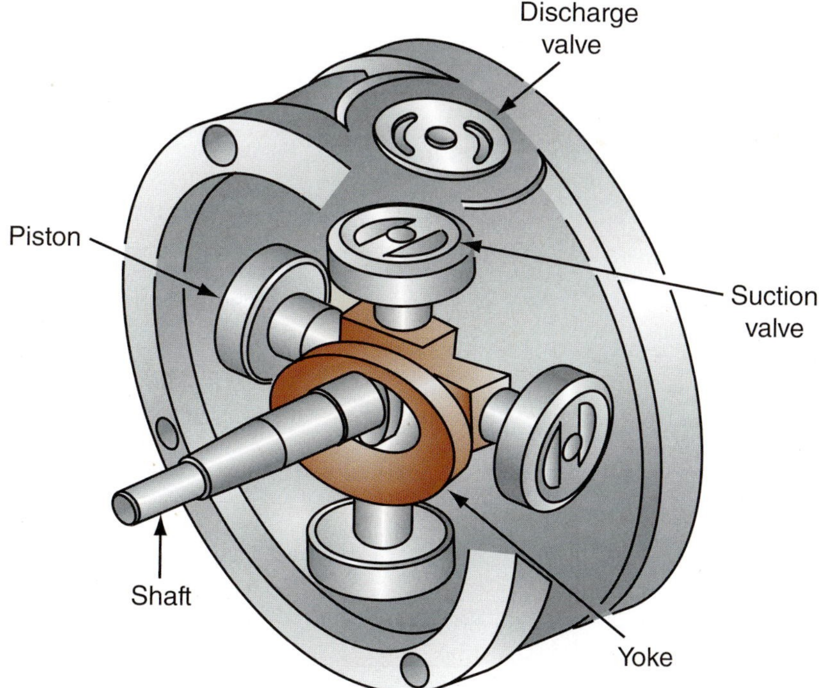

FIGURE 9-24 **A typical scotch yoke compressor details.**

VARIABLE DISPLACEMENT COMPRESSORS

Harrison first introduced a variable displacement compressor in 1985. A variable displacement compressor is designed to match any automotive air-conditioning load demand under all conditions. This is accomplished by varying the displacement of the compressor by changing the stroke (displacement) of the pistons. The axially oriented pistons are driven by a variable-angle wobble (swash) plate. The angle of the wobble plate is changed by a bellows-activated control valve located at the rear of the compressor in the suction port (low-pressure) side and is opened and closed in response to changes in low-side pressure (Figure 9-25). The internal pressure of the crankcase is controlled by the operation of the control valve. The angle of the wobble plate is determined by the pressure differential between the crankcase's internal pressure and the piston cylinder pressure.

When air-conditioning demand is high, with an increased heat load, the refrigerant pressure on the low side will increase. In this situation, the suction pressure will be above the control point, and the control valve bellows will compress to open the low-pressure side valve and close the high-pressure side valve, which will maintain a bleed from the compressor crankcase to the suction side (Figure 9-26). In this case, the crank-case's internal pressure will be equal to the pressure on the suction port side. The internal pressure in the cylinder will be greater than the crankcase pressure, and the wobble plate will be at the maximum angle, providing greatest piston travel (stroke) and displacement (Figure 9-27).

Operation of the control valve is dependent on differential pressure, known as delta p (Δp).

Conversely, when the air-conditioning demand is low, such as when the passenger compartment or ambient air temperature is low or during high-speed driving, the suction pressure will also be low. In this situation, the suction pressure will be below the control point, and the control valve bellows will expand to close the low-pressure side valve and open the high-pressure side (discharge port) valve, which will bleed high pressure into the compressor crankcase (Figure 9-28). In this case, the crankcase internal pressure will trigger a pressure differential between the internal pressure in the cylinder and the crankcase, and the wobble plate will move to its minimum angle, providing the least piston travel (stroke) and displacement (Figure 9-29).

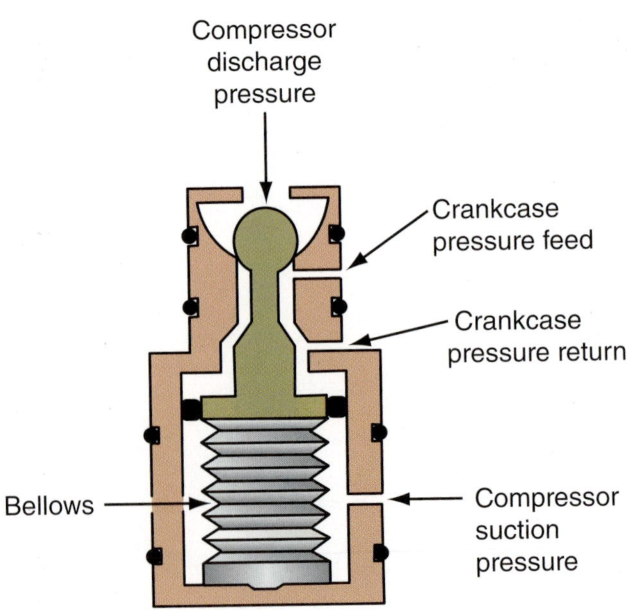

FIGURE 9-25 **Variable displacement compressor control valve assembly.**

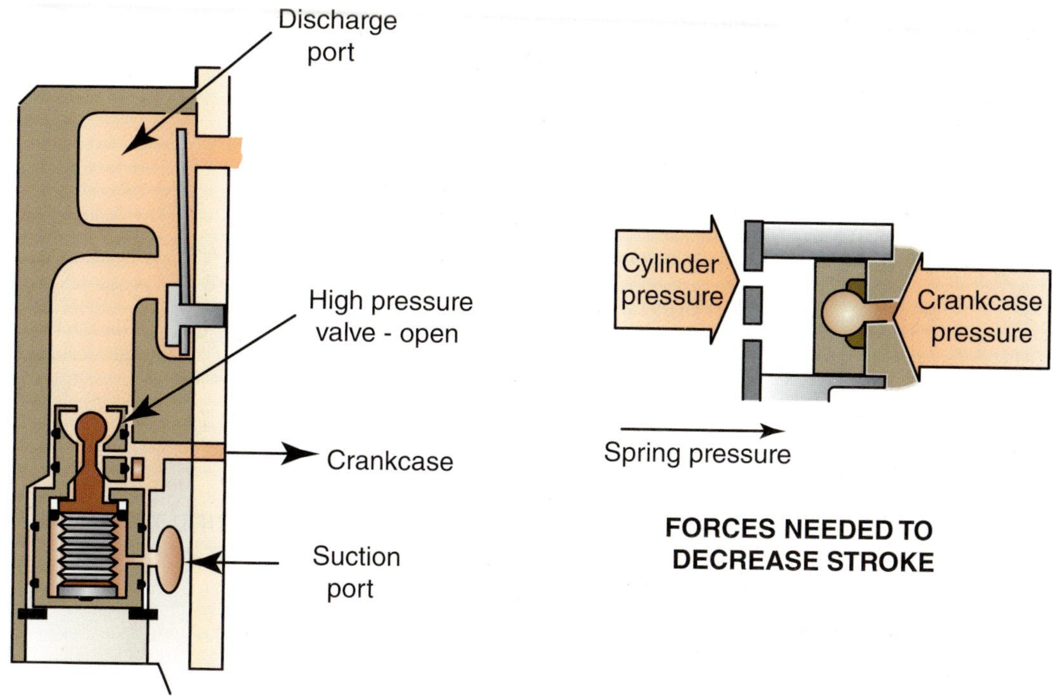

FIGURE 9-26 **Variable displacement compressor control valve in position to lower compressor crankcase pressure and provide maximum wobble plate displacement angle.**

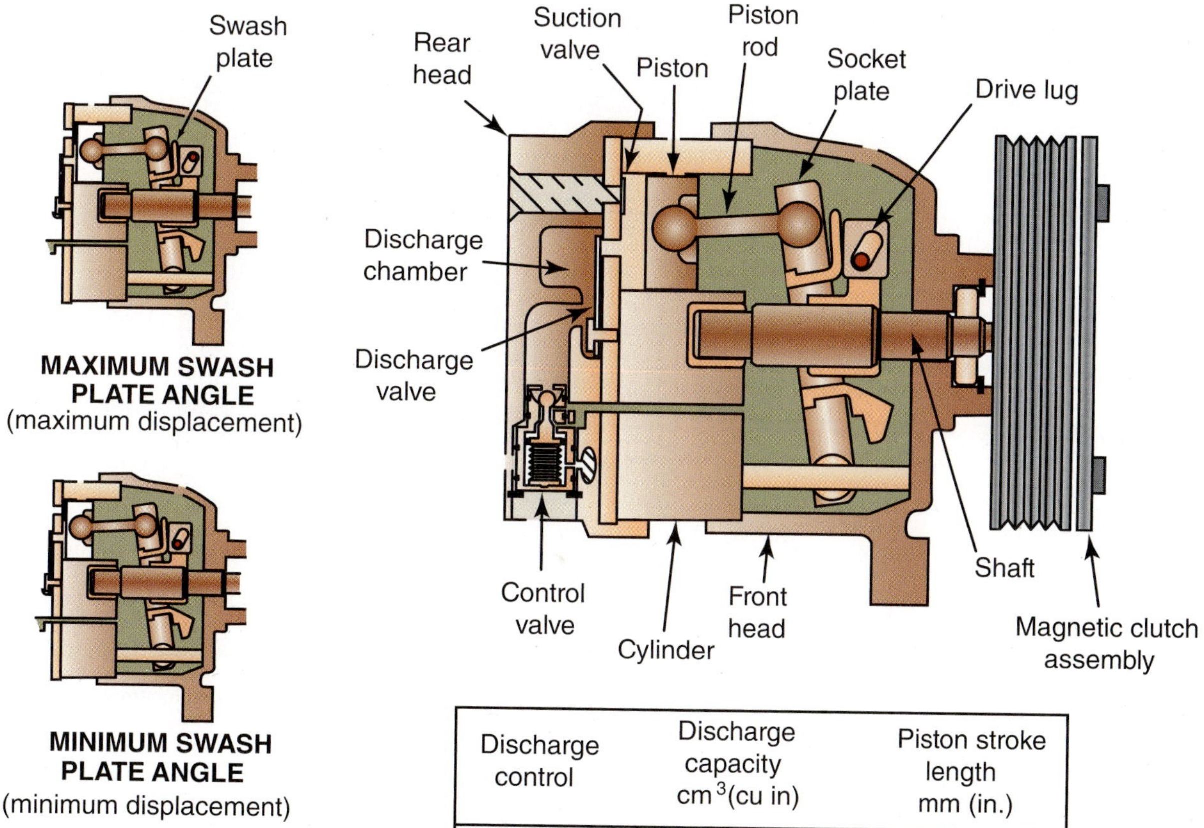

Discharge control	Discharge capacity cm^3(cu in)	Piston stroke length mm (in.)
Minimum	10.5 (0.641)	1.6 (0.053)
Maximum	184 (11.228)	28.6 (1.126)

FIGURE 9-27 **Variable displacement compressor at maximum displacement.**

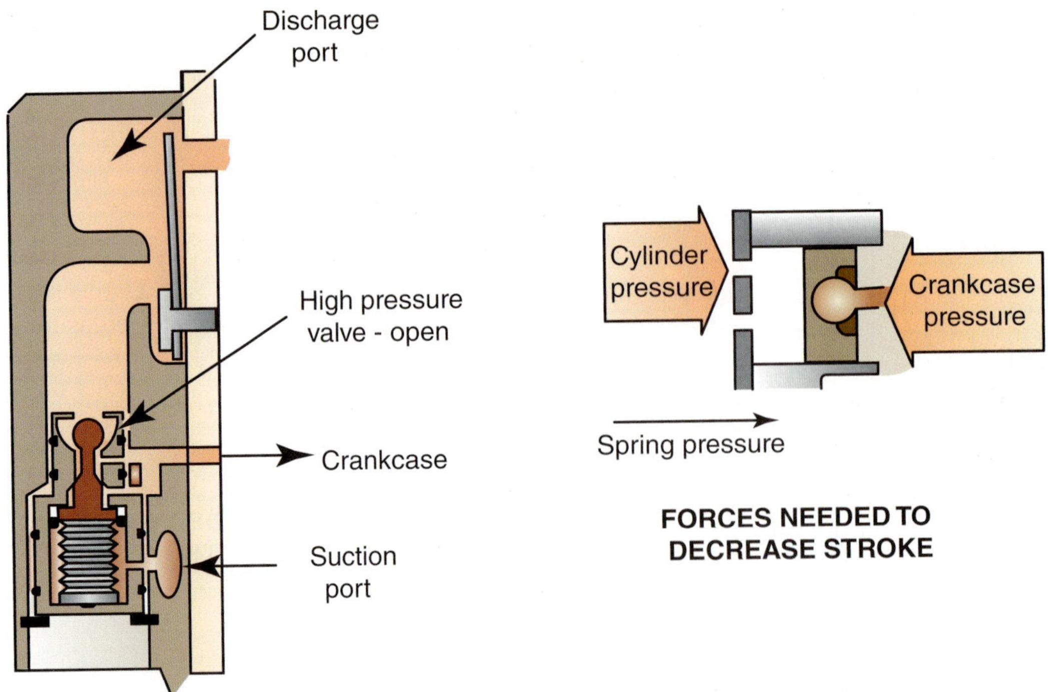

FIGURE 9-28 **Variable displacement compressor control valve in position to increase compressor crankcase pressure and provide minimum wobble plate displacement angle.**

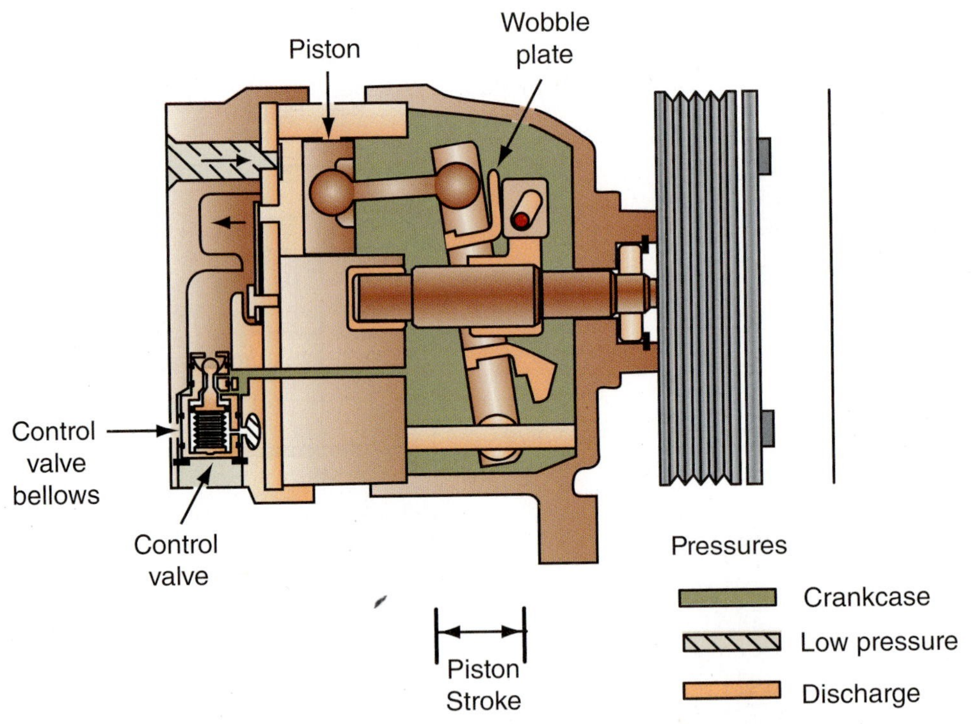

FIGURE 9-29 **Variable displacement compressor at minimum displacement.**

The angle of the wobble plate is actually controlled by a force balance on the pistons. Only a slight increase of the crankcase-suction pressure differential is required to create a force on the pistons sufficient to result in a movement of the wobble plate.

Temperature, then, is maintained by varying the capacity of the compressor, not by cycling the clutch on and off. This action provides a more uniform method of temperature control and at the same time eliminates some of the noise problems associated with a cycling clutch system.

Electronic Variable Displacement Compressor

An electronic variable displacement compressor design is based on a swash plate design, but it is able to change piston stroke in response to the cooling capacity required (Figure 9-30). The piston stroke is changed by the tilt of the swash (wobble) plate, which changes refrigerant discharge volume. A typical discharge volume for an electronic variable displacement compressor varies from 0.885 to 11.228 cu. in. (14.5 to 184 cm^3). It is similar in design to the V-5 compressor but uses an electronic high-pressure control valve pulse–width–modulated solenoid, which replaces the conventional control valve and is often used with an automatic temperature control climate control system.

The electronic control valve solenoid's magnetic coil receives a duty cycle signal from the refrigerant control system air-conditioning amplifier. The duty cycle signal regulates the amperage applied to control the high-pressure control valve lift amount. By changing the high-pressure valve lift, the high-pressure electronic control valve is able to vary compressor volume by routing more or less pressure to the rear chamber of the compressor, which in turn changes the angle of the wobble plate. The compressor output can be varied from 1 to 100 percent. With this wide range of outputs it optimizes the air-conditioning system's efficiency while lowering CO_2 emissions, improving fuel economy, and improving engine performance.

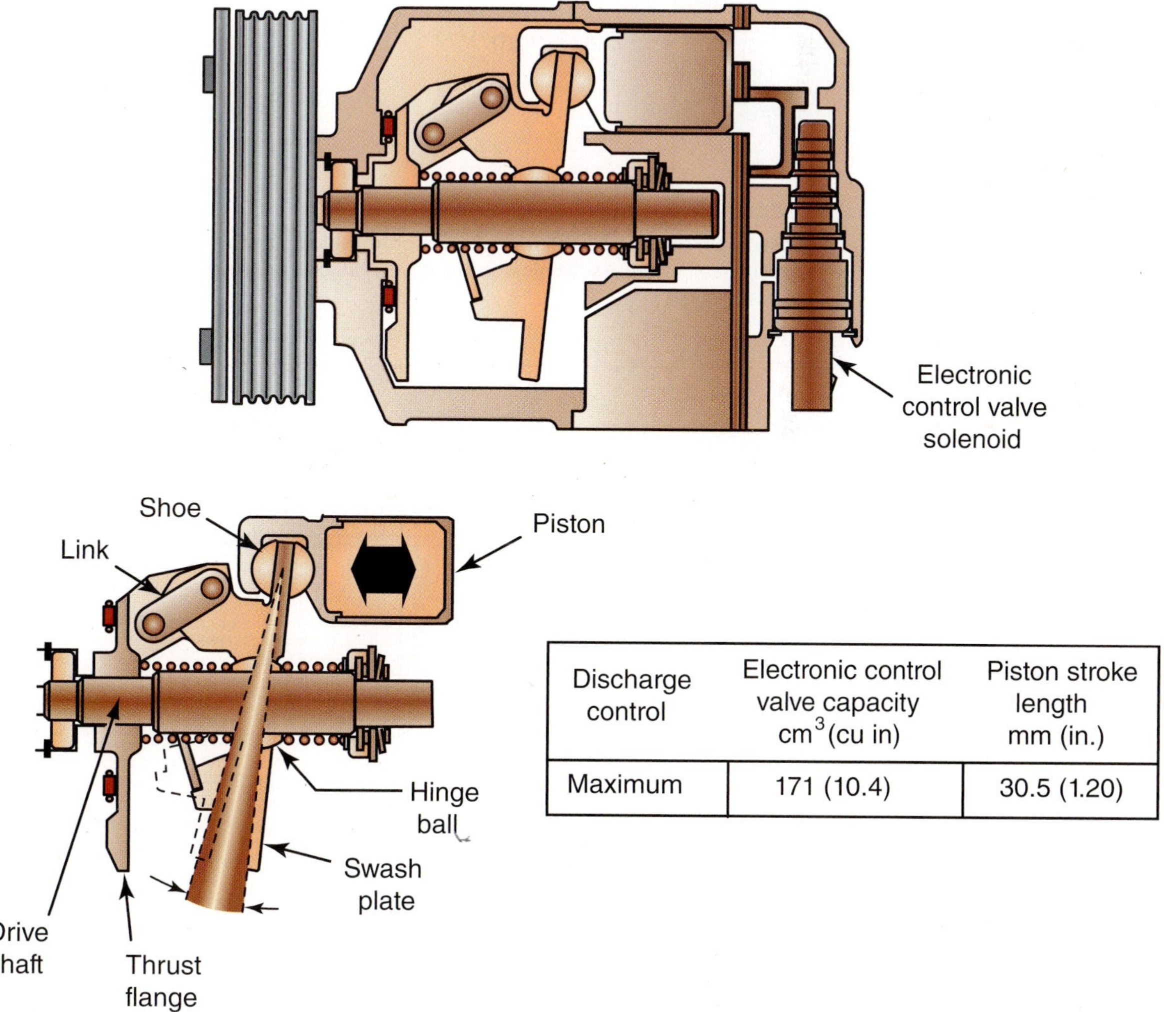

Discharge control	Electronic control valve capacity cm^3 (cu in)	Piston stroke length mm (in.)
Maximum	171 (10.4)	30.5 (1.20)

FIGURE 9-30 **The variable displacement compressor piston stroke is changed by the tilt of the swash (wobble) plate, which changes refrigerant discharge volume based on a pulse width–modulated signal to the electronic control valve.**

A broken discharge valve plate will set up a mini-vibration sufficient to "shake" the compressor loose enough to cause belt slippage.

For maximum cooling, the high-pressure control valve solenoid is closed by the magnetic field (amperage) applied to the electronic coil at the electronic control valve solenoid (Figure 9-31). The climate control module increases the duty cycle (on time) during high heat load conditions based on evaporator temperature sensor data and other system inputs (Figure 9-32). This creates a pressure balance between inside the crankcase (by lowering crankcase pressure) and the suction line, causing the swash plate to move to the maximum stroke position.

To reduce compressor output, the electrical signal to the electronic control valve solenoid is turned off and the high pressure control valve is opened by spring force (Figure 9-33). The climate control module decreases the duty cycle (on time) during low heat load conditions based on evaporator temperature sensor data and other system inputs (Figure 9-34). At this point, the suction-line pressure is still low (maximum output), which enables the suction

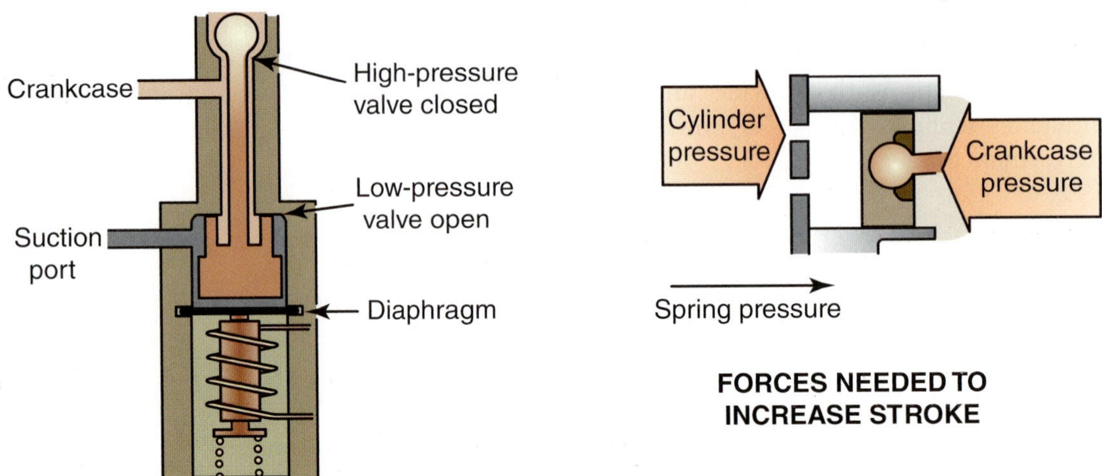

FIGURE 9-31 **For maximum cooling, the high-pressure control value solenoid is closed by the magnetic field applied to the electronic coil at the electronic control valve solenoid, opening the pressure passage from the suction port to the crankcase.**

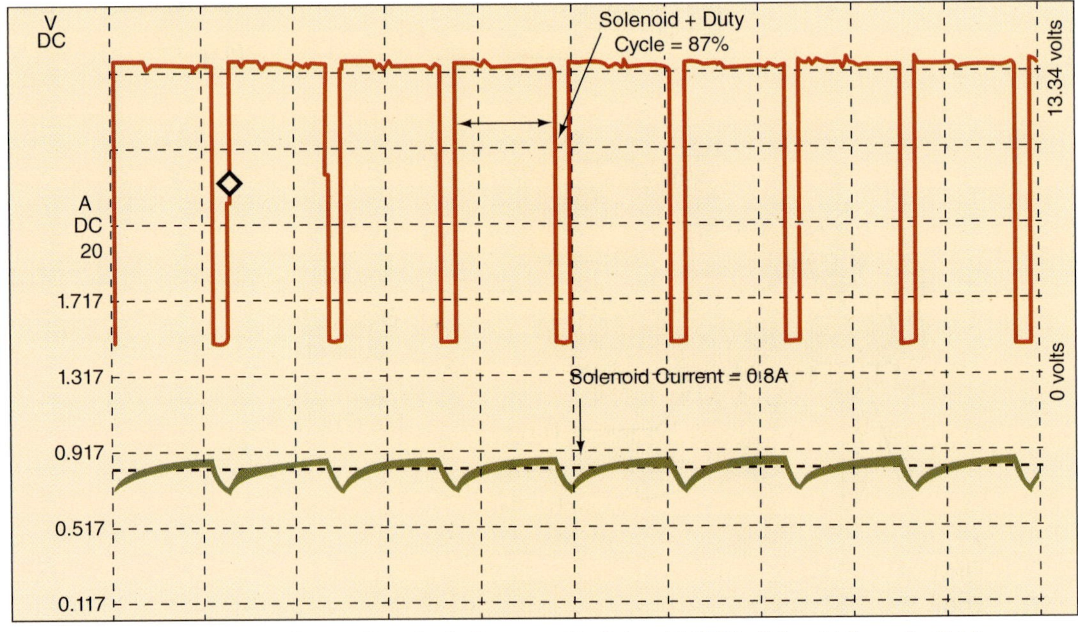

FIGURE 9-32 **As the duty cycle is increased to the variable displacement compressor, control solenoid amperage increases, creating a strong magnetic field opening the solenoid.**

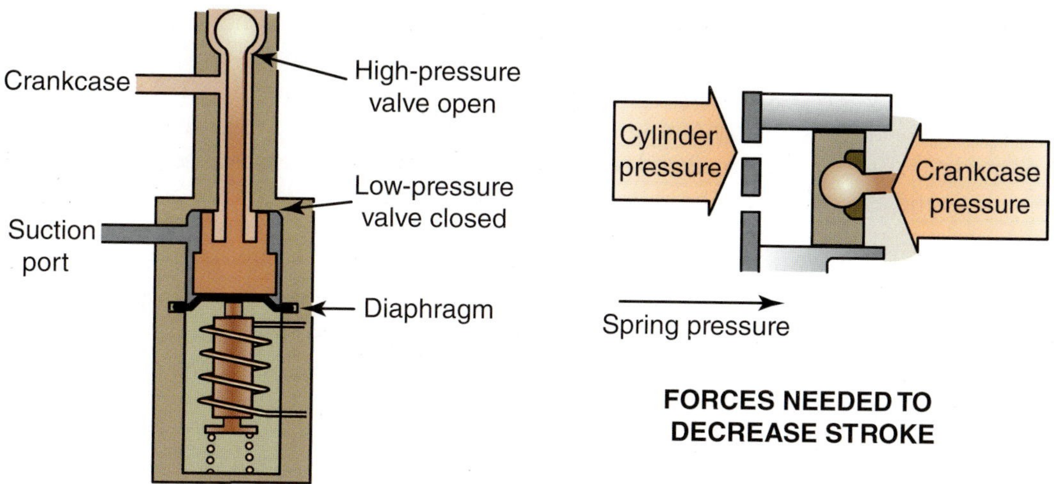

FIGURE 9-33 To reduce compressor output, the high-pressure control valve solenoid is turned off. This allows discharge pressure to enter the crankcase.

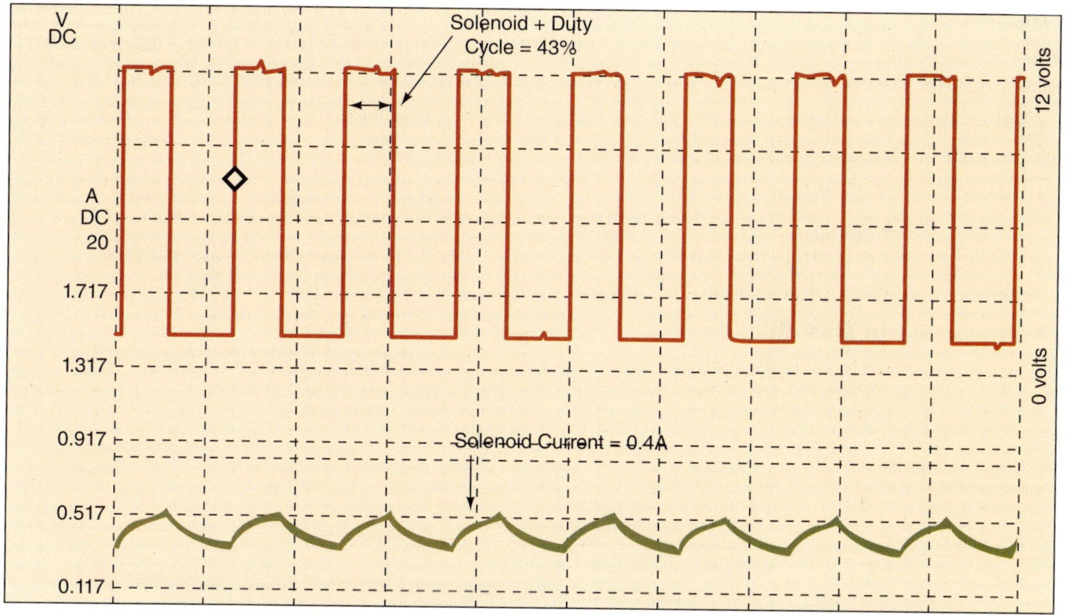

Low/Medium Heat Load – Medium Duty Cycle Command to Solenoid (43%). Solenoid Currect = 0.4A

FIGURE 9-34 As the duty cycle is decreased to the variable displacement compressor, control solenoid amperage also decreases, creating a weak magnetic field and allowing the solenoid to close.

port to close and the discharge port to open. This in turn allows high pressure to enter the crankcase. This increased crankcase pressure is applied to the swash plate, creating a pressure difference between the pistons and the crankcase, which changes the angle of the swash plate and reduces displacement.

During diagnosis, if the compressor is not producing adequate pressure after the system refrigerant has been recovered and recharged with the proper amount of refrigerant, be sure to check that the climate control module is sending the correct duty cycle signal to the compressor. An unplugged solenoid will cause the compressor to default to minimum displacement. Always check for both powertrain and climate control system stored trouble codes, which could inhibit proper system operation. Inaccurate climate control sensor inputs such as inaccurate evaporator, ambient or cabin temperature sensor readings, or pressure sensors could cause the wrong displacement command to be sent.

On belt-driven applications there is no need for an electromagnetic clutch, but some design applications do use one. The compressor output can be lowered to almost zero output pressure when there is no command for air conditioning. The compressor shaft turns continuously whenever the engine is running, so lubrication is critical. It is essential that the system be properly charged with both oil and refrigerant at all times to maintain adequate compressor lubrication and avoid failure. The clutchless system is designed to keep the oil charge circulation going even when the air conditioning is turned off. There is a vibration damper built into the compressor pulley to absorb engine torque fluctuations. The pulley is also designed with a spoke limiter mechanism attached to the compressor drive shaft that will break away in the event the compressor locks up, allowing the pulley to free-wheel and continue to rotate. This is a safety feature that allows the compressor to be driven by a common drive belt without the fear of belt system failure due to compressor failure.

Diagnosing Problems and Making Repairs

Shop Manual
Chapter 9, page 340

Broken discharge valves in compressors (Figure 9-35) are not uncommon. Broken suction valves and piston rings are less often encountered, but lead to the same diagnosis. Broken valves or rings are easier to diagnose in one- and two-cylinder reciprocating compressors than in multicylinder compressors. The manifold and gauge set is the diagnostic tool most often used to determine valve plate condition. The first indication of failure is a higher than normal low-side (suction) pressure accompanied by a lower than normal high-side (head) pressure.

Valve and ring failures, however, are not as easily diagnosed in four-, five-, and six-cylinder compressors. The first indication of valve or ring failure in these compressors is that the belt(s) will not remain tightened. One defective discharge valve plate in a six-cylinder compressor, for example, sets up a vibration that, when not otherwise detected, literally shakes the belt(s) loose. This is true regardless of how well the adjustment provisions are tightened.

Many simple compressor repairs are usually a routine service provided by the automotive air-conditioning technician. These repairs include checking and adding oil, replacing the crankshaft seal, and, in many units, replacing the valve plate assembly. More complex repairs are often "shopped out" to a specialty shop that has the facilities for semi-mass-rebuilding procedures. Because of the general high cost of labor, one-on-one compressor rebuilding is not usually economically feasible.

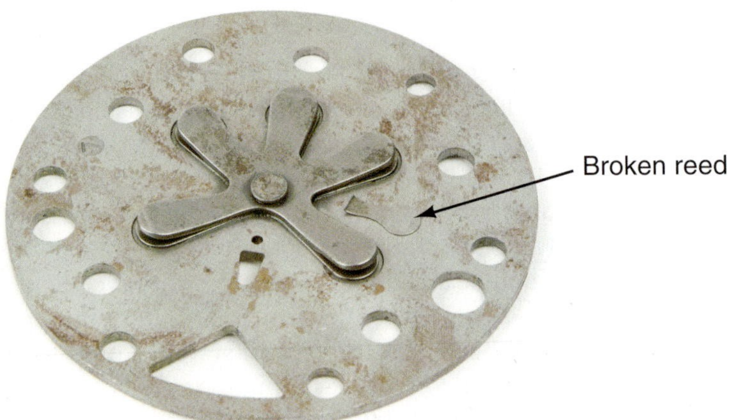

FIGURE 9-35 **Broken reed in valve plate.**

Compressor Failure

Shop Manual
Chapter 9, page 343

Compressor failure, representing almost 30 percent of all vehicle air-conditioning system repairs, is the leading cause of system failure according to a survey conducted in the late 1990s by the Mobile Air Conditioning Society Worldwide. The principal cause of compressor failure was found to be leaks, followed by internal mechanical problems. Clutch problems were the least common reasons for compressor failure. It should be noted that repeat compressor failures are often caused by a restricted condenser that went undiagnosed after the initial compressor failure. Many of today's honeycomb design condensors cannot be flushed and should be replaced if a restriction is suspected or a catastrophic compressor failure has occurred with particles present in the compressor oil.

The survey revealed that over half of the reported compressor failures were to R-134a air-conditioning systems in vehicles that had been retrofitted to avoid using R-12 refrigerant. This is probably because many older compressors designed for R-12 simply will not withstand the rigors of R-134a's higher operating pressures. The lack of proper lubrication is also implicated as a problem with vehicle air-conditioning system compressors and is generally due to not properly changing the lubricant during retrofit procedures or not checking lubricant during repair procedures.

Most compressors are designed to function with a compression ratio between 5:1 and 7:1. One may make a quick check of a compressor's operating compression ratio by dividing high-side psia by low-side psia. Note that these are absolute pressures, so one must add 15 to both low-side and high-side gauge readings. The formula is:

$$\text{Compression Ratio} = \frac{\text{High-Side Pressure} + 15}{\text{Low-Side Pressure} + 15}$$

For example, assume that the low-side pressure is 30 psig and the high-side pressure is 220 psig. When 15 is added to these values, they become 45 psia and 235 psia, hence:

$$\frac{235}{45} = 5.2$$

Pressure ratios above 7.5:1 can cause early compressor failure because of the added load on bearings, pistons, and seals. Also, higher operating temperatures generated by the higher operating pressures can cause lubrication breakdown, which results in harmful deposits on the compressor's internal assembly.

Electric Motor-Driven Compressors

Shop Manual
Chapter 9, page 372

Electric motor–driven air-conditioning compressors used on current production hybrid vehicles enable air-conditioning operation even when the internal combustion engine is not running. The system uses a humidity sensor to improve the efficiency of passenger compartment dehumidification. The electric air-conditioning compressor is driven by an integrated electric motor (Figure 9-36). The compressor motor runs on three-phase alternating current (AC) power at 200+ volts supplied by the hybrid AC inverter.

The electric compressor is a scroll design compressor driven by a brushless electric motor. The basic construction of the compressor is the same as a standard belt-driven scroll-type compressor. Special compressor oil is used to insulate the compressor housing from the high voltage components inside the compressor. Always refer to the manufacturer's specifications for proper oil selection; currently 2004 and later model Toyota electric compressors use ND11 compressor oil with a high insulation value. Because the compressor electric motor is powered by over 200V AC, special high-voltage circuit safety precautions must be followed. Always read and understand manufacturer high-voltage safety procedures prior to working on or near any high-voltage circuits or components.

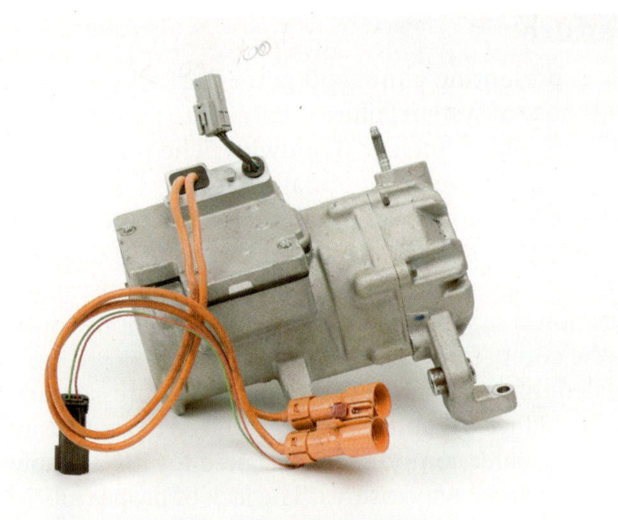

FIGURE 9-36 **Electric air-conditioning compressor is driven by an integrated 200-volt electric motor.**

The compressor utilizes an oil separator that reduces the circulation of compressor oil through the system. As the refrigerant gas and oil leave the compressor discharge port, they flow around a cylindrical pipe in the oil separator. The centrifugal force that is created by the rotation of the refrigerant gas and oil causes the oil to separate out due to the difference in specific gravity between the two. The lighter refrigerant gas passes out the discharge port, while the heavier compressor oil is discharged into an oil storage reservoir. From the reservoir, the oil travels back to the inside of the compressor and the process is repeated. Other than the compressor, the rest of the air-conditioning system operates and is diagnosed in the same way as in a conventional system.

Never work on the high-voltage systems of a hybrid electric vehicle until you have completed a hybrid electric training program.

The following are basic hybrid electric vehicle warnings that should be adhered to:

- Never assume that a hybrid electric vehicle is shut off, because they are silent.
- Always verify that the Ready indicator is off.
- Remove the service plug and wait at least 5 minutes before touching any of the high-voltage wires, connectors, or components.
- Always wear insulated electrical (lineman) gloves before coming in contact with any high-voltage system wiring or components, to prevent electrocution.
- *Never* damage, cut, or open any *orange* high-voltage power cables or components.
- Always test circuit for voltage with a digital multimeter to ensure that a 0 volt reading is obtained before touching any high-voltage terminals.
- Special electrically insulating refrigerant oil is used to keep voltage from conducting from the electric drive motor to the compressor case. Toyota specifies ND-11 compressor oil. Always refer to the specific manufacturer's recommendations for proper compressor oil usage for the application in question. Some manufacturers warn that a diagnostic trouble code may be set or a complete hybrid electric system shutdown could occur due to the addition of even a small amount of the wrong lubricating oil being added to a system. They also warn that if the wrong lubricant is added and circulated through the system that all major refrigerant system components (evaporator, compressor, condenser, storage container, etc.) must be replaced.

Honda hybrid electric vehicles use a dual-scroll air-conditioning compressor design. It is actually two compressors in one. It has a conventional belt-driven 75 cc scroll compressor and a smaller 15 cc scroll compressor driven by a 3-phase high voltage electric motor. Depending on conditions and the demand on the air-conditioning system the compressor can be switched from belt drive to electric drive or engage both. During idle stops the air-conditioning compressor operates in electric-only mode unless cooling demand is high and the internal combustion motor is operating or if the high voltage battery pack state of charge is low in which case the internal combustion engine would be commanded on.

It is predicted that eventually the belt-driven air-conditioning compressor on all platforms will be replaced by a high-voltage electric motor–driven compressor, which is more reliable and less prone to leakage. But this will have to wait until high-voltage power generation is available on more platforms.

Stretch to Fit Belts

Ford, General Motors, and Subaru began using stretch to fit belts to drive their air-conditioning compressors and other single-drive belt accessories on some models beginning in 2008. Stretch to fit belts were designed by Gates, a global supplier of OE belts and hoses, as an alternative to adjustable micro-V belts. The belt is very similar in appearance to a conventional micro-V belt. The difference is the use of a reinforcing cord made of polyamide material, which is more elastic than traditional aramid or polyester fibers used in traditional designs that makes them self-tensioning. The polyamide fibers and elastic backing compound allow the belt to stretch and then retract to maintain proper tension throughout its lifetime without the need for a manual or automatic belt tensioner. The micro-V belt ribbed drive surface is still made with EPDM for long service life that resists cracking. Out of the package they are shorter than their actual working length, but once installed they automatically tension, providing optimum load-carrying capacity. A stretch to fit belt may be identified by looking at the part number; if the last letters are "SF," as in K030195SF (Figure 9-37), then it is a stretch to fit

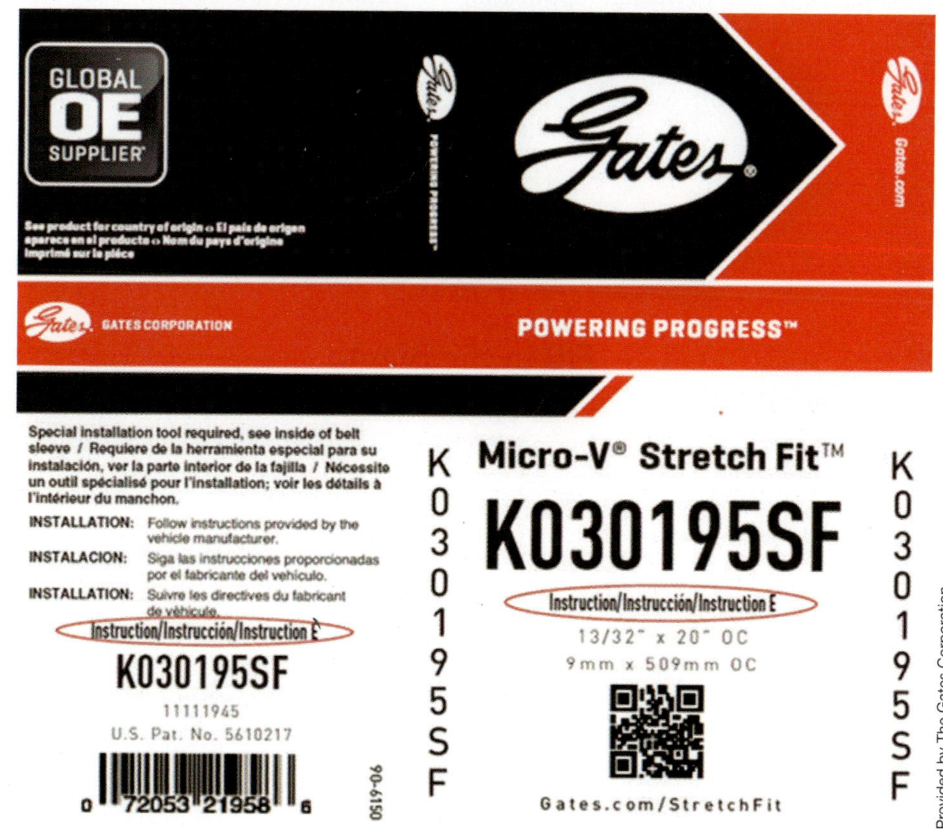

FIGURE 9-37 **A stretch to fit belt can be identified by looking at the last two letters of the part number; if the number ends in "SF" then it is a stretch to fit belt.**

belt requiring special service instruction for replacement. General Motors has indicated that these are one-time use belts and once removed should be replaced, meaning that if you find it necessary to remove an air-conditioning compressor driven by a stretch to fit belt, the belt must be replaced to maintain optimal load-carrying capacity.

Terms to Know

Auxiliary
Axial plate
Compression
Crankshaft
Discharge valve
Exhaust
High pressure
Intake
Low pressure
Rotary
Scroll
Serpentine
Suction
Suction valve
Swash plate
Wobble plate
Zener clamping diode

SUMMARY

- Reciprocating compressors have a piston or pistons that draw low-pressure heated refrigerant vapor into a chamber, increases its heat content and pressure, and "pumps" it out as a high-pressure high-temperature vapor.
- A scroll compressor draws low-pressure heated refrigerant vapor through its suction port into a continuously moving scroll where the vapor's pressure and temperature are increased and then forced out through its discharge port as a high-pressure high-temperature vapor.
- In a rotary compressor, a rotating vane draws in low-pressure heated refrigerant vapor through the suction port and increases its temperature and pressure before forcing it out through the discharge port. It is discharged as a high-temperature high-pressure vapor.
- An electromagnetic clutch is used to engage and disengage (turn on and off) a compressor, as desired, in present applications of automotive air-conditioning systems.

REVIEW QUESTIONS

Short-Answer Essays

1. Describe the operating principles of a reciprocating compressor.
2. Why is low pressure important?
3. What are other terms used to describe an axial plate?
4. What type of refrigerant compressor designs are used on hybrid electric vehicles?
5. Explain the operation of an electronic variable displacement compressor.
6. Describe the function of a variable displacement compressor.
7. Describe the operating principles of a rotary compressor.
8. What is the purpose of a magnetic clutch:
   a. In a fixed displacement compressor system?
   b. In a variable displacement compressor system?
9. What are the two primary functions of a compressor?
10. Why is the scroll compressor considered to be the most efficient?

Fill in the Blanks

1. In-car temperature is controlled in a variable displacement compressor system by varying the capacity of the ______________, not by cycling the ______________ on and off.
2. Compression in a scroll compressor is achieved by the interaction of a orbiting ______________ and a stationary ______________.
3. A ______________ compressor is designed to match any automotive air conditioning ______________ demand under all conditions.
4. Electric motor–driven air conditioning ______________ used on current production hybrid vehicles enable air-conditioning operation even when the ______________ is not running.

5. A ______________ compressor has only one cylinder with a compression stroke (output) for every ______________ of rotation.

6. A single belt driving all accessories is often called a ______________ drive system.

7. The compressor serves two important functions, to create a ______________ condition within the system, and it ______________ refrigerant vapor from a low pressure to a ______________, thereby increasing its ______________.

8. An electromagnetic clutch is used to turn the ______________ on and off.

9. If a compressor clutch ______________ is set too loose, the clutch may not engage or it may slip, if too tight the clutch may ______________.

10. A ______________ ______________ compressor is designed to match any automotive air-conditioning load demand under all conditions.

Multiple Choice

1. The electronic variable displacement compressor is based on what type of compressor design?
   - A. Rotary compressor design
   - B. Swash plate compressor design
   - C. Scroll compressor design
   - D. Vane compressor design

2. Electric motor–driven air-conditioning compressors used on hybrid vehicles are based on what compressor design?
   - A. Wobble plate compressor design
   - B. Swash plate compressor design
   - C. Scroll compressor design
   - D. Rotary compressor design

3. An air-conditioning compressor clutch failed due to overheating. What is the most likely cause of the failure?
   - A. A low refrigerant charge
   - B. A faulty compressor clutch diode
   - C. High resistance in the compressor
   - D. A loose drive belt clutch circuit

4. The compressor pumps refrigerant as a
   - A. liquid.
   - B. liquid that changes to a vapor.
   - C. vapor that changes to a liquid.
   - D. vapor.

5. The methods used to drive a compressor are being discussed:

   Technician A says that automotive air-conditioner compressors have an electromagnetic clutch.

   Technician B says that a serpentine belt system is often used to drive a compressor clutch.

   Who is correct?
   - A. A only
   - B. B only
   - C. Both A and B
   - D. Neither A nor B

6. A technician finds an air-conditioning clutch will not engage. Upon initial testing the compressor is receiving both proper power and ground at the compressor clutch electrical connector; in addition the compressor input shaft turns freely by hand. Which of the following is the most likely cause?
   - A. A faulty pressure cycling switch.
   - B. Low refrigerant level.
   - C. A locked-up compressor.
   - D. Compressor clutch air gap too large.

7. All of the following statements about variable displacement compressors are true, except:
   - A. Variable displacement compressors control evaporator pressure by varying their pumping capacity.
   - B. Variable displacement compressors have a cycling clutch.
   - C. Variable displacement compressors do not have a cycling clutch.
   - D. Variable displacement compressor piston stroke is adjusted by an internal pressure valve.

8. Reciprocating compressors are being discussed.

   Technician A says that the suction valve is opened to allow refrigerant vapor to enter due to a differential in pressure above and below the valve.

   Technician B says that the discharge valve is opened to allow refrigerant vapor to leave due to a differential in pressure above and below the valve.

   Who is correct?
   - A. A only
   - B. B only
   - C. Both A and B
   - D. Neither A nor B

9. The function of a scroll compressor is being discussed.

   Technician A says that its function is to create a low pressure condition in the system.

   Technician B says that its function is to increase the temperature and pressure of refrigerant vapor.

   Who is correct?
   - A. A only
   - B. B only
   - C. Both A and B
   - D. Neither A nor B

10. The following are automotive air-conditioning system compressor types, except:
    - A. Rotary
    - B. Swash plate
    - C. Centrifugal
    - D. Scroll

Chapter 10

Case and Duct Systems

Upon Completion and Review of this Chapter, you should be able to:

- Identify types of case/duct systems.
- Discuss air distribution through the case/duct system.
- Understand the airflow through the case/duct system for defrost mode, heat mode, and cool mode.
- Understand and control unpleasant HVAC odor.
- Identify the need for and location of cabin air filters.
- Understand Mode Door Actuator operation: cable, vacuum, and electric.

The heater coolant flow control valve is generally found outside the case/duct system for ease of service.

The evaporator core is often easier to service in a split case system.

Hybrid means that no particular system is in mind—that the illustration is representative of any system.

Mode is the manner or state of existence of something, for example, heat or cool, open or close.

Introduction

This chapter is intended to provide a basic understanding of the automotive heater/air-conditioner case/duct system for factory-installed heater/air-conditioning systems. The system discussed in this unit should be considered typical. It is not representative of any particular automotive case/duct system. A typical automotive heater/air-conditioner case/duct system (Figure 10-1), at first glance, may seem to be a complicated maze of passages and doors. Actually, it is much simpler than it first appears (Figure 10-2).

The case/duct system serves two purposes. First, it houses the heater core and the air-conditioner evaporator. Second, it directs fresh or recirculated conditioned air into the vehicle. This air is directed through selected components into the passenger compartment via selected outlet provisions, such as panel registers, a floor outlet, and defroster outlets.

The supply air may be either fresh (outside) or recirculated (in-car) air, depending upon the system **mode** selected. After air is heated or cooled (conditioned), it is delivered to either the floor outlet, dash (panel) outlets, or the **defrost** (windshield) outlets.

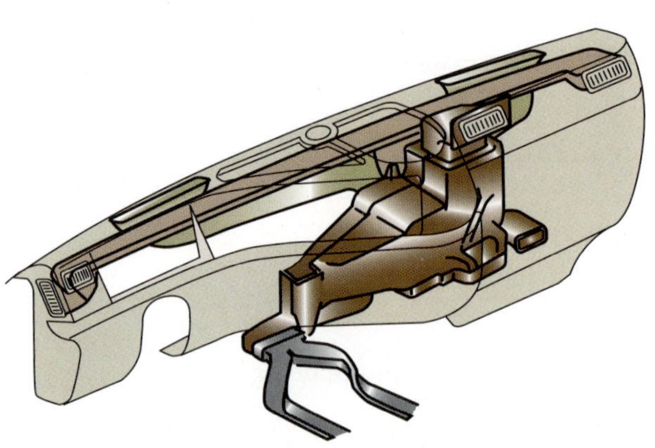

FIGURE 10-1 **A typical case and duct system may seem like a maze at first glance.**

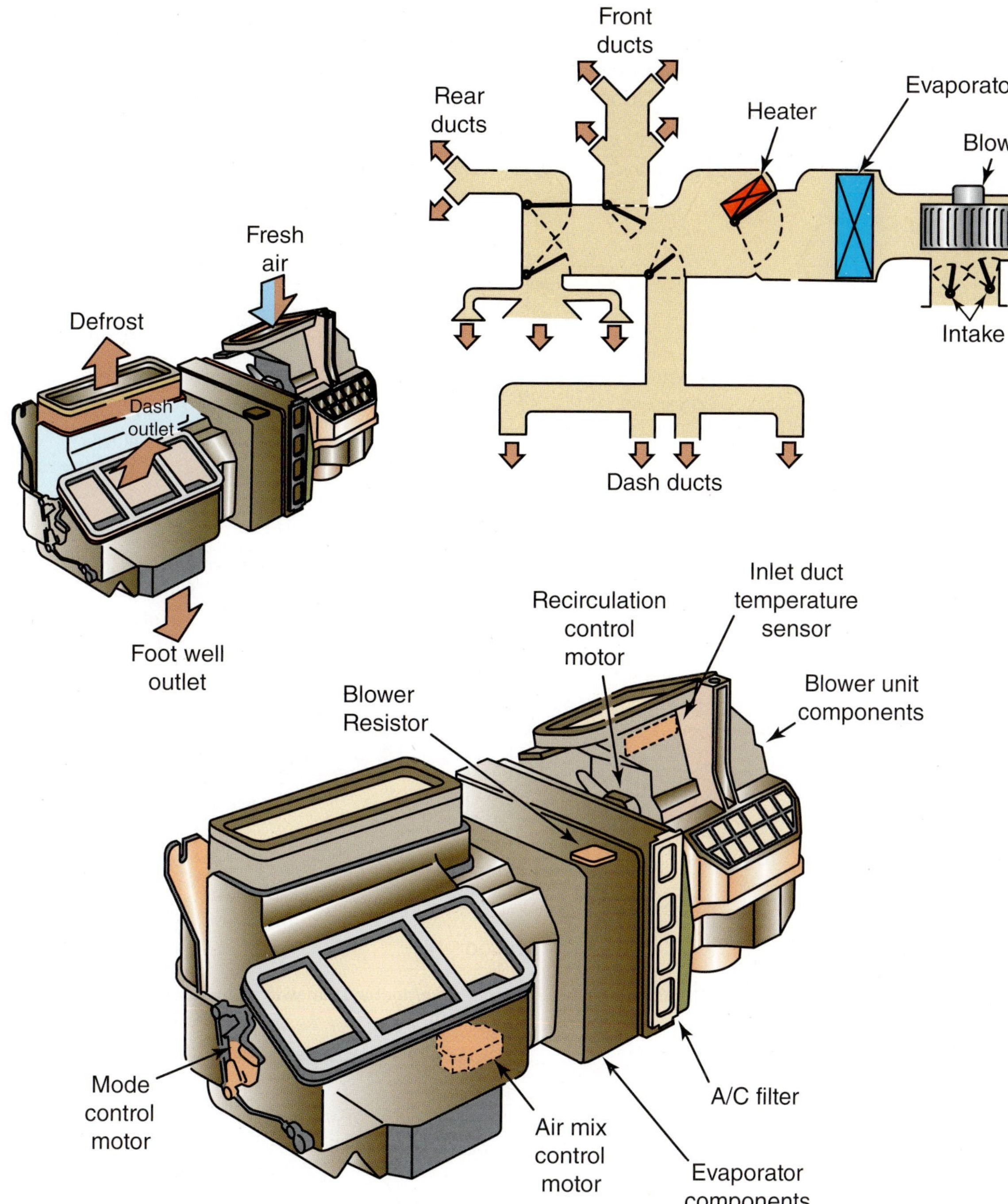

FIGURE 10-2 **A typical case and duct system showing both exterior and interior view.**

To **defrost** is to remove frost. It also refers to the vent position where warm outlet air is directed to the windshield.

Two basic types of case assemblies are used to house the heater core and air-conditioner evaporator: the independent case assembly and the split case assembly.

The independent case, which is used on compact and small cars, may have an upstream blower (Figure 10-3) or a downstream blower (Figure 10-4). Either an upstream integral blower (Figure 10-5) or an independent blower (Figure 10-6) is used on split case systems. The split case system, which is used on larger cars, is located on both sides of the engine firewall. The independent case system is usually located under the dash, on the inside of the firewall.

For simplicity of understanding, a typical hybrid case/duct system is illustrated in this unit. Also, for the purposes of explanation, this system is theoretically divided into three Shop

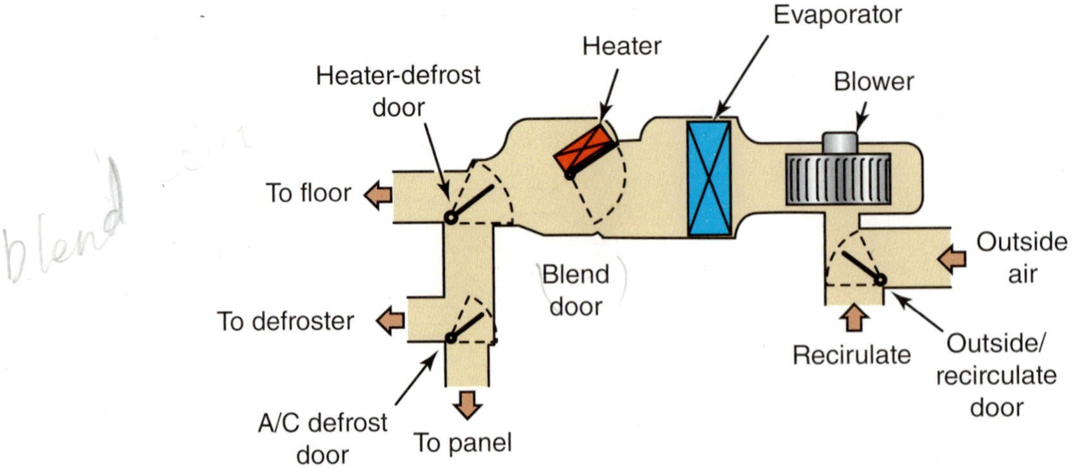

FIGURE 10-3 **An independent case/duct system with an upstream blower.**

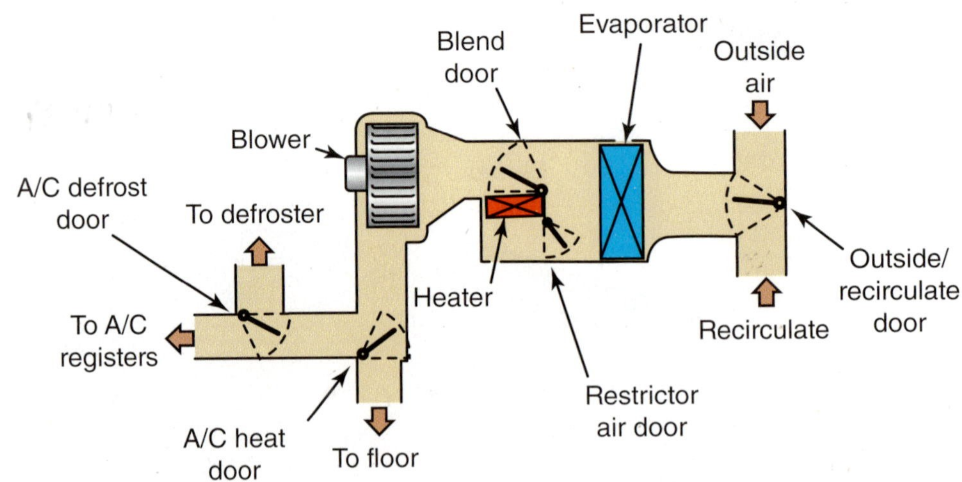

FIGURE 10-4 **An independent case/duct system with a downstream blower.**

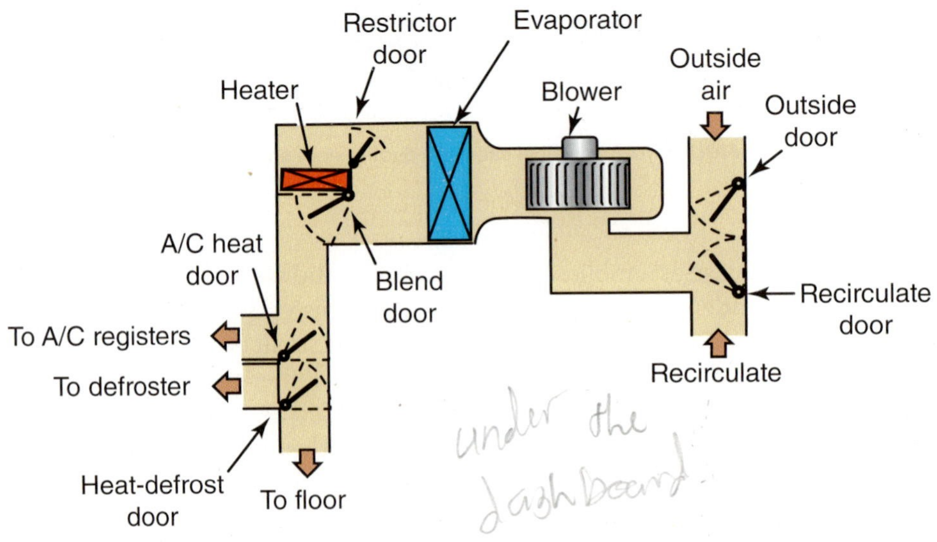

FIGURE 10-5 **A split case system with an upstream blower.**

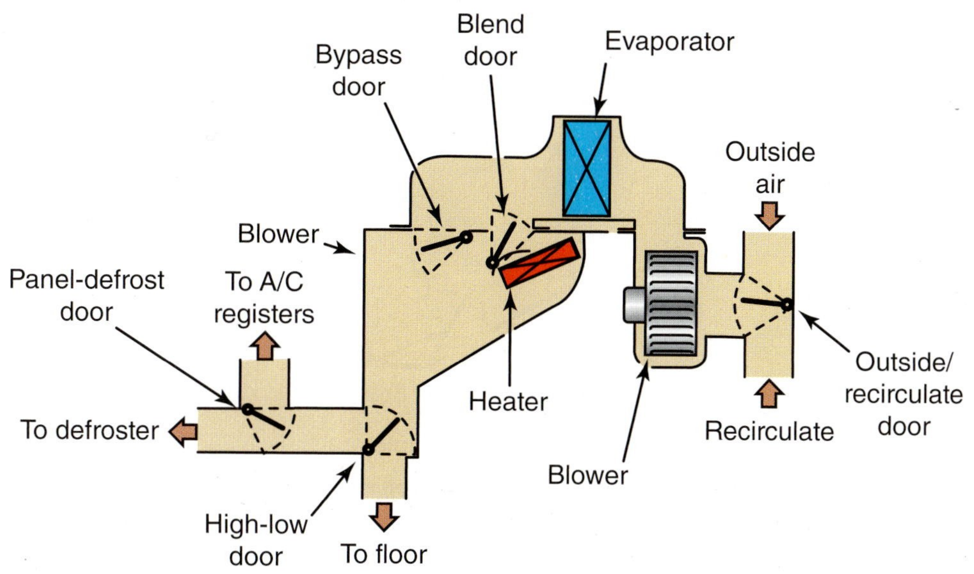

FIGURE 10-6 **A split case system with an independent blower.**

Shop Manual
Chapter 10, page 397

The **plenum** is an area filled with air at a pressure that is slightly higher or lower than the surrounding air pressure, such as the chamber just before the blower motor.

Manual sections (Figure 10-7): the air intake, heater core and air-conditioning evaporator (**plenum**), and air distribution.

Each will be studied, first individually, then as a complete system. Remember, however, that this discussion is of factory-installed or original equipment manufacturer (OEM) installation.

This test also assumes that the heating system is in proper working order.

There are many considerations involved in the design of an automotive air-conditioning system and the volume of airflow requirements of the case and duct system. The interior passenger compartment soak temperatures of the vehicle are affected by many factors. Temperatures are affected by both exterior and interior surface colors and tint versus nontint windows. Heat load on the air-conditioning system varies with the number of passengers in the vehicles, due to both body heat and breath temperatures, which are typically 40°F–60°F above ambient air temperatures. The temperature of the interior surfaces due to the radiant heat of the sun can range from 50°F–100°F above ambient air temperatures. All of these factors and more affect the overall performance of the heating and air-conditioning system.

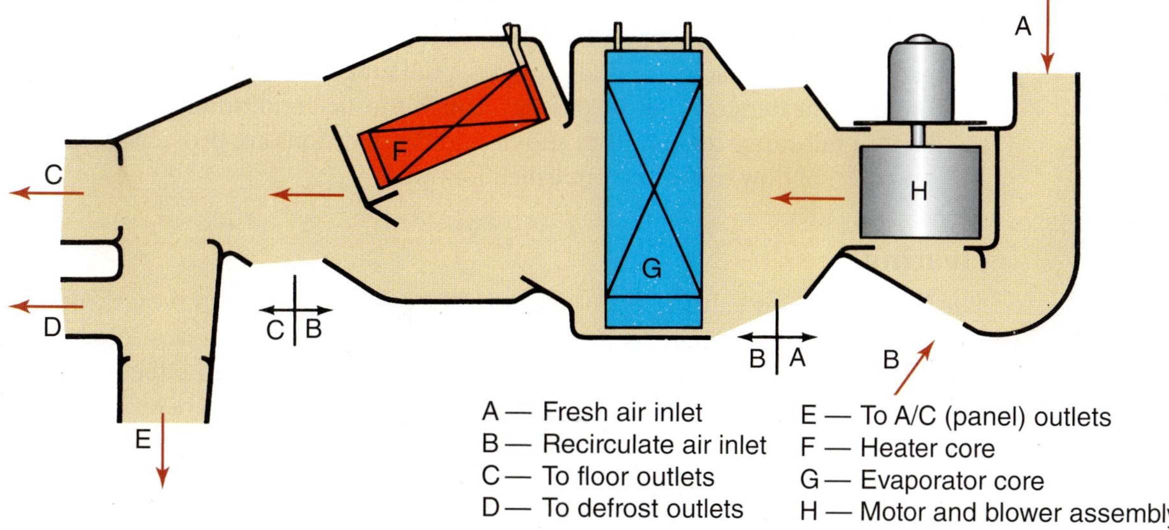

FIGURE 10-7 **Three sections, right to left: (A) air intake section; (B) plenum section; and (C) air distribution section.**

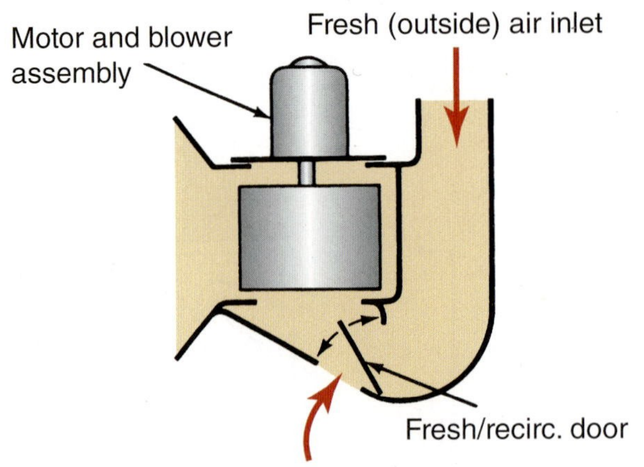

FIGURE 10-8 **Air intake (inlet) section.**

Shop Manual
Chapter 10, page 398

Understanding how the entire system functions is essential to proper diagnosis of the system as a whole.

Air Intake

To **recirculate** is to reuse, to circulate over and over again.

Fresh air intake is generally through vents provided just in front of the windshield, often hidden by the hood cowl.

The air intake or inlet section (Figure 10-8) consists of a fresh (outside) air inlet, a **recirculate** (inside) air inlet, and a fresh-recirculate blend-air door. The outlet of this section is to the blower inlet. The fresh air inlet provides the system with fresh outside air; the recirculate air inlet provides a recirculated in-car air.

The position of the fresh-recirculate door depends on the system mode. Generally, in all modes except maximum cooling (MAX A/C), the air supply is from the outside ambient. In MAX A/C, the air supply is from the inside (recirculated). Even in the MAX A/C mode, some systems provide for up to 20 percent fresh air. This is to provide for a slightly positive in-vehicle pressure. A slightly positive pressure must be maintained inside the vehicle to prevent the possibility of the entrance of dangerous exhaust gases that could produce a hazardous, if not lethal, in-vehicle atmosphere when all the windows are tightly closed.

Core Section

The **blend door** is a door in the duct system that controls temperature by blending heated air and cool outside air.

A **Bowden cable** is a wire cable inside a metal or rubber housing used to regulate a valve or to control a remote device.

The core section, more appropriately called the plenum section, is the center section of the system. It consists of the heater core, the air-conditioning evaporator, and a **blend door**. The blend door may be operated by a **Bowden cable** or, on many systems today, it is operated by an electric or vacuum actuator and provides a full range control of the airflow either through or around the heater core. All air passes through the air-conditioning evaporator. It is in this section that full-range temperature and humidity conditions are provided for in-car comfort. A description of how this is accomplished follows.

Heating

Shop Manual
Chapter 10, page 402

The heater coolant valve is open to allow hot engine coolant to flow through the heater core. Cool outside fresh air is heated as it passes through the heater core. In the heating mode, the air conditioner is not operational; therefore, it has no effect on temperature as the air first passes through the evaporator. The desired temperature level is achieved by the position of the blend door. This allows a percentage of the cooler outside air to bypass the heater core, tempering the heated air. The heated and cool air are blended in the plenum to provide the desired temperature and humidity level before passing to the air distribution section.

Cooling

If all other conditions are correct, the compressor is turned on in the cooling mode. In maximum cooling (MAX A/C), recirculated air passes through the air-conditioner evaporator and is then directed back into the vehicle. In other than MAX A/C, fresh outside air passes through the air-conditioning evaporator and is cooled before delivery into the vehicle.

The desired in-vehicle temperature level is achieved by the position of the blend door. The blend door allows a percentage of cooled air to pass through the heater core to be reheated. The cooled air passing through the evaporator and the reheated air passing through the heated core are blended in the plenum to provide the desired temperature level. This tempered air is then directed to the air distribution section.

Some may refer to the core section as the mixing section; it is in this section that heated and cooled air are mixed.

The coolant flow control may allow from partial to full flow of coolant based on the temperature selected by the operator.

Distribution Section

The air distribution section directs conditioned air to be discharged to floor outlets, defrost outlets, or dash panel outlets. Depending on the position of the **mode doors**, conditioned air may be delivered to any combination of these outlets. There are two mode (blend) doors in the air distribution section: the HI/LO door and the DEF/AC door. The HI/LO door provides 0–100 percent full-range conditioned air outlet control to the HI (dash) and LO (floor) outlets. The DEF/AC door provides conditioned air outlet control either to the defrost (windshield) outlets or to the dash panel outlets.

A full flow of refrigerant in the evaporator core provides maximum cooling at all times.

Combined Case

The combined case/duct system provides full-range control of air circulation through the heater core and air-conditioner evaporator. Figure 10-9 shows 100 percent recirculated air through the air-conditioner evaporator and out through the panel outlets.

This may typically represent mode and blend door positions when maximum cooling (MAX A/C) is selected during high in-car ambient temperature conditions.

Figure 10-10 shows 100 percent fresh air circulation through the heater core and out through the floor outlets. This may typically represent the mode and blend door positions when heat is selected during low in-car ambient temperature conditions. A variation (Figure 10-11) shows some of the heated air diverted to the defrost outlets. This would be the typical application to clear the windshield of fog or light icing conditions.

The bi-level control is sometimes referred to as the HI/LO door or control.

Mode doors are diverters within the duct system for directing air to various locations.

Even on Recirculate, up to 20 percent fresh air may be brought in to maintain a positive in-car pressure.

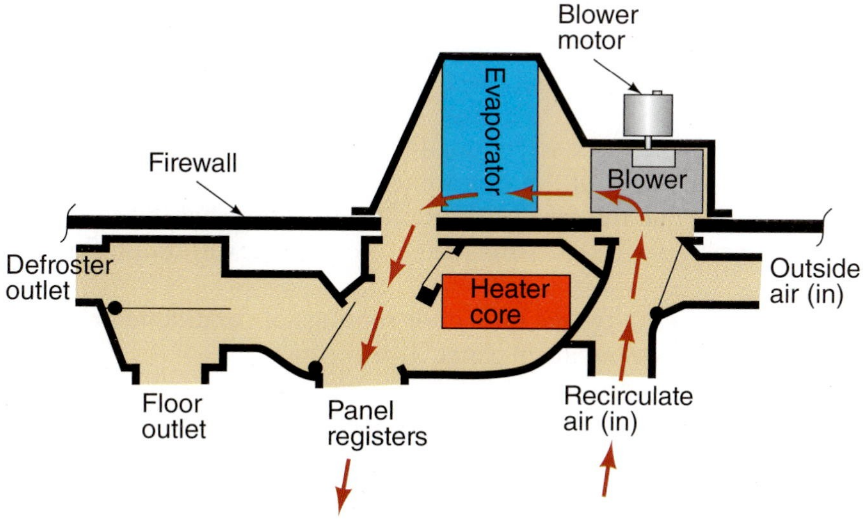

FIGURE 10-9 **All recirculated air through the evaporator and out the panel registers (outlets).**

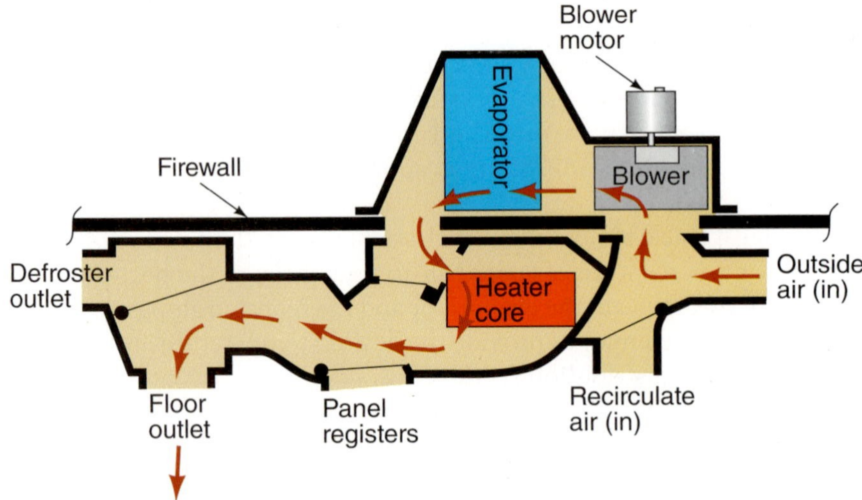

FIGURE 10-10 **All fresh air flows through the heater core and out the floor outlets. Though air flows through the evaporator, the compressor is not running and there is no cooling effect.**

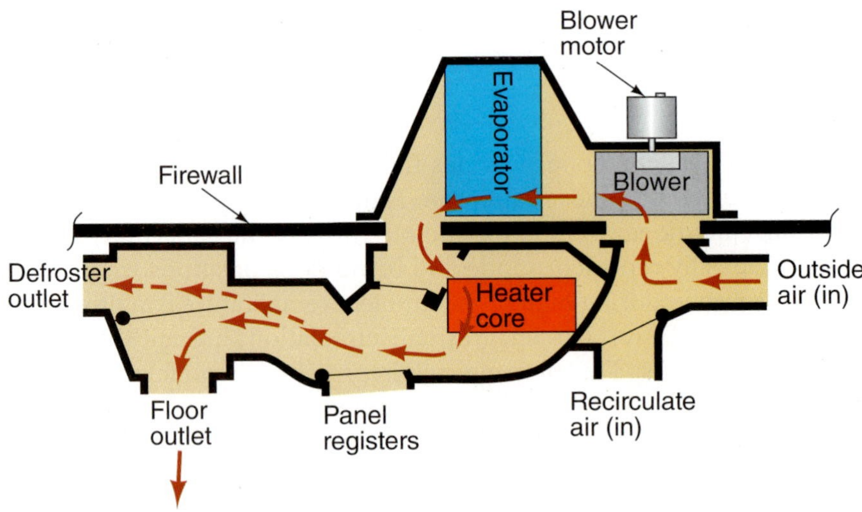

FIGURE 10-11 **The same condition as illustrated in Figure 10-10, but with some air diverted to the defroster outlet.**

Shop Manual
Chapter 10, page 412

Bi-level is a condition whereby air is delivered to two levels in the vehicle, generally the floor and dash outlets.

In MAX cooling, the heater coolant control valve is closed if the system is equipped with one.

Air Delivery

In addition to OFF, there are six selections for the condition of the air to be provided to the passenger compartment of the car. Some may require recirculated air, and others may require fresh air. While the select conditions may differ slightly from one car model to another, they typically are MAX, panel, panel/floor (**bi-level**), floor, floor/defrost, and defrost.

Following are some of the typical duct door routings of conditioned air for the various selections available at the driver control panel.

MAX

In MAX (maximum) cooling (Figure 10-12), the compressor is running and the outside/recirculate air door is closed to ambient air. Flow is from in-car air, through the evaporator, and out through the panel registers. Bi-level, which will provide some air to the floor outlet (Figure 10-13), may be selected.

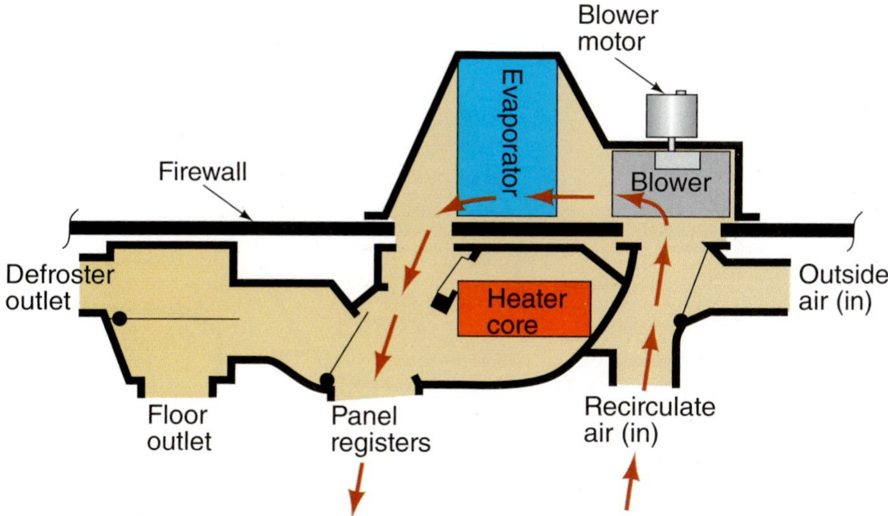

FIGURE 10-12 **In MAX cooling, airflow is from in-vehicle, through the evaporator, and out the panel outlets.**

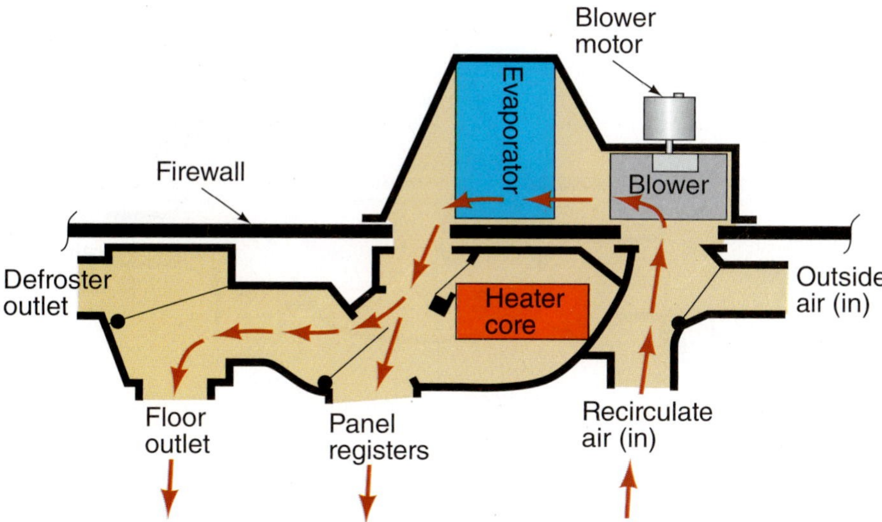

FIGURE 10-13 **MAX cooling with BI-LEVEL selected.**

If MAX heating (Figure 10-14) is selected, the compressor is not running and the heater coolant valve is open, if the system is equipped with one. Airflow is from in-car, through the evaporator and heater core, and out the floor outlet. If bi-level is selected (Figure 10-15), some air is directed to the panel registers. A small amount of air in either condition is directed to the windshield to prevent fogging.

Panel (Norm)

If normal (panel) cooling is selected, the air-conditioner compressor is running. Airflow is from outside ambient, through the evaporator, and out the panel registers (Figure 10-16). For humidity control, some air may be directed through the heater core (Figure 10-17) as well.

The "panel" registers are those visible on the dash assembly.

If normal (panel) heating is selected, the air-conditioning compressor is not running and the coolant control valve is open, if the system is equipped with one. Airflow is from the outside ambient, through the heater core, and out the floor outlets (Figure 10-18).

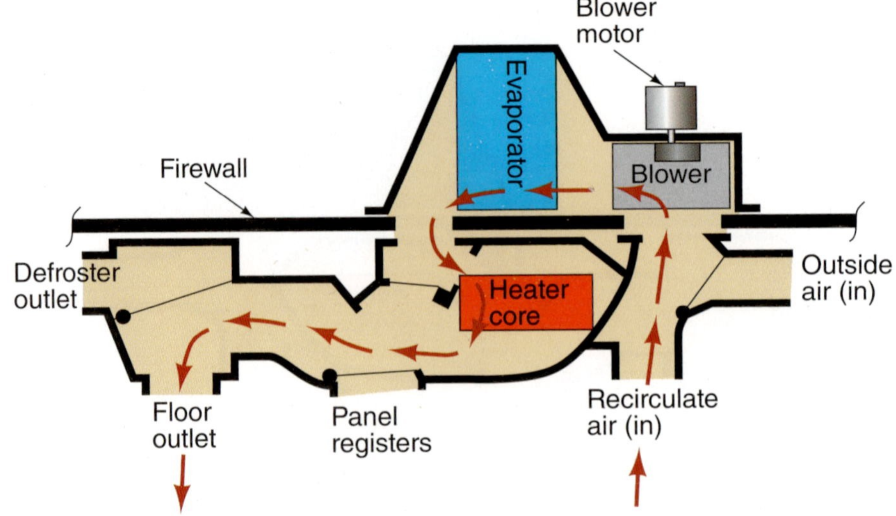

FIGURE 10-14 **Airflow when MAX heating is selected.**

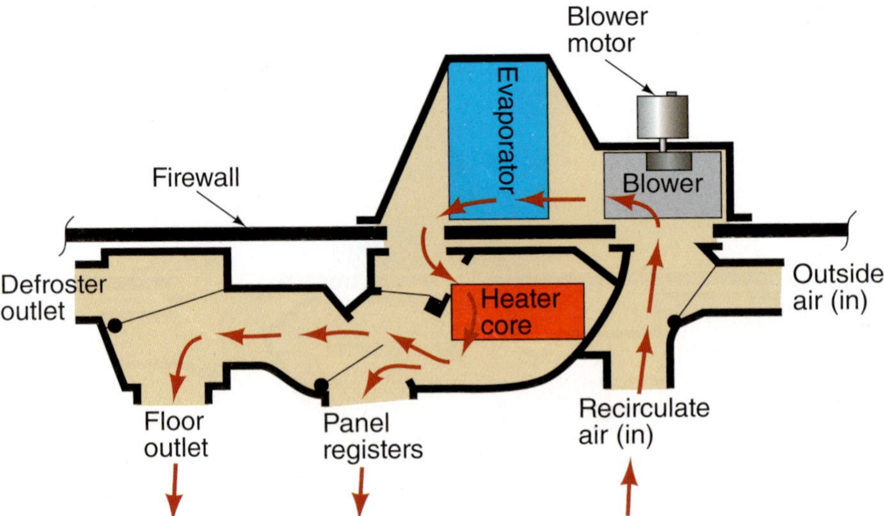

FIGURE 10-15 **MAX heating with BI-LEVEL selected.**

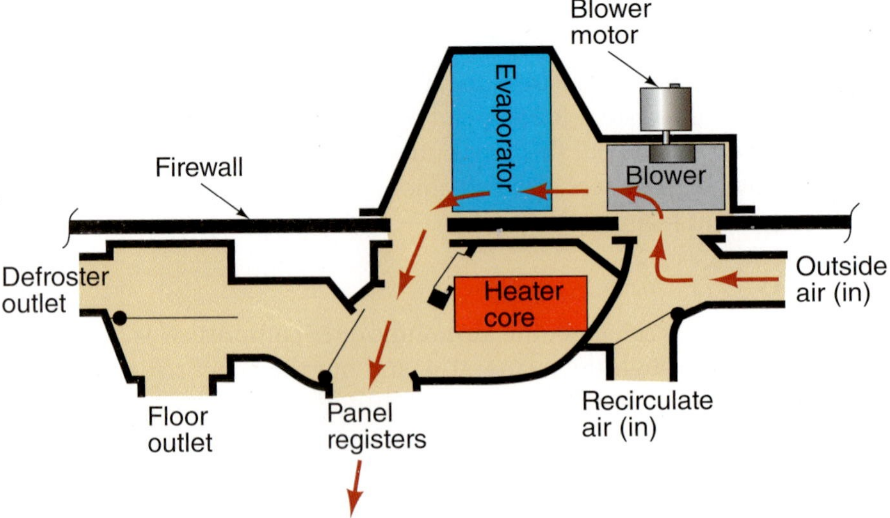

FIGURE 10-16 **Normal cooling (air conditioning) is selected.**

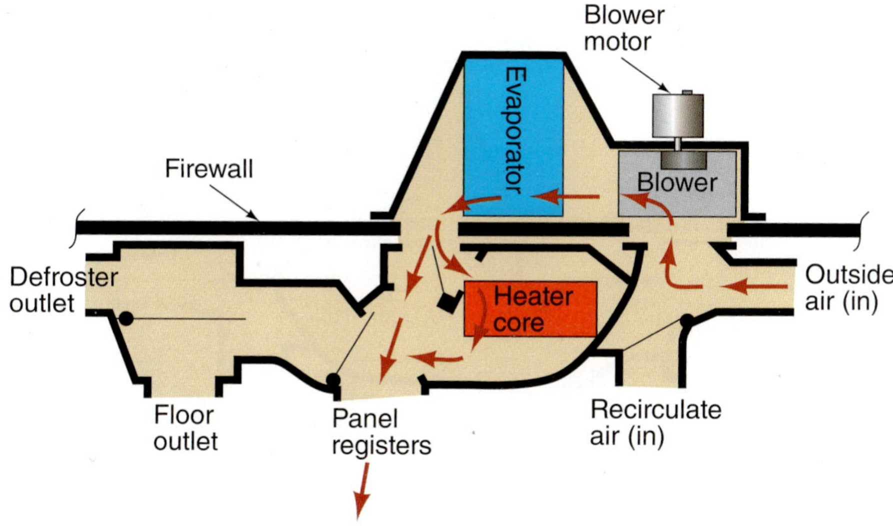

FIGURE 10-17 **Airflow when humidity control is required with normal cooling.**

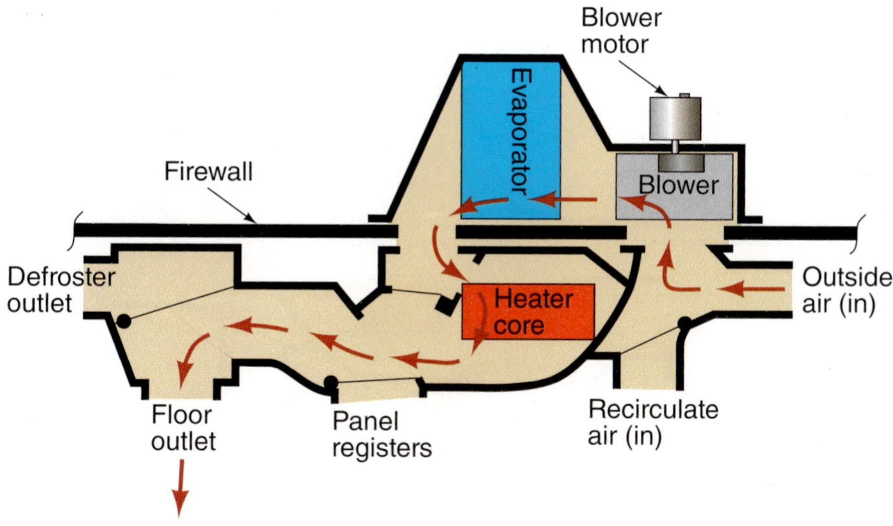

FIGURE 10-18 **Airflow when normal heating is selected.**

Floor outlets are not generally visible from the seated position.

In either condition, cooling or heating, a small amount of conditioned air is directed to the defrost outlets as an aid to prevent windshield fogging (Figure 10-19).

ATC is a recognized acronym for automatic temperature control.

Panel/Floor (Bi-level)

Bi-level air in the cooling mode can be selected (Figure 10-20) to provide some conditioned air to the floor outlet. Bi-level air may also be selected in the heating mode to provide some air through the panel registers (Figure 10-21).

The bi-level setting simply means that conditioned air may be provided at two outlets, panel and floor, as desired by some vehicle occupants. In some systems, this is referred to as HI-LO or PNL/FLR. This condition is similar in operation to **MIX**.

MIX is a term used to describe the bi-level or HI/LO mode position.

Vent

The **vent** brings in unconditioned, ambient air when neither heating nor cooling is desired. The compressor is not running, and the heater coolant valve is not open if the system is equipped with one. Air passage is from ambient air through the heater or evaporator core to

To **vent** is to introduce fresh outside air into the vehicle.

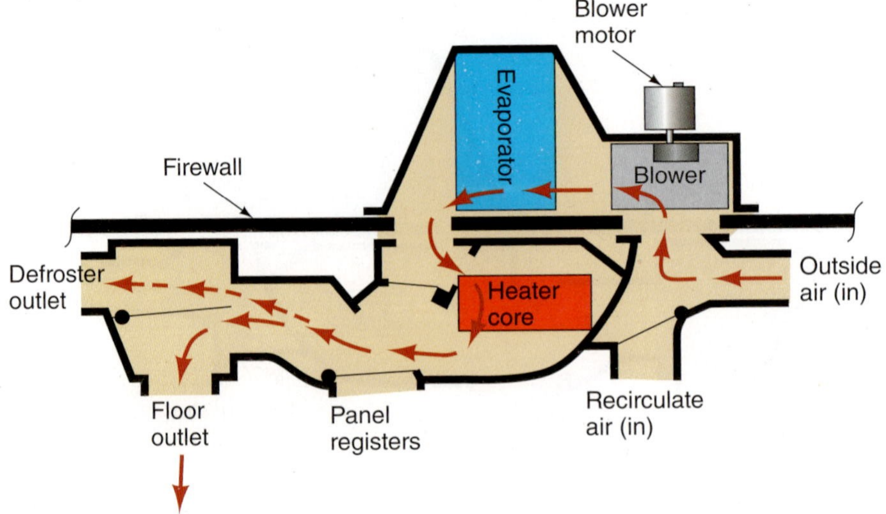

FIGURE 10-19 **Some conditioned air is directed to the defroster outlets to prevent windshield fogging.**

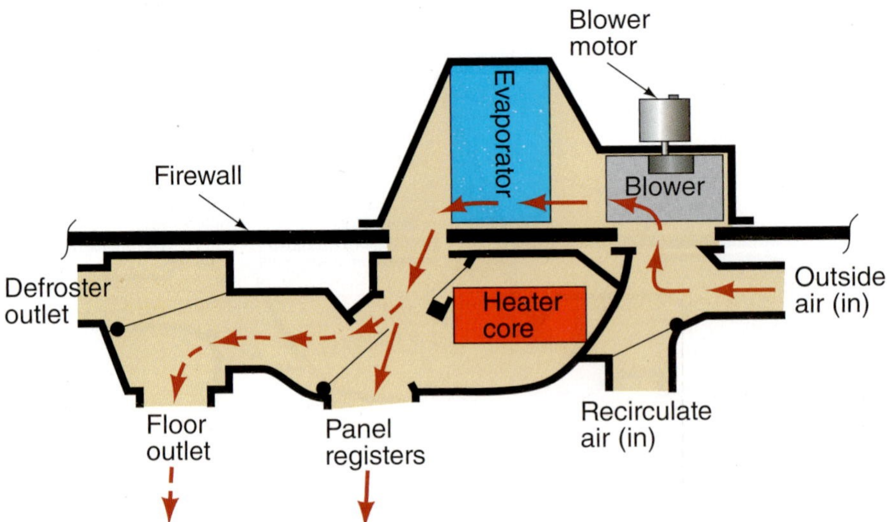

FIGURE 10-20 **Airflow in the cooling mode when BI-LEVEL is selected.**

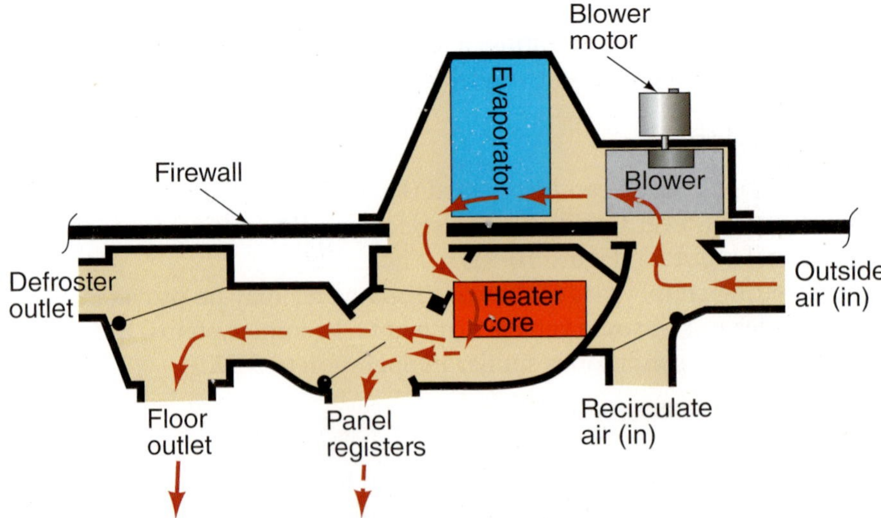

FIGURE 10-21 **Airflow in the heating mode when BI-LEVEL is selected.**

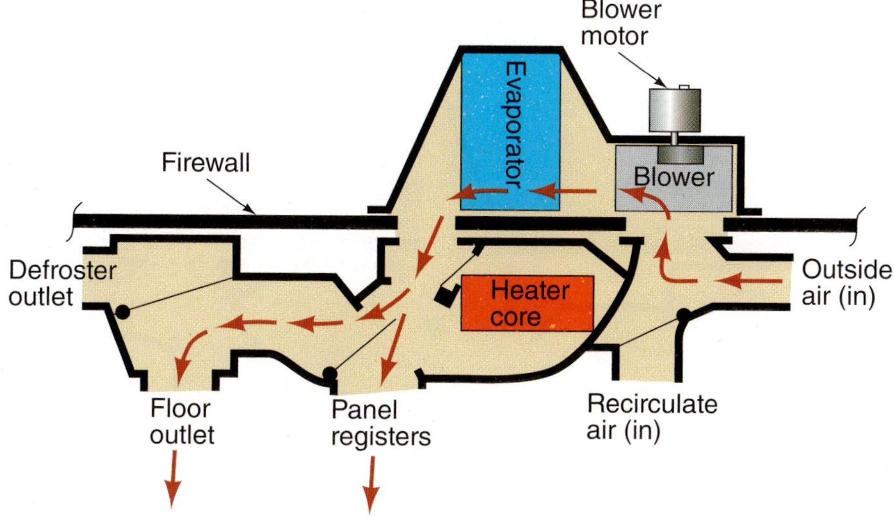

FIGURE 10-22 **Airflow when VENT is selected in the bi-level condition.**

the selected outlets—floor outlets and panel registers. Figure 10-22 shows the vent setting selected with bi-level air delivery.

Shop Manual
Chapter 10, page 413

Heat/Cool

A temperature control is generally provided to select in-car temperature. There are two methods: manual/semiautomatic and automatic (Figure 10-23). Temperature and mode are manually selected in the manual/semiautomatic temperature control (**SATC**) system.

SATC stands for semiautomatic temperature control.

In the automatic temperature control (**ATC**) system, the selected temperature and mode are a fully automatic function of a digital microprocessor. The microprocessor compares data

ATC stands for automatic temperature control.

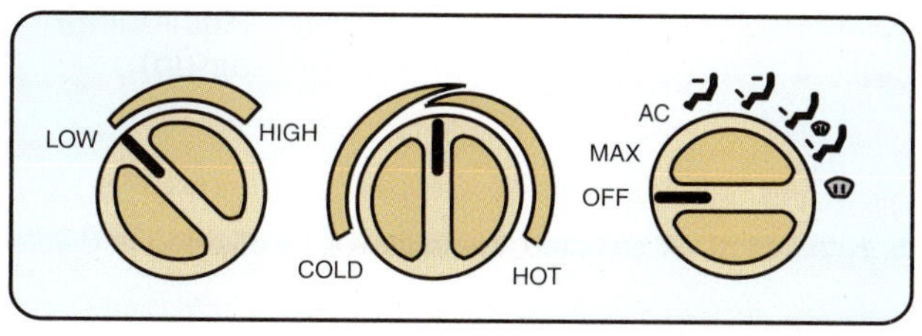

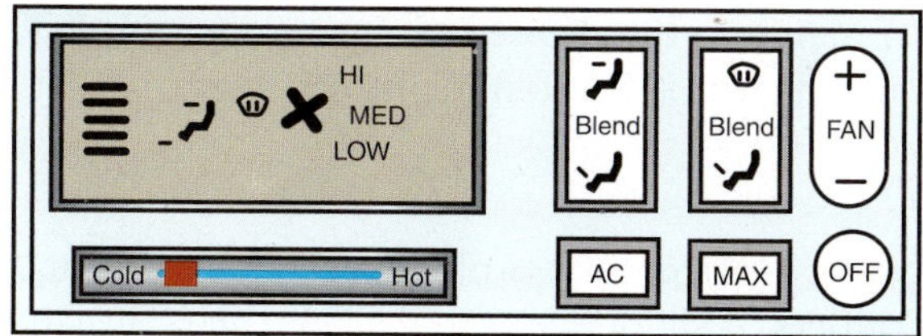

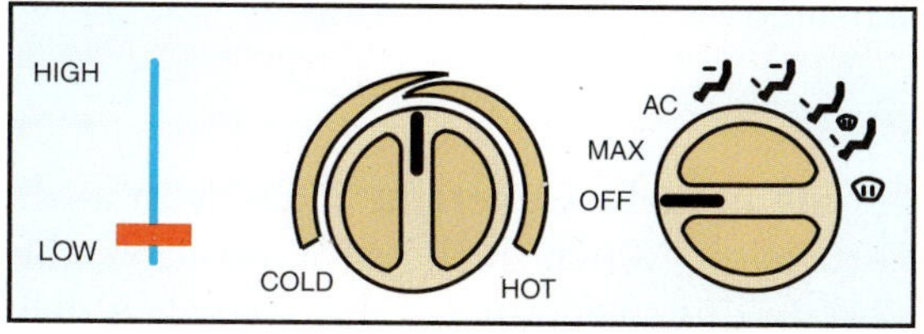

FIGURE 10-23 **Typical air-conditioner/heater controls.**

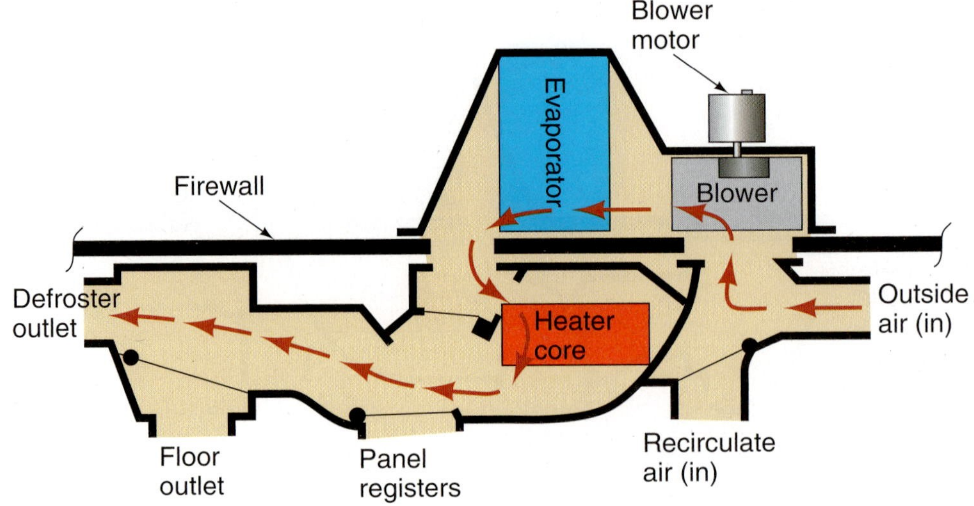

FIGURE 10-24 **Airflow when DEFROST is selected.**

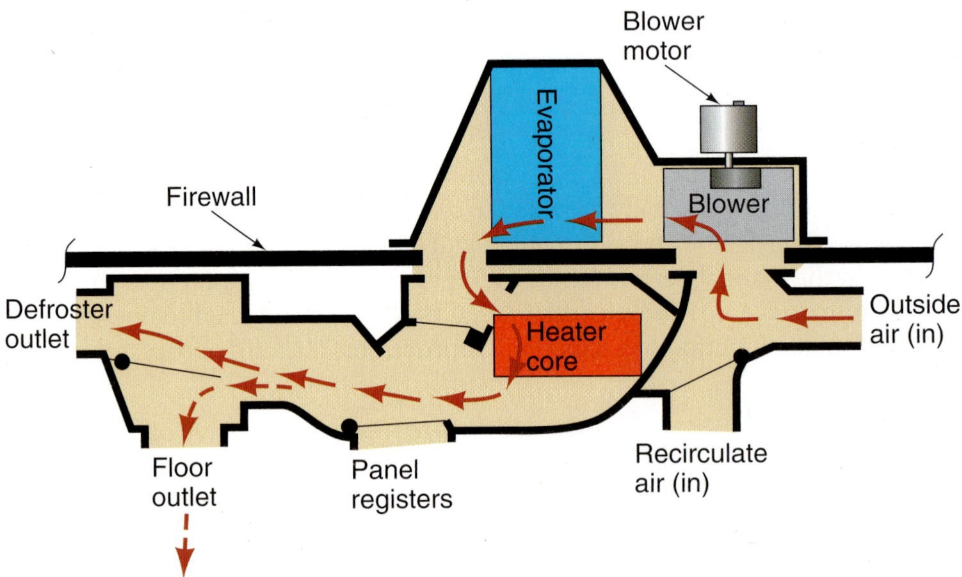

FIGURE 10-25 **With DEFROST selected, some air will be diverted to the floor outlets.**

from designated sensors to maintain the desired in-car temperature. These sensors will be discussed in greater detail in Chapter 11, System Controls.

Defrost

In the defrost position (Figure 10-24), outside ambient air passes through the heater core and is directed to the defroster outlets. A slight amount of heated air is directed to the floor outlets (Figure 10-25). If the outside ambient air temperature is above 50°F (10°C), the compressor may operate to temper the heated air for humidity control.

MIX

MIX may be selected on some models. Generally, those with a MIX select do not have a bi-level (HI-LO) select provision. When in the MIX position, the floor/defroster door opens halfway (Figure 10-26). In this selection, conditioned air is delivered to the floor and defrost outlets. In-car temperature is controlled by adjusting the temperature control lever.

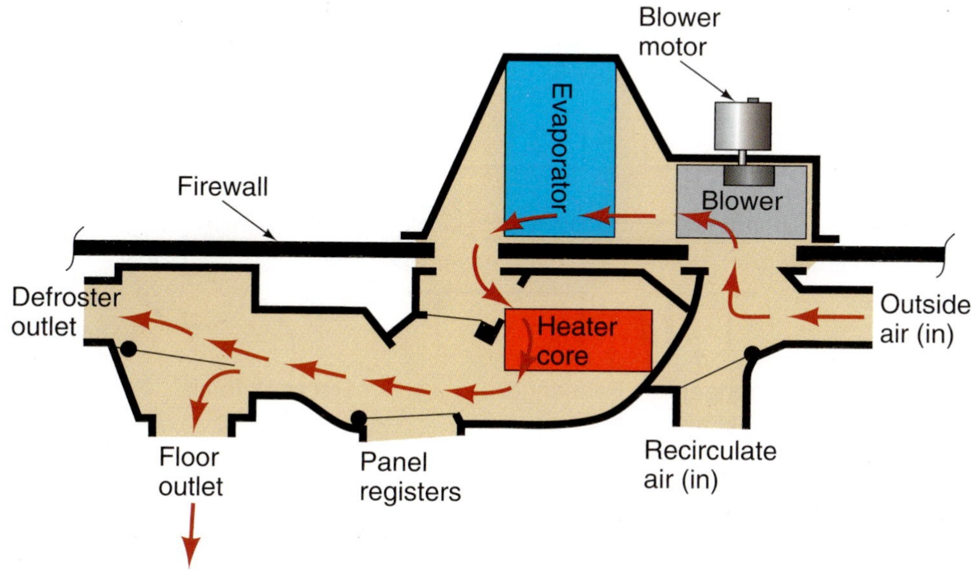

FIGURE 10-26 **Airflow when MIX or BI-LEVEL is selected.**

Other settings for MIX are HI/LO and BI-LEVEL.

The compressor will operate if in-car temperature conditions warrant. The compressor will also operate if outside ambient air temperature is above 50°F (10°C) as an aid for in-car humidity control.

Dual-Zone Duct System

The dual-zone duct system found in some cars has a separate driver- and passenger-side duct system (Figure 10-27). Both sides have a defrost/air-conditioning outlet door and a heater/floor outlet door that operate together. The passenger has control of the passenger-side temperature door only.

The dual-zone duct air distribution system can be controlled by either a manual or automatic climate control system and has a separate temperature control for the passenger. The passenger may adjust the temperature of the air at the outlet vents on the passenger side in an automatic climate control system only within the limits set by the driver—generally up to 30°F (16.7°C) cooler or warmer than that selected by the driver. Passenger control on a manual climate control system is not usually restricted by the driver's temperature selection.

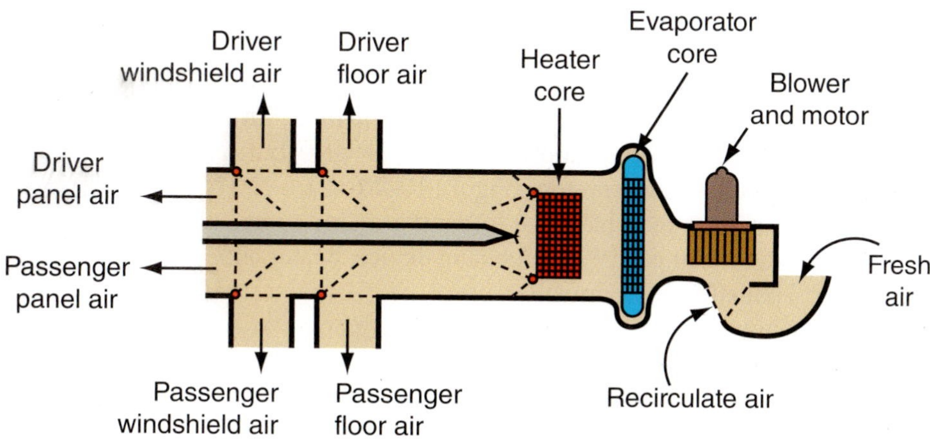

FIGURE 10-27 **A typical dual-duct system.**

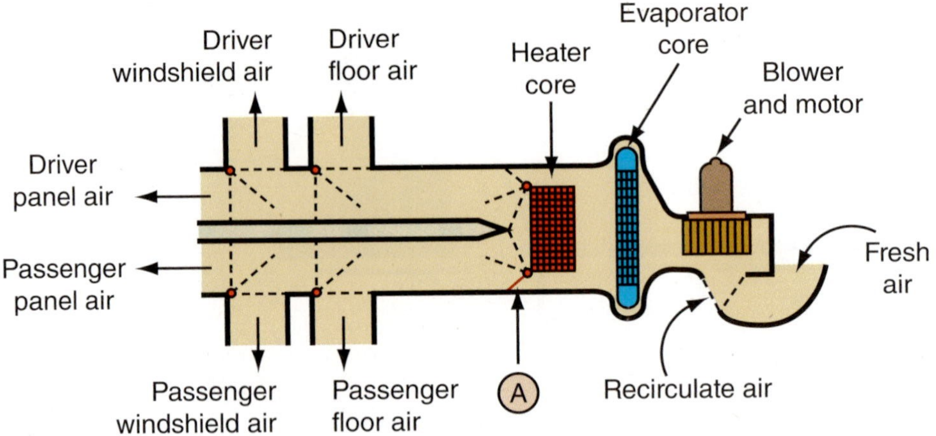

FIGURE 10-28 **A typical dual-zone duct system with passenger-side full hot selected (A).**

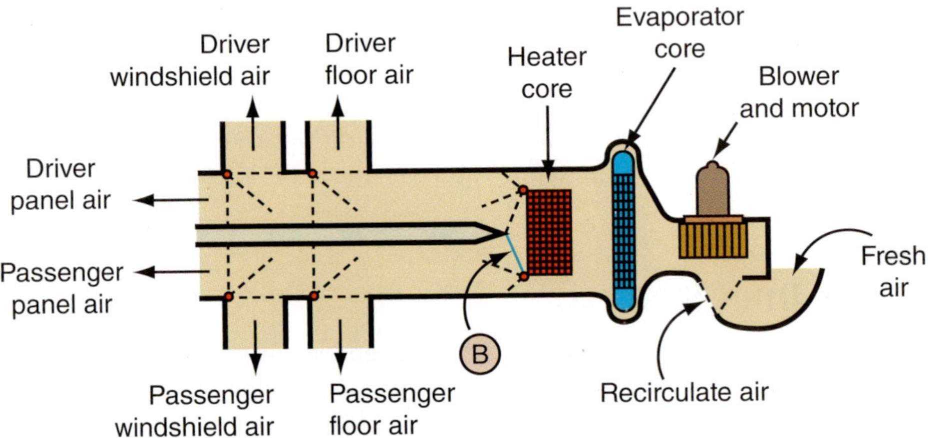

FIGURE 10-29 **A typical dual-zone duct system with passenger-side full cold selected (B).**

The passenger manually controls the position of the passenger-side temperature air door, controlling the discharge temperature of the passenger-side air outlets between full hot (Figure 10-28) and full cold (Figure 10-29). The actual passenger-side temperature depends on the general operation of the system. The passenger controls do not engage or disengage the compressor, change blower speed, or reposition the passenger mode door.

Rear Heat/Cool System

Some trucks and vans may be equipped with a rear air distribution system to provide rear heating, cooling, or a combination of both. The rear air distribution system is often referred to as an auxiliary air-conditioning system (Figure 10-30).

Depending on design, it may have the following major components: blower and motor, temperature door, evaporator core with metering device, heater core with flow control, outlet mode door, control panel(s), and controller.

The rear auxiliary system that provides only heating or cooling does not require an outlet or temperature mode door. The heat-only control panel has a blower speed control accessible to the rear-seat passengers. The rear blower master control for the cooling-only system is generally in the front control panel. The switch in the REAR position permits the rear blower switch to select the speed of the rear blower.

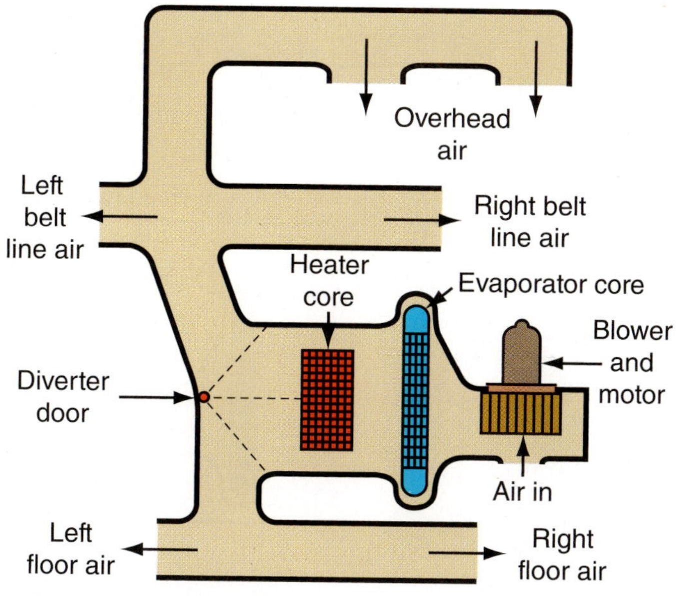

FIGURE 10-30 **A typical rear (auxillary) heat-cool system duct.**

Systems that provide both heating and cooling have an outlet mode door to direct outlet air to the upper or lower vents (Figures 10-31 and 10-32). Some systems may have a temperature door controlling outlet air temperature, while others are controlled by the front master control. The heat/cool system generally has both front and rear control panels for controlling the rear air distribution system. The control panels allow selection of the blower speed, the mode door, and in some systems the temperature door position.

Because the rear system is connected in parallel to the front system, the rear controls cannot override the master controls, such as to energize or de-energize the compressor or heater control valve. The rear system can only provide cooling or heating when cooling or heating is selected in the front system.

It should be noted that in some minivans rear heat/cool temperature control is determined by the driver's temperature control setting and only offers heating or cooling to the rear compartment. This is considered a two-zone system (front driver/passenger separate zones), with

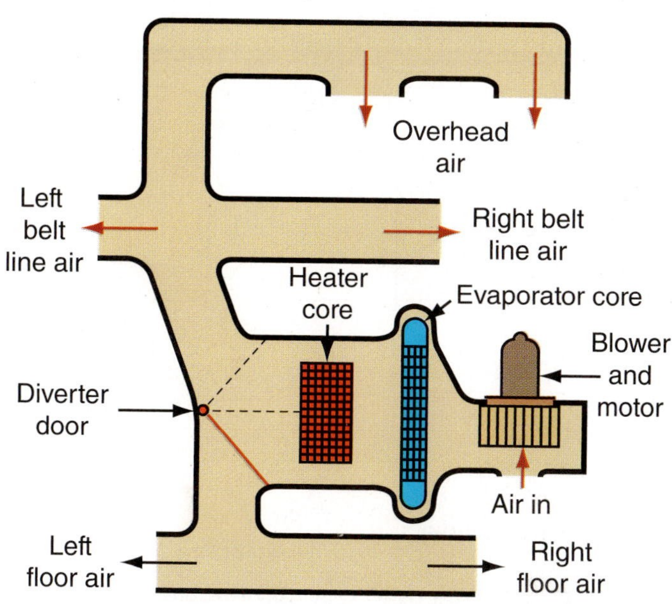

FIGURE 10-31 **Air diverted to upper vents.**

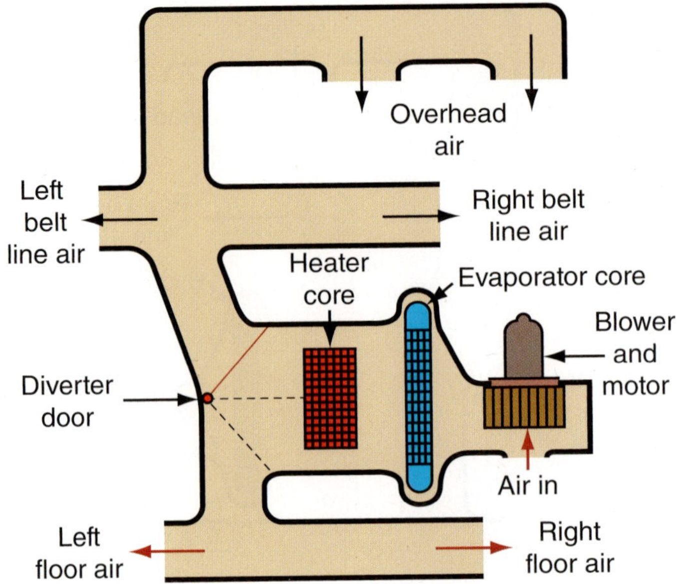

FIGURE 10-32 **Air diverted to lower (floor) vents.**

the rear compartment only able to control the fan speed in order to control the temperature. The driver must slide the temperature control past the 75 percent point toward full hot position in order for the rear compartment to be heated or past the 25 percent full cold position in order for the rear compartment to be cooled (Figure 10-33). If the midpoint on the driver's temperature control is selected, the rear compartment will receive full heating or full cooling; depending on the last position selected by the driver, no temperature blending is available to the rear compartment. True three-zone systems have a rear HVAC duct system with full blend features, as depicted in Figures 10-30, 10-31, and 10-32.

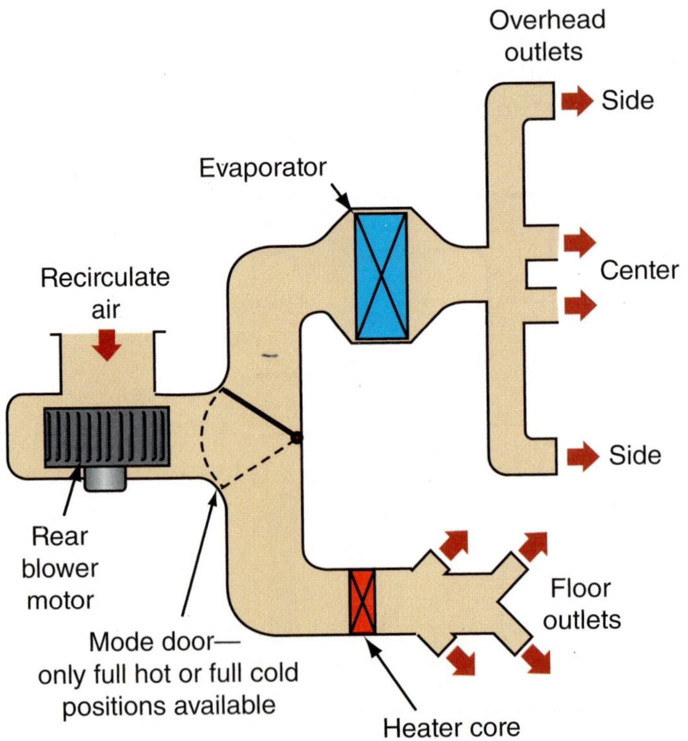

FIGURE 10-33 **Typical rear HVAC system on two-zone system with driver in control of temperature control.**

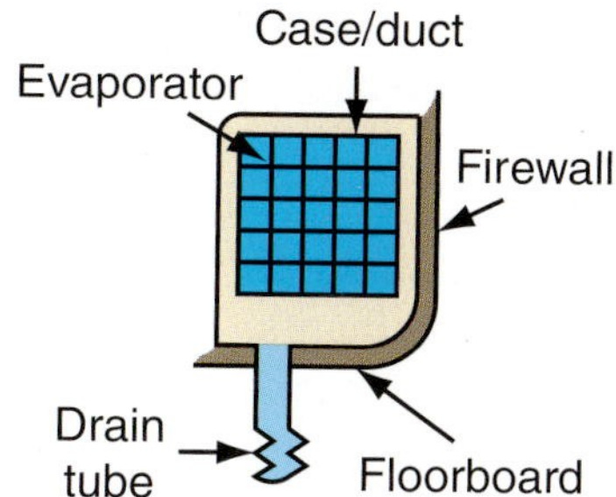

FIGURE 10-34 **An evaporator drain extends through the floorboard of a vehicle.**

FIGURE 10-35 **An evaporator drain hole hose extending from the bottom of a vehicle.**

Evaporator Drain

Moisture extracted from the air in the evaporator is readily expelled from the evaporator case through a drain tube that extends through the floorboard of the vehicle. To prevent insects from entering the evaporator through this tube, it has a molded, accordion shape (Figure 10-34). The weight of the moisture, as it collects, overcomes the rigidness of the tube closure and allows the water to pass.

Over time, however, airborne debris and dust may restrict the tube to the point that causes water to back up inside the case. This is sometimes evidenced by water droplets exiting the dash outlets with the air or dripping onto the floor mats in the passenger compartment. There have even been reports of drivers getting a cold, wet right foot when making a hard right turn.

If this becomes a problem, it is necessary to clean out the drain tube to ensure that it will open to allow water to pass. This is best done from under the vehicle (Figure 10-35). For obvious reasons, do not stand immediately under the tube while cleaning it.

Odor Control

During the normal operation of the passenger comfort heating/ventilation/air-conditioning (**HVAC**) system, moisture can accumulate in the ductwork and collect on the evaporator core surface. Under normal operation, most of this moisture will drain out of the case drain, causing no problems. Air that enters the passenger compartment contains microscopic contaminants and bacteria that will stick to the moisture on the evaporator and case. During periods of high humidity or when the recirculation mode is used extensively, this becomes an ideal

HVAC stands for heating/ventilation/air conditioning and is generally used when referring to the system as a whole.

Shop Manual
Chapter 10, page 407

environment for odor-causing mold, bacteria, and mildew to grow. When these environmental conditions arise, a musty odor develops and becomes very pronounced when the HVAC system is first turned on.

The industry has developed commercially available treatments to combat this problem by chemically coating the evaporator core and duct with an antimicrobial deodorizer and disinfectant product to eradicate these contaminants. These products generally offer protection for three cooling seasons or more under normal air-conditioning use. They are applied using a siphon-type sprayer (Figure 10-36).

In addition to chemically treating the HVAC system, the case drain vent must be checked to be sure that it is not plugged and will properly drain the system of excess moisture. Some manufacturers have also integrated a feature into their HVAC systems that leaves the blower motor on for a brief period of time after the air conditioner is turned off to allow the evaporator core and ducts to dry even after the vehicle has been turned off. There are also kits

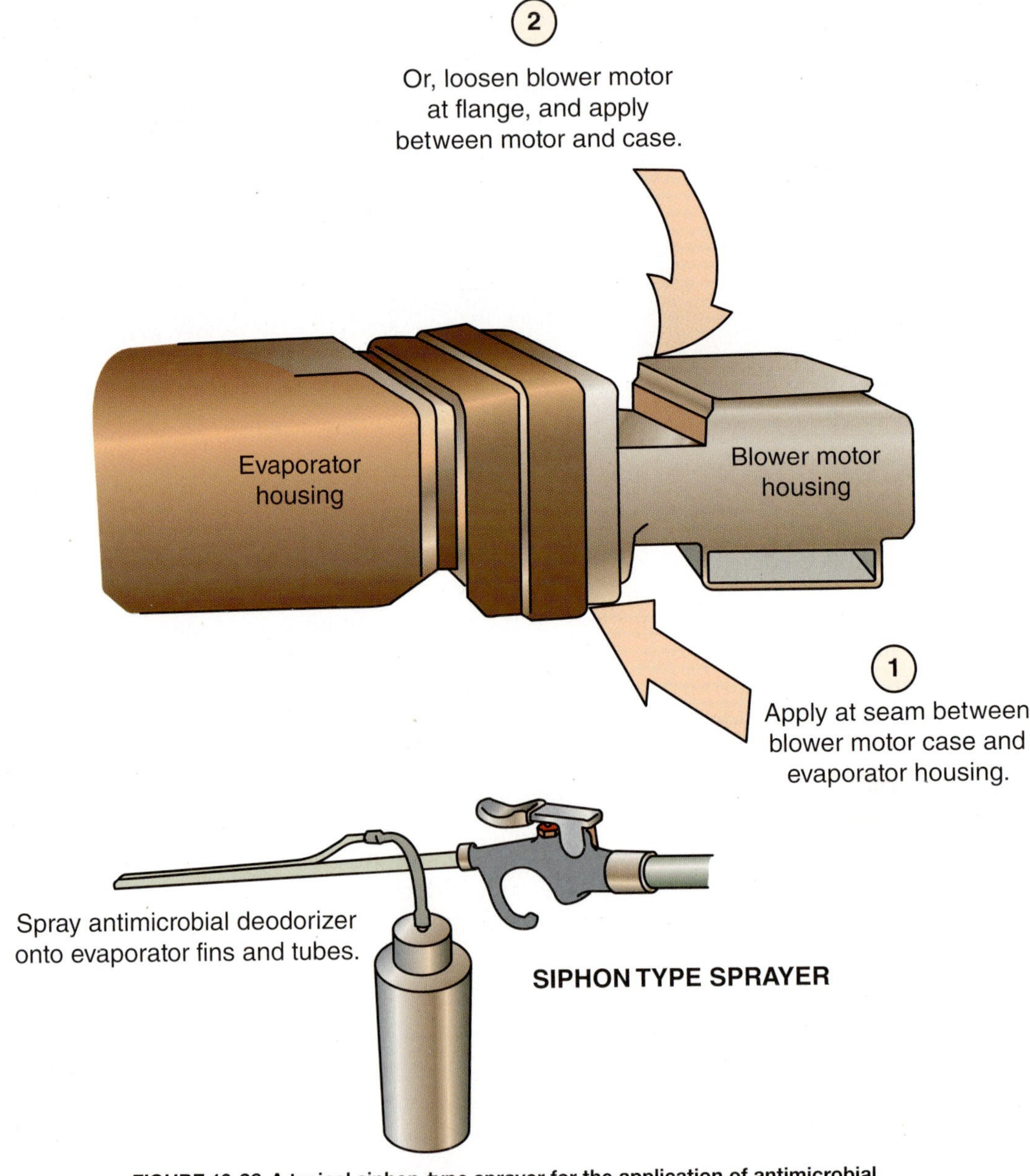

FIGURE 10-36 **A typical siphon-type sprayer for the application of antimicrobial deodorizer to eliminate musty odors.**

available to add this delay timer feature to other vehicles. Some manufacturers have sent out service bulletins recommending the installation of a blower delay feature on troublesome HVAC systems and have developed specific kits for specific models. General Motors calls their system Electronic Evaporator Dryer (EED), and it is not polarity sensitive, which means it may be installed on any GM vehicle without the need for a special harness. The EED pulses the blower motor in 10-second bursts, and it also incorporates an ambient temperature sensor that automatically disables the EED if the ambient temperature drops below 60°F (15.56°C) due to low levels of microbial growth at low temperatures.

In addition, many makes and models of vehicles today offer cabin air filters to trap these microscopic contaminants and bacteria before they enter the HVAC system. The next section will discuss this topic in more detail.

Cabin Air Filters

The cabin air filter is another feature of many vehicles today. It is a passenger compartment filter medium installed in the duct system to filter out pollen and dust particles, which would otherwise enter the interior of the vehicle. The introduction of the air filters in the automotive air-conditioning system of domestic vehicles has been slow. The first occurrence was found in the 1938 Nash; the next occurrence was not until more than 35 years later in Oldsmobile's 1974 Toronado and 88. They were first introduced across major car lines in European vehicles during the mid-1980s but have since become a feature on many vehicles produced both in the United States and abroad. It is expected that the popularity of cabin filters will grow, and some estimates indicate that 85 percent of the cars and light trucks sold in the United States contain one or more cabin filters.

As a vehicle travels down the road or is sitting in traffic, the air outside the vehicle (which may contain high levels of dust and pollen as well as other impurities) is drawn in through the fresh air intake system even when the blower motor is not on. The cabin filter is placed in the fresh air intake ductwork (Figure 10-37) and is designed to reduce pollens, bacteria, dust, exhaust gases, and mold spores, as well as other tiny airborne allergens that may enter a vehicle's ventilation system. Mold spores are the main contributor to the musty, stale smell that may be emitted from the ventilation system.

Shop Manual
Chapter 10, page 414

Most cabin air filters have the ability to remove up to 95 percent of all particles that are 3 microns or larger. There are two main filter designs used today: the particle filter and the absorption filter. The particle filter is designed to remove solid particles larger than 3 microns (less than one millionth of an inch in diameter) such as dust, pollen, soot, and mold spores. The filter element is made of a special paper or nonwoven microfiber fleece. The filter material may also be electrostatically charged to improve its efficiency. The absorption filter is designed to remove odor-causing particles and gases. It uses activated charcoal to remove these particles. Activated charcoal is a carbon substance that has been treated with oxygen to open up millions of tiny pores, which increases its surface area. As the impurities pass by the

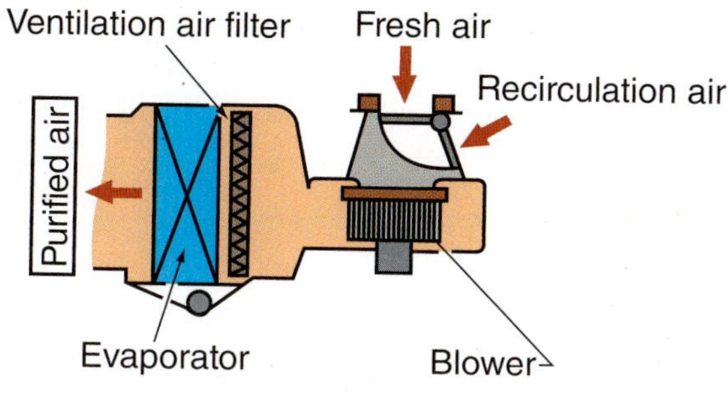

FIGURE 10-37 **Many vehicles have a cabin air filter.**

activated charcoal's surface, they are attached to the charcoal surface and trapped, much like a magnet attracts and holds metal particles. Because activated charcoal attracts and holds impurities, it will eventually become saturated and require replacement. The combination filter (or two-stage filter) combines both the particle filter and the absorption filter into one assembly.

The cabin air filter is part of routine maintenance and is generally changed every 15,000 miles (24,000 km), but you should refer to specific vehicle service schedules for manufacturers' recommendations. If the filters are not serviced regularly, they will eventually cause an airflow restriction as they become clogged. The cabin air filter is generally located at the fresh air intake under the hood (Figure 10-38) or under the dash (Figure 10-39) on the passenger side of the vehicle. Consult the manufacturer's service information for exact locations and service procedures. A clogged filter can create an air pressure drop, placing a greater demand on the blower motor and, perhaps, leading to an early failure. Because it restricts airflow, a clogged filter will also affect air-conditioning, heating, and defroster performance.

Some systems that are equipped with a cabin air filter are also equipped with an airborne pollutant (air-quality) sensor (Figure 10-40). It is located in the intake air plenum chamber and is tasked with detecting pollutants in the ambient air. The sensor signal is used by the climate control module for the automatic air recirculation mode. In this mode if the sensor detects pollutants in the fresh air intake, it will automatically switch to the air recirculation mode. These systems are used in conjunction with a combination cabin air filter that combines a particle filter (dust and pollen) and activated charcoal to cleanse the air of some chemical contamination. The sensor is designed to detect oxidizable and reducible gases such as carbon monoxide and nitrous oxides and as such is not designed to detect all odor-causing agents. Once the pollution concentrations drop below a calibration threshold for the system, the fresh air mode door is opened to once again to draw fresh air from outside of the vehicle into the passenger compartment.

The sensor determines the average pollutant concentrations and sends this information as a digital square-wave signal to the climate control module (Figure 10-41). During initial system activation phase, when automatic mode is first turned on, the sensor takes approximately 30 seconds to initialize. The climate control module uses this information to determine when and for how long it should select the recirculation mode. The control module only selects recirculation mode during peak pollution levels and only for a set period of time depending

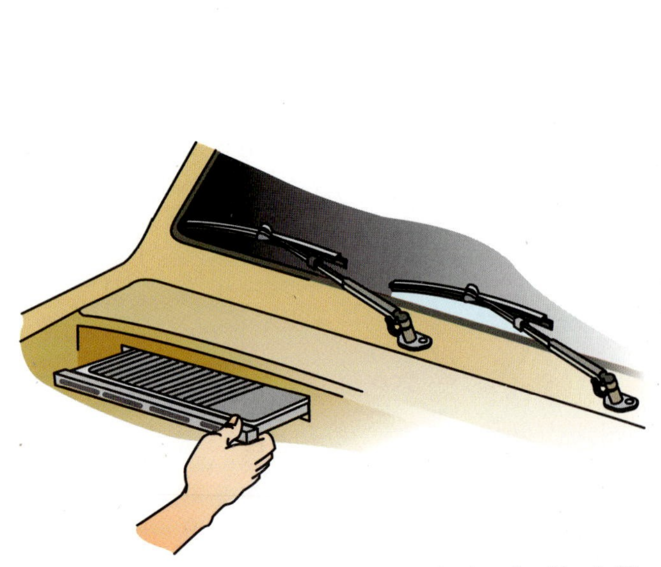

FIGURE 10-38 **Typical location of an under-hood cabin air filter.**

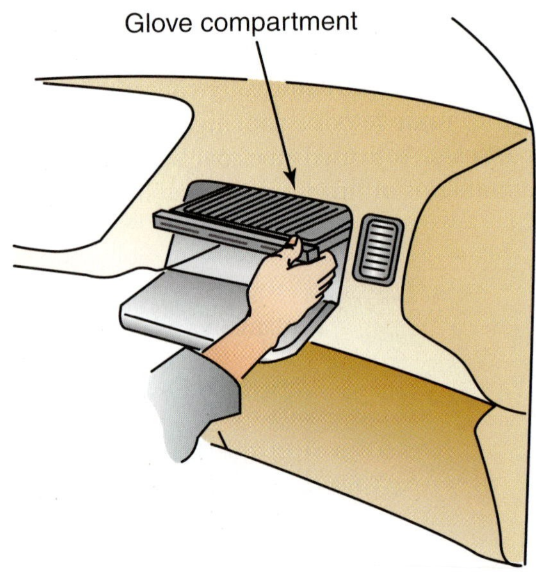

FIGURE 10-39 **Typical location of an under-dash cabin air filter.**

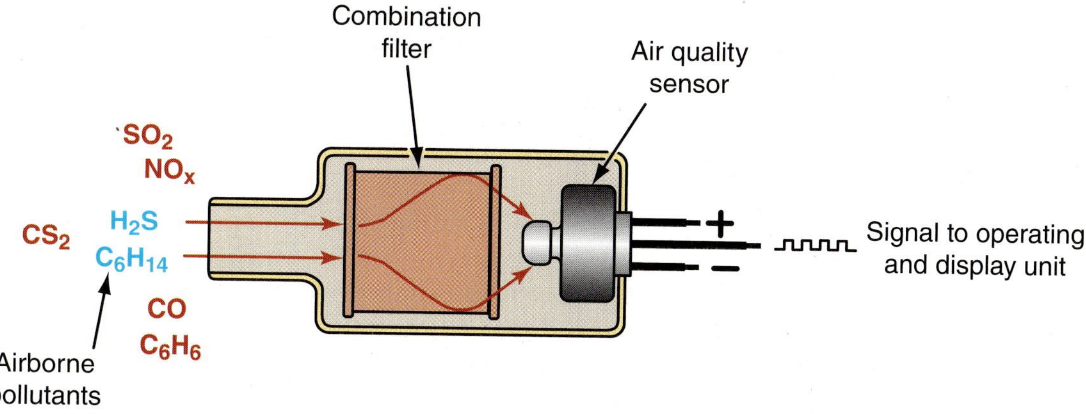

FIGURE 10-40 **Some automatic temperature control systems are equipped with and airborne pollutant sensor in the fresh air intake plenum and are used to automatically activate the recirculation mode.**

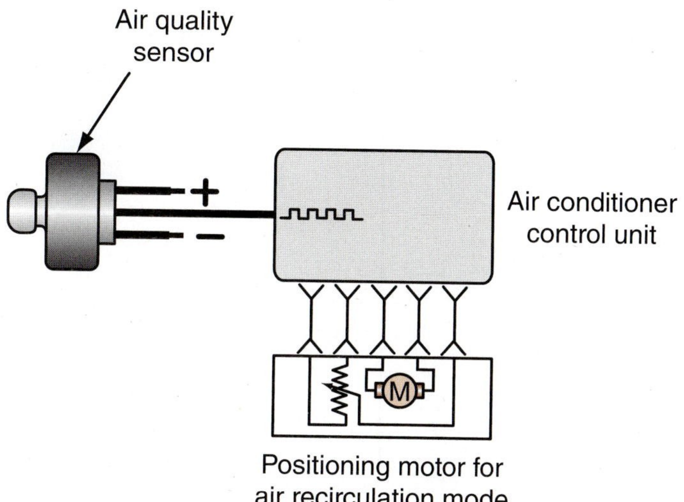

Ambient temperature	Air pollution level	Air recirculation
> +2˚C	Low rise	Yes min. 25 sec.
> +2˚C	Low	No
+2˚C - –5˚C	Higher rise	Yes
< –5˚C	Higher rise	max.15 sec.
ECON mode compressor off		max.15 sec.
Defrost mode		No
Warm-up phase of sensor appox. 30 sec.		No

FIGURE 10-41 **The airborne pollutant (air quality) sensor sends a digital square- wave signal to the climate control (air conditioning) module if airborne pollutants are detected.**

on ambient air temperature and air pollution levels. The system will not remain in the recirculation mode for extended periods of time in heavily polluted areas. This automatic function is not available in all mode ranges, such as defrost mode or during the warm-up phase of the sensor, or in manual mode. Many systems will also switch to the recirculation mode when the wash/wipe system is activated to clear the windshield.

The sensor is a mixed oxide sensor (oxide mixed with tin or tungsten) utilizing semiconductor technology with platinum and palladium as a catalyst and an operating temperature of 662°F (350°C), but with a very low power consumption of 0.5 watts. The material in the mixed oxide sensor changes its electrical properties (resistance) when it comes in contact with reducible or oxidizable gases (Figure 10-42). From an internal operational standpoint, if the internal resistance of the sensor rises, oxidizable gases are present, and if the resistance falls, reducible gases are present. Because of the nature of the chemical and physical properties of a mixed oxide sensor it is able to detect both gases simultaneously.

The system is automatically calibrating and self-learning and if the sensor fails, the automatic recirculation mode is no longer available and a trouble fault code is set.

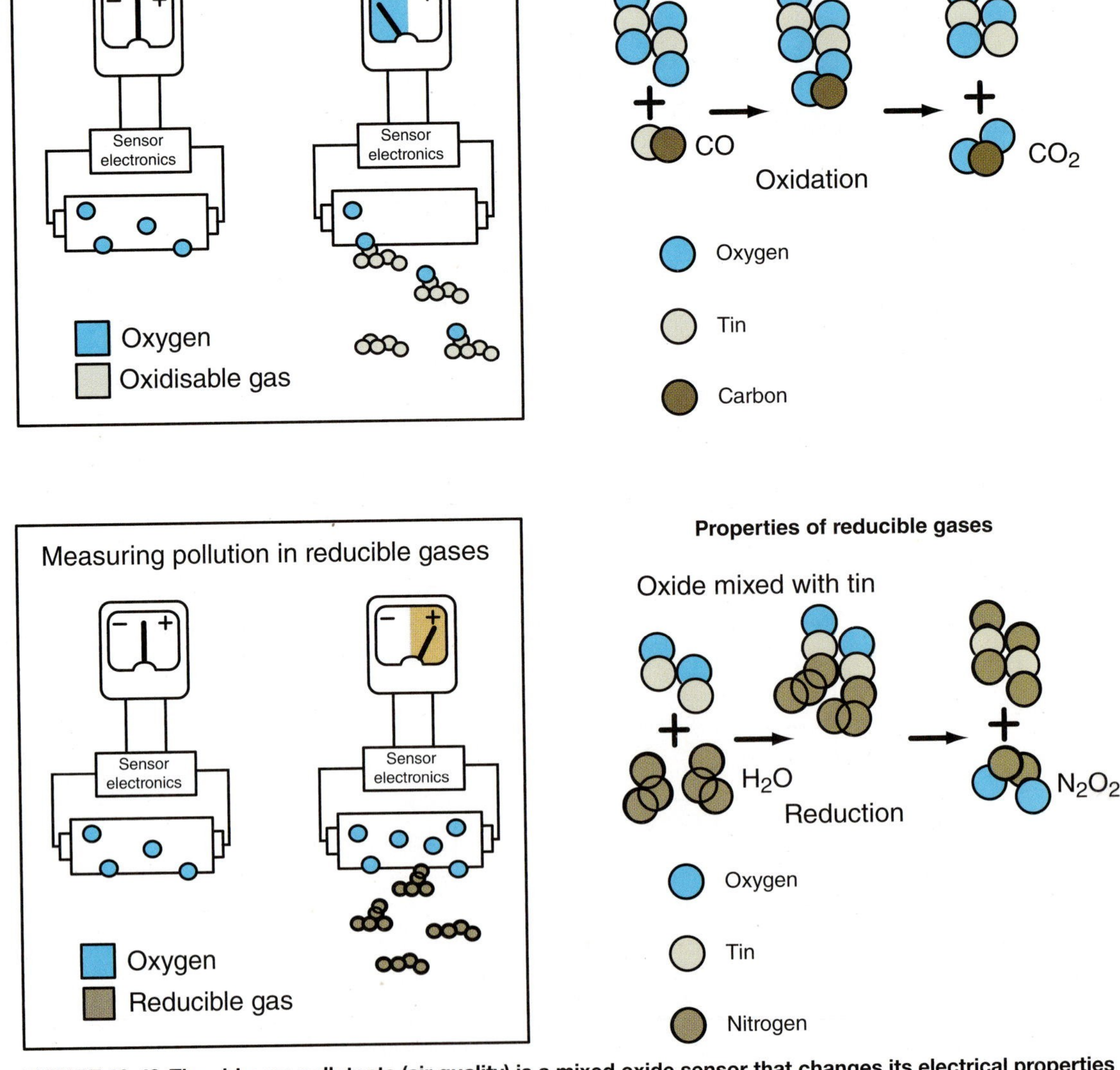

FIGURE 10-42 **The airborne pollutants (air quality) is a mixed oxide sensor that changes its electrical properties (resistance) when it comes in contact with reducible or oxisable gases.**

The Air Door Control System

The air door control system uses vacuum-operated or electrically powered motors to position the air doors, also referred to as mode doors, to provide the desired in-vehicle air delivery conditions. These generally include OFF, MAX, VENT, BI-LEVEL, HTR, BLEND, and DEF. The airflow pattern for each of these conditions is given in Chapter 8 of the Classroom Manual.

There are basically three types of control systems: vacuum, rotary vacuum, and vacuum solenoid and electric motor.

Vacuum Control

In the vacuum control system, vacuum actuators, also called vacuum motors, are used to position the air doors and valves. This system relies on a vacuum signal from either the engine manifold or an onboard vacuum pump. Vacuum is applied to the selected actuator by a rotary vacuum valve (RVV) or a vacuum solenoid in the main control panel.

Rotary Vacuum Valve

When a rotary vacuum valve system (Figure 10-43) is used to control air doors and valves, the master control head selector rotates a vacuum switch that aligns the vacuum passages in the valve to direct a vacuum signal to the appropriate vacuum actuator(s) for the mode selected.

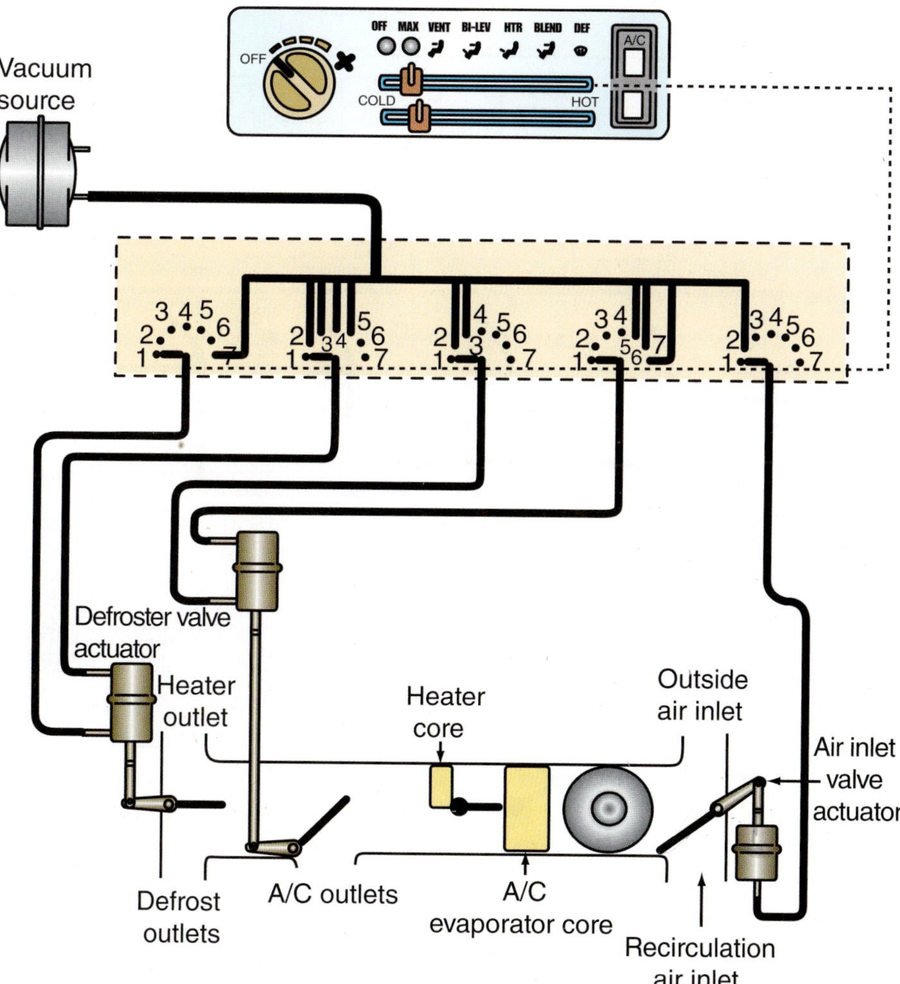

FIGURE 10-43 **Typical rotary vacuum valve system.**

Vacuum Solenoids

When vacuum solenoids (Figure 10-44) are used to control doors and valves, the master control head selector contains electrical circuits that provide a ground path for the selected vacuum solenoid. The selected (energized) solenoid allows vacuum to be applied to the selected actuator.

An automatic air distribution system often uses vacuum solenoids that are located inside the programmer to control the position of the mode doors. The programmer controls the electrical ground side of the solenoids to establish a ground path to the selected vacuum solenoid. When the ground path is removed, the vacuum actuator is allowed to vent.

Electric Actuator Motors

Many vehicles today use electric actuator motors to control air distribution mode doors and temperature blend doors. Electric actuators may be used alone or in combination with cable or vacuum control, with some doors operated by electric actuators and others controlled by cables. There are several types in use as electric mode door actuators. The two-position type either fully opens or fully closes a mode door; the fresh air/recirculation door is often this type. Another type is the variable-position actuator, which can position the mode door at any point from fully open to fully closed; temperature blend doors are typically of this design.

The manual control head (HVAC dash control assembly) on many of these systems does not directly control the actuators. The manual control head is instead wired to the body control module (BCM) or a separate HVAC control module that is connected to the actuators

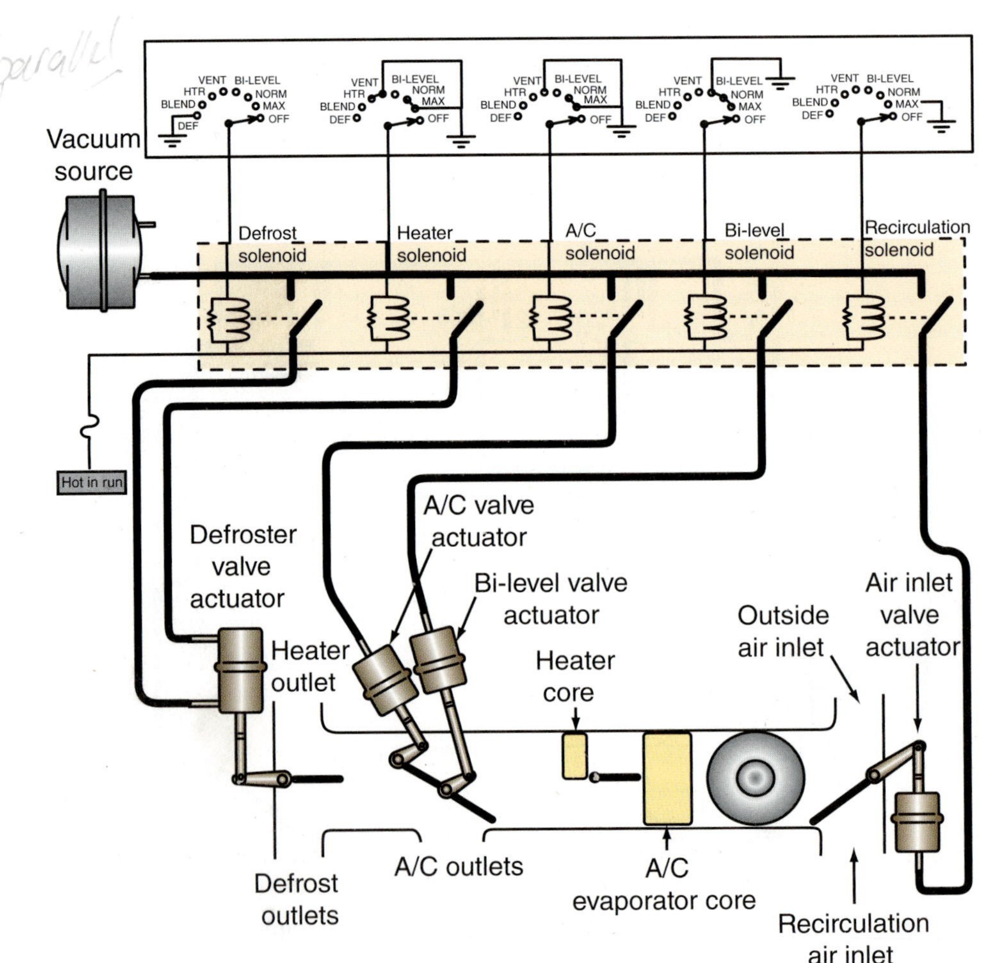

FIGURE 10-44 **A typical vacuum solenoid control.**

and controls their position in response to inputs from the control head. Many of these systems use a multiplexed switch in the control head to control air distribution mode door position, meaning they only use one wire to communicate to the control module by using a different resistance value in the switch for each function. The control module interprets this information by dropping a voltage through the circuit (Figure 10-45). Most temperature selectors use a potentiometer to command the temperature blend door.

Mode door actuators all perform the same function. They position the mode doors based on driver input to the control head assembly. Electric mode door actuators may be five-wire, three-wire, or two-wire controlled. Both the two-wire and the five-wire actuators (Figure 10-46) generally use a driver circuit in the control module to control their movement. The control module will supply 12 volts to one driver circuit and ground the other driver circuit, thus giving bidirectional control depending on the polarity of the two wires to move the motor in one direction or the other. Which driver is negative and which is positive controls the rotational direction of the motor. When both sides of the actuator motor are power or ground, the motor stops (it is electrically balanced). The control module determines door position through feedback circuits. On the two-wire actuator, the control module counts the actuator commutator pulses to determine door position. On the five-wire actuator, the control module determines mode door position through the potentiometer feedback voltage signal, which is built into the actuator assemblies. The three-wire actuator (Figure 10-47) generally has an external 12-volt supply and ground. Control of the actuator is through a logic module built in to the actuator—in essence, a smart motor. The three-wire actuators are

Shop Manual
Chapter 10, page 415

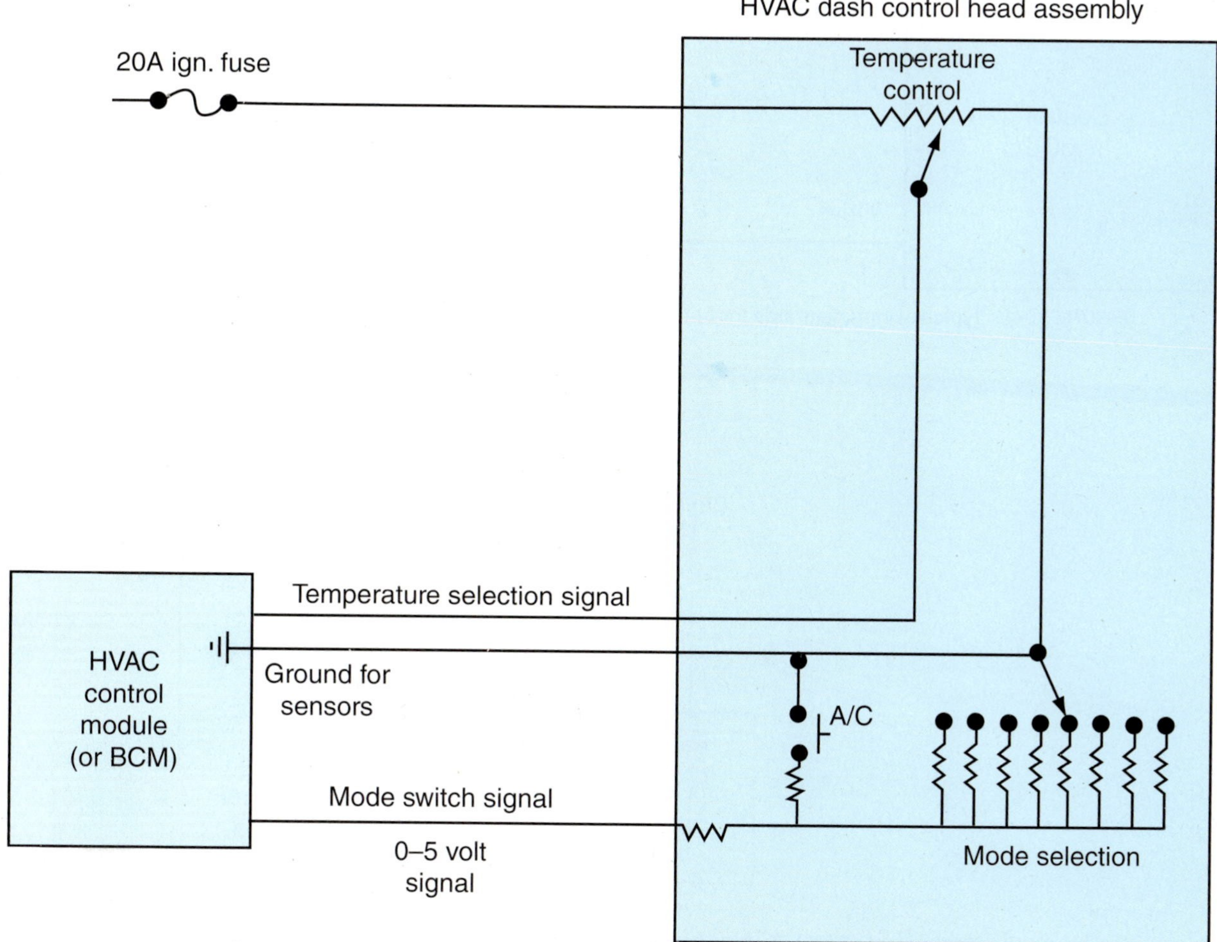

FIGURE 10-45 **A typical wiring schematic for a multiplexed HVAC control system.**

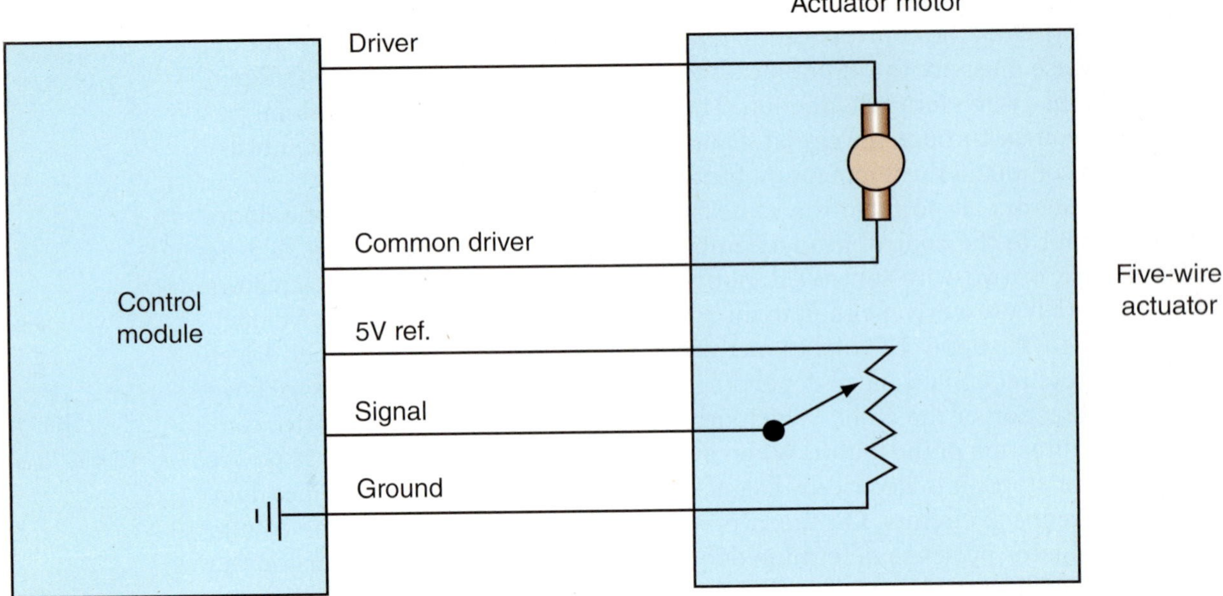

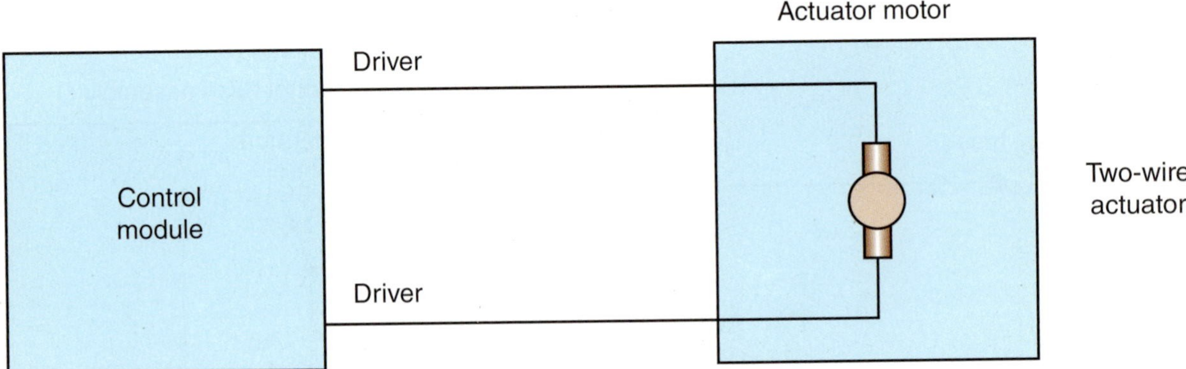

FIGURE 10-46 **Typical wiring diagrams for both five-wire and two-wire bidirectional mode door actuators.**

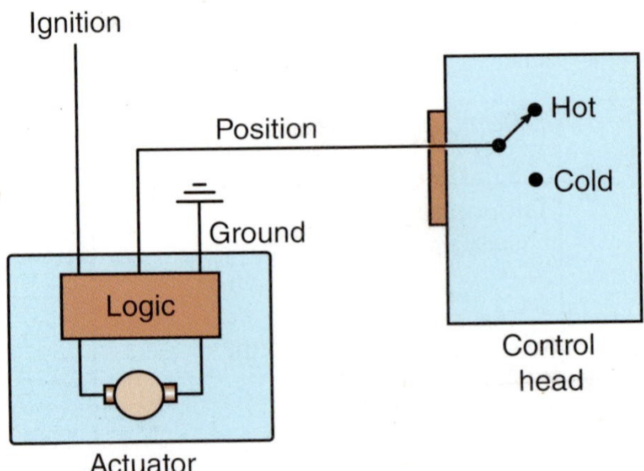

FIGURE 10-47 **Typical mode door motor schematic for a three-wire actuator motor.**

self-calibrated. Remember when checking a three-wire actuator that the 12-volt power and ground circuits are constants, and the input line from the control head is the position control (command) circuit. Electric mode door actuators are not adjustable and must be replaced if faulty. It is also necessary to initialize a calibration procedure in order for the control module to relearn mode door position if the actuator or control module is replaced or otherwise loses its memory. Consult the manufacturer's service information for exact procedures on recalibrating the systems.

Control System Faults

An inoperative vacuum motor could be due to a loss of vacuum signal at the appropriate time. This could be the fault of the vacuum source, vacuum switch, check valve, reserve tank, hose, restrictor, or vacuum motor. To determine the cause, disconnect the suspected vacuum motor (Figure 10-48) and substitute another vacuum source, such as the vacuum pump. If the motor is inoperative, it should be replaced. If it is operative, check further for the source of the problem.

If an inoperative fresh air door or mode door is the problem, a fault in the vacuum control system is again indicated. Most older systems use vacuum motors with a vacuum selector valve at the control head to control the operation of these doors. Some vehicles have vacuum motors controlled by electric solenoids, whereas others use electric motors at the doors. There are also systems in which all the mode doors have electric motor control (Figures 10-46 and 10-47).

Generally, a vacuum system problem can be traced to a cut, kinked, crimped, or disconnected vacuum hose. A faulty selector valve, vacuum actuator, storage tank, or check valve may also be the problem.

In an electric solenoid or electric motor system, an electrical system defect may be responsible for improper mode door operation. Because these systems function electrically, it is wise to consult the appropriate manufacturer's service manual for specific troubleshooting procedures. One must be extremely careful when troubleshooting electrical systems under the dash. Just one improper test point could cause serious damage to one of the onboard computers.

The area under the dash is cramped and congested—filled with wires, vacuum hoses, and ducts, as well as various electrical and mechanical components and assemblies. It is not, therefore, easy to gain access to any of the components, especially those associated with the vacuum control system. The control panel on many vehicles, however, may be pulled out far

FIGURE 10-48 **A typical vacuum motor.**

enough from the dash to gain access. Extreme caution must be exercised when gaining access to any under-dash component. Failure to do so could trigger the air bag restraint system.

Visual Inspection

When addressing a customer complaint for poor or insufficient heating or cooling, the first step is to make a visual inspection. The following should be included in the inspection:

- **Is the case and ductwork sound?** Check for cracks or broken or disconnected ducts.
- **Are the vacuum hoses sound?** Check for disconnected, split, damaged, or kinked vacuum hoses.
- **Is the airflow restricted?** Check to ensure that mode doors are opening. Check for leaves or other debris that may block airflow, such as at the fresh air inlet screen located at the base of the windshield. Is the cabin filter clean, if so equipped?
- **Are the cables secure?** Check for loose, broken, or binding mode door control cables.

> **AUTHOR'S NOTE: When checking electric actuator circuits, use only high-impedance multimeters. Never use a test light that could overload the circuits in the microprocessor. Always follow the manufacturer's recommended diagnostic procedures.**

A BIT OF HISTORY

No company has an exclusive patent on HFC-134a, and the industry-wide availability of it allowed for the rapid introduction of R-134a as the industry replacement for R-12, which began in 1992.

Mode Door Adjustment

A cable or vacuum actuator is used to position one or more of the mode doors in the duct system. The cable-operated system (Figure 10-49) consists of a steel cable encased in a plastic, nylon, or steel housing. It is used to connect the mode door to the control panel. Adjustments are made on the mode door end of the cable by repositioning the cable housing in its mounting bracket. The cable is usually held in place with a clip or a retainer secured in place with a hex-head screw.

The only adjustment possible for a vacuum actuator (Figure 10-50) is in the linkage, if there are adjustment provisions. If the problem proves to be a defective vacuum motor, however, it must be replaced. First, ensure that there is a vacuum signal at the vacuum motor indicating that the vacuum system is sound. More information on troubleshooting and servicing of the vacuum-operated and electrically operated actuators can be found in Chapter 11 of this manual as well as in Chapters 10 and 11 of the Shop Manual.

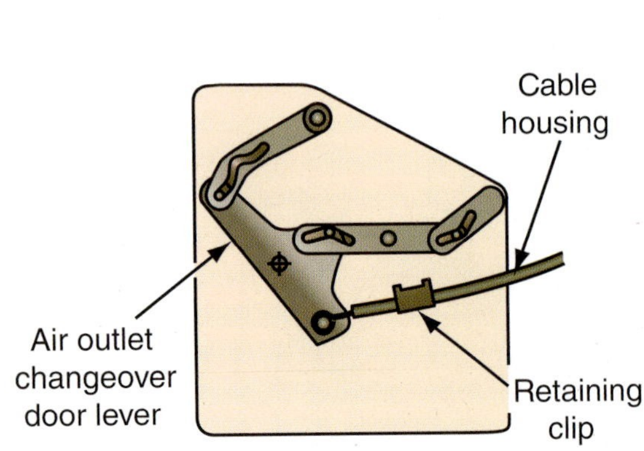

FIGURE 10-49 A typical cable-operated mode door.

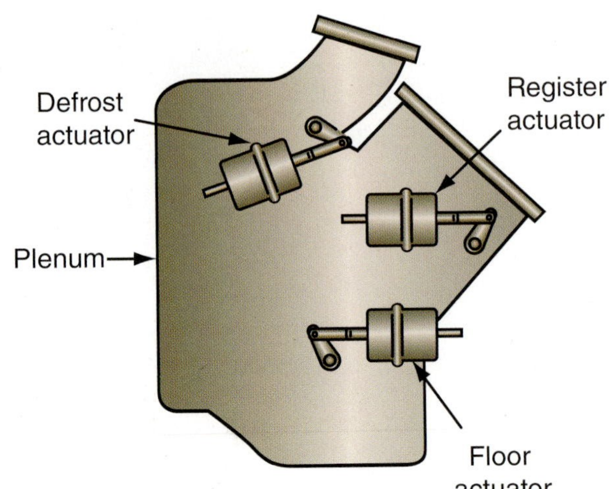

FIGURE 10-50 A typical vacuum-operated mode door.

SUMMARY

- There are many variations of mode and blend door positions, as well as many case/duct system designs.
- Doors may be electric, vacuum, or cable operated.
- Some doors are either fully opened or fully closed; others are infinitely variable.
- Because of the many different applications and methods of control (Figures 10-51 and 10-52), it is necessary to consult a particular manufacturer's manual for specifications and testing procedures.

Terms to Know

Automatic Temperature Control (ATC)
Bi-level
Blend door
Bowden cable
Defrost
Heating/ventilation/air conditioning (HVAC)
MIX
Mode
Mode doors
Plenum
Recirculate
Semiautomatic temperature control (SATC)
Vent

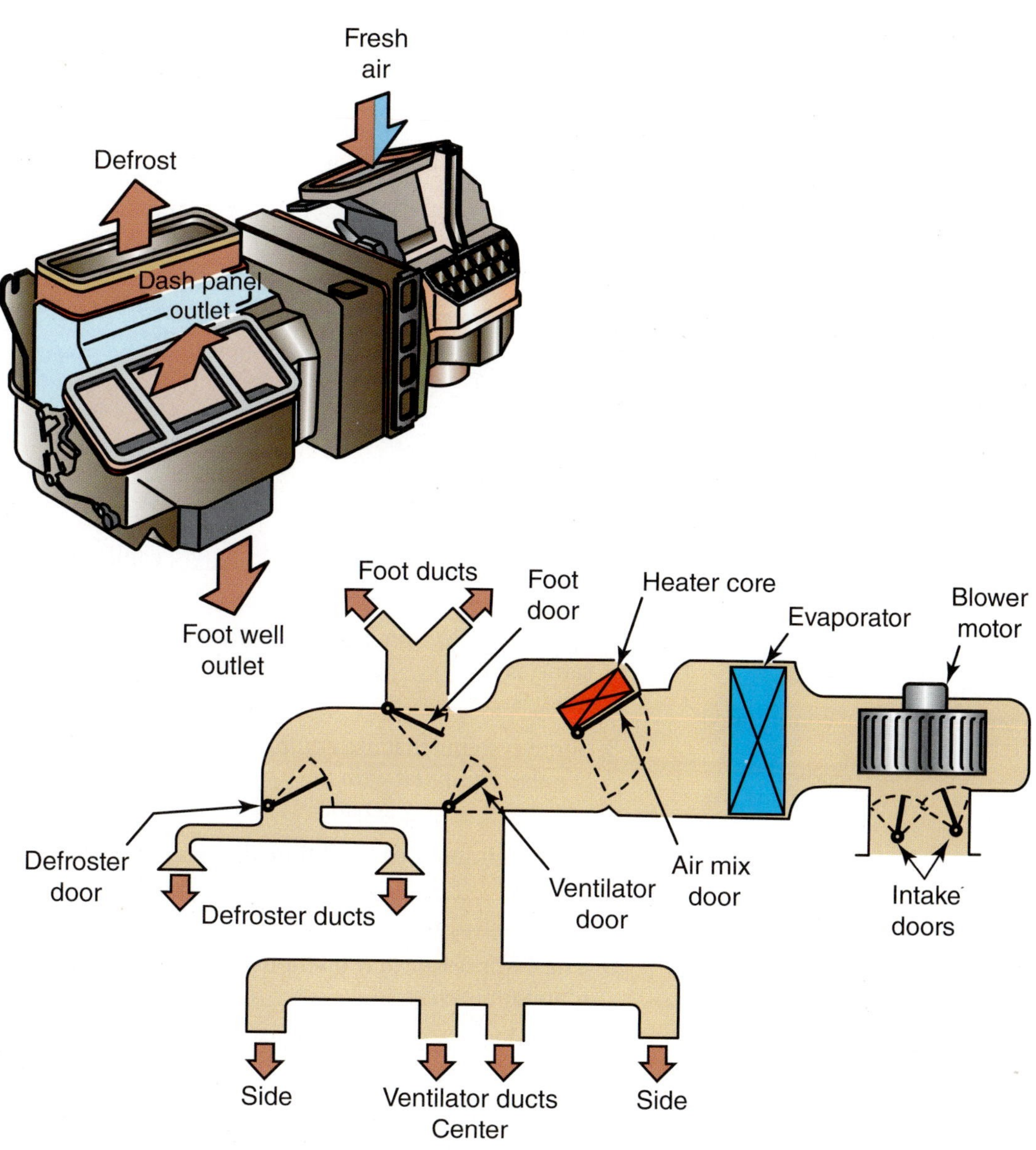

FIGURE 10-51 **One type of case and duct system. Compare with Figure 10-52.**

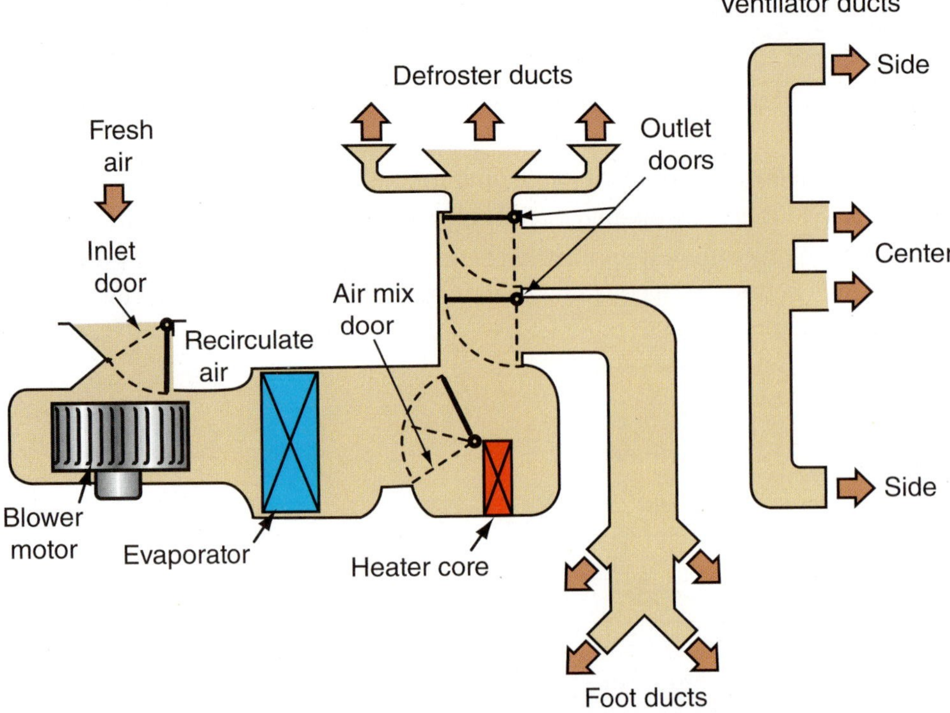

FIGURE 10-52 **Another type of case/duct system. Compare with Figure 10-51.**

REVIEW QUESTIONS

Short-Answer Essays

1. What is one of the two purposes of the case/duct system?
2. Why do some case and duct systems develop a musty odor?
3. Where is the air taken from that is to be "conditioned" for MAX cooling?
4. What is the purpose of maintaining a slightly positive pressure in the vehicle's interior?
5. Explain how the programmer controls the movement of the electronic actuated blend door.
6. How are pollen and dust particles stopped from entering the passenger compartment?
7. Where is air directed in the distribution section?
8. Describe the airflow through the case/duct system when DEFROST is selected.
9. How does the blend door create the desired in-vehicle temperature and how is the air tempered to maintain a desired humidity?
10. Define the term *bi-level*.

Fill in the Blanks

1. Mode doors may be ______________, cable, or ______________ operated.
2. The MIX selection on some models is basically the same as the ______________ selection on other models.
3. Some systems that are equipped with a cabin air filter are also equipped with an ______________ airborne ______________ sensor.
4. A small amount of conditioned air may be directed to the ______________ outlets to prevent windshield ______________.
5. The compressor may operate in the heat mode to help maintain in-vehicle ______________.
6. Cooled air is generally delivered into the passenger compartment through the ______________ vents.
7. Heated air is generally delivered into the passenger compartment through the ______________.
8. The ______________ duct system found in some cars has a separate driver- and passenger-side ______________ system.

9. There are two types of blower system: an upstream blower and a _______________.

10. The case system provides a _______________ for the components, whereas the duct system provides a for the airflow.

Multiple Choice

1. A vehicle is brought into the shop with a complaint of weak airflow from all the dash vents. Which of the following is the most likely cause?
   A. A blend door stuck in the heat position
   B. The recirculation mode door stuck in the recirculation position
   C. A dirty cabin air filter
   D. A mode door stuck in the floor vent position

2. The rear evaporator duct is cooling properly on a vehicle with a rear heating and cooling system, but the front evaporator duct temperature is too warm. What is the most likely cause of this condition?
   A. An undercharged refrigerant system
   B. A misadjusted front blend door
   C. An overcharged refrigerant system
   D. A stuck open rear thermal expansion valve

3. When the air-conditioning system is first turned on, a musty odor comes from the vents. What is the most likely cause for this?
   A. An undercharged refrigerant system
   B. A misadjusted front blend door
   C. Mold formation on the evaporator core
   D. Ice formation on the evaporator core

4. The following statements about a typical system set to MAX cooling are true, except:
   A. The compressor clutch is engaged.
   B. The blower motor is running.
   C. The outside/recirculate door is in recirculate position.
   D. Return air is from vehicle exterior.

5. During normal air-conditioning system operation, the position of the system mode doors:
   A. Is fully opened.
   B. Is fully closed.
   C. Is partially opened (or closed).
   D. Depends on the mode selected.

6. In the MIX/BILEVEL position, conditioned air is delivered to the _______________ outlet.
   A. Panel
   B. Defrost
   C. Floor
   D. Both B and C

7. An inoperative vacuum motor could be caused by all of the following, except:
   A. Defective vacuum switch
   B. Kinked vacuum hose
   C. Mode door cable misadjustment
   D. Leaking vacuum reserve tank

8. The airflow path in the case/duct system illustrated in Figure 10-52 is being discussed:
   Technician A says that airflow is also through the heater core in the cooling mode.
   Technician B says that airflow is also through the evaporator in the heating mode.
   Who is correct?
   A. A only
   B. B only
   C. Both A and B
   D. Neither A nor B

9. *Technician A* says that up to 20 percent of fresh air provides a means to maintain a positive in-vehicle pressure when the windows are closed.
   Technician B says this positive pressure is necessary to provide a proper balance of air pressure within the air delivery system.
   Who is correct?
   A. A only
   B. Both A and B
   C. B only
   D. Neither A nor B

10. In the case and duct system the air distribution mode doors may be operated by all of the following except:
    A. Vacuum motors
    B. Cable
    C. Electrically powered motors
    D. Potentiometers

Chapter 11

System Controls

Upon Completion and Review of this Chapter, you should be able to:

- Understand the requirements of fuses and circuit breakers for electrical circuit protection.
- Understand the role of the manual master control assembly in a HVAC system.
- Discuss the operation and function of the temperature control thermostat.
- Explain the role of the low- and high-pressure cutoff switches.
- Compare the function of gauges versus lamps for engine coolant temperature.
- Discuss vacuum control devices and understand their function as well as vacuum system diagrams.
- Discuss and compare the differences of automatic temperature control systems that use various pressure- and temperature-actuated controllers and recognize the components.
- Explain the operation and role of refrigerant pressure sensors.
- Discuss the operation of various climate control system input sensors.
- Discuss the role of the automotive scan tool in adding in the diagnosis of HVAC system failures.
- Discuss the various body and network codes associated with the operation of the HVAC system.
- Describe the operation of the Controller Area Network (CAN) protocol and climate control local area network (LAN) systems.
- Discuss the operation and role of heated and climate controlled seats (CCS).

A **fuse** is an electrical device used to protect a circuit against accidental overload or unit malfunction.

A **circuit breaker** is a bimetallic electrical device used to protect a circuit against accidental overload or unit malfunction. It automatically resets once it cools down.

Introduction

The automotive air-conditioning control system can be as simple as that found in aftermarket installations or as complex as that found in computer-controlled automatic factory-installed systems. The simple system usually consists of a master on/off switch, blower control, thermostat, blower motor, clutch coil, and **fuse** or **circuit breaker**.

Note in the schematic (Figure 11-1) that only one wire is shown from the battery. The other side of the battery, as well as the blower motor and clutch coil, terminate to ground. The vehicle chassis, body, and all metal parts are *common* (ground) in 12-volt, direct-current (DC) automotive electrical systems. A separate ground wire circuit is not required unless the car has fiberglass or other nonconducting body components. An electrical symbol (Figure 11-2) is used to indicate a ground connection. Examples of other wiring diagram symbols are shown in Figure 11-3.

The schematic for a factory-installed heater/air-conditioner control system is more complex (Figure 11-4). Actually, the schematic in this illustration has been condensed so that

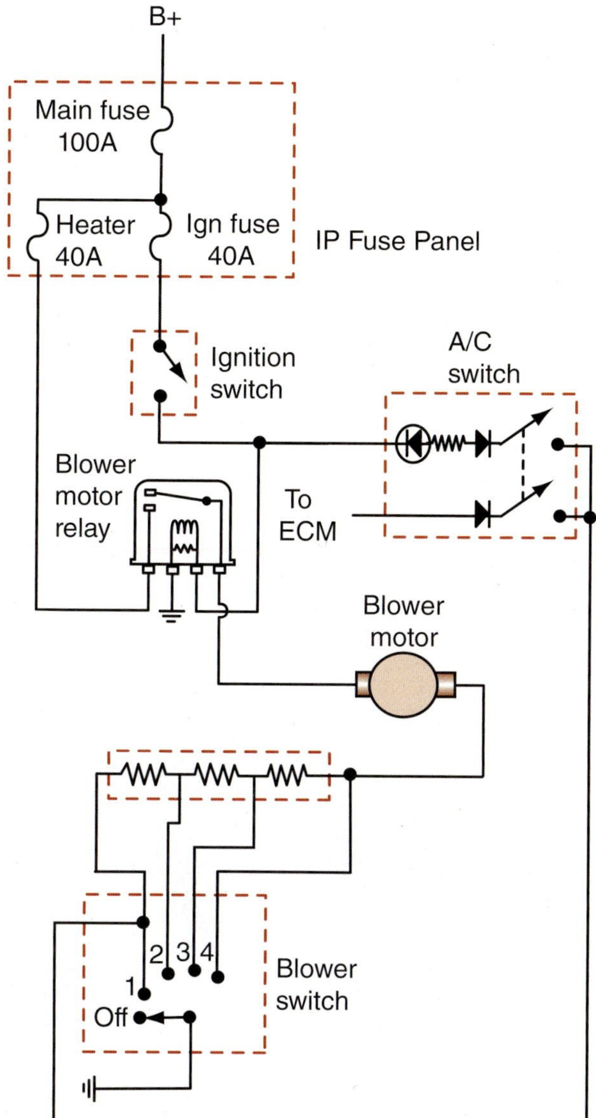

FIGURE 11-1 **A typical air-conditioner/heater system schematic.**

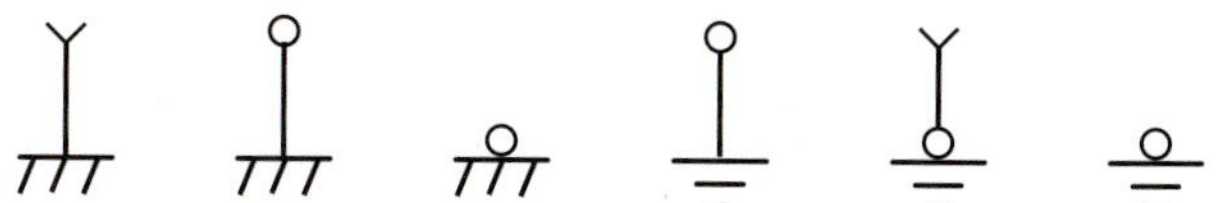

FIGURE 11-2 **Electrical symbols used to identify a ground connection.**

it can be shown on one page. Many schematics of factory-installed heater/air-conditioning systems require several pages in the shop service manual for illustration.

The air-conditioning and heater electrical systems are integral and often share fuses or circuit breakers. The blower motor serves both the heater and air conditioner in factory-installed systems. Electrical circuits associated with the heating and cooling system, such as those used to warn of engine overheating conditions, may also be a part of the electrical system schematic.

SYMBOLS USED IN WIRING DIAGRAMS			
+	Positive		Temperature switch
—	Negative		Diode
	Ground		Zener diode
	Fuse		Motor
	Circuit breaker	C101	Connector 101
	Condenser	→	Male connector
Ω	Ohm		Female connector
	Fixed value resistor		Splice
	Variable resistor	S101	Splice number
	Series resistors		Thermal element
	Coil		Multiple connectors
	Open contacts	88:88	Digital readout
	Closed contacts		Single filament bulb
	Closed switch		Dual filament bulb
	Open switch		Light emitting diode
	Ganged switch (N.O.)		Thermistor
	Single-pole double-throw switch		PNP bi-polar transistor
	Momentary contact switch		NPN bi-polar transistor
	Pressure switch		Gauge
	Battery		Wire Crossing

FIGURE 11-3 **Typical schematic symbols.**

MANUAL MASTER CONTROL

The **master control** is the primary or main control.

The manual **master control** (Figure 11-5) generally includes the blower speed control provisions. The variable (infinite) speed control, also known as a **rheostat** (Figure 11-6), is generally found on aftermarket systems. Also used are four- or five-position blower speed controls.

A **rheostat** is a wirewound variable resistor with one input and one output wire.

The four-speed five-position control (Figure 11-7) has three resistors to provide selected blower motor speed. The first position is OFF. The second position, LOW, supplies current to the motor through all three resistors. This provides maximum reduced voltage for low-speed operation. The next position, (M1), supplies current to the motor through two resistors to provide medium-low speed. Current is supplied to the motor through one resistor in the fourth position, (M2), to provide medium-high blower speed. The fifth position, HIGH, supplies full battery voltage to the motor to provide high-speed operation of the blower.

A multiwound motor is also referred to as a tapped motor.

Some multiposition blower speed control switches (Figure 11-8) do not have resistors. These switches provide full battery voltage to either of several windings in the motor.

The blower motor speed control may be a part of the master ON/OFF control (Figure 11-9). The blower switch, in either ON position, provides full battery voltage to the control thermostat. In this arrangement, the compressor clutch does not engage unless the blower motor is running.

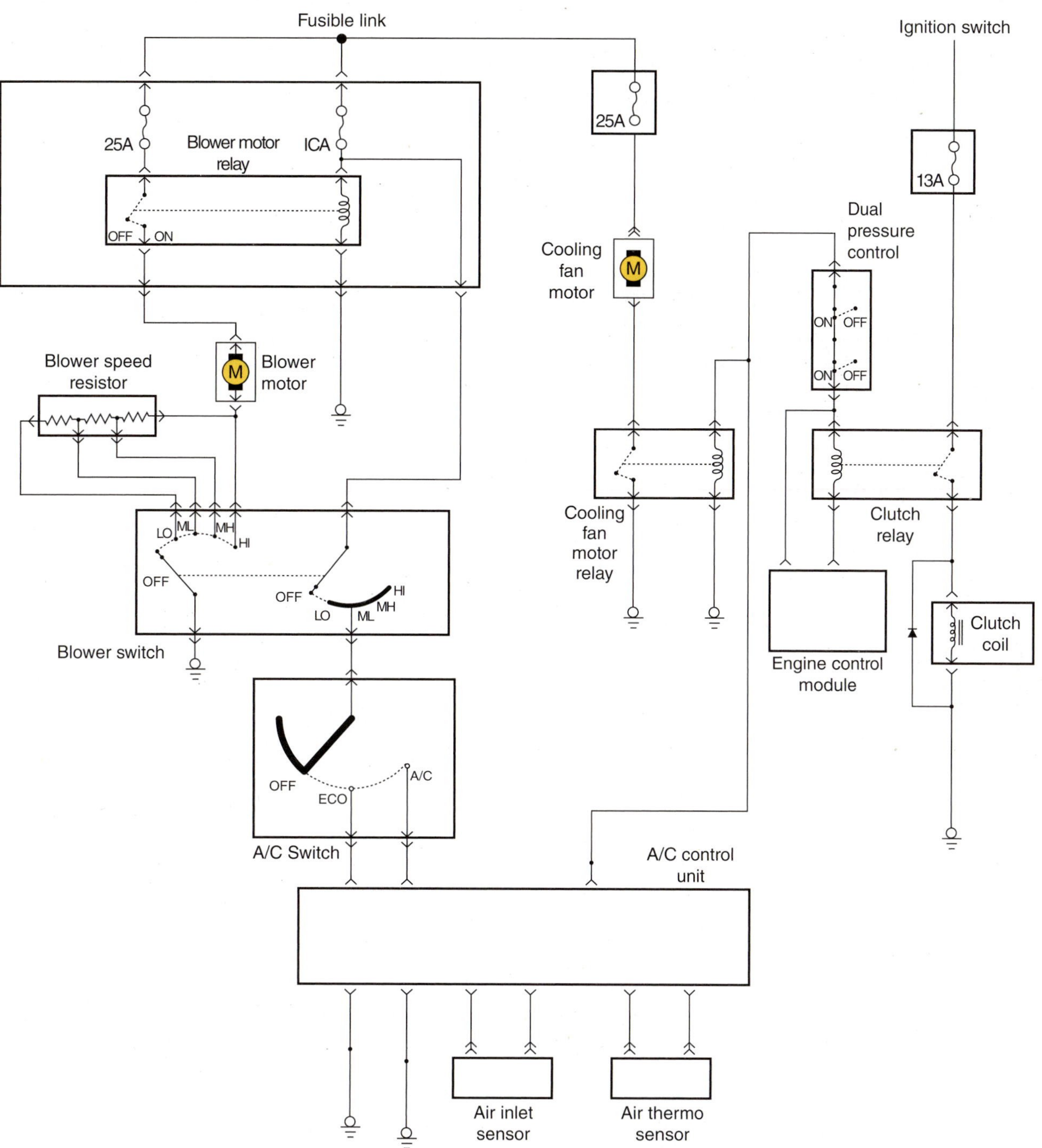

FIGURE 11-4 A typical climate control system electrical diagram.

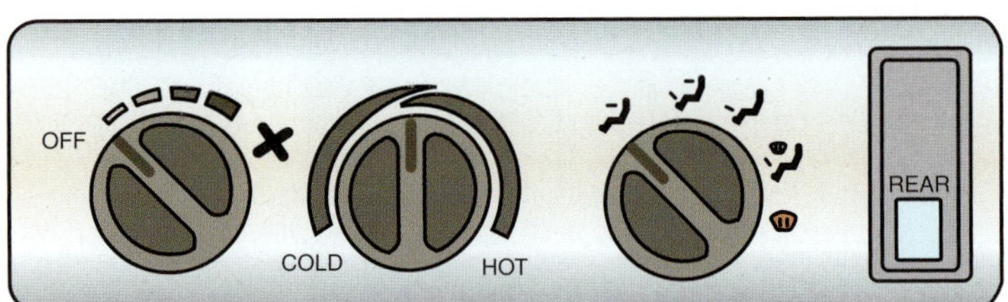

Heater control

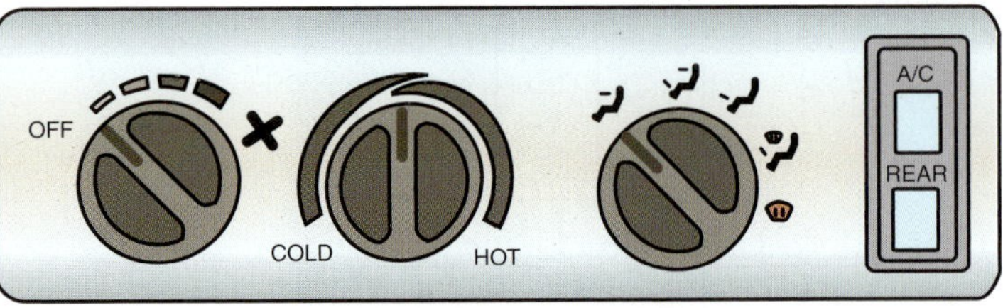

Manual A/C control

FIGURE 11-5 **Typical master controls.**

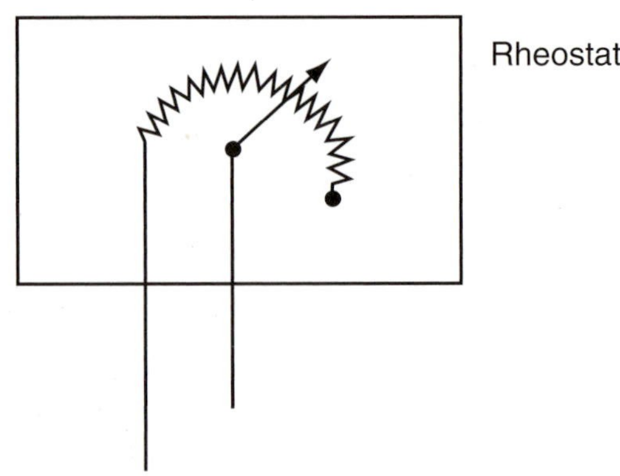

FIGURE 11-6 **A rheostat is a two-wire variable resistor with no feedback signal wire, common on dashlight dimmer circuits.**

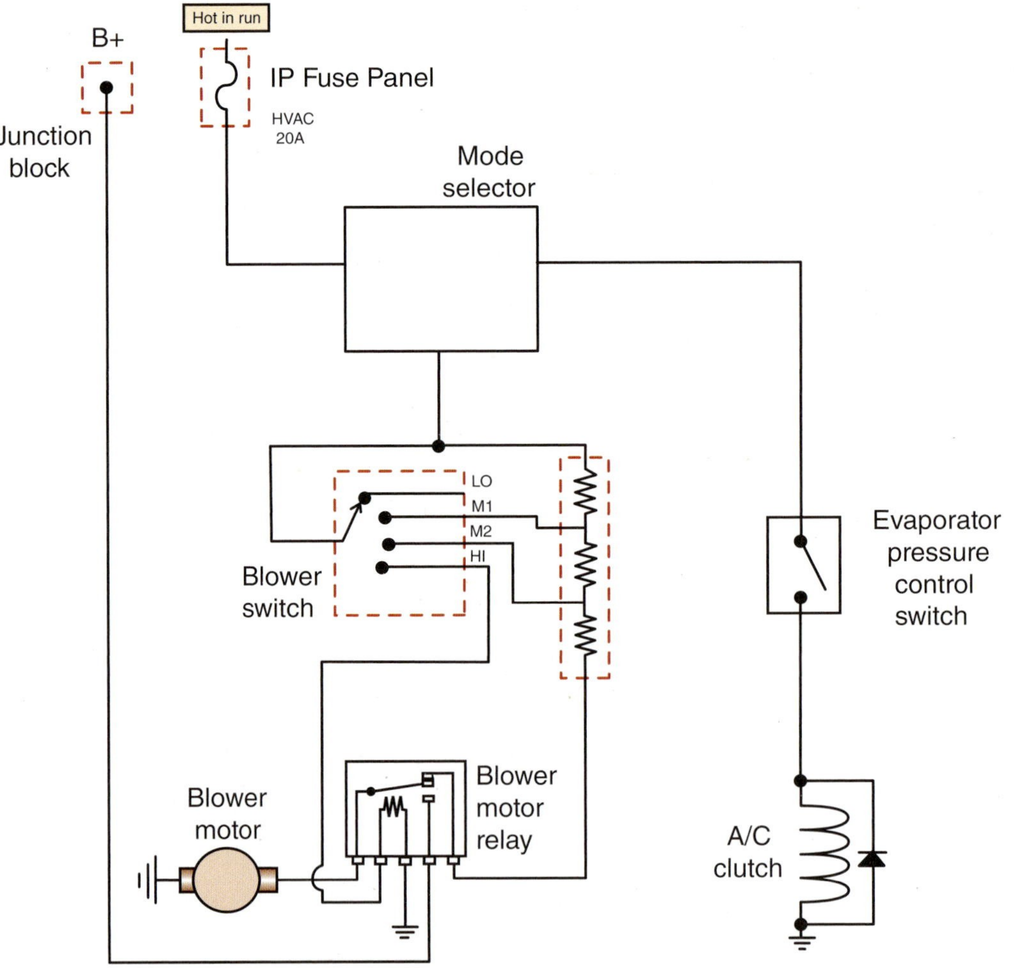

FIGURE 11-7 **Wiring diagram for typical multispaced blower motor.**

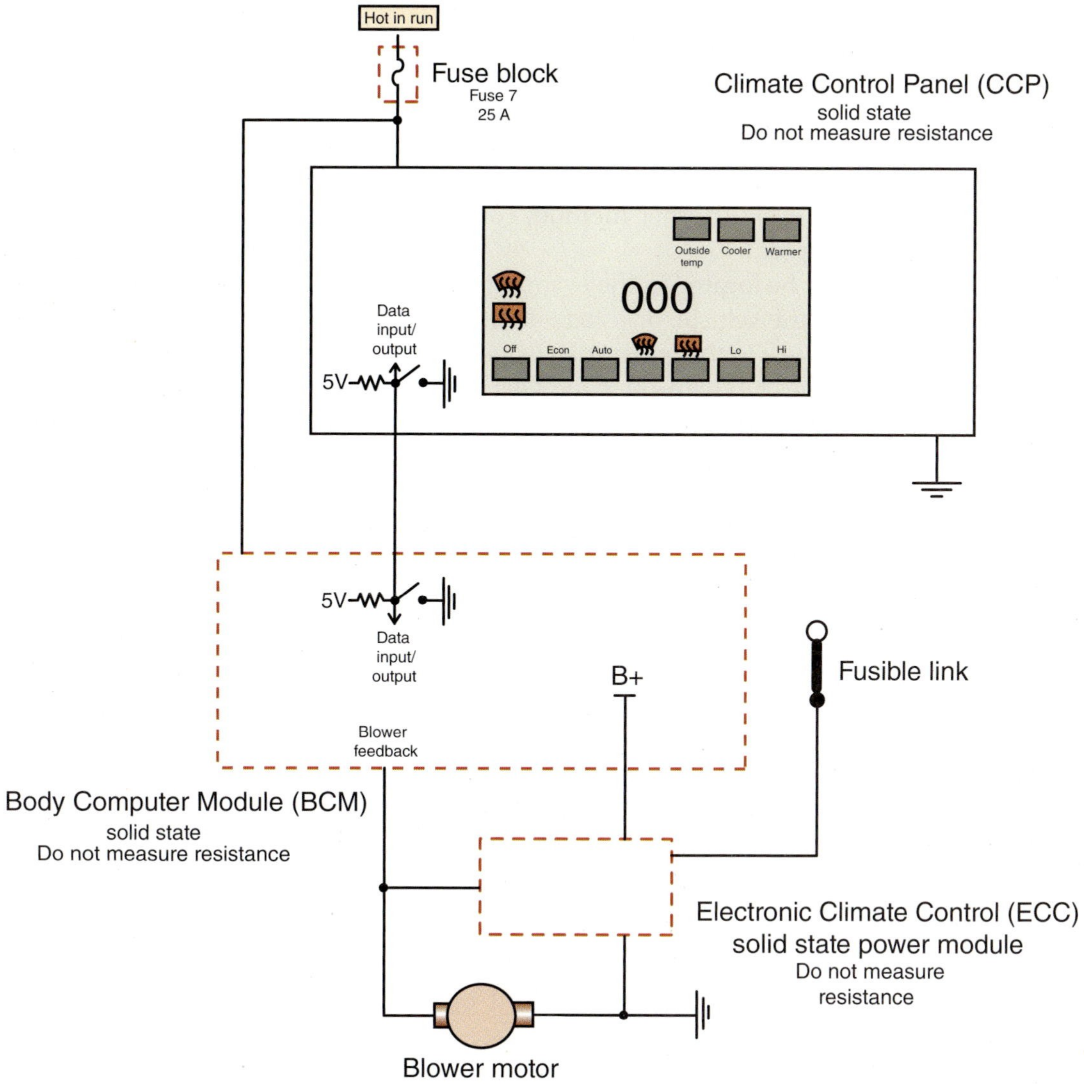
Hot in run
Fuse block
Fuse 7
25 A
Climate Control Panel (CCP)
solid state
Do not measure resistance
Outside temp
Cooler
Warmer
000
Off
Econ
Auto
Lo
Hi
Data input/ output
5V
5V
Data input/ output
B+
Fusible link
Blower feedback
Body Computer Module (BCM)
solid state
Do not measure resistance
Electronic Climate Control (ECC)
solid state power module
Do not measure resistance
Blower motor

FIGURE 11-8 **Wiring schematic showing a pulse width modulated motor control circuit and feedback circuit to the BCM.**

A/C
FRONT
AUTO
REAR
SYNC
OFF
REAR
REAR

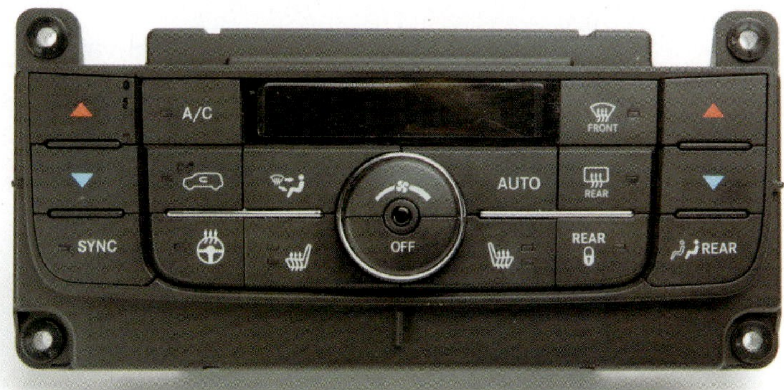

FIGURE 11-9 **A master control panel.**

Shop Manual
Chapter 11,
page 448

Some thermostat capillary tubes are inserted into a well provided in the suction line immediately after the evaporator.

The capillary tube is often filled with the same fluid used in the system, either R-12 or R-134a.

Thermostat

An electromagnetic clutch is used on the compressor of all automotive air-conditioning systems to turn it on when cooling is desired and off when cooling is not desired. The clutch is often used to provide a means of in-vehicle temperature control. One way to accomplish this is to control the compressor operation with a temperature-sensitive switch known as a thermostat (Figure 11-10). Another method, control with a pressure-sensitive switch, is covered later in this chapter.

The thermostat may be located in the evaporator, where it senses the temperature of the air being delivered into the vehicle, or it may be mounted in such a manner that its remote bulb is immersed into a well in the outlet tube (suction line) of the evaporator (Figure 11-11). The thermostat cycles the clutch on-off at the selected setting. This in turn controls the average in-vehicle temperature.

The thermostat senses the temperature of the evaporator core air or of the vapor leaving the evaporator. A temperature above that selected closes the thermostat contacts to provide current to the clutch coil. The clutch is energized and the compressor operates. A temperature

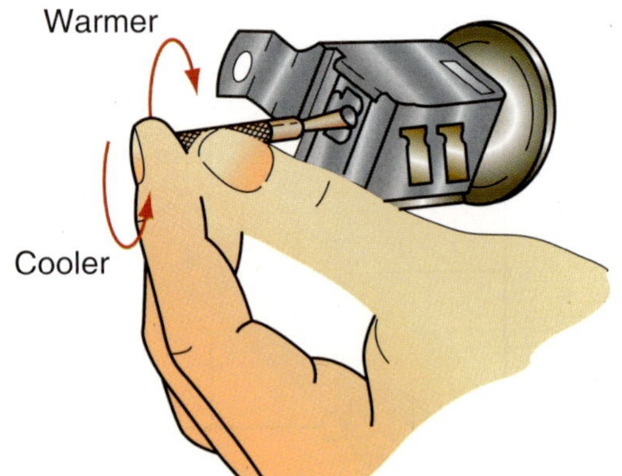

FIGURE 11-10 Typical thermostat adjustment.

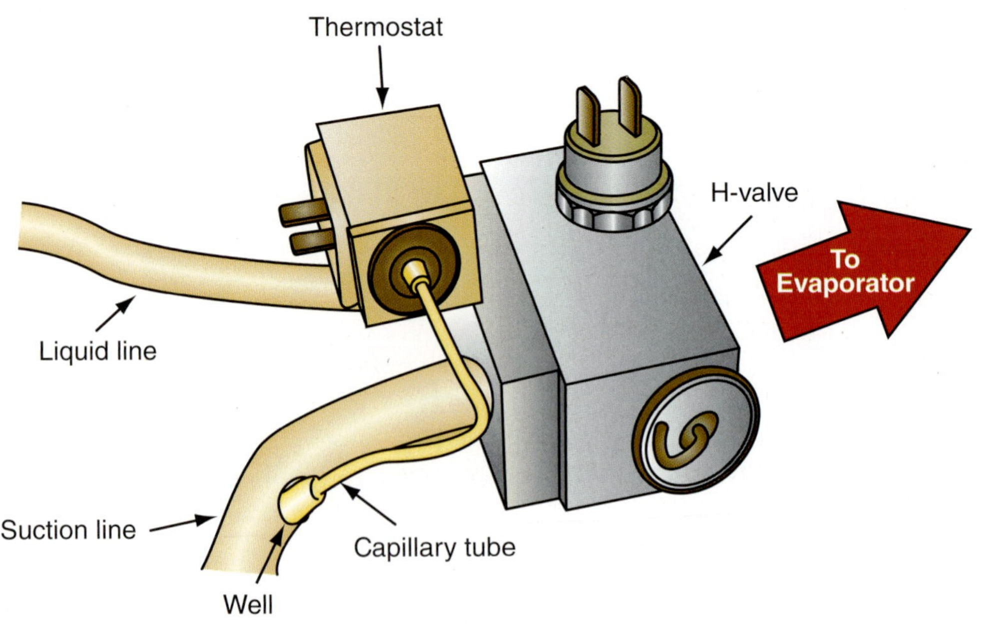

FIGURE 11-11 A typical thermostat location on an H-valve.

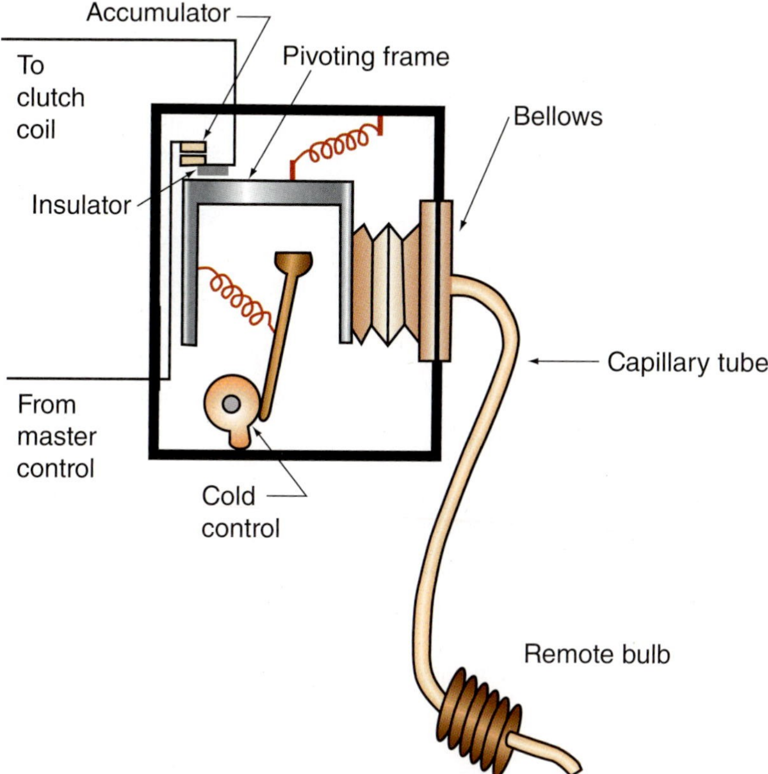

FIGURE 11-12 **Thermostat capillary tube shown with points open.**

at or below that selected opens the thermostat switch to interrupt current to the clutch coil. The clutch then de-energizes and compressor operation stops.

Thermostats generally have an OFF position so that the clutch can be turned off regardless of the temperature. In this way, the blower motor can be operated without a refrigerating effect.

Thermostat Construction

A capillary tube connected to the thermostat is filled with a temperature-sensitive fluid or vapor. The capillary is attached to a bellows within the thermostat (Figure 11-12). This **bellows**, in turn, is attached to a swinging frame assembly. Two electrical contact points are provided. One contact is fastened to the swinging frame through an **insulator**, and the other electrical contact is fastened to the body of the unit, again through an insulator.

A **bellows** is an accordion-type chamber that expands and contracts as its interior pressure is increased or decreased to create a mechanical action, such as in a thermostatic expansion valve.

An **insulator** is a nonconductive material, such as the covering on electrical wire.

Thermostat Operation

As the inert gas in the capillary tube expands, a pressure is exerted on the bellows. The bellows, in turn, closes the electrical contacts (Figure 11-13). The temperature selection is provided by a cam that is connected to the swinging frame via a shaft to an external control knob. When the knob is turned clockwise, the spring tension is increased against the bellows. If more pressure is required to overcome the increased spring tension, more heat is necessary. Because it is heat that is being removed from the evaporator, a lower temperature is required to open the points. On a temperature rise, the heat again exerts pressure on the bellows to close the points and allow for cooling.

A second spring in the thermostat regulates the temperature interval that the points are open. This interval usually represents a temperature rise (delta T or Δ_T) of 12°F (6.6°C), providing sufficient time for the evaporator to defrost.

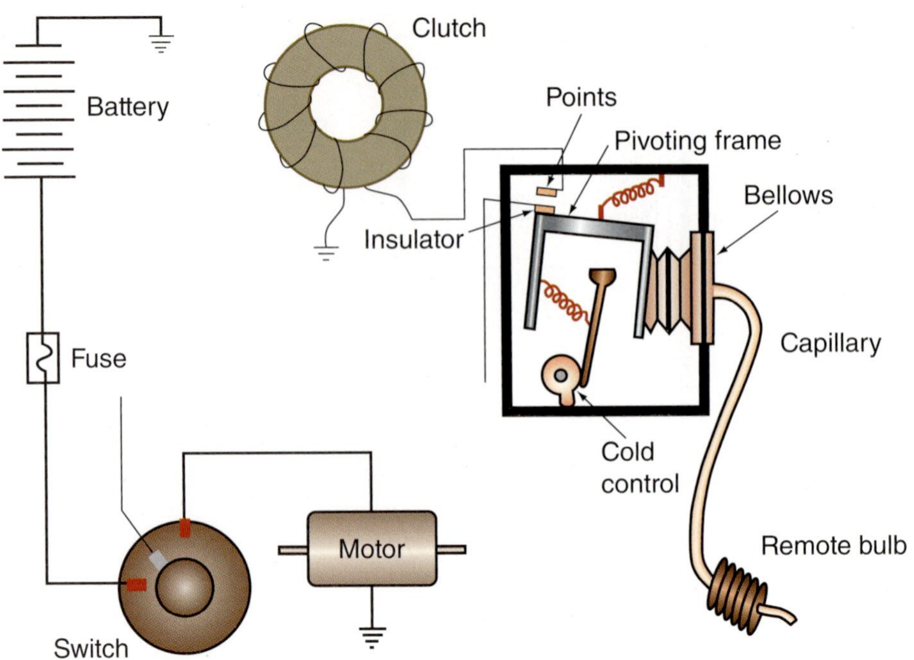

FIGURE 11-13 **Thermostat operation: points closed.**

Thermostat Installation and Handling

As with any device containing a capillary tube, care must be exercised when handling a thermostat. There should be no sharp bends or kinks in the capillary. When a bend must be made, it should be no sharper than can be formed around the end of a thumb.

For best results, the end of the capillary tube should be inserted into the evaporator core between the fins to a depth of about 1 in. (25.4 mm). The capillary should not be inserted all the way through the fins because it may interfere with the blowers, which are often mounted behind the core. If a remote bulb prevents insertion, the remote bulb should be fastened against the evaporator core (Figure 11-14).

If the capillary is damaged and has lost its charge of inert gas for any reason, the thermostat must be replaced. When there is no fluid in the capillary, the unit has no ON cycle. The capillary cannot be recharged using standard equipment. Many thermostats are adjustable.

FIGURE 11-14 **Thermostat remote bulb inserted into the evaporator core.**

Pressure Cutoff Switch

Some systems have a low- and high-pressure cutoff switch as part of the clutch circuit. These switches, which are normally closed (nc), are sensitive to system pressure and open in the event of abnormally low or high pressure. This in turn interrupts electrical current to the clutch coil to stop the compressor. The low pressure switch serves two purposes on some systems: temperature control and system protection.

Some pressure switches (Figure 11-15) can be replaced without having to remove the refrigerant from the system. Others, however, require that the refrigerant be recovered before the old switch can be removed. Those that do not require refrigerant removal have a valve depressor located inside the threaded end of the pressure switch (Figure 11-16). This pin presses on the Schrader-type valve stem as the switch is screwed on and allows system pressure to be expressed on the switch.

Shop Manual
Chapter 11,
page 455

The clutch-cycling low-pressure switch is often found on the accumulator.

A pressure cycling switch may open (close) on low or high pressure depending on application.

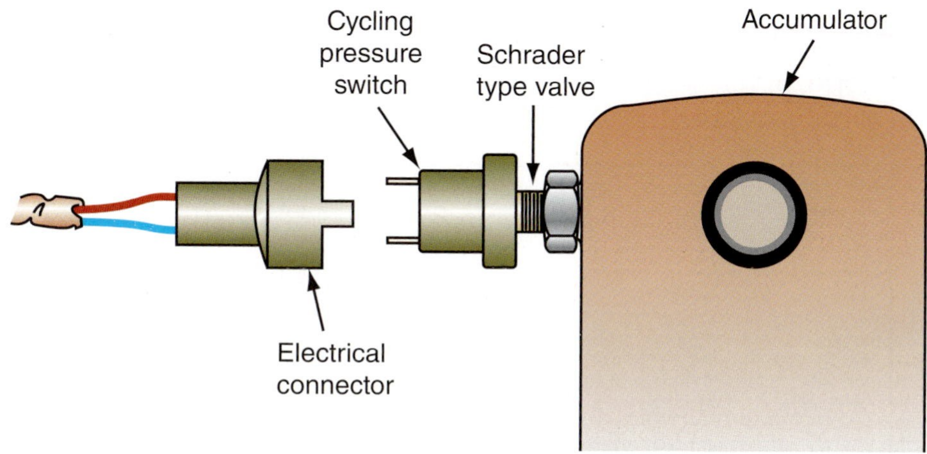

FIGURE 11-15 **Some pressure switches can be replaced without removing the refrigerant.**

FIGURE 11-16 **A valve depressor located inside the threaded portion of the pressure switch.**

Low-Pressure Switch

A **pressure switch** is an electrical switch that is actuated by a predetermined low or high pressure. A pressure switch is generally used for system protection.

Many cycling clutch orifice tube systems and fixed orifice tube cycling clutch systems use a **pressure switch** instead of a thermostat. The pressure cycling switch, which is mounted on the accumulator, senses low-side pressure to cycle the clutch off at about 24−26 psig (165−179 kPa) and cuts back in at about 40−42 psig (276−290 kPa). These pressures correspond to temperatures of 24°F−27°F (−4.4°C−−2.8°C) and 43°F−45°F (6.1°C−7.2°C), respectively, for R-12. They correspond to 27.5°F−30°F (−2.5°C−−1.1°C) and 45°F−47°F (7.2°C−8.3°C), respectively, for R-134a. This maintains a cold evaporator while controlling for freeze-up. The pressure−temperature relationships for R-12 and R-134a in the evaporating range is given in Figure 11-17.

The low-pressure cutoff switch (Figure 11-18) may be found in the system anywhere between the evaporator inlet and the compressor inlet. In the event of an abnormally low pressure of 8–10 psig (55.2–68.9 kPa), the switch will open to stop the compressor. This

TEMPERATURE		PRESSURE			
°F	°C	CFC-12		HFC-134a	
		psig	kPa	psig	kPa
20	−6.7	21.0	144.8	18.4	126.9
21	−6.1	21.7	149.6	19.2	132.4
22	−5.6	22.4	154.4	19.9	137.2
23	−5.0	23.2	160.0	20.6	142.0
24	−4.4	23.9	164.8	21.4	147.6
25	−3.9	24.6	169.6	22.0	151.7
26	−3.3	25.4	175.1	22.9	157.9
27	−2.8	26.1	180.0	23.7	163.4
28	−2.2	26.9	185.5	24.5	168.9
29	−1.7	27.7	191.0	25.3	174.4
30	−1.1	28.4	195.8	26.1	180.0
31	−0.6	29.2	201.3	26.9	185.5
32	0.0	30.1	207.5	27.8	191.7
33	0.6	30.9	213.1	28.7	197.9
34	1.1	31.7	218.6	29.5	203.4
35	1.7	32.6	224.8	30.4	209.6
36	2.2	33.4	230.3	31.3	215.8
37	2.8	34.3	236.5	32.2	220.0
38	3.3	35.2	242.7	33.2	228.9
39	3.9	36.1	248.9	34.1	235.1
40	4.4	37.0	255.1	35.1	242.0
41	5.0	37.9	261.3	36.0	248.2
42	5.6	38.8	267.5	37.0	255.1
43	6.1	39.8	274.4	38.0	262.0
44	6.7	40.7	280.6	39.0	268.9
45	7.2	41.7	287.5	40.1	276.5
46	7.8	42.6	293.7	41.1	283.4
47	8.3	43.6	300.6	42.2	291.0
48	8.9	44.6	307.5	43.3	298.6
49	9.4	45.7	315.1	44.4	306.1
50	10.0	46.7	322.0	45.5	313.7

FIGURE 11-17 **Low-side pressure–temperature chart for R-12 and R-134a, English and metric.**

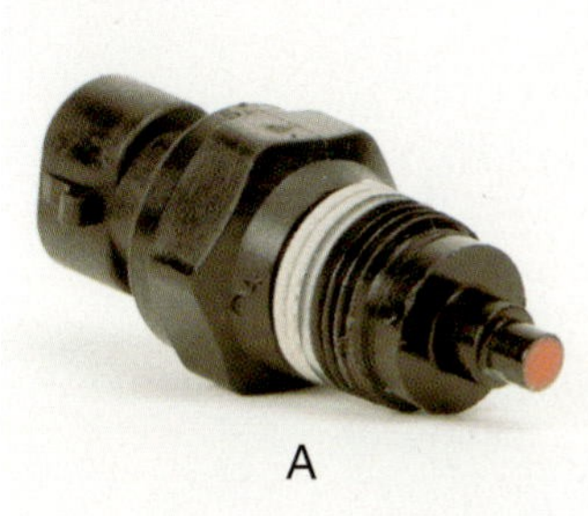

FIGURE 11-18 **Low-pressure cutoff switch: (A) male thread; and (B) female thread.**

prevents further reduction of system pressure to protect the system from the possible entrance of air or moisture, as would be the case with a low-side leak.

High-Pressure Switch

The high-pressure cutoff switch is found on some General Motors, Chrysler, and Subaru car lines. This switch is found in the system anywhere between the compressor outlet and the condenser inlet. In the case of an abnormally high pressure of 300–500 psig (2,068.5–3,447.5 kPa), the switch will open, stopping compressor action. This prevents system damage or rupture that may be caused by a further increase in pressure.

The high-pressure switch is normally closed (nc) and opens if the air-conditioning system pressure exceeds 425–435 psig (2,930–2,953 kPa). It closes when the system pressure drops to below 200 psig (1,379 kPa). This switch provides for system safety if, for any reason, pressures exceed safe limits. Unlike the low-pressure switch, the high-pressure switch does not provide data to the microprocessor. This switch is usually in series with the compressor clutch circuit.

Conditions such as a defective condenser fan motor may cause excessive high-side pressure.

Compressor Discharge Pressure Switch

Many factory systems use a compressor discharge pressure switch to disengage the compressor clutch electrical circuit if the refrigerant charge in the system is not adequate enough to provide sufficient circulation. The compressor discharge pressure switch is also called a no-charge switch, ambient low-temperature switch, or a low-pressure cutoff switch. The switch is designed to open electrically to shut off the compressor when high-side system pressure drops below 37 psig (255 kPa). This switch also performs the secondary function of an outside ambient air temperature sensor. When outside ambient air temperature falls below 40°F (4.4°C), the reduced corresponding refrigerant pressure, 36.9 psig (254.4 kPa), keeps the switch open.

The compressor discharge pressure switch is actually a low-pressure switch.

The compressor discharge pressure switch (Figure 11-19), which is located in the compressor housing or the high-pressure discharge line from the compressor or receiver-drier, cannot be repaired. If it fails in service, it must be replaced with a new unit. Its function is to protect the compressor. That function should not be defeated, such as by bypassing it with a jumper wire.

Factory-Installed Wiring

The electrical schematic illustrated earlier in Figure 11-4 is typical of any of the hundreds that illustrate the wiring of factory-installed air-conditioning (cooling and heating) systems. When servicing a particular system, it is necessary to consult the appropriate service manual for specific information and schematic details. As previously discussed, it must be noted that various

FIGURE 11-19 **A compressor discharge pressure switch.**

methods of temperature control are used: thermostat, pressure control, variable displacement compressor, and blend air (warm and cool).

Coolant Temperature Warning System

Shop Manual
Chapter 11, page 456

When operating, the automotive air-conditioning system places a high demand and an additional heat load on the engine cooling system. The condenser is located upstream (in front) of the radiator. Air intended to remove heat from the engine coolant in the radiator first passes through the condenser.

A malfunctioning air conditioner will often affect engine coolant temperature. Conversely, an overheated engine will affect air-conditioning performance. To monitor the engine coolant condition, a dash light or a dash gauge (Figure 11-20) is used. Either type has a sending unit located in the engine coolant system.

Most telltale lamp systems have one lamp: HOT. The telltale lamp is often referred to as an "idiot light."

Lamps

There are two types of engine coolant lamp systems: the one-lamp system and the two-lamp system. The one-lamp system (Figure 11-21) warns that the engine has overheated and that immediate attention is required.

FIGURE 11-20 **A typical engine coolant temperature gauge.**

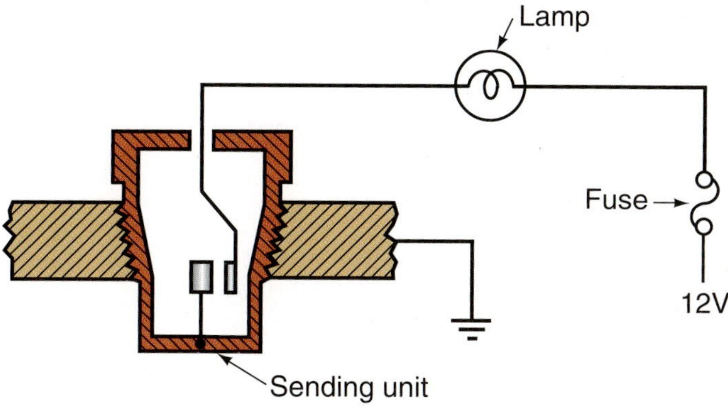

FIGURE 11-21 **A one-lamp coolant warming system.**

The two-lamp system has one lamp to indicate cold and another lamp to indicate hot. In the two-lamp system, the cold switch is closed until the engine coolant temperature reaches its normal operating temperature, usually about 180°F (82.2°C). In both the one- and two-lamp systems, the hot contacts of the sending unit close when engine coolant temperature reaches about 250°F (121.1°C). The actual temperature at which this switch closes depends upon the engine design.

The main disadvantage of the telltale light system is obvious: the hot lamp generally is not illuminated until after there is a problem. This is probably why the telltale light is sometimes referred to as an "idiot light."

Gauges

The coolant temperature gauge system (Figure 11-22) consists of two parts: the dash (gauge) unit and the engine (sending) unit. The sending unit contains a sintered material known as a thermistor. A thermistor changes resistance in relation to its temperature.

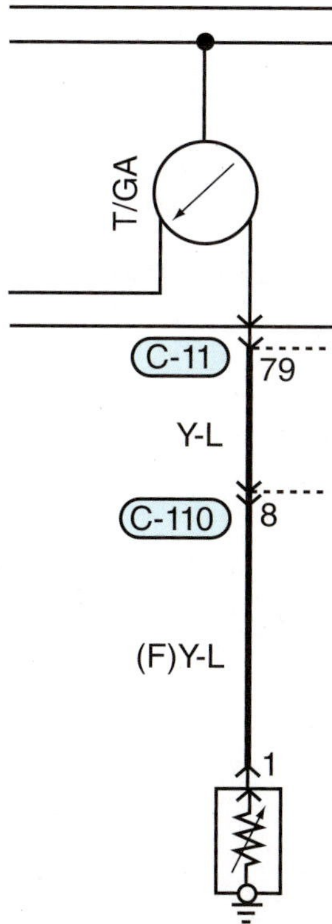

FIGURE 11-22 **Engine coolant temperature gauge schematic.**

The higher the temperature of a thermistor, the lower its resistance.

This material, which is sealed in a metal bulb, is screwed into a coolant passage of the engine. It has a high resistance when cold and a low resistance when hot.

The varying resistance of the sending unit regulates the amount of current passing through the coil of the dash gauge and moves the pointer accordingly.

Control Devices

Negative refers to any pressure less than atmospheric; *positive* refers to any pressure above atmospheric.

Many controls used in automotive air-conditioning systems are either negative (vacuum) or positive (pressure) actuated. Others are mechanically, electrically, or electronically actuated.

Vacuum Circuits

To understand vacuum circuits, it is first essential that the term vacuum be defined and understood. Vacuum is defined as a space that is devoid of matter. Because all things contain matter in some form, it would seem that there is no such thing as a vacuum.

For all practical purposes, therefore, a vacuum is better thought of as a portion of space that is partially devoid of matter. For a clearer understanding, consider that a vacuum is a space in which pressure is below atmospheric pressure. A good example of a vacuum is demonstrated by a person drinking through a straw (Figure 11-23). As the person sucks on the straw, a slight vacuum is created in the straw. Atmospheric pressure, which is greater than the vacuum pressure, is exerted against the surface of the liquid. This difference in pressure, known as **delta P (Δp)**, forces the liquid up the straw.

Delta P (Δp) is a term used when referring to a difference in pressure.

Atmospheric Pressure. Atmospheric pressure at sea level is 14.696 psia (101.328 kPa absolute). For all practical purposes, this value is usually rounded off to 14.7 psia (101.4 kPa absolute) or 15 psia (103.4 kPa absolute). At sea level, then, a pressure of 14 psia (96.5 kPa absolute) is a vacuum. Traditionally, English system vacuum pressure values are given in inches of mercury (in. Hg).

Atmospheric pressure is reduced by about 0.4912 psia (3.3868 kPa) per 1,000 ft. (304.8 m) of elevation.

FIGURE 11-23 A person drinking through a straw: (A) vacuum pressure; and (B) atmospheric pressure.

Vacuum Terms. Most automotive manufacturers' manuals give vacuum value requirements and specifications using the term *inches* only. In this manual, references to vacuum values are given in the English and metric absolute scales of pressure. The conversion chart in Figure 11-24 may be used as an aid for comparison of inches to psia and kPa absolute.

Vacuum-Operated Devices

Many vacuum-operated devices, such as heater coolant valves and mode doors, are activated with a **vacuum pot**, also called a vacuum motor or vacuum power unit.

A **vacuum pot** is a device designed to provide mechanical action by the use of a vacuum signal.

Single-Chamber Pot. The exertion (force) of atmospheric pressure on one side of a diaphragm causes the diaphragm to move toward the lower (vacuum) pressure side (Figure 11-25). This moves the device that is to be controlled through a lever, arm, or rod linkage.

Dual-Chamber Pots. Dual-chamber vacuum pots (motors) operate below atmospheric pressure based on a pressure differential (Ap) from one side to the other. A higher pressure on either side will move the diaphragm to the side with a lower pressure. This provides a push or pull effect on the vacuum pot.

A vacuum pot is also called a vacuum motor or vacuum power unit.

Changes in atmospheric pressure do not affect the operation of dual-chamber vacuum motors.

Inches of Mercury (in. Hg)	Pounds per Square Inch Absolute (psia)	Kilopascals Absolute (kPa absolute)
28.98	0.5	3.45
27.96	1.0	6.89
26.94	1.5	10.34
25.92	2.0	13.79
24.90	2.5	17.24
23.88	3.0	20.68
22.86	3.5	24.13
21.83	4.0	27.58
20.81	4.5	30.03
19.79	5.0	34.47
18.77	5.5	37.92
17.75	6.0	41.37
16.73	6.5	44.82
18.71	7.0	48.26
14.69	7.5	51.71
13.67	8.0	55.16
12.65	8.5	58.61
11.63	9.0	62.05
10.61	9.5	65.50
9.59	10.0	68.95
8.57	10.5	72.40
7.54	11.0	75.84
6.52	11.5	79.29
5.50	12.0	82.74
4.48	12.5	86.19
3.46	13.0	89.63
2.44	13.5	93.08
1.42	14.0	96.53
0.40	14.5	99.98
	15.0	103.42

FIGURE 11-24 **Conversion chart: absolute scale versus atmospheric scale.**

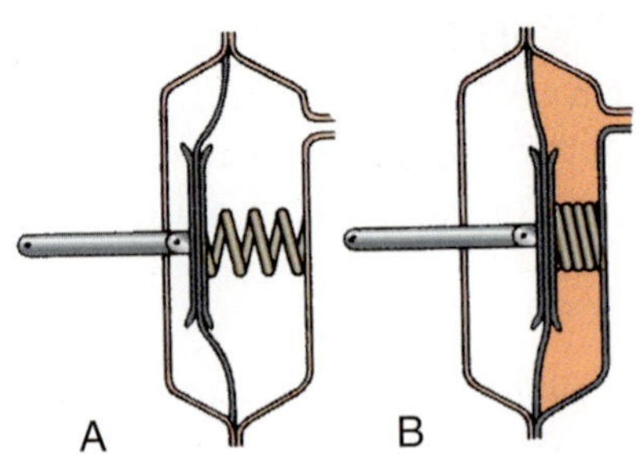

FIGURE 11-25 **Movement of a single-chamber vacuum motor: (A) no vacuum applied; and (B) full vacuum applied.**

Vacuum Source

When running, the automobile engine provides a ready source of vacuum. This source is usually taken off the intake manifold and routed to the various components through small-diameter synthetic rubber, plastic, or nylon hoses. The engine vacuum supply source can vary from 0.01 in. Hg (14.7 psia or 101.4 kPa absolute) to 20 in. Hg (4.89 psia or 33.7 kPa absolute), or more. Actual vacuum conditions depend on certain engine conditions. The reason for the vacuum variation is not important in this discussion. It is important, however, to be aware that engine vacuum does vary.

The vacuum reserve tank and check valve prevent erratic operation of the vacuum motors.

Because of this vacuum variation, a reserve tank and check valve are used (Figure 11-26). This combination of devices provides the means to maintain maximum vacuum values to properly operate air-conditioning and heater vacuum controls under all engine operating conditions. It should be noted that more than one reserve tank or check valve may be found in the air-conditioning and heating system vacuum circuit. It may also be noted that the vacuum system may serve other components, such as the power brake booster.

Shop Manual
Chapter 11, page 457

Reserve Tank. Vacuum reserve tanks are provided in a variety of sizes and shapes. Some early tanks, which are made of metal, resemble a large juice can (Figure 11-27). Others (Figure 11-28) are made of plastic and resemble a sphere. Vacuum reserve tanks generally require no maintenance, but they sometimes develop pinhole-sized leaks due to exposure to the elements. When a vacuum tank is found to be leaking, it may be repaired. If the reservoir fails to hold a vacuum, the mode doors may operate sporadically or not at all. The default or normal position for the mode door when no vacuum is applied is typically in the defrost mode, supplying air to the windshield on many vehicles. This is done for safety so that the windshield will stay clear even with a system failure.

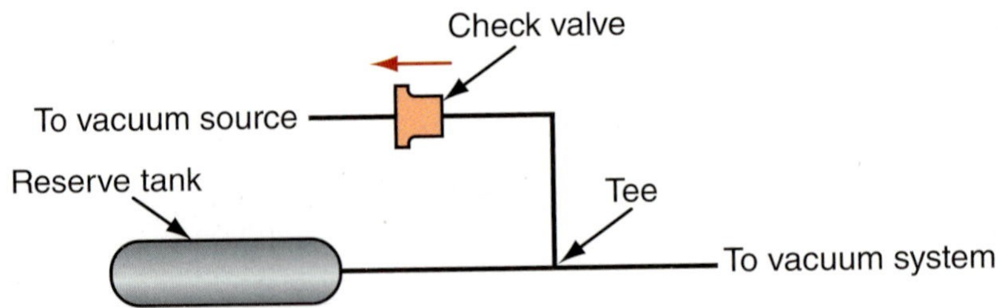

FIGURE 11-26 **Reserve vacuum tank and check valve.**

FIGURE 11-27 **An early vacuum reserve tank.**

FIGURE 11-28 **A plastic vacuum reserve tank.**

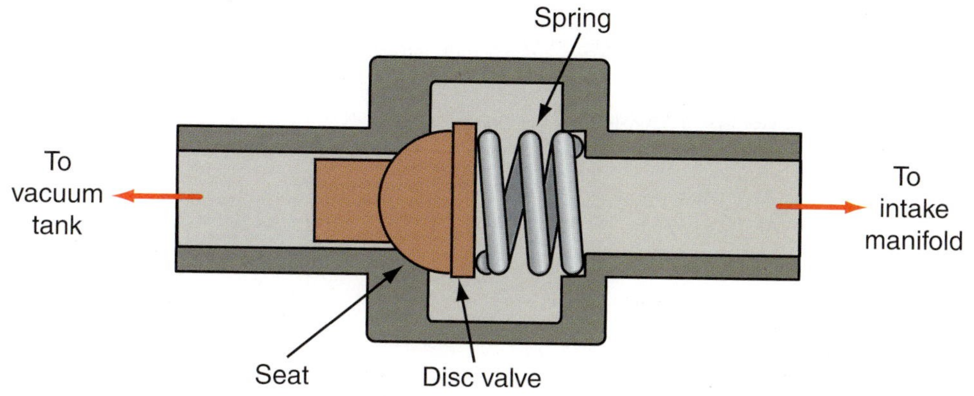

FIGURE 11-29 **A typical vacuum check valve.**

Vacuum signal legend includes "pv" for partial vacuum, "nv" for no vacuum, and "v" or "fv" for full vacuum.

Shop Manual
Chapter 11, page 410

Check Valve. Many types and styles of check valves are used in the automotive vacuum circuit. Essentially, a check valve (Figure 11-29) allows the flow of a fluid or vapor in one direction and blocks the flow in the opposite direction. Many systems use a check valve between the engine manifold supply vacuum hose and the reservoir tank. This ensures a consistent supply of vacuum and a source of vacuum if manifold vacuum decreases (such as under a load).

Restrictor. Some vacuum systems have a restrictor to provide a delay or to slow the operation of a device. Restrictors have a small orifice that sometimes becomes clogged with lint or other airborne debris. Attempts to clean a restrictor usually prove unsuccessful, and replacement is suggested. To test a restrictor, simply use a vacuum pump and gauge setup.

A BIT OF HISTORY

In 1964, Cadillac introduced the first automatic air-conditioning climate control system. This was one of the most significant advancements in available options for the luxury car market.

Vacuum System Diagrams

The vacuum system diagram in Figure 11-30 must be considered to be typical only because it is a composite of one of hundreds of variations. The manufacturer's vacuum system diagram for a specific year/model car must be followed. Basically, the vacuum system is used to open, close, or position the heater coolant valve and mode doors to achieve a desired preselected temperature and humidity level. Pressure differentials provide a source of power to perform the mechanical movement of devices.

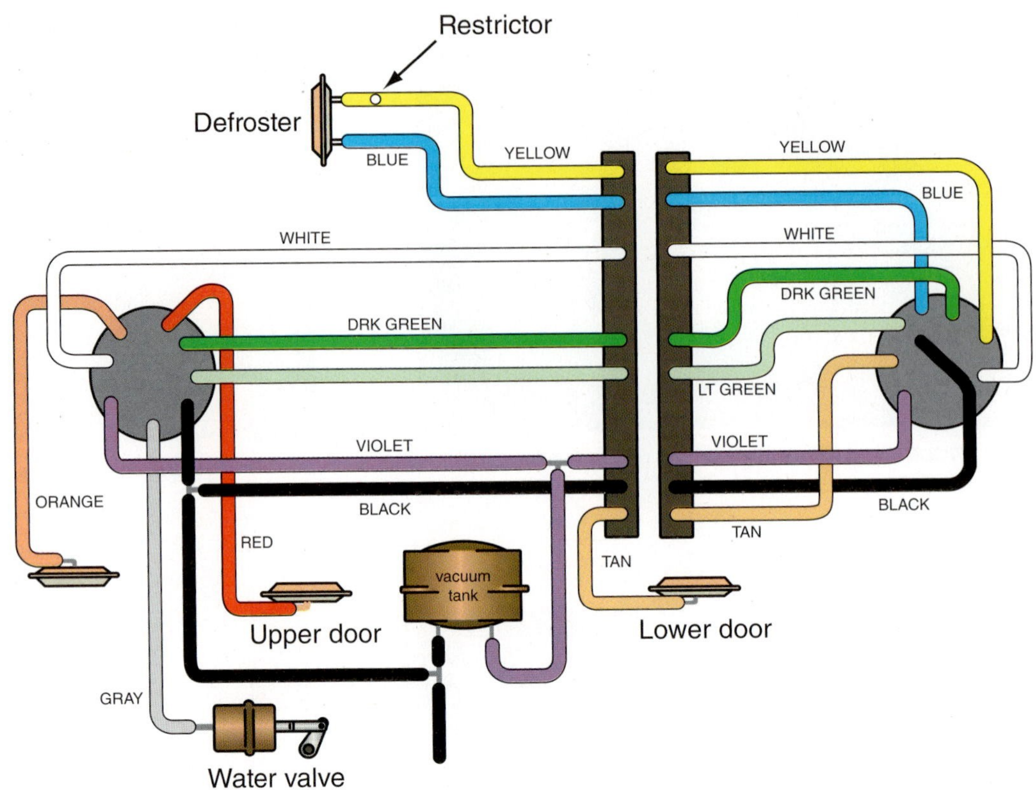

FIGURE 11-30 **A typical vacuum system diagram.**

Shop Manual
Chapter 11,
page 458

The average person is comfortable at a temperature of 78°F to 80°F with a relative humidity (RH) of 45–50 percent.

Automatic Temperature Controls

Many different types of semiautomatic and automatic temperature control systems are used today—so many that it is not possible to cover each system individually in this manual. Systems are modified or changed from year to year and from car model to car model.

Many solid-state components are so sensitive that even the 1.5-volt battery used in an analog ohmmeter may destroy them. A digital ohmmeter, then, must be used whenever a manufacturer's specifications suggest that component resistance measurements be taken. Some components and circuits are so sensitive to outside influence, however, that some schematics are labeled "Do not measure resistance." Heed this caution when it is noted to avoid damage to delicate electronic components.

Though systems differ in many respects, all are designed to provide in-car temperature and humidity conditions at a preset level (within system limitations), regardless of the temperature conditions outside the car (Figure 11-31). The temperature control also functions to hold the relative humidity within the car to a healthy level and to prevent window fogging.

For example, if the desired temperature is 75°F (23.89°C), the automatic control system will maintain an in-car environment of 75°F (23.89°C) at 45–55 percent humidity, regardless of the outside weather conditions.

In even the hottest weather, a properly operating system can rapidly cool the automobile interior to the predetermined temperature (75°F or 23.89°C). The degree of cooling then cycles to maintain the desired temperature level. In mild weather conditions, the passenger compartment can be held to this same predetermined temperature (75°F or 23.89°C) without resetting or changing the control.

During cold weather, the system rapidly heats the passenger compartment to the predetermined 75°F (23.89°C) level and then automatically maintains this temperature level.

Automatic climate control systems have a range of functions from FULL AUTO mode to semi auto and manual override control. To select FULL AUTO mode, the AUTO button on the climate control panel is selected. On some systems, the fan control knob must be set to the AUTO position. Cabin temperature is determined by the temperature control dial or increase/decrease temperature buttons (Figure 11-31). The system is designed to select the appropriate mix of volume (fan speed) and temperature of cooled or heated air in order to get the passenger cabin to the temperature selected as quickly as possible. On very cold days, the system may not run the blower motor until the heater core has reached a preset temperature, to avoid blowing frigid air on the occupants. The exception to this is if defrost is selected, in which case the blower motor will run to direct air to the windshield.

On many systems, if the temperature selected is the lowest limit or the highest limit, the system will produce maximum cooling or heating respectively. At these positions, the system will no longer regulate interior temperature, but instead produce maximum output. The system only regulates interior temperature when the temperature control knob is set between the maximum and minimum limits.

When the AUTO mode is selected, the automatic climate control system allows the vehicle occupant semiautomatic mode control of various functions manually, while still maintaining preset cabin temperature. These features vary among manufacturers but generally allow for driver- and passenger-side temperature control and air discharge selection, as an example. Often selecting recirculation or changing fan speed will disengage FULL AUTO control.

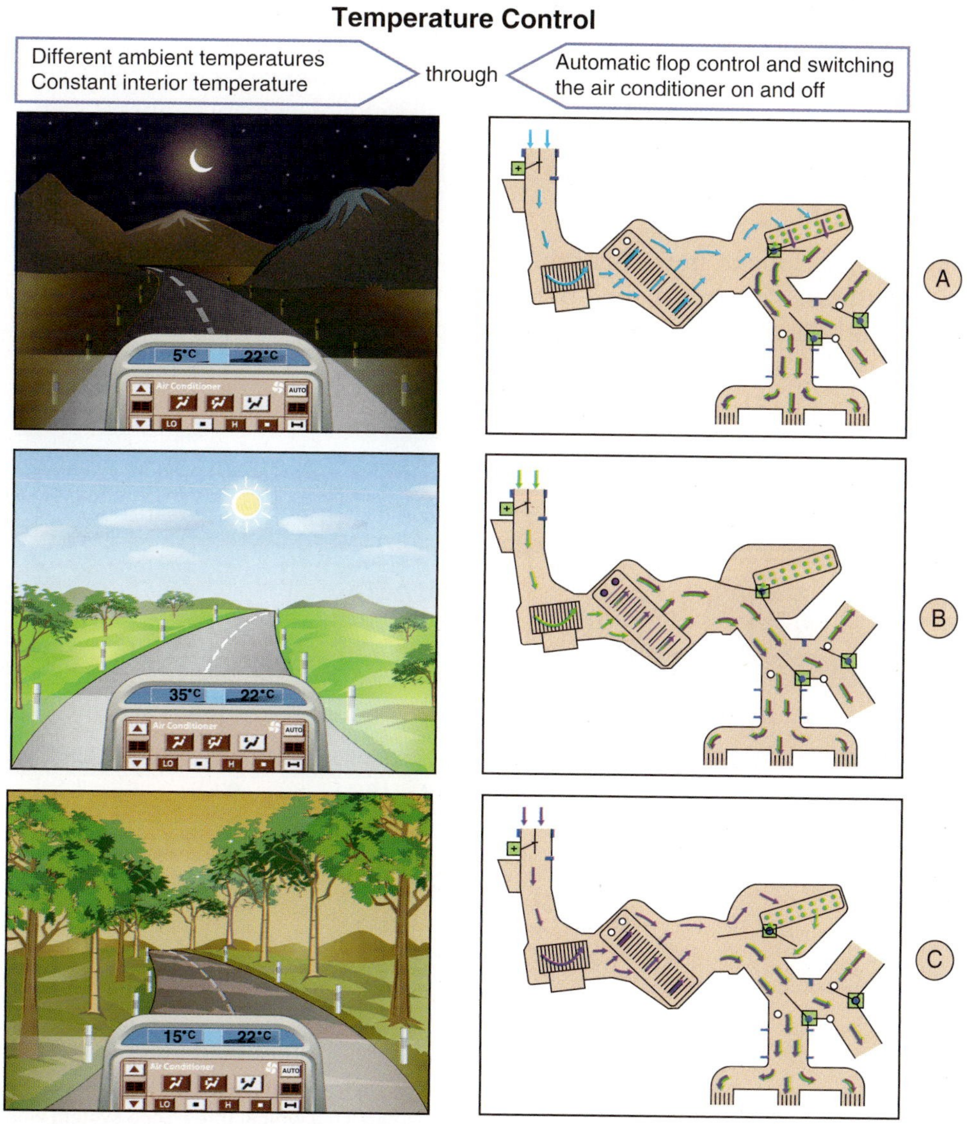

Temperature Control

Different ambient temperatures
Constant interior temperature
through
Automatic flop control and switching the air conditioner on and off

A

B

C

FIGURE 11-31 **The AUTO feature on the climate control head allows the driver to select and maintain a desired temperature setting.**

The intent of this text is to give an overall understanding of the components of the various systems, not to cover any particular system in detail. These components include but are not limited to:

- Coolant temperature sensor
- In-car temperature sensor
- Outside temperature sensor
- High-side temperature switch
- Low-side temperature switch
- Evaporator thermistor
- Low-pressure switch
- High-pressure switch
- Vehicle speed sensor
- Throttle position sensor
- Sun load sensor
- Power steering cutout switch

Many automotive electronic temperature control systems have self-diagnostic test provisions whereby an onboard microprocessor-controlled subsystem will display a code. This code (number, letter, or alphanumeric) is displayed to tell the technician the cause of the malfunction. Some systems also display a code to indicate which computer detected the malfunction. The manufacturer's specifications must be followed to identify the malfunction display codes.

It is possible for the air-conditioning system to malfunction even though self-check testing indicates there are no problems. It is then necessary to follow a manufacturer's step-by-step procedure to troubleshoot and check the system.

In-car comfort may dictate a temperature setting other than that determined to be ideal for the average person.

Control Panel

The control panel is found in the instrument panel at a convenient location for both driver and front-seat passenger access. Two types of control panels (Figure 11-32) may be found: manual and push-button. All serve the same purpose: to provide operator input control for

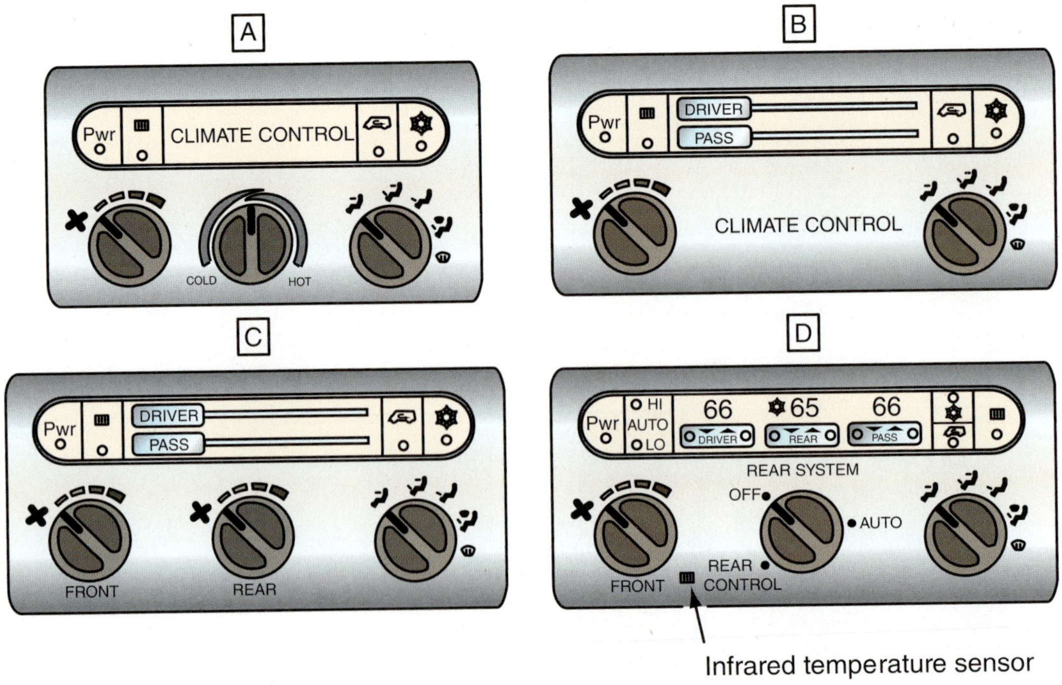

FIGURE 11-32 **Examples of typical control panels for HVAC systems: (A) a typical single-one system; (B) a typical zone system; (C) a typical three-zone system; and (D) a typical three-zone system with automatic temperature control (ATC).**

the air-conditioning and heating system. Some control panels have features that other panels do not have, such as provisions to display in-car and outside air temperature in English or metric units.

Provisions are made on the control panel for operator selection of an in-car temperature, generally between 65°F (47.2°C) and 85°F (56.6°C) in one-degree increments. Some have an override feature that provides for a setting of either 60°F (42.2°C) or 90°F (72.2°C). Either of these two settings will override all in-car temperature control circuits to provide maximum cooling or heating conditions.

A microprocessor is usually located in the control head to input data to the **programmer**, based on operator-selected conditions. When the ignition switch is turned off, a memory circuit will remember the previous setting. These conditions will be restored each time the ignition switch is turned on. If the battery is disconnected, however, the memory circuit is cleared and must be reprogrammed.

A **programmer** is the part of an automatic temperature control system that controls the blower speed, air mix doors, and vacuum or electrical actuators.

Some **dual systems** also have a control panel in the rear of the vehicle (Figure 11-33) for the comfort and convenience of the rear-seat passengers.

Dual systems usually refers to systems with two evaporators in an air-conditioning system, one in the front and one in the rear of the vehicle, driven off a single compressor and condenser system.

A mini-microprocessor is found in the control head to input temperature and humidity data selected by the operator to the programmer. Most electronic temperature control heads have provisions for self-testing, known as onboard diagnostics (OBD). This system provides a number, letter, or alphanumeric code to give the technician information relative to the problem. Manufacturers' charts must be consulted to interpret a particular code. For example, Ford's "09" or "88," a no-trouble code, corresponds to General Motors' "70." Another example, code "14," indicates "control head defective" in a Ford system, whereas "36" indicates "ATC head communications failure" in a Chrysler system.

Note too that some electronic climate control (ECC) programmers (Figure 11-34) have an "ECC diagnostic connector" provision for the connection of an external readout. Many others use the onboard OBD II 16 pin-data link connector (DLC) to access diagnostic information, codes, and system data streams.

Comfort of those in the passenger compartment is maintained by mixing cooled and ambient or heated air in the plenum section of the heater/air-conditioning duct system. In the AUTO mode, the operator sets the desired comfort level, often humidity as well as temperature. Both the quality as well as the quantity of air delivered to the passenger compartment is then controlled automatically. The blower speed and air delivery can, however, be manually controlled if desired.

The control panel in Figure 11-32D is typical for an automatic temperature control (ATC) system. The control panel may be used to select a predetermined temperature level that will automatically be maintained at all times. If desired, the operator can override the automatic

FIGURE 11-33 **A typical rear control panel.**

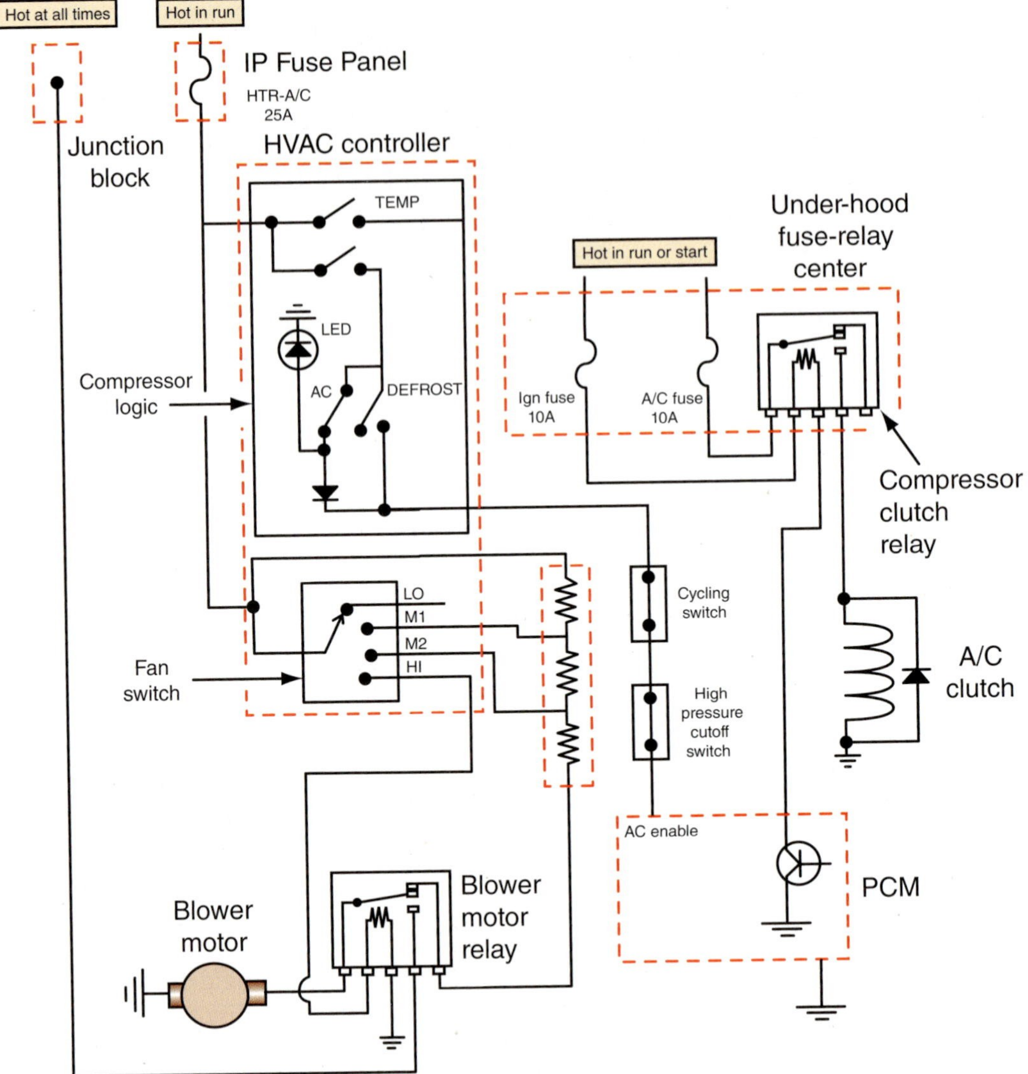

FIGURE 11-34 **A typical electronic climate control system schematic.**

provisions of the control head by selecting MAX A/C or MAX heat. Manual modes are also available for selecting BI-LEVEL, DEF, VENT, and DEFOG operation. Check the heater and A/C function test chart (Figure 11-35) for the proper system response for the various control settings.

Electronic Temperature Control Systems

Many types of electronic temperature control systems are in use. The flowcharts shown in Figure 11-36 illustrate two typical systems. The following information relates to many of the components found in an electronic temperature control system. Not all components, however, are found in all systems.

High-Side Temperature Switch

The high-side temperature switch is located in the air-conditioning system liquid line between the condenser outlet and the orifice tube inlet (Figure 11-37). Though it is a temperature-sensing device, it provides air-conditioner system pressure data to the processor. System temperature is determined by system pressure based on the temperature–pressure relationship of the refrigerant.

STEP	CONTROL SETTINGS				SYSTEM RESPONSE			
	Mode control	Temp control	Fan control	Blower speed	Heater outlet	A/C outlets	Defrost outlets	Remarks
1	OFF	60	Does not function	OFF	No airflow	No airflow	No airflow	A
2	AUTO	60	LO	LO	Min. airflow	Airflow	No airflow	A
3	AUTO	60	LO to HI	LO to HI	Small airflow	Airflow	No airflow	D
4	BI-LEVEL	60	LO to HI	LO to HI	Airflow	Airflow	Small airflow	A,D
5	AUTO	90	HI	HI	Airflow	No airflow	Small airflow	A,B,C,D
6	DEF FRT	90	HI	HI	Small airflow	No airflow	Airflow	A,D
7	DEF REAR	90	Does not change system response					A

REMARKS:

A. The word "AUTO" must appear in L/H upper corner of display when in automatic mode.
Mode arrows in display must indicate flow from appropriate outlet.
LEDs must light above mode buttons when selected.

B. Listen for air noise reduction as recirculation door closes.

C. During transition from air-conditioner outlets to heater outlets, there will be a period of time when one half of the air will be directed from the defroster outlets.

D. Check for airflow at side window defogger outlets in all modes but OFF.

FIGURE 11-35 **A typical heater and air-conditioning function test chart.**

Low-Side Temperature Switch

The low-side temperature switch is located in the air-conditioning system line between the orifice tube outlet and the condenser inlet. Its purpose is to sense low-side refrigerant pressure and to provide this information to the microprocessor.

Refrigerant Pressure Sensors

Many air-conditioning systems today are equipped with a refrigerant pressure sensor, also called a pressure transducer, on both the high-pressure and low-pressure sides of the refrigerant system. The pressure sensors provide the climate control module, body control module (BCM), and ECM/PCM with refrigerant pressure data. The pressure sensor is an electromechanical sensor and it appears physically similar to other refrigerant pressure cutoff switches, with the exception that it has three wires connected to it. Like many other computer sensors, the three circuits are the 5-volt reference circuit, a ground circuit, and a signal circuit, all connected to the control module (Figure 11-38). The typical voltage signals are 0.1 volts at 0 psig to 4.9 volts at 450 psig or above. With the use of a scan tool, both high-side and low-side refrigerant system pressures can be monitored without the need for a manifold and gauge set. One possible system code that can be stored related to this sensor is P0530—air-conditioner pressure sensor circuit.

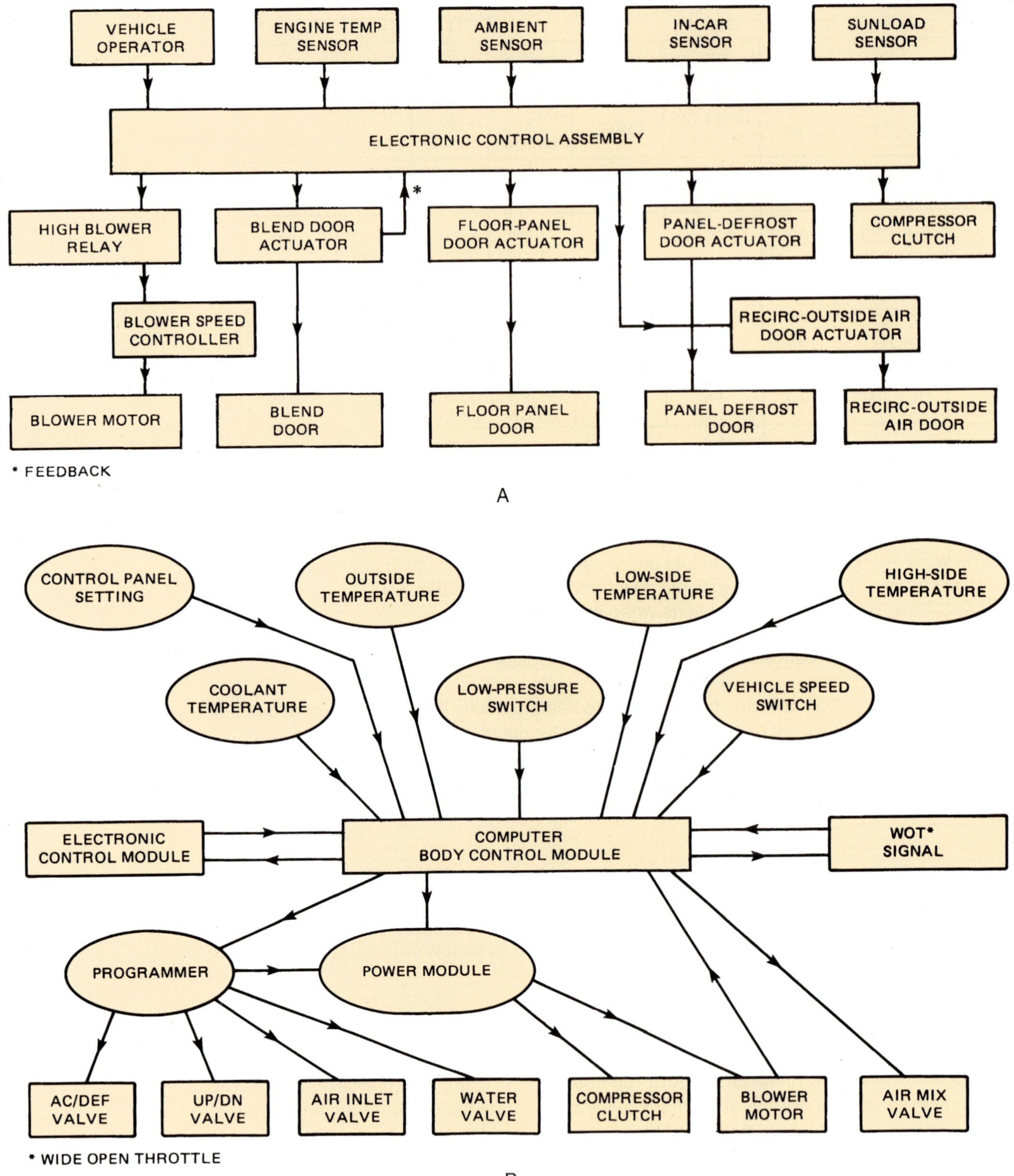

FIGURE 11-36 **Electronic temperature control flowcharts with five inputs (A) and nine outputs (B).**

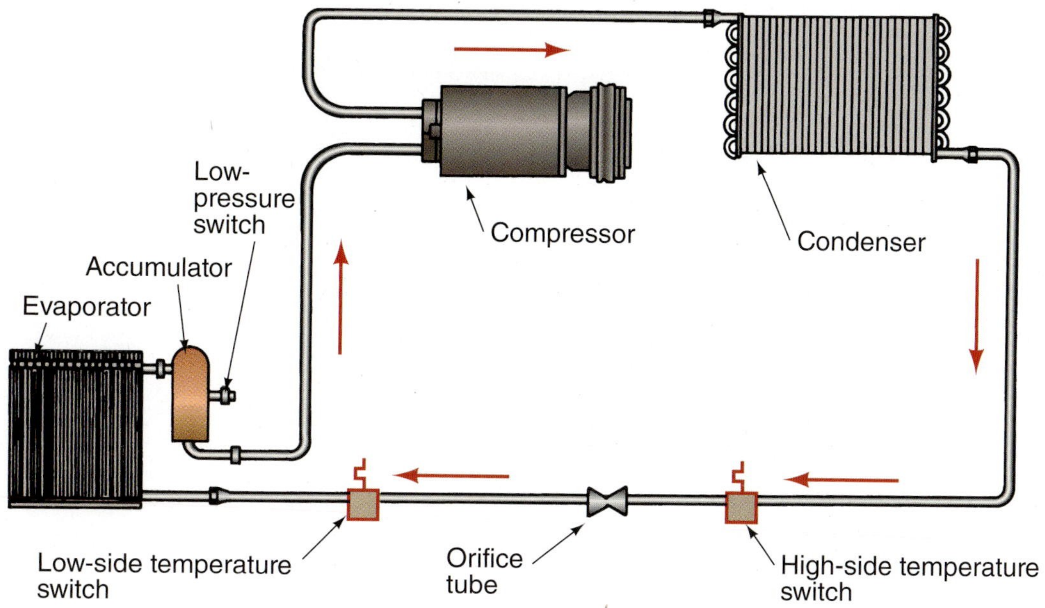

FIGURE 11-37 **Location of the low- and high-side temperature switch.**

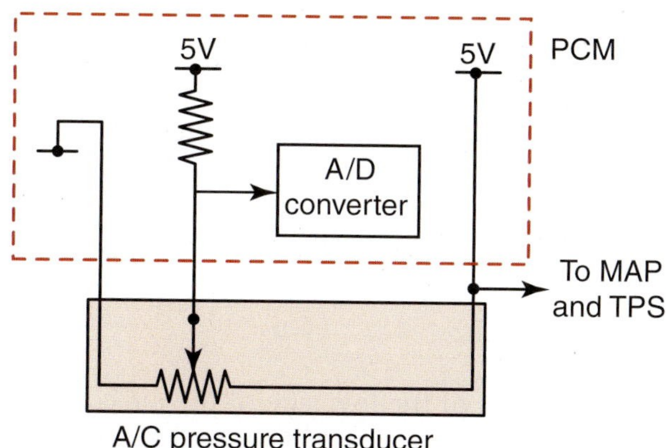

FIGURE 11-38 **The AC pressure transducer monitor high- and low-side pressure in the air-conditioning system and sends this data to the ECM/PCM or the BCM and climate control module, depending on system design.**

Shop Manual
Chapter 11,
page 468

Refrigerant pressure sensors also protect the air-conditioning system from excessively high or low pressures. If the high-pressure sensor detects excessively high pressure on the high side of the system, this voltage information is sent to the ECM/PCM or climate control module. If a high-pressure condition is detected that is above manufacturer specifications, which are typically in the range of 431 psig (2972 kPa), the control module turns off the air-conditioner compressor clutch relay until pressure drops to a safe level. Many systems will also disable the compressor if discharge pressure is too low, typically below 29 psig (203 kPa).

As was discussed earlier, the refrigerant system is also protected by a high-pressure relief valve that is located in the rear cylinder head of the compressor (Figure 11-39). If pressure in the refrigerant system reaches an unusually high level that will cause serious damage to system components (generally above 500 psig), the pressure release valve will open, releasing refrigerant into the atmosphere.

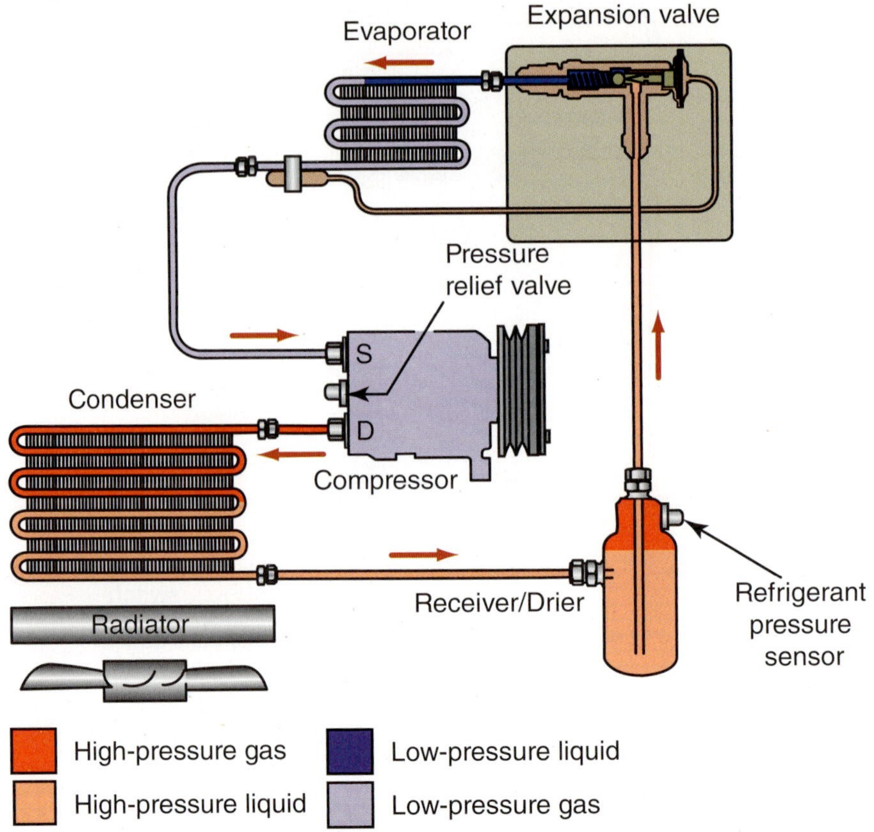

FIGURE 11-39 **The high-pressure relief valve is located at the rear of the compressor and the high-pressure sensor is located on the high-side line between the compressor discharge port and the restriction device.**

Conditions such as a loss of refrigerant may cause abnormally low low-side pressures.

Low-Pressure Switch

The low-pressure switch is located in the low side of the air-conditioning system, usually on the accumulator (Figure 11-40). This normally closed (nc) switch opens when system low-side pressure drops below 2–8 psig (13.8–55.2 kPa). An open low-pressure switch signals the microprocessor to disengage the compressor clutch circuit to prevent compressor operation during low-pressure conditions. Low-pressure conditions may result due to a loss of refrigerant or a clogged metering device.

Pressure Cycling Switch

The pressure cycling switch is found on some systems. It is used as a means of temperature control by opening and closing the electrical circuit to the compressor clutch coil. On cycling clutch systems, this switch usually opens at a low pressure of 25–26 psig (172.4–179.3 kPa) and closes at a high pressure of 46–48 psig (317.2–331 kPa). On some systems, this switch may be in line with the compressor clutch coil. On other systems, it may send data to the microprocessor to turn the compressor on and off.

Input Sensors

A sensor is a general name given to a transducer, which is short for "transfer indicator." In automotive terms, a sensor is a device that is capable of sensing a change in pressure, temperature, or other controlled variables. The climate control module uses input sensor output voltage data (Figure 11-41) to determine actuator commands that will be required to maintain a selected temperature (Figure 11-42). Climate control system input sensors are generally either a thermistor or a type of diode (in other words, photovoltaic diode).

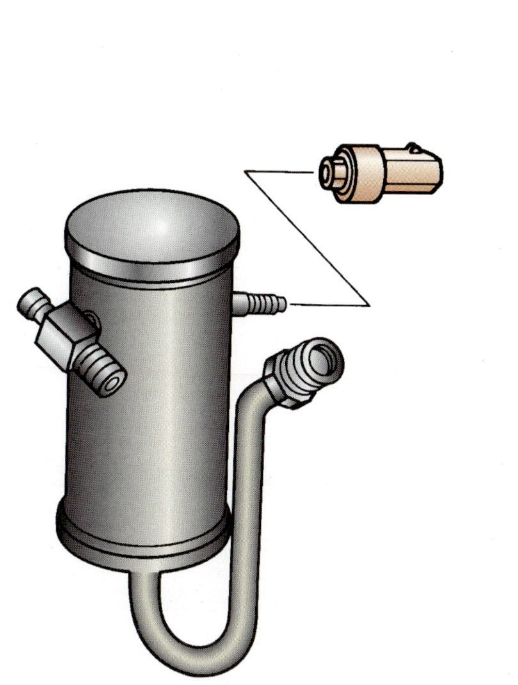

FIGURE 11-40 The low-pressure switch is usually located on the accumulator.

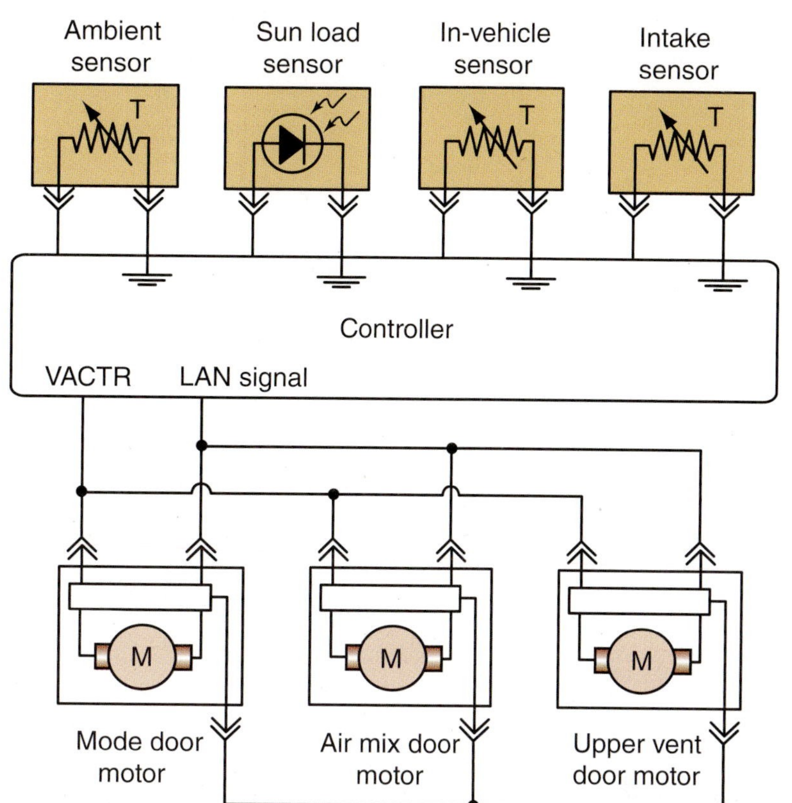

FIGURE 11-41 The most common types of climate control sensors used as either the thermistor or the diode. These input sensors supply data directly to the climate control module or BCM.

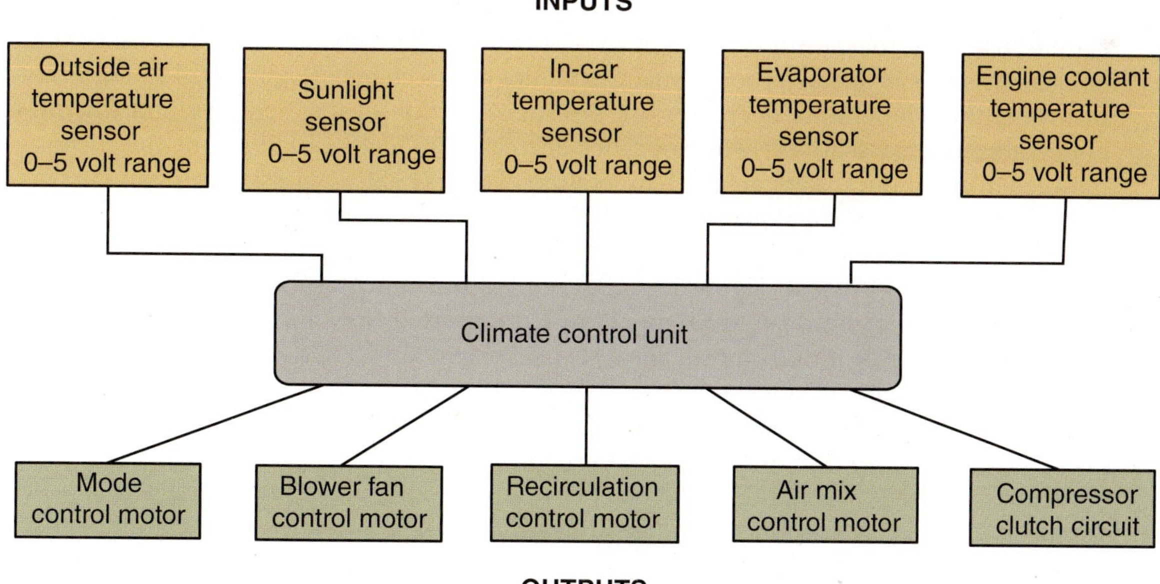

FIGURE 11-42 The climate control module logic uses input sensor information to determine output actuator control to maintain selected cabin temperature.

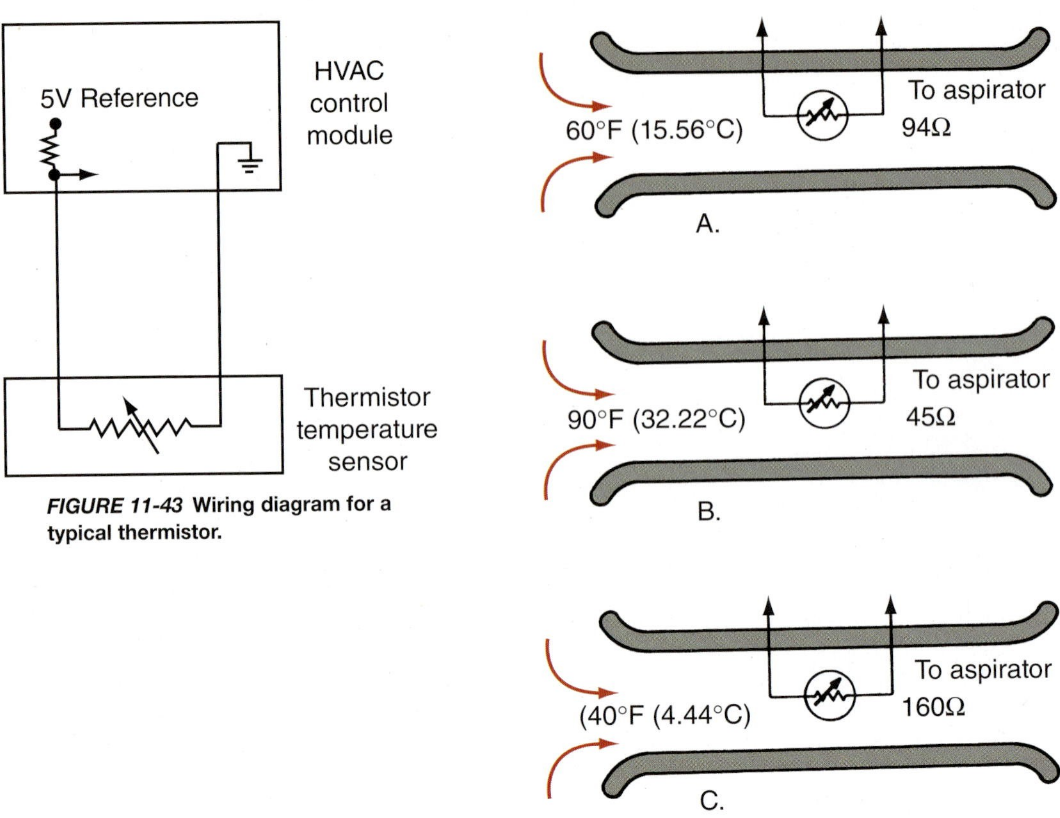

FIGURE 11-43 **Wiring diagram for a typical thermistor.**

FIGURE 11-44 **The resistance of a thermistor changes as temperature changes.**

Although they may vary in physical appearance, all thermistors have the same general operating characteristics (Figure 11-43)—that is, they are extremely sensitive to slight changes in temperature. A thermistor may be one of either two designs, a negative temperature coefficient (NTC) or positive temperature coefficient (PTC) design. As a rule, climate control systems use NTC thermistor designs. In an NTC thermistor, a change in the resistance value of each sensor is inversely proportional to a temperature change (Figure 11-44). For example, when the temperature decreases, the resistance of the sensor increases; and when the temperature increases, the sensor resistance decreases. In Figure 10-59A, one thermistor is installed in an air duct. With air at a temperature of 60°F (15.56°C) passing through the duct, the resistance value of the thermistor is 94 ohms. Referring to the resistor value given in the chart (Figure 11-45), this thermistor is currently reading correctly. If the temperature in the duct is 90°F (32.22°C), as in Figure 11-44B, then the resistance of the thermistor decreases to about 45 ohms. If, however, the temperature is decreased to 40°F (4.44°C), the thermistor resistance is increased to 160 ohms (Figure 11-44C). Even though this value is not on the chart, it appears the thermistor is reacting as designed.

A graph of individual sensor values at various temperatures (Figure 11-46) may be compared with the examples given to this point. Note that each sensor has a different value for a particular temperature. Though all NTC thermistors react to temperature changes in the same manner, the specific resistance value for a given temperature varies among sensor types and specific vehicle applications. Always refer to specific vehicle information when trying to determine proper thermistor operation.

Shop Manual
Chapter 11,
page 462

Sun Load Sensor

The sun load sensor (Figure 11-47) is usually found atop the dashboard, adjacent to one of the radio speaker grilles. The sun load sensor is a photovoltaic diode that sends an appropriate

Temperature* °F	°C	Resistance Ohms
50	10.0	120
51	10.6	117.5
52	11.1	115
53	11.7	112.5
54	12.2	110
55	12.8	107
56	13.3	104
57	13.9	101.5
58	14.4	99
59	15.0	96.5
60	15.6	94
61	16.1	92.5
62	16.7	91
63	17.2	89
64	17.8	87
65	18.3	85

Temperature* °F	°C	Resistance Ohms
66	18.9	83
67	19.4	81
68	20.0	79
69	20.6	77
70	21.1	75
71	21.7	73.5
72	22.2	72
73	22.8	70.5
74	23.3	69
75	23.9	67.5
76	24.4	66
77	25.0	64.5
78	25.6	63
79	26.1	61.5
80	26.7	60
81	27.2	58.5

*Temperature of ambient air passing across the thermistor

FIGURE 11-45 **Thermistor values.**

signal to the microprocessor to aid in regulating the in-car temperature. The sun load sensor can also be found under the defrost grille at about the center of the windshield. It is a thermistor that is sensitive to the heat load of the sun on the vehicle. As light level increases, the resistance of the photodiode increases.

The BCM compares the sun load values with in-car temperature values to determine how much cooling is required in order to maintain selected in-vehicle temperature conditions (Figure 11-48).

When sunlight intensity is high, the control module will automatically increase blower fan speed and increase the volume of air flowing to the dash discharge vents. This in turn improves comfort by preventing the passengers from feeling hotter from the effects of direct sunlight increasing the temperature of the upper portion of the body.

Outside Temperature Sensor

The outside temperature sensor (OTC), also called an ambient temperature sensor (ATS), is a negative temperature coefficient thermistor in a protective housing located behind either the front bumper or grille (Figure 11-49). As with an NTC thermistor, when the temperature of the sensor increases, the resistance of the sensor will decrease (Figure 11-50). The climate control module passes a 5-volt reference through the thermistor and measures current flow. The purpose of the OTC is to sense outside ambient temperature conditions to provide data to the microprocessor.

Shop Manual
Chapter 11, page 466

Data from the OTC is processed by the BCM and is displayed on the electronic climate control (ECC). Through a rather complicated process, the OTC provides information regarding outside ambient temperature that is essential for the proper operation of an electronic automatic temperature control (EACT) system.

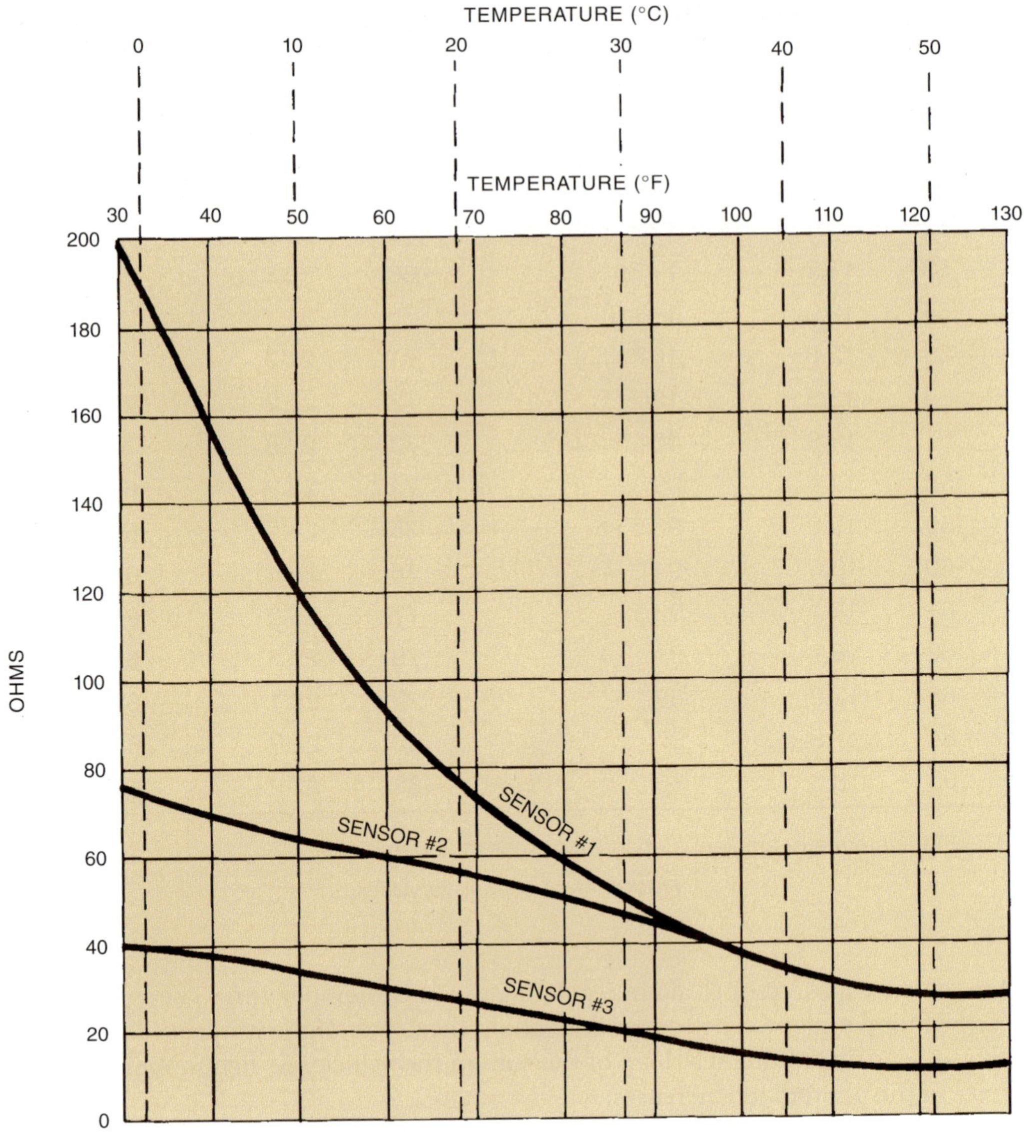

TEMPERATURE (°C)
0
10
20
30
40
50
TEMPERATURE (°F)
30
40
50
60
70
80
90
100
110
120
130
OHMS
200
180
160
140
120
100
80
60
40
20
0
SENSOR #1
SENSOR #2
SENSOR #3

FIGURE 11-46 **Individual sensor values graphed.**

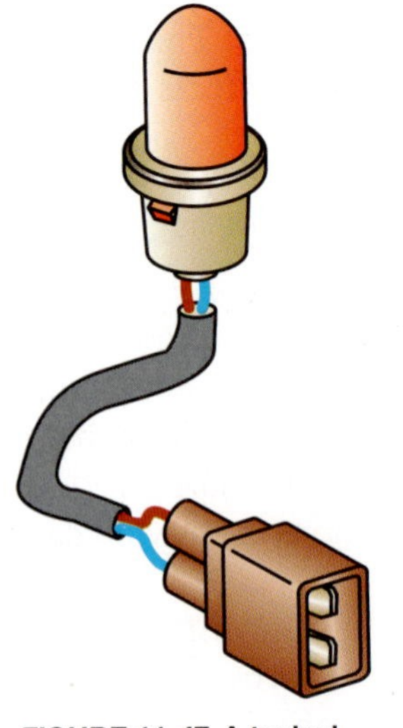

FIGURE 11-47 **A typical sun load sensor.**

5 volt reference
Automatic Temperature Control module
Sun load sensor
Photodiode

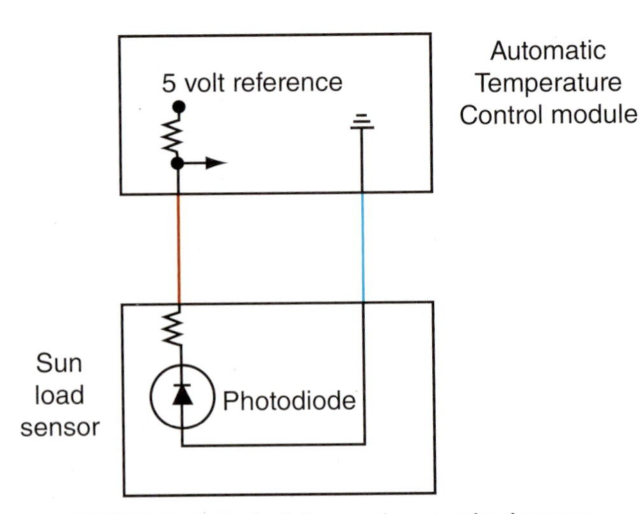

FIGURE 11-48 **Typical diagram for a sun load sensor.**

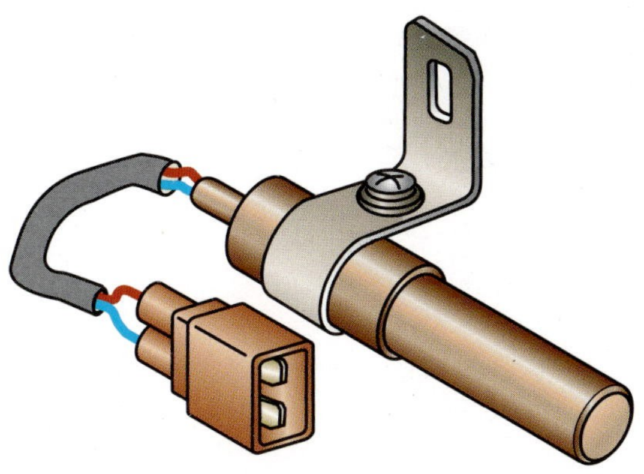

FIGURE 11-49 **The outside temperature sensor is located behind the grille.**

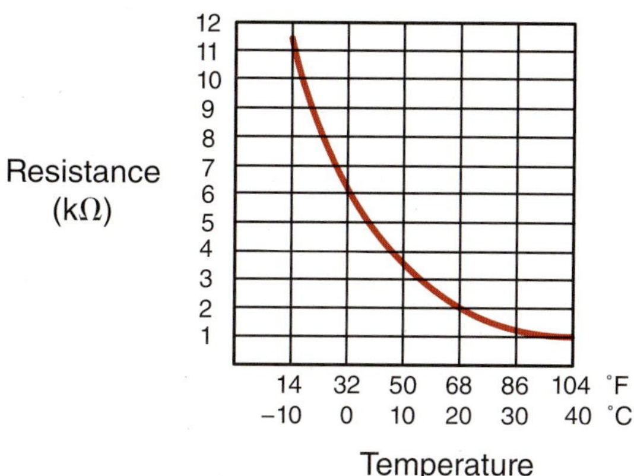

FIGURE 11-50 **Both the OTC and the in-car temperature sensor are negative temperature coefficient thermistors that decrease in resistance with an increase in temperature.**

This sensor circuit has several programmed memory features to prevent false ambient temperature data input during periods of low-speed driving or when stopped close behind another vehicle, such as when waiting for a traffic signal. If ambient air temperature is below the minimum preprogrammed level, the control module will not allow the air-conditioning compressor clutch to engage. Air conditioning is not generally required at temperatures below 50°F (10°C). This is due to the low relative moisture content contained in air at that temperature, even at the saturation point.

An **aspirator** is a device that uses suction to move air, accomplished by a differential in air pressure.

In-Car Temperature Sensor

Shop Manual
Chapter 11, page 467

The in-car temperature sensor, also called an in-vehicle sensor (Figure 11-51), is located in a tubular device called an **aspirator**. A small amount of in-car air is drawn through

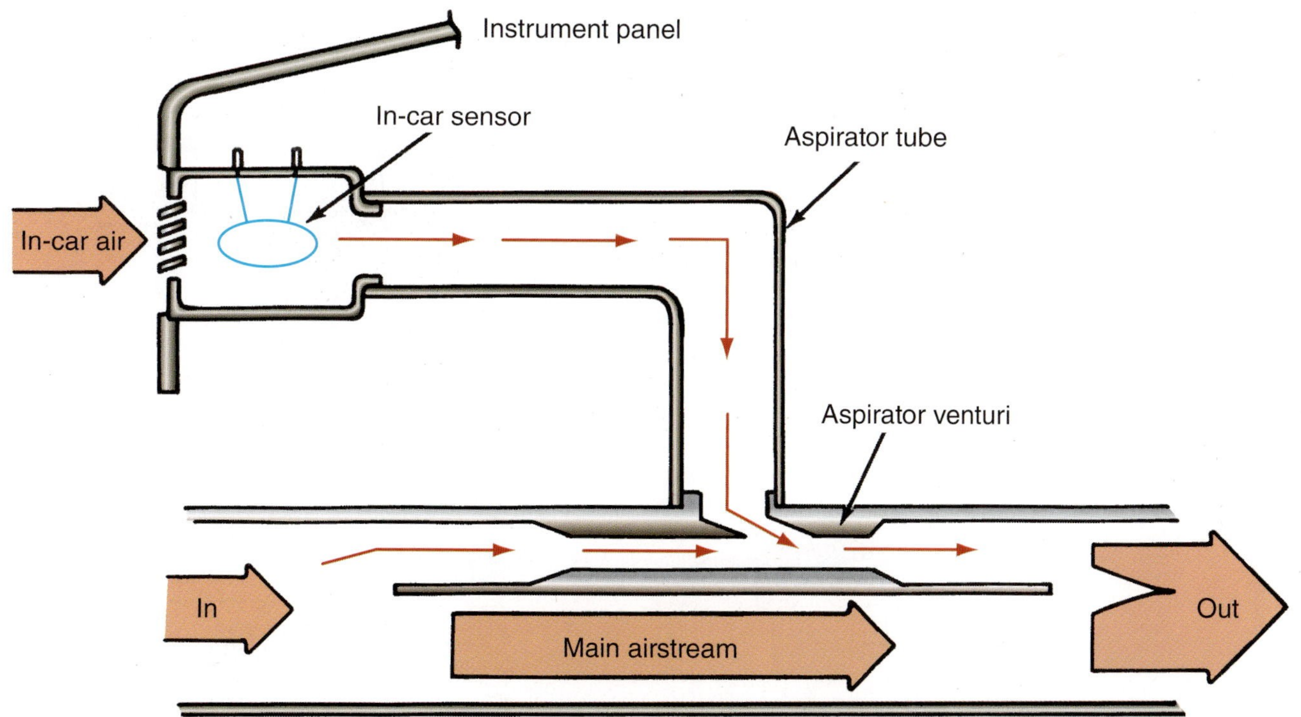

FIGURE 11-51 **A typical in-car temperature sensor and aspirator assembly.**

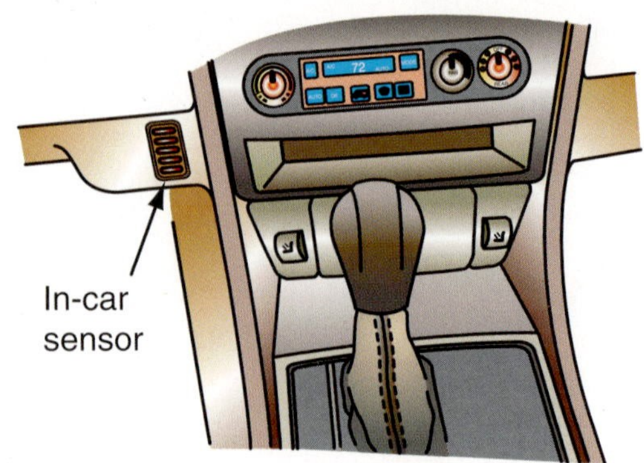

FIGURE 11-52 **The in-car temperature sensor may be located in a separate louvered housing on the dashboard or climate control panel.**

the aspirator across the in-car sensor to provide average in-car temperature data to the microprocessor.

The aspirator is a small duct system that is designed to cause a small amount of in-car air to pass through it. The main airstream causes a low pressure (suction) at the inlet end of the aspirator. This causes in-car air to be drawn into the in-car sensor plenum. The in-car sensor, located in the plenum, is continuously exposed to average in-car air to monitor the in-car air temperature.

Like the OTC, the in-car temperature sensor is also a negative temperature coefficient thermistor. Some in-car temperature sensors use a small electric fan positioned behind the thermistor to draw in-car air across sensor (Figure 11-52) to maintain accurate in-car temperature readings.

Evaporator Temperature Sensor

Shop Manual
Chapter 11, page 461

The evaporator temperature sensor (Figure 11-53), also called a fin temperature sensor, is used on many systems to control evaporator temperature. It is located at the evaporator and is fitted in between the cooling fins to measure and control evaporator core temperature. The evaporator temperature sensor is another negative temperature coefficient (NTC) thermistor.

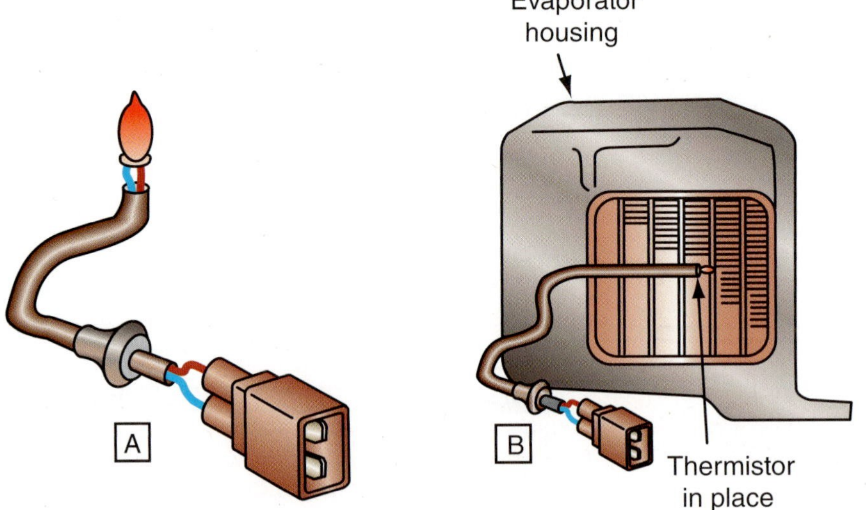

FIGURE 11-53 **A typical evaporator thermistor (A) in position; and (B) between evaporator cooling fins.**

Like all other NTC thermistors, when the temperature of the sensor increases, the resistance of the sensor will decrease. The climate control module passes a 5-volt reference through the thermistor and measures current flow. The purpose of the evaporator temperature sensor is to sense evaporator core temperature conditions and provide the data to the controller. By monitoring the core temperature, the microprocessor can determine whether the compressor clutch should be turned on or off to add additional refrigerant temperature regulation. The compressor is turned off when the evaporator core temperature drops to 34°F (1.1°C). This prevents the formation of frost and ice on the fins of the evaporator.

On electronically controlled variable displacement compressors, the output volume of the air-conditioner compressor can be controlled by pulse width modulation of the compressor electronic pressure control valve. In this manner, the compressor output can be varied to complement the actions taken by the thermal expansion valve to maintain the evaporator core at the optimal temperature range for current system demands.

Infrared Temperature Sensor

Some vehicles with automatic temperature control are equipped with an infrared temperature sensor (ITS) to determine passenger compartment temperature. It may be located in the temperature control assembly or near the dash discharge outlets at the center of the dash. Infrared temperature sensors measure surface temperatures rather than air temperature and thus can adjust passenger compartment temperatures based on the perceived temperature (caused by ultraviolet radiation from the sun or evaporative heat loss) of the passenger instead of actual air temperature. The temperature control module interprets the data received by the ITS and evaporator temperature sensor to adjust blower speed and the amount of refrigerant flowing through the evaporator core to maintain the selected temperature level of the passenger compartment. If an infrared sensor is found to be defective, it must be replaced as an assembly.

Shop Manual
Chapter 11, page 464

Air Humidity Sensor

Information from the air humidity sensor is used by some hybrid electric platforms to reduce air-conditioning compressor load when in-car humidity levels are low. On some platforms, the sensor is located at the base of the rearview mirror and integrates an air humidity sensor, windshield temperature sensor, and interior temperature sensor. This information is used to permit adaptive control of the automatic defrost function.

The air humidity sensor integrates a temperature sensor since determining air humidity is dependent on air temperature. The ability of air to hold moisture is directly proportional to its temperature. The humidity sensor uses a capacitive thin-layer film to measure moisture content of the air. The capacitor uses a special dielectric material that absorbs water vapor. The water absorbed by the capacitance material changes the electrical properties and thus the capacitance of the capacitor (Figure 11-54). By measuring the capacitance and converting this to a voltage signal, air humidity can be measured.

Coolant Temperature Sensor

The engine coolant temperature (ECT) sensor is an NTC thermistor (Figure 11-55) that is located in an engine coolant passage. It provides engine coolant temperature information to various control modules including the climate control module ECM/PCM and BCM.

The control unit supplies a 5-volt reference voltage to the sensor, and as the engine temperature increases, the sensor resistance will decrease. As sensor resistance decreases, it creates an easier path to ground, which results in a lower voltage observed at the control module. Based on the information received from the ECT sensor, the climate control module can increase or decrease blower fan operation on some vehicle models. As an example, on a vehicle with FULL AUTO mode climate control, the blower fan will not operate if engine coolant temperature is too low.

Shop Manual
Chapter 11, page 465

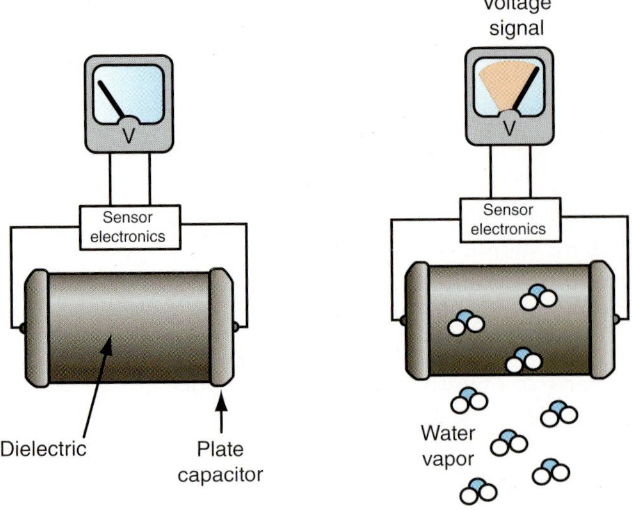

FIGURE 11-54 **The humidity sensor uses a capacitive thin-layer film to measure moisture content of the air. The water absorbed by the capacitance material changes the electrical properties and thus the capacitance of the capacitor.**

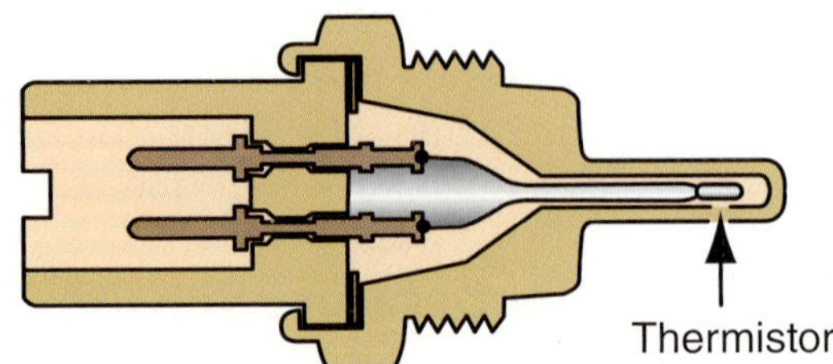

FIGURE 11-55 **The ECT sensor is an NTC thermistor located in an engine coolant passage.**

This sensor also provides input information to other onboard computers to provide data for fuel enrichment, ignition timing, exhaust gas recirculate operation, canister purge control, idle speed, and closed-loop fuel control. A defective coolant temperature sensor will cause poor engine performance, which will probably be evident before poor air-conditioning performance is noticed.

Vehicle Speed Sensor

The vehicle speed sensor is a pulse generator that is usually located at the transmission output shaft. It provides actual vehicle speed data to the microprocessor as well as other subsystems, such as the electronic control module (ECM).

Throttle Position Sensor

The throttle position sensor is actually a potentiometer with a voltage input from the processor. The processor, then, determines throttle position based on the return voltage signal. At the wide-open throttle (WOT) position, the compressor clutch is disengaged to provide maximum power for acceleration. This device is often called the WOT sensor and is most often found on diesel engine–equipped vehicles.

Heater Turn-On Switch

The heater turn-on switch is usually a bimetallic snap-action switch found in the coolant stream of the engine. Its purpose is to prevent blower operation when engine coolant temperature is below 118°F–122°F (48.9°C–50°C), if heat is selected.

If cooling is selected, the programmer will override this switch to provide immediate blower operation, regardless of engine coolant temperature.

Brake Booster Vacuum Switch

The brake booster vacuum switch is a low-pressure switch.

The brake booster vacuum switch is found on some cars. Its purpose is to disengage the air-conditioning compressor whenever braking requires maximum effort. This switch, which is usually in series with the compressor clutch electrical circuit, does not provide data to the microprocessor.

Power Steering Cutoff Switch

The power steering cutoff switch, which is found on some cars, is used to disengage the air-conditioning compressor whenever power steering requires maximum effort. This switch, on some cars, is in series with the compressor control relay and does not provide data to the programmer. On other applications, this switch is in the electronic control module and provides feedback data to the microprocessor.

The power steering cutoff switch is a high-pressure switch

Analyzing Sensor Input Information

The following is a simplified description of how the electronic climate control system uses input sensor information to control outputs in the FULL AUTO mode automatic climate control mode.

The climate control module must analyze data from the in-car temperature sensor, the outside ambient air temperature sensor (Figure 11-56), and the temperature level selected at the control panel before calculating the best position for the recirculation door to optimize system performance.

When the climate control module calculates where discharge air should be directed and at what volume, it must analyze data from the in-car temperature sensor, the sun load sensor (Figure 11-57), and the temperature level selected at the control panel.

The climate control module then figures the best position for the mode door control motors to optimize system passenger comfort.

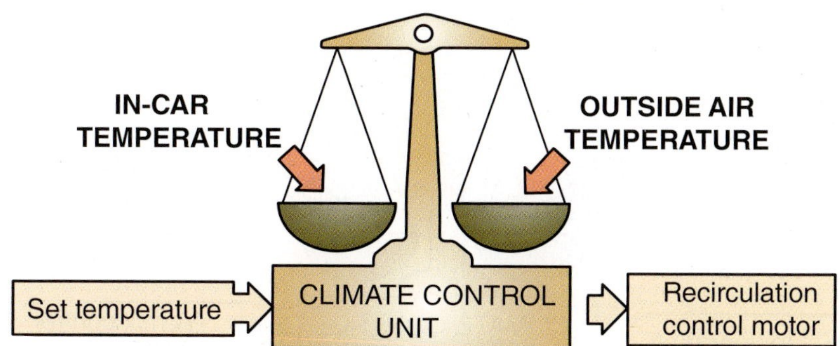

FIGURE 11-56 **The climate control module will weigh various inputs when determining the position for the recirculation mode control door.**

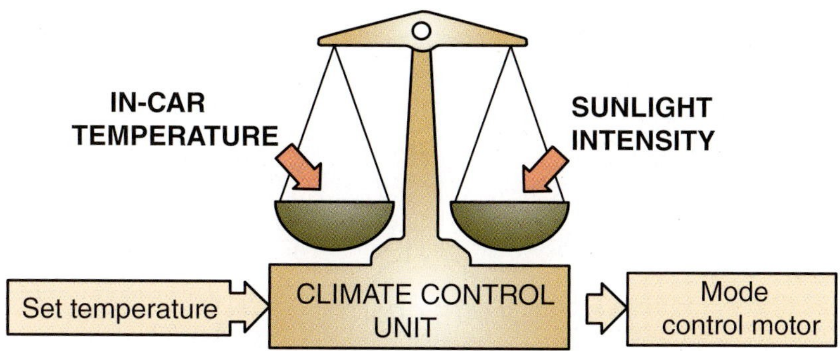

FIGURE 11-57 **The climate control module will weigh various inputs when determining the position for the mode door control motors.**

When the weather is cooler and both heating and dehumidification may be required, the climate control module will determine whether hotter or cooler air is needed in the passenger compartment to maintain the selected preset temperature. It must analyze data from the in-car temperature sensor, the outside ambient air temperature sensor (Figure 11-58), and the temperature level selected at the control panel. The climate control module then calculates the best position for the air mix control motor and heater valve, if equipped to optimize system passenger comfort.

At other times, the control module determines the appropriate speed for the blower motor to achieve both a high level of passenger comfort and system performance (Figure 11-59). This can be a complicated process, and the control module must interpret information from the sun load sensor, in-car temperature sensor, evaporator temperature, engine coolant temperature, and, as always, the desired passenger compartment temperature selected.

During varying heat load conditions, the climate control module may need to cycle the compressor clutch off to maintain evaporator core temperature and to avoid frost or ice forming on the evaporator cooling fins (Figure 11-60). The control module analyzes data from the in-car temperature sensor, selected passenger compartment temperature, and the evaporator temperature sensor to determine whether the air-conditioning compressor clutch should be engaged or disengaged.

The climate control system logic is much more complex than can be fully explained in the examples given here, and often many of the described conditions are occurring simultaneously. These brief examples should improve your understanding of the system complexity and how decisions are arrived at.

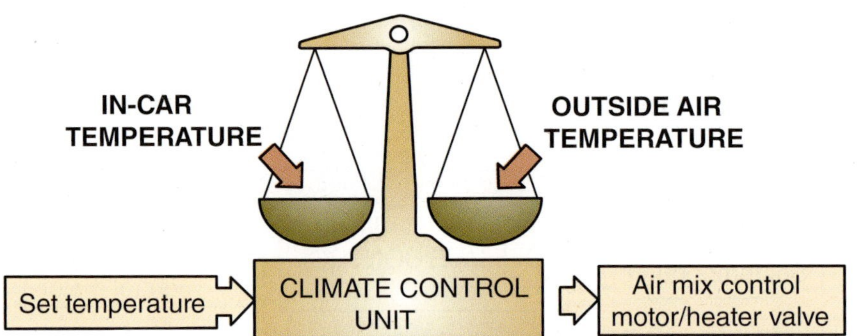

FIGURE 11-58 The climate control module will determine the best position for the air mix door, based on both inside and outside temperatures.

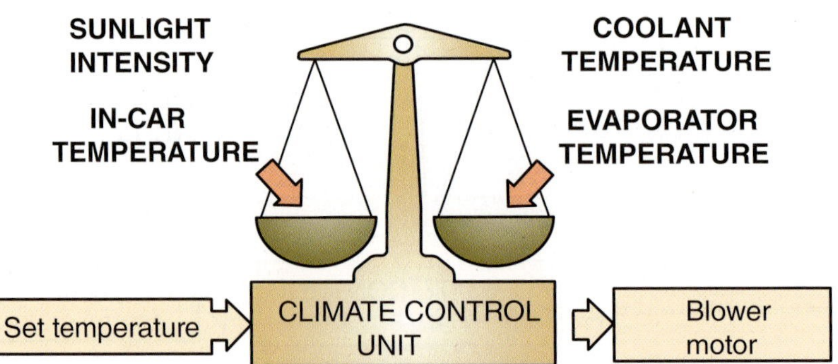

FIGURE 11-59 The climate control module will determine the best blower motor speed, based on many inputs, to achieve both a high level of passenger comfort and system performance.

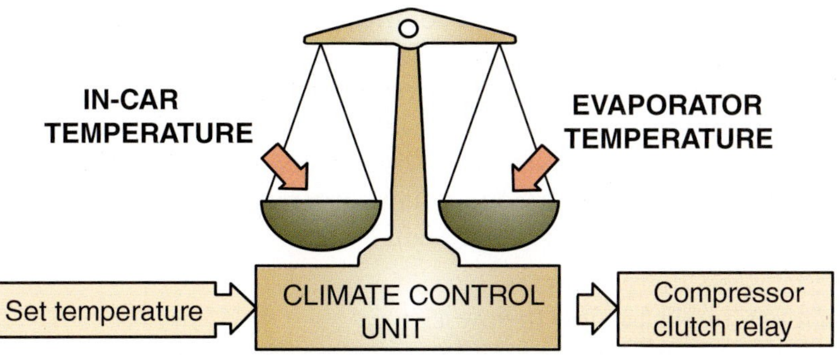

FIGURE 11-60 **The climate control module at times decides whether the air-conditioning compressor clutch should be engaged or disengaged.**

Scan Tool

A scan tool (Figure 11-61) is a microprocessor that is designed to communicate with the vehicle's computer. When connected to the computer through diagnostic connectors, a scan tool accesses diagnostic trouble codes (DTCs), runs tests to check system operations, and monitors the activity of the system. Both trouble codes and test results are displayed on a light-emitting diode (LED) screen or are printed out on the scanner printer.

Today, most scan tools have large screens with many lines of information and graphing capabilities. Many may be interfaced with a laptop or desktop computer for increased data display and graphing, as well as enabling service technicians to store information and create their own data and waveform library. Most scan tools can store the test data in a random access memory (RAM) that can be accessed by a printer, personal computer, or an engine analyzer to retrieve the information.

The average person is comfortable at 78°F–80°F (25.6°C–26.7°C) at a relative humidity (RH) of 45–50 percent.

Trouble codes set by the computer help the technician identify the cause of the problem. Most diagnostic work on computer control systems should be based on a description of symptoms to help locate any technical service bulletins that refer to the problem. One can also use the symptom description to locate the appropriate troubleshooting sequence in the manufacturer's service manuals.

Since 1996 and the introduction of OBD II and a standard 16-pin data link connector (DLC), most climate control systems can be accessed through this same connector assembly with a generic OBD II scan tool. Manufacturer-specific and many enhanced generic scan tools also have the ability not only to retrieve data and trouble codes from the climate control module or BCM but also to allow the technician to activate output devices like the

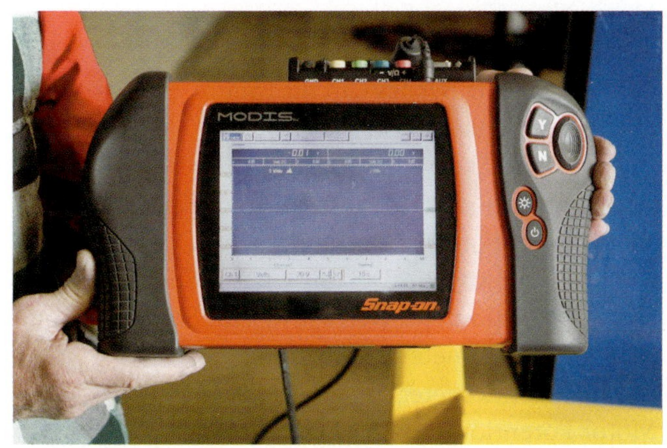

FIGURE 11-61 **A typical handheld scan tool with graphing features.**

air-conditioner compressor clutch when a command is received from the scan tool. This is generally referred to as bidirectional control. The scan tool can also read refrigerant system high-side and low-side operating pressures on systems equipped with pressure transducers.

If the air-conditioning system compressor will not engage, a scan tool can be used to verify whether an air conditioner request signal is present when the HVAC control switch is set to AC or DEFROST. Sometimes the air-conditioner compressor clutch will not be commanded on if any of the following conditions are present:

Shop Manual
Chapter 11,
page 485

- Battery voltage is less than 10.5 volts.
- Intake air temperature is too low (below 45°F).
- Engine coolant temperature is too high (greater than 260°F).
- Throttle angle is above 90 percent.
- Refrigerant system high-side pressure is above 431 psig (2972 kPa).
- Refrigerant system discharge pressure is below 29 psig (203 kPa).

It should be noted that these are only examples and that specific vehicle information should be consulted.

The following is a list of OBD II codes relevant to HVAC and climate control system diagnostics:

Generic OBD-II Body Codes:

Shop Manual
Chapter 11,
page 486

B1200—Climate control push-button circuit failure
B1239—Airflow Blend Door Driver Circuit Failure
B1242—Airflow Recirculation Door Driver Circuit Failure
B1249—Blend Door Failure
B1250—Air Temperature Internal Sensor Circuit Failure
B1251—Air Temperature Internal Sensor Circuit Open
B1252—Air Temperature Internal Sensor Circuit Short to Battery Power
B1253—Air Temperature Internal Sensor Circuit Short to Ground
B1254—Air Temperature External Sensor Circuit Failure
B1255—Air Temperature External Sensor Circuit Open
B1256—Air Temperature External Sensor Circuit Short to Battery Power
B1257—Air Temperature External Sensor Circuit Short to Ground
B1258—Solar Radiation Sensor Circuit Failure
B1259—Solar Radiation Sensor Circuit Open
B1260—Solar Radiation Sensor Circuit Short to Battery
B1261—Solar Radiation Sensor Circuit Short to Ground
B1262—Servo Motor Defrost Circuit Failure
B1263—Servo Motor Vent Circuit Failure
B1264—Servo Motor Foot Circuit Failure
B1265—Servo Motor Cool Air Bypass Circuit Failure
B1266—Servo Motor Air Intake Left Circuit Failure
B1267—Servo Motor Air Intake Right Circuit Failure
B1268—Servo Motor Potentiometer Defrost Circuit Failure
B1269—Servo Motor Potentiometer Defrost Circuit Open
B1270—Servo Motor Potentiometer Defrost Circuit Short to Battery
B1271—Servo Motor Potentiometer Defrost Circuit Short to Ground
B1273—Servo Motor Potentiometer Vent Circuit Failure
B1273—Servo Motor Potentiometer Vent Circuit Open
B1274—Servo Motor Potentiometer Vent Circuit Short to Battery
B1275—Servo Motor Potentiometer Vent Circuit Short to Ground
B1276—Servo Motor Potentiometer Foot Circuit Failure

B1277—Servo Motor Potentiometer Foot Circuit Open
B1278—Servo Motor Potentiometer Foot Circuit Short to Battery
B1279—Servo Motor Potentiometer Foot Circuit Short to Ground
B1280—Servo Motor Potentiometer Cool Air Circuit Failure
B1281—Servo Motor Potentiometer Cool Air Circuit Open
B1282—Servo Motor Potentiometer Cool Air Circuit Short to Battery
B1283—Servo Motor Potentiometer Cool Air Circuit Short to Ground
B1284—Servo Motor Potentiometer Air Intake Left Circuit Failure
B1285—Servo Motor Potentiometer Air Intake Left Circuit Open
B1286—Servo Motor Potentiometer Air Intake Left Circuit Short to Battery
B1287—Servo Motor Potentiometer Air Intake Left Circuit Short to Ground
B1288—Servo Motor Potentiometer Air Intake Right Circuit Failure
B1289—Servo Motor Potentiometer Air Intake Right Circuit Open
B1290—Servo Motor Potentiometer Air Intake Right Circuit Short to Battery
B1291—Servo Motor Potentiometer Air Intake Right Circuit Short to Ground
B1849—Climate Control Temperature Differential Circuit Failure
B1850—Climate Control Temperature Differential Circuit Open
B1851—Climate Control Temperature Differential Circuit Short to Battery
B1852—Climate Control Temperature Differential Circuit Short to Ground
B1853—Climate Control Air Temperature Internal Sensor Motor Circuit Failure
B1854—Climate Control Air Temperature Internal Sensor Motor Circuit Open
B1855—Climate Control Air Temperature Internal Sensor Motor Circuit Short to Battery
B1856—Climate Control Air Temperature Internal Sensor Motor Circuit Short to Ground
B1857—Climate Control On/Off Switch Circuit Failure
B1858—Climate Control A/C Pressure Switch Circuit Failure
B1859—Climate Control A/C Pressure Switch Circuit Open
B1860—Climate Control A/C Pressure Switch Circuit Short to Battery
B1861—Climate Control A/C Pressure Switch Circuit Short to Ground
B1862—Climate Control A/C Lock Sensor Failure
B1946—Climate Control A/C Post Evaporator Sensor Circuit Failure
B1947—Climate Control A/C Post Evaporator Sensor Circuit Short to Ground
B1948—Climate Control Water Temperature Sensor Circuit Failure
B1949—Climate Control Water Temperature Sensor Circuit Short to Ground
B1966—A/C Post Heater Sensor Circuit Failure
B1967—A/C Post Heater Sensor Circuit Short to Ground
B1968—A/C Water Pump Detection Circuit Failure
B1969—A/C Clutch Magnetic Control Circuit Failure
B2175—A/C Request Signal Circuit Short to Ground
B2380—Heater Coolant Temp Sensor Circuit Short to Ground
B2381—Heater Coolant Temp Sensor Circuit Open
B2428—A/C Post Heater Sensor 2 Circuit Failure
B2429—A/C Post Heater Sensor 2 Circuit Short to Ground
B2513—Blower (Fan) Circuit Failure
B2514—Blower (Fan) Circuit Short to Battery Power
B2515—Heater Blower Relay Circuit Failure
B2516—Blower Control Circuit Failure
B2518—Compressor Over Temperature Fault
B2606—A/C Temperature Sensor Out of Range

Generic OBD-II Network Codes:

U0164—Lost Communications with HVAC Control Module
U0165—Lost Communications with HVAC Control Module, rear
U0324—Software Incompatibility with HVAC Control Module
U0325—Software Incompatibility with Auxiliary Heater Control Module
U0422—Invalid Data Received from Body Control Module
U0424—Invalid Data Received from HVAC Control Module
U0425—Invalid Data Received from Auxiliary Heater Control Module

Though this list is extensive, it is not intended to be comprehensive. Always consult specific manufacturer service information to aid in the diagnostic process. Using a scan tool and following the published diagnostic steps for a specific trouble code will improve your ability to repair HVAC system failures. Though scan tools are a vital tool, not all system failures have set diagnostic trouble codes. You must become familiar with circuit wiring diagrams and the network communication process on the vehicle you are working on. In addition, you must become comfortable in the use of digital multimeters and all the functions available on them. A thorough understanding of automotive electrical systems and electronics is required in today's world of advanced electrical control devices, along with knowledge of specific system operation.

Electronic HVAC Control Module Self-Diagnosis

Some older climate control systems have the ability to perform self-diagnostics without the aid of a scan tool. These systems can run function tests and enter sensor/actuator relearn procedures. In addition, these systems can display trouble codes either in the form of flash coding LEDs on system control buttons or by displaying the code on the climate control screen.

Shop Manual
Chapter 11, page 477

To access trouble codes, the system must enter the self-diagnosis mode. The retrieval system varies among manufacturers and even varies between models and years, making it necessary to consult vehicle-specific service information to enter the self-diagnosis mode. In general, the electronic control panel will enter the self-diagnosis mode by pressing a series of buttons on the control panel in a specific sequence. Once the self-diagnosis mode is entered, the control panel will display the code or flash an LED on one of the control panel buttons, with any specific codes stored. In addition, some systems will perform a self-test on all output devices and cycle the HVAC system through all available mode cycles (in other words, operate all mode control doors in the ventilation system).

The following is an example of a portion of one manufacturer's sequence required to access the Self-Diagnostic Mode (Figure 11-62):

1. Turn the ignition switch ON.
2. Within 10 seconds of turning the ignition switch on, press and hold the HVAC control panel OFF button for at least 5 seconds.
3. Check inputs from each sensor circuit for opens or shorts:
   - **a.** Does code 20 appear on the display?
     - **i.** Yes—Go to step 4.
     - **ii.** No—Go to step 13.
4. Check Advance function:
   - **a.** Push the driver's side temperature control UP switch.
   - **b.** Advance to step 5—Mode and intake door motor position switch check.
     - **i.** Yes—Go to step 6.
     - **ii.** No—Replace Multifunction control panel assembly (switch malfunction).
5. Check Return function:
   - **a.** Push the driver's side temperature control DOWN switch.

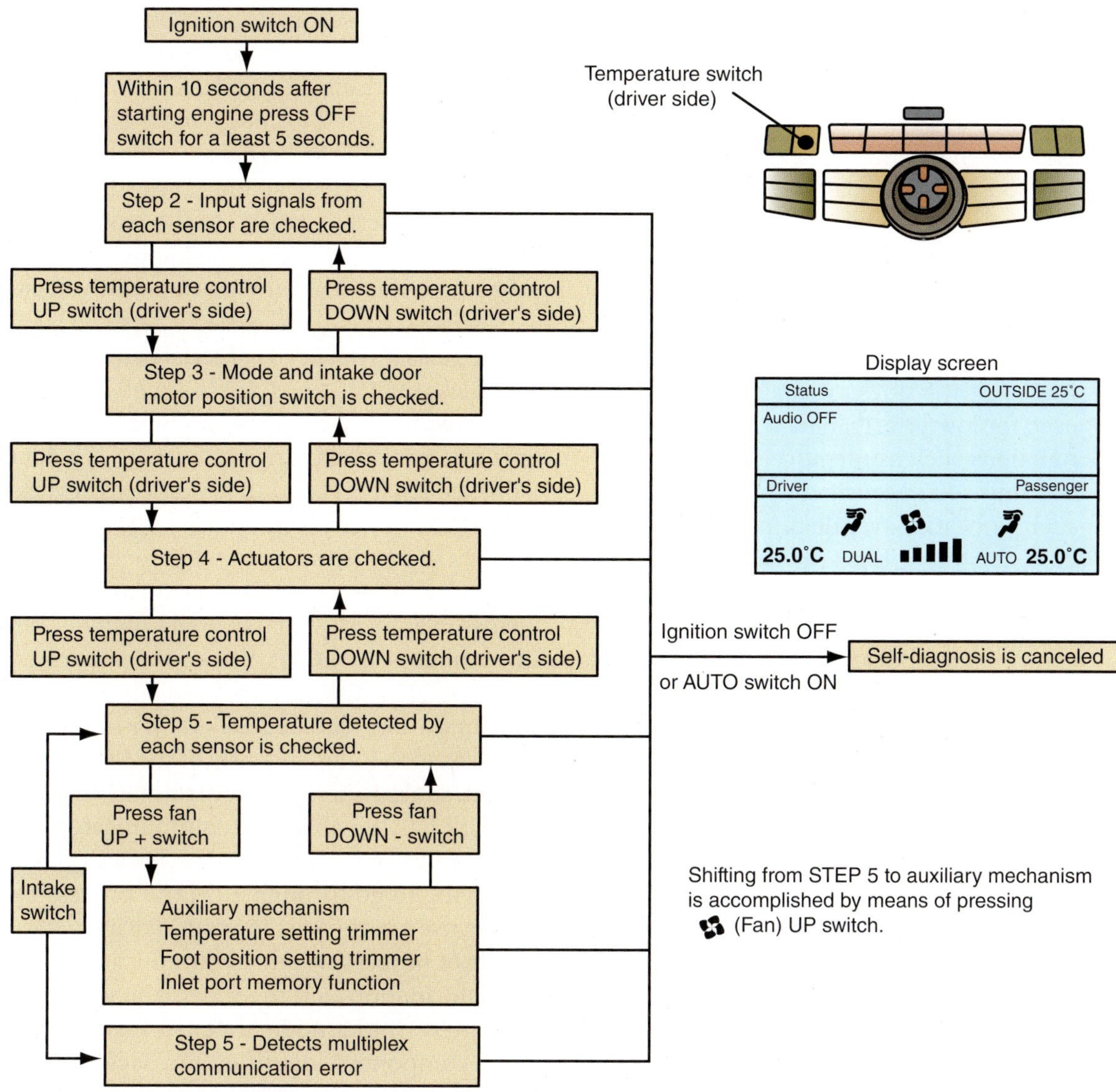

FIGURE 11-62 **An example of a manufacturer's sequence to access the Self-Diagnostic Mode for the electronic climate control system.**

- **b.** Return to step 3—Inputs from each sensor circuit are checked for opens or shorts.
  - **i.** Yes—Push the driver's side temperature control UP switch and advance to step 6 again: Mode and intake door motor position switch check.
  - **ii.** No—Malfunctioning control panel assembly (switch) or control module.

6. Check Mode and intake door motor position switch:
   - **a.** Does code 30 appear on the display?
     - **i.** Yes—Go to step 14.
     - **ii.** No—Go to step 7. This sequence continues for several more steps, but the general idea of how the systems work should be apparent.

Controller Area Network

If you are going to work on an automatic climate control system, it is necessary to understand the network on which they communicate. The **Controller Area Network (CAN)** protocol is the latest serial bus communication network used on OBD II systems and offers real-time

Controller Area Network (CAN) A high-speed serial bus communication network. The CAN protocol has been standardized by the International Standards Organization (ISO) as ISO 11898 standard for high-speed and ISO 11519 for low-speed data transfer.

Starting in the 2008 model year, the universal communication method or "language" is used for sharing of information between modules on a vehicle network or with a scan tool.

control. The speed of data transmission is expressed in bits per second (bps). The high-speed version, which can operate at 1 megabit per second (Mbps) and is used for power train management systems, performs in virtually real-time data rate transfer speeds. The low-speed version can operate at 125 kilobits per second (Kbps) and is used for body control modules and passenger comfort features. Although the prefix kilo usually indicates a multiplier value of 1,000, in a serial data stream a kilobyte has a value of 1,024 bytes of data. This is the mathematical result of a base two numbering system (ones and zeros) carried to the tenth place. Additionally, a megabyte has a value of 1,048,576 bytes of data, which is one kilobyte (1,024) squared. The CAN system has allowed for improved communication with onboard vehicle systems and is a true multiplexed network.

The CAN communication line is divided into three classes or speeds of serial data transfer. Class A is the slowest transmission rate with speeds less than 10 kbps. Class A networks are used for low priority data transmission; generally related to noncritical body control module functions such as memory seats. Class B networks are mid-speed range networks with data transmission speeds between 10 kbps and 125 kbps; generally related to less critical devices such as heating, ventilation, and air conditioning (HVAC); advanced lighting systems; and dash clusters. Class C networks have the fastest data transmission rate with speeds up to 1 Mbps. Class C networks are the most expensive to produce and are used for "mission critical" data transmission that flow at real-time speeds. Examples of Class C data include fuel control and ABS activation activity. The data link connector (DLC) is also connected to the Class C network for improved onboard diagnostics.

CAN enables the use of enhanced diagnostics and more detailed DTCs. With CAN, a scan tool is capable of communicating directly with sensors, independently of the PCM. The CAN protocol uses smart sensors. Each component contains its own control unit (microprocessor) called a "node." Each node on the network has the ability to communicate over a twisted pair of wires or a single wire, called a data bus, with all the other nodes on the network (bidirectional communication) without having to go through a central processing unit (Figure 11-63), unlike other multiplexed systems used in the past for data sharing. Every component on the network is independently capable of processing and communicating data over a common transmission line. Nodes transmit information (messages) with an identifier that prioritizes the message. The messages transmitted from a node are a package of data bits, which include a beginning of message signal, component identifier, message (sensor output signal), and an end of message signal. Because this is a bidirectional communication network, the control module receiving the data will send a signal back that the information was received. In order for this sophisticated communication protocol to function, the data transmission package must be a set size (number of bits) and format, and the information order must be consistent for all devices.

When multiple nodes need to send data simultaneously to the control module, the node will first see whether the data bus is busy. The system uses collision detection similar to that of an Ethernet system. But unlike Ethernet, the CAN system can handle high data transmission rates. In essence, the node is looking into traffic to see whether a higher priority node should be allowed to pass. Each CAN node on the network will have its own network unique identifier code, and nodes may be grouped based on function. The data message is then transmitted with its unique identifying code onto the network. Each node and control module on the network will perform an acceptance test of the transmission to determine whether it is relative based on its identifier. Relative information is processed and nonrelative information is ignored. Then the system segments transmissions based on the priority identifier (Figure 11-64) of the data package. The priority is determined by the unique number of the identifier, with lower number identifiers having higher priority. This guarantees higher priority node identifier messages access to the network, and lower priority node messages will be automatically retransmitted in the next available bus cycle based on priority.

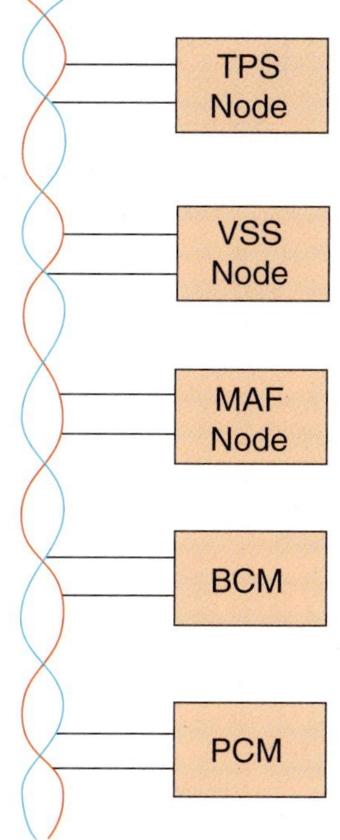

FIGURE 11-63 **A twisted pair data bus network with all nodes communicating on a shared line.**

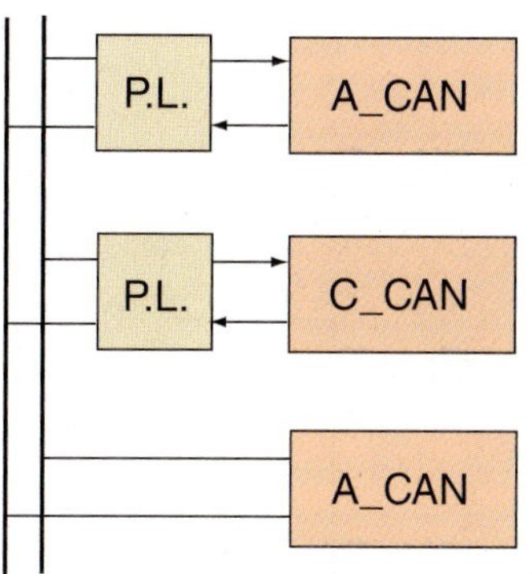

FIGURE 11-64 **Each node on the network has its own address and priority identifier. When sending data simultaneously, the CAN system segments data transmissions based on the priority identifier.**

Because the CAN protocol technology allows for many nodes on one set of wiring, the overall vehicle wiring harness size is greatly reduced. A twisted pair wired network contains a CAN H (+) and a CAN L (−) wire. The CAN bus is a differential bus system where the data signal from the CAN H (+) wire is a mirror image of the CAN L (−) network wire (Figure 11-65). The combination of the twisted pair network wiring combined with the differential bus data eliminates the effect of EMF noise on the data transmission. Multiple networks on the vehicle can be linked together by gateways if necessary (Figure 11-66). Class C high-speed data flows on one network, Class B mid-speed data flows on a second network, while Class A low-speed data flows on a third network. As an example, the coolant

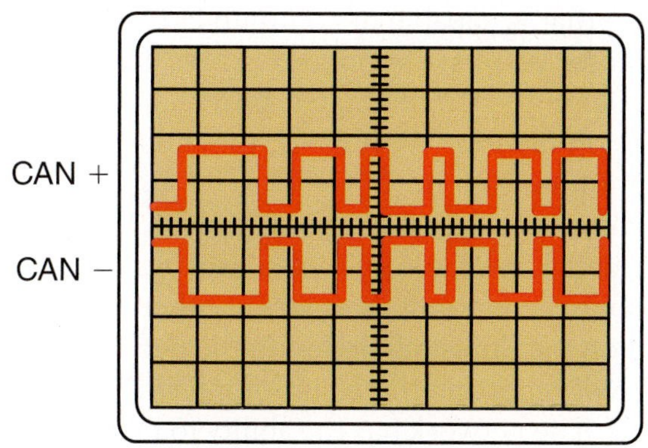

FIGURE 11-65 **Waveform of the CAN H (+) and of the CAN L (−) twisted pair data bus network.**

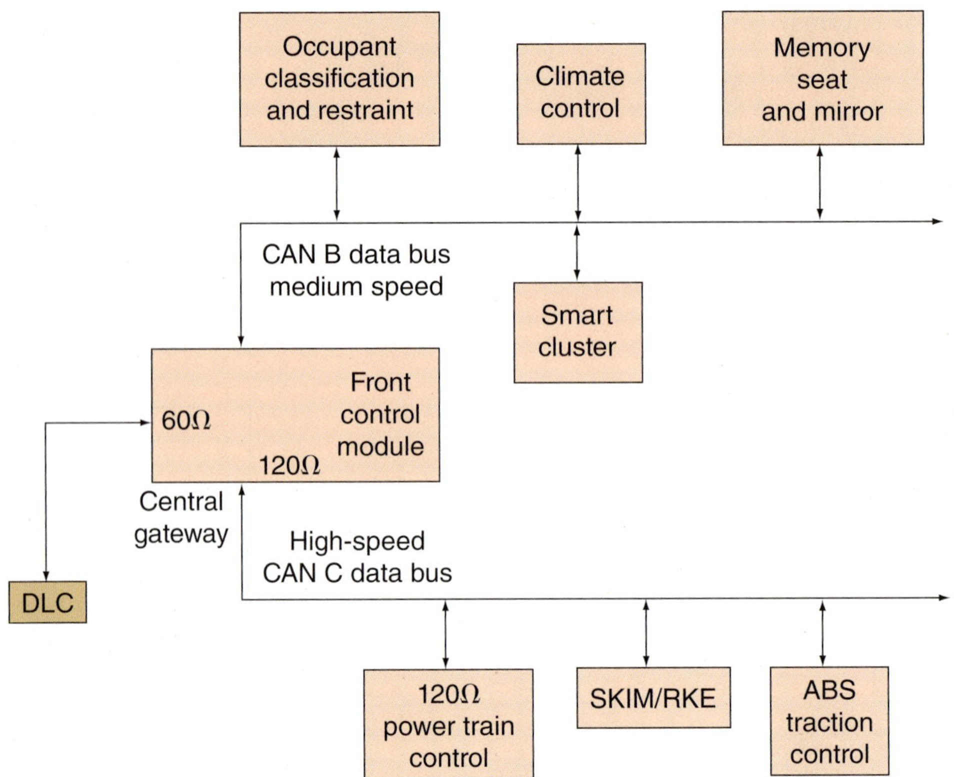

FIGURE 11-66 **An example of a typical CAN B network and a CAN C network.**

temperature (CT) sensor will place its data on the network data bus, allowing any control module on the network direct access to the information without the need for one control module (for example, the BCM) requesting the information from another control module (for example, the PCM). The BCM has direct access to information without having to request it from the PCM.

In 1996, the Environmental Protection Agency (EPA) specified that all vehicles be able to transmit generic scan tool data. However, proprietary data, any data other than P0 codes and data streams, were free to use any other protocol the manufacturer chose. The CAN PCM will still transmit data to the data link connector (DLC) in SAE's generic scan tool protocol, as specified by the EPA for generic scan tool data communication, such as generic DTCs. But, in order to access all the functions available, you will need to have a scan tool that is compatible with CAN if that is the vehicle's network operation system. Unless you have purchased a scan tool since the turn of the century (2000), it will be necessary to upgrade your existing scan tool or replace it with one that is capable of communicating with CAN. The EPA emission regulations for the 2008 model year have specified CAN as the new scan tool communications protocol for all vehicles sold in the United States, providing the repair technician more data for troubleshooting emission failures. With CAN, the industry finally has a single standard for onboard diagnostic communication.

The CAN protocol still allows access to the typical DTC information and data streams, but with enhanced DTC detail. A scan tool is also capable of bidirectional communication directly with a smart sensor or actuator node as well as other control modules on the network. In addition, flash calibration for almost all nodes on the network will become commonplace. A smart sensor is capable of reporting the result of internal voltage drops, opens, grounds, and other self-test features. The network has the ability to take faulty sensors off-line and can self-diagnose the difference between a faulty device and a faulty circuit.

Climate Control Network

The climate control network on a CAN system communicates on a medium-speed CAN B data bus network with data transmission speeds between 10 kbps and 125 kbps. The climate control system uses a small local area network (LAN) between the climate control module and the various mode door and air mix door motors in the system (Figure 11-67). The climate control module and motors are connected by a data transmission line and a power supply line. The data transmission line transmits initial compulsory start signal, component addresses, motor opening angle signals, error-checking messages, and motor stop signals. A local control unit (LCU) is integrated into each mode door motor, air mix door motor, recirculation door motor, and even the climate control module (Figure 11-68). The LCU is responsible for the following network functions: component address, data transmission, motor opening angle signals, motor stop and drive decision, opening angle sensor, and comparison data.

The climate control module receives data from each of the system sensors (Figure 11-69) and based on this information will send open/close commands out to the LCUs of the various mode door motors (nodes). Each node is responsible for reading its respective signal according to the address code (Figure 11-70). Next, each node sends position information back to the control module LCU, where the information is compared to the command signal (Figure 11-71). Subsequently, the climate control module can select hot/cold, defrost/vent, fresh air/recirculated air, or any combinations required, and verify that the command was followed.

On some vehicles, if the vehicle battery has been disconnected, the climate control module will engage the air-conditioning compressor clutch. A relearn procedure must be performed for all the mode door motors with a scan tool or, on some models, through their self-diagnostic procedure. A scan tool has become a necessary tool for many vehicle diagnostic procedures. Also it is more important than ever to follow manufacturers' diagnostic steps in the service information published.

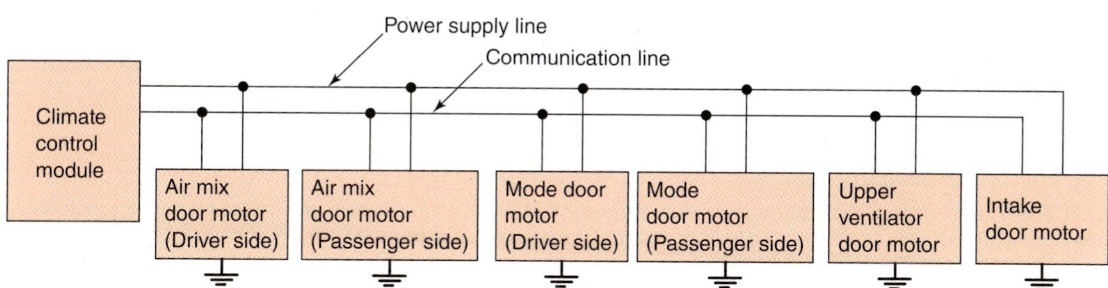

FIGURE 11-67 **The climate control system uses a small local area network (LAN) between the climate control module and the various mode door and air mix door motors in the system.**

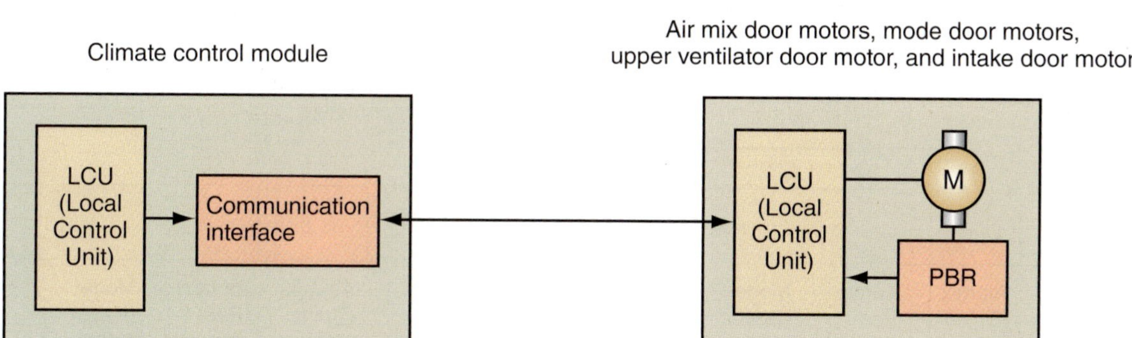

FIGURE 11-68 **A local control unit (LCU) is integrated into each mode door motor, air mix door motor, recirculation door motor, and even the climate control module.**

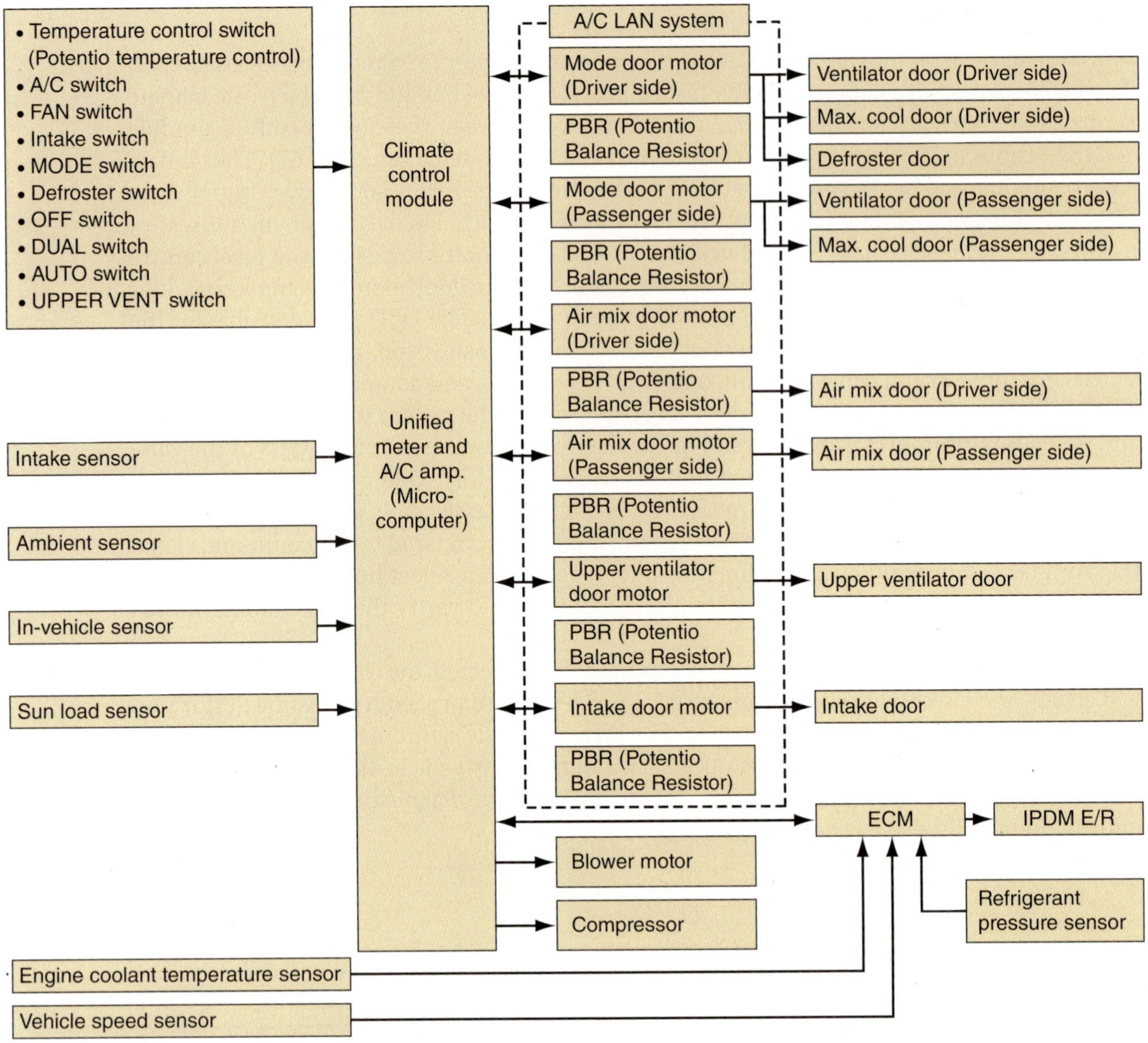

• Temperature control switch (Potentio temperature control)
• A/C switch
• FAN switch
• Intake switch
• MODE switch
• Defroster switch
• OFF switch
• DUAL switch
• AUTO switch
• UPPER VENT switch
Intake sensor
Ambient sensor
In-vehicle sensor
Sun load sensor
Climate control module
Unified meter and A/C amp. (Micro-computer)
A/C LAN system
Mode door motor (Driver side)
PBR (Potentio Balance Resistor)
Mode door motor (Passenger side)
PBR (Potentio Balance Resistor)
Air mix door motor (Driver side)
PBR (Potentio Balance Resistor)
Air mix door motor (Passenger side)
PBR (Potentio Balance Resistor)
Upper ventilator door motor
PBR (Potentio Balance Resistor)
Intake door motor
PBR (Potentio Balance Resistor)
Ventilator door (Driver side)
Max. cool door (Driver side)
Defroster door
Ventilator door (Passenger side)
Max. cool door (Passenger side)
Air mix door (Driver side)
Air mix door (Passenger side)
Upper ventilator door
Intake door
ECM
IPDM E/R
Blower motor
Compressor
Refrigerant pressure sensor
Engine coolant temperature sensor
Vehicle speed sensor

FIGURE 11-69 **The climate control module receives data from each of the system sensors and based on this information will send open/close commands out to the LCUs of the various mode door motors.**

Transmission data form
Start
Address
Opening angle data
Error check
Stop signal
Transmitted from unified meter and A/C Amp.
Transmitted from door motors
Air mix (Driver side)
0.1 sec
Air mix (Pas-senger side)
Mode (Driver side)
Mode (Pas-senger side)
Upper ventilator
Intake
Air mix (Driver side)
Air mix (Pas-senger side)
Mode (Driver side)

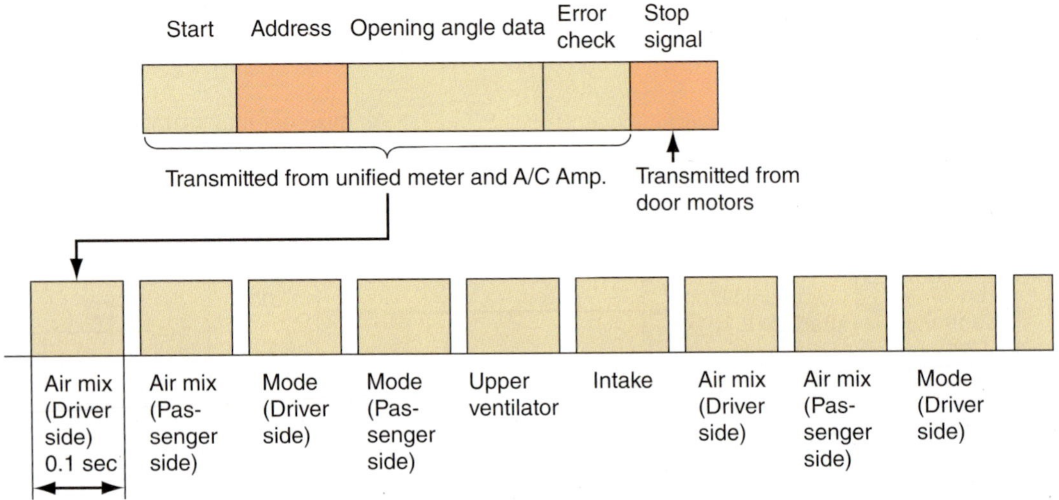

FIGURE 11-70 **The LCU communication data stream contains initial compulsory start signal, component addresses, motor opening angle signals, error-checking messages, and motor stop signals.**

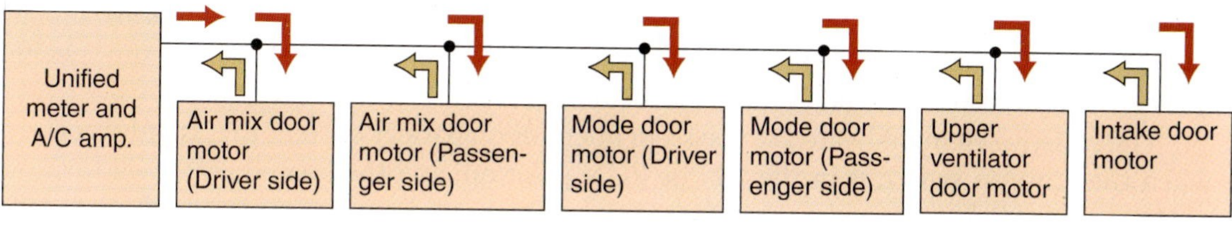

FIGURE 11-71 **Each node sends position information back to the control module LCU, where the information is compared to the command signal.**

Heated and Climate Controlled Seating

Though not part of the HVAC system on vehicles, both heated seats and climate controlled seats (CCS) are part of the overall passenger compartment comfort package on many vehicles. Seat surface temperature is an important factor for overall customer satisfaction of the entire vehicle climate control system. The average vehicle HVAC system can adjust the cabin air temperature to the occupant's comfort zone in about 12 to 15 minutes when outside temperature conditions are extreme, hot or cold. Studies have shown that when vehicles are only driven for short distances, the HVAC system does not have enough time to reach optimum performance levels to meet the needs of the passenger compartment comfort level selected. To meet these demands, manufacturers have developed both heated seats and climate controlled seating.

Heated seats are generally only offered for the two front passenger seats and are controlled independently from two separate switches typically located on the center console or on the instrument panel center stack (Figure 11-72). The most common design uses two resistive heating elements per seat (Figure 11-73). One is located in the seat back and the other is located in the bottom seat cushion. When electrical current passes through the heated seat element, the resistance of the wire used in the element converts the electrical energy into heat energy. The seats are controlled by a heated seat module that contains the control logic and software for the system. On current systems, the module communicates on the Controller Area Network (CAN) data bus. The following discussion covers the more complicated CAN

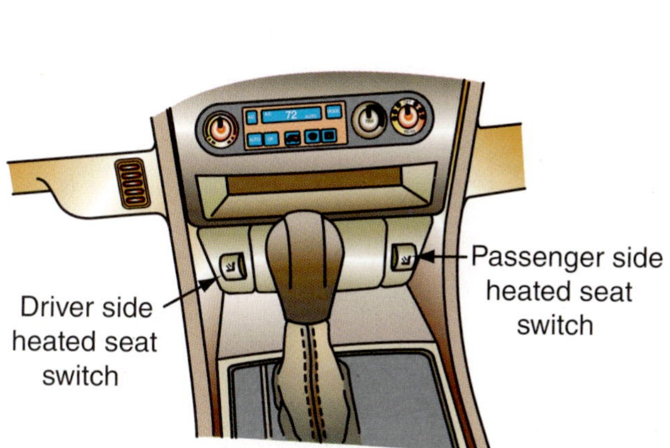

FIGURE 11-72 **Front passenger heated seats are controlled independently from two separate switches typically located on the center console or on the instrument panel center stack.**

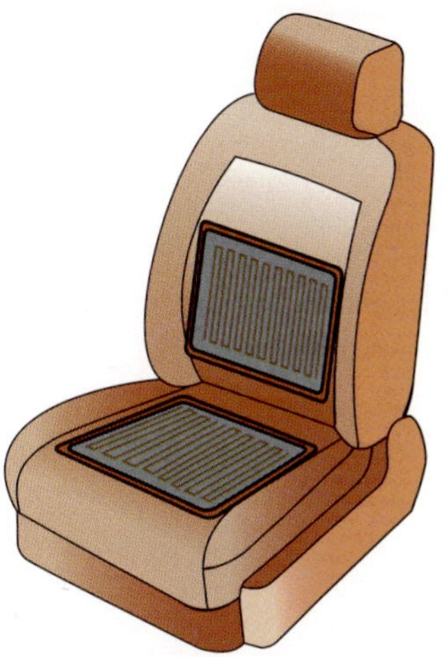

FIGURE 11-73 **The most common design uses two resistive heating elements, one in the seat cushion and one in the back.**

data bus systems; it should be noted that although the basic design of the systems are similar among manufacturers, specific system operation and diagnosis information for the specific make, model, and year should always be consulted.

The resistive element heated seat system operates on battery current that is only received when the ignition switch is in the ON or RUN position, and the heated seats will turn off anytime the switch is moved from this position. The heated seat control module responds to messages sent from both the heated seat switches and the ignition switch status by controlling integrated solid-state relays for the 12-volt output to the heating elements located in each seat. These switches' inputs send a resistive signal to a controller node, which in turn sends a signal by the CAN data bus to the heated seat module, signaling the module to either energize or de-energize the heating element in the respective seat(s). The individual heated seat switches generally have two settings: one for low temperature (100.4°F or 38°C) and one for high temperature (107.6°F or 42°C). In addition, the switches contain an amber LED for both high/low heat setting to indicate to the passenger which setting has been selected. Some systems will supply a boosted heat level for the first few minutes of operation to quickly bring the seat up to temperature when the high temperature switch position is selected and then drop back to normal high-temperature current flow after this initial boost period. Most systems will also automatically shut off after operating for extended periods of time, generally two hours. The system will also shut off if an open or short is detected in the heating element system.

The resistive element in the heated seat cushion and back consists of a carbon fiber element with multiple circuits wired in parallel with one another (Figure 11-74). If one or more circuits in the element develop a malfunction, the other circuits will continue to operate and provide heat, though a dead spot (no heat) in the cushion may develop. The carbon fiber element is captured between the leather seat cover and the cushion assembly. The heated seat element cannot be repaired and must be replaced as an assembly if found to be damaged or inoperative.

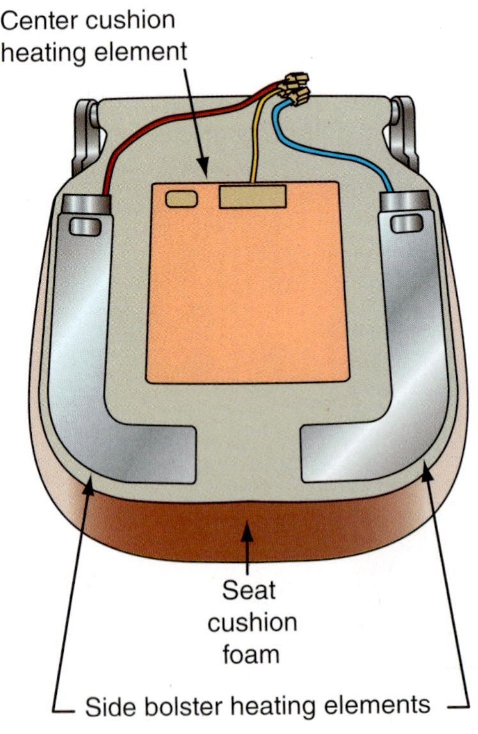

FIGURE 11-74 **Heated seat resistive elements.**

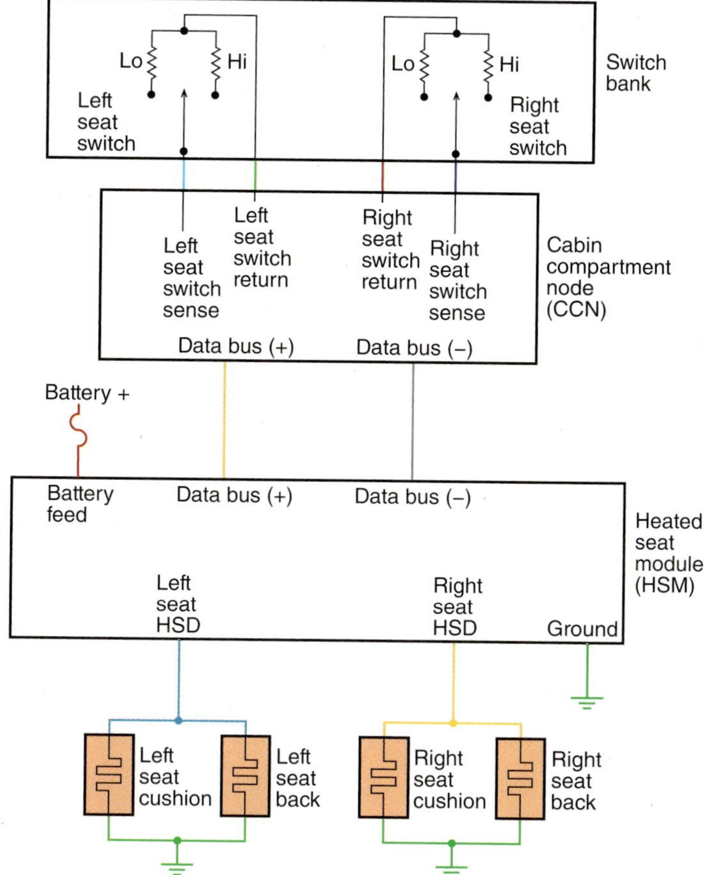

FIGURE 11-75 **Typical wiring diagram for a resistor element heated seat.**

To properly diagnose these systems, a scan tool and the appropriate service information for the specific make, model, and year of the vehicle is required (Figure 11-75). These systems are equipped with a low/high voltage cutoff feature and will shut down if the vehicle voltage drops below 11.1 volts or rises above 15.5 volts as a general rule. Always verify the proper system voltage and that the battery is fully charged prior to any diagnosis of the heated seat system.

Heated seats have been around for many years, but climate controlled seating is a relatively new addition to the comfort package. Some early design systems utilized a resistive heating element to heat the seat and cooled air from the vehicle air-conditioning system to cool the seat. Most current production climate controlled seating is self-contained (Figure 11-76) and does not use a standard resistive heating element to heat the seat cushion or rely on the vehicle air-conditioning system. Instead, both the passenger seat heating and cooling modes draw air in from the passenger cabin, and a solid-state thermoelectric device (TED) heat pump rapidly converts electric current into heat or cool air. Then the integrated blower pumps the air through the seat cushions (Figure 11-77). The Amerigon's Climate Control Seat ™ (CCS™) system is completely independent of the vehicle's HVAC system and is one of the most popular self-contained systems being used by most automotive manufacturers in the United States, Japan, and Europe. Each seat has independent electrical controls and greatly improves passenger comfort by focusing the cooling/heating directly on the passenger, independent of cabin air temperature.

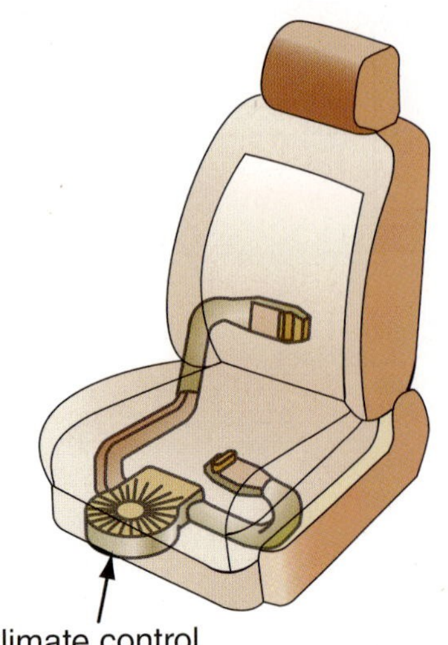

FIGURE 11-76 Climate controlled seat with integrated heating and cooling unit as well as self-contained air ducts and blower assembly.

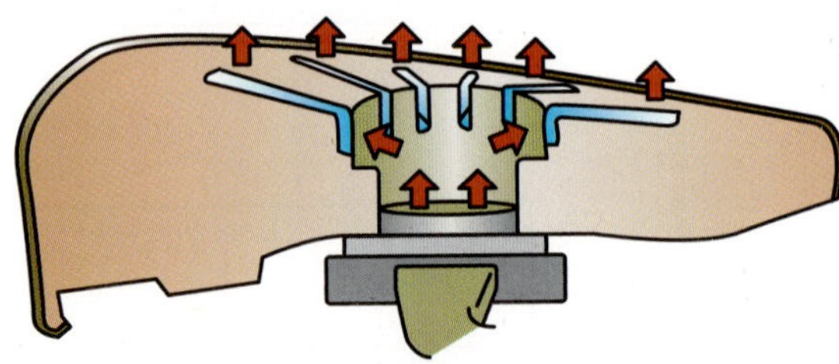

FIGURE 11-77 Air that has been either heated or cooled, based on driver command, is pumped through the passenger in the seat cushion.

Terms to Know

Aspirator
Bellows
Circuit breaker
Controller Area Network (CAN)
Delta P (Ap)
Dual systems
Fuse
Insulator
Master control
Pressure switch
Programmer
Rheostat
Vacuum pot

SUMMARY

- Many of the components of an automatic temperature control system are covered in this chapter.
- Because of the complexity of the automatic control system and its number of variations, it is essential that manufacturers' specifications, manuals, and schematics be consulted for any specific year/make/model car to be serviced.

REVIEW QUESTIONS

Short-Answer Essay

1. What is the purpose of the clutch diode?
2. Explain what is meant by this sentence: "The change in resistance value of each sensor is inversely proportional to a temperature change."
3. What is the relationship of the aspirator to the in-car temperature sensor?
4. Compare the difference between a temperature-controlled switch and a pressure-controlled switch.
5. Describe a sun load sensor.
6. On an automatic temperature control system what occurs if the temperature selected is the lowest limit or the highest limit?
7. Describe an air humidity sensor and its relationship to air temperature.
8. Describe the evaporator temperature sensor and the outside temperature sensor and how they are similar.
9. Describe the term input sensor.
10. Briefly describe the climate control network on a CAN bus system.

Fill in the Blanks

1. The climate control network on a ______________ system communicates on a medium-speed ______________ ______________ data bus network.

2. The control module analyzes data from the ______________ ______________ ______________, selected passenger compartment temperature, and the evaporator temperature sensor to determine whether the air-conditioning compressor clutch should be ______________ ______________ ______________.

3. The climate control module uses input sensor ______________ ______________ ______________ ______________ to determine actuator commands that will be required to maintain selected temperature.

4. On an automatic climate control system with a ______________ ______________ sensor when sunlight intensity is ______________, the control module will automatically ______________ blower fan speed and increase the volume of air flowing to the dash discharge vents.

5. The air ______________ sensor integrates a temperature sensor since determining air humidity is dependent on air ______________.

6. The FOTCC system uses a ______________ sensitive compressor cycling switch instead of a ______________ sensitive switch.

7. The outside temperature sensor (OTC), also called an ______________ ______________ ______________, is a ______________ temperature coefficient thermistor.

8. Some in-car temperature sensors use a small ______________ ______________ positioned behind the thermistor to draw in-car air across the sensor to maintain accurate in-car ______________ readings.

9. An automatic air-conditioning system should provide selected in-car ______________ and ______________ at all times.

10. Many air-conditioning systems today are equipped with a refrigerant pressure sensor, also called a ______________ ______________, on both the ______________ pressure side and the ______________-pressure side of the refrigerant system.

Multiple Choice

1. Which of the following is the best description of a pressure switch designated NC?
   A. It is held closed during normal pressure conditions.
   B. It is held closed during extreme low-pressure conditions.
   C. It is held closed during extreme high-pressure conditions.
   D. It is held open during extreme high-pressure conditions.

2. The temperature fluctuates between hot and cold on a vehicle equipped with an automatic temperature control system. Which of the following is the most likely cause?
   A. An open blower motor control relay
   B. A low refrigerant charge level
   C. A faulty coolant temperature sensor
   D. A disconnected in-car temperature sensor aspirator tube

3. All of the following statements are correct, *except*:
   A. A low-pressure cutoff switch opens at a predetermined low pressure.
   B. Atmospheric pressure at sea level is 14.696 psia.
   C. A clutch coil resistor is used to prevent high voltage spikes.
   D. A sensor is an electrical input device that may be used to sense temperature or pressure.

4. A vehicle's automatic temperature control system does not hold the temperature set on the temperature control assembly. What operation should the technician perform first?
   A. Change the in-car temperature sensor.
   B. Evacuate and recharge the refrigerant system.
   C. Refer to the appropriate diagnostic fault chart.
   D. Replace the mix door actuator.

5. The blower motor does not run when the HEAT mode is first selected and the engine is cold on a vehicle that is equipped with an automatic temperature control system. Which of the following is the most likely cause?
   A. The ambient air temperature is open.
   B. The system is functioning normally.
   C. There is an excessive voltage drop at the blower motor.
   D. The coolant temperature sensor is out of range.

6. When the air-conditioning switch is turned to the OFF position, the compressor clutch stays engaged. Which of the following is the most likely cause?
   A. The compressor clutch air gap is set too wide.
   B. An electronic thermostat is out of specification.
   C. A pressure switch is stuck closed.
   D. A compressor clutch relay is stuck in the closed position.

7. The blower motor on a four-speed switched system only functions on the high-speed setting. Which of the following is the most likely cause?
   A. The current draw at the blower motor is too high.
   B. The resistor block assembly is faulty.
   C. There is an open blower motor switch.
   D. The blower motor ground is faulty.

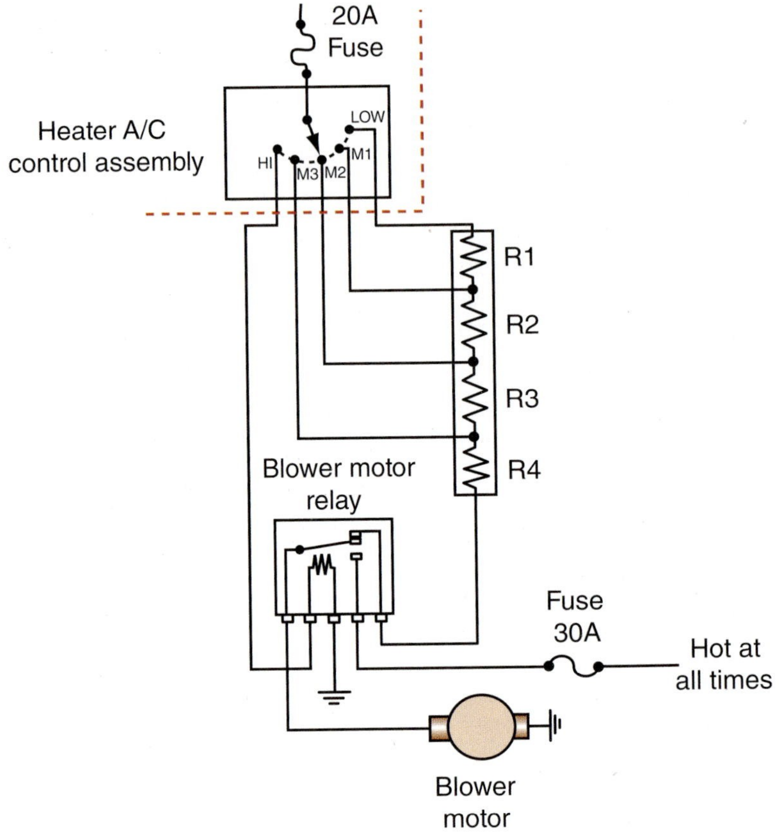

8. The blower motor in the above figure has blower motor speeds M2, M3, and High, but does not run in either LOW or M1. Which of the faults listed below could be the cause?
   A. Faulty heater A/C control assembly
   B. Faulty resistor block
   C. Faulty blower motor relay
   D. Faulty blower motor

9. The following may be used to maintain in-vehicle temperature, except:
   A. Evaporator thermistor
   B. Variable displacement compressor
   C. Mass airflow sensor
   D. Pressure cycling switch

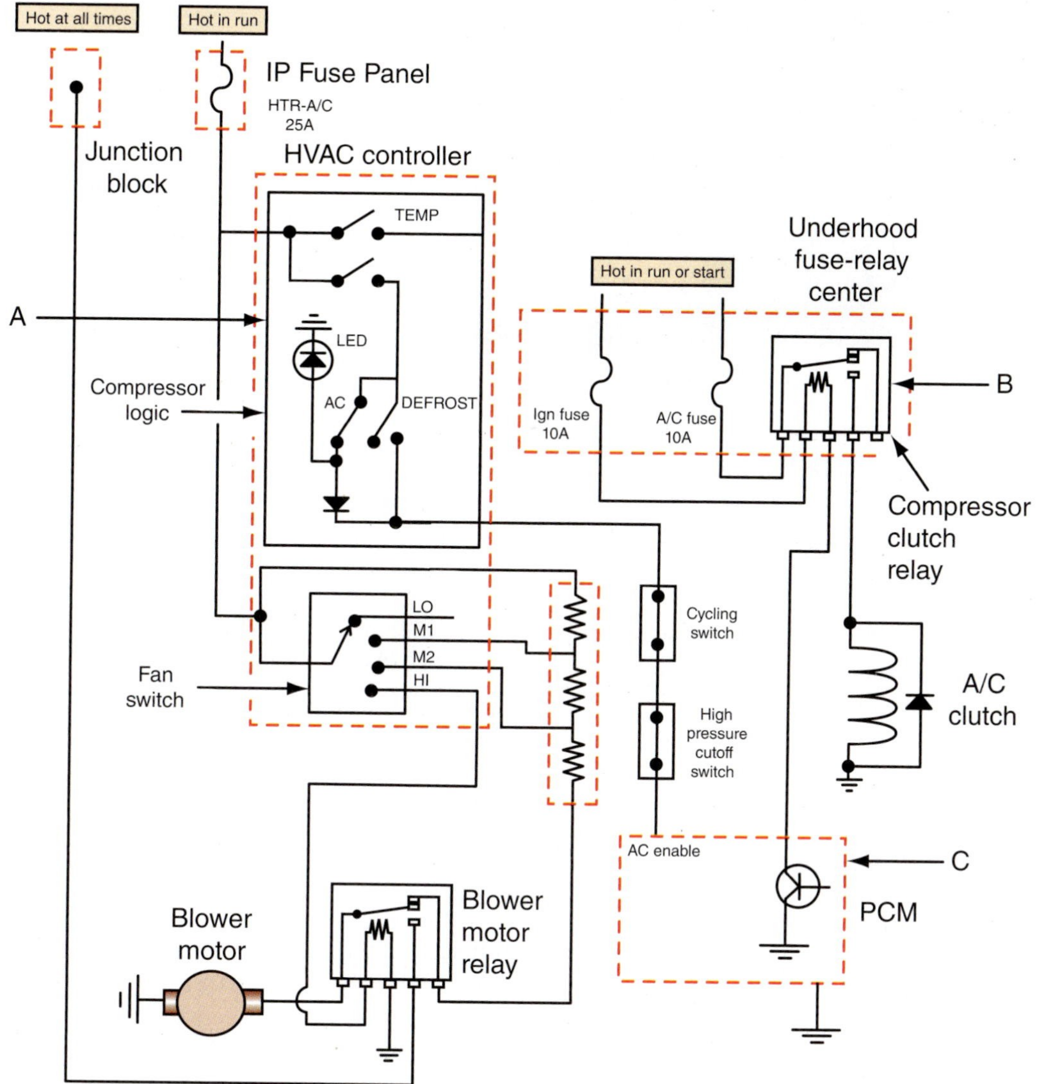

10. In the above figure if a ground is not supplied by the PCM at point "C," what effect would this have on the circuit shown?

    A. A/C compressor clutch would always be engaged.
    B. The blower motor would not have high speed.
    C. The blower motor would not be turn on.
    D. A/C compressor clutch would not be turn on.

Chapter 12

RETROFIT AND FUTURE TRENDS (R-12 TO R-134A)

UPON COMPLETION AND REVIEW OF THIS CHAPTER, YOU SHOULD BE ABLE TO:

- Discuss CO_2 (R-744) as a refrigerant and why some are still holding out hope for it in the future.
- Discuss the various refrigerants approved to replace R-12.
- Identify the refrigerant approved to replace R-12 in automotive air-conditioning systems.
- Understand the problems associated with contaminated refrigerant.
- Compare components used in R-134a systems with those used in R-12 systems.

Retrofit is the process of modifying equipment that is already in service by installing updated parts and materials made available after the time of original manufacturing.

Alternative refrigerant is a refrigerant that can be used to replace an existing refrigerant, such as ozone-friendly R-134a that is used to replace ozone-depleting R-12.

INTRODUCTION

In this chapter, we will first discuss future trends and changes that are taking place or may take place within the refrigerant industry. Only a few years ago, it looked like the industry was going to abandon R-134a in favor of a refrigerant with a lower global warming potential (GWP). In Europe, the switch to R-1234yf has already been taking place. Daimler is one of the manufacturers that is still dedicating research and development money into R-744 CO_2 refrigerant systems, while other European manufacturers are looking at alternative refrigerants similar in system performance to R-134a that can compete with R-1234yf. At the time of this writing, Daimler has decided to continue producing new vehicle platforms using R-134a in violation of the EU requirement to switch to a low-GWP refrigerant beginning in 2013 and fully abandon R-134a by 2017. Daimler contests that their testing indicates that R-1234yf is a flammable refrigerant and as such is unsafe in their vehicle platforms. Ultimately this will be decided in the European court system. In addition, some say do not count R-134a out in the U.S. market due to improvements in system designs, smaller sizes, and decreased lifetime leakage rates. It appears R-134a will be replaced by a refrigerant with a GWP number below 150 by the early 2020s. Some are also predicting that as the industry eventually shifts to fully electric vehicle–powered platforms that gas-based air-conditioning systems will be replaced by a fully electric gasless air-conditioning system, which is yet to be developed for automotive commercial applications. But be sure, whatever the changes are we will be ready to service them when the time comes.

When speaking about an automotive air-conditioning system, the term ***retrofit*** is used to describe the process of converting an R-12 system to one using an **alternative refrigerant**. In this text, it will be assumed that the conversion refrigerant is R-134a because the automotive industry worldwide chose this refrigerant to be the replacement in new, as well as retrofitted, automotive air-conditioning systems. Due to the current age of vehicles (pre-1994) that require retrofitting, it is becoming a less prevalent procedure. At some point, retrofitting will be limited to vehicles under consideration for restoration or those luxury platforms that retain substantial market value. Regardless of what happens in the future we will not be retrofitting either an R-12 system or R134A to R-1234yf refrigerant. In fact this is not allowed by the EPA. R-134A will continue to be available for servicing systems into the future regardless of the popularity of R-1234yf.

Most automobile manufacturers have developed retrofit kits and procedures for some of their late 1980 through early 1990 model vehicles. These kits and information are intended to provide the best level of performance with R-134a refrigerant, with no regard for costs. Vehicle owners, however, generally do not want to pay the high cost for a retrofit and are therefore often the target of the independent technician offering an inexpensive solution—the "magic bullet," so to speak.

NEW REFRIGERANT SYSTEMS ON THE HORIZON

Air-conditioning systems, which were once considered a luxury, are now considered standard equipment on most passenger vehicles. However, with increased popularity came environmental hazards. We have already discussed the environmental hazards and dangers of an R-12 system, but there are also dangers associated with R-134a. R-134a is not an ozone-depleting refrigerant, but it is a greenhouse gas. The concern in the world community has shifted from the threat of ozone-depleting chemicals to a concern over global warming. In 1997, the U.S. government decided not to sign the Kyoto Protocol, an international agreement that, among other things, set reduction quotas for the production of greenhouse gases that contribute to global warming. The U.S. government did agree with the principle of the agreement and decided to implement a voluntary U.S. policy to limit the production of greenhouse gases.

Global warming has become a more pronounced concern, but as of the time of printing this textbook, the United States had not passed any legislation to phase out or ban the use of R-134a refrigerant but is pressuring the industry by incentivizing manufacturers to choose a lower GWP refrigerant with the use of carbon credits. With that said there is considerable research taking place to find an alternative to R-134a that will be more environmentally friendly and still offer the passenger comfort levels currently achievable at a reasonable cost and dependability. The front-runner and the one that has gained the most widespread acceptance is HFO-1234yf (**R-1234yf**), though some are still holding out long-term hopes for CO_2 (**R-744**) systems, but that appears doubtful at least for now.

R-1234yf is a low-GWP refrigerant developed by Honeywell and Dupont.

R-744 is the refrigerant gas designation given to carbon dioxide (CO_2).

CO_2 (R-744) Refrigerant Systems

Another alternative refrigerant that once held widespread hopes for being the next refrigerant due to its low GWP rating of 1 is carbon dioxide, CO_2 (R-744), a naturally occurring gas. Though most research has been shelved for now, further discussion is warranted to understand why the industry makes the choices that it does and other options that are available though not currently likely.

A CO_2 system is similar to today's systems, but the operating pressures are extremely high, seven to ten times greater than for R-134a systems. Currently, the efficiency of the overall system is much lower than R-134a systems. Less-efficient systems require more power to perform the same job, and because we still rely on the internal combustion engine, the overall benefits of the CO_2 system are negated. As research continues, R-744 systems are becoming more efficient.

Carbon dioxide shows potential as a refrigerant because the properties of the gas make it ideal for small portable refrigeration systems. Although technically CO_2 is a greenhouse gas, its release into the atmosphere is harmless because there is enough CO_2 occurring naturally or obtained through natural chemical reaction that no new CO_2 would have to be created. The new R-744 refrigerant systems could greatly benefit automotive manufacturers, who could, for the first time since R-12 systems, offer a system that cools better than the current R-134a systems and poses no environmental hazard that could affect costs through government regulations, but high costs negate this benefit.

One large hurdle that automotive manufacturers must first overcome is the extreme line pressure that the system must operate at. The average high-side pressure will be in the range of 2,000 psig (13789.5 kPa), which means system components like the compressor and lines must be strengthened. On the other hand, the new R-744 systems require about half the refrigerant capacity of current R-134a systems, with better cooling results.

Adiabatic expansion is a process that occurs without the loss or gain of heat.

There are some differences in the R-744 system compared to the R-134a system. The R-744 system (Figure 12-1) has a gas cooler instead of a condenser to reduce the temperature of the R-744 gas but not condense it. Instead, some of the gas condenses as it passes through the expansion valve as a result of **adiabatic expansion**. Further cooling occurs by exchanging heat with the inner heat exchanger, which is between the gas cooler and evaporator on the low side of the system.

The R-744 system can also serve as a source of heat for the passenger compartment through the use of an integrated heat pump (Figure 12-2), a special heat exchanger that uses the heat created in the air-conditioning system to provide heat for the passenger compartment. This will be a very useful system for both passenger compartment heating and cooling, especially on electric-powered or fuel cell–powered vehicles (which have little or no waste heat). In addition, with smaller, more fuel-efficient internal combustion engines, the need for a supplemental heater is also growing, especially in the compact diesel market. This system will enable both heating and cooling simultaneously by regulating refrigerant gas flow through both expansion valves in the system. One expansion valve will regulate refrigerant flow to the evaporator for cooling, while the second expansion valve will regulate refrigerant flow to the interior gas cooler (heater core) for heating.

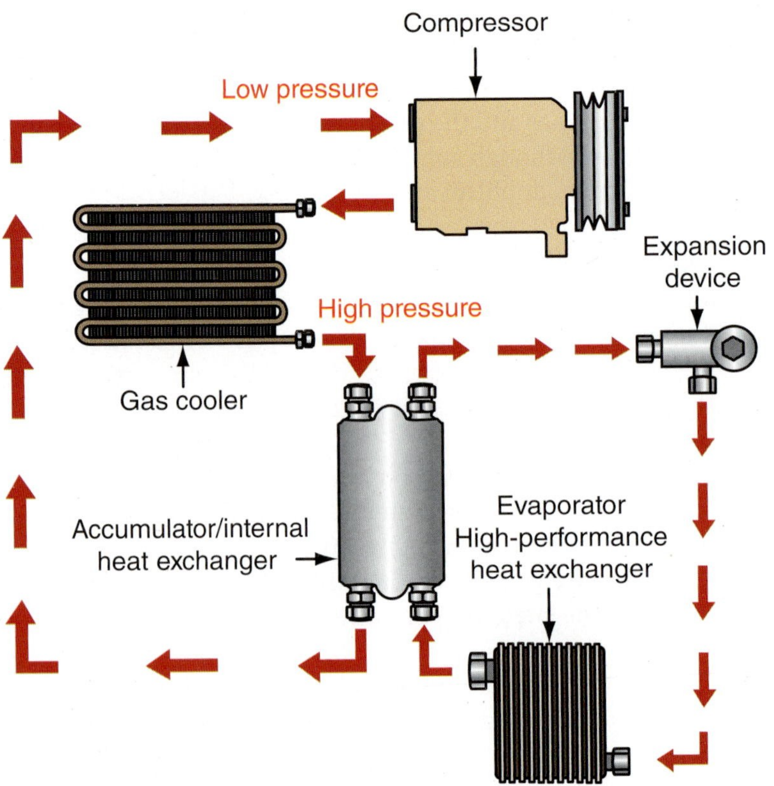

FIGURE 12-1 A typical R-744 (CO_2) refrigerant system.

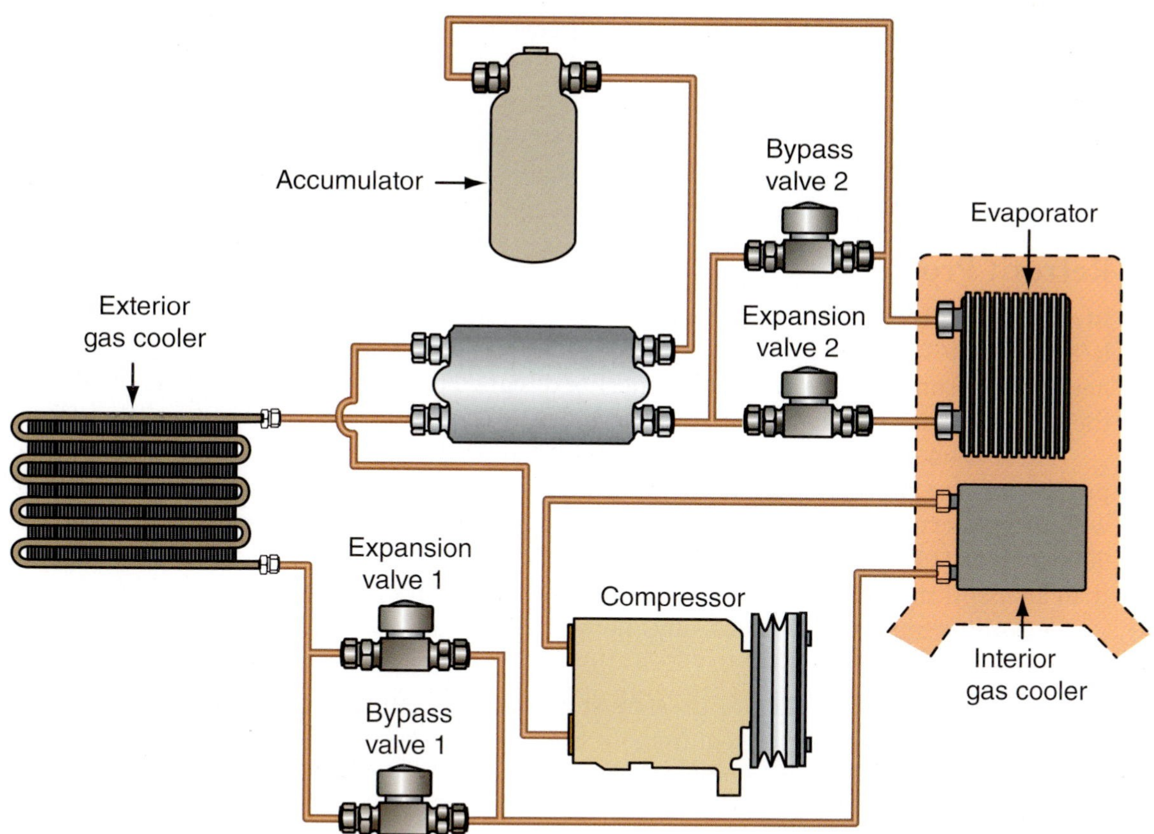

FIGURE 12-2 **An R-744 system with heat pump for passenger compartment heating.**

As for the near future, we will see more electronics integrated into the automotive air-conditioning system. The use of computer-controlled variable displacement compressors will increase, and electric-driven compressors will become commonplace on vehicles with start-stop technology (idle shutoff systems). There will also be an increased use of electronic expansion valves and orifice tubes. The future looks bright and exciting for the automotive air-conditioning industry.

The Inexpensive Retrofit

The procedures for an inexpensive retrofit are relatively simple and generally do not require major component replacements. The process usually only requires removal of the R-12 refrigerant, new fittings, new label, and the addition of the proper lubricant. For many, this simple inexpensive retrofit will provide the owner with an air-conditioning system performance that is comparable to the former R-12 system. Even if the retrofit results in slightly reduced performance, it is usually sufficient for customer satisfaction.

The EPA has an ongoing program intended to educate car owners on matters concerning retrofit options. Many car owners, however, rely on their service technician for their education. When recommending a retrofit to a customer, the "three C's" should be discussed: cost, climate, and components.

Cost. What is the value of the car? How long will the customer continue to drive it? Is it a refrigerant leaker or is this the first time the air-conditioning system has been serviced? How much is the customer willing to spend?

Climate. Does the customer live in the North and require minimal air-conditioning performance because the car is only used for occasional short pleasure trips, or does the customer live in the South and need maximum performance because the car is used five or six days a week for business?

Components. Are the air-conditioning system components in good working order? Are they compatible with the new refrigerant, R-134a? Are there any indications of leaking hoses, restrictions in the system, or a noisy compressor? If not operational, did the system cool satisfactorily when it was last working? If R-12 system performance is no more than marginally satisfactory, retrofitting will not make it better. To the contrary, owners should be prepared for a slight reduction in system performance.

Retrofit Problems

In older cars, it is often necessary that worn air-conditioning system components be replaced. R-134a operates at a higher pressure than R-12 and will put additional stress on system components. Older, somewhat worn components not designed for R-134a service may not withstand the higher pressures and are more likely to fail.

There is no such thing as a universal retrofit kit that can be purchased nor is there a set procedure for a technician to follow that will ensure a successful retrofit for every car. Even within a given vehicle model, the retrofit requirements will vary. For example, a vehicle driven 90,000 miles (144,810 kilometers) in southern Florida may require a more extensive retrofit than an identical car driven, say, 25,000 miles (40,225 kilometers) in northern Minnesota.

According to EPA regulations, any alternate refrigerant used to replace R-12 requires the following:

- Unique service fittings (Figure 12-3) must be used on both the high side as well as the low side of the system. This requirement is intended to reduce the likelihood of **cross-contamination** of the air-conditioning system or the repair facility's refrigeration service equipment.
- Use of the new refrigerant must be noted on a uniquely colored label (Figure 12-4) to distinguish the type refrigerant and lubrication used in the system.
- All R-12 must be properly removed from the system before filling the system with an alternative refrigerant.
- To prevent release of refrigerant to the atmosphere, a high-pressure compressor shut-off switch must be installed on any system equipped with a pressure relief device.
- Separate, dedicated EPA-approved equipment must be used to recover R-12 refrigerant from the system.
- Barrier hoses must be used with alternative refrigerant blends that contain HCFC-22.

Cross-contamination is when one refrigerant is contaminated with another. This usually occurs due to improper or incomplete service procedures.

FIGURE 12-3 **An R-134a system uses unique hose fittings.**

FIGURE 12-4 **A label identifies the type of refrigerant in the system.**

The Replacement Refrigerant of Choice

Several refrigerants in addition to R-134a are now listed by the EPA as acceptable for motor vehicle air conditioner (MVAC) use under their SNAP plan. Others are under SNAP review. The SNAP program tests and evaluates substitute refrigerants for their effect on human health and the environment. SNAP does not test and evaluate refrigerants for performance or durability. Except for R-134a, no refrigerant has been endorsed by vehicle manufacturers for use in MVACs. While some alternate refrigerants are being marketed as "drop-ins," there is by definition no such thing as a refrigerant that can literally be "dropped in" on top of existing R-12 in a system.

The current refrigerant of choice—R-134a—is considered to be one of the safest refrigerants based on toxicity data. Extensive tests indicate that R-134a does not pose cancer or birth defect hazards, is not corrosive on steel, aluminum, or copper samples, and is not flammable at ambient temperatures at atmospheric pressure. Service equipment and vehicle air-conditioning systems, however, should not be pressure or leak tested using compressed air. Some mixtures of air and R-134a have been known to be combustible at elevated pressures.

As with any other chemical, R-134a should be handled with respect; work in a well-ventilated area, wear adequate personal protection, avoid open flames, and do not inhale any vapor.

System Charge

The amount of R-134a charged into the system should initially be 80–90 percent of the charge of R-12. Most manufacturers provide guidelines regarding the amount of R-134a to be used.

Lubricants

The mineral oil used with R-12 cannot be adequately transported through the system by R-134a. Most, but not all, automobile manufacturers chose polyalkaline glycol (PAG) lubricants for use in new and retrofitted air-conditioning systems charged with R-134a. PAGs are very hygroscopic; they draw water from the atmosphere when exposed to open air. Some specialists choose to use polyol ester (POE) lubricants (Figure 12-5), believing that PAG's hygroscopic nature limits its lubricating ability and causes corrosion in a system. Although it is less hygroscopic than PAG, care must still be taken with POE to ensure that excess moisture does not enter the system.

Personal protection such as PVC-coated gloves or barrier creams and OSHA-approved safety goggles should be used when handling these lubricants. Prolonged skin contact or

FIGURE 12-5 **A small container of POE lubricant.**

eye contact can cause irritations such as stinging and burning. Avoid breathing any vapors produced by these lubricants, and only use them in a well-ventilated area. They should be stored in tightly sealed containers to prevent contamination by humidity and to ensure that the vapors do not escape.

Flushing. The amount of mineral oil that can remain in a system after retrofitting without affecting performance is still being debated. The technician should always remove as much of the mineral oil as possible, however. Removal may require draining components such as the compressor and accumulator. Tests have shown that any residual R-12 remaining in the system will not have a significant effect on system performance. If the vehicle manufacturer does not recommend flushing the system during the retrofit procedure, it can be assumed that flushing is not necessary.

Hoses and O-Rings. Tests have shown that lubricant used in an automotive air-conditioning system is absorbed into the hose to create a natural barrier to R-134a permeation. In most cases, R-12 nonbarrier hoses will perform well for R-134a service, provided they are in good condition. Any replacement hose, however, should be of the barrier type.

If the fittings were not disturbed during retrofit, replacing them should not be necessary. Most retrofit instructions suggest lubricating replacement green or blue R-134a O-rings with mineral oil to provide protection because the mineral oil also provides a natural barrier.

Compressors. Most compressors that function satisfactorily in an R-12 system will continue to function after retrofitting an R-134a system. When a compressor is first operated with R-12, a thin film of metal chloride forms on bearing surfaces to serve as an antiwear agent. This protection continues even after the system has been retrofitted to R-134a. This may explain why new R-12 compressors often fail when installed in an R-134a system without the benefit of a break-in period with R-12.

Viton® is a registered trademark of Dupont Dow Elastomers.

Some older compressors have seals made of Viton® that are not compatible with R-134a or the new lubricants and must be replaced. Also, any compressor that is not in good working order should be replaced during the retrofit procedure with one designed for R-134a service.

Desiccants. R-12 systems often use silica gel or a desiccant designated XH-5, while R-134a systems use either XH-7 or XH-9. Some recommend replacement during the retrofit procedure of the accumulator or receiver-drier to one having XH-7 or XH-9 desiccant. It is generally agreed, however, that the accumulator or receiver-drier should be replaced if the vehicle has over 70,000 miles (112,630 kilometers), is five years or more old, or is opened up for major repair.

Condensers and Evaporators. It is generally accepted that if an R-12 system is operating within the manufacturer's specifications, there may be no need to replace the condenser or evaporator. The higher vapor pressures associated with R-134a, however, may result in lost condenser capacity. When planning a retrofit, the technician should consider the airflow and condenser design.

A pusher-type cooling fan mounted in front of the condenser often has improved the performance of a retrofitted air-conditioning system. Bent, misshapen, or improperly positioned airflow dams and deflectors also affect performance. Hood seal kits are often recommended for retrofit procedures.

Pressure Cutout Switch. Systems not equipped with a high-pressure cutout switch should have one installed to prevent damage to air-conditioning system parts and to prevent refrigerant emissions. The high-pressure cutout switch will disengage the compressor clutch during high-pressure conditions, thereby reducing the possibility of venting refrigerant and engine cooling system overheating.

Metering Devices. Orifice tubes, thermostatic expansion valves, pressure cycling switches, or other pressure controls may have to be changed during the course of a retrofit.

With the exception of R-134a, all approved alternate refrigerants are blends; they contain two or more refrigerants. In addition to R-134a, the following alternate refrigerants are available. One must be cautioned that not all are approved by the EPA for use in MVAC and MVAC-like appliances. Some are considered by the EPA to be dangerous, and heavy penalties are imposed on those who use them. The EPA makes no exceptions and its rules are simple: Use it—get caught—pay the penalty. There are no excuses. There are one or more questions in the certification exam that test whether the technician "knows the law." If there are any doubts, call the Stratospheric Ozone Hotline and ask. Their toll-free number is (800) 296-1996.

Other Refrigerants

The Mobile Air Conditioning Society (MACS) has warned on many occasions that several refrigerant products are being offered as substitutes for R-12. Many of these refrigerants contain butane (R-600), ethane (R-170), or propane (R-290). Although they are all refrigerants, they are also very flammable materials.

By the close of 1993, 13 states and the District of Columbia had established laws that prohibited the use of any flammable refrigerant in mobile air-conditioning equipment. The first states to enact the law were Arkansas, Connecticut, Idaho, Indiana, Kansas, Louisiana, Maryland, North Dakota, Oklahoma, Texas, Utah, Virginia, and Washington.

In early 1994, Florida was first to pass a law to make it illegal to use any flammable refrigerant in an automobile air-conditioning system. It is now a violation of federal law to use any refrigerant, flammable or otherwise, in a mobile air-conditioning system if it has not been approved by a department of the EPA known as the SNAP program (*www.epa.gov/ozone/snap/*). Currently, there are five refrigerants that are not approved. Although there are 10 refrigerants that are approved for use, only one, R-134a, has been universally accepted by the automotive industry. All refrigerants used in a mobile air-conditioning system must have unique fittings and be identified by labels. There are also requirements for compressor high-pressure cutoff switches to prevent venting to the atmosphere.

Everyone is looking for the "magic bullet," a drop-in replacement for R-12. So far, it does not exist. MACS warns:

- Use only R-12 in an R-12-equipped system.
- Use only R-134a in an R-134a-equipped system.
- Follow retrofit procedures to use R-134a in an R-12 system.
- Do not use refrigerants that contain a toxic substance.
- Do not use a refrigerant that contains a flammable substance.
- The use of unauthorized refrigerants will void the manufacturer's warranties.
- Talk to your customer about prior automotive air-conditioning service. Take no chances with health and safety. Use extreme caution if an unknown refrigerant has been introduced into the system.
- Protect yourself, your equipment, and your refrigerant. Use a refrigerant identifier on every job.

Substitute Refrigerants

Five of the 10 substitute refrigerants found acceptable for automotive use by the EPA contain HCFC-22 as a main component. The use conditions for these refrigerants—R-406/GHG/McCOOL, GHG-X4/Autofrost/Chil-it, Hot Shot/Kar Kool, GHG-HP, and GHG-X5—in addition to unique fittings, labels, and compressor shutoff switch, require barrier hoses. The other five refrigerants accepted by the EPA—R-134a, FRIGC FR-12, Free Zone RB-276, Ikon-12, and Freeze-12—have the same use conditions except they do not require barrier hoses.

Currently, with the exception of R-134a, no vehicle manufacturer approves the use of any of these refrigerants for use in any of their air-conditioning systems as a substitute refrigerant for R-12.

There are three refrigerants at present that are not acceptable to the EPA due to their flammability. These refrigerants are OZ-12, HC-12a, and Duracool-12. Also, refrigerant R-176 is not acceptable because it contains R-12, and R-405A is unacceptable because of its potential association with global warming and high stratospheric lifetime.

The EPA last accepted a substitute refrigerant in mid-1997. It must be noted that the EPA accepts refrigerant for use in certain applications, such as MVACs. However, this does not mean that the EPA recommends or otherwise endorses any particular refrigerant for any particular use. The agency, however, does recognize that R-134a is currently the accepted refrigerant for vehicle use by the industry.

Dedicated service and storage equipment is required by the EPA for each type of refrigerant used in a service facility. For the average facility, that means two systems: one for R-12 and one for R-134a. If one decides to service vehicles using, for example, FREEZE-12, a third set of service and storage equipment must be purchased for use. This is required even though FREEZE-12 contains 80 percent R-134a. The other 20 percent contains HCFC-142b, and it would contaminate the R-134a equipment.

For the latest updates and information on refrigerant approval or any other stratospheric ozone issue, one may contact the EPA. Contact information, toll-free numbers, FAX numbers, and Web site addresses are found in the Appendix.

Freeze 12

Freeze 12 (Figure 12-6)—a blend of 80 percent R-134a and 20 percent HCFC-142b—is acceptable for automotive use subject to having proper fittings, labeling, and a compressor shutoff switch. It is not a drop-in replacement for R-12 or R-134a. The high-side service port must be 3/8-24 right-hand thread and the low-side service port must be 716-20 right-hand thread. The label background color is required to be yellow.

Free Zone/RB-276

Free Zone/RB-276 (Figure 12-7)—a blend of 79 percent R-134a, 19 percent HCFC-142b, and 2 percent lubricant—is acceptable for automotive use subject to having proper fittings, labeling, and a compressor shutoff switch. It is not a drop-in replacement for R-12 or R-134a. The high-side service port must be ½-13 right-hand thread and the low-side service port must be $\frac{9}{16}$-18 right-hand thread. The label background color is required to be light green.

Hot Shot/Kar Kool

Hot Shot—a blend of 50 percent HCFC-22, 39 percent HR-124, 9.5 percent HCFC-142b, and 1.5 percent R-600a—is acceptable for automotive use subject to having proper fittings, labeling, barrier hoses, and a compressor shutoff switch. It is not a drop-in replacement for R-12 or R-134a. Although this refrigerant contains **hydrocarbons** (R-600a, Isobutane), it is not flammable as blended. The high-side service port must be $\frac{5}{8}$-18 left-hand thread and the low-side service port must be $\frac{5}{8}$-18 right-hand thread. The label background color is required to be medium blue.

Hydrocarbons are organic compounds containing only hydrogen (H) and carbon (C).

GHG-HP

This refrigerant is a blend of 65 percent HCFC-22, 31 percent HCFC-142b, and 4 percent R-600a. It is acceptable for automotive use subject to having proper fittings, labeling, barrier hoses, and a compressor shutoff switch. Although it contains hydrocarbons (R-600a, Isobutane), it is not considered flammable as blended. It is not a drop-in replacement for R-12 or R-134a. The required fitting sizes and label background color were undetermined at the time of this writing. Contact the EPA for this information.

FIGURE 12-6 Typical Freeze 12 containers.

FIGURE 12-7 A Free Zone/RB-276 refrigerant cylinder.

GHG-X4/Autofrost/Chil-It

This refrigerant—a blend of 51 percent HCFC-22, 28.5 percent HCFC-124, 16.5 percent HCFC-142b, and 4 percent R-600a—is acceptable for automotive use subject to having proper fittings, labeling, barrier hoses, and a compressor shutoff switch. Although it contains hydrocarbon (R-600a, Isobutane), it is not flammable as blended. It is not a drop-in replacement for R-12 or R-134a. The high-side service port must be 0.305-32 right-hand thread and the low-side service port must be 0.368-26 right-hand thread. The label background color is required to be red.

GHG-X5

GHG-X5—a blend of 41 percent HCFC-22, 15 percent HCFC-142b, 40 percent HFC-227ea, and 4 percent R-600a—is acceptable for automotive use subject to having proper fittings, labeling, barrier hoses, and a compressor shutoff switch. This refrigerant contains 4 percent Isobutane, a hydrocarbon, but it is not considered flammable as blended. It is not a drop-in replacement for R-12 or R-134a. The high-side service port must be ½-20 left-hand thread and the low-side service port must be 916-18 left-hand thread. The label background color is required to be orange.

R-406A/GHG

This refrigerant is a blend of 55 percent HCFC-22, 41 percent HCFC-142b, and 4 percent R-600a. It is acceptable for automotive use subject to having proper fittings, labeling, barrier hoses, and a compressor shutoff switch. It is not considered flammable as blended, although it contains Isobutane (R600a), a hydrocarbon. It is not a drop-in replacement for R-12 or R-134a. The high-side service port must be 0.305-32 left-hand thread and the low-side service port must be 0.368-26 left-hand thread. The label background color is required to be black.

Ikon-12

This refrigerant was approved for automotive air-conditioning system use in mid-1996. The manufacturer, Ikon Corporation, claims that the composition of this refrigerant is confidential business information. Requirements relating to fitting sizes and label color are not developed at the time of this writing. Nor is it yet known whether barrier hoses are required. Contact the EPA or the manufacturer for more information.

FRIGC FR-12

This refrigerant is a blend of 39 percent HR-124, 59 percent R-134a, and 2 percent R-600. It is acceptable for automotive use subject to having proper fittings, labeling, and a compressor shutoff switch. It is not considered flammable as blended, though it contains a hydrocarbon, Butane (R-600). It is not a drop-in replacement for R-12 or R-134a. The high-side and low-side service ports must be a quick disconnect type, but they are different from the R-134a service ports. The label background color is required to be grey.

OZ-12®

This refrigerant, a hydrocarbon Blend A, is not SNAP-approved by the EPA. The agency claims that it contains a flammable blend of hydrocarbons and that insufficient data was submitted to demonstrate its safety.

R-176

This refrigerant contains R-12, HCFC-22, and HCFC-142b. It is not SNAP-approved by the EPA, which claims that it is not appropriate to use an R-12 blend as an R-12 substitute.

HC-12a®

This refrigerant, a hydrocarbon Blend B, is not SNAP-approved by the EPA, which claims that it contains a flammable blend of hydrocarbons and that insufficient data was submitted to demonstrate its safety.

Duracool 12a

This refrigerant is not SNAP-approved by the EPA. It is identical to HC-12a® in composition, but it is produced by a different manufacturer.

R-405A

This refrigerant is not SNAP-approved by the EPA because it contains perfluorocarbons, which are implicated in global warming.

MT-31

This blend proposed as an R-12 substitute is not approved by the EPA for use in any application because of the toxicity of one of its components.

The Do-It-Yourselfer

Although the Clean Air Act (CAA) amendments prevent the sale of small containers of R-12 to the general public, it is not unusual to find small cans of R-134a (Figure 12-8) on the shelves of automotive parts shops. Nor is it at all uncommon for do-it-yourselfers (DIYer) to somehow acquire refrigerant and install it in their personal vehicles. If asked by friends to purchase refrigerant for them, be aware that you may be subject to the wrath of the EPA for doing so, and act accordingly. Some refrigerants are flammable under certain conditions. Refrigerants are not compatible with each other and will contaminate the system.

Water (H_2O), a refrigerant, is unwanted in an air-conditioning system.

FIGURE 12-8 **A small container and can tap of both R-12 and R-134a; notice the different style of can taps.**

Contaminated Refrigerant

Mixing two or more different refrigerants in an air-conditioning system contaminates the refrigerant. The refrigerant is contaminated in that it is no longer "pure" and will not react chemically and physically as intended. Not only will the system not function properly, if it functions at all, but also contaminated refrigerant can damage expensive equipment, such as a recovery/recycle unit.

Do not put any additional refrigerant in a recovery cylinder if the present date is five years or more past the test date stamped on the cylinder shoulder or collar. There are no exceptions to the rule that recovery cylinders must be inspected every five years. There is no "grace period." Using a recovery cylinder beyond the reinspection date can result in heavy penalties.

If there is any doubt as to the purity of the refrigerant in the vehicle, do not service the air-conditioning system unless you are properly equipped.

With the transition to CFC-free air-conditioning systems, the likelihood of cross-mixing refrigerants is a growing concern. Different refrigerants, as well as their lubricants, are not compatible and should not be mixed. It is possible, however, for the wrong refrigerant to be mistakenly charged into an air-conditioning system or for refrigerants to be mixed in the same recovery tank. Also, because recovery/recycling equipment is generally designed for a particular refrigerant, inadvertent mixing can cause damage to the equipment.

A refrigerant identifier tester is far superior to pressure–temperature comparisons because, at certain temperatures, the pressures of R-12 and R-134a are too similar to differentiate with a standard gauge. This is easily noted in the chart shown in Figure 12-9. For example, at 90°F (32.2°C), both 95 percent R-12 and 95 percent R-134a have about the same pressure—111 and 112 psig, respectively. Given that this chart is accurate to plus or minus 2 percent, there is really no way of determining which type refrigerant is in the air-conditioning system or tank. Also, because other substitute refrigerants and blends may have been introduced into the automotive air-conditioning system, they can contaminate a system or tank and may not be detected by the pressure–temperature method. A refrigerant identifier would conclude the refrigerant in our example to be UNKNOWN. The purity of refrigerants has been set by SAE purity standards for both R-12 and R-134a. The purity standard for recycled R-12 is J1991 and the specified limits are 15 parts per million (ppm) by weight for water, 4,000 ppm by weight for refrigerant oil, and 330 ppm by weight for noncondensable gases (air). The purity standard for recycled R-134a is

The thermometer should be placed in an area where it can "sense" free air.

AMB TEMP		R-12/R-134a PERCENT BY WEIGHT										
°F	°C	100/0	98/2	95/5	90/10	75/25	50/50	25/75	10/90	5/95	2/98	0/100
65	18.3	64	67	71	74	83	84	78	73	70	67	64
70	21.1	70	74	79	82	90	92	87	81	77	74	71
75	23.9	77	81	85	91	99	101	96	89	85	83	79
80	26.7	84	88	93	99	107	110	105	98	95	92	87
85	29.4	92	96	101	108	116	120	114	106	103	100	95
90	32.2	100	105	111	116	125	130	125	116	112	109	104
95	35.0	108	114	119	126	135	140	135	126	122	119	114
100	37.8	117	123	127	135	145	151	145	136	133	130	124
105	40.6	127	132	138	146	158	164	159	149	144	141	135
110	43.3	136	142	147	156	170	176	173	164	157	152	146
115	46.1	147	152	159	166	183	192	184	175	168	163	158
120	48.9	158	164	170	177	195	205	196	187	181	176	171

CFC-12/HFC-134a Cross-Contamination Chart. All pressures are given in psig. For kPa, multiply psig by 6.895. For example, 100 percent R-12 at 95 °F (35 °C) is 108 psig or 744.7 kPa.

FIGURE 12-9 **Temperature–pressure chart of R-12 and R-134a mixed refrigerants.**

J2099, and the specified limits are 15 parts per million (ppm) by weight for water, 500 ppm by weight for refrigerant oil, and 150 ppm by weight for noncondensable gases (air). Refrigerant should test at least 98 percent pure when tested with a purity tester. If the refrigerant is less than 98 percent pure, it should be considered contaminated refrigerant and treated as such.

Proper Equipment

Recovery cylinders must be inspected every five years.

Being properly equipped means that you have access to and use "recovery only" equipment (Figure 12-10) that meets SAE's J2209 standards. You must also have proper recovery cylinders that meet rigid U.S. Department of Transportation (DOT) specifications. These cylinders should be marked "CONTAMINATED REFRIGERANT" for identification. Only contaminated refrigerant should be recovered into this cylinder.

Disposable cylinders, known as DOT 39s (Figure 12-11), must not be used for recovered refrigerant. Federal law prohibits refilling these cylinders.

FIGURE 12-10 **A typical refrigerant recovery unit.**

FIGURE 12-11 **Disposable R-134a cylinder.**

Do Not Take Chances

If there is any doubt about the purity of the refrigerant, question the customer. Ask the customer such questions as:

- When is the last time the system was serviced?
- Who worked on it?
- What parts were replaced?
- Do you have a copy of the work order?

Remember, under the CAA, anyone performing repairs for consideration (pay) to a motor vehicle air-conditioning system must be certified, must use recovery/recycle equipment, and must comply with all rules and regulations.

Be very suspicious if the customer's response to "Who worked on it?" is something like, "Well, my neighbor works at a shop and he did it over the weekend as a favor to me." You should then ask, "Where?" Chances are the customer will reply, "At my home."

The good-intentioned neighbor may have simply been trying to do a favor. Whatever the reason, he could have contaminated the system. After all, the air conditioner still does not work. If it worked, your customer would not have brought it to you for service.

Do not take a chance. Test a sample of the refrigerant with a purity tester as covered in Chapter 7 of this manual and Chapter 7 of the Shop Manual. If a purity tester is not available and there is even the slightest doubt, turn the vehicle away. An alternative is to keep the vehicle overnight, a period of 12 hours or more, and check for refrigerant purity according to a temperature–pressure chart before attempting repairs.

If in doubt, perform the purity test.

Purity Test

A determination of the purity of the refrigerant in the vehicle is possible while allowing for reasonable inaccuracies of the gauge, the thermometer, and the reader. After a 12-hour period, the pressure should nearly match that expected for any given temperature if the refrigerant is pure.

There are other factors to be considered, however, when testing refrigerants. For example, if there is air in the system, an accurate reading may not be noted. If there is any doubt, do not run the risk of contaminating a good tank of refrigerant.

Disposal of Contaminated Refrigerant

Contaminated refrigerant may be reclaimed to ARI-700-88 standards, or it may be destroyed by fire. This is usually accomplished at an off-site reclamation facility that is equipped to handle such problems. Remember, however, that it is your responsibility to legally *dispose* of contaminated refrigerant.

Make no attempt to destroy refrigerant without the proper equipment.

Use of Alternate Refrigerants

Many new alternative refrigerants marketed for use in motor vehicle air-conditioning systems are being touted by their manufacturers and distributors. Whether employed by a nationwide repair chain or a one-person service facility, the technician should take the time to determine how well an alternative refrigerant will perform and whether it may pose any problems for customers or raise liability issues.

Health and the Environment

The EPA's SNAP program determines what risks to human health and the environment are posed by refrigerant alternatives. The EPA evaluates the alternative refrigerant's ozone-depleting potential (ODP), GWP, flammability, and toxicity. The SNAP evaluation, however, does not determine whether the alternate refrigerant will provide adequate performance or whether it will be compatible with the components of the system.

Use Conditions

The EPA places conditions or restrictions on how an alternative can be used. Under SNAP, for example, an R-12 substitute requires the use of a new label and new fittings unique to the alternative. There are no exemptions to the rule for do-it-yourself (DIY) mechanics.

Because of the vast range of equipment types and designs, the EPA does not issue retrofit procedures. The manufacturer of the system is the best source of information about how well a given substitute will perform. Additionally, one must determine whether charging a system with a particular "new" refrigerant will void any manufacturer's warranty.

Clean Air Act (CAA)

The CAA requires that the EPA establish standards for recovery, recycling, and reclamation of refrigerants, including alternatives, accepted under SNAP. If standards have not been published by the EPA for a particular alternative, they may be under development. Ensure that the refrigerant manufacturer intends to work with the EPA to develop uniform methods for extraction, recycling, and reclamation.

Standards

The Air Conditioning and Refrigeration Institute (ARI), a manufacturers' trade association, develops standards for the industry. ARI's standard 700 specifies acceptable levels of refrigerant purity for R-12, as well as for certain refrigerant blends. The purpose of the standard is to enable end users to evaluate and accept or reject any refrigerant.

The American Society of Heating, Refrigerating and Air-Conditioning Engineers (ASH-RAE), a trade association, sets many of the standards and guidelines to provide a uniform method of rating refrigerants for toxicity and flammability and to assign refrigerant numbers. In fact, before ARI determines that its standard 700 should apply to a particular refrigerant, it must receive a classification from ASH-RAE. However, ASH-RAE classification is not required for SNAP acceptability.

Flammability

Both ASHRAE and the EPA evaluate refrigerants for flammability. The EPA requires that a new refrigerant be analyzed according to a test of the American Society of Testing Materials (ASTM). This test determines the concentrations in the air at which a substance is flammable at normal atmospheric pressure. Some hydrocarbons, for example, ignite at concentrations as low as 2 percent by volume. If a blend contains a flammable component, the EPA requires leak testing to ensure that the blend does not change and become flammable.

If a system is charged with an alternative refrigerant that later becomes unavailable, the system may have to be retrofitted again, a service the customer may feel is unfair and be unwilling to pay for.

Grace Period

A refrigerant manufacturer must submit information on a new refrigerant for SNAP review at least 90 days before marketing. The CAA, however, does not prohibit the sale and use of that refrigerant after the 90-day period. If the agency is still engaged in its review after 90 days, the refrigerant can be sold and used even though it is not formally approved. The EPA may later determine that the refrigerant is unacceptable, and you may be stuck with an inventory of refrigerant that cannot be legally used.

Use versus Sale

The CAA granted to the EPA authority to regulate the use of alternative refrigerants, not the sale of them. If, for example, the EPA determines that an alternative is unacceptable for automotive service, it is still legal to sell it to the automotive trade. Using it in a customer's air-conditioning system, however, is considered illegal, and the technician who serviced the air-conditioning system may be fined $25,000 and have to serve up to five years in prison.

Retrofit Components

Following is an overview of some of the problems and conditions associated with components, listed in no particular order, when retrofitting an automotive air-conditioning system from R-12 to R-134a refrigerant.

Saddle clamp access valves have been used to gain access to domestic "hermetic" air conditioners for many years.

Access Valves

There is a distinct difference between the access valves used on R-134a systems (Figure 12-12) and those used on R-12 systems (Figure 12-13). Adapters (Figure 12-14) are available that are to be used on R-12 fittings during retrofit procedures to make them compatible with R-134a equipment. A special adapter, called a saddle clamp access valve (Figure 12-15), is available for installation where space does not permit the R-134a adapter to convert the R-12 valve.

Shop Manual
Chapter 12, page 518

Accumulator

Accumulators (Figure 12-16) in R-12 systems typically have a desiccant designated as XH5. This desiccant is not compatible with R-134a refrigerant. The desiccant to be used in R-134a systems is designated XH7 or XH9. This desiccant is found in accumulators and receivers designated for R-134a service. General Motors and Ford do not recommend that their accumulator be changed because the desiccant used is compatible with R-134a. Both XH7 and XH9 desiccants are compatible with R-12 as well as R-134a.

If a clutch cycling pressure switch (CCPS) is to be changed, however, the accumulator may have to be replaced to accommodate the metric threads found on the switch. In some retrofit packages, an adapter may be included for English-to-metric thread conversion.

Do not attempt to screw a metric fitting into an English fitting.

Compressor

Compressors are being redesigned to withstand the slight increase in pressures associated with R-134a. Most compressor rebuilders are also incorporating these design changes into their rebuilding procedures. When purchasing a new or rebuilt compressor for an R-134a system, make sure that it has been identified for that application.

It is not recommended that a compressor be replaced as a matter of course for retrofitting. The compressor should only be replaced if it is defective. Do not replace a compressor simply because the system is being retrofitted.

FIGURE 12-12 **Access valve port on an R-134a system.**

FIGURE 12-13 **Access valve port on an R-12 system hose.**

FIGURE 12-14 **Adapter fittings for retrofitting R-12 access valve fittings over to R-134a access valve fittings.**

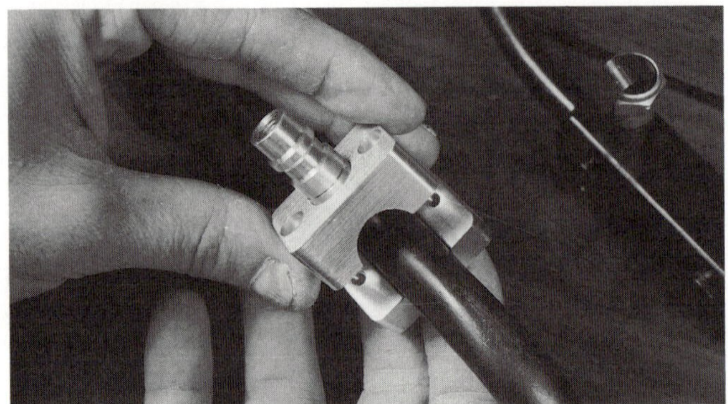

FIGURE 12-15 **A saddle clamp access valve.**

FIGURE 12-16 **A typical accumulator.**

Engine performance is often affected by the increased load created by the air conditioner.

Condenser

To change any part of an original design is to change the performance of the equipment. This may be especially true for the condenser. The engine cooling system may also be affected by the slight increase in pressure (and temperature) of the condensing R-134a refrigerant. A dam may be considered to reduce the problems. Some manufacturers recommend replacing the condenser assembly during the retrofit procedure with one designed for use with R-134a, or system performance may suffer after retrofitting. Always refer to the manufacturer's recommendations prior to performing retrofit procedures or quoting the cost of the service to your customer.

Dams. A dam, loosely identified, is the sealing provision located between the radiator and condenser that helps to direct ambient and ram air through both components. It is critical that all condenser and radiator seals be in place. All holes, regardless of how small, that could allow air to bypass either component should be blocked off to ensure maximum airflow.

Do not overload the circuit by adding a second motor to the original relay.

In some installations, the condenser will be changed. Because the mounting space is limited, this usually means a condenser with more fin area or fins per inch (FPI). A higher rpm cooling fan motor may be used to replace the original motor. In other cases, a second motor and fan, a pusher type, may be placed in front of the condenser. The idea is to improve or increase airflow to remove more heat.

If a fan and motor are added, a relay should also be added. This is to ensure that the electrical system is not overloaded. The coil of the relay may be wired in with the compressor clutch circuit (Figure 12-17) to ensure that the fan is running when the air-conditioning system is turned on. An in-line fuse is included to protect the circuit.

Evaporator

The evaporator is not replaced unless it is found to be leaking. There have been no problems reported when using R-12 evaporators for an R-134a retrofit.

Minor changes are necessary for evaporators designated for use with R-134a refrigerant to accommodate the slightly higher pressure that may be expected.

Hoses

Generally, hoses (Figure 12-18) used for automotive air-conditioning service in 1989 and later year/model vehicles need not be replaced when retrofitting from R-12 to R-134a. The exception is if the hose is found to be leaking during retrofit procedures.

O-Rings and Seals

Although O-rings made of epichlorohydrin and designated for R-12 service are not compatible with R-134a, it is not recommended that they be replaced when retrofitting an air-conditioning system. The exception is if the fitting is found to be leaking. In that case, use only O-rings and seals (Figure 12-19) designated for the refrigerant being used in the system. Generally, O-rings and seals for R-12 systems are black. Unfortunately, some manufacturers prefer black R-134a O-rings and seals as well. Many, however, color code the O-rings and seals designated for R-134a service. When in doubt, use color-coded neo prene or HSN/HNBR O-rings or seals; they are also compatible with R-12.

If O-rings are to be replaced, use components designated for use with R-134a.

Metering Devices

Metering devices should not be changed as a matter of practice when retrofitting a system. There are two types of metering devices used in the modern automotive air-conditioning system: the thermostatic expansion valve (TXV) and the fixed orifice tube (FOT).

Superheat is heat that is added after the refrigerant has vaporized.

Thermostatic Expansion Valve. The TXV (Figure 12-20) does not have to be replaced when retrofitting a system from R-12 to R-134a. If, however, a TXV is found to be defective, it should be replaced with a model designed for use with the system refrigerant. An R-12 TXV used in an R-134a system will result in higher superheat and improved overall evaporator temperature. An R-134a valve used in an R-12 system will have reduced superheat and will not perform as well. Because superheat has a direct effect on performance, it is not advisable that the superheat be allowed to increase more than 3°F (1.7°C) over that of the operating R-12 system.

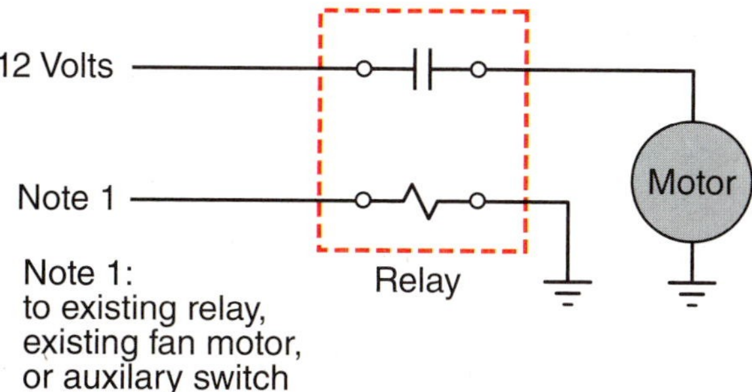

FIGURE 12-17 **Schematic for adding an auxiliary condenser fan motor.**

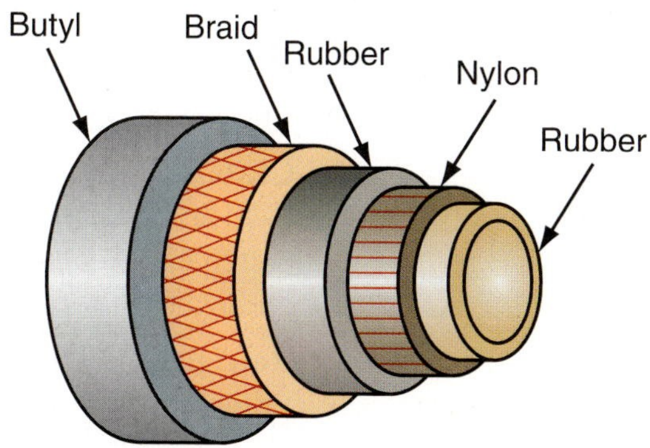

FIGURE 12-18 **Construction detail of a barrier hose.**

FIGURE 12-19 **Typical air conditioner O-rings and seals.**

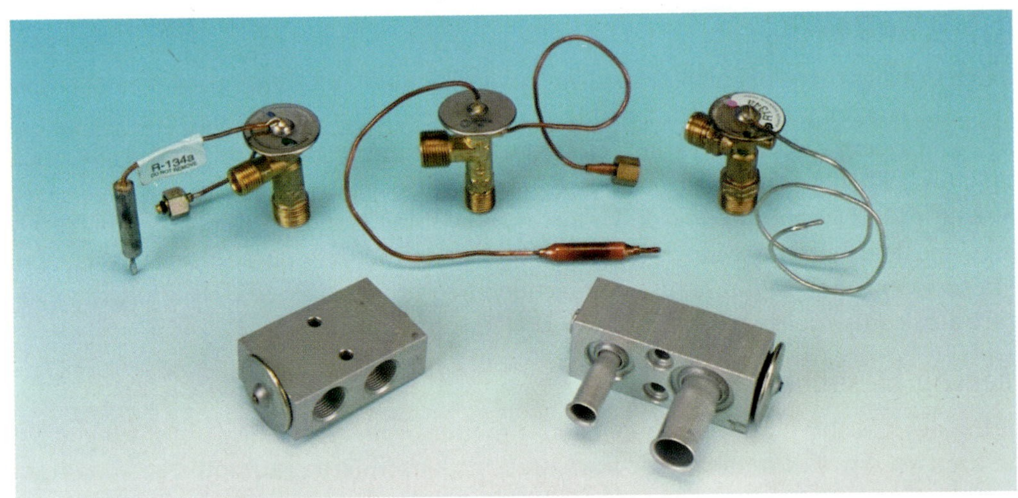

FIGURE 12-20 **Typical thermostatic expansion valves.**

If a new TXV is required, use one that is designed for the specific refrigerant in the system. As a rule of thumb: R-12 valves should not be used on automobiles originally equipped with R-134a systems and, conversely, R-134a valves should not be used on R-12 systems.

Orifice Tube. With the exception of one automobile manufacturer, it is not recommended that the orifice tube (Figure 12-21) be replaced when retrofitting an air-conditioning system. Volvo recommends changing the orifice tube to one that has a 0.002 in. (0.0508 mm) or smaller orifice. If it is changed, however, a slight increase in high-side pressure may be noted.

Pressure Switch

Do not bypass system protective switches.

Either or both of two switches may be recommended for change during some retrofit procedures. These switches are the CCPS and the refrigerant containment device. A brief description follows.

Clutch Cycling Pressure Switch. The CCPS (Figure 12-22) may be changed for some R-134a retrofits. The difference is that the R-134a switches are calibrated for slightly lower clutch cycling pressures. Also, the mounting threads are metric to prevent the connection of an English-thread R-12 switch in an R-134a system.

Refrigerant Containment Device. This device, which is new for 1994 and later model/ year vehicles, may also be included in some retrofit kits for earlier model year vehicles. The refrigerant containment device includes models for single- and dual-function refrigerant containment switches and the air conditioner high-pressure transducer.

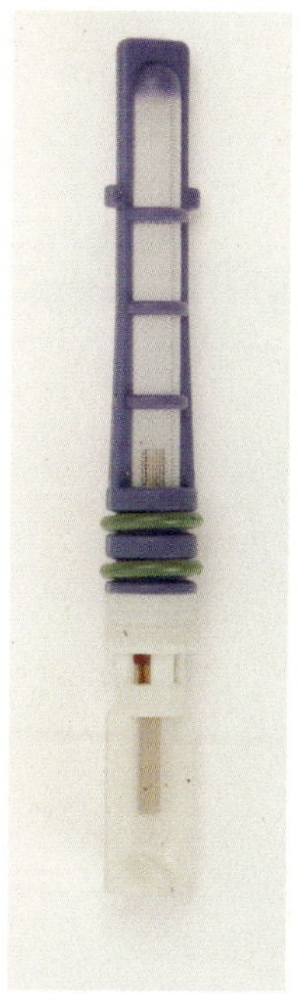

FIGURE 12-21 **An orifice tube.**

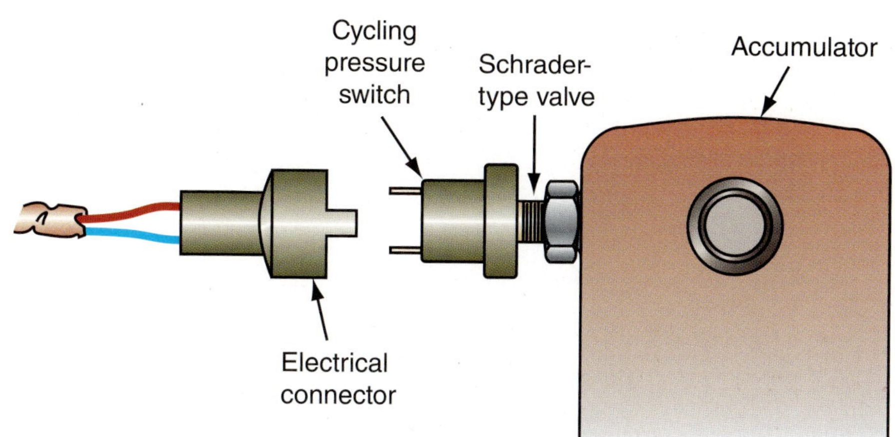

FIGURE 12-22 **A cycling clutch pressure switch.**

Each have their specific applications. The single switch is in general use, controlling the compressor clutch; the dual switch also includes provisions to control the condenser fan. The transducer is in some solid-state temperature control systems.

High-Pressure Switch

The EPA requires the installation of a high-pressure cutout switch, also called a refrigerant containment switch. Its purpose is to interrupt the clutch coil circuit, thereby stopping the compressor before high-side pressure reaches the point at which it would open the high-pressure relief valve and release refrigerant into the environment.

One such switch designed for retrofit (Figure 12-23) is included with a tee-fitting service valve and is actually a dual-pressure switch. It opens the compressor clutch circuit at a high pressure of about 390–400 psig (2,689–2,758 kPa) and closes the circuit at about 310–315 psig (2,137–2,172 kPa). For low-pressure protection, the switch opens the clutch circuit at a low pressure of about 28 psig (193 kPa), a condition that would not exist on the high side of the system unless the refrigerant had leaked out. This then would prevent the compressor from running when the air-conditioning system is first turned on.

Receiver-Drier

The receiver-drier (Figure 12-24) used in R-12 systems typically has XH5 desiccant. This desiccant is not compatible with R-134a refrigerant. To be sure, the receiver-drier should be replaced during retrofit procedures with a unit designated for R-134a service and PAG or ester lubricants. This desiccant, designated XH7 or XH9, is also compatible with R-12 refrigerant and mineral oils.

A system has an accumulator or a receiver-drier, not both.

Retrofit Labeling Requirements

Once the original refrigerant has been replaced with a SNAP-approved alternative refrigerant, label (Figure 12-25) must be affixed over the old label. The label must contain detailed information about the retrofit installation. The label's background color is unique for each SNAP refrigerant (Figure 12-26). The label must show:

- Name and address of the installer who performed the retrofit
- The refrigerant name and identification number (i.e., R-134a)

- The refrigerant charge amount installed
- The date the retrofit was performed
- The type of refrigerant oil installed and the part number
- The amount of refrigerant lubricant (oil) installed

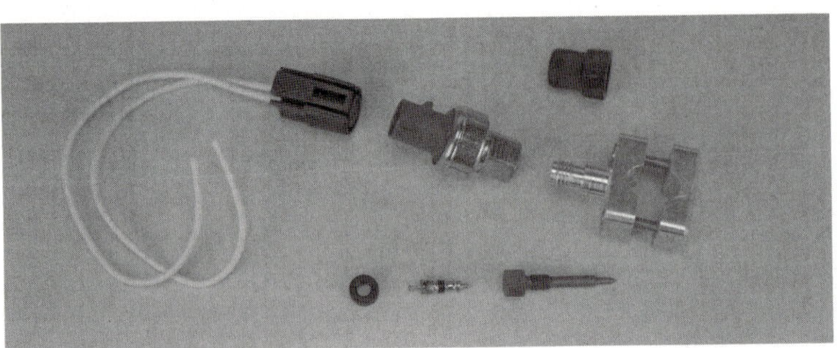

FIGURE 12-23 **A high-pressure cutout switch with saddle clamp.**

FIGURE 12-24 **A typical receiver-drier.**

NOTICE: RETROFITTED TO R-134a

RETROFIT PROCEDURE PERFORMED TO SAE J1661
USE ONLY R-134a REFRIGERANT AND SYNTHETIC
OIL TYPE: ____1____ PN: ____2____ OR
EQUIVALENT, OR A/C SYSTEM WILL BE DAMAGED

REFRIGERANT CHARGE/AMOUNT: ____3____
LUBRICANT AMOUNT: ____4____ PAG ☐ ESTER ☐ 5

RETROFITTER NAME: ____6____ DATE: ____7____
ADDRESS: ____8____
CITY: ____9____ STATE: ____10____ ZIP: ____11____

1 Type: manufacturer of oil (Saturn, GM, Union Carbide, and so forth).

2 PN: Part number assigned by manufacturer.

3 Refrigerant charge/amount: Quantity of charge installed.

4 Lubricant amount: Quantity of oil installed (indicate ounces, cc, or mL).

5 Kind of oil installed (check either PAG or ESTER).

6 Retrofitter name: Name of facility that performed the retrofit.

7 Date: Date retrofit is performed.

8 Address: Address of facility that performed the retrofit.

9 City: City in which the facility is located.

10 State: State in which the facility is located.

11 Zip: Zip code of the facility.

FIGURE 12-25 **A typical retrofit label.**

Refrigerant	Background
CFC-12	White
HFC-134a	Sky blue
Freeze 12	Yellow
Free Zone / RB-276	Light green
Hot Shot	Medium blue
GHG-X4	Red
R-406A	Black
GHG-X5	Orange
GHG-HP	Not yet developed*
Ikon-12 / Ikon A	Not yet developed*
FRIGC FR-12	Gray
SP34E	Tan
R-426A (RS-24, new formulation)	Gold
R-420A	Dark green (PMS #347)

FIGURE 12-26 **A unique background color label is required for each SNAP-approved refrigerant.**

Flushing is the process of removing solid particles, such as metal flakes or dirt.

SYSTEM FLUSHING

Little is written about flushing because "the jury is still out" on the subject. Some claim that flushing is necessary to "clean" an air-conditioning system, whereas others claim flushing causes more harm than good. There is no need to **flush** an air-conditioning system to remove refrigerant. Refrigerant removal is accomplished with the use of a proper refrigerant recovery machine. Flushing, then, is only performed in an attempt to clean the system of excess debris and lubricant. Actually, during the flushing procedure, a great deal of the debris will be caught in the screens of the metering device and dehydrator (receiver or accumulator). Also, the lubricant will be trapped in the bottom tank of the evaporator and dehydrator. Little, if anything, will be removed by flushing if the individual components are not removed from the vehicle and flushed individually.

General Motors (GM), as a rule, does not recommend flushing an air-conditioning system. There are but two possible exceptions to the rule: a lubricant overfill or lubricant contamination. Even then, GM recommends removing and draining the accumulator in an effort to remove the lubricant before considering flushing the air-conditioning system. If flushing is determined to be necessary, the only flushing chemical approved by GM is the same refrigerant that the air-conditioning system was originally charged with.

Accordingly, flushing procedures are not specifically covered in this manual. If it is found, however, that an automotive air-conditioning system needs flushing, one should follow the specific instructions included with the flushing equipment. Several flushing systems are available. Robinair, for example, has air-conditioning flushing kits (Figure 12-27), which include an accessory that connects to their recovery/recycle machines and uses recovered refrigerant as a flushing agent.

A BIT OF HISTORY

The time is at hand to seriously consider the retrofit market for automotive air-conditioning systems. According to a report made by the International Trade Commission, the agency that tracks refrigerant production, only about half as much R-12 was produced in 1994 as in 1993. They also caution us that production of all CFCs ended January 1, 1996. That means that the only R-12 now available for automotive use is that which has been recovered and recycled or was produced prior to production being banned by the Clean Air Act.

AN INDUSTRY STUDY

A paper sponsored by Elf Atochem, a leading manufacturer of refrigerants, was presented in the winter of 1994 at the International CFC and Halon Alternatives Conference. Written by a staff engineer and a senior technician at Elf Atochem's fluoro chemicals research and

FIGURE 12-27 **Typical flush system.**

development center, the paper revealed that retrofitting is simpler than originally believed. The report was the result of a study of a fleet of 17 employee-volunteered vehicles retrofitted in 1993 to R-134a and polyalkaline glycol (PAG) lubricant. In early 1994, 20 more vehicles were retrofitted with R-134a and polyol ester (POE) lubricant. Only refrigerant and lubricant were replaced; no system components were replaced. Some were power flushed to remove as much of the mineral oil as possible; others were simply drained and refilled. The study involved a random selection of both domestic and imported cars and light trucks.

Regardless of the procedure, flush or no flush, there was little or no noticeable difference in the performance of any of the vehicles. There were only two reported failures; both lost their complete R-134a charge due to O-ring failure. Another vehicle lost 6 percent of its charge of R-134a due to a leak. The worst "leaker" in the study was actually a control vehicle that had not been retrofitted at all. This vehicle lost 38 percent of its R-12 charge due to a leak.

AUTHOR'S NOTE: If after retrofitting a system from R-12 to R-134a the high-side head pressure is excessive, the installation of an auxiliary cooling fan will help lower this pressure. The auxiliary fan should be wired so that it will run continuously when air conditioning is selected.

SUMMARY

- The concern in the world community has shifted from the threat of ozone-depleting chemicals to a concern over global warming.
- In the United States and Asia, manufacturers are not required to phase-out R-134a.
- Some are still holding out long-term hopes for CO_2 (R-744) systems, but that appears doubtful at least for now.
- Follow appropriate manufacturer's retrofit procedures.
- After retrofit, appropriate decals must be affixed to identify the type of refrigerant in the system.
- A light blue (the industry color code for R-134a) decal is placed over the current R-12 decal (Figure 12-25).
- A yellow (the color used for caution) decal may be placed around the hoses at the service fittings.

TERMS TO KNOW

Adiabatic expansion
Alternative refrigerant
Cross-contamination
Flushing
Global Warming Potential (GWP)
Hydrocarbons
Retrofit
R-744

REVIEW QUESTIONS

Short-Answer Essay

1. Describe the meaning of the term retrofit.
2. What is the meaning of the term global warming potential?
3. What did the European Union mandate beginning January 1, 2011?
4. Describe the proper disposal of contaminated refrigerant.
5. Are vehicles manufactured or sold in the United States required to phase-out R-134a refrigerant systems?
6. Describe the difference in application for an XH9 drier compared to an XH5 drier.
7. CO_2 is a greenhouse gas, so how could it be used in an air-conditioning system as an environmentally friendly alternative?
8. What is an important consideration for the condenser during retrofit procedures?
9. Describe the conditions under which hoses and O-rings should be replaced during retrofit procedures.
10. Identify and describe either of the components that are included on R-134a systems but are not found on R-12 systems.

Fill in the Blanks

1. The EPA's ______________ ______________ ______________ program determines what risks to human health and the environment are posed by refrigerant alternatives.
2. Mixing two or more different refrigerants in an air-conditioning system ______________ the refrigerant.
3. If a hose is found to be ______________ during retrofitting it should be ______________.
4. ______________ ______________ is a process that occurs without the loss or gain of heat.
5. ______________ ______________ ______________ is when one refrigerant is contaminated with another.
6. R-12 receiver-driers containing ______________ desiccant must be replaced with receiver-driers containing ______________ or ______________ desiccant when retrofitting to R-134a.
7. ______________ is the refrigerant gas designation given to carbon dioxide (CO_2).
8. Another alternative refrigerant that once held widespread hopes for being the next refrigerant due to its low GWP rating of 1 is ______________, a naturally occurring gas.
9. ______________ ______________, known as DOT 39s, must not be used for ______________.______________
10. MACS is an acronym for ______________ ______________ ______________

Multiple Choice

1. All of the following alternate refrigerants are approved for automotive use, *except*:

A. Freeze-12
B. FRIGC FR-12
C. Ikon-12
D. Duracool 12a

2. Replacement of hoses during retrofit procedures is being discussed.

   Technician A says hoses need not be replaced unless they are the old-style nonbarrier type.

   Technician B says hoses need not be replaced unless they are equipped with the old-style O-rings and fittings.

   Who is correct?

   A. A only
   B. Both A and B
   C. B only
   D. Neither A nor B

3. Refrigerant contamination is being discussed.

   Technician A says that R-12 with a 5 percent trace of R-22 is considered contaminated.

   Technician B says that R-12 with a 5 percent trace of R-134a is considered contaminated.

   Who is correct?

   A. A only
   B. B only
   C. Both A and B
   D. Neither A nor B

4. The concern in the world community has shifted from the threat of ozone-depleting chemicals to a concern over which of the following?

   A. Global cooling
   B. Global warming
   C. Global weather
   D. Global disaster

5. *Technician A* says that a DOT 39 cylinder may be used to temporarily store contaminated refrigerant.

   Technician B says that, if they are used, DOT 39s must be marked "CONTAMINATED REFRIGERANT" for identification.

   Who is correct?

   A. A only
   B. B only
   C. Both A and B
   D. Neither A nor B

6. Which of the following desiccants should be used with HFC-134a refrigerant?

   A. XH5
   B. XH7
   C. XH9
   D. Both B and C

7. A refrigerant retrofit label must contain all of the following *except:*

   A. Name and address of the installer who performed the retrofit
   B. The refrigerant charge amount installed
   C. The amount of R-12 removed from the system
   D. The type of refrigerant oil installed and the part number

8. *Technician A* says that a red label is required to identify R-134a in an automotive air-conditioning system.

   Technician B says that a yellow label is used to identify R-12 in an automotive air-conditioning system.

   Who is correct?

   A. A only
   B. B only
   C. Both A and B
   D. Neither A nor B

9. The retrofit label's background color for R-134a refrigerant is:

   A. White
   B. Tan
   C. Royal blue
   D. Light blue

10. *Technician A* says that O-rings designated for R-12 service need not be changed for retrofit.

    Technician B says that O-rings designated for R-134a service may also be used for R-12 service.

    Who is correct?

    A. A only
    B. B only
    C. Both A and B
    D. Neither A nor B

GLOSSARY
GLOSARIO

Note: **Terms are highlighted in color**, followed by **Spanish translation in bold**.

Absolute zero The complete absence of heat, believed to be −459°F (−273.15°C). This is shown as 0 degrees on the Rankine and Kelvin temperature scales.

Cero absoluto Ausencia completa de calor, lo cual se cree ser −459°F (−273.15°C). Se indica como 0 degrees en las escalas de temperatura Rankine y Kelvin.

Absorb To take in or to suck up; to become a part of itself.

Absorber Admitir o aspirar; llegar a ser una parte de sí mismo.

Accumulator A tank-like vessel located at the outlet (tailpipe) of the evaporator to receive all of the refrigerant that leaves the evaporator. This device is constructed to ensure that no liquid refrigerant enters the compressor.

Acumulador Recipiente parecido a un tanque ubicado en la salida (tubo de escape) del evaporador para recibir todo el refrigerante que sale del evaporador. Dicho dispositivo está diseñado de modo que asegure que el refrigerante líquido no entre en el compresor.

Additive A substance added to another substance, expected to increase its quality or performance. Antifreeze, for example, is an additive that may be added to water (H_2O) to raise its boiling point and lower its freezing point.

Aditivo Sustancia que se agrega a otra sustancia para mejorar su cali-dad o rendimiento. Por ejemplo, el anticongelante es un aditivo que puede agregarse al agua (H_2O) para elevar su punto de ebullición y disminuir su punto de congelación.

Adiabatic expansion The process of expansion without the gain or loss of heat.

Expansión adiabiático El proceso del expansión sin un gano o pérdida del calor.

Adsorb To take up and hold a thin layer of vapor or liquid molecules on the surface of a solid substance.

Adsorber Recoger y retener una capa delgada de moléculas de vapor o líquido en la superficie de una sustancia sólida.

Aftermarket A term generally given to a device or accessory that is added to a vehicle by the dealer after original manufacturer, such as an air-conditioning system.

Postmercado Término dado generalmente a un dispositivo o accesoria que el distribuidor de automóviles agrega al vehículo después de la fabricación original, como por ejemplo un sistema de acondicionamiento de aire.

Air conditioning (A/C) The process of adjusting and regulating by heating or refrigerating; the quality, quantity, temperature, humidity, and circulation of air in a space or enclosure; to condition the air.

Acondicionamiento de aire Proceso de ajustar y regular al calentar o enfriar, la calidad, cantidad, temperatura, humedad, y circulación de aire en un espacio o encerramiento; para acondicionar el aire.

Allotrope A structurally different form of an element. For example, though different in structure, the properties of graphite and diamond are the same as the element carbon (C).

Aliotropo Una forma de un elemento que es diferente en estructura. Por ejemplo, aunque son diferentes en estructura, las propiedades del grafito y diamante son las mismas del elemento carbón (C).

Alternative refrigerant A refrigerant that can be used to replace an existing refrigerant, such as ozone friendly R-134a that is used to replace ozone-depleting R-12.

Refrierante alternativa Un refrigerante que se puede usar para reem-plazar un refrigerante actual, tal como el R-134 que no agota el ozono del medio ambiente para reemplazar el R-12 que sí lo agota.

Ambient All around, surrounding, or encompassing.

Amblente El area circundante o los alrededores.

Ampere (A) Ampere or Amperage is the unit for measuring current. One ampere equals a current flow of 6.28 × 1018 electrons per second.

Amperio (A) el amperio o amperaje es la unidad de medición de la corriente. Un amperio equivale a un flujo de corriente de 6,28 × 1018 electrones por segundo.

Antifreeze A commercially available additive solution used to increase the boiling temperature and reduce the freezing temperature of engine coolant. A solution of 50 percent water and 50 percent antifreeze is suggested for year-round protection.

Anticongelante Solución aditiva comercialmente disponible utilizada para elevar la temperatura de ebullición y disminuir la temperatura de congelación del enfriador del motor. Se sugiere una solución de un 50 por ciento de agua y un 50 por ciento de anticongelante para proveer protección durante todo el año.

Arid Dry.

Arido Seco.

Aspirator A device that uses suction to move air, accomplished by a differential in air pressure.

Aspirador Dispositivo que aspira para mover el aire; dicho movimiento se debe a una diferencia en la presión del aire.

ATC Abbreviation for automatic temperature control.

ATC Abreviatura de control automático de temperatura.

Atmosphere Air.

Atmosfera El aire.

Atom The smallest possible particle of matter.

Atomo Partícula más pequeña de materia.

Automatic A self-regulating system or device that adjusts to variables of a predetermined condition.

Automático Sistema o dispositivo con regulación automática que se adjusta a estadores variables de una condición predeterminada.

Auxiliary A backup component or system. The rear evaporator in a dual air-conditioning system is often referred to as an "auxiliary evaporator."

Auxiliar Un componente o sistema auxiliar. El evaporador trasero en un sistema del aire acondicionado muchas veces se refiere como un evaporador auxiliar.

Average A single value that represents the median.

Promedio Un valor único que representa la mediana.

Axial Pertaining to an axis; a pivot point.

Axial Perteneciente a un eje; punto de giro.

Axial plate That part of an automotive air-conditioner compressor piston assembly that rotates as a part of the driven shaft.

Placa axial La parte del conjunto del pistón de un compresor del acondi-cionador de aire automotriz que gira como parte del árbol mandado.

Barrier A term given to something that stands in the way, separates, keeps apart, or restricts; an obstruction.

Barrera Término dado a algo que impide, separa, mantiene separado, o restringe; una obstrucción.

Bellows An accordion-type chamber that expands and contracts as its interior pressure is increased or decreased to create a mechanical action, such as in a thermostatic expansion valve.

Fuelles Cámara en forma de acordeón que se dilata y se contrae cuando su presión interior se aumenta o se disminuye para crear una acción mecánica, como por ejemplo en una válvula de expansión termostática.

Bi-level A condition whereby air is delivered at two levels in the vehicle, generally to the floor and dash outlets.

Bilevel Condición por medio de la cual se envía el aire a dos niveles en el vehículo, generalmente al piso y a las salidas del tablero de instrumentos.

Blend air door A door in the duct system that controls temperature by blending heated and cooled air.

Puerta de aire mezclado Puerta en el sistema de conductos que regula la temperatura al mezclar el aire calentado y enfriádo.

Blend door See Blend air door.

Puerta de mezcla Ver Blend air door [Puerta del aire mezclado].

Block valve A type of thermostatic expansion valve utilizing an internal sensing bulb.

Válvula de bloque Tipo de válvula de expansión termostática que utiliza una bombilla sensora interior.

Bowden cable A wire cable inside a metal or rubber housing used to regulate a valve or control from a remote place.

Cable Bowden Cable trenzado de alambre que está envuelto por una cubierta de metal o de caucho y que se utiliza para regular una válvula o regulador desde un sitio a distancia.

British thermal unit (Btu) A measure of heat energy; one Btu is the amount of heat necessary to raise one pound of water 1°F.

Unidad térmica británica (Btu) Medida de energía calorífica; un Btu es igual al calor necesario para elevar la temperatura de una libra de agua en 1°F.

Butane A colorless gas (C_4H_{10}) that is used as a fuel.

Butano Gas incoloro (C_4H_{10}) utilizado como combustible.

Bypass A passage or hose that directs coolant around thermostat when thermostat is closed and back to the water pump.

Desviado Un pasaje o manguera que dirige el refrigerante alrededor del termostato cuando está cerrado y lo dirige a la bomba de agua.

Capillary tube A tube with a calibrated inside diameter and length used to control the flow of refrigerant. In automotive air-conditioning systems, the tube connecting the remote bulb to the expansion valve or to the thermostat is called the capillary tube.

Tubo capilar Tubo cuyo diámetro y longitud interiores son calibrados; se utiliza para regular el flujo de refrigerante. En sistemas automotrices para el acondicionamiento de aire, el tubo que conecta la bombilla a distancia a la válvula de expansión o al termostato se llama el tubo capilar.

Case A thing used to hold, cover, or contain something, such as the evaporator and heater cores of an air-conditioning system.

Caja Cosa utilizada para guardar, cubrir, o contener algo, como por ejemplo los núcleos del evaporador y del calentador de un sistema de acondicionamiento de aire.

CCOT Cycling clutch orifice tube.

CCOT Tubo de onificio del embrague con funcionamiento cíclico.

CCPS Cycling clutch pressure switch.

CCPS Autómata manométrico del embrague con funcionamiento cíclico.

CDPS Compressor discharge pressure switch.

CDPS Autómata manométrico de discarga del compresor.

Centrifugal Moving away from the center or axis; to develop a force that is progressively away or outward from the axis.

Centrífugo Moviéndose hace afuera del centro o el eje; desarollar una fuerza que se mueva progresivamente hacia afuera del eje.

Centrifugal impeller Rotating water pump impeller that uses centrifugal force to force water drawn in from the center outward to the outlet passage.

Impulsor centrífugo Un impulsor giratorio de la bomba de agua que usa la fuerza centrífuga para forzar el agua proveniente del centro hacia afuera al pasaje de salida.

Centrifugal pump A type of pump used to circulate coolant by centrifugal force in an automotive cooling system.

Bomba centrifuga Tipo de bomba utilizada para la circulación del enfriante por medio de la fuerza centrífuga en un sistema automotriz para el acondicionamiento de aire.

CFCs See Chlorofluorocarbon.

CFCs Véase Clorofluorocarbono.

Change of state Rearrangement of the molecular structure of matter as it changes between any two of the three physical states: solid, liquid, or gas.

Cambio de estado Reordenamiento de la estructura molecular de materia al cambiarse entre cualquier de los tres estados físicos: sólido, líquido, o gas.

Check valve A one-way valve that only allows flow in one direction and restricts flow in the opposite direction.

Válvula de retención Una válvula de una vía que solo permite fluir en una dirección y restringe el flujo de la dirección opuesta.

Chlorine (Cl) A poisonous greenish-yellow gas used in some refrigerants and known to be harmful to the ozone (O_3) layer.

Cloro (Cl) Gas venenoso de color verdusco amarillo utilizado en algunos refrigerantes; conocido como una sustancia nociva al ozono (O_3).

Chlorofluorocarbon (CFC) A manufactured compound used in refrigerants such as R-12, more accurately designated CFC-12.

Clorofluorocarbono Compuesto sintético utilizado en refrigerantes como por ejemplo el R-12; designado con más precisión como CFC-12.

Circuit breaker A bimetallic device used instead of a fuse to protect a circuit.

Disyunto Dispositivo bimetálico utilizado en vez de un fusible para la protección de un circuito.

Clean Air Act (CAA) A Title IV amendment signed into law in 1990, which established national policy relative to the reduction and elimination of ozone-depleting substances.

Ley para Aire Limpio Enmienda Título IV firmado y aprobado en 1990 que estableció la política nacional relacionada con la reducción y eliminación de sustancias que agotan el ozono.

Clutch An electromechanical device used to engage and disengage the compressor in an automotive air-conditioning system.

Embrague Un dispositivo electromecánico que sirve para embragar y desembragar el compressor en un sistema del aire acondicionado automotivo.

Clutch diode A diode placed across the clutch coil to prevent unwanted electrical spikes as the clutch is engaged and disengaged.

Diodo de embrague Diodo que cruza la bobina del embrague para evitar impulsos afilados eléctricos no deseados al engranarse y desen-granarse el embrague.

Clutch field Consists of many windings of wire and is fastened to the front of the compressor. Current applied to the field sets up a magnetic field that pulls the armature in to engage the clutch.

Campo del embrague Consiste de muchos devanados de alambre y se fija a la parte delantera del compresor. La corriente aplicada al campo produce un campo magnético que tira la armadura para engranar el embrague.

Cold The absence of heat.

Frío Ausencia de calor.

Collector A tank located in the radiator to collect coolant.

Colector Tanque ubicado en el radiador para acumular enfriante.

Compression The act of reducing volume by pressure.

Compresión Acción de disminuir el volúmen por efectos de la presión.

Compression stroke That part of the compressor piston that travels from the bottom of its stroke to the top of its stroke.

Carrera de compresión Parte del movimiento del pistón del compre-sor desde la posición inferior de su carrera hasta la posición superior de su carrera.

Compressor A component of the refrigeration system that pumps refrigerant and increases the pressure of the refrigerant vapor.

Compresor Componente del sistema de refrigeración que bombea el refrigerante y eleva la presión del vapor del mismo.

Condenser The component of a refrigeration system in which refrigerant vapor is changed to a liquid by the removal of heat.

Condensador Componente de un sistema de refrigeración en el que el vapor del refrigerante se convierte en un líquido debido a la eliminación de calor.

Conduction The transmission of heat through a solid.

Coducción Transferencia de calor a través de un sólido.

Contaminated Not being of pure form. Refrigerant is considered contaminated when it contains greater than 2 percent of one or more other gases.

Contaminado El no ser de una forma pura. El refrigerante se considera ser contaminado cuando contiene uno o más gases en una cantidad de 2 por ciento o más.

Contamination A matter rendered unpure due to the production of foreign matter.

Contaminación Una materia hecha impura debido a la introducción de materia extraña.

Control thermostat A temperature-actuated electrical switch used to cycle the compressor clutch on and off, thereby controlling the air-conditioning system temperature.

Termostato de control Un interruptor eléctrico actuado por la temperatura que sirve en los ciclos de apagado y prendido del embrague del compresor, así controlando la temperatura del sistema del aire acondicionado.

Control valve A mechanical, pneumatic, or electric valve used to control the flow of coolant into the heater core.

Válvula de regulación Válvula mecánica, neumática, o eléctrica utilizada para regular el flujo de enfriante al núcleo del calentador.

Controller Area Network (CAN) A high-speed serial bus communication network. The CAN protocol has been standardized by the International Standards Organization (ISO) as ISO 11898 standard for high-speed and ISO 11519 for low-speed data transfer.

Red de área de controlador (CAN) red de comunicaciones de bus serial de alta velocidad. El protocolo CAN ha sido estandarizado por la Organización Internacional de Estandarización (ISO) como la norma ISO 11898 para la transferencia de datos a alta velocidad y la norma ISO 11519 para la transferencia de datos a baja velocidad.

Convection The transfer of heat by the circulation of a vapor or liquid.

Convección Transferencia de calor mediante de la circulación de un vapor o líquido.

Coolant pump A term often used when referring to a water pump.

Bomba del enfriante Término utilizado con frecuencia al referirse a una bomba de agua.

Crankshaft That part of a reciprocating compressor on which the wobble plate or connecting rods are attached to provide for an up-down or to-fro piston action.

Cigueñal Parte de un compresor recíproco sobre la cual se fijan las bielas a la placa oscilante para permitir el movimiento de arriba abajo o el de un lado para otro.

Cross-contamination Contamination that can occur when a system is retrofitted to an alternative refrigerant and the entire original refrigerant is not removed or when one piece of equipment is used for more than one type of refrigerant.

Contaminación cruzada La contaminación que puede ocurrir cuando un sistema se equipa después de fabricación para aceptar un refrigerante alternativo y no se remueva todo el refrigerante original o cuando una pieza del equipo se usa para más de un tipo del refrigerante.

Cycle It is one complete on-off occurrence. Including the total on time and off time. Cycles occurrences are measured in Hertz (Hz).

Ciclo una incidencia completa de encendido apagado. Incluido el tiempo total de encendido y el de apagado. Los ciclos se miden en Hertz (Hz).

Cycling clutch A clutch that is turned on/off to control temperature.

Embrague con funcionamiento cíclico Embrague que se pone en marcha y se apaga para regular la temperatura.

Cylinder A circular tube-like opening in a compressor block or casting in which the piston moves up and down or back and forth; a circular drum used to store refrigerant.

Cilindro Apertura circular parecida a un tubo en un bloque del compresor o una pieza en los que el pistón se mueve de arriba abajo o de un lado a otro; un tambor circular utilizado para el almacenaje de refrigerante.

DC See Direct current (DC).

CC Ver Direct current [Corriente continua (cc)].

Deep-tissue temperature The sub-surface temperature, such as the internal temperature of the human body of 98.6°F.

Temperatura interna La temperatura bajo la superficie, tal como la temperatura interna del cuerpo humano que es el 98.6°F.

Defrost To remove frost.

Decongelar Deshelar.

Deice To remove ice or heavy frost.

Deshelar Derretir hielo o una gran cantidad de escarcha.

Delta P A term used when referring to a difference in pressure.

P delta Término utilizado al referirse a una diferencia en presión.

Delta T A term used when referring to a difference in temperature.

T delta Término utilizado al referirse a una diferencia en temperatura.

Desiccant A drying agent used in refrigeration systems to remove excess moisture. The deciccant is located in the receiver-drier or accumulator.

Desecante Agente secador utilizado en sistemas de refrigeración para eliminar un exceso de humedad. El desecante está ubicado en el receptor/secador o en el acumulador.

Diode An electrical check valve. Current flows only in one direction through a diode.

Diodo Válvula eléctrica de retención. La corriente fluye en una sola dirección a través de un diodo.

Direct current (DC) A type of electrical power used in mobile applications. A unidirectional current of substantially constant value.

Corriente continua (cc) Tipo de potencia eléctrica utilizada en circunstancias cuando el objeto va a ser móvil. Una corriente de un solo sentido de un valor substancialmente constante.

Discharge Bleeding some or all of the refrigerant from a system by opening a valve or connection and permitting the refrigerant to escape slowly.

Descarga Desangramiento de una porción o de todo el refrigerante de un sistema al abrir una válvula o conexión y permitir su escape gradual.

Discharge line Connects the compressor outlet to the condenser inlet.

Línea de descarga Conecta la salida del compresor a la entrada del condensador.

Discharge stroke See Compression stroke.

Carrera de descarga Ver Compression stroke [Carrera de compresión].

Discharge valve The outlet valve.

Válvula de descarga La válvula de salida.

Disposal Get rid of something.

Eliminación Eliminar algo.

Distilled water One hundred percent pure H_2O.

Agua destilada El H_2O cien por ciento puro.

DIY Do-it-yourself.

"DIY" Expresión en jerga que significa que una persona hace algo por su propia cuenta.

Dobson unit (DU) A measure of ozone density level, named after Gordon Dobson, a British meteorologist who was the inventor of the measuring device (called a spectrophotometer).

Unedad Dobson (DU) Una medida al nivel de densidad del ozono, nombrado por Gordon Dobson, un meteorologista inglés que fue el inventor de un dispositivo de medida (llamado el espectrofotómetro).

DOT U.S. Department of Transportation.

DOT Departamento de Transportes de los Estados Unidos deAmerica.

Drier A device containing desiccant; a drier is placed in the liquid line to absorb moisture in the system.

Secador Dispositivo que contiene un desecante; se ubica un secador en la línea de líquido para absorber la humedad presente en el sistema.

Dual systems Two systems.

Sistemas dobles Dos sistemas.

Duty cycle The length of time that an output actuator is energized during one complete on-off cycle, it is a measurement of pulse-width modulation.

Ciclo de servicio el período de tiempo en que un impulsor de salida está energizado durante un ciclo de encendido-apagado completo. Es una medición de la modulación de ancho de pulso.

Electrolysis The decomposition of a compound caused by the action of an electric current passing through it.

Electrólisis La decomposición de un compuesto causada por la acción de un corriente eléctrica que pasa por en medio.

Electromagnet A soft iron core surrounded by a coil of wire that will temporarily become a magnet when an electrical current is passed through it.

Electroimán Un núcleo de hierro blando rodeado por una bobina de alambre que se convierte brevemente en un imán cuando es atrave-sado por un corriente eléctrico.

Electromagnetic clutch An electrically controlled device used to start and stop compressor action.

Embrague electromagnético Dispositivo controlado electrónicamentey utilizado para arrancar y detener la acción del compresor.

Ethylene glycol A colorless liquid used in the production of antifreeze $HOCH_2CH_2OH$.

Glicol etileno Un líquido sín color que se usa en la producción del anticongelante $HOCH_2CH_2OH$.

Evacuate To create a vacuum within a system to remove all air and moisture.

Evacuar Dejar un vacío dentro de un sistema para eliminar todo aire y humedad.

Evaporation The changing of a liquid to a vapor while picking up heat.

Evaporación La conversión de un líquido en vapor al acumular el calor.

Evaporator The component of an air-conditioning system that conditions the air.

Evaporador Componente en un sistema de acondicionamiento de aire que acondiciona el aire.

Exhaust A pipe through which used gases or vapors pass. See also Discharge.

Escape Tubo por el cual pasan los gases or vapores gastados. Ver tambien Discharge [Descarga].

Expansion tank An auxiliary tank that is usually connected to the inlet tank or a radiator to provide additional storage space for heated coolant; often called a coolant recovery tank.

Tanque de expansión Tanque auxiliar que normalmente se conecta al tanque de entrada o a un radiador para proveer almacenaje adicional del enfriante calentado. Llamado con frecuencia tanque para la recuperación de enfriante.

Expansion tube A metering device used at the inlet of some evaporators to control the flow of liquid refrigerant into the evaporator core. Also see Fixed orifice tube (FOT).

Tubo de expansión Dispositivo para la dosificación y utilizado a la entrada de algunos evaporadores para regular el flujo de refrigerante líquido dentro del núcleo del evaporador. Ver tambien Fixed orifice tube (FOT) [Tubo de orificio fijo].

Expansion valve A term often used when referring to a thermostatic expansion valve.

Válvula de expansión Un término que se usa comunmente para referirse a una válvula de expansión termoestática.

Fan A device that has two or more blades attached to the shaft of a motor. The fan is mounted in the evaporator and causes air to pass over the evaporator. A fan is also a device that is mounted on the water pump and has four or more blades that cause air to pass through the radiator and condenser.

Ventilador Dispositivo provisto de dos aletas o más fijadas al arbol de un motor. El ventilador está montado en el evaporador y hace que el aire pase sobre el evaporador. Un ventilador también puede ser un dispositivo montado en la bomba de agua y que tiene cuatro aletas o más que hacen que el aire pase por el radiador y el condensador.

Fan clutch A device used on engine-driven fans to limit their terminal speed, reduce power requirements, and lower noise levels.

Embrague de ventilador Dispositivo utilizado en ventiladores acciona-dos por motores para limitar su velocidad terminal, disminuir requisitos de potencia, y bajar los niveles de ruino.

Field coil See Clutch Field and Electromagnet.

Bobina del campo Ver Clutch field [Campo del embrague] y Electromagnetic [Electroimán].

Fixed orifice tube (FOT) A refrigerant metering device used at the inlet of evaporators to control the flow of liquid refrigerant allowed to enter the evaporator.

Tubo de orificio fijo Dispositivo para la dosificación de refrigerante utilizado a la entrada de los evaporadores para regular el flujo de refrigerante líquido permitido entrar en el evaporador.

Flooded See Flooding.

Inundado Ver Flooding [Inundación].

Flooding A condition caused by too much liquid refrigerant being metered into the evaporator.

Inundación Condición ocasionada por una cantidad excesiva de refrigerante líquido dosificado al evaporador.

Flush To remove solid particles such as metal flakes or dirt. Refrigerant passages are purged with a clean dry gas such as nitrogen (N).

Limpiar por inundación Remover las partículas sólidas, como por ejemplo escamas metálicas o polvo. Se purgan los pasajes de refrigerante con un gas limpio y seco, como por ejemplo el nitrógeno (N).

FOTCC Fixed orifice tube cycling clutch.

FOTCC Tubo de orificio fijo del embrague con funcionamiento cíclico.

FPI Feet per inch or fins per inch.

FPI Pies por pulgada o aletas por pulgada.

Fuse An electrical device used to protect a circuit against accidental overload or unit malfunction.

Fusible Dispositivo eléctrico utilizado para proteger un circuito contra una sobrecarga imprevista o una disfunción de la unidad.

Gas A state of matter. A vapor that has no particles or droplets of liquid.

Gas Estado de materia. Vapor desprovisto de partículas o gotitas de líquido.

Gauge A device used to measure pressure or force scaled in English or metric values.

Calibrador Dispositivo utilizado para medir la presión o fuerza; provisto de una escala en valores ingleses y/o métricos.

Global warming The gradual warming of the earth's atmosphere due to the greenhouse effect. See Greenhouse effect.

Calentamiento mundial Calentamiento gradual de la atmósfera de la Tierra debido al efecto de invernadero. Ver Greenhouse effect [Efecto de invernadero].

Global warming potential (GWP) Global warming potential is an index number that is an estimate of how much a given mass of a gas will contribute to global warming compared to the same mass of carbon dioxide, where carbon dioxide is given the number 1.

Potencial de calentamiento global (GWP) el potencial de calentamiento global es un índice numérico que es un cálculo estimativo de cuánto contribuirá una masa dada de gas al calentamiento global comparada con la misma masa de dióxido de carbono, donde el dióxido de carbono recibe el número 1.

Greenhouse effect A greenhouse is warmed because glass allows the sun's radiant heat to enter but prevents radiant heat from leaving. Global warming is caused by some gases in the atmosphere that act like greenhouse glass; hence, the term greenhouse effect.

Efecto de invernadero Se calienta un invernadero porque el vidrio permite la entrada del calor radiante del sol pero impide la salida del calor radiante de la Tierra. El calentamiento mundial es ocasionado por algunos gases en la atmósfera que actuan como el vidrio de un invernadero; por eso, se utiliza el término efecto de invernadero.

Halide Any compound of a halogen with another element such as refrigerant.

Halogenuro Cualquier compuesto de un halogenuro y otro elemento, como por ejemplo el refrigerante.

Halogen Refers to any of the five chemical elements—astatine (At), bromine (Br), chlorine (Cl) fluorine (F), and iodine (I)—that may be found in some refrigerants.

Halógeno Se refiere a cualquier de los cinco elementos químicos—astatinio (At), bromo (Br), cloro (Cl), fluoro (F), y yodo (I)—que pueden estar presentes en algunos refrigerantes.

Head pressure Pressure of the refrigerant from the discharge reed valve through the lines and condenser to the expansion valve orifice.

Altura piezométrica Presión del refrigerante de la válvula de la lámina de descarga a través de las líneas y el condensador al orificio de la válvula de expansión.

Heat Energy; any temperature above absolute zero.

Calor Energía; cualquier temperatura superior al cero absoluto.

Heater core A heat exchanger that extracts the heat from coolant warmed by the engine to heat the passenger compartment.

Núcleo de calor Un cambiador de calor que extrae el calor del refrigerante calentado por el motor para calentar el compartimento del pasajero.

Heat load The load imposed on an air conditioner due to ambient temperature, humidity, and other factors that may produce unwanted heat.

Carga de calor Carga impuesta sobre un acondicionador de aire debido a la temperatura ambiente, humedad, y otros factores que pueden producir calor no deseado.

HFC See Hydrofluorocarbon.

HFC Ver Hidrofluorocarbono.

High pressure A relative term to describe excessive refrigerant pressure in the high side of an air-conditioning system.

Alta presión Un término relativo que describe una presión excesiva del refrigerante en el lado alto de un sistema de aire acondicionado.

High-pressure cutoff switch An electrical switch that is activated by a predetermined high pressure. The switch opens a circuit during high-pressure periods.

Interruptor de cierre de alta presión Interruptor eléctrico que es accionado por una alta presión predeterminada. El interruptor abre un circuito durante períodos de alta presión.

High-pressure switch See High-pressure cutoff switch.

Autómata manométrico de alta presión Ver High-pressure cutoff switch [Interruptor de cierre de alta presión].

High side That part of an air-conditioning system extending from the compressor outlet to the metering device inlet.

Lado alto Esa parte de un sistema de aire acondicionado que se extiende de la salida del compresor a la entrada del dispositivo medidor.

HL/LO A term often used to refer to bi-level.

HI/LO (alto/bajo) Término utilizado con frecuencia para referirse a binivel.

Hot gas line A line that carries hot gas such as the discharge line from the compressor to the condenser.

Línea de gas caliente Línea que conduce el gas caliente, como por ejemplo la línea de descarga, desde el compresor hasta el condensador.

Humid Damp.

Húmedo Que contiene humedad.

Humidity See Moisture. See also Relative humidity.

Humedad Ver Moisture [Humedad]. Ver también Relative humidity [Humedad relativa].

HVAC The abbreviation used for heating ventilation and air conditioning.

HVAC La abreviación que se usa para la ventilación del calor y el aire acondicionado.

H-valve An expansion valve with all parts contained within that is used on some Chrysler and Ford lines.

Válvula H Válvula de expansión que contiene todas las partes dentro de sí misma; se utiliza dicha válvula en algunos modelos de vehículos de las compañias automotrices Chrysler y Ford.

Hybrid organic acid technology (HOAT) G-05 extended-life coolants are ethylene-glycol Glysantin-based formula refrigerants that are low in silicate, have low pH, and are phosphate-free. HOAT coolants use both organic and inorganic carbon-based additives for long life protection.

Tecnología de ácido orgánico híbrido (HOAT) Los refrigerantes de vida extendida G-05 son refrigerantes cuyo base es una fórmula de etilenglicol Glysantin con bajo silicato, bajo pH y sin fosfato. Los refrigerantes HOAT utilizan aditivos de carbono orgánicos e inorgánicos para protección de larga duración.

Hydrocarbon An organic compound containing only hydrogen (H) and carbon (C).

Hidrocarbono Compuesto orgánico que contiene solo el hidrógeno (H) y el carbono (C).

Hydrochloric acid A corrosive acid produced when water and R-12 are mixed, as within an automotive air-conditioning system.

Ácido hidroclórico Ácido corrosivo producido cuando se mezcla el agua con el R-12, como por ejemplo dentro de un sistema automotriz para el acondicionamiento de aire.

Hydrogen The lightest of all known substances; it is colorless, odorless, and flammable (H).

Hidrógeno La más ligera de todas las substancias conocidas que es sin color, sin olor y es inflamable.

Hydrolysis The chemical reaction with water whereby a substance is changed into one or more other substances.

Hidrólisis La reacción química con el agua en el cual una sustancia se cambia a una o más substancias.

Hydrofluorocarbon (HFC) A compound consisting of hydrogen, fluorine, and carbon. HFCs are a class of replacement refrigerants for CFC refrigerants and are not ozone depleting.

Hidrofluorocarbono (HFC) Un compuesto de hidrogeno, fluór y carbono. Los HFC pertenecen a una clase de refrigerantes que reem-plazan los refrigerantes CFC y no reducen el ozono.

Hydrostatic pressure The pressure exerted by a fluid.

Presión hidrostática Presión ejercida por un fluído.

Hygroscopic Readily absorbing and retaining moisture.

Higroscópico Que absorbe y conserve facilmente la humedad.

Idiot light Slang term often used for engine coolant and oil pressure warning lights; also called telltale light.

Luz para idiotas Término en jerga utilizado con frecuencia para las luces de advertencia de enfriante de motor y/o de presión de aceite.

Impeller A rotating member with fins or blades used to move liquid, for example, the rotating part of a water pump.

Impulsor El elemento rotativo provisto de aletas utilizadas para mover líquido, p.e., la parte giratoria de una bomba de agua.

Infrared It is the invisible light rays just beyond the red end of the visible spectrum and have a penetrating heating effect.

Infrarrojo son los rayos de luz invisible que se encuentran justo después del extremo rojo del espectro visible y tienen un efecto de calor penetrante.

Inject To insert, usually by force or pressure.

Inyectar Insertar, normalmente por medio de la fuerza o presión.

Insulator A nonconductor such as the covering of a wire (electrical) or a tube (thermal).

Aislador Elemento no conductor, como por ejemplo la cubierta de un alambre (eléctrico) o un tubo (térmico).

Intake See Suction.

Toma Ver Suction [Succión].

kiloPascal absolute See kPa absolute.

KiloPascal absoluto Ver kPa absolute [kPa absoluto].

kPa An abbreviation for the metric pressure measure "kilopascal," sometimes written "kiloPascal," equivalent to 0.145 psi on the English scale.

kPa Una abreviación de la medida métrica de presión "kilopascal" que a veces se escribe "kilopascal", equivalente a 0.145 libra por pulgada cuadrada en la gama inglesa.

kPa absolute A metric unit of measure for pressure measured from absolute zero.

kPa absoluto Unidad métrica de medida para presión medida del cero absoluto.

Latent heat The amount of heat required to cause a change of state of a substance without changing its temperature.

Calor latente La cantidad de calor requerida para ocasionar un cambio de estado de una sustancia sin cambiar su temperatura.

Liquid line The line connecting the drier outlet with the expansion valve inlet. The line from the condenser outlet to the drier inlet is sometimes called a liquid line.

Línea de líquido Línea que conecta la salida del secador con la entrada de la válvula de expansión. La línea de la salida del condensador a la entrada del secador a veces se llama una línea de líquido.

Low pressure A relative term to describe below-normal pressure in the low side of an air-conditioning system.

Baja presión Un término relativo para describir una presión bajo lo normal en el lado bajo de un sistema del aire acondicionado.

Low-pressure cutoff switch An electrical switch that is activated by a predetermined low pressure. This switch opens a circuit during certain low-pressure periods.

Interruptor de cierre de baja presión Interruptor eléctrico que es accionado por una baja presión predeterminada. Dicho interruptor abre un circuito durante ciertos períodos de baja presión.

Low-pressure switch See Low-pressure cutoff switch.

Autómata manométrico de baja presión Ver Low-pressure cutoff switch [Interruptor de cierre de baja presión].

Low side That part of the air-conditioning system extending from the inlet of the evaporator metering device to the inlet of the compressor.

Lado bajo Esa parte del sistema del aire acondicionado que se extiende de la entrada del dispositivo medidora del evaporador a la entrada del compresor.

Low-side service valve A device located on the suction side of the compressor that allows the service technician to check low-side pressures or perform other necessary service operations.

Válvula de servicio del lado de baja presión Dispositivo ubicado en el lado de succión del compresor; dicha válvula permite que el mecánico verifique las presiones en el lado de baja presion o que lleve a cabo otras funciones de servicio necesarias.

Lubricant A substance, more commonly referred to as "oil," thought of as a petroleum-based product that is used to coat moving parts to reduce friction between them. The term generally refers to the new synthetic lubricants, such as polyalkaline glycol (PAG) and polyol ester (POE) that are used with HFC refrigerants.

Lubricante Una substancia, comunmente referido como "aceite" que se suele definir como un producto a base de petroleo que se usa para cubrir las partes en movimiento para reducir la fricción entre ellas. El término generalmente se refiere a los lubricantes nuevos sintéticos, tal como el glicol polialcalino (PAG) y el poliol ester (POE) que se usan con los refrigerantes HFC.

Malfunction Failure to work or perform.

Disfunción Dejar de funcionar correctamente.

Master control A primary or main control.

Regulador maestro Control primario o principal.

Matter Anything that occupies space and possesses mass. All things in nature are composed of matter.

Materia Todo lo que ocupe espacio o tenga masa. Todas las cosas en la naturaleza se componen de materia.

Metering device A device for metering the proper amount of refrigerant into an evaporator. The two types for automotive air-conditioning system service are thermostatic expansion valve (TXV) and orifice tube (OT).

Dispositivo medidora Un dispositivo para medir la cantidad adecuada del refrigerante entrando al evaporador. Los dos tipos en el servicio del sistema aire acondicionador automotivo son la válvula de expansión termostático (TXV) y el tubo del orificio (OT).

Miscible Another word for mixable.

Miscible Otra palabra que significa que se puede mezclar.

MIX A term often used when referring to HI/LO.

MEZCIA. Término utilizado con frecuencia al referirse a HI/LO (alto/bajo).

Mode Manner or state of existence of a thing; for example, hot or cool.

Modo Manera o estado de existencia de una cosa; p.e., calor o fresco.

Mode door A diverter door within the duct system for directing air through the heater and evaporator core.

Puerta de modo Puerta desviadora dentro del sistema de conductos para conducir el aire a través del núcleo del calentador y/o del evaporador.

Moisture Droplets of water in the air; humidity, dampness, or wetness.

Humedad Gotitas de agua en el aire.

Molecule Two or more atoms chemically bound together.

Molécula Dos o más átomos quimicamente ligados.

Negative Minus, less than zero. The ground (–) side of a battery or DC electrical circuit. A pressure below atmospheric; a vacuum.

Negativo Menos de cero. El lado puesto a tierra (–) de un acumulador o corriente eléctrica de corriente continua. Una presión inferior a la de la atmósfera; un vacío.

Nitrogen (N) An odorless, tasteless, and colorless element that composes over 80 percent of the atmosphere and is essential for all animal and plant life.

Nitrógeno (N) Elemento inodoro, insípido e incoloro que compone mas de un 80 por ciento de la atmósfera y es esencial para toda vida vegetal y animal.

Normally closed (nc) A switch or device that is closed in its relaxed (normal) position.

Normalmente cerrado Conmutador o dispositivo que está cerrado en su posición relajada (normal).

Normally open (no) A switch or device that is open in its relaxed (normal) position.

Normalmente cerrado Conmutador o dispositivo que está cerrado en su posición relajada (normal).

Ohm (Ω) The unit used to measure the amount of electrical resistance in a circuit or an electrical device. One ohm is the amount of resistance present when one volt pushed one ampere through a circuit or device.

Ohmio (Ω) la unidad utilizada para medir la cantidad de resistencia eléctrica de un circuito o dispositivo eléctrico. Un ohmio es la cantidad de resistencia presente cuando un voltio empuja un amperio a través de un circuito o dispositivo.

Ohm's law Defines the relationship between voltage, resistance, and current.

Ley de Ohmio define la relación entre la tensión (o voltaje), la resistencia y la corriente.

Organic acid technology The term used to describe the chemical additive package in extended life coolant.

Tecnología de ácidos orgánicos El término que se usa para describir el paquete de aditivos de químicas del refrigerante de larga vida.

Orifice A small hole of calibrated dimensions for metering fluid or gas in exact proportions.

Orificio Un hoyo pequeño de dimensiones calibradas para medir los flúidos o los gases en proporciones exactas.

Orifice tube See Expansion tube and Fixed orifice tube (FOT).

Tubo de orificio Ver Expansion tube [Tubo de expansion] y Fixed orifice tube [Tubo de orificio fojo (FOT)].

Overcooling A general term used if the engine does not reach design operating temperature in a predetermined time period, such as would be the case if the thermostat were removed.

Sobreenfriamiento Término general utilizado si el motor no alcanza la temperatura de functionamiento de diseño dentro de un período pre-determinado de tiempo, como por ejemplo si se removara el ter-mostato.

Overflow tank Another term used when referring to the coolant recovery tank that allows for both coolant expansion and contraction from the cooling system of the engine.

Tanque de derrame Otro término que se usa cuando se refiere al tanque de rescate del refrigerante que permite la expansión y contracción del refrigerante debido al sistema refrigerante del motor.

Overheating A general term used if the engine exceeds design operating temperature.

Sobrecalentamiento Término general utilizado si el motor excede la temperatura de funcionamiento de diseño.

Oxygen (O) An odorless, tasteless, colorless element that forms about one-fifth of our atmosphere. An essential element for animal and plant life.

Oxígeno (O) Elemento inodoro, insipido e incoloro que forma más o menos la quinta parte de nuestra atmósfera. Es un elemento esencial para toda vida vegetal y animal.

Ozone (O_3) An unstable pale-blue gas with a penetrating odor; it is an allotropic form of oxygen (O) that is usually formed by a silent electrical discharge in the air.

Ozono (O_3) Gas inestable de color azul palido que tiene un olor pene-trante; es una forma alotrópica de oxígeno (O) normalmente pro-ducido por una descarga eléctrica silenciosa en el aire.

Ozone depletion The reduction of the ozone layer due to contamination, such as the release of CFC refrigerants into the atmosphere.

Agotamiento del ozono La reducción de la capa de ozono debido a la contaminación, tal como los refrigerantes CFC a la atmósfera.

Performance The way in which something functions.

Ejecución La manera en la cual algo funciona.

Perspiration The salty fluid secreted by sweat glands through the pores of the skin.

Sudor Fluído salado secretado por las glándulas sudoríparas a través de los poros de la piel.

Pitch Set at a particular degree or angle.

Inclinación Puesto a un grado o ángulo específico.

Plenum See Plenum chamber.

Pleno Ver Plenum chamber [Cámara impelente].

Plenum chamber An area filled with air at a pressure that is slightly higher than the surrounding air pressure, such as the chamber just before the blower motor.

Cámara impelente Área en la cual existe una condición de sobrepresión, como por ejemplo la cámara justo enfrente del motor del soplador.

PM Preventive maintenance.

PM Mantenimiento preventativo.

POE An abbreviation for the synthetic lubricant polyol ester.

POE Abreviatura del lubrificante sintético poliolester.

Positive The hot (+) side of a battery or electrical circuit. Also, a pressure above atmospheric.

Positivo El lado cargado (+) de un acumulador o circuito eléctrico. También una presión superior a la de la atmósfera.

Power train control module Generally abbreviated PCM, it is the computer system for the engine management and emission systems.

Módulo de control del trén motriz Generalmente abreviado de PCM y es el sistema de computador de los sistemas de regulación del motor y emisiones.

Predetermined The preset parameters used.

Predeterminado Los parámetros prescritos que se usan.

Pressure cap A radiator cap that increases the pressure of the cooling system and allows higher operating temperatures.

Tapa de presión Tapa de radiador que eleva la presión del sistema de enfriamiento y permite un funcionamiento a temperaturas mas altas.

Pressure switch An electrical switch that is actuated by a predetermined low or high pressure. A pressure switch is generally used for system protection.

Autómata manométrico Interruptor eléctrico accionado por una alta o baja presión predeterminada. Generalmente se utiliza un autómata manométrico para la protección del sistema.

Price leader An item that a merchant may sell at cost or near cost to attract customers.

Artículo en venta Un artículo que un negociante puede vender en costo o casi en costo para atraer a la clientela.

Programmer The part of an automatic temperature control system that controls the blower speed, air mix doors, and vacuum diaphragms.

Programador Parte de un sistema de regulación automática de temperatura que regula la velocidad del soplador, puertas de mezcla de aire, y diafragmas al vacío.

Propylene glycol A colorless liquid used in the production of antifreeze, $C_3H_8O_2$; it is considered a low toxicity antifreeze.

Propileno glicol Un líquido sin color que se usa en la producción del anticongelante, $C_3H_8O_2$, y que se considera un anticongelante de baja toxicidad.

Pulse width modulation (PWM) On-off duty cycling of a component. The period of time for each cycle does not change; only the amount of on time in each cycle changes. The length of time in milliseconds that an actuator is energized.

Modulación por ancho de pulsos (PWM) El ciclo útil de un componente de conexión. El período de tiempo de cada ciclo no cambia; Sólo la cantidad de tiempo en cada ciclo cambia. La longitud de tiempo en milisegundos que se energiza un actuador.

Pump The compressor. Also refers to the vacuum pump.

Bomba El compresor. Se refiere tambien a la bomba de vacío.

Pump down See Evacuate.

Vaciar Ver Evacuate [Evacuar].

Pure Not mixed with anything else.

Puro Quo no se ha mezclado con ninguna otra cosa.

R-12 An abbreviation to identify the chlorofluorocarbon family of ozone-depleting refrigerants. R-12, or CFC-12, was a popular refrigerant for automotive air-conditioning systems service until it was phased out of production.

R-12 Una abreviación para identificar la familia de los refrigerantes clo-rofluorocarburo que agotan la capa de ozono. El R-12, o CFC-12, fue un refrigerante de sistemas de aire acondicionado automotivos muy popular hasta que su producción fue restringida.

R-1234yf is a low GWP refrigerant developed by Honeywell and Dupont.

R-1234yf es un refrigerante con bajo potencial de calentamiento global (GWP) desarrollado por Honeywell y DuPont.

R-134a An abbreviation to identify the ozone-friendly hydrofluorocarbon family of refrigerants. R-134a, or HFC-134a, is the refrigerant of choice for replacing R-12 in automotive air-conditioning systems.

R-134a Una abreviación para identificar la familia de los refrigerantes clorofluorocarburo que no agotan la capa de ozono. El R-134a, o HFC-134a, es el refrigerante que más se usa para reemplazar el R-12 en los sistemas de aire acondicionado automotivo.

R-744 The trade name for carbon dioxide (CO_2) gas, it is considered to be one of the alternative refrigerant gases that will be used for refrigerant systems.

R-744 El nombre registrado del gas carbónico (CO_2), que se considera ser uno de los gases refrigerantes alternativos que serán usados para los sistemas refrigerantes.

Radiation The transfer of heat without heating the medium through which it is transmitted.

Radiación La transferencia del calor sin calentar el medio por el cual se está transmitiendo.

Radiator A coolant-to-air heat exchanger. The device that removes heat from coolant passing through it.

Radiador Intercambiador de calor del enfriante al aire. El dipositivo que remueve calor del enfriante que pasa por él.

Ram air The term used to describe the flow of air striking the front of a vehicle as it travels in a forward direction.

Aire admitido en marcha El término que se usa para describir el flujo del aire que golpea la parte delantera del vehículo mientras que éste viaja en un movimiento delantera.

RCD Refrigerant containment device.

RCD Dipositivo para contener refrigerante.

Receiver A container for the storage of liquid refrigerant.

Receptor Recipiente para el almacenaje de refrigerante líquido.

Receiver-dehydrator A combination container for the storage of liquid refrigerant and a desiccant.

Receptor-deshidratador Recipiente de combinación para el almacenaje del refrigerante líquido y un desecante.

Receiver-drier See Receiver-dehydrator.

Receptor-secador Ver Receiver-dehydrator [Receptor/deshidratador].

Reciprocating To move to and fro, fore and aft, or up and down.

Movimiento alternativo Moverse de un lado para otro, de atrás para adelante, o de arriba para abajo.

Reciprocating piston(s) A compressor assembly that uses the back and forth movement of a piston to cause a rotary motion of the compressor crankshaft.

Pistones recíprocos Una asamblea de compresor que usa el movimiento oscilante de un pistón para causar un movimiento rotario del cigueñal del compresor.

Recirculate To reuse. To circulate a fluid or vapor over and over again.

Recircular Utilizar de nuevo. Hacer circular un fluído o vapor repetida-mente.

Reclaim To process used refrigerant to new product specifications by means that may include distillation. This process requires that a chemical analysis of the refrigerant be performed to determine that appropriate product specifications are met. This term implies the use of equipment for processes and procedures usually available only at a reprocessing facility.

Recuperar Procesar refrigerante gastado a nuevas especificaciones para productos por un medio que puede incluir la distilación. Este proceso require que se realice un análisis químico del refrigerante para determinar si se pueden cumplir con las especificaciones apropiadas para dicho producto. Este término implica la utilización de equipo para procesos y procedimientos normalmente disponibles solo en una instalación de reprocesamiento.

Recover To remove refrigerant in any condition from a system and to store it in an external container without necessarily testing or processing it in any way.

Recobrar Remover refrigerante en cualquier condición de un sistema y almacenarlo en un recipiente externo sin necesariamente probarlo o procesarlo.

Recovery cylinder A recovery cylinder for R-12 and R-134a must meet DOT specifications 4BA-300. These cylinders are characterized by a combined liquid/vapor valve located at the top. A dip tube is used to feed liquid refrigerant from the bottom so it can be dispensed without inverting the cylinder. A recovery cylinder should be painted gray with a yellow shoulder.

Cilindro de recuperación Un cilindro de recuperación para R-12 y/o R-134a tiene que cumplir con las especificaciones 4BA-300 del DOT. Se caracterizan estos cilindros por una válvula combinada de líquido y vapor ubicada en la parte superior. Se utiliza un tubo probador para alimentar el refrigerante líquido desde la parte inferior para que pueda dispensarse sin invertir el cilindro. Un cilindro de recuperación debe ser pintado el color gris con un resalto amarillo.

Recovery tank See Recovery cylinder and Expansion tank.

Tanque de recuperación Ver Recovery cylinder [Cilindro de recuperación] y Expansion tank [Tanque de expansión].

Recycle To clean refrigerant for reuse by oil separation and to pass through other devices such as filter-driers to reduce moisture, acidity, and particulate matter. Recycling applies to procedures usually accomplished in the repair shop or at a local service facility.

Reciclar Limpiar el refrigerante para ser utilizado de nuevo por medio de la separación de aceite y del pasaje a través de otros dispositivos, como por ejemplo filtro-secadores, para disminuir la humedad, acidez, y materia partícula. El reciclamiento se aplica a los procedimientos normalmente realizados en el taller de reparación o en la instalación de servicio local.

Reed valve The leaves of steel located on the valve plate of a compressor. The suction reed valve opens to admit refrigerant on the intake stroke of the compressor and closes to block refrigerant flow on the exhaust stroke. The discharge reed valve, on the other hand, is closed to block refrigerant flow on the intake stroke and opens to expel refrigerant on the exhaust stroke.

Válvula de lámina vibrante Las láminas del acero ubicados en la placa de la válvula del compresor. La válvula succión de lámina abre para admitir al refrigerante en la carrera de entrada del compresor y cierra para bloquear el flujo del refrigerante en la carrera de escape. La válvula de descarga de lámina, en cambio, es cerrada para bloquear el flujo del refrigerante en la carrera de entrada y abre para expeler el refrigerante en la carrera de escape.

Refrigerant The chemical compound used in a refrigeration system to produce the desired cooling.

Refrigerante Compuesto químico utilizado en un sistema de refrigeración para producir el enfriamiento deseado.

Refrigeration To use an apparatus to cool, keep cool, chill, and keep chilled under controlled conditions by natural or mechanical means as an aid to ensuring personal safety and comfort. To cool the air by removing some of its heat content. The removal of heat by mechanical means.

Refrigeración Utilizar un aparato para enfriar y mantener el frío bajo condiciones controladas por medios naturales o mecánicos para ayudar a asegurar la seguridad y comodidad personales. Enfríe el aire removiendo una porción de su contenido de calor. La remoción del calor por medios mecánicos.

Relative humidity The actual moisture content of the air in relation to the total moisture that the air can hold at a given temperature.

Humedad relativa Contenido verdadero de humedad del aire en relación a la humedad total que el aire puede mantener a una temperatura dada.

Remote bulb A sensing device connected to the expansion valve by a capillary tube. This device senses the tailpipe temperature and transmits pressure to the expansion valve for its proper operation.

Bombilla a distancia Dispositivo sensor conectado a la válvula de expansión por un tubo capilar. Este dispositivo siente la temperatura del tubo de escape y transmite presión a la válvula de expansión para su funcionamiento correcto.

Reserve tank See Overflow tank.

Tanque de derrame Ver Overflow tank (Tanque de derrame).

Restriction A blockage in the air-conditioning system caused by a pinched or crimped line, foreign matter, or moisture freeze-up.

Limitación Bloqueo en el sistema de acondicionamiento de aire ocasionado por una línea pellizcada o arrugada, una materia extraña, o la congelación de humedad.

Restrictor Decreases the flow of a liquid or gas.

Limitador Disminuye el flujo de un líquido o un gas.

Retrofit To modify equipment that is already in service using parts and materials made available after the time of original manufacture.

Retromodificación Modificar el equipo que ya está en servicio usando las partes y/o los materiales disponibles después del tiempo de la fabricación original.

Reverse flow Direction opposite that of which is considered the standard direction. In engine cooling systems, reverse flow systems circulate coolant first through the cylinder head(s) and then through the engine blow.

Flujo en reverso La dirección opuesta de la que se considera la dirección normal. En los sistemas de refrigeración automotivos los sistemas de flujo en reversa circulan el refrigerante primero por la(s) culata(s) de cilindro y luego por el bloque motor.

Rheostat A wire-wound variable resistor used to control blower motor speed.

Reostato Resistor variable devanado con alambre utilizado para regular la velocidad del motor del soplador.

Rotary The turning motion around an axis.

Rotario El movimiento de girar alrededor de un eje.

Saddle valve A two-part accessory valve that may be clamped around the metal part of a system hose to provide access to the air-conditioning system for service.

Válvula de silleta Válvula de accesorio de dos partes que puede fijarse con una abrazadera a la parte metálica de una manguera del sistema para proveer acceso al sistema de acondicionamiento de aire para llevar a cabo el servicio.

SAE Society of Automotive Engineers.

SAE Sociedad de Ingenieros Automotrices.

SATC Semiautomatic temperature control.

SATC Regulador semi automático de temperatura.

Screen A metal mesh located in the receiver, expansion valve, and compressor inlet to prevent particles of dirt from circulating through the system.

Cribadora Malla metálica ubicada en el receptor, la válvula de expansión, y la entrada del compresor para evitar que las partículas de polvo se circulen a través del sistema.

Scroll A spiral, rolled, or convoluted form.

Arollado Una forma de espiral, de enrollado o convoluto.

Sensible heat Heat that causes a change in the temperature of a substance but does not change the state of the substance.

Calor sensible Calor que ocasiona un cambio de temperatura de una sustancia pero que no cambia el estado de dicha sustancia.

Sensor A temperature-sensitive unit such as a remote bulb or thermistor. See Remote bulb and Thermistor.

Sensor Unidad sensible a la temperatura, como por ejemplo una bombilla a distancia o termistor. Ver Remote bulb [Bombilla a distancia] y Thermistor [Termistor].

Serpentine Refers to the circuitous and twisted or winding path taken by one drive belt used to turn numerous pulleys.

Serpentina Refiere a la senda indirecto y girando or torciéndose que toma una correa de impulso para girar poleas numerosas.

Serpentine belt A flat or V-groove belt that winds through all of the engine accessories to drive them off the crankshaft pulley.

Correa serpentina Correa plana o con ranuras en V que atraviesa todos los accesorios del motor para forzarlos fuera de la polea del cigueñal.

Service port A fitting found on some control devices and in the low-and high-sides of an air-conditioning system, used to gain access into the system for diagnostics and service procedures.

Puerta de servicio Un montaje que se encuentra en algunos dispositivos de control y en los lados de baja y alta presión de un sistema de aire acondicionado que sirve para dar acceso al sistema para efectuar los procedimientos de diagnóstico y/o el servicio.

Service valve See Service port.

Valvula de servicio Ver Service Port [Orificio de servicio].

Shroud A duct-like cover to ensure that maximum airflow is directed over the engine by the engine-driven fan assembly.

Gualdera Cubierta parecida a un conducto para asegurar que un flujo máximo de aire es conducido sobre el moto por el conjunto del ventilador accionado por el motor.

SNAP Acronym used for the EPA's Significant New Alternatives Policy program, which reviews alternatives to CFC-12 (R-12) refrigerant.

SNAP Una sigla del programa Poliza de Alternativas Nuevas Significantes de la EPA (Agencia de Protección del Medio Ambiente) que revisa las alternativas del refrigerante para el CFC-12 (R-12).

Snapshot A feature of OBD II that shows, on various scanners, the conditions that the vehicle was operating under when a particular trouble code was set. For example, the vehicle was at 225°F, ambient temperature was 55°F, throttle position was part throttle at 1.45 volts, rpm was 1,450, brake was off, transmission was in third gear with torque converter unlocked, air-conditioning system was off, and so on.

Instantáneo Una característica del OBD II que muestra, en varios detectores, las condiciones bajo las cuales operaba el vehículo cuando se registró un código de fallo. Por ejemplo, el vehículo regis-traba 225°F, la temperatura del ambiente era el 55°F, la posición del regulador estaba en una posición parcial de 1.45 voltios, el rpm era 1,450, el freno estaba desenganchada, la transmisión estaba en la ter-cera velocidad con el convertidor del par desenclavado, el sistema de acondicionador de aire estaba apagado, y etcétera.

Specific heat The quantity of heat required to change one pound of a substance by 1°F.

Calor específico Cantidad de calor requerida para cambiar una libra de una sustancia en un grado Fahrenheit.

Stabilize To keep from fluctuating.

Estabilizar Prevenir las fluctuaciones.

Starved Refers to a condition whereby too little refrigerant is metered into the evaporator.

Falta de refrigerante Se refiere a una condición en la cual no se dosifica la cantidad suficiente de refrigerante al evaporador.

Strainer See Screen.

Colador Ver Screen [Cribadora].

Stroke The distance a piston travels from its lowest point to its highest point.

Carrera Distancia que un pistón viaja desde el punto más bajo hasta el mas alto.

Suction A negative force or pressure.

Succión Fuerza o presión negativa.

Suction line The line connecting the evaporator outlet to the compressor inlet.

Conducto de succión Línea que conecta la salida del evaporador a la entrada del compresor.

Suction pressure Compressor inlet pressure. Reflects the pressure of the system on the low side.

Presión de succión La presión de la entrada del compresor. Refleja la presión del sistema del lado de baja presión.

Suction service valve See Low-side service valve.

Válvula de succión de servicio Ver Low-side service valve [Válvula de servicio del lado de baja presión].

Suction valve The low-side service valve is often referred to as the suction valve.

Válvula de succión La válvula de servicio del lado bajo de presión suele referirse como la válvula de succión.

Superheat Adding heat intensity to a gas after the complete evaporation of a liquid.

Sobrecalentar El agregar intensidad calorífica a un gas después de la evaporación completa de un líquido.

Surface temperature The inner temperature of a body, such as water.

Temperatura de la superficie La temperatura interior de un cuerpo, tal como el agua.

Swash plate A type of concentric plate found on some compressor crankshafts used to move the pistons to and fro or back and forth.

Placa oscilante Tipo de placa concéntrica ubicada en algunos cigueñales de compresor y utilizada para mover los pistones de un lado para otro o de un lado a otro.

Telltale light A dash lamp to indicate a malfunction such as low oil pressure or overheating; also called an "idiot light."

Luz indicadora Una lámpara en el tablero de instrumentos para indicar una disfunción, como por ejemplo la baja presión del aceite o el sobrecalentamiento; tambien ilamada "luz para idiotas".

Temperature Heat intensity measured on a thermometer.

Temperatura Intensidad calorífica medida con un termómetro.

Temperature–pressure relationship The relationship that exists, in the English system of measure, of the similarities between temperature and pressure readings of refrigerants R-12 and R-134a in a refrigeration system.

Relación temperatura-presión La relación que existe en el sistema de medida inglés de las similaridades entre las lecturas de temperatura y presión de los refrigerantes R-12 y R-134a en el sistema de refrigeración.

Temperature switch A switch actuated by a change in temperature at a predetermined point.

Interruptor de temperatura Interruptor accionado por un cambio de temperatura a un punto predeterminado.

Thermistor A temperature-sensing resistor that has the ability to change values with a change in temperature.

Termistor Resistor sensible a temperatura que tiene la capacidad de cambiar valores al ocurrir un cambio de temperatura.

Thermostat A device used to cycle the clutch to control the rate of refrigerant flow as a means of temperature control. The driver has control over the temperature desired.

Termostato Dispositivo utilizado para ciclar el embrague para regular la proporción del flujo de refrigerante como medio de regulación de temperatura. El conductor puede regular la temperatura deseada.

Thermostatic expansion valve The component of a refrigeration system that regulates the rate of flow of refrigerant into the evaporator as governed by the action of the remote bulb-sensing tailpipe temperatures.

Válvula de expansión termostática Componente de un sistema de refrigeración que regula la proporción del flujo de refrigerante en el evaporador, lo cual es controlado por la acción de la bombilla a distancia que siente las temperaturas del tubo de escape.

Toxicity Toxic or poisonous quality.

Toxicidad Calidad tóxica o venenosa.

Ultraviolet (UV) radiation The invisible rays from the sun that have damaging effects on the earth. Ultraviolet radiation causes sunburns.

Radiación ultravioleta Rayos invisibles del sol que tienen efectos dañosos en la tierra. La radiación ultravioleta es la causa de la insolacion.

Vacuum Any pressure below atmospheric pressure.

Vacío Cualquier presión inferior a la de la atmósfera.

Vacuum motor A device designed to provide mechanical action by the use of a vacuum signal.

Motor de vacío Dispositivo diseñado para proveer acción mecánica por medio del uso de una señal de vacío.

Vacuum pot See Vacuum motor.

Olla de vacio Ver Vacuum motor [Motor de vacío].

Vacuum signal Level of vacuum received.

Señal de vacío El nivel del vacío que se ha recibido.

Vapor See Gas.

Vapor Ver Gas.

Variable displacement To change the displacement of a compressor by changing the stroke of the piston(s).

Desplazamiento variable Cambiar el desplazamiento de un compressor al cambiar la carrera del pistón o de los pistones.

Vent A condition whereby fresh outside air may be introduced into the vehicle.

Ventilación Condición por medio de la cual el aire fresco exterior puede introducirse al vehículo.

Virgin Newly manufactured or produced, not previously used.

Virgen Fabricado o producido últimamente; no utilizado previamente.

Visual inspection An inspection by sight as opposed to smell, hearing, or touch.

Inspección visual Una inspección usando el sentido de vista en vez del olfato, el oído o el tacto.

Voltage (V) The difference or potential that indicates an excess of electrons at the end of the circuit farthest from the electromotive force. It is the electrical pressure that causes electrons to move through a circuit. One volt is the amount of electrical pressure required to move one amp through one ohm of resistance.

Tensión o voltaje (V) la diferencia o potencial que indica un exceso de electrones en el extremo del circuito más alejado de la fuerza electromotriz. Es la presión eléctrica que provoca que los electrones se muevan a través de un circuito. Un voltio es la cantidad de presión eléctrica necesaria para mover un amperio a través de un ohmio de resistencia.

Wobble plate An offset plate that is secured to the main shaft and moves the piston(s) to and fro.

Placa oscilante Placa de desviación que se fija al árbol principal y mueve el pistón o los pistones de un lado para otro.

Zener clamping diode A one-way electrical gate with a threshold voltage used to suppress damaging voltage spikes to integrated circuits.

Diodo estabilizador zener Una entrada eléctrica de una vía con un umbral de voltaje que se usa para suprimir los picos dañosos de voltaje en los circuítos integrados.

INDEX

C

D

E

F

G

H

I

K

L

M

N

O

P

Q

R

S

T

U

V

W

Z